Application of Geographic Information Systems in Hydrology and Water Resources Management

SOME OTHER TITLES PUBLISHED BY
The International Association of Hydrological Sciences (IAHS)

Hydrological Applications of Remote Sensing and Remote Data Transmission. Proceedings of a symposium held during the IUGG Assembly, Hamburg, August 1983
edited by B. E. Goodison
Publ.no.145 (1985), price $48
ISBN 0-947571-10-8

Conjunctive Water Use. Proceedings of a symposium held during the Second Scientific Assembly of IAHS, Budapest, Hungary, July 1986
edited by S. M. Gorelick
Publ.no.156 (1986), price $48
ISBN 0-947571-65-5

Hydrologic Applications of Space Technology. Proceedings of the Cocoa Beach Workshop, August 1985
edited by A. I. Johnson
Publ.no.160 (1986), price $45
ISBN 0-947571-85-X

Groundwater Monitoring and Management. Proceedings of the Dresden Symposium, March 1987
edited by G. P. Jones
Publ.no.173 (1990), price $55
ISBN 0-947571-51-5

Consequences of Spatial Variability in Aquifer Properties and Data Limitations for Groundwater Modelling Practice. Report prepared by a Working Group of the IAHS International Commission on Groundwater
by Adrian Peck, Steven Gorelick, Ghislain de Marsily, Stephen Foster & Vladimir Kovalevsky
Publ.no.175 (1988), price $45
ISBN 0-947571-61-2

Remote Data Transmission. Proceedings of a workshop held during the IUGG Assembly, Vancouver, August 1987
edited by A. I. Johnson & R. W. Paulson
Publ.no.178 (1989), price $30
ISBN 0-947571-81-7

New Directions for Surface Water Modelling. Proceedings of a symposium held during the Third IAHS Scientific Assembly, Baltimore, Maryland, May 1989
edited by M. L. Kavvas
Publ.no.181 (1989), price $50
ISBN 0-947571-96-5

Groundwater Contamination. Proceedings of a symposium held during the Third IAHS Scientific Assembly, Baltimore, Maryland, May 1989
edited by L. M. Abriola
Publ.no.185 (1989), price $40
ISBN 0-946571-17-5

Remote Sensing and Large-Scale Global Processes. Proceedings of a symposium held during the Third IAHS Scientific Assembly, Baltimore, Maryland, May 1989
edited by A. Rango
Publ.no.186 (1989), price $40
ISBN 0-947571-22-1

Groundwater Management: Quantity and Quality. Proceedings of the Benidorm Symposium, October 1989
edited by A. Sahuquillo, J. Andreu & T. O'Donnell
Publ.no.188 (1989), price $60
ISBN 0-947571-32-9

ModelCARE 90: Calibration and Reliability in Groundwater Modelling. Proceedings of the symposium held in The Hague, September 1990
edited by Karel Kovar
Publ.no.195 (1990), price $55
ISBN 0-947571-67-1

Groundwater Contamination Risk Assessment: A Guide to Understanding and Managing Uncertainties. Report prepared by a Working Group of the International Commission on Groundwater
by Eric Reichard, Carl Cranor, Robert Raucher & Giovanni Zapponi
Publ.no.196 (1990), price $40
ISBN 0-947571-72-8

The Hydrological Basis for Water Resources Management. Proceedings of the Beijing Symposium, October 1990
edited by Uri Shamir & Chen Jiaqi
Publ.no.197 (1990), price $60
ISBN 0-947571-77-9

Hydrology for the Water Management of Large River Basins. Proceedings of a symposium held during the IUGG Assembly, Vienna, August 1991
edited by F. H. M. van der Ven, D. Gutknecht, D. P. Loucks & K. Salewicz
Publ.no.201 (1991), price $55
ISBN 0-947571-97-3

Hydrological Basis of Ecologically Sound Management of Soil and Groundwater. Proceedings of a symposium held during the IUGG Assembly, Vienna, August 1991
edited by H. P. Nachtnebel & K. Kovar
Publ.no.202 (1991), price $55
ISBN 0-947571-03-5

Please send orders and enquiries to:

IAHS Press, Institute of Hydrology
Wallingford, Oxfordshire OX10 8BB, UK
Telephone: +44 491 38800; telex: 849365 hydrol g
Fax: +44 491 32256

Office of the Treasurer IAHS, 2000 Florida
Avenue NW, Washington, DC 20009, USA
Telephone: +1 202 4626900; telex: 7108229300;
Fax: +1 202 3280566

Please send credit card orders (VISA, ACCESS, MASTERCARD, EUROCARD) and IAHS membership orders to the Wallingford address only. A catalogue of publications may be obtained free of charge from either of the above addresses.

Application of Geographic Information Systems in Hydrology and Water Resources Management

Edited by

research for
man and environment

K. KOVAR
*Secretary of the IAHS International Commission on Groundwater
National Institute of Public Health and Environmental Protection
(RIVM), PO Box 1, 3720 BA Bilthoven, The Netherlands*

H.P. NACHTNEBEL
*Institut für Wasserwirtschaft, Hydrologie und konstruktiven Wasserbau
Universität für Bodenkultur (BOKU)
Nussdorfer Lände 11, A-1190 Vienna, Austria*

Proceedings of an international conference held in Vienna, Austria, from 19 to 22 April 1993. This conference was jointly organized by:

the International Commission on Groundwater of the International Association of Hydrological Sciences (IAHS)

the United Nations Educational, Scientific and Cultural Organization (UNESCO) – as a contribution to subprogramme M-2-3 of UNESCO's IHP-IV

Universität für Bodenkultur, Vienna

The conference was sponsored and supported by: The International Association of Hydrogeologists (IAH), the International Ground Water Modeling Center (IGWMC), the International Institute for Applied Systems Analysis, Austria (IIASA), the Association of Ground Water Scientists and Engineers, USA (AGWSE), the American Society of Testing and Materials (ASTM); the International Committee on Remote Sensing and Data Transmission of IAHS (ICRSDT).

IAHS Publication No. 211

Published by the International Association of Hydrological Sciences.

IAHS Press, Institute of Hydrology, Wallingford, Oxfordshire OX10 8BB, UK.

IAHS Publication No. 211.

ISBN 0-947571-48-5

IAHS is indebted to the employers of the Editors for their invaluable support and the services provided that enabled the Editors to function effectively and efficiently. Without this support the proceedings of HydroGIS 93 would not have been pre-published.

The designations employed and the presentation of material throughout the publication do not imply the expression of any opinion whatsoever on the part of IAHS concerning the legal status of any country, territory, city or area or of its authorities, or concerning the delimitation of its frontiers or boundaries.

The use of trade, firm, or corporate names in the publication is for the information and convenience of the reader. Such use does not constitute an official endorsement or approval by IAHS of any product or service to the exclusion of others that may be suitable.

The Conveners would like to express their thanks to all who assisted in organizing the conference. They especially would like to thank the members of the Scientific Advisory Committee who were:

Dr G. Barrocu (Italy)
Dr M. Brilly (Slovenia)
Dr M.J. Clark (UK)
Dr K. Fedra (Austria)
Dr T. Givone (France)
Mr A.I. Johnson (USA)
Dr S. Kaden (Germany)
Dr A. Kilchenmann (Germany)
Dr A. Kraus (Austria)
Dr D.P. Loucks (USA)
Dr H.J. Scholten (The Netherlands)
Dr S.P. Simonovic (Canada)
Dr K.M. Strzepek (USA)

The French papers were checked and corrected by Dr J.A. Rodier whose service is acknowledged with gratitude. Possible typing errors in these two papers are the responsibility of the Editors.

It is highly appreciated that Dr K. Fedra took responsibility to organize a computer workshop during the conference to provide an opportunity for participants to demonstrate their software applications.

The Conveners of the conference would also like to express their thanks for financial support obtained from the following companies and institutions:
    Environmental Systems Research Institute (ESRI, Redlands, California) contributed towards the cost of printing the colour illustrations in the proceedings
    Datamed GmbH (Vienna, Austria), vendor of ARC/INFO
    Austrian Airlines, official carrier for the conference
    Austrian Ministry of Science
    Raiffeisen Landesbank Niederösterreich, Vienna

Our special thanks go also to the members of the Organizing Committee, especially to Dr J. Fürst and Dipl.Ing. G. Reichel, both from the Institut für Wasserwirtschaft, Hydrologie und konstruktiven Wasserbau at the Universität für Bodenkultur, Vienna, Austria.

Finally, the Editors wish to acknowledge the conference authors for their patience and cooperation during the editing process.

The camera-ready copy for the papers was prepared by the Editors, and assembled and finished at IAHS Press, Institute of Hydrology, Wallingford, Oxfordshire, UK, by Penny Kisby.

Printed in The Netherlands by Krips Repro, Meppel.

Preface

Hydrological modelling and water resources management are primarily related to spatial and dynamic processes. Until recently, the complexity of spatially distributed hydrological data sets prevented detailed modelling but stimulated the application of spatially averaged models. The fast progress in computer technology has promoted the development of spatial data base systems, of efficient object-related map manipulation routines and of high resolution visualization techniques. In general, these tools are integrated into Geographic Information Systems from which many software packages are commercially available, differing only in their capabilities of additional utilities, in hardware requirements and in their user friendliness.

In the last few years, many GIS meetings have been held and the majority of the topics addressed were in the field of regional planning, geography, geodesy and environmental management. Although these experiences are quite valuable for hydrologists and water resources systems engineers, additional specific requirements have to be considered for water related applications.

The objectives of the Conference on the Application of Geographic Information Systems in Hydrology and Water Resources Management were to exchange experiences in the application of GIS and to identify research needs with respect to the specific requirements of hydrology and water resources related management. The remarkable interest of the scientific community involved in water related issues was reflected by the submission of more than 180 abstracts. The 69 papers included in this volume are grouped under the following topics:
- Decision support and expert systems
- Methodological aspects and application of GIS in remote sensing
- Digital terrain models and GIS
- Application of GIS in three- and four-dimensional problems
- Coupling GIS with hydrological models
- Application of GIS in water and environmental management
- Application of GIS in surface water systems
- Application of GIS in groundwater systems

It is the opinion of the Conveners that this volume is a contribution to the evaluation of the benefits and drawbacks of the application of GIS in hydrological sciences. Further, another intention was to identify specific requirements of the hydrological community to achieve successful application of GIS.

The interest and support of UNESCO in the conference are greatly appreciated. The conference was a contribution to the International Hydrological Programme (1990-1995), in particular to subprogramme M-2-3: *The Use of Geographic Information Systems in Hydrological and Water Resources Studies.*

The Conference Conveners:

H.P. Nachtnebel
Institut für Wasserwirtschaft, Hydrologie und konstruktiven Wasserbau
Universität für Bodenkultur (BOKU)
Nussdorfer Lände 11, A-1190 Vienna, Austria

K. Kovar
Secretary of the IAHS International Commission on Groundwater
National Institute of Public Health and Environmental Protection (RIVM)
PO Box 1, 3720 BA Bilthoven, The Netherlands

Contents

Preface *by H.P. Nachtnebel & K. Kovar* — v

1 Decision Support and Expert Systems

Application of fuzzy sets in ecohydrological expert modelling
Wim J. Droesen & Luc H.W.T. Geelen — 3

Application of GIS in Decision Support Systems for groundwater management *J. Fürst, G. Girstmair & H.P. Nachtnebel* — 13

An expert system approach of integrating hydrological database, models and GIS: application of the RAISON System *D.C.L. Lam & D.A. Swayne* — 23

Application of ILWIS to decision support in watershed management: case study of the Komering river basin, Indonesia *A.M.J. Meijerink, C.M. Mannaerts, H.A. de Brouwer & C.R. Valenzuela* — 35

Hydrological modelling and GIS: the Sandia Environmental Decision Support System *Steven P. Frysinger, Richard P. Thomas & Alva M. Parsons* — 45

An integrated information system for environmental hydraulics using Smallworld GIS *P. Ruland, U. Arnold & G. Rouvé* — 51

Flood control management by integrating GIS with expert systems: Winnipeg City case study *S.P. Simonovic* — 61

2 Methodological Aspects and Application of GIS in Remote Sensing

Relevance of Geographic Information Systems of landscape characteristics for hydroclimatological models *R. Avissar* — 75

Process, scale and constraints to hydrological modelling in GIS
R.B. Grayson, G. Blöschl, R.D. Barling & I.D. Moore — 83

High resolution satellite imagery and GIS as a dynamic tool in groundwater exploration in a semi-arid area *Per Gustafsson* — 93

GIS application to water resources management in the land planning context: a methodological proposal *F. Jemma* — 101

Environmental modelling and GIS: dealing with spatial continuity
Karen K. Kemp — 107

Application of GIS for study of groundwater flow in arid regions using remote sensing data *Yu.L. Obyedkov & I.S. Zektser* — 117

Application of GIS and remote sensing in hydrology *G.A. Schultz*   127

3 Digital Terrain Models and GIS

Integrating a physically based hydrological model with GRASS
S. Chairat & J.W. Delleur   143

Matching standard GIS packages with urban storm drainage simulation software *J. Elgy, C. Maksimovic & D. Prodanovic*   151

Modelling the spatial variability of hydrological processes using GIS
I.D. Moore, J.C. Gallant, L. Guerra & J.D. Kalma   161

Hydrologically oriented GIS and application to rainfall-runoff distributed modelling: case study of the Arno basin *P. La Barbera, L. Lanza & F. Siccardi*   171

Developing a spatially distributed unit hydrograph by using GIS
David R. Maidment   181

Raster based modelling of watersheds and flow accumulation
Broder Merkel & Barbara Sperling   193

Hydrological terrain features derived from a pyramid raster structure
Wolfgang Rieger   201

TOPMODEL as an application module within WIS
Renata Romanowicz, Keith Beven, Jim Freer & Roger Moore   211

4 Application of GIS in Three- and Four-Dimensional Problems

Development of a three-dimensional hydrogeological framework model for the Death Valley region, southern Nevada and California, USA
Claudia Faunt, Frank D'Agnese & A. Keith Turner   227

Integrated three-dimensional geoscientific information system (GSIS) technologies for groundwater and contaminant modelling
Thomas R. Fisher   235

Integration of three-dimensional groundwater modelling techniques with multi-dimensional GIS *Judith Schenk, Keith Kirk & Eileen Poeter*   243

Integration of space and time in GIS applications: one step further to model reality? *Bart Kusse, Colin Gray & Henk J. Scholten*   251

Database development and GIS application in support of groundwater management: case study at the Austrian-Bohemian Border *M. Stibitz, Z. Patzelt & J. Wolfbauer*   263

Deep waste injection, Western Canada: analysis of data and requirements for numerical simulation *J.R. Underschultz & A.T. Lytviak*   271

5   Coupling GIS with Hydrological Models

Multi-media: an integrated approach combining GIS within the
framework of the Water Act 1989 *Christine H. Ashton &
David Simmons* — 281

Groundwater modelling and GIS: integrating MICRO-FEM and ILWIS
A. Biesheuvel & C.J. Hemker — 289

Models, GIS, and expert systems: integrated water resources models
K. Fedra — 297

Application of a GIS for simulating hydrological responses in
developing regions *Stefan W. Kienzle* — 309

Hydrological modelling within GIS: an integrated approach *Neil Stuart
& Christopher Stocks* — 319

6   Application of GIS in Water and Environmental Management

A national groundwater model combined with a GIS for water
management in The Netherlands *W.J. De Lange &
J.L. Van der Meij* — 333

The use of GIS techniques to assess discharge consents and abstraction
licences *N.J. Bonvoisin & R.V. Moore* — 345

Un système d'information géographique adapté à l'évaluation de la
pollution agricole diffuse *D. Cluis & E. Quentin* — 355

GIS and multivariate ecological analysis: an input to integrated river
channel management *A.M. Gurnell, D. Simmons, P.J. Edwards,
J. Ball, J. Feaver, A. McLellan & G. Ogle* — 363

The importance of GIS in regional geohydrological studies
J.H. Hoogendoorn, W. Van der Linden & C.B.M. Te Stroet — 375

GIS in water-related environmental planning and management:
problems and solutions *Stefan O. Kaden* — 385

Interactive multi-media, GIS, and water resources simulation
Daniel P. Loucks — 399

A GIS-based hierarchical simulation model for assessing the impacts of
large dam projects *R.A. McDonnell & W.D. Macmillan* — 409

Application of GIS to water management in a developing region
J.J. Olivier & D.R. McPherson — 417

GIS structure for the Nile River forecast project *John C. Schaake,
Robert M. Ragan & Edward J. Vanblargan* — 427

Groundwater system modelling and management using Geoscientific
Information Systems *K.E. Kolm & J.S. Downey* — 433

National water resources management planning based on GIS *G. Keser & J.J. Bogardi* — 439

7 Application of GIS in Surface Water Systems

Data constraints on GIS application development for water resource management *M.J. Clark* — 451

Validation of the ANSWERS catchment model for runoff and soil erosion simulation in catchments in The Netherlands and the United Kingdom *A.P.J. De Roo* — 465

Exemple d'utilisation d'un SIG pour la gestion des données d'un modèle hydrologique à mailles carrées *F. Delclaux & G. Boyer* — 475

Mapping and management of flood plains *P. Farissier & P. Givone* — 485

Application of GIS in modelling winter orographic precipitation, Gunnison River Basin, Colorado, USA *L.E. Hay, W.A. Battaglin, M.D. Branson & G.H. Leavesley* — 491

Linking sediment and nutrient export models with a geographic information system *E. Klaghofer, W. Birnbaum & W. Summer* — 501

RHINEFLOW: an integrated GIS water balance model for the river Rhine *W.P.A. Van Deursen & J.C.J. Kwadijk* — 507

Application of GIS in determining sources and loads of pollutants transported by regional rivers into The Netherlands *A. Molenaar, W. Bleuten, M.J. Zeylmans & N.F.M. Van Leeuwen* — 519

Flood simulation assisted by a GIS *I. Muzik & C. Chang* — 531

"APS on GIS": an operational forest hydrological GIS *D.W. Van der Zel & F.V. Rabe* — 541

Spatially oriented surface water hydrological modelling and GIS *M. Brilly, M. Smith & A. Vidmar* — 547

Identification of heavy metal concentrations in surface waters through coupling of GIS and hydrochemical models *T.J. Kern & J.D. Stednick* — 559

8 Application of GIS in Groundwater Systems

Application of GIS for aquifer vulnerability evaluation *G. Barrocu & G. Biallo* — 571

Development and application of a groundwater model integrated in the GIS GRASS *O. Batelaan, F. De Smedt, M.N. Otero Valle & W. Huybrechts* — 581

Use of volume modelling techniques to estimate agricultural chemical mass in groundwater, Minnesota, USA *William A. Battaglin* — 591

Application of a geographic information system in analyzing the occurrence of atrazine in groundwater of the mid-continental United States *M.R. Burkart & D.W. Kolpin*   601

EGIS, a geohydrological information system *F. Deckers*   611

The management of groundwater resources in the Salinas Valley, California, by computer *Philip Hall & Matthew Zidar*   621

Mapping procedures for assessing groundwater vulnerability to nitrate and pesticides *G. Sokol, Ch. Leibundgut, K.P. Schulz & W. Weinzierl*   631

Development of the GIS-based "RIVM National Groundwater Model for The Netherlands (LGM)" *R. Lieste, K. Kovar, J.G.W. Verlouw & J.B.S. Gan*   641

Application of geographical information systems to support groundwater modelling *H.P. Nachtnebel, J. Fürst & H. Holzmann*   653

Groundwater modelling using GIS at the Amsterdam water supply *T.N. Olsthoorn, P.T.W.J. Kamps & W.J. Droesen*   665

Evaluation of a new integrated software platform for data storage and analysis pertinent to hazardous waste site investigations *M.A. Sen & B. Kelk*   675

Application of water balance model to western Saudi Arabia and use of GIS in future *Ali Ü. Sorman, Yakup Basmaci & Kamil Eren*   685

1 Decision Support and Expert Systems

Application of fuzzy sets in ecohydrological expert modelling

WIM J. DROESEN
Wageningen Agricultural University, Dept. of Landsurveying and Remote Sensing, P.O. Box 339, 6700 AH Wageningen, The Netherlands

LUC H.W.T. GEELEN
Amsterdam Water Supply, Vogelenzangseweg 21, 2114 BA Vogelenzang, The Netherlands

Abstract A fuzzy ecohydrological model is presented. The model consists of fuzzy sets, fuzzy relations and some fuzzy aggregation operators. The sets and relations are defined by experts and represent their knowledge of the ecohydrological system. Essentially fuzzy sets and fuzzy relations allow the modelling of such a complex system into a limited set of decision rules by introducing vagueness. The degree of vagueness reflects the uncertainty involved. The benefits of fuzzy ecohydrological modelling over crisp modelling are presented using data of a test site in the catchment area of the Amsterdam Water Supply.

INTRODUCTION

The Amsterdam Water Supply withdraws groundwater from a catchment area in the dunes bordering the North Sea. Due to these activities most of the wet dune slacks dessicated and lost their specific ecological values. A study has been conducted to look at the possibilities to regenerate these former wet slacks by reduction or reallocation of groundwater catchment, if necessary supported by nature management.

In Fig. 1 a diagram of the ecohydrological model is presented. The hydrological system can be seen as the key factor for directing the development of the ecological system. Complementary nature management can be applied either to enhance the ecological effects due to changes in the hydrological situation or to undo negative effects of the latter. Central to the research was the development of a model to estimate nutrient availability for the vegetation dependent, on amongst others, soil moisture content. The latter model is the subject of this paper. The hydrological submodel is presented elsewhere in this volume (Olsthoorn *et al.*, 1993).

Due to a lack of data a statistical nutrient availability model was not feasible. However, over the years experts gained adequate knowledge about the ecohydrological system. This knowledge is formalized in "if <condition> then <hypothesis>" decision rules. The total set of rules and the definition of the conditional and hypothetical classes form a logical model.

The evaluation of the results raised questions about the data quality. There was a need for a quality measure to accompany the data. By redefining

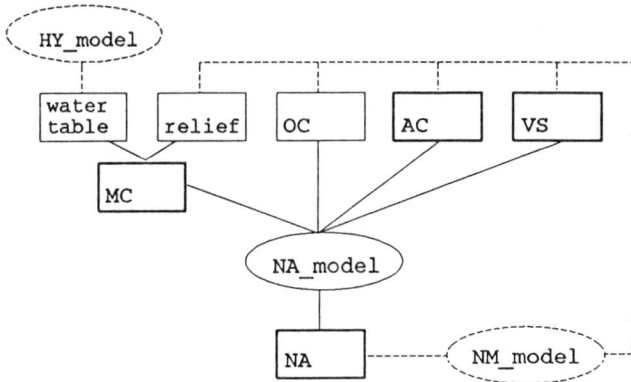

Fig. 1 Diagram of the ecohydrological model containing a hydrological (HY), a nutrient availability (NA) and a nature management (NM) submodel. The unbroken lines of the diagram show the model parts to be handled in the paper. Abbreviations: MC=moisture content, OC=organic matter content, AC=acidity, VS=vegetation structure, and NA=nutrient availability.

the crisp expert model in fuzzy sets and fuzzy relations the latter is achieved. In this paper the fuzzification of the original crisp model is outlined. Recently in GIS literature two papers appeared on the implementation of a fuzzy data model in a relational database management system (Wang *et al.*, 1990; Kollias & Voliotis, 1991). We used the Oracle RDBMS for data modelling and the GIS Genamap for preparing input data and map production. We will not deal with the hydrological and ecological justification of the model. A full description of the model is presented by Koerselman *et al.* (1992).

GENERAL DESCRIPTION OF THE BOOLEAN LOGICAL MODEL

Four attributes are chosen to construct a feature space enabling a rough discrimination between different ecological systems:
(a)   moisture content MC;
(b)   acidity AC;
(c)   nutrient availability NA of the topsoil;
(d)   vegetation structure VS.

Moisture content and nutrient availability are highly discriminating attributes because together with light they form the primary production factors for vegetation. The irradiance is not used as an attribute for it is spatially constant. Acidity is added to the feature space for it affects the species composition, as does vegetation structure.

Figure 1 shows how these variables are derived. Moisture content of the topsoil is estimated by the depth of the groundwater level below terrain surface. Acidity and organic matter content can be easily measured and are expressed in pH-values and weight per area, respectively. Finally nutrient availability is estimated by an expert model.

In the feature space classes are defined by defining thresholds on the attribute axes. For instance the variable MC is represented by three ordinal classes:

$$MC = \{low\_mc, medium\_mc, high\_mc\} \qquad (1)$$

Nutrient availability in sandy dunes is largely governed by processes dependent on the following three variables: (a) moisture content, (b) acidity and (c) nutrient supply captured by the organic matter. For every combination of these conditional classes a hypothetical nutrient availability class is indicated. Because each variable is represented by 3 classes (3x3x3=) 27 decision rules are required. Together they form the nutrient availability model NAM. The following rule is given as example:

$$\text{IF MC = medium AND OC = low AND AC = medium AND VS = herbaceous}$$
$$\text{THEN NA = very low} \qquad (2)$$

FUZZY EXPERT MODEL

Due to the complexity of ecohydrological processes the application of a set of simple Boolean decision rules does not reflect the expert's uncertainty. The uncertainty seems to be threefold (Molenaar & Janssen, 1991). Uncertainty appears with respect to the definition of classes (or sets); e.g. what is a high moisture content? Secondly, the inference formalized in production rules is not error free; How sure is the relation between condition and hypothesis? Finally, when we apply this rule to a specific terrain element a third source of uncertainty is the observation accuracy of the element. The first two types of uncertainty are addressed in this paper and handled by fuzzy sets and fuzzy relations respectively (Zimmermann, 1985).

Uncertainty and fuzzy sets

In classical Boole's set theory classes are defined by membership functions MF for a specific variable x, which take as values only 1 or 0. A membership value of 1 means "belongs to" the set or class, while a value of 0 means that it does not.

Contrary to the MF^B of a crisp set, a fuzzy set is defined by a continuous function $MF^F(x)$ taking values between 0 and 1. The generalization of crisp sets to fuzzy sets enables the use of partial memberships. The membership value indicates the degree of compatibility of an attribute value with the set. This compatibility measure is not necessarily a probability.

We use class labels to represent fuzzy sets. These labels are values of the variable of interest. Consider, for example, the variable moisture content MC with the three classes; low_mc, medium_mc and high_mc. The intension of the classes is defined by a membership function on the variable "depth of groundwater" indicated by x. The variable MC is thus represented by three membership grades, one for each fuzzy class. Variables defined in this way are called fuzzy variables (Klir & Folger, 1988).

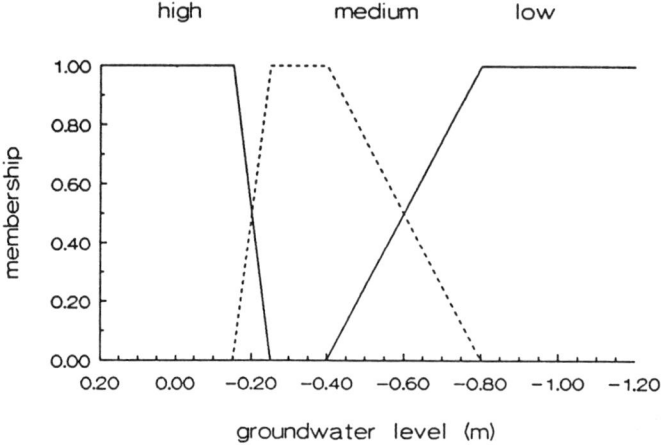

Fig. 2 Membership functions of the classes of the fuzzy variable moisture content.

In Fig. 2 the fuzzy variable moisture content is presented. The Boolean equivalent is defined by exact boundaries at groundwater levels of 0.20 and 0.60 m below the surface. The membership functions stem from the subjective decision of a team of experts. The vagueness of the classes is governed by the range of overlap between classes.

In ecohydrology the definition of fuzzy classes usually resembles much better the experts perception of the character of data. Why have classes a fuzzy character? The vagueness seems to proceed from the context dependency of the class definitions. Consider again, as an example, moisture content. The relation between depth of groundwater and the topsoil moisture content is not simple. In order to make a good estimation of the moisture content more variables and their mutual relations need to be modelled. Moreover mutual dependencies between classes of different fuzzy variables exist, whereas these dependencies are hard to model. For instance, a high organic matter content usually results in a higher moisture content.

Crisp data reflect nothing of this complexity. By adapting the degree of fuzziness of the sets we are able to express this complexity even in the case of relatively little data. Although seemingly less precise than crisp data, fuzzy data are in fact a more adequate representation of reality (Klir & Folger, 1988).

Fuzzy relations

Besides the definition of fuzzy variables expert knowledge is formalized in decision rules. These rules represent the presence or absence of association between the conditional and hypothetical classes (or sets). When the presence of association is indicated with 1 and absence with 0, this is called a crisp relation. In general practice an expert is not equally certain of all production rules. By indicating the strength of association by a membership grade a fuzzy relation is obtained.

The nutrient availability model (NAM) is a fuzzy relation $R^{NAM}(MC,AC,OC,NA)$ being a subset of the Cartesian product of the four distinguished fuzzy variables. The subset is defined by a membership function, which is estimated by a team of experts. Because each fuzzy variable contains 3 classes except for nutrient availability having 5 classes, the fuzzy relation contains a subset of the (3x3x3x5=) 135 ordered tuples. Thus a 5-tuple can be defined consisting of 4 classes and a membership function value.

Fuzzy operations

By proper fuzzy operations a membership value is to be calculated for the joint occurrence of classes in the feature space for each terrain element (i,j). This means the derivation of the membership function for the fuzzy relation on the Cartesian product of the fuzzy variables moisture content, acidity and nutrient availability $R_{(i,j)}(MC,AC,NA)$.

A full description of the subsequent fuzzy operations is omitted. We confine with giving the fuzzy equivalents of the applied Boolean AND and OR operators. These are the MIN and MAX operator, respectively.

DESCRIPTION OF THE DATA

One former dune slack is selected to demonstrate the functionality of the model. In this area, which measures 15 ha, soil data are obtained by field sampling. Water levels are estimated by a hydrological model. The conditions for the model calculations are specified in an hydrological management scenario. We arbitrarily chose a scenario that resulted in a simulated phreatic surface yielding a full range of soil moisture conditions.

By interpolation between the irregular point observations estimations are obtained for all variables in a regular grid. In order to satisfactorily represent the spatial variability of especially the relief, a cell size of 10x10 m was chosen. In Table 1 the statistics calculated over all gridcells are presented. All variables vary considerably throughout the area and reflect properly the diversity common within dune slacks.

Table 1 Statistics of the estimated grid cell values.

variable		mean	sd	min	max
terrain height (h)	m	4.89	0.60	3.83	9.56
water level (wl)	m	4.58	0.58	4.05	5.87
h - wl	m	0.31	0.49	-0.63	3.69
organic matter	kg m^{-2}	5.30	2.04	1.70	10.15
pH	pH	4.12	0.94	3.55	6.26

RESULTS

The result of the crisp model typically consists of a single outcome per terrain element, i.e. grid cell. The fuzzy model yields a vector of membership values for all classes and combination of classes in the feature space. This vector provides the opportunity to create more than one view on the fuzzy data model.

Figure 3a shows a map of the membership value of the most possible class combination for each grid cell. This map clearly shows that for extended areas the uncertainty is considerably high. The difference between the most and second most possible (Fig. 3b) combination of classes shows the relative uncertainty in a terrain element.

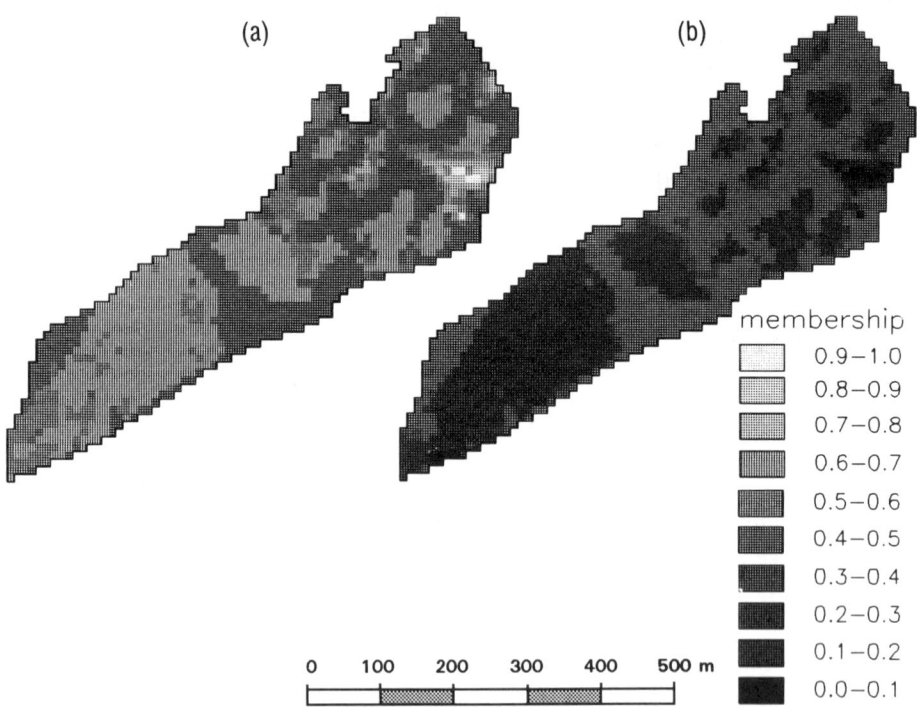

Fig. 3 Membership values of the first (left) and second (right) most possible class combination. The difference between the membership values indicates the relative uncertainty in a terrain element.

The ecologists of the Amsterdam Water Supply were particularly interested in specific class combinations having much potential from an ecological point of view. The class combination high moisture content, high acidity and low nutrient availability is taken as an example (Fig. 4). The crisp model predicts this particular combination for only 11 terrain elements. In the fuzzy data model all locations with a non-zero membership value indicate that there is a possibility for an interesting ecosystem to develop. From the point of view of nature management this allows attention to be paid to those areas and the application of adequate management measures to enhance their ecological potential. Clearly some potential interesting areas would be overlooked in the crisp case.

So far the queries yielded a result for each terrain element. In Table 2 the aggregated information of the testsite is summarized. For convenience only the classes occupying more than 0.1% of the test site are presented. The estimated areas can differ considerably between the two methods, up to a maximum of 24 600 m². It is also remarkable that the areal extent of states which are not present in the crisp model result, can be quite high, i.e h_mc, h_ac and m_na, and l_mc, h_ac and m_na.

By summing all the minimum areal estimates per state, i.e the minimum

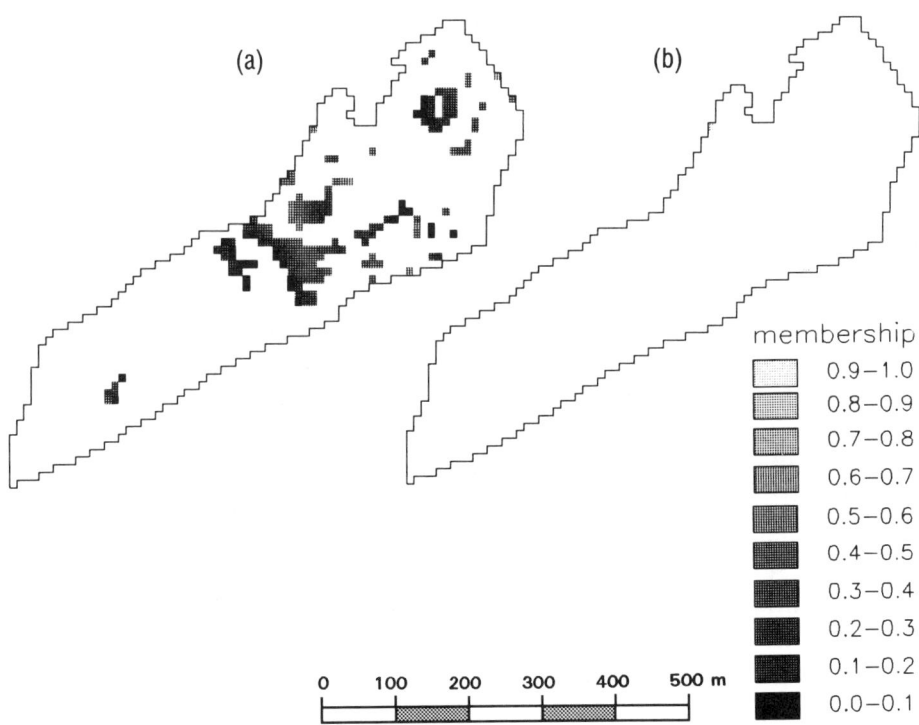

Fig. 4 Membership values for the class combination high moisture content, high acidity and low nutrient availabilty in the fuzzy (left) and crisp (right) model. The terrain elements with a zero membership value are left blank.

Table 2 Areal estimates for the class combinations obtained by the fuzzy en crisp model. Only the class combinations occupying more than 1% of the test site are presented:

class			area (100 m^2)		
MC	AC	NA	fuzzy	crisp	diff.
h	h	h	438.1	684.1	-246.0
h	h	m	199.5	0.0	199.5
h	h	l	32.9	11.0	21.9
m	h	h	14.1	10.1	4.0
m	m	m	17.6	33.2	-15.6
m	m	l	16.1	0.0	-16.1
m	l	m	133.7	319.0	-185.3
m	l	l	135.6	18.0	117.6
m	l	vl	57.5	44.0	13.5
l	m	m	15.2	23.7	-8.5
l	l	m	124.9	0.0	124.9
l	l	l	125.0	242.0	-117.0
l	l	vl	88.2	94.0	-5.8

of the fuzzy and crisp model result, the total area is obtained, which is correspondingly allocated to states by both models. This area amounts 90 090 m^2, which is 60% of the total area. The remaining 40% of the area is differently distributed over the states by the models.

CONCLUDING REMARKS

(a) Fuzzy sets and fuzzy relations appear to be useful concepts to describe uncertainty in logical expert modelling. The results of the fuzzy model give more adequate information per terrain element. The same holds for spatially aggregated information;

(b) the subjectively defined membership functions must not primarily be judged as weak model parts, but more as a best attempt to express uncertainty. Uncertainty is undoubtedly present in a model of a complex ecosystem;

(c) recently Burrough & Heuvelink (1992) presented a study on the sensitivity of fuzzy and crisp logical modelling to uncertain input data. Their results suggest that the fuzzy model outperforms its crisp counterpart. We intend to carry out a comparable study for our data;

(d) one more task for the authors is the development of a decision support model to allocate nature management measures. The expectation is that the latter model will prove to be more useful when based on a fuzzy data model.

REFERENCES

Burrough, P.A. & Heuvelink, G.B.M. (1992) The sensitivity of Boolean and continuous (fuzzy) logical modelling to uncertain data. *Proc. of 3rd European conf. on GIS*, Munich,1032-1041.

Klir, G.J. & Folger, T.A. (1988) Fuzzy sets, uncertainty, and information. Prentice-Hall Int., London.

Koerselman, W., Doing, H., Geelen, L.H.W.T., Leltz, G.M. & Nijssen, E.M. (1992) Nutrient availability of dune slacks in relationship to environmental conditions and external inputs. KIWA, Nieuwegein (in Dutch).

Kollias, V.J. & Voliotis, A. (1991) Fuzzy reasoning in the development of geographical information systems. FRSIS: a prototype soil information system with fuzzy retrieval capabilities. *Int. J. GIS* 5(2), 209-223.

Molenaar, M. & Janssen, L.L.F. (1991) Integrated processing of remotely sensed and geographic data for land inventory purposes. ASPRS/NCGIA Symposium on Integration of RS and GIS, Baltimore, USA.

Olsthoorn, T.N., Kamps, P.T.W.J. & Droesen, W.J. (1993) Groundwater modelling using GIS by the Amsterdam Water Supply. In: *Application of GIS in Hydrology and Water Resources* (ed. by K.Kovar & H.P.Nachtnebel) (Proc. Vienna conf., April 1993). IAHS Publ. no. 211.

Wang, F., Hall, G.B.& Subaryono (1990) Fuzzy information and processing in conventional GIS software: database design and application. *Int. J. GIS* 4(3), 261-283.

Zimmermann, H.J. (1985) Fuzzy set theory - and its applications.Kluwer-Nijhoff Publ., Dordrecht.

Application of GIS in Decision Support Systems for groundwater management

J. FÜRST, G. GIRSTMAIR & H.P. NACHTNEBEL
Institut für Wasserwirtschaft, Hydrologie und konstruktiven Wasserbau, Universität für Bodenkultur, Nussdorfer Lände 11, A-1190 Wien, Austria

Abstract Based on the experiences from two case studies, this paper describes the application of GIS in Decision Support Systems (DSS) for groundwater management. A DSS concept of model based scenario analysis can be efficiently supported by GIS components. The most important and beneficial contributions of GIS to DSS are recognized for the spatial database, spatial data analysis and map display. In both of the discussed prototypes, the underlying GIS (ARC/INFO and GRASS) had deficits in support of 3D data. Some disadvantages are also in the sometimes tedious procedures of interfacing GIS tools and data with groundwater models, and the poor performance of interactive graphical tasks.

INTRODUCTION

Geographical Information Systems (GIS) attained a prominent position among the computer tools for decision support of problems with a spatial dimension. Especially in the context of environmental protection numerous applications of GIS are reported (e.g., Pillmann & Jaeschke, 1990; Aiken, 1992). In many cases the "decision support" is simply information retrieval, filtering and display.

For groundwater management, computer applications ranging from databases, statistical analysis, time series analysis, geostatistics to numerical groundwater models and graphics programs are used since many years. Incompatibilities and inconsistencies between these tools that were primarily not developed for groundwater management restricted their efficient application.

Numerical groundwater models are successfully used for analyzing groundwater flow and transport (Bachmat *et al.*, 1985). While giving accurate results and being numerically efficient, they require a remarkable effort to prepare the input and to interpret the output. User friendly interfaces were a step to handle these models more easily. This did not resolve the incompatibilities to other necessary tools for groundwater decision support, because it was still driven by the requirements of the groundwater model.

The variety of tasks and already available partial solutions lead to a data driven, modular concept of decision support software. All the required tools are combined in a common user friendly interface and access the same data. This paper demonstrates, that GIS are a reasonable basis for efficient and economic development of DSS for groundwater management.

REQUIREMENTS ON DECISION SUPPORT SYSTEMS FOR GROUNDWATER MANAGEMENT

Typical actions with impacts on regional groundwater systems and thus requiring groundwater management are:
(a) groundwater pumping for domestic, industrial and agricultural supply;
(b) hydraulic engineering constructions such as hydro power plants, river regulations, dams, reservoirs;
(c) other actions influencing the regional water cycle: roads and highways, tunnels, waste dump sites, sewer systems, etc.

While general objectives like maximizing benefits and minimizing the negative impacts are often easily postulated, it is difficult to formulate direct search strategies for an "optimal" solution.

The decision problems are usually neither well structured nor unambiguous because of multiple objectives and criteria. Uncertainty in the data, subjective and socio-political aspects are further reasons that there is no unique, objective way to find a "best" solution (Fedra & Reitsma, 1990). The huge number of variables induced by the necessary spatial discretization of input, state and output of a groundwater system imposes another limit on the solution by straightforward systems analysis approaches.

Scenario analysis proved to be a reasonable concept to find acceptable solutions. The reaction of the groundwater system to a range of different decision alternatives is investigated under different assumptions of aquifer properties, boundary and hydrological conditions.

If a responsible decision maker can do the analysis, he will investigate a feasible set of decision alternatives and iteratively find an acceptable solution. While this solution is not guaranteed to be optimal, it is likely to be robust and highly acceptable, because the impacts of the decisions can be easily compared and the sensitivity to any of the assumptions of aquifer properties, boundary and hydrological conditions can as well be verified.

A set of tools that support quick and comprehensive generation, simulation, analysis and comparison of scenarios is a prerequisite for successful scenario analysis as close as possible to the decision makers. Model based DSS are a recent attempt to meet these requirements. They couple specialized models for parts of the problem with the database. A friendly, interactive user interface makes them accessible at the level of the decision maker (Bonczek et al., 1981; Sol, 1983).

The key term of modern DSS is information management. Their central component is its database. All the functions of the DSS access this database and add and/or retrieve information. This holds for a simple time series plotter as well as for a complex numerical groundwater model. So the database is "the model" on which the functions (including numerical groundwater "models") provide different views and perform experiments.

Database

A large amount of different categories of data forms the model of a groundwater system in the computer.

Time invariant spatial data: Primarily the hydrogeological properties of the aquifer (geological layers, hydraulic conductivity, porosity), but also maps of land use, valuable habitats or regional planning topics belong to this category. Ideally, we need to know all this information as continuous functions of space. However, in most cases only point (e.g., from boreholes) or line related data (e.g., seismic profiles) are available.

Time series data: For analysis of the historic behaviour of a groundwater system, groundwater levels, solute concentrations, discharges and meteorological variables are measured at several observation points. These variables are a function of space and time. The density of observation points varies and is often very low. Also the observation interval is usually not constant and observation times are not the same on different points.

We must distinguish original (observed) data from computed data. Another important aspect is that also the origin, method of measurement, reliability and similar attributes are available.

The database has to handle time invariant spatial data, time series data and textual information or expertise. The organization of the spatial data must provide the following access methods:
(a) selection of objects by identification, coordinates, and/or attribute values
(b) selection by topological criteria (neighbours, nearest points)

The time series database must provide selection of observation points by name and position, by time criteria and by value. A special case is the selection of a value at a given time for a set of observation points.

Textual information must be incorporated to capture expertise and to document the important steps of analysis.

Functions

The functions of a DSS for groundwater management can be grouped in three categories: information, scenario generation and decision support.

Information functions are the tools for data acquisition, data analysis and display. Data acquisition tools cover the tasks of map digitizing, data entry and conversion and import of external data.

Data analysis functions provide specific information about the state of the groundwater system as represented in the database. They include time series analysis, spatial analysis, neighbourhood analysis, area statistics, cluster analysis and others.

Scenario generation and simulation are the key functions. The main goal of building a computer model of a groundwater system is to provide a facility for specifying decision alternatives and simulate their impact in the model. Numerical groundwater models are frequently applied, powerful tools to study the dynamic behaviour of a regional groundwater system under various conditions.

User interface

A large database and an extensive and complex set of tools makes a friendly user interface a precondition for successful use. The users want convenient and safe ways to submit their questions to the DSS and they want fast, intuitive presentation of the results (Loucks *et al.*, 1985). State of the art technology for this purpose are graphical user interfaces and extensive use of interactive computer graphics.

APPLICATION OF GIS

In this section the experiences with the development of two prototypes of DSS are discussed.

ARCGW is based on the vector based GIS ARC/INFO. The development approach in ARCGW was to extend the GIS with groundwater specific modules and then to wrap up those groundwater tools and useful ARC/INFO functions in a menu driven user interface that appears to the user as the DSS. It was developed for conducting environmental impact analysis of hydropower schemes on the Danube.

LGW (Logical Groundwater Workstation) is a more independent system with a relational database system, a graphical user interface and some components of the raster based GIS GRASS. The DSS is used by the federal water authority to identify the impacts of conflicting projects on a major Austrian groundwater basin and to assess the degree of interference of the projects.

GIS based spatial database

In ARCGW the spatial database is completely designed as a set of ARC/INFO point, line and polygon coverages and triangulated irregular networks (TIN). Original data are distinguished from working data (Fig. 1). There is also a complete representation of the geometry, boundary conditions, aquifer parameters and computed results of the integrated finite difference groundwater model created in ARC/INFO.

Before the database is ready for scenario analysis, the conceptual model of the groundwater system has to be derived. We have to construct spatial estimates of the aquifer parameters from scarce point (e.g., boreholes) or line (e.g., seismic profiles) information. While general purpose GIS supply rather simple spatial interpolation techniques, adequate geostatistic methods like kriging from other sources had to be accessed. Similarly, the available data structures did not meet the requirements of 3D surface modelling. Only the

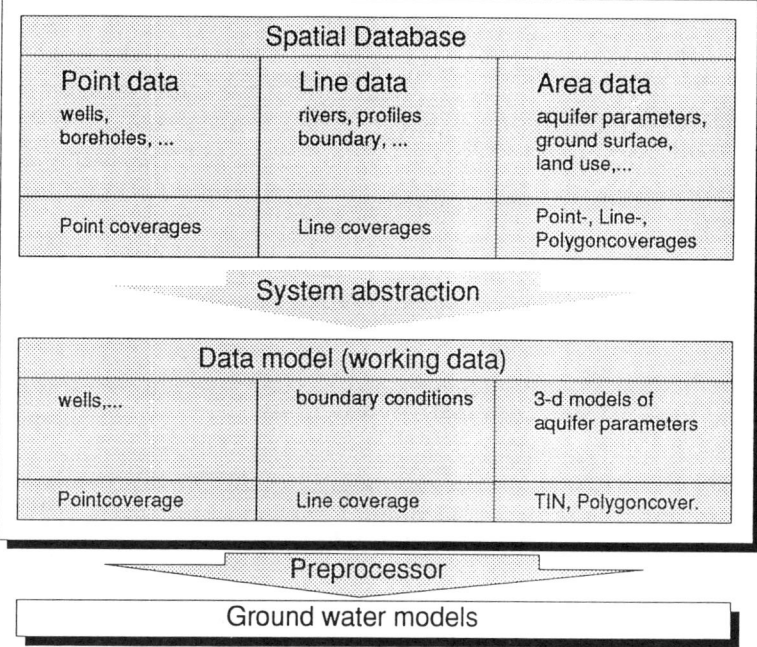

Fig. 1 Schematic representation of the spatial database in ARCGW.

latest releases of GIS packages begin to incorporate advanced interpolation techniques and 3D support.

In LGW the primary database is implemented in a relational database system. Simple spatial selection and export tools exist to create point-, vector- or raster maps in the raster based GIS GRASS. Data are only temporarily kept in GRASS when GIS functions need to be utilized.

GIS tools used for building the database

The database is doubtlessly the key element of a DSS and unfortunately the most expensive one. There is no need that the users of a DSS perform the data acquisition themselves and the necessary functions need not be part of the DSS. However, an integrated GIS based solution has several advantages:

(a) digitizing, import, export, cleaning or building of topologies are readily available basic components of GIS;

(b) a minimum set of data acquisition tools is desirable in any DSS for updating the database;

(c) operators of the DSS can gain considerable insight into the behavior of the groundwater system and get a good impression of the quality and reliability of data, if they are involved in the data acquisition;

(d) if desired, two different user interfaces for DSS and data acquisition can be created with little additional effort and without compatibility problems;

(e) with GIS being increasingly used by consultants and public authorities, there is hope that data can be exchanged on electronic media or via computer networks. Major GIS packages have the advantage of supporting export and import to and from each other.

In ARCGW ARC/INFO functions are used for digitizing hardcopy maps. Where data are available in tabular format, they are converted to standard ASCII files suitable for ARC/INFO map generation. An AML (Arc Macro Language) menu interface can simplify the most frequently needed tasks.

In LGW GRASS digitizing and map import/export tools are available as well.

GIS tools for data analysis and display

To get a conceptual understanding of the considered groundwater system we must analyse the data and display various aspects. Some tasks that are efficiently supported by GIS are described below.

Contour maps Contour maps contain very comprehensive information about a 2D variable. While the lines facilitate easy reading of the displayed variable's value at any point, also the direction and magnitude of its gradient can be seen. Two similar variables (e.g., two groundwater tables) can easily be compared by displaying their contour lines simultaneously on one map.

Colour coded maps These maps can very efficiently convey information about a 2D phenomenon. With an adequate selection of the color palette, assignment of colors to values of the variable and the number of colors, interesting effects can be achieved. A gradually varying palette with many colors and a linear assignment to the values is well suited for displaying topography, geologic layers etc. The information is similar to contour maps.

A palette of two colours, e.g. green and signal red can intuitively signal areas where certain constraints, e.g. a toxic concentration, are violated.

Transparent overlay over e.g. scanned topographical maps could improve readability and acceptance considerably. However, this feature is not directly supported by the two GIS we used. Another limitation comes from the available hardcopy facilities.

On bit-mapped graphics displays the efficiency can further be improved by dynamic legends. By a mouse click in a field of the legend the area of the corresponding category is immediately highlighted. Optionally, the size and percentage of this category are also displayed. Further, the exact value of the variable at the position of the mouse cursor is continuously displayed. This feature was not available in the used GIS packages but was developed for the graphical postprocessor of the groundwater model in LGW.

Cross sections To understand the geological shape of an aquifer sections along arbitrary profiles are helpful. The even more effectfull fence diagrams (perspective view of several cross sections) are so far mainly a feature of surface modelling packages. Neither ARC/INFO nor GRASS offered suitable cross section tools. Thus, an interactive cross section tool was developed for the LGW postprocessor, which allows to draw an arbitrary line on screen and immediately pops up the corresponding cross section with ground surface, bottom and top of aquifer as well as selected groundwater heads.

Area statistics Area statistics are a standard feature of any GIS. In groundwater management, e.g., a table of the areas of significant drawdown could help to compare alternatives.

Map calculation Arithmetic and Boolean operations on maps are a well known characteristic of GIS. These operations require complex geometric intersection in vector based GIS like ARC/INFO, but are easily implemented in raster based GIS. For example, the transmissivity is easily computed as the product of saturated thickness of the aquifer and hydraulic conductivity. Four maps are required to compute the transmissivity map: hydraulic conductivity, bottom and top of aquifer and a map of groundwater heads.

Thiessen polygons (Voronoi diagram) Thiessen polygons can be a simple method of regionalization of point related data. This method is popular for regionalization of precipitation measurements. An estimate of available groundwater volume can easily be computed from a Thiessen map of groundwater heads, storage coefficient and saturated thickness.

Rule based combination of maps GRASS has an expert system module that can evaluate a set of rules with raster maps as objects and generate a new raster map with the conclusion. Also for GRASS, Buehler & Wright (1990) describe a Bayesian reasoning shell for land management that also could find applications in groundwater management.

GIS support for scenario generation and simulation

A finite difference 2D groundwater flow model is used in both, ARCGW and LGW to compute groundwater flow. This model is integrated into the DSS via pre- and postprocessors. The preprocessor is the "command board" of the numerical model. Its main features are interactive graphical design of the model grid, definition of boundaries using available maps in the GIS, and automatic interpolation of the aquifer parameters for the designed grid. In ARCGW, this required direct access to ARC/INFO coverages using the ARC/INFO subroutine libraries (Fürst *et al.*, 1987). In LGW, pre- and postprocessors access data in a hierarchical file system for better performance.

In ARCGW, the postprocessor simply converts the format of the model's output and passes selected results to the DSS database for further analysis and display. Thus, the results of the model also can be compared with other information in the database.

The LGW postprocessor is a specifically developed module for the interpretation and comparison of scenarios. The most important raster GIS functions were implemented for the finite difference grid. As data are read from binary files and are kept in memory for different views, a high speed of execution is achieved. Currently, colour coded maps with dynamic legends, contour maps, cross sections, vector maps and time series for unsteady scenarios are supported. For a given grid, arithmetic operations are possible between aquifer parameters and computed results.

For background maps, GRASS vector and point maps can be displayed. The postprocessor was implemented in C language, with an OSF/Motif graphical user interface and all graphics based on the X-Window System.

GIS support for the user interface

In ARCGW the complete user interface was developed with the Arc Macro Language AML (Fürst *et al.*, 1989). AML also allows the integration of system and user developed external programs. The main advantage of AML is the easy customization and extension of user interfaces even for the end user. While offering just the required subset of GIS and groundwater functions through the interface, access to the complete GIS is always possible for the experienced user.

In LGW, some of the major modules have their own OSF/Motif interface, others are implemented as UNIX command line tools and are accessed via the GRASS add on utility Xgen, a script based application generator to drive UNIX command lines via an OSF/Motif interface.

CONCLUSIONS

In the development and application of two DSS for groundwater management GIS provided a considerable part. If the software architecture of the GIS is sufficiently open, integration of GIS into DSS has advantages compared to using less powerful components and tools for the DSS.

The raster based GIS has the disadvantage that the minimum resolution of a raster map has to be decided when it is created. Thus, if for a given area analysis needs to be done in different scales (a regional and a local), we have either huge dimensions of the raster or information might get lost.

GIS operations are performed faster on a raster, which is important for an interactive tool.

The availability of GIS source code, or at least a linkable library for direct data access is essential for a reasonable performance of interactive programs. Interfaces using file export techniques often have bad performance and are only acceptable for tasks of lower priority.

Support of 3D data elements was not satisfactory in both, ARC/INFO and GRASS. Especially, features of digital terrain models like non-convex boundaries and break lines were missed.

A higher degree of modularity would avoid significant overhead, when only part of the GIS needs to be integrated.

REFERENCES

Aiken, R. (editor), (1992) Education and Society. *Information Processing 92*, Volume II. Elsevier Science Publishers B.V. (North-Holland).
Bachmat, Y., Andrews, B., Holtz, D. & Sebastian, S. (1985) Groundwater Management: The Use of Numerical Models. AGU, *Water Resources Monographs*. Vol. 5, 2nd Edn.
Bonczek, R.H., Holsapple, C.W.& Whinston, A.B. (1981) *Foundations of Decision Support Systems*. Academic Press, New York.
Buehler, K.A. & Wright, J.R. (1990) B-infer: A Bayesian reasoning shell for land management - Part I: Structure and function. To appear in: *J. Comput. in Civil Engng*, ASCE.
Fedra, K. & Reitsma, R. (1990) Decision support and geographical information systems. In: *Geographical Information Systems for Urban and Regional Planning (ed. by H.J. Scholten & J.C.H. Stillwell)*. Kluwer Academic Publishers, 177-188.
Fürst, J., Haider, S. & Nachtnebel, H.P. (1987) ARC/INFO as a tool in groundwater systems analysis. *ESRI European User Conference*, Kranzberg, Germany, 7/1-12.
Fürst, J., Haider, S. & Nachtnebel, H.P. (1989) Improved groundwater modeling by coupling a finite difference groundwater model with a geographic information system. *Workshop on Computer-Based Information Systems in Environmental Protection and Public Administration, Envirotech Vienna 1989*, 123-132.
Loucks, D.P., Kindler, J. & Fedra, K. (1985) Interactive water resources modeling and model use: An overview. *Wat. Resour. Res.* **21**(2), 95-102.
Pillmann, W. & Jaeschke, A. (eds) (1990) *Proceedings of the 5th Symposium on Informatics for Environmental Protection*, Vienna 1990, Informatik-Fachberichte 256. Springer-Verlag.
Sol, H.H. (editor) (1983) Processes and tools for decision support. *Proc. of the joint IFIP WG 8.3 / IIASA Working Conference on Processes and Tools for Decision Support*. North Holland Publ.

An expert system approach of integrating hydrological database, models and GIS: application of the RAISON System

D.C.L. LAM
National Water Research Institute, Burlington, Ontario, Canada L7R 4A6

D.A. SWAYNE
Department of Computing and Information Science., University of Guelph, Guelph, Ontario, Canada N1G 2W1

Abstract Integration of database, models and GIS requires an intelligent system. An expert system approach is proposed in which logical inference and integrative design techniques are used to handle both fundamental operations and special cases and to provide open architecture programming capabilities among database, model, GIS and expert systems. Two examples are given to illustrate the basic and advanced applications.

INTRODUCTION

Geographical Information Systems (GIS) have gained much attention in many areas that require spatial description and manipulation of information. For hydrological sciences, GIS applications are needed for map information, regional planning, geological and geographical overlay analyses. These applications are also timely in the sense that modern computers, especially microcomputers and workstations, are now capable of running these GIS's and produce high quality coloured outputs that were considered impossible a decade ago.

However, not all research problems can be solved by GIS's. The expectations of what GIS can do are often too high. As it stands, GIS is only a graphical tool, not a fully developed intelligent system. Many hydrological problems require the incorporation of other techniques. For example, the integration of temporal and spatial data require not only a GIS but also methodologies for trend analysis and computer animation. Raster-to-vector conversion claimed by many GIS's as a standard feature is problematic and requires a high level of manual guidance, if not artificial intelligence. Many GIS's can be linked to other software such as databases, spreadsheet and graphics, on an ad hoc basis rather than on a generic basis. Few offer a seamless interface and arbitrary exit and re-entry points for large scale repetitive processing. Except for some satellite or radar data, almost every set of data used in GIS is derived through some form of interpolation. As soon as GIS users begin to treat interpolated data as actual data, the question of spatial uncertainty and error is conveniently forgotten. Only a few GIS's offer functionalities or programming capabilities to handle spatial fuzziness intelligently (Wang *et al.*, 1990).

From the GIS user's and developer's point of view, it is natural to expect that future GIS's will incorporate intelligent interfaces and technologies. One may build such an intelligent system as a subsystem of a GIS, with the GIS remaining as the focus. However, one may also develop the intelligent system as the focus by itself, so that GIS becomes one of the many modules that make up the system. The other modules may include database, knowledge base, text documents, language parser, trend analysis, models, expert systems and computer graphics. It sounds rather academic, if not trivial, that one should differentiate at the outset the GIS from the intelligent system as the focus. The implication for the system development could lead to very different far reaching design problems and user interfaces. This paper describes an example of the intelligent expert system approach (as opposed to the GIS approach), emphasizing its design philosophy and domain of application for hydrological research.

THE EXPERT SYSTEM APPROACH: THE RAISON SYSTEM AS AN EXAMPLE

The expert system or knowledge-based system approach is concerned with the storage, retrieval and manipulation of knowledge. Here, knowledge (in a broad sense) is meant to include GIS maps, numerical data values, expert rules, regulatory guidelines, model choices, parameter selection, logical algorithms, statistical methods and other forms of information. With the advent of computers (microcomputers and workstations), information technologies now exist that can store, retrieve and manipulate various forms of the knowledge under one intelligent system (e.g. RAISON, see Lam & Swayne, 1991). Figure 1 shows the design of such a system, with the intelligent central command serving the various knowledge modules. The intelligent interface consists of the logical unit, i.e. the so-called inference engine (Lam & Swayne, 1991), and the linkage menus that enable one form of knowledge (e.g. GIS) to interact with another (e.g. model parameter selection).

Specifically, we have developed a knowledge-based system RAISON (Regional Analysis by Intelligent Systems ON microcomputers) that follows the design features given in Fig.1. It is operational on PC computers (386's or 486's) and is currently being restructured for workstations. The RAISON System consists of a hierarchy network of program packages (Fig. 2) designed for different applications: basic, special, expert systems and programming language. In the basic version, there are four components (GIS, database, spreadsheet and analysis) which were original programmes written in the "C" and "C++" languages specifically for RAISON to ensure seamless linkages among all components. Data can be retrieved according to conditions set by the user (e.g. for summer only or when pH < 6) from the database to the spreadsheet. The spreadsheet offers standard calculations as well as additional functions for environmental applications including graphics. Statistical analysis

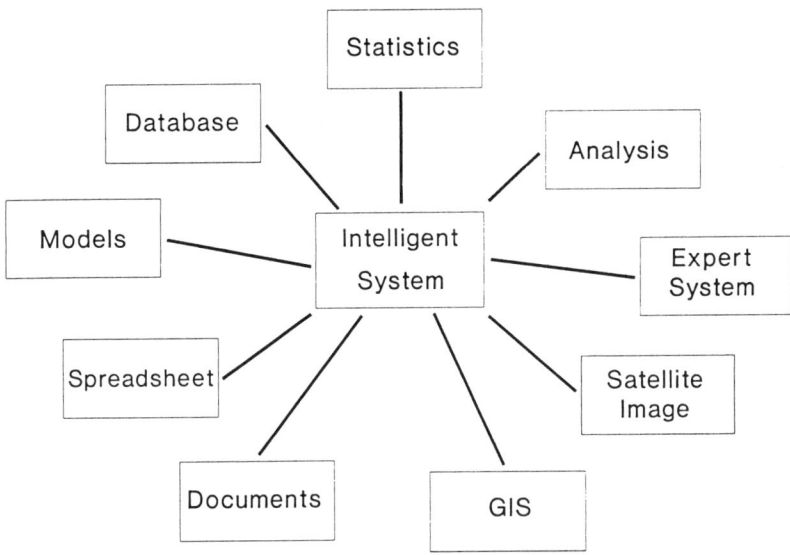

Fig. 1 The expert system approach to the design of an intelligent system for integrating GIS, database, models, expert systems, documents, statistics and other components.

(e.g. normal and lognormal distributions, or time series) can be assessed from the spreadsheet and the graphics (e.g. box plot or time plot) can be displayed on GIS maps. Search for keywords or phrases from a site document can be used to highlight those sites found with such keywords on the GIS map. These and other functionalities are fundamental building blocks for the intelligent interfaces between GIS and other components. While these components are all internal to and written specially for RAISON, each component can import and export files with other GIS and database software.

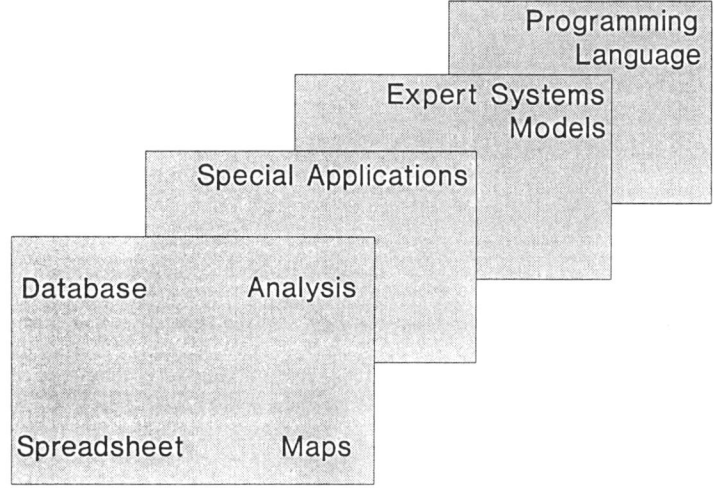

Fig. 2 The hierarchy of RAISON software: basic, special, expert system and programming language versions.

The next level of RAISON software pertains to special applications (Fig. 2). The design of the RAISON System calls for a generic system. However, from time to time, special applications require specific functions to be developed. For hydrologists and limnologists, a contouring package has been designed to handle both horizontal and vertical cross-sections. For example, by selecting stations on land or in the lake along a chained line, the contour of a given variable (temperature) can be plotted for the vertical section along the line defined by these selected points. Horizontal contours can be constructed with a number of interpolation methods (e.g. the kriging method). Special constraints are allowed for interpolation over general or preselected areas (e.g., drainage basins). The interpolation can be based on a coarse grid which tends to generate homogeneous contours (e.g. for air pollution maps), or can be based directly on the pixel cells on the screen to capture heterogeneous details (e.g. for soil maps). These special functions are necessary for many hydrological and environmental applications and are designed to bridge the gaps between what ordinary GIS can offer and what hydrologists actually require.

The third level of RAISON programmes consists of the logical inference or expert system component (Fig.2). A decade ago, most expert systems were programmed in PROLOG or LISP and limited to a few applications (e.g. medical diagnostics). Recently, efficient methods have been developed and can be written in "C" or "C++" that can be directly linked to database, GIS and other systems. The expert system component in RAISON is written in "C++" and is based on the Induction Dichotomy "ID3" decision tree algorithm (Quinlan, 1986). It offers both forward chaining and the backward chaining sequences of activating the rules. For forward chaining, one proceeds from hypotheses to reach conclusions; for backward chaining, one starts from assumed conclusions to see if hypotheses can be established. The rules, typically tens or hundreds of them, are entered via the RAISON spreadsheet. The ID3 algorithm converts the rules into a near-optimal tree structure (Quinlan, 1986) which enables a query structure to be set up so that, instead of following all the tens and hundreds of rules, a few queries will quickly lead to a conclusion. During the firing of these rules, inputs from other RAISON components (e.g. a slope estimated from a GIS map) can be received or the results can cause action in other components (e.g. model outputs on a GIS map). In this way, the expert system becomes the intelligent system (Fig. 1) that manipulates information using a combination of logical inference and scientific knowledge.

The most advanced version in the hierarchy of RAISON software is the one with the RAISON Programming Language (RPL). The RPL is an open architecture programming language that was developed specifically for RAISON to customize an application, i.e. to assess directly all components in the first three levels (Fig. 2) and to create new interfaces or functions required by the application. The RPL is a simple language which is quite similar to the BASIC Language (RPL is also an interpreter). It can be used to write simple and complicated functions (e.g., for the database or the GIS) or to program mathematical models as required. Alternately, existing models written in other languages (e.g. FORTRAN) can be run outside of RAISON and the model results stored in a database assessable by RAISON for subsequent presentation using the GIS and graphics capabilities. One can develop an expert system application for selecting and running the appropriate model for a given set of data and condition using RPL (Lam *et al.*, 1989). Indeed, a number of innovative combination of database-GIS-model-expert system applications can be made with the advanced version. For example, in entering the expert rules, instead of inputting the value or description of the attributes (e.g. high, low or medium), the rule entry could be an RPL program name. When the rule containing this program name is fired, the program will be run to assess a GIS map for information or to activate an interface prompting for user input. Through this approach, the number of ways of integrating database, GIS and models is almost limitless (e.g. to run a model within an expert system or to call another expert system from one expert system, etc.).

EXAMPLES OF APPLICATIONS

The design concept of a hierarchy of intelligent software, with generic fundamental building blocks as well as special application tools, is central to developing an intelligent system using GIS, modelling and expert system technologies. The open architecture of the RAISON Programming Language offers the versatility needed to adapt from one area of application to another area. RAISON has been used in a number of environmental and hydrological applications including acid rain (Lam *et al.*, 1989), mine effluent (Bowen *et al.*, 1991), drinking water and health (Swayne *et al.*, 1992), hydrological modelling (Bobba *et al.*, 1992) and water quality modelling (Booty *et al.*, 1992). The following are two recent examples.

Query system for drinking water wells

In collaboration with the Ontario Ministry of Environment, we have developed a query information system for drinking water wells, using the RAISON System. In the prototype development, information from the Essex County in Southwestern Ontario, Canada, is featured. This system has been implemented for PC microcomputers and is currently tested by the Ontario Ministry of Environment for application in other counties. Figure 3 shows some of the results from the prototype currently being developed on the IRIS Workstation of Silicon Graphics, Inc. As in the microcomputer version, the system consists of GIS, spreadsheet, database, expert system, models and graphics assessable through special icons (Fig. 3). The query system was developed to answer enquiries (mainly from telephone) on information pertaining to existing drinking water wells. To guide the information retrieval, the user can select

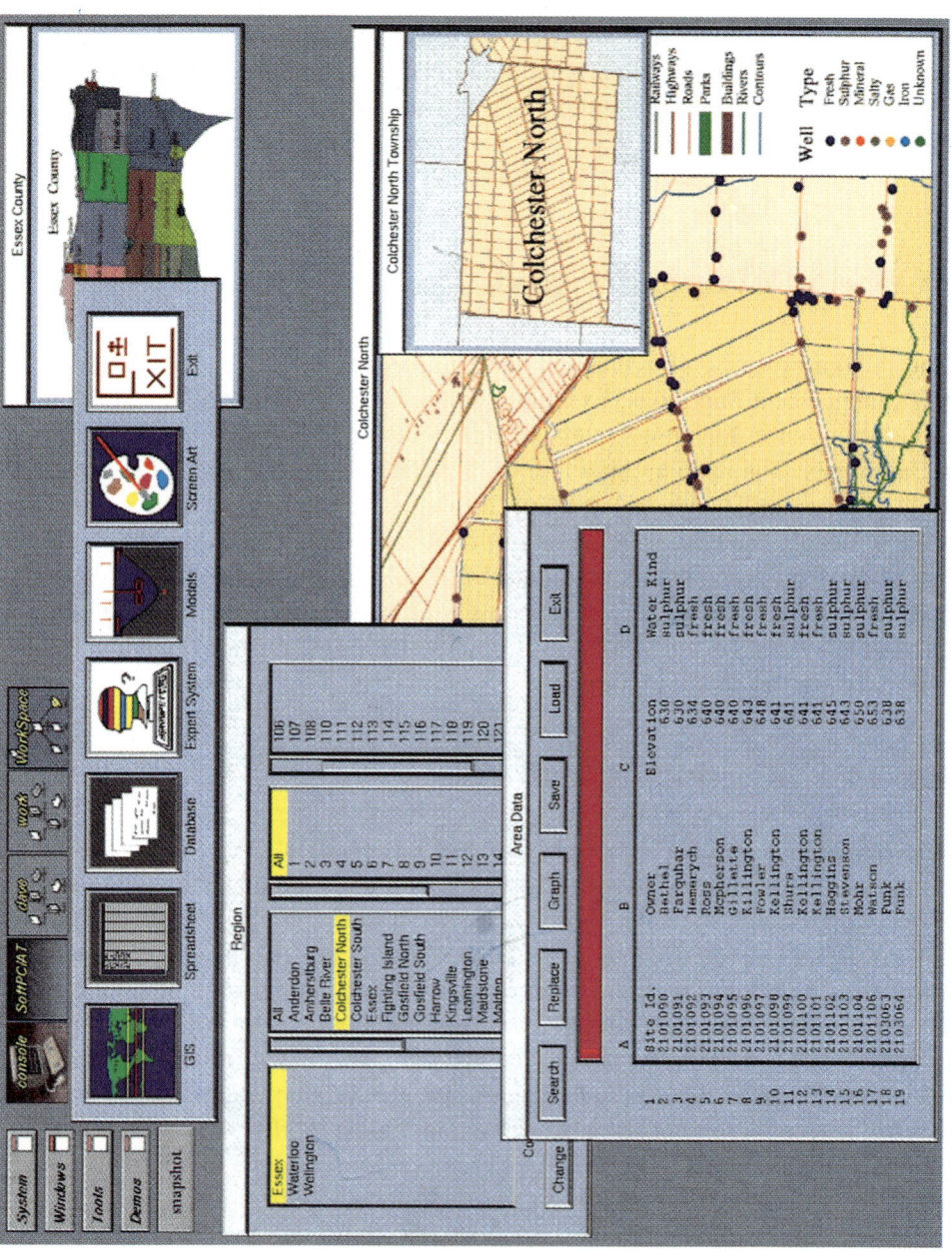

Fig. 3 The Essex County drinking water well query system, clockwise from top: icons for assessing systems, County map showing townships, Colchester North Township with concessions and lots (insert) and landmarks (zoom), area data and region menu.

from a region menu (Fig. 3) the County (e.g. Essex), the Township (e.g. Colchester North), the Concession Number and the Lot Number, and so on, in telescoping order of regional divisions and subdivisions. Alternately, if the enquirer is not certain about the names and numbers of these regional divisions, guidance by GIS maps is provided to the user. By zooming into the correct county and township, landmarks such as railway roads, buildings, rivers, etc. (Fig. 3) can be shown on the maps to help locate the well in question. When the well is located, either through the region menu or the GIS maps, information (e.g. elevation, water kind) concerning the well can be retrieved from the area data (Fig. 3) stored in the database. Sometimes, it is necessary to retrieve the whole file of area data (Fig. 3) for a concession or a township to look for or to compare the information of adjacent wells. This type of operation can be easily implemented with the basic version (Fig. 2) of the RAISON System.

The information stored in the query system can be used to generate regional information. For example, the elevation data for all the wells in Essex County can be retrieved from the database, analyzed for statistical distributions and then plotted on the GIS map according to a colouring scheme (Fig. 4). Contouring schemes for the elevation with continuity constraints at the land-water boundary along the lakefront have been experimented in the special version of the RAISON System. An example of this type of horizontal contours for well elevation is shown in Fig. 4. The query system can be also used to retrieve data within any polygon drawn by the user with a mouse on the map. For example, the data retrieved in the spreadsheet shown in Fig.4 pertain to the polygon AB drawn on the map for Colchester North Township in Fig. 4. Based on the soil type and sampling depth data retrieved, RAISON was able to produce the vertical cross section view (Fig. 4) of the soil type profile along AB.

Landfill sites advisory system

The selection of landfill sites is subject to environmental, structural, social and other constraints. Rules for following guidelines and satisfying these constraints have been obtained from experts and entered into the expert system via the RAISON spreadsheet (Lam & Swayne, 1992). Part of the rule base can be seen in Fig. 5 in the decision tree structure given by the ID3 algorithm. As an example, the log file (session.log) shown in Fig. 5 indicated that in the forward chaining mode, the user had chose the "*" , i.e. "don't care", description for contamination. As shown in the decision tree, this choice led to a second query which was on the rate of slope. The user could then answer the query by selecting the appropriate condition (e.g. rate-of-slope > 15) if it was known. Then the next query on the length of slope would be prompted. In the example, the answer for the slope length was chosen as greater than or equal to 500, for which case a conclusion was reached. The recommendation or result was to use dikes and berms as highlighted in the decision tree and shown in the session log (Fig. 5).

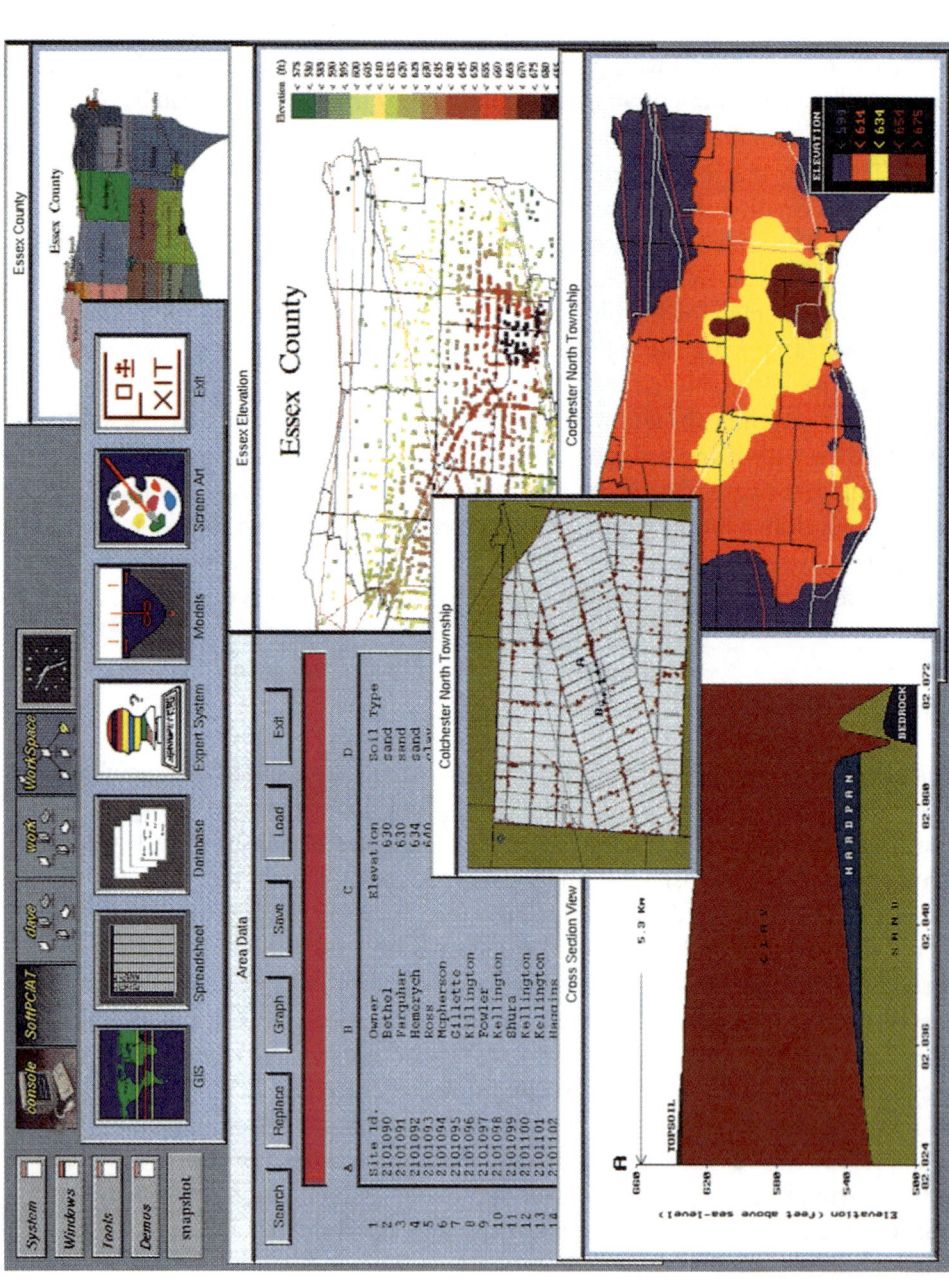

Fig. 4 Well Query System (continued), clockwise from top: system icons, county map, individual well colourized by elevation classes, elevation contours, vertical section profile along AB (insert) and data associated with AB.

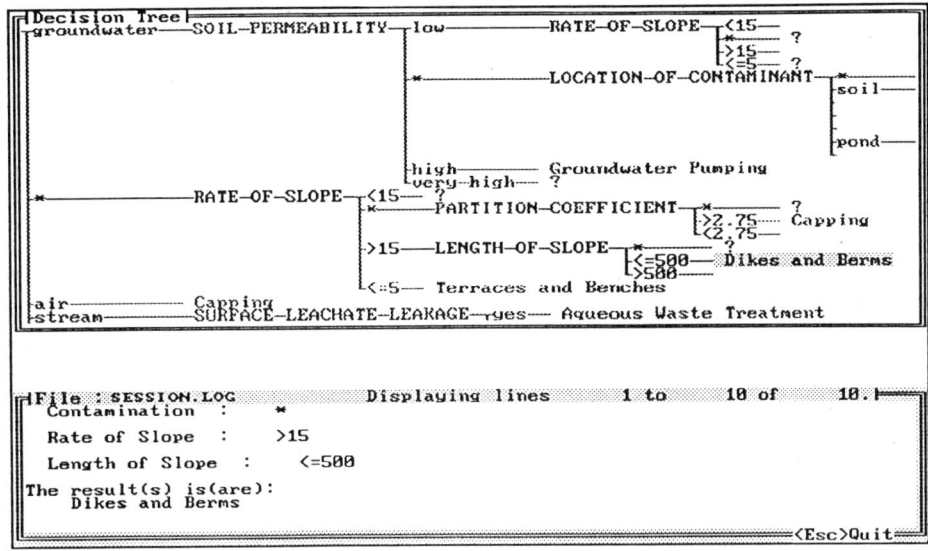

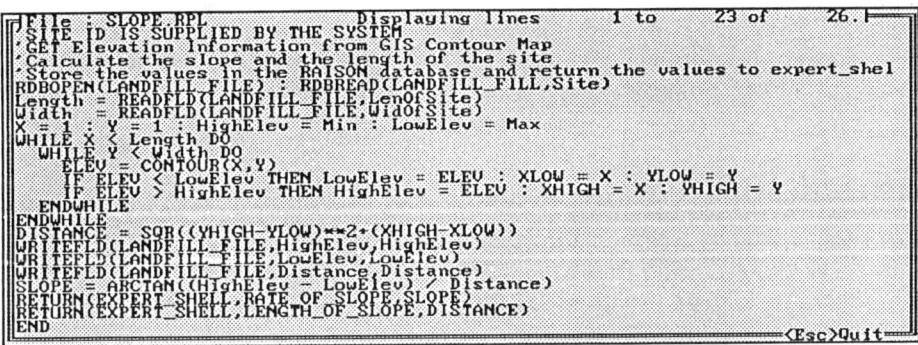

Fig. 5 Landfill site advisory system, from top: decision tree (partial); query session log file showing queries and conclusion; and an example of an RPL (to calculate rate and length of slope from GIS data)

Alternately, if the slope information was not readily available, a calculation procedure or model may be needed. To do so, the procedure was written in RPL and its name, e.g. "slope.rpl", was input as part of the rule base. When this portion of rule was activated, the RPL program would be run. Figure 5 shows the slope.rpl program that calculated the rate of the slope as well as the length of the slope of the chosen site. Typically this was carried out as a scenario game playing exercise in which the user was asked to place the proposed dump site on a GIS map (Fig. 6). When this rule was fired, the slope.rpl program was run and the elevation contour map information for the chosen site would be used to calculate the rate of slope and length of slope. The results were then checked against the appropriate decision tree path (e.g. rate-of-slope > 15) and the next query would be prompted, and so on. This intelligent automation of that part of the rules that require GIS inputs or calculations is an example of the types of integration among GIS, models and expert system that the advanced version of RAISON can provide.

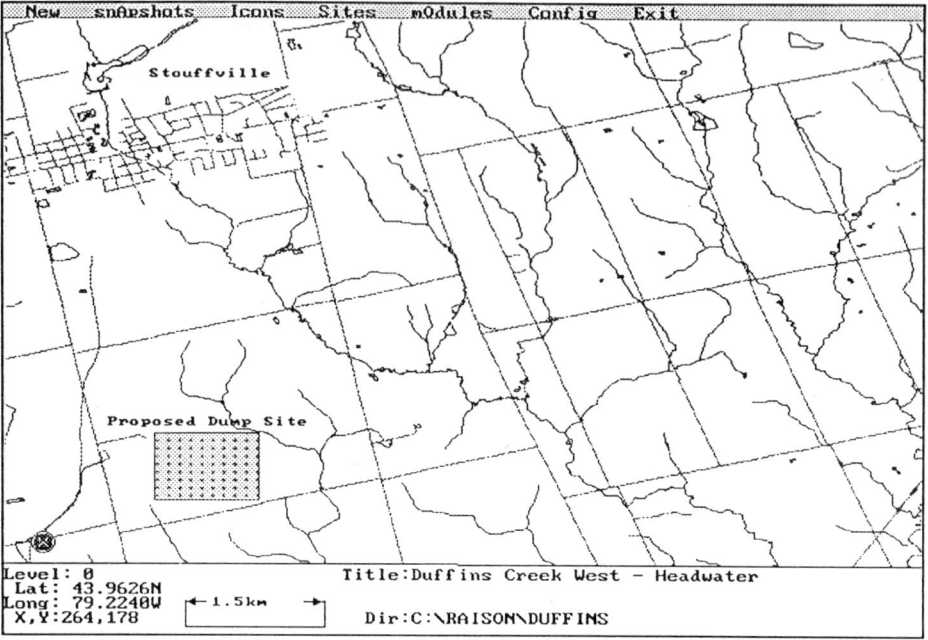

Fig. 6 Proposed dump site placed on a RAISON map.

FUTURE WORK

There are many ways in which database, models, GIS and expert system can be combined to achieve practical solutions for hydrological and environmental problems. For example, by delegating subtasks to simple models or procedures, and by allowing for nested expert systems (i.e. expert system that containing rules that calls for other expert systems), one can design efficient systems that can generate new information as a result of combining both logical inference and conventional scientific methods. Future work on uncertainty of data and the fuzziness representation in GIS and other database may need this type of combination of heuristic and deterministic methodologies.

Acknowledgements This study was supported in part by the Artificial Intelligent Research and Development Fund administered by Industry, Science and Technology Canada. The authors are indebted to J. Kerby, R. Ruddock, I. Wong, J. Stoery and A. Storey for assisting system development. The authors are also thankful to S. Singer and T. Chang, M.O.E., for advice and support.

REFERENCES

Bobba, A.G., Lam, D.C.L., Kay, D. & Ullah, W. (1992) Interfacing a hydrological model with the RAISON Expert System. *Wat. Resour. Management* **6**, 25-34.

Booty, W.G., Lam, D.C.L., Bobba, A.G. Wong, I., & Bowen, G.S. (1992) An expert system for water quality modelling. *J. Environ. Monitor. & Assessment* **23**, 1-18

Bowen, G.S.,Lam, D.C.L., Wong, I., Swayne, D.A. & Storey, J. (1991) A microcomputer-based system for data management and impact assessment of mine effluent. *Proc. Int. conf. Abatement of acidic mine drainage*, Montreal, Sept. 16-18, 1991. 271-290.

Lam, D.C.L., Swayne, D.A., Storey, J. & Fraser, A.S. (1989) Watershed acidification models using the knowledge-based systems approach. *Ecol. Modelling* **47**, 131-152.

Lam, D.C.L. & Swayne, D.A. (1991) Integrating database, spreadsheet, graphics, GIS, statistics, simulation models and expert systems: experiences with the RAISON system on microcomputers. *NATO ASI Series*, Vol. G 26. Springer-Verlag Publ., Heidelberg, 429-459.

Lam, D.C.L. & Swayne, D.A. (1992) Some experiences in applying the RAISON Expert System to environmental problems. In: *Advances in Artificial Intelligence - Theory & Application*, (ed. by J.W. Brahan & G.E. Lasker), Int. Inst. Advanced Studies in Systems Research & Cybernetics, Baden-Baden, Germany, 59-64.

Quinlan, J.R. (1986) Induction of decision trees. *Machine Learning* **1**, 81-106.

Swayne, D.A., Wang, C.W., Storey, J., Lam, D.C.L. & Kerby, J.P. (1992) An expert system for assessing safety and security of heterogeneous public water resources. *J. Environ. Monitor. & Assessment* **23**, 51-70.

Wang, F., Hall, G.B. & Subryono (1990) Fuzzy information representation and processing in conventional GIS software: database design and application. *Int. J. GIS* **4**, 261-283.

Application of ILWIS to decision support in watershed management: case study of the Komering river basin, Indonesia

A.M.J. MEIJERINK, C.M. MANNAERTS, H.A. DE BROUWER & C.R. VALENZUELA

International Institute of Aerospace Surveys and Earth Sciences (ITC),350 Boulevard 1945, P.O.Box 6, 7500 AA Enschede, The Netherlands

Abstract Aerospace and terrain data were coupled to a model and decision rule base in a GIS to assess the extent of environmental degradation in a large tropical catchment. The GIS was used for various modelling approaches and applied here to erosion - sedimentation hazard assessment in order to determine the present and near future land degradation and specifically the sediment source areas. Use of complex mapping units in GIS are discussed. Together with a ecological land suitability modelling, trend scenarios were worked out for increased coffee production in the headwaters areas and for alluvial downstream damage. The various management scenarios permitted to draw the attention of policy- and decision makers to the interrelationships between economic productivity and environmental degradation in this large tropical catchment.

INTRODUCTION

Watershed management involves a problem diagnosis in a catchment with emphasis on erosion and sedimentation, and the planning of remedial action to alleviate problem issues. Land use planning forms a key activity if man plays an important role, as is the case in the large Komering catchment in South Sumatra. The central role of GIS in the study is illustrated in Fig.1. In the confrontation between conservation and -mainly agricultural- development, scenarios are run within the GIS, which can support decision making for watershed management. Such scenarios require bio-physical as well as socio-economic data bases. Data requirements for conventional GIS processing generally pose a major problem in data scarce areas such as large tropical catchments. Therefore specific approaches had to be developed. The transformation of the available data into information makes use of a "decision rule base", as is shown in Fig.1. This can be thought of as appropriate combination rules but also of models and the conditions for their application and data quality requirements. Scenarios created with the GIS have to be evaluated, both in terms of their accuracy and outside criteria, which may be of financial or political nature. Several future scenarios can be simulated i.e., continued trend, reduced, changing or forced trend, all focusing around the question what irreversible natural resource degradation is incurred when remedial actions or policies are implemented or not. The continued trend scenario for example deals with the spatial forecasting of environmental degradation when present activities and conditions are maintained and no remedial action is taken. This pertains to the downstream damages in the

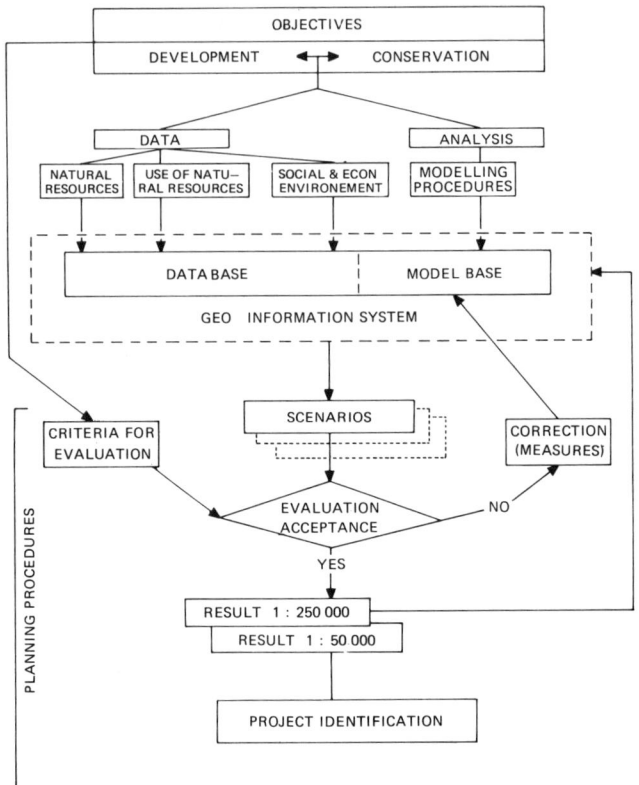

Fig. 1 The role of geographic information systems in planning.

extensive, rice production decline in the alluvial zones, as well as to the land degradation upstream, where certain parts have turned into fully unproductive lands. Trend scenarios, made with the GIS, more easily attract the attention of the decision makers at various levels. Remedial action scenarios (reducing, forced trends) can be worked out using various policy or management options. A number of aspects however remain evident; (i) the multidisciplinary nature of the problem requires the sharing of the data bases and the joint development of the rule base. This requires a certain synergism among the various data bases and spatial and physical modelling approaches; (ii) scenarios have to be created through interaction between the researchers and the authorities.

PROBLEM DIAGNOSIS

In the alluvial downstream part of the Komering river basin, a change in the sediment regime of the river over the last decades has caused much damage. The river bed raised by deposition of sands which has led to impediment of the drainage from the backswamps to the river (Meijerink *et al.*, 1988). Rice is cultivated in the backswamps and planting occurs in stages as the floodwaters recede, starting at the lower part of the natural levees and ending at the lower parts of the backswamps. Infrared aerial photography depicts very clearly the stages of rice planting and growth and can thus be used for an accurate delineation of the relative heights. From this information, a fairly accurate relative DTM (digital elevation or terrain model) can be obtained, which can be transformed to a DTM with georeferenced control points from height data from the contour map. Because of the impeded drainage, the lowest backswamps have turned into permanent swamps, which are extending in size as could be mapped from comparison of old photography and more recent SPOT satellite data. Apart from the damage to thousands of hectares of rice cultivation, the widening and shallowing up of the river has caused extensive river bank erosion with much damage to infrastructures. By using the relative DTM compiled in the GIS as well as the interpretations from sequential imagery, a prognosis of the present and near future damage could be made. Single theme maps of the damages, with estimated costs, presented to the decision makers, largely improved their willingness to seriously consider the problems in the upper catchment.

The large upper catchment (about 4000 km^2) shows large spatial variability in rainfall, topography, geology, geomorphology, soils, and vegetation. An important cash crop is coffee. From old aerial photography and three multispectral satellite images of different years (Landsat 1, SPOT and Landsat 4) the severe decrease of the primary rainforest could be mapped. Much effort was spend in digital image classification techniques, but the best results were obtained by manually digitizing and interpreting a multitemporal colour coded image. The reasons for the failure of automated classifications were the spectral intensity effects caused by the relief and the variations in composition and density of the rainforest. It turns out that the forest dwindled from some 60% in pre-war times to less than 15% in more recent times. Concerning the upstream erosion and downstream sedimentation hazard, a first problem was to locate the source areas of the increased bedload of the river. In certain parts of the catchment active landsliding takes place, but not all landslides cause an increase in the bedload. Soil degradation by sheet and rill erosion had to be assessed because of its contribution to the sediment yield affecting a planned reservoir as well as the fact that these processes in some parts have led to unproductive lands. For the planning of remedial measures, relocation of the farmers to areas in the catchment less susceptible to soil degradation and local intensification of the agriculture had to be considered, requiring land evaluation. We however concentrate here on the hydrology, in particular on the interrelationships between erosion and sedimentation processes.

DATA BASE DEVELOPMENT FOR COMPLEX LARGE TROPICAL WATERSHEDS

The knowledge and methodology of capturing complex field data by stereo interpretation of aerial photographs, developed during the fifties and sixties by either geologists (Ray, 1960; Miller, 1961), soil scientists (Goossen, 1967) and geomorphologists (Verstappen, 1977) seem to have escaped the attention of many hydrologists. They made use of the natural associations of landscape parameters in defining units, which have a complex attribute list. It is not difficult to prepare a data structure for such units to be used in GIS evaluations (Meijerink, 1988). One may view these efforts as precursors of object oriented approaches. Complex units can be used in the simulation modelling; either by extracting single attribute values if so desired, or by addressing the properties of the complex units in relational statements. A second important observation is that physical degradation processes can be attributed to the complex terrain units (e.g., types of erosion, intensity and frequency). In conventional GIS approaches, one is supposed to arrive at such information only after many combinations. While working with DTM's and other earth science related data planes, the natural associations between geology, geomorphology soils, and even vegetation are re-discovered. However, few attempted to work with complex terrain units, which are perhaps a to high level of fuzziness or abstraction. In the case of the Komering catchment an engineering hydrological approach, using only single attribute values in numerical models, meets with great difficulties and the same is true for the regionalized variable approach. In the first place nearly all of the required conventional input is lacking, and the effort of obtaining them through surveys would be fully out of proportion. Hydrometric data for model calibration would have to be incorporated. Worse is that the state of the art modelling, for example in erosion modelling, is inadequate in this region, as will be explained below.

The data bases, compiled for the upper Komering relied heavily on aerial photo-interpretation and satellite image processing, which was fully incorporated in ILWIS. They consisted of: (i) a metric topographic data base, which contained all available topographic data and constituted the georeference for the other databases; (ii) a terrain data base consisting of non-hierarchical complex terrain mapping units (TMU), based on geology and genetic geomorphological criteria, but containing statistical parametric data and the soil associations. Sub-units are described in the database, but their location within the main units is unknown, because of the scale of survey; (iii) a land cover data base consisting of units of different complexity (structure, phenology, etc.) with a pointer to farming systems; (iv) a socio-economic data base, including farming system data.

THE MODEL BASE

The lack of specific data for the upper catchments in many tropical countries places much restriction on the choice of models, their calibration and validation. Moreover, at the level where funding is decided, little priority is usually given to some in-depth research proposals, as it is assumed that the "state of the art" technology is adequate for supporting the overall decisions concerning the management of the watersheds, which are the responsibilities of line ministries and not of research organizations. Thus, the scientist has little alternative but to adjust to the situation. A problem arises when the practical application of models, even the simple ones, do not produce reliable results.

A comparison was made between the results of two erosion prediction models, i.e., the USLE model and the procedure described by Morgan *et al.* (1982), which has also been tested in Malaysia. The results are shown in Fig. 2 and need little comment. The ULSE model was applied using regionally tested rainfall erosivity factors, based on published research in Indonesia (e.g. Bols, 1979), as well as soil erodibility- and cover factors derived from runoff plots (Coster, 1938) and erosion plot data (Kurnia, 1986; Gintings, 1981). The results of the model of Morgan *et al.* (1982) were field tested against detailed field descriptions of erosion, which included rills, root exposures, pedestals, interviews with farmers and so on. Results indicate that this model could be conditionally used to replace the time consuming field observations. In the Komering watershed, sediment gauging was done on two catchments with a size of a few hundred square kilometres, draining mainly the sandy Ranau tuffs and andesitic volcanics. The resulting sedigraphs were very difficult to interpret and could not be used for estimation of long term sediment yields or extrapolation to other catchments. Obviously, much more in-depth study is required. The downstream problems are essentially caused by bed load derived from fast geological erosion (landslides of various types, gullying) and not by sheet and rill erosion which produce predominantly suspended load. The effects of fast geological erosion could not be modelled numerically and therefore an alternative method was used.

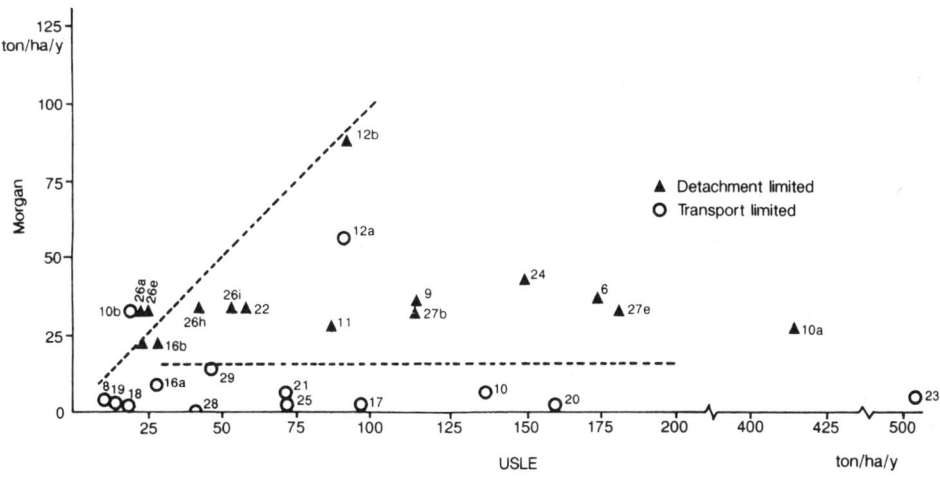

Fig. 2 Comparison between Morgan's soil erosion assessment procedure and the unadjusted USLE model estimates for the sample sites in the Komering study watershed.

DEVELOPMENT OF THE RULE BASE

Because of the difficulties sketched above, the following actions were undertaken:
(a) adjust the models with local field data from sample sites;
(b) specify conditions and selection criteria for application of erosion model type to certain parts of the catchment.

Still, as no satisfactory results could expected from the model base alone for the entire watershed (Mannaerts & Nurdin, 1988), a number of decision rules for the direct classification of relative erosion rates using the complex terrain and cover units were developed, based on empirical expert survey knowledge. The model and rule base were combined to complete the erosion scenario of the entire watershed. This hybrid approach is illustrated in Fig. 3. Where the procedure of Morgan (op.cit.) could not be applied, it was first decided to assess relative erosion intensity by using a regionally adapted USLE approach. Because of the high correlation between monthly rainfall and the rainfall erosivity index of the USLE, the annual rainfall could be used for the erosivity map by interpolation. An estimate of the soil erodibility was made in the relational data base by ranking the soil associations of the TMU's against

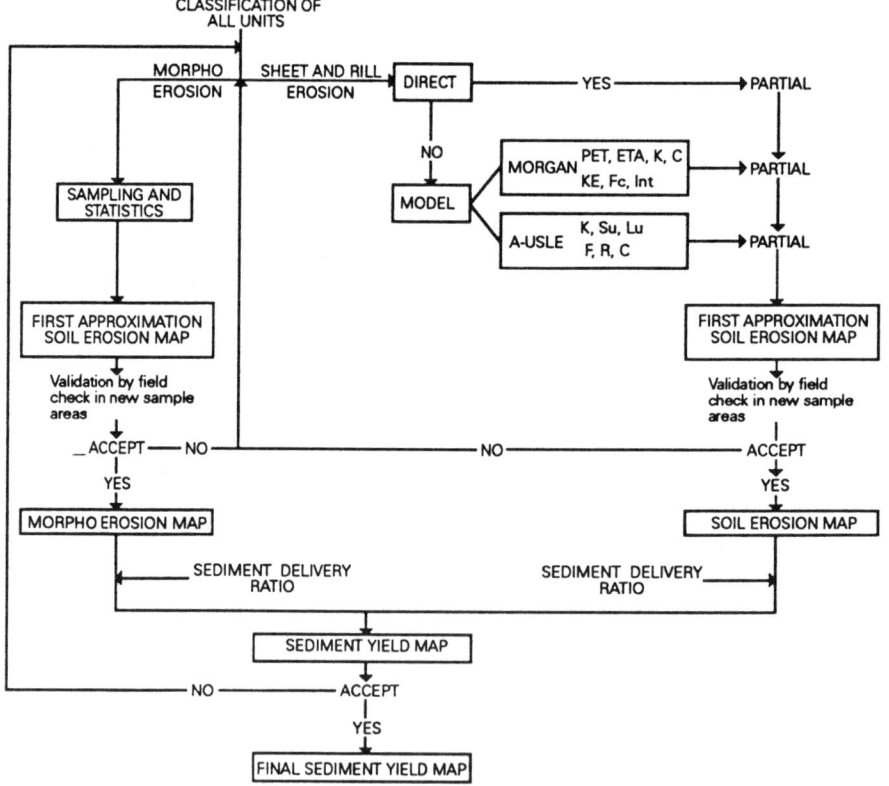

Fig. 3 Flow chart illustrating the hybrid soil erosion hazard assessment as applied to the Komering watershed.

Table 1 Slope length and gradient functions for erosion assessment with the modified USLE method.

Parent material	Slope functions	
	slope length	slope gradient
Andesitic tuff	$E=(L/22.1)^{0.30}$	$E=4.268*\sin S^{0.60}*\cos S$
Stratovolcano	$E=(L/22.1)^{0.31}$	$E=6.432*\sin S^{0.77}*\cos S$
Ranau tuff	$E=(L/22.1)^{0.30}$	$E=6.589*\sin S^{0.75}*\cos S$
Basalt	$E=(L/22.1)^{0.30}$	$E=8.801*\sin S^{0.90}*\cos S$
Tert.sediment	$E=(L/22.1)^{0.70}$	$E=6.432*\sin S^{0.77}*\cos S$

results of field tests. Considerable model improvement was obtained by using the slope functions shown in Table 1, which have been developed from numerous measurements of soil depositions behind trees or bushes in downslope direction. Two locally derived vegetation cover sub-factors were also developed and applied. Furthermore, a slope shape factor and a sediment delivery ratio, based on photo-interpretation and field checks, were added. The multiplicative algorithm was then applied to the terrain and cover mapping units and the interpolated rainfall erosivity map. This approach, although far from perfect, is an improvement on simple multiplication of individual pixel values, as is usually done with a GIS, ignoring thereby any catenary influence, complex terrain units and even Wischmeier's warning (Wischmeier & Smith, 1976) for the misuse of the USLE model. The adjustments were made by iterations against the set of erosion estimates of sample sites covering various geomorphological, soil and cover combinations in the watershed. While doing this, it turned out that the extreme relative classes "low erosion" (i.e. < 2.5 t ha^{-1} year^{-1}) or the class "very high erosion" (>30 t ha^{-1} year^{-1}) could be classified directly by means of a set of decision rules based on relational statements.

RELATIONAL MODELLING USING COMPLEX UNITS

With knowledge of the processes occurring in the watershed and the terrain and cover units in the relational data base, it is possible to formulate conditional rules for classifying hydrologic characteristics. The rule may pertain to a single data layer. For example, under secondary forest less than 10 years old, and old abandoned coffee plantations in the catchment, nowhere sheet and rill erosion was observed, because of the extreme dense vegetation. This is also substantiated by experimental evidence on erosion plots in Indonesia. Similarly, no erosion occurs on non-dissected fluvial terraces and certain tuff plateaux. More often, more data layers are addressed in one rule. For example: " there is no or little overland flow, and thus no sheet erosion, on the andesitic volcanoes above 800 m elevation (metric database), where andosols are found (terrain or tmu-database), which retain their high infiltration and resistance to splash and entrainment for several years after deforestation (cover database). If erosion occurs in these units, it is at the base of convex slopes (tmu database) where minor slumping may occur because of high moisture conditions. This form of mass wasting contributes only to the sediment yield if the drainage is incised (tmu database)." The areas which are responsible for the increased bedload were defined by rules concerning the degree of dissection and internal relief, in relation to the deforestation. Statistical data on landslide densities and some individual landslide modelling supported the formulation of the rules. The above knowledge can be implemented in a GIS in various ways, partly in the data base in the descriptions of the complex units, and if logical combinations are required, by one or more two-dimensional tables. The GIS used (ILWIS) is capable of handling relational operators in the spatial domain. The relational rules have the advantage that only rules with some certainty are implemented. From this follows that only parts of the watershed are classified, and in other parts other solutions have to be found, to be merged at a later stage.

PLANNING SCENARIOS

In order to present options to the decision makers, the GIS turned out to be a versatile tool, once the data bases and the model base became available. Numerous scenarios of the present and near future situation could be generated and adjusted. Continued trend scenarios for downstream damage were made, revealing large losses of fertile lowland irrigation areas if no remedial actions were taken urgently. A number of changing trend scenarios were worked out for increased coffee production in the upper part of the Komering basin. Therefore, the results of the crop suitability modelling, using the numerical LECS model of Wood & Dent. (1983), modified and adjusted for the study area by Elbersen *et al.* (1988) were merged with the hybrid degradational erosion modelling to form an ecological suitability map for coffee. This product was combined with land availability maps and accessibility maps to make a forecast where the largest pressure of the land occurs, also for the near future situations with increases in population. Other scenarios show the areas where, after creating an infrastructure, farmers from the deforested areas with the high sediment production and degradation hazard could be relocated. The subcatchments and sites where large sediment trapping structures are required could be spatially located, but not the estimate of the expected filling-up rates for which further research was required and justified in view of the financial expenditure. A provisional estimate of the sediment yield of the entire upper watershed was made, which effected the design of the planned dam and reservoir in the catchment.

CONCLUSIONS

The role of ILWIS for decision support in watershed management is analyzed. Lack of quantitative data in this tropical environment made it necessary to develop the terrain mapping unit concept and use these spatial entities as central core objects within a relational data structure. A number of critical areas were pinpointed by remote sensing of the Komering basin area using aerial photographs and multispectral scanner data of the Landsat and Spot satellites. The combination of hydrological, soils, vegetation and small scale agro-economical field surveys led to the formulation of a set of decision rules, relating combinations of attributes of both the terrain unit and vegetation cover databases to the intensity of water and soil resource degradation phenomena. This rule base was in the first instance used to verify conventional erosion modelling techniques applied to the catchment. Secondly, the rules were used to map geological and morpho-erosion processes, the latter could not be assessed by standard prediction methods. The diagnosis of the main problem issues was then done by linking the decision rule base to the relational modelling capabilities of ILWIS. The more object-oriented terrain unit approach made cartographic modelling, respectively in the spatial (e.g., using remotely sensed and digital terrain modelling outputs) and in the tabular attribute database environment possible. A scenario-driven analysis framework was furthermore implemented. This permitted the effects, future trends and the spatial extent of the present or proposed land and water management options in the affected areas of the Komering basin to be evaluated more precisely. It is shown that GIS, when reinforced by remotely sensed data, a relational data structure and field surveys, can be a powerful support tool for solving environmental problems in water resources management.

REFERENCES

Beasley, D.B. & Huggins, L.F. (1981) ANSWERS users manual. Agr.Eng.Dept. Purdue Univ. Indiana, USA.
Bols, P. (1979) Contribution to the study of surface runoff and erosion of Java. Ph.D.thesis, Univ. of Gent, Belgium.
Coster, C. (1938) Surface runoff and erosion on Java. Tectona **31** (9/10), 613-728.
Elbersen, G.W.W., Ismangun, D.S., Sutaatmadja,A. & Solihin, A.A. (1988) Small scale soil survey and automated land evaluation. ITC J. 1988-1, 51-59.
Gintings, S.A.N. (1981) Surface runoff and erosion under continuous coffee and forest in S.Lampung). Laporan no.399 (in Malay Indonesian), BPH (Forest Research Institute).
Goossen, D. (1967) Aerial photo interpretation in soil Survey. Fao, Rome, Soils Bulletin no.6.
Kurnia, U., Abdurrachman, A. & Sukmana, S. (1986) Comparison of two methods is assessing the soil erodibility factor of selected soils in Indonesia. Pemb. Pen. Tanah dan Pupuk. no.5, 33-36
Mannaerts, C.M. & Nurdin, E. (1988) Introduction of ILWIS/GIS elements in RTL-RLKT design planning in Indonesia. Lokakary Pemgembangan - Ilwis Workshop, 8-10 August 1988, Palembang, Indonesia, (Min. of Forestry, Indonesia).
Miller, V.C. (1961) Photogeology. McGraw Hill Book Co. N.Y.
Meijerink, A.M.J. (1988) Data acquisition and data capture through terrain mapping units. ITC J. 1988-1, 23-44.
Meijerink, A.M.J., Van Wijngaarden, W. & Amier Asrun, S. (1988) Downstream damage caused by upstream land degradation in the Komering Basin. ITC Journal 1988-1, 96-108.
Morgan, R.P.C., Hatch, T. & Wam Harun, S.W. (1982) A simple procedure for assessing soil erosion risk: a case study for Malaysia. Zeitschrift fuer Geomorphologie, Suppl. Band 44, 69-89.
Ray, R.G. (1960) aerial photographs in geologic interpretation and mapping. U.S. Geol. Surv. Prof. Paper no. 373.
Verstappen, H.Th.(1977) Remote Sensing in Geomorphology. Elsevier Sci. Publ. Co.

Wischmeier, W.H. & Smith, D.D. (1976) The use and misuse of the universal soil loss equation. J. *Soil and Wat. Cons.* **31** (1), 5-9.

Wood, S.R & Dent, F.J. (1983) LECS Methodology. Ministry of Agriculture, Government of Indonesia/FAO- AGOF/INS/78/006, Manual 5.

Hydrological modelling and GIS: the Sandia Environmental Decision Support System

STEVEN P. FRYSINGER
AT&T Bell Laboratories, Holmdel, New Jersey 07733, USA

RICHARD P. THOMAS
Science Applications International Corporation, c/o Sandia National Laboratories, Albuquerque, New Mexico 87122, USA

ALVA M. PARSONS
Sandia National Laboratories, Albuquerque, New Mexico 87122, USA

Abstract The Sandia Environmental Decision Support System (SEDSS) is a family of workstation applications being developed jointly by Sandia National Laboratories and AT&T Bell Laboratories. Sharing a common user interface philosophy and software architecture, the members of the SEDSS family are designed to facilitate human decision making with respect to the hydrological aspects of hazardous waste management. The Monitor Well Network Designer (MWND), the first member of the SEDSS family, is designed to satisfy the United States' Resource Conservation and Recovery Act (RCRA) groundwater monitoring regulations. RCRA requires that a hazardous waste management facility have a groundwater monitoring system consisting of at least one upgradient and three downgradient wells. These regulations are subjective, making it difficult for the owner/operator and the regulators to resolve whether a monitor well network satisfies the regulatory requirements. To minimize this subjectivity, Sandia developed a probabilistic approach which defines and quantifies the performance measures to evaluate monitor well network effectiveness (Parsons & Davis, 1991). MWND implements this strategy, integrating a highly interactive graphical user interface, various modelling and simulation processes, and the GRASS Geographical Information System (GIS) to yield an Environmental Decision Support System (EDSS) which is easily used by both regulators and operators to evaluate assumptions about a site's hydrological characteristics and negotiate the design of a monitor well network.

INTRODUCTION

Environmental protection is, if nothing else, an information-intensive pursuit. Environmental decision making requires consideration of large numbers of variables from such disparate disciplines as hydrology, chemistry, and economics. In support of the complex analysis tasks facing environmental decision makers, a class of computer workstations, Decision Support Systems (DSS), has been applied to environmental problems. The Sandia Environmental Decision Support System (SEDSS) is a family of workstations being developed jointly by Sandia National Laboratories and AT&T Bell Laboratories, and is representative of this technology. Sharing a common user interface philosophy and software architecture, the members of the SEDSS family are designed to facilitate human decision making with respect to the hydrological aspects of hazardous waste management. The first member of the SEDSS family, the Monitor Well Network Designer (MWND), helps both site managers and regulators determine the number and placement of monitor wells required around a hazardous waste site pursuant to the United States' Resource Conservation and Recovery Act (RCRA).

DISCUSSION

The information required to manage the environmental impacts of human activity is multidisciplinary, involving interactions between a great many disciplines, and it is usually uncertain. While this uncertainty is often ignored, the prudent decision maker will instead want to explore the implications of the uncertainty, in order to better understand the likely ramifications of various decision alternatives. Environmental information is also inherently spatial in nature. Beyond the trivial observation that the environment itself is a spatial entity, one must consider that most variables in a given environmental management decision can be considered a function of space, whether one is considering the physical transport of contaminants or the extent of health or economic risks associated with a remediation activity.

Information systems for environmental management

Traditional information systems have found wide use in environmental management. These range from Relational Data Base Management Systems (RDBMS) and spreadsheets through Computer-Aided Drawing (CAD) systems.
    More recently, Geographical Information Systems (GIS) have begun to play a significant role in environmental information management, offering generic spatial data analysis and management capabilities to the environmental manager knowledgable enough to apply them Environmental Decision Support Systems (EDSS) have recently sprung from this GIS foundation and are just beginning to evolve into mainstream management tools.

What is an environmental decision support system?

An Environmental Decision Support System is a computer-based system supporting human decision makers in the environmental management arena. Unlike GISs, which are generic kits of tools, EDSSs are focused on a particular class of problem, and are usually designed to support a particular category of decision maker.
    EDSSs often integrate monitoring data, providing a coherent view of these data and enhancing human comprehension of them. Through the use of

modelling and simulation, they support the analysis of candidate decision alternatives, and help the analyst to evaluate the uncertainty inherent in the problem at hand.

Since GISs provide a generic platform for analysis of spatial data, while EDSSs provide analytical procedures targeting specific problems (using both spatial and aspatial data), an EDSS may clearly be supported by an underlying GIS. The balance of this paper will be devoted to an example of an EDSS which is supported by the GRASS GIS.

Case study: the Sandia Monitor Well Network Designer

Sandia National Laboratories, as the operator of chemical waste facilities regulated under the Resource Conservation and Recovery Act (RCRA), needs to determine the number and location of monitoring wells to be installed around these sites. RCRA specifies a minimum of one upgradient background well and three downgradient monitor wells, but it does not provide a quantitative basis for determining the actual number and placement of wells to be drilled and monitored. As a result, Sandia (like other waste site operators) is often compelled by their regulators to install more and more wells, without benefit of a rational placement strategy or knowledge aforehand of the total number of wells which will be required.

In response to this circumstance, Sandia undertook the use of mathematical models of the site's geohydrology to determine likely flow paths, and place wells accordingly (Parsons & Davis, 1991). Using computational models of flow in the unsaturated and saturated zones, and applying Monte Carlo simulation techniques to sample the uncertain parameters of their conceptual model of the site's hydrology and geology, it is possible to evaluate the performance of an actual or proposed monitoring well network, and to determine where one might best locate (additional) wells around the site in order to achieve a desired probability of plume detection (should a leak occur). This analytical methodology employs a scoring approach, taking the plume realizations generated by Monte Carlo simulation as representative of the population of all possible plumes. For each plume realization, the plume is considered detected if it intersects at least one well in the monitoring network (which can be composed of either actual wells, hypothetical wells, or some combination of the two). The fraction of plume realizations so detected is then taken as the probability of detection of any plume by the network. Conventional optimization methods can then be employed to determine combinations of well locations which achieve a specified threshold detection probability with the minimum number of wells.

Unfortunately, any such modelling effort requires the development of a rather extensive set of assumptions about the hydrology and geology of the site, and uncertainties about these assumptions are not easily conveyed to the regulator when presenting modelling results. Furthermore, a credibility issue arises when the regulator cannot themselves manipulate the parameters of the conceptual model in order to evaluate the sensitivity of the modelling solution.

The solution to these difficulties, represented by the MWND EDSS, was to develop an interactive geohydrological modelling platform to share with the regulators so they can control the assumptions behind modelling operations, and make their own decisions based upon actual data. Since the cost of monitoring wells is significant, MWND serves an important function in helping to ensure that contaminant plumes are likely to be detected while minimizing the number of wells installed to achieve the desired detection goal. MWND allows the analyst to interactively manipulate the parameters of the hydrological conceptual model (including consideration of uncertainty), use various computational methods to simulate contaminant plumes, invoke linear programming optimization algorithms to suggest well locations which meet the desired detection probability, and use interactive visualization techniques to conduct exploratory data analysis. Since it offers a "what if" capability similar to spreadsheets, MWND allows analysts who traditionally distrust modelling results to develop their own results based upon their own conceptual model, and to interactively test the sensitivity of these results to changes in the conceptual model.

Design objectives

A platform for environmental decision making must maximize utility, ease of use, extensibility, and portability.

MWND's utility derives from the fact that it integrates access to empirical monitoring data, environmental models, simulation and optimization codes, spatial and aspatial data management, and graphical representation of data. Designed for use by professionals without computer expertise, MWND is a highly interactive, mouse-driven workstation incorporating field-proven human-computer interface principles. The human interaction paradigm for MWND was derived through human factors task analysis, focusing on the specific needs of hydrologists seeking to understand and evaluate well placement options.

To simplify evolution to new applications, the user interface o MWND was, as much as possible, designed for generic quantitative environmental problems, with elements specific to the monitor well network partitioned in such a way as to facilitate replacement. Since MWND integrates computational modelling programs as discrete processes, the software architecture encourages the modification or replacement of the computational aspects of the EDSS. This helps to alleviate a philosophical obstacle in well network negotiations: agreement on the appropriate computational models.

To enhance portability, MWND is hosted on Unix^{TM} workstation platforms, and its windowing is achieved through the public-domain X-Window System. Spatial data management and display are accomplished through the public-domain GRASS GIS, which is both widely available and also facilitates data compatibility with other common existing GIS databases.

Integration with GRASS

MWND, like its Sandia EDSS siblings, is integrated with GRASS in such a way that the user perceives themselves to be using a single software application, at the same time that the GIS and the control aspects of the EDSS are compartmentalized as much as possible. An X-based Graphical User Interface (GUI) control process runs concurrently with the GRASS monitor, which has been modified to be positioned in the appropriate place on the screen with window "decorations" removed.

Non-animated graphics are generated through the GRASS monitor: animated graphics are generated by the control program on top of the GRASS display window. GRASS functions are used to handle such operations as region selection, data layer management, and raster generation. In some of MWND's younger siblings, GRASS's more sophisticated analytical functions are also tapped.

Limits of GIS in this role

In the course of integrating the GRASS GIS, there were several limiting factors which had either to be repaired, taking advantage of the fact that source code is provided with GRASS, or worked around in some other way.

An example of a simple modification may be found in the GRASS monitor, the process which displays spatial data in an X-Window. The original GRASS monitor offers no non-interactive way to control its initial position or size. We therefore had to modify the monitor code slightly to permit the use of environment variables to control this, as well as to cause the usual X-Window decorations (such as the header bar) to be omitted.

A more complicated difficulty involves the use of colours for animation. GRASS, like most GISs, is more like a painter's canvas than an animator's movie screen. Thus, it provides no mechanism for real-time movement of graphical objects. Since such animation is crucial to a highly interactive EDSS, animation was implemented "on top" of the GRASS monitor's window, taking advantage of the fact that X allows one process to acquire access to a window created by another process. However, since GRASS uses all 8 bits of the standard X-Window's colour table, the only way to draw and erase animated objects without disturbing the underlying GRASS image was to draw the objects in an "XOR" colour, a colour defined, for each pixel, as the complement of the underlying pixel. With this definition, an object is erased by redrawing it recomplementing each pixel and thus restoring the original colour to that pixel. While this has the advantage of causing all portions of the animated object to be a different colour from its surround most of the time, it has the disadvantage of eliminating the use of colour as a means for distinguishing different animated objects. One as yet untested solution to this may be to modify GRASS so that it does not use the entire 8-bit colour table.

The most fundamental limitation of GIS in EDSS applications, however, is the two-dimensional nature of the conventional GIS data model. While current GIS technology provides "2.5D" capabilities, such as perspective surface plots, there is no true three dimensional raster structure available in common GISs. It is possible to simulate, for data storage purposes, a 3D raster by using multiple layers of 2D rasters, but access to data so stored is clumsy at best, and analysis can only be performed by tools custom written for the purpose.

CONCLUSION

We have, through the Monitor Well Network Designer, demonstrated the concept of highly integrated environmental models and GIS, taking advantage of many of the features of the GRASS GIS. On this foundation, it will be possible to expand the Sandia Environmental Decision Support System family at greatly reduced cost and with consistent, friendly human interfaces. Ultimately, we hope that this leads to better environmental management decisions and corresponding improvements in environmental stewardship.

Acknowledgements The MWND EDSS was developed by a team of Sandia and Bell Labs scientists and engineers. Besides the authors, this team included Dave Copperman and Joe Levantino (of Bell Labs) and Steve Conrad, Roger Cox, Paul Davis, Jim McCord, and Tony Zimmerman (of Sandia). The US Army Construction Engineering Research Laboratory has contributed substantially to this effort through their development of GRASS and support of its users.

REFERENCES

Parsons, A.M. & Davis, P.A. (1991) A proposed strategy for assessing compliance with the RCRA groundwater monitoring regulations. In: *Current Practices in Ground Water and Vadose Zone Investigations* (ed. by D.M. Nielsen & M.N. Sara). ASTM STP 1118, American Society for Testing and Materials, Philadelphia, USA.

An integrated information system for environmental hydraulics using Smallworld GIS

P. RULAND
Institute of Hydraulic Engineering and Water Resources Development, Aachen University of Technology, Mies-van-der-Rohestr. 1, 5100 Aachen, Germany

U. ARNOLD
IKW Consultants for Public and Environmental Affairs, Teltower Damm 52, 1000 Berlin 37, Germany

G. ROUVÉ
Institute of Hydraulic Engineering and Water Resources Development, Aachen University of Technology, Mies-van-der-Rohestr. 1, 5100 Aachen, Germany

Abstract The fusion of powerful data management, interactive graphics, and sophisticated simulation models into a single, versatile and easy to use information system has been a challenge of ever growing appeal. Within this paper an object-oriented solution approach is presented for the context of Environmental hydraulics. Key components are: the object-oriented Smallworld GIS, finite element flow and transport models, and special object types necessary to incorporate the finite element simulation tools into the GIS structure. A first prototype of this integrated information system was developed for risk assessment and restoration planning of polluted sediments in rivers, lakes and reservoirs. Promising application results could be obtained in a case study at the river Agger in Germany.

INTRODUCTION

In old industrialized regions river systems embody small reservoirs and lakes forming dead zones and sediment traps. During the time of growth a large number of mud deposits developed into hazardous mud sites due to considerable amounts of adsorbed pollutants of any kind. Thus, these mud sites can be seen as the river´s long term memory of its own pollution history.

Today, these deposits are identified as one of the major risk to the water quality and the whole aquatic environment. In case of extreme hydraulic conditions such as careless weir operations or huge floods the deposits can be destabilized. In first consequence the destabilized mud load usually causes a strong BOD/COD increase leading to critically low oxygen concentrations downstream which may be lethal for the fish population.

Moreover, spreading and sedimentation of heavily polluted mud from "old" layers may even cause a long term change of the biochemical conditions due to reinitialized chemical reactions between pollutants and the water constituents.

A simple solution to this problem could be dredging and depositing the mud on a safe site. However, this is a rather costly method and appropriate sites are hardly available. To decide about measures to be taken with these mud deposits a large amount of information and many different aspects have to be taken into account.

First, a family of simulation models is needed to model the physical processes involved in sediment transport. Given certain hydraulic stress conditions such as flooding or depletion of a reservoir these models can evaluate the risk of erosion. For the case study a system of depth averaged simulation models is used (Ruland & Rouvé, 1992; Ruland, 1993). Their operation is rather complex and requires to a great extent the handling of spatial data. GIS can be efficiently used to support this task and to reduce the danger of data mismanagement.

Secondly, in addition to simulating the sediment transport a lot of information has to be taken into account such as field data concerning the actual distribution of sediment pollutants, material properties, geographical, geological and hydrological/hydraulic boundary conditions or data describing the ecological sensitivity and economic use of the river system concerned. In many cases, the results of "what if" scenarios, simulations and multi-thematic risk assessments have to be condensed and translated into clear statements and presentations suitable for decision maker with non technical background.

This paper presents the description of a customized GIS which suits the above outlined needs.

CUSTOMIZATION OF SMALLWORLD GIS

The following chapters discusses some of the special features the object oriented Smallworld GIS possesses. The way the simulation models are incorporated into the GIS is outlined and several other means of presenting and analysing information are discussed. In order to understand the customization some general remarks about the GIS have to be made.

General layout of Smallworld GIS

Smallworld GIS is a Unix based software running on all modern workstations. Its Graphics is based on X-windows libraries such as OpenLook or Motiv. It utilizes a version managed relational database to store all data and can also access commercially available Databases such as Oracle and Ingres.

The key feature of Smallworld GIS is that it is based on an object oriented language called Magik. The Magik system comprizes three parts, a compiler for the Magik source code, a database manager to operate the relational database, and a window manager for the connection to the low level graphics provided by the window system.

The Magic language includes objects which operate the window system such as the class "frame" and "canvas" forming together a drawable window on the screen. With its graphics toolkit Magik overrides the slight differences (for example between OpenLook and Motiv) in order to create window system independent code.

The complete GIS is implemented by means of object classes and methods in Magik. The user can change these objects or add new to adopt the system to specific needs. It is beyond the scope of this paper to explain more about object oriented programming, however, a good introduction can be found in Budd (1990).

Because of the underlying object oriented language it is natural that Smallworld uses an object oriented approach to model the world. This means that thematic separation of geographic entities is done using a vehicle called "real world object" (rwo). Basically this is again a Magik object but with additional behaviour, such that it includes geometric attributes and that its data is stored in the relational database. Figure 1 shows the layout of a sample object capable of representing a house on the map. Thematic separation is achieved by operating the visibility of objects and their geometric attributes.

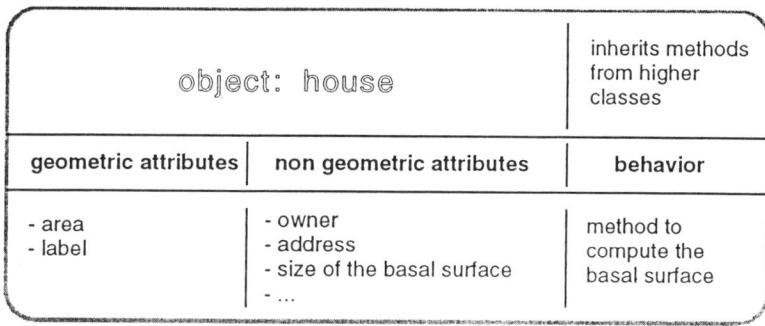

Fig. 1 Layout of the sample real world object "house".

The topology is modelled by means of geometric and topologic objects connected to the rwo´s. Figure 2 gives a view on the different Magik objects used for this purpose. Geometric entities for the rwo´s can be "point", "chain" and "area" representing a point, a line or an area on the map surface. Beneath these objects is a level of topologic objects maintaining topologic relationships between the geometric objects. The topologic objects are responsible for the spatial query functions of the system such as "contained by", "overlay" and so on.

Specific real world objects

With Smallworld GIS the user creates geographic objects like house, land parcel or street by means of Magik. Only a few lines of code are needed for the implementation that includes roughly the following steps:
(a) by defining a prototype object of the desired rwo the object class begins to live in the Magik object space;

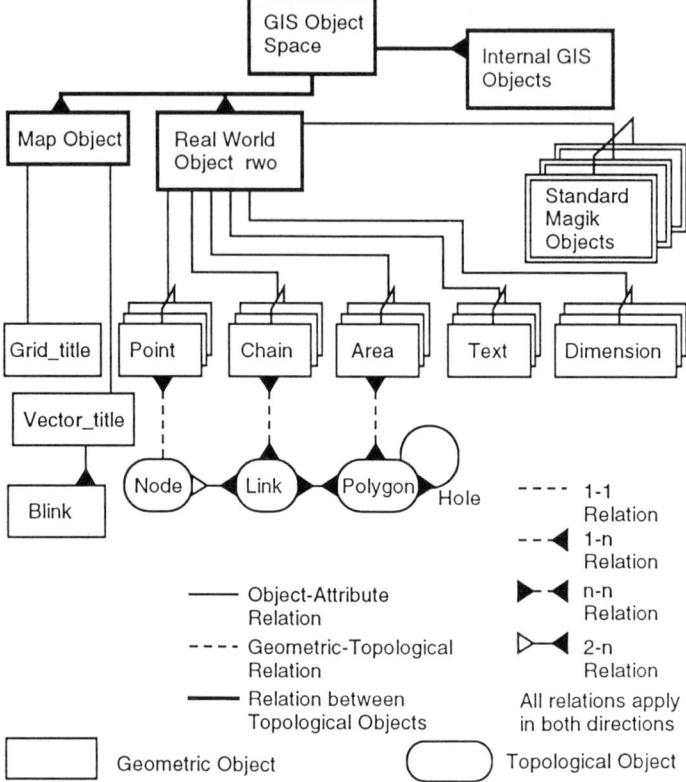

Fig.2 Overview about different Magik objects used in the Smallworld System (Smallworld, 1992).

(b) next a table is to be created in the relational database. The columns of this table coincide with the slots of the Magik object;
(c) finally the geometric attributes of the rwo´s are to be defined. The system inserts appropriate tables into the database to maintain their geometric attributes and topology.

The standard real world objects have methods being inherited from an upper class. These methods include all necessary behaviour needed for the function of the GIS. For example each rwo has its own editor. It is used for operational function such as the creation of new instances (inserting a new record in the table and digitizing the geometry on the screen) as well as the visualization of their data. At any time after having performed step (a) the user can define methods belonging to that object class that add certain behaviour or override existing functions. To a great extent, this has been done for the following objects.

Rwo "fe_element" and "fe_node" Within the system of mathematical models to simulate sediment transport the hydrodynamic and advection-diffusion model are based on partial differential equations. They are solved by the Finite Element method, hence it is essential to have finite elements and their nodes in the GIS.

There are two main purposes the objects fe_element and fe_node are needed for. First they are used to construct the mesh on the base of a map of a river, lake or reservoir. Second they carry all simulation data and provide the means to visualize these data in an appropriate way. The construction of the map is either done by screen digitizing or by using a grid generator that gets its input from a coarse mesh of elements.

The rwo "data_section" is an object with the geometric attribute chain (line). Over the area of a Finite Element mesh an arbitrary line can be digitized. On request (by a button added to the editor of the rwo data_section) the object queries the topological database for fe_elements lying beneath the digitized line. Be means of these fe_elements the simulation data (i. e. velocity, concentration and bathemetry) is interpolated along this line and in a pop-up window the simulation data is drawn in a cross sectional view (Fig. 3).

The rwo "sediment_quality" and "water_quality" These two types of rwo's have the only purpose of presenting information being connected to the water or sediment quality of a specific lake or reservoir. The rwo's have the geometric attribute point. Its graphical appearance on the map is a symbol (for example the letters "SQ"). The user can query this symbol and gets the information contained in the slots. Additionally there is the possibility to automatically read in a text file with more information on the quality. On mouse click this is shown in a scrollable popup window.

The rwo "river_section" is in its first place a normal GIS rwo with the geometric attribute line representing a part of a river. The rwo contains among others two slots with the water quality in that river section and the medium discharge. The rwo owns a method that draws around the river a band the colour of which depends on the water quality and the width depends on medium discharge.

In addition to these special rwo's a lot of other exist behaving as normal GIS objects such as house, land parcels, lakes, reservoirs, hydro power plants etc..

Customization of GIS functions and graphics

The customization of the GIS not only adds special behaviour to the rwo's but also inserts new menus and panels to the standard GIS. Again, these submenus and panels are based on Magic objects. The main window operated by an object of the class graphic_system is told to include the additional buttons and on request instances of the object classes representing a certain submenu are created.

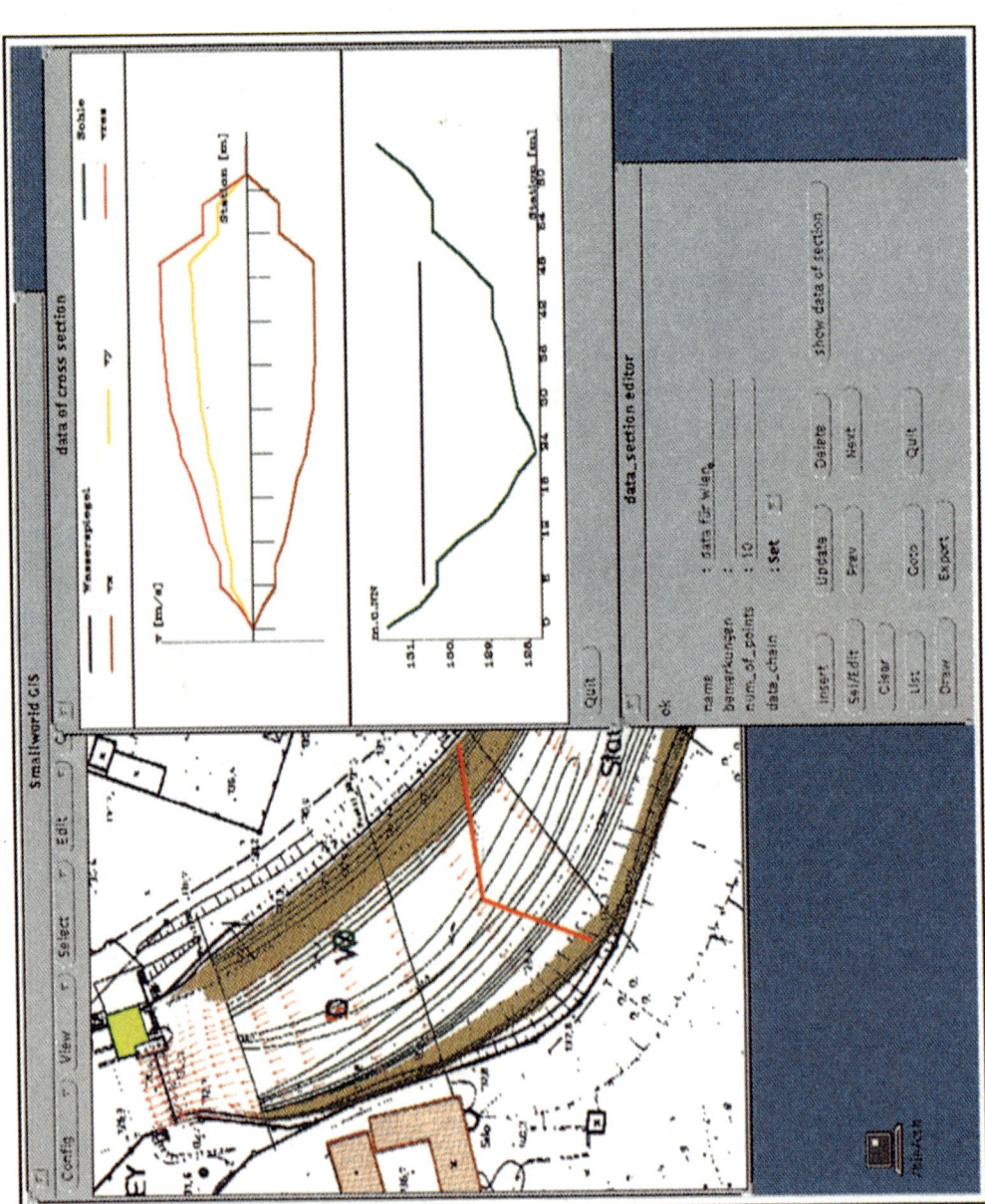

Fig. 3 Diagram of simulation data of an arbitrary cross section (red line in the reservoir) in a drawable pop up window.

Smallworld GIS handles groups of objects (as they result for example from spatial queries or predicate searches in the database) in an object called set. These sets are essential as they provide a mean to group objects and hand them from submenu to submenu. Some of the features described here rely on these sets.

A good example for a customized menu is the menu that supports the construction of a finite element mesh. It handles different kinds of sets containing fe_elements and fe_nodes, provides topology check of the latter and enables communication with the simulation program.

Managing and visualizing simulation data The simulation models produce data which can be stored in either the fe_element or the fe_node rwo's. As these can only contain the data of one simulation run a special database table was created that stores efficiently an unlimited number of runs. The user can select out of a list a certain run to be visualized. These data are inserted into the elements and nodes. Both rwo types have specific drawing methods that query for the graphical visualization chosen. Each time a fe_node or fe_element is drawn it sends a request (message) to its draw method.

Currently three types of graphic visualization of the mesh are implemented. To edit the mesh and to select correct material types the elements are drawn with a colour coding for the material type. For the flow data the nodes are drawn with an arrow representing the velocity and elements fallen dry get a yellow colour (see Fig. 5, discussed later). Finally the concentration of suspended sediments can be drawn with a colour coding of the elements. The result of this procedure is a thematic map.

Other means of graphical presentation Because of the graphics toolkit there are in fact no limitations on the way information is presented in a textual or graphical way. The possibilities range from cartographic elements on the drawing surface such as histograms or pie charts to the creation of drawable and writable subwindows for special purposes.

CASE STUDY: RIVER AGGER IN GERMANY

The river Agger is one of the tributaries of the river Rhine. It embodies a chain of shallow reservoirs that are used for power generation. All of them contain mud deposits being polluted with heavy metals and hydrocarbonates. Under normal operation the mud deposits are not destabilized, but for maintenance of the weirs and turbines the water level of the reservoir has to be lowered from time to time to a certain extent. These depletions and also floods are the hydraulic stress conditions for mud mobilization.

Within the case study, reservoir "Haus Ley" was chosen as an example to evaluate the risk of erosion by means of the model family. Figure 4 shows this reservoir with its finite element mesh and the distribution of material constants.

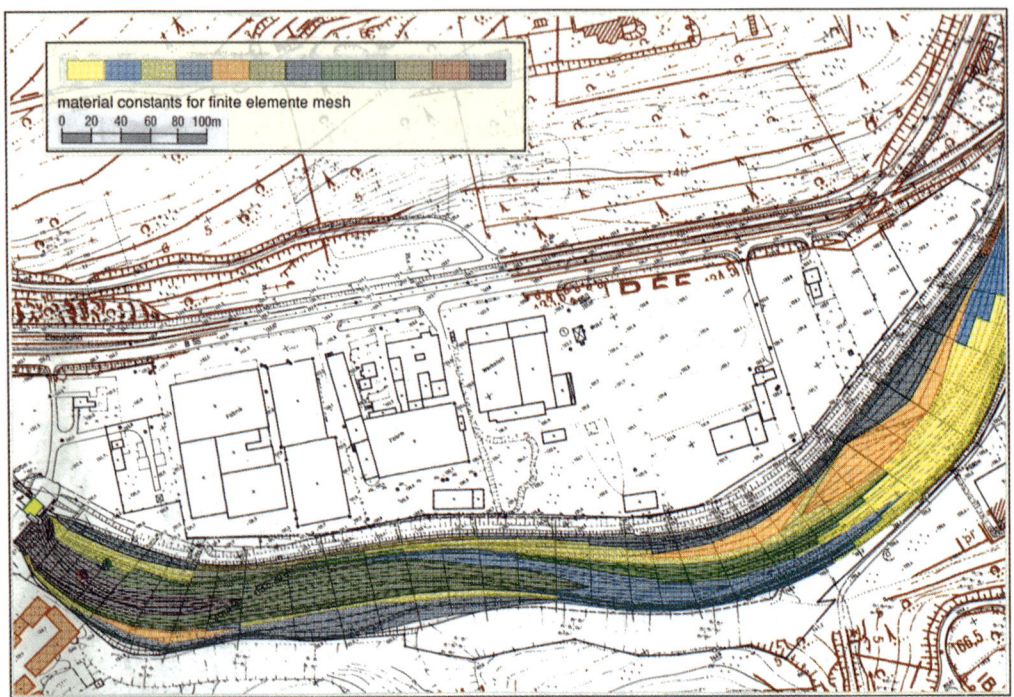

Fig. 4 Reservoir "Haus Ley" with the finite element mesh used by the simulation models.

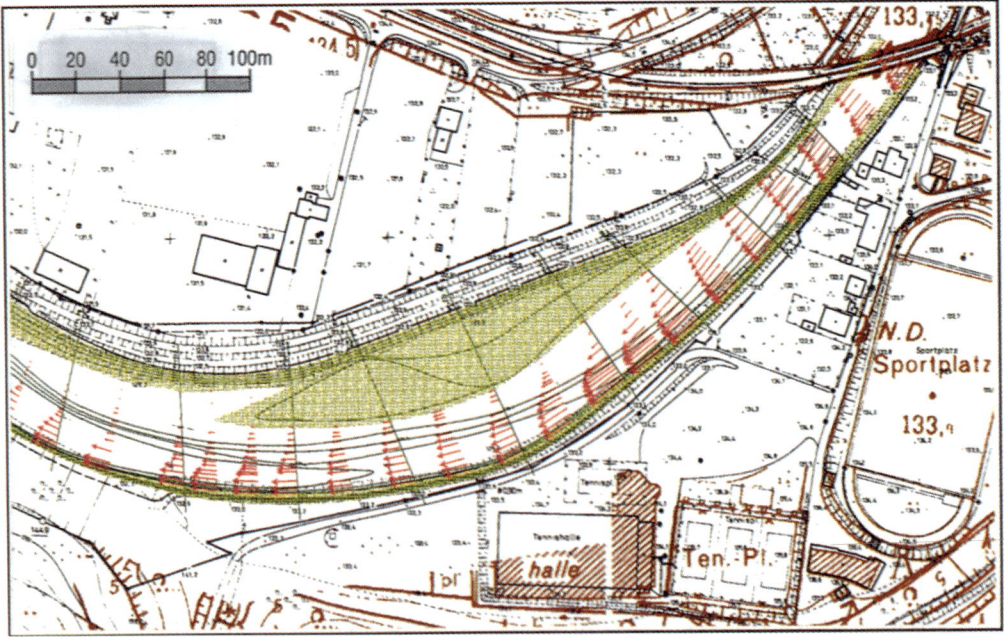

Fig. 5 Part of the reservoir "Haus Ley" showing the distribution of flow velocities in the partly depleted reservoir.

An impression of the distribution of flow velocities in the partly depleted reservoir is given in Fig. 5. It shows the entrance area of the reservoir (inflow of the river). Due to the lowered water level the reservoir has turned back into a river with relatively high flow velocities.

The results of the investigation on the erosion risk performed so far showed that erosion can occur under some circumstances. Because of the uncertainty of the material parameters of the sediments more sophisticated measurements on the sediment are carried out. Currently the discussion of what to do with these deposits goes on. The information system is permanently improved to serve as an information basis for this discussion.

FINAL REMARKS

There are two main reasons for integrating simulation models into a GIS instead of creating a simulation specific graphical user interface. First, a GIS already provides large amount of well developed and highly useful tools to manage and present all different kinds of data and not only those data needed for and created by simulation. Secondly, the given data management framework of a GIS, where simulation models are just a subject of data browsing methods to be incorporated, guarantees the consistency of the information model and thus leads to a higher level of extendibility of integrated systems.

Due to the object oriented layout of Smallworld GIS the integration could even be more straight forward. The capability of objects, i. e. active data structures, to incorporate and activate object specific methods (procedures) and to communicate with other objects, turned out to be a fundamental advantage in creating an interactive graphics system surface. User defined object behaviour for special purposes such as mesh digitizing, model initialization or model operation control could be inserted relatively easily without reprogramming the whole information system. Thinking in terms of objects, their functions and behaviour starting with the basic components proceeding to complex groups of objects and ending with compound sets for the whole system facilitated the design of a whole customized information system for the given problem and application domain.

Still there are major tasks left, since the object-oriented architecture is limited to the GIS part of the system. Ever higher capabilities are thinkable if a restructuring of the simulation models to object-oriented architecture can be achieved.

Acknowledgements The study is funded by the Ministry for Science and Research of the state North Rhine-Westphalia, Germany. Additionally it was sponsored by the German company "Moselkraftwerke Andernach GmbH" which operates the power plants at the river Agger. Their help in providing necessary data is also acknowledged.

REFERENCES

Budd, T. A. (1990) An Introduction to Object Oriented Programming. Oregon State University, Oregon, USA.
Ruland, P. (1993) Numerische Simulation des Feststofftransports unter Verwendung eines objektorientierten geographischen informationssystems (Numerical simulation of sediment transport using a GIS). Mitteilungen des Instituts für Wasserbau und Wasserwirtschaft, RWTH Aachen, Germany.
Ruland, P.& Rouvé, G. (1992) Risk assessment of sediment erosion in river basins. Fifth International Symposium on River Sedimentation, Karlsruhe 1992, Germany, 357-364.
Smallworld (1992) Smallworld GIS Customisation Overview. Smallworld Systems GmbH, Europaring 60, Ratingen, Germany.

Flood control management by integrating GIS with expert systems: Winnipeg City case study

S.P. SIMONOVIC
Department of Civil Engineering, University of Manitoba, Winnipeg, Canada R3T 2N2

Abstract A prototype decision support system (DSS) which would assist in the determination of optimal planning and operating policy decisions for an urban flood control system is herein presented. Development of integrated software tools has been reinforced in recent years by the increase in the amount of computer power available to the everyday user and also by the emergence of several technologies such as geographic information systems (GIS) and expert systems (ES) which have been successfully applied to various aspects of water resources engineering. Some of the issues addressed in the development of the system include the effective integration of GIS, ES and optimization techniques in a decision support framework, and also the inter-communication between these modules. The DSS presented is capable of guiding the user through various aspects of the flood management problem such as assisting interactively in entering input into a model for expected annual damage calculation, providing useful help screens and interfacing with a geographic database to provide default values and visual interpretation of data. Preferred solutions can be determined through multi-objective analysis and a visual comprehension of the outcomes can be displayed using the GIS. The user is also assisted in generating possible scenarios by an expert system which recommends flood management options based on expert opinion. A case study of the City of Winnipeg flood control system is presented to illustrate the use of the DSS.

INTRODUCTION

The issue of making optimal planning and policy decisions in flood-control management is one that has received continued attention in recent years. One main focus in attempts to deal with floods has been to try to obtain a better understanding of the cause-effect components of the physical process. Flood control, as many other problems in water resources, is ill-structured and the very nature of the problem calls for some form of subjective evaluation or informed judgement, in addition to the use of numerical models for its resolution. Since floods affect different groups of people, another problem is that they may not agree on what should be the best solution to a prevailing problem.

Number of different approaches have been used to resolve the issues involved with flood control planning and management. They include interactive involvement of users, or affected parties, in the decision making process (Johnson, 1986), and the use of an expert systems approach (Savic & Simonovic, 1991) to bring a high level of expert opinion into the decision-making process. Recent trends in the solution of ill-defined water resource problems have been to aggregate several models, procedural and heuristic, into integrated software tools. Such collections have been called decision support systems (Sprague & Carlson, 1982). They have been used in water resources (Loucks & daCosta, 1991; Simonovic & Grahovac, 1991) to enhance the interactive modelling process and to aid in providing a more comprehensive understanding of the physical systems being addressed.

DSS approach focuses on the interaction and interface between the user or analyst and the data, and models and computer (the human-model-machine interface). There are several important issues to be addressed in the process of developing a computer-aided DSS in order to increase the likelihood that implementation will be successful. Initially, formulation of the problem should be carried out jointly by both the developer and the user of the system. The system should take into account not just the technical elements of the project, but the institutional as well. By carefully testing a DSS throughout development and implementation, the developer should be able to make sure that the results produced are realistic and the solutions recommended are feasible. In developing any DSS it is recommended to use a modular approach which allows flexibility to ensure successful implementation. One aspect of flexibility is in being able to update the DSS to take advantage of new modelling techniques, new computer technology, and new data as they become available.

The system developed combines the use of optimization techniques and other numerical models with GIS, to facilitate spatial decision support; ES and engineering expertise to facilitate intelligent decision support (Simonovic & Grahovac, 1991) in the problem domain; and the use of database management software to facilitate model data input and data management.

FLOOD MANAGEMENT PROCEDURE

The DSS provides support for flood management following procedure presented in Fig. 1. The first part of the procedure comprises the survey of all information related to the flood plain, including location and structure data. The process of generating the initial state usually includes: conducting a field survey of the region; geographic data input; and updating the relational database. This data is used in second part of the procedure to develop aggregated elevation-damage curves for each reach. Data preparation is the second important task supported by the DSS. An expert system component assists in the input of data into the models to be used in flood damage analysis. Third part of the procedure is generation of flood management alternatives. The selection of type of structural and/or non-structural flood control measures depends on factors such as the physical and economic feasibility of the proposed measure, the stage of the expected flood, the amount of flood warning time, and the velocity and duration of the expected flood. The flood management process ends with the disaster analysis.

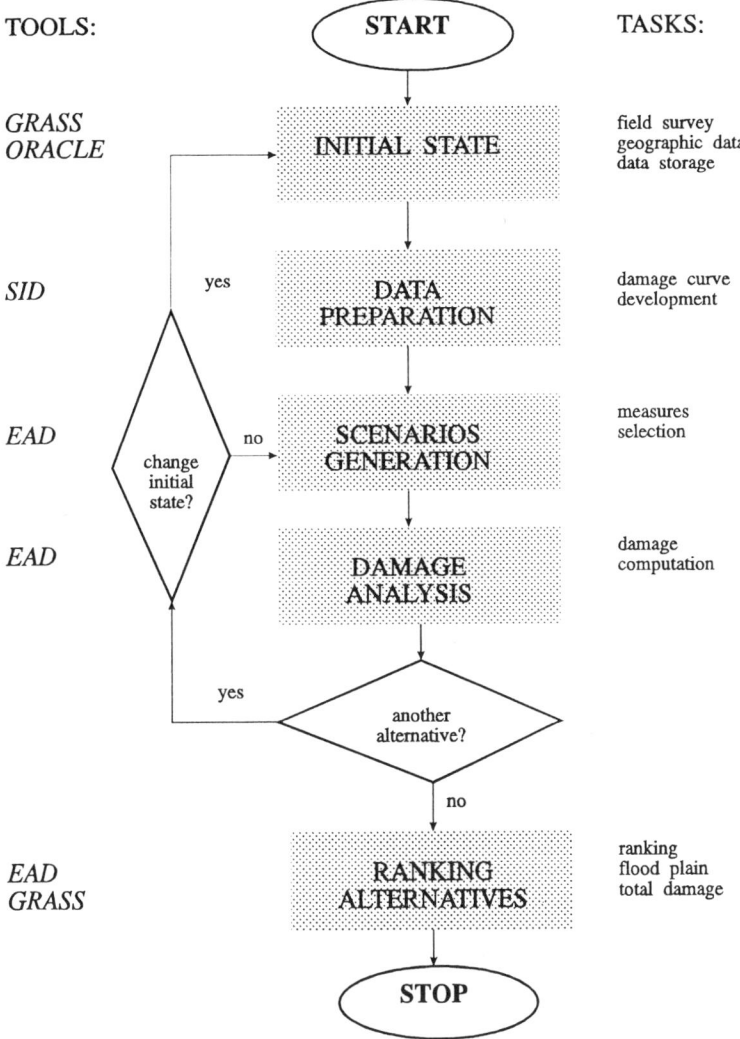

Fig. 1 Diagram of the flood management procedure.

Initial state

A field survey is normally carried out to obtain detailed information on the characteristics of study area. They include: type; cost; and ownership of structures in the region. Data from this process is stored in the study area database.

    A major aspect in the use of any mathematical model is data acquisition and pre-processing for its effective use. Advances in the field of remote sensing technology have made geographic and other data more readily available as input to GIS and hydrological models (Neumann *et al.* 1990). Geographic data such as land-use, land cover and elevation can be extracted from areal photographs and satellite pictures, or it may be input from ordinary maps by the use of a scanner or a digitizer. Data may consist of several map layers, each containing features which contribute towards the complete description of the study area. A typical cross section of map layers which make up a study area are:

    Layer 1 Designated Area (e.g., airstrips, golf courses);
    Layer 2 Building (e.g., fire stations, houses, schools);
    Layer 3 Structure (e.g., storage tanks, smoke stacks, antennas);
    Layer 4 Roadway/Railway (e.g., roads, bridges, rail lines, tunnels);
    Layer 5 Utility (e.g., electrical poles, water towers);
    Layer 6 Hydrography (e.g., rivers, ditches, dykes, ponds);
    Layer 7 Hypsography (e.g., contours, water levels);
    Layer 8 Land Cover (e.g., fields, orchards, trees); and
    Layer 9 Textual (e.g., road names, cities, villages).

GIS packages facilitate the task of data input and management by extracting data from these map layers and assist in generating a complete inventory of the flood plain.

    The relational database tool is used to store and manipulate the parameters required by the mathematical model for structure inventory analysis. Some examples are data such as the location of each building in the flood plain, the owner, the cost of structure, the damage category of each structure, etc. For the effective performance, an effective link between the GIS and the relational database tool is necessary. Simple example of the SQL command such as:

        UPDATE St.Adolphe
        SET V1FS = 50000
        WHERE IBLDG = 365;

performs the task of assigning a value of $50 000 to building number 365.

    The existing conditions in the flood plain can be efficiently documented by using the information from the GIS and field survey. The completely updated database can be output to a file, making such information available for use by the structure inventory analysis model.

Data preparation

Data preparation task uses initial state data and process them for the mathematical models to be used in flood damage analysis. This task is assisted by an expert system (ES) module through a sequential consultation designed to maintain the focus of the user on the necessary activities within each step of the process.

    Data preparation procedure is following a hierarchical scheme which improves the comprehension of the user towards the analysis procedure as a whole and helps in selecting different alternatives to be examined. The rules stored in the ES's knowledge base govern the interactive process of data preparation. The system basically inquires whether any non-structural flood control measures will be applied to structures in the flood plain. Three options are provided in the DSS namely, raise-to-target-elevation analysis, flood proofing analysis and relocation of structures analysis.

The process continues with inquiring information related to the flood plain such as the number of damage categories, the number of damage reaches and the different damage functions that may be applied to a particular category. The damage function for a particular structure represents the manner in which the elevation-damage relationship is defined for that particular category or structure. The damage function may be represented as an elevation-percent-damage relationship, or as an elevation-direct-damage relationship.

Using this hierarchically organized consultation process the user can generate several different scenarios and obtain the elevation-damage relationships for them.

Generation of flood management alternatives

This task includes the selection of the type of structural or non-structural flood control measures to be used in a given situation. Structural flood control measures (such as reservoirs, dykes and dams) are normally implemented over a regional scale, while non-structural measures are generally implemented to individual or small groups of structures.

The optimal choice of implementation of structural or/and non-structural measures is based on the physical parameters and also on engineering expertise. Expertise is required to take into account the human factor which may lead to some unexpected situations.

Disaster analysis

A flood management process ends with the disaster analysis. Flood damage analysis for alternative flood control plans is performed within this task. Again, a systematic hierarchical consultation supported by an ES module is used to conduct the analysis. Basically two types of information is inquired from the user: (a) a period of analysis; and (b) number of different flooding scenarios to be used. Data may be input in three different formats, exceedance-frequency values, flow values, or stage values. All the data received from the user are written to data files which are transformed to be used in the flood damage analysis of alternative flood control plans.

There are several measures that are computed in order to determine the consequences of a particular flooding event or a series of events. They are: expected annual damage; specific event flood damage; area flooded by a specific event; and individual structure damage.

ARCHITECTURE OF THE DECISION SUPPORT SYSTEM

Traditionally, the architecture of DSSs has been following one of three main approaches: functional approach which is decision process oriented; a tool based approach; and a combined approach which utilizes beneficial features of previous two (Sprague & Carlson, 1982). The latter approach, known as the combined architecture, is modular and adaptive. An adaptation of the combined approach has been derived for the development of the DSS for flood control management. Proposed architectural design of DSS is illustrated in Fig. 2. In general, a DSS is made of four main components, the system manager, database module, modelbase module and the display module. The system manager is the component which integrates the other parts of the system and also controls all the processes within the DSS. The database component is the module that handles data storage manipulation and extraction. It also provides

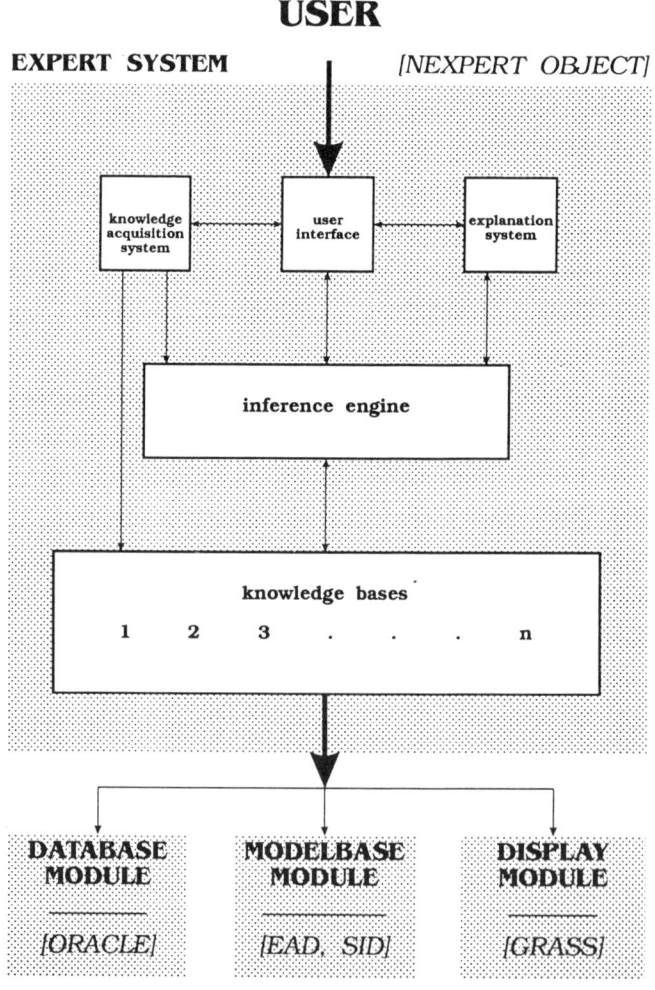

Fig. 2 Architectural design of the DSS for flood control management.

facilities for updating and modifying data used in the process of flood control management. The modelbase component of the system is the part which contains the different process models pertinent to the flood control management. Display module is introduced to enhance the input of data into the models, provide an efficient user interface, and aid in the visual comprehension of spatial information used by the models for flood control management.

The system manager

The central component of a DSS is the system manager which controls the various parts of the system. In the development of DSS for flood control management the system manager has been designed using ES technology.

The use of an ES shell as the central part of the DSS provides an easy way to incorporate heuristics and rules-of-thumb into the consultation in the areas where engineering expertise may benefit the user. DSS for flood control management contains number of knowledge bases designed to perform different tasks. Each of these knowledge bases has access directly to the other modules of the DSS, e.g. the database management subsystem, the modelbase subsystem and the display subsystem.

The main role of the ES is overall system coordination. Different modules are linked together using the knowledge specific to the flood control management process. Order in which the knowledge bases are activated is controlled with the main knowledge base. Specific characteristics of the problem and types of measures used in the flood damage analysis are used to lead the process. The inference engine of the ES shell is the actual component which determines the search strategy and order in which the available knowledge will be used. This engine activates in turn, the knowledge bases that need to be fired and determines the order of firing. Beside coordinating role, the ES provides assistance in data preparation, running different mathematical models, and presenting the results in the graphical form using GIS.

The database module

This part of the system facilitates data storage and manipulation. It has been found to be important, in the development of large applications programs which utilize large amounts of data and have several programs communicating with each other, to have a module which has the capability of providing efficient data storage, handling and manipulation services. A relational database manager named ORACLE has been selected for the development of the database module of DSS for flood control management.

ES developed is fully integrated with ORACLE through a database bridge. Structured Query Language (SQL) is used to transfer the instructions provided by the ES into the ORACLE. Data from other components of the system are loaded into ORACLE and manipulated using SQL commands to be output in the format required by pertinent models for flood damage analysis.

The modelbase component

This component of the DSS contains mathematical models which are accessed by the ES to perform procedural and analytical tasks. For the flood control management, this module is comprised of software for: (a) structure inventory damage (SID) analysis;and (b) computing the average annual damage or flood damage from a particular flood event (EAD).

The SID program (US Army Corps of Engineers, 1989a) is part of the flood damage analysis package. It is designed to assist in the systematic collection, management, and processing of data related to structures subject to flooding. Its main function is to generate elevation-damage functions by damage categories and reaches. SID can also perform an analysis of non-structural flood control measures.

The EAD program (US Army Corps of Engineers, 1989b) is developed to assist in economic evaluation of flood plain management plans. The program can compute flood damage in three modes: (a) damage for specific event; (b) expected annual damage for specific year or years; and (c) equivalent annual flood damage. Damage calculation can be performed for several damage categories such as urban, agricultural and residential. Expected annual damage can be calculated for the conditions existing in the past or expected in the future.

The display module

The major role of this component of the DSS is to improve the comprehension of the spatial and time-dependent information utilized in flood control management. This component aids the decision maker in analyzing the results of various alternatives thus leading to an optimal choice of flood protection.

Expert system shell used a built-in user interface (called session controller) which has been utilized as adequate user-machine interface in the DSS for flood control management. Most of the geographical data and the results of the analysis are presented using GIS named GRASS. The Geographic Resource Analysis Support System (GRASS) (US Army Corps of Engineers, 1991) is a public domain GIS. It is an integrated set of programs designed to provide digitizing, image processing, map production and GIS capabilities to users. GRASS has been used within the DSS to facilitate the task of data input and management by providing the routines for analyzing and extracting data from the map layers listed in the previous section of the paper. Using GRASS a complete inventory of the flood plain can be generated. Area flooded by a specific event and individual structure damage are also presented using GRASS. The elevation corresponding to the particular event is input to ES and the flooded area corresponding to this stage is displayed on the elevation map. The GRASS is then used to generate a report on the total area taken up by the displayed area, and this is then multiplied by actual area of each cell to arrive at the actual flooded area. Individual structure damage is determined by comparing the flood stage for particular flood and the ground elevation of the buildings in the flood plain. If flooding is found to have occurred, then the damage done to the structure is computed using the damage function for the particular structure and the cost of the structure. The total damage data is then displayed using GRASS.

WINNIPEG CITY FLOOD MANAGEMENT: CASE STUDY

The city of Winnipeg in Manitoba, Canada and surrounding areas along the banks of the Red River have a history of flooding. Floods have caused much damage and destruction (1950, 1970, 1974, 1979). In recent years several flood control measures were put in place: (a) dyking system along the Red, Assiniboine and Seine rivers thorough the city (after 1950); (b) the Red River floodway (1968); (c) the Portage diversion (1970); and (d) the Shellmouth reservoir (1972). The operation of the flood control system maintains upstream water levels to their natural levels. Therefore, the flood hazard of areas upstream of the Red River floodway is not reduced. During the 1974 and 1979 events, residents upstream of the floodway incurred heavy damages.

The area selected for testing the DSS covers 10x13 km^2 just upstream of the inlet control structure and Red River floodway. The main community in the region is St Adolphe. The main occupation of people leaving in the area is farming. The average cost of buildings in the area is between $80 000 and $50 000. West dyke, which is a part of the flood control system runs from north to south through the western part of the area. Buildings located west of this dyke are protected to a height of 787.4 ft above sea level. The DSS is used to generate and evaluate suitable flood control measures for the area.

Data for the analysis

GRASS GIS software was employed for pre-processing the two main map layers to be used in the analysis: (a) the elevation map; and (b) the buildings map. The elevation map for the study area contained elevations that ranged from 705.4 ft at the surface of the Red river to 820.2 ft at the top of the Inlet Control structure. The contour intervals obtained after running the interpolation program was 3.3 ft. In processing the building map, a single point was marked for every structure or building. By overlaying this new map on the elevation map the elevation corresponding to these points is determined.

The SID program was used to determine the elevation-damage relationship for the existing conditions in the region and for a number of flood

control measures, economically and physically feasible to be applied. The flood control measures investigated are:
(1) existing conditions;
(2) raising individual structures to a height of 3 ft;
(3) raising individual structures to a height of 5 ft;
(4) flood proofing individual structures to a height of 3 ft;
(5) flood proofing individual structures to a height of 5 ft;
(6) relocation of structures to above the 754.0 ft;
(7) dyking along the Red River to the 725.4 ft level (20 ft dyke); and
(8) dyking along the Red River to the 745.4 ft level (40 ft dyke).

The SID consultation was performed for each of the eight options, and the output from different alternatives were saved in a data storage to be used by the damage model EAD.

Two particular damage analyses were performed: (a) damage for a specific flood event: and (b) expected annual damage over a 50 year period. Three separate floods were selected for the specific event analysis: 1950, 1970 and 1974. The data used for the computation of the expected annual damage are: (1) period of analysis 50 years, (2) study year 1992, (3) base year 1995, (4) dollar year 1992, (5) affluence factor 1.2, (6) discount rate 5%, and (7) price level index 1.0.

Average annual costs as a percentage of structure value data were obtained using US Federal Insurance Agency estimate costs of implementing specific flood control measures.

Results of the analysis

Main results of damage analysis are shown in Table 1. Alternative 6 (relocation of structures above a height of 754.0 ft) is the measure that reduces damages the most in all the flood events. It is interesting to point out that raising structures proved to be more effective in reducing damages than flood proofing to similar heights. The expected annual damage computation generated the

Table 1 Results of damage analysis.

Measure	Damage for specific event			Expected annual damage			Equivalent annual damage
	1950	1970	1974	2004	2024	2044	
1	2571	1319	2212	389	525	707	435.6
2	1614	980	1446	269	363	488	300.9
3	1261	779	1137	208	281	378	232.9
4	2479	1288	2149	355	477	644	396
5	1673	1134	1533	301	406	547	336.8
6	865	0	592	44	60	81	49.9
7	2571	1319	2212	379	510	688	423.9
8	2571	1319	2212	294	395	532	382.2

same results. Alternative 6 is the measure that has the lowest value of equivalent annual damage. Average annual costs of implementing the chosen flood control measures to the community as a whole were also computed. Flood proofing options would be the least costly to implement and the relocation of structures and the Red River dyking options the most expensive.

Spatial interpretation

Use of the GIS software provides an effective way of displaying and manipulating the results of the analysis. Research presented in this paper is investigating several alternative ways in which the GIS can provide visual support to the planning process.

Showing the flooded areas To examine the consequences of flooding from specific events, flooded areas resulting from different measures can be displayed both in 2- and 3-dimensional overlay on an elevation map layer.

Displaying damage by elevation For each point in the area and for every level of elevation, a display of the resultant damage can be provided. This display is useful in locating areas at high risk of sustaining major damages due to a particular flood at a particular elevation. In deriving these maps, the damage function relationship for each structure is utilized to obtain elevation-damage relationships. GRASS Ginfer program is then applied to generate a new map which is an aggregation of the damages incurred by each building in the reach at a particular elevation. Following rule is an example of the procedure used for displaying damage by elevation:

IFMAP	flooded area	725 ft
ANDIFMAP	elevation	705 ft
ANDIFMAP	buildings	1
THENMAPHYP	result	50000

The interpretation of this rule is: if the three maps (flooded area, buildings and elevation) are used and there is a value representing 725 ft in the map layer "flooded area" and there is a value representing 705 ft in the "elevation" map and if there is a cell representing category 1 in the "building" map, then assign the value of $50 000 to this cell in a new map "result". Value of $50 000 represents the damage to a building submerged to 20 ft (725 - 705 ft).

FUTURE WORK

Decision support system has been developed to provide comprehensive support in the domain of flood control management. Emphasis of the development was on integrating ES, GIS, database manager, and analytical models into an efficient tool. ES technology helped in the management of the overall process of analysis and provided support for strong interaction between the user and the machine. GIS played a key role in addressing the spatial aspects of the problem domain. The use of a relational database to handle the storage and retrieval of data facilitated the data management process. Example study area contained 1213 buildings. With a larger and more built up area the data requirements would become large and use of an efficient tool for their management will be essential for the successful analysis.

Several directions have been identified for the future system enhancement: (a) incorporation of knowledge bases containing engineering expertise to be applied in the selection of the type of non-structural measures; (b) inclusion of the model for hydrological computations of water surface elevation; and (c) incorporation of animation to simulate flood wave movement along the flood plain, and sound effects which will be used to indicate the flood magnitude.

Acknowledgements Programming work on the development of the DSS presented in the paper has been done by my former graduate student K.A.Agyare. Financial support for the research has been provided by Natural Sciences and Engineering Research Council of Canada through the research grant OGP0004416.

REFERENCES

Johnson, L.E. (1986) Water resources management decision support systems. *ASCE J. Wat. Resour. Planning and Management* **112**(3), 308-325.

Loucks, D.P. & daCosta, J.R. (eds) (1991) Decision support systems: Water resources planning. NATO ASI Series, Springer-Verlag, New York.

Neumann, P., Fett, W. & Schultz, G.A. (eds) (1990) A geographic information system as a database for distributed hydrological models. In: *Proceedings of the international symposium on remote sensing in water resources*, IAH, The Netherlands.

Savic, D.A. & Simonovic, S.P. (1991) An interactive approach to selection and use of single multipurpose reservoir models. *Wat. Resour. Res.* **27**(10), 2509-2521.

Simonovic, S.P. & Grahovac, J. (1991) Evolution of decision support system for reservoir operations: Manitoba Hydro case study. In: *Decision Support Systems: Water Resources Planning* (ed. by D.P. Loucks & J.R. daCosta). NATO ASI Series, Springer-Verlag, New York, 485-526.

Sprague, R.H.& Carlson, E.D. (1982) Building effective decision support systems. Prentice Hall, New Jersey.

US Army Corps of Engineers (1989a) Structure inventory of damage analysis (SID) User's manual. Hydrologic Engineering Centre, Davis, California.

US Army Corps of Engineers (1989b) Expected annual flood damage (EAD) computation User's manual. Hydrologic Engineering Centre, Davis, California.

US Army Corps of Engineers (1991) GRASS version 4 User's Manual. Construction Engineering Research Laboratory, Champaign, Illinois,

2 Methodological Aspects and Application of GIS in Remote Sensing

Relevance of Geographic Information Systems of landscape characteristics for hydroclimatological models

R. AVISSAR
Department of Meteorology and Physical Oceanography, Rutgers University, New Brunswick, New Jersey 08903, USA

Abstract An analysis with the Fourier Amplitude Sensitivity Test (FAST) and a sophisticated scheme of land surface processes for hydroclimatological models indicated that plant stomatal conductance and land surface roughness are the two landscape characteristics of greatest importance, in densely vegetated surfaces, for hydroclimatological models. In bare and sparsely vegetated surface, soil surface wetness, instead of stomatal conductance, becomes the predominant characteristic. The relation between these landscape characteristics and land surface energy fluxes is strongly nonlinear. As a result, an estimation of the mean landscape characteristics at the resolvable scale of the hydroclimatological models is not sufficient. Higher statistical moments of these characteristics are required. The organization of these characteristics at the global scale presents an important challenge for the Geographic Information Systems (GIS) scientific community.

INTRODUCTION

The Earth's surface absorbs over 70% of the energy absorbed into the climate system, and many physical, chemical, and biological processes take place there. Of particular importance are the exchanges of mass (notably water), momentum, and energy between the surface and the atmosphere. Also, the bulk of the biosphere lives there and plays a major role in many biogeochemical cycles, some of which may be affecting the chemical composition of the atmosphere and thereby the climate. The production and transport of atmospheric pollutants from various human activities is but one aspect of this question. Other natural processes also contribute to the modification of the radiative properties of the atmosphere: for example, most aerosols originate at the surface (Avissar & Verstraete, 1990).

The importance of Earth's surface processes in climate and weather predictions has been demonstrated with both general circulation models (GCM's), which simulate the entire Earth's atmosphere with a horizontal resolution (i.e., the horizontal domain represented by one element in the numerical grid) of 100 000 to 250 000 km^2 (e.g., Shukla & Mintz, 1982; Dickinson & Henderson-Sellers, 1988; Henderson-Sellers *et al.*, 1988; Shukla *et al.*, 1990), and mesoscale atmospheric models, which simulate regional and local atmospheric phenomena with typical horizontal resolutions of 25 to 2500 km^2 (e.g., Mahfouf *et al.*, 1987; Segal *et al.*, 1988; Avissar & Pielke, 1989; Pielke & Avissar, 1990; Avissar, 1992a; Avissar & Chen, 1992a; Avissar & Chen, 1992b).

For instance, Shukla *et al.* (1990) showed with a GCM that turning the

Amazonian tropical forests into agriculture could reduce precipitation and lengthen the dry season in this region. As a result, serious ecological implications would be expected. However, Avissar & Chen (1992b) demonstrated, with a mesoscale model, that local (patchy) deforestation (as occuring in the Amazonian tropical forest) can produce local atmospheric circulations stronger than sea breezes, and eventually produce convective clouds and precipitation. These apparently contradictory results simulated by the two types of model emphasize not only the impact of the Earth's surface on atmospheric processes, but also the importance of scaling in land-atmosphere interactions studies.

These global and regional processes are due mainly to the fact that the input of solar and atmospheric radiation at the Earth's surface is redistributed very differently on dry and wet land. Indeed, on bare dry land, the absorption of this energy results in a relatively strong heating of the surface, which usually generates high sensible and soil heat fluxes. Typically, the soil heat flux represents 10% to 20% of the sensible heat flux. In that case, there is no evaporation (i.e., no latent heat flux). On wet land, however, the incoming radiation is used mainly for evaporation, and the heat flux transferred into the atmosphere and conducted into the soil is usually much smaller than the latent heat flux. When the ground is covered by a dense vegetation, water is extracted mostly from the plant root zone by transpiration. In that case, latent heat flux is dominant even if the soil surface is dry, but as long as there is enough water available in the plant root zone and plants are not under stress conditions. As a result, the characteristics of the atmosphere above dry land and wet (vegetated) land are significantly different (e.g., Avissar, 1992a).

Hence, it is widely agreed upon that the parameterization of the Earth's surface is one of the more important aspects of hydroclimatological models, and that it should be treated with extreme care to ensure successful simulations with these models. Several parameterizations of the land surface have been suggested for application in these models. As recently reviewed and discussed by Avissar & Verstraete (1990), these parameterizations have improved from the prescription of surface potential temperature as a periodic heating function (e.g., Neumann & Mahrer, 1971; Pielke, 1974; Mahrer & Pielke, 1976) to more realistic formulations based on the solution of energy budget equations applied to the soil surface (e.g., Pielke & Mahrer, 1975) and, when present, vegetation layers (e.g., McCumber, 1980; Yamada, 1982; Sellers *et al.*, 1986; Dickinson *et al.*, 1986; Avissar & Mahrer, 1988).

Yet the benefits gained by introducing more realistic land surface parameterizations in these models is counterbalanced by the need to provide a larger number of landscape characteristics, which often are not readily available. Consequently, an important contribution of GIS to this type of modeling would be to organize those characteristics that have the most significant impact on hydrological processes at the appropriate spatial and temporal resolutions.

The aim of the present paper is to identify applications and requirements of geographic information systems (GIS) for landscape characteristics. For this purpose, first an evaluation of which landscape characteristics are of greatest importance for hydroclimatological models is described. Then, the impact of spatial variability of these characteristics on hydroclimatological processes is discussed. Finally, requirements of GIS of landscape characteristics for hydroclimatological models are pointed out.

WHICH LANDSCAPE CHARACTERISTICS ARE IMPORTANT?

In order to estimate which land surface characteristics are of greatest importance for hydroclimatological models, Collins & Avissar (1992) used the Fourier Amplitude Sensitivity Test (FAST) suggested by Cukier *et al.* (1973), in conjunction with the Land-Atmosphere Interactions Dynamics (LAID), a land surface scheme for atmospheric models developed by Avissar & Mahrer (1988). This land surface scheme is conceptually similar to the Biosphere-Atmosphere Transfer Scheme (BATS) developed by Dickinson *et al.* (1986) or the Simple Biosphere Model (SiB) suggested by Sellers *et al.* (1986) since it is based on the same physical principles and the solution of energy budget equations for soil surface and vegetation (yet some of the mechanisms are parameterized differently in these schemes).

FAST determines the relative contribution of the distribution of individual input parameters (assuming that their exact value is unknown) to the variance of land surface energy fluxes. By simultaneously varying all parameters according to their individual probability density functions, the number of computations needed is very much reduced by this technique. For instance, in a traditional sensitivity analysis, if ten values are used within the ranges of ten input parameters, a total of 10^{10} (i.e., 10 billion) model runs would be needed to cover all possible combinations of the surface characteristics. Such an analysis would be computationally and time prohibitive. For comparison, only 411 runs are required with FAST to produce an objective analysis.

Ten input parameters were selected for this analysis: soil albedo (α_G), soil emissivity (ε_G), soil thermal conductivity (λ), vegetation albedo (α_v), vegetation emissivity (ε_v), vegetation extinction coefficient (κ_v), soil surface wetness (s_w), relative stomatal conductance (g_{sc}), surface roughness (z_0), and leaf area index (l). These parameters, were prescribed by continuous probability distributions, over a given range of possible values. The maximum and minimum values were obtained from the literature (Dickinson, 1983; Pielke, 1984; Goudriaan, 1988) and are summarized in Table 1. Note that in some cases, values beyond the range found in the literature were used as maxima and minima to be sure of completeness. For the vegetation extinction coefficient, a maximum and a minimum were chosen to be the highest and lowest possible values obtained by the method of estimation of κ_v devised by

Table 1 Land surface parameters used as input to the FAST. For the analysis, values were chosen within the range between maximum and minimum values.

Parameter	Minimum	Maximum
soil albedo (α_G)	0.05	0.95
soil emissivity (ε_G)	0.80	0.995
soil thermal conductivity (λ)	0.05 W m^{-1} K^{-1}	3.0 W m^{-1} K^{-1}
vegetation albedo (α_v)	0.10	0.30
vegetation emissivity (ε_v)	0.90	0.99
vegetation extinction coefficient (κ_v)	0.30	2.30
soil surface wetness (s_w)	0.0	1.0
relative stomatal conductance (g_{sc})	0.0	1.0
surface roughness (z_0)	2x10^{-5} m	6.0 m
leaf area index (l)	0.0	10

Goudriaan (1988). Theoretically, κ_v can range between 0 and ∞ for specific leaf orientations and solar zenith angles. In nature, however, leaf orientation is somewhat random; therefore, a minimum of 0.29 and a maximum of 2.26 are sufficient for the characterization of the extinction of solar radiation by a plant canopy (Goudriaan, 1988).

Note, g_{sc}, s_w, and λ are usually linked to the subsurface soil water potential in the LAID parameterization. Profiles with depth of soil temperature and soil water can be determined, and g_{sc} is a function of soil water potential in the root zone. This was not done for the present study, and these parameters are independent within the model, as required by the FAST analysis.

Assuming various probability density functions (pdf's) of land surface characteristics under a broad range of atmospheric conditions, this analysis demonstrated that the land surface parameters that affect the most land surface energy fluxes are, in vegetated land, stomatal conductance and surface roughness. In bare land, soil-surface wetness (instead of stomatal conductance) and surface roughness were the dominant parameters. Albedo and leaf area index (where vegetation completely covers the ground surface) have also some impact on the fluxes, but to a lesser extent. Clearly, varying the leaf area index from a value of zero (i.e., bare soil) to a value greater than one can strongly affect the fluxes, especially when the soil surface is dry but there is enough water in the plant root zone for transpiration. Thus, in that case, leaf area index has a strong impact on the fluxes.

It must be mentioned that the relative importance of these characteristics depends on atmospheric forcing conditions (i.e., radiation, temperature, humidity, and wind speed). For instance, stomatal conductance (or soil-surface wetness in bare land) plays a major role under unstable atmospheric conditions, while surface roughness tends to be the most important characteristic under neutral atmospheric conditions. This implies that these five parameters (i.e., stomatal conductance, soil-surface wetness, surface roughness, albedo, and leaf area index) need to be considered carefully in atmospheric models.

IMPORTANCE OF SPATIAL VARIABILITY

For organizing landscape characteristics in an appropriate GIS format, it is important to know not only which characteristics are the most significant for hydroclimatological models, but also how their variability found at the resolvable scale of the model is expected to affect the simulated processes. For this purpose, Li & Avissar (1992) studied the impact of variability of these five land surface characteristics on the turbulent heat fluxes near the ground surface, with a statistical-dynamical land surface scheme developed by Avissar (1992a). With this scheme, land surface characteristics are represented by a pdf rather than by a single representative value (as done in current state-of-the-art schemes such as BATS, SiB, or LAID).

Eleven different pdf's were considered for each one of the five land surface characteristics: five beta distributions, three log-normal distributions, two normal distributions, and one rectangular distribution. They are shown in Fig. 1. All possible combinations of the pdf's of land surface characteristics and a broad range of atmospheric conditions were considered. A comparison of land surface energy fluxes integrated over a horizontal domain represented by one grid element in a GCM with either the pdf's or the corresponding mean characteristics indicated that, under unstable atmospheric conditions, stomatal conductance and leaf area index variability have the most significant effect on spatially integrated energy fluxes from vegetated land. In bare land, soil-surface wetness, instead of stomatal conductance, strongly affected the surface fluxes. Under neutral and stable atmospheric conditions, surface roughness appeared to have the predominant effect on the surface fluxes.

Differences as large as 150, 135, 105, and 100 W m^{-2} were found between the heat fluxes calculated with the pdf's or the mean leaf area index, stomatal conductance, soil-surface wetness, and surface roughness, respectively, emphasizing the importance of accounting for spatial variability of these characteristics in atmospheric models. The spatial variability of albedo created differences of not more than 20 W m^{-2}, indicating a relatively linear relation between this characteristic and land surface energy fluxes. In general, Li & Avissar (1992) found that the more positively skewed the pdf, the largest the difference between the heat fluxes calculated with the pdf or the mean. Avissar (1992b) and Otte *et al.* (1992) demonstrated that a large spatial variability of stomatal conductance is measured in otherwise homogeneous canopies. This variability was characterized by a lognormal pdf (e.g., Fig. 2) that remained relatively stable during the entire measuring period (about two month during summer time in New Jersey). Interestingly, among the various pdf's investigated by Li & Avissar (1992), lognormal distributions of stomatal

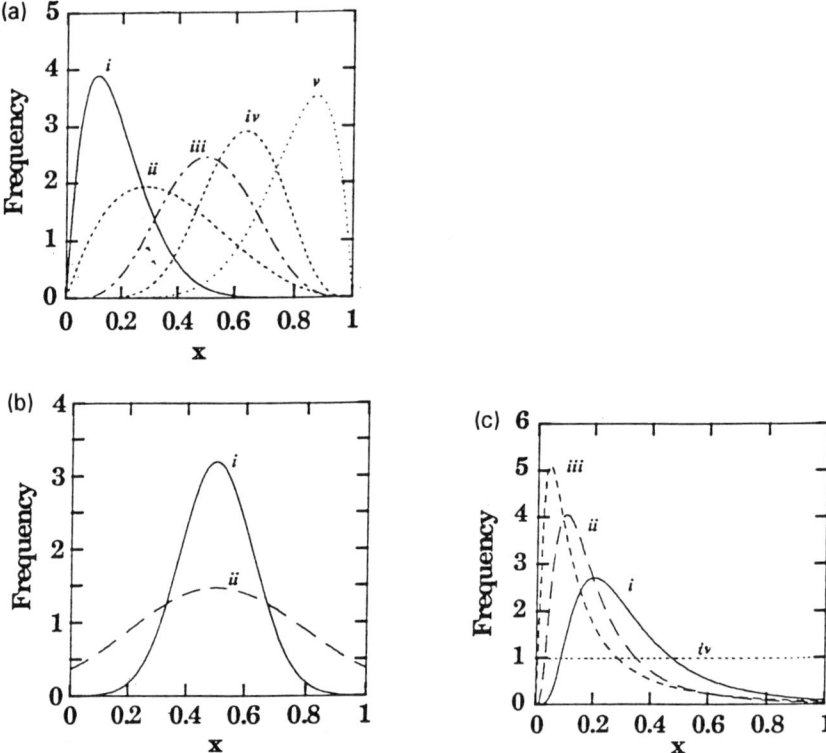

Fig. 1 Distributions used to represent spatial variability of different land surface characteristics (a) beta distributions (p=2 and q=9 in Curve (i); p=2 and q=3.5 in Curve (ii); p=5 and q=5 in Curve (iii); p=8 and q=5 in Curve (iv); and p=8 and q=2 in Curve (v)); (b) normal distributions (mean is 0.5 in both curves, and standard deviation is 0.125 in Curve (i) and 0.3 in Curve (ii)); and (c) log-normal distributions (mean is 0.35, 0.23, and 0.23, and standard deviation is 0.235, 0.198, and 0.34 in Curve (i), Curve (ii), and Curve (iii), respectively) and rectangular distribution (Curve (iv)). From Li & Avissar (1992).

conductance affects relatively strongly the difference between the heat fluxes calculated with the pdf or the mean.

CONCLUSIONS

Water availability for evapotranspiration at the land surface has a considerable impact on the entire atmospheric planetary boundary layer (PBL). High temperature, low humidity, and a well-developed PBL are obtained over dry land. On the contrary, low temperature, high humidity, and a shallow PBL are obtained over wet (vegetated) land.

The analysis described in this paper indicated that five land surface

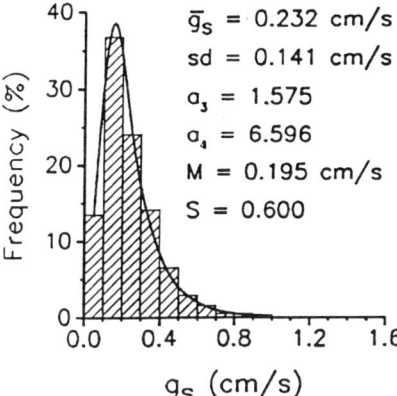

Fig. 2 Histogram of stomatal conductances (g_s) based on 1150 observations collected in a homogeneous potato field in New Jersey, during the summer of 1989. Adapted from Avissar (1992b).

characteristics (stomatal conductance, soil surface wetness, surface roughness, leaf area index, and albedo) must be described as accurately as possible in atmospheric models. While using a mean value of the albedo does not affect too much the land surface energy fluxes, the spatial variability (i.e., the higher statistical moments) of the four other characteristics should be considered. The organization of these characteristics at the global scale is an important challenge for the GIS scientific community.

Acknowledgement This research was supported by the US Department of Energy under Grant DE-FG02-92ER61453, by the National Science Foundation (NSF) under Grants ATM-9016562 and EAR-9105059, and by the National Aeronautics and Space Administration under Grant NAGW-2658.

REFERENCES

Avissar, R. (1992a) Conceptual aspects of a statistical-dynamical approach to represent landscape subgrid-scale heterogeneities in atmospheric models. *J. Geophys. Res.* **97**, 2729-2742.

Avissar, R. (1992b) Observations of leaf stomatal conductance at the canopy scale: an atmospheric modeling perspective. *Bound.-Layer Meteor.*, in press.

Avissar, R. & Chen, F. (1992a) Potential use of the mesoscale kinetic energy (MKE) for the parametrization of subgrid-scale (mesoscale) processes in large-scale atmospheric models. *J. Atmos. Sci.*, accepted for publication.

Avissar, R. & Chen, F. (1992b) An approach to represent mesoscale (subgrid-scale) fluxes in GCM's demonstrated with simulations of local deforestation in Amazonia. *Hydrol. Sci. J.*, submitted.

Avissar, R. & Mahrer, Y. (1988) Mapping frost-sensitive areas with a three-dimensional local scale numerical model. Part I : Physical and numerical aspects. *J. Appl. Meteor.* **27**, 400-413.

Avissar, R. & Pielke, R.A. (1989) A parameterization of heterogeneous land-surface for atmospheric numerical models and its impact on regional meteorology. *Mon. Wea. Rev.* **117**, 2113-2136.

Avissar, R. & Verstraete, M.M. (1990) The representation of continental surface processes in atmospheric models. *Reviews of Geophysics* **28**, 35-52.

Collins, D. & Avissar, R. (1992) An evaluation with the Fourier Amplitude Sensitivity Test (FAST) of which land-surface parameters are of greatest importance for atmospheric modelling. *J. Climate*, accepted for publication.

Cukier, R.I., Fortuin, C.M., Shuler, K.E., Petschek, A.G. & Schaibly, J.H. (1973) Study of the sensitivity of coupled reaction systems to uncertainties in rate coefficients. I. Theory. *J. Chem. Phys.* **59**, 3873-3878.

Dickinson, R.E. (1983) Land surface processes and climate - surface albedos and energy balance. *Adv. Geophysics* **25**, 305-353.

Dickinson, R.E. & Henderson-Sellers, A. (1988) Modelling tropical deforestation, a study of GCM land-surface parametrizations, *Quart. J. R. Met. Soc.* **114**, 439-462.

Dickinson, R.E., Henderson-Sellers, A., Kennedy, P.J. & Wilson, M.F. (1986) Biosphere-Atmosphere Transfer Scheme (BATS) for the NCAR Community Climate Model. *NCAR Tech. Note NCAR/TN-275+STR*, Natl. Cent. for Atmos. Res., Boulder, Colorado, USA.

Goudriaan, J. (1988) Bare bones of leaf-angle distribution in radiation models for canopy photosynthesis and energy exchange. *Agric. For. Meteor.* **43**, 155-169.

Henderson-Sellers, A., Dickinson, R.E. & Wilson, M.F. (1988) Tropical deforestation, important processes for climate models, *Climatic Change* **13**, 43-67.

Li, B. & Avissar, R. (1992) The impact of spatial variability of land-surface characteristics on land-surface heat fluxes. *J. Climate*, submitted.

Mahfouf, J.F., Richard, E. & Mascart, P. (1987) The influence of soil and vegetation on the development of mesoscale circulation. *J. Clim. Appl. Meteorol.* **26**, 1483-1495.

Mahrer, Y. & Pielke, R.A. (1976) Numerical simulation of the air flow over Barbados. *Mon. Wea. Rev.* **104**, 1392-1402.

McCumber, M.C. (1980) A Numerical Simulation of the Influence of Heat and Moisture Fluxes upon Mesoscale Circulations. Ph.D. Dissertation, University of Virginia, Charlottesville.

Neumann, J. & Mahrer, Y. (1971) A theoretical study of the land and sea breeze circulations. *J. Atmos. Sci.* **28**, 532-542.

Otte, M.J., Naot, O. & R. Avissar (1992) The impact of stomatal conductance variability on land-surface energy fluxes. *Bound.-Layer Meteorol.*, in preparation.

Pielke, R.A. (1974) A three-dimensional numerical model of the sea breezes over south Florida. *Mon. Wea. Rev.* **102**, 115-139.

Pielke, R.A. (1984) Mesoscale Meteorological Modeling. Academic Press, New York.

Pielke, R.A. & Avissar, R. (1990) Influence of landscape structure on local and regional climate. *Landscape Ecology* **4**, 133-155.

Pielke, R.A. & Mahrer, Y. (1975) Technique to represent the heated-planetary boundary layer in mesoscale models with coarse vertical resolution. *J. Atmos. Sci.* **32**, 2288-2308.

Segal, M., Avissar, R., McCumber, M. & Pielke, R.A. (1988) Evaluation of vegetation effects on the generation and modification of mesoscale circulations. *J. Atmos. Sci.* **45**, 2268-2292.

Sellers, P.J., Mintz, Y., Sud, Y.C. & Dalcher, A. (1986) A simple biosphere model (SiB) for use within general circulation models. *J. Atmos. Sci.* **43**, 505-531.

Shukla, J. & Mintz, Y. (1982) Influence of Land-Surface Evapotranspiration on the Earth's Climate. *Science* **215**, 1498-1501.

Shukla, J., Nobre, C. & Sellers, P. (1990) Amazon deforestation and climate change. *Science* **247**, 1322-1325.

Yamada, T. (1982) A numerical model simulation of turbulent airflow in and above canopy. *J. Meteorol. Soc. Japan* **60**, 439-454.

Process, scale and constraints to hydrological modelling in GIS

R.B. GRAYSON
Centre for Environmental Applied Hydrology, University of Melbourne, Parkville, 3052, Australia

G. BLÖSCHL
Institut für Hydraulik, Gewässerkunde und Wasserwirtschaft, Technische Universität Wien, Karlsplatz 13/223, 1040 Vienna, Austria

R.D. BARLING
Centre for Environmental Applied Hydrology, University of Melbourne, Parkville, 3052, Australia

I.D. MOORE
Centre for Resource and Environmental Studies, The Australian National University, Canberra City, ACT 2601, Australia

Abstract Distributed parameter hydrological models attempt to represent the spatial behaviour needed to address integrated catchment management issues and are, in principle, compatible with the structure of GIS. Unfortunately these models have been developed for research catchments that are orders of magnitude smaller than management areas. This difference in scale requires simplifying assumptions that undermine the validity of the original model. Furthermore, the fundamental premise of the original models for use as predictive tools is questioned. The incorporation of hydrological models with GIS introduces additional problems. There is a perception that GIS can "generate information" via interpolation but this is only possible under certain specific circumstances. Also, the sophisticated graphics and data handling features of GIS can be used to seduce the user into an unrealistic sense of model accuracy. Alternative approaches to management issues are needed and these require a fundamental shift in thinking away from quantitative answers and toward a greater reliance on simple spatial modelling combined with qualitative reasoning. This approach is consistent with the availability of data and with our current ability to represent the system.

INTRODUCTION

In the past, the management of surface water resources has concentrated on reservoir analysis, flood forecasting and the control of industrial and urban point source pollution. With the increased pressure on natural resources and the growing awareness of environmental issues, the focus has shifted towards the integrated management of catchments. Now, the targeting of land-use strategies, non-point-source pollution control and the identification of the spatial movement of nutrients and sediments are critical management needs. While lumped hydrological models were adequate for addressing the earlier problems, the recent management needs have stimulated the development of new modelling approaches that explicitly take into account the spatial distribution of catchment properties.

Geographic Information Systems (GIS) are now widely being used for storing and visualizing spatially distributed catchment data. Distributed parameter models are being used to model spatially distributed processes which are, in principle, compatible with the data structure of GIS. These distributed models generally have been developed for small research catchments, while managerial decisions are needed for basins several orders of magnitudes larger. On the other hand, many of the more traditional lumped models are well suited to large catchments, but are clearly not capable of providing distributed information.

This paper expands on the dilemma. Starting with the technical aspects of upscaling from research to management sized catchments, we then discuss some fundamental issues in distributed hydrological modelling. The perception of hydrological information in the context of GIS is addressed and we proceed to ethical issues related to combining GIS and hydrological models. Finally, we ask the question: "Is there a way out of this dilemma?"

THE *SCALE* OF THE PROBLEM

Here, we take the advantages of distributed parameter rainfall runoff models at "face value" and explore some of the issues that arise when these models are applied to large scale management areas.

Distributed parameter hydrological models such as SHE (Abbott *et al.*, 1986), IHDM (Beven *et al.*, 1987), ANSWERS (Beasley *et al.*, 1980) and THALES (Grayson *et al.*, 1992a) have been designed to represent the processes of runoff in a theoretically sound manner. They are based on the solution of equations that represent the physical processes leading to runoff and the subsequent routing of the runoff through the landscape. By explicitly representing topography and utilizing a distributed parameter structure, the spatial variability of such catchment features as soils, vegetation, surface roughness and rainfall can be represented. Furthermore, by solving equations that are based on an understanding of the small scale processes, the parameters used in such models have a physical meaning and can, in principle, be measured in the field. This philosophy of modelling has been thoroughly presented in the past (e.g., Freeze & Harlan, 1969).

The algorithms in the distributed parameter model are often based on an understanding of processes at the scale of laboratory soil columns or small plots where such characteristics as saturated hydraulic conductivity and the soil water retention curve are well defined. When applied to the research catchments used for model verification, these characteristics are measured at numerous points in the field and it is assumed that within a particular model element, the parameter distribution can be replaced by a constant. Loosely speaking, this is the "continuum" assumption.

So what happens to model parameterization when the element size increases from the order of hundreds of square metres (as is the case when applied to research catchments) to the order of square kilometres (as is the case when applied to problems at the "management" scale)?

"Effective" values

As element size increases, the assumption that a measured value of some model parameter such as saturated hydraulic conductivity (K_s) is representative of the value for the whole element becomes more and more questionable. This is because the spatial variability in the parameter is too great to be represented by a single value. In principle it might be possible to take enough measurements to determine the distribution for the parameter but in practice, the required number of measurements is too great to be feasible (e.g., Loague & Gander, 1990). It must therefore be assumed that parameter values are "effective" values that result in the observed input/output relationship for the particular element but are not the manifestation of a physically measurable quantity.

Klemes (1986) considered this problem in relation to measurement of hydrological variables and stated:

"It also seems obvious that the search for new measurement methods that would yield areal distributions, or at least reliable areal totals or averages, of hydrologic variables such as precipitation, evapotranspiration, and soil moisture would be a much better investment for hydrology than the continuous pursuit of a perfect massage that would squeeze the nonexistent information out of the few poor anaemic point measurements ....".

The same comment applies to establishing parameter values for the algorithms in distributed parameter models. At present, the methodology for estimating "effective" parameters is incomplete and it is necessary to calibrate the important model parameters using measured data. This then introduces some additional problems associated with the type of data we commonly have available for model calibration (namely catchment outflow hydrographs) and the distributed nature of the models we are testing.

In a recent application of a physically-based, distributed parameter model to a 7 ha, agricultural catchment at Wagga Wagga, New South Wales, Australia, Grayson *et al.* (1992a, b) illustrated that it was possible to obtain similar fits of outflow hydrographs for different combinations of model parameters, but the distributed values of flow depth and velocity were vastly different. The parameters were "effective" in that they represented the input/output relationships, but the internal estimates of flow from the model were simply values that, at the scale of integration assumed by the model, provided satisfactory estimates of catchment outflow but were not true reflections of the catchment behaviour.

The problem of different parameter combinations yielding similar output has been recognized for some time (e.g., Philip, 1975; Anderson & Rogers, 1987). This points to the importance of analysing distributed model behaviour, rather than an integrated value such as runoff, when assessing the performance of distributed parameter models (e.g., Blöschl *et al.*, 1991; Moore & Grayson, 1991).

Threshold processes

An additional problem that arises when the element scale becomes too large is that the continuum assumption may break down. This occurs when, within an element, threshold changes in runoff response occur, such as when surface runoff is generated from saturated source areas within an element. This process can occur on areas much smaller than the element size of a "management" scale model.

It is possible to use distribution functions for parameters such as K_s in order to simulate the effect of such processes on element runoff response (e.g., Beven & Kirkby, 1979), however, these are only useful if the spatial characteristics of runoff are not required. The Representative Elementary Area (REA) concept of Wood *et al.* (1988) is the extension of this argument. They define a critical area for the simulation of catchment outflow. For catchments larger than the REA, it is not necessary to explicitly represent the internal pattern of hydrological behaviour. When the hydrological model is being used to "drive" a model of upland erosion, knowledge of these patterns is crucial to the representation of the erosion process. A distribution function approach enables the surface runoff hydrograph to be represented but does not indicate the areas from which the runoff is produced.

It should be noted from this discussion that the scale at which the continuum assumption is valid becomes a function of the dominant hydrological processes operating in the catchment and the type of output information required by the user. If the user is interested only in catchment outflow, rather than upland erosion, large elements might be used but in this case it is likely that much simpler models would be more appropriate (Beven, 1989). Hence, if the potential of physically based distributed parameter models is to be realized, the element size would need to be small enough for valid application of the continuum assumption for the process of interest.

Tipping the *scales*

The preceding discussion is based on the assumption that at the scale of a small research catchment, the fundamental premises of physically-based, distributed parameter models are valid. There is a growing body of literature that disputes this contention, further undermining the "conceptual soundness" of incorporating distributed parameter models into a GIS.

Physically based, distributed parameter models have been shown to be useful for the synthesis and interpretation of detailed data and for hypothesis testing but for the purpose of prediction, many authors have argued that they must be used with a great deal of caution, if at all (e.g., James & Burges, 1982; Klemes, 1982; Beven, 1983, 1989; Anderson & Rogers, 1987). This is because the difficulties in scaling occur not only between the research catchment and management area scales but also between the laboratory and research catchment scales. Authors such as Philip (1972, 1975), Klemes (1983), Hillel (1986), Anderson & Rogers (1987), Beven (1989), and Grayson *et al.* (1992a,b) have indicated that even at the research catchment scale, the algorithms used to represent hydrological processes may not be valid and even when they are, their parameterization is uncertain.

So not only are there problems associated with the large scale of management areas compared to research catchments but also of the fundamental premises of the original models.

In 1975 Philip, referring to the problems of interpreting parameter values from catchment outflow, stated that :

"........the investigator who believes in the physical reality of the parametric values he infers does so at his (or her) own peril."

Based on the preceding discussion this could be rephrased :

"........the investigator who combines distributed parameter hydrological models with GIS does so at his (or her) own peril."

GENERATING INFORMATION

Historically, GIS were developed for storing, retrieving and visualizing georeferenced information (Heatwole *et al.*, 1987). Early applications consisted primarily of mapping tasks and the efficiency of GIS in this area is widely recognized. When combining a GIS with a distributed parameter hydrological model, the mapping capabilities of the GIS greatly reduce the processing time for data preparation and presentation. This is sometimes referred to as a decision support system (Fürst, 1991). But in fact, these features neither enhance nor diminish the fundamental applicability of the hydrological model, i.e., the GIS is "hydrologically neutral" to the inclusion of a distributed parameter hydrological model. However, in many cases, the ease of data manipulation and interpolation within the GIS environment can encourage further misuse of hydrological models because a notion is held by some that the GIS can be used to "generate" hydrological information (e.g., Vandenbroucke & Orshoven, 1991).

In looking at the problem of data generation in more detail, we consider the hypothetical example of a distributed parameter hydrological model used for simulating surface and subsurface flow. Digital elevation data are used to define the element network and hydrological point data (e.g., precipitation and soil characteristics) provide the input for the model simulations. Some would argue that many of the scale related problems outlined in previous sections could be overcome simply by using a smaller element size. This is particularly tempting, since in most cases the spatial variation of topography is known at a much finer resolution than any hydrological data. Figure 1 hypothesizes the change of "hydrological information content" with changing element size (i.e., spatial resolution) for the above example. "Information content" relates to the upper limit of information that a model can yield from a particular data set.

The relationship in Fig. 1 implies that once the element size increases beyond the scale of the measurements (e.g., the spacing of soil samples), there is a decline in the amount of information provided by the model. This is a generally accepted notion and is caused by lumping of subgrid information. Conversely, as the element size is reduced, two possible scenarios are envisaged. In the first, topography is assumed to be a useful surrogate for the dominant hydrological processes so a finer spatial resolution increases the information content. This is the basis of terrain index methods such as those proposed by Moore *et al.* (1991). In the second case, hydrological processes are dominated by other factors such as geological structure, soil characteristics or preferential flow paths which are not related to topography. Thus, a finer spatial resolution does not increase the level of hydrological information. The information content can only be increased by interpolation if underlying relationships are present and can be defined.

For topography based interpolations, it is therefore vital for modellers (and users) to check the assumption that topography is related to the dominant hydrological processes in the particular management situation. The modeller who fails to recognize these subtleties will misinterpret the results and place greater confidence in the spatial variability of model output than its underlying performance justifies.

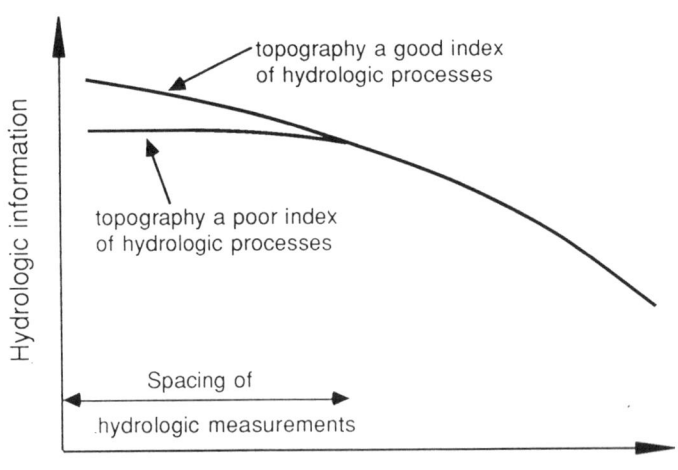

Fig. 1 Conceptual representation of the relationship between hydrologic information and element size for topography based interpolations.

ADDITIONAL DANGERS OF THE GIS ENVIRONMENT

In the previous section we made the point that, in theory, GIS are "hydrologically neutral" to combination with distributed parameter hydrological models. In practice this is not necessarily the case and issues of professional responsibility and ethics of model use arise. For example, when the sophisticated graphic capabilities of GIS are used to visualize the spatial predictions from a hydrological model, the results can be seductively convincing and conceal subtleties. Of course visualization is extremely useful but the modeller must ensure that this capability is used to reveal what is known rather than to conceal the assumptions and uncertainty underlying the coloured image.

In fact, the combination of user friendly GIS with complex hydrological models has created two serious problems (Heatwole *et al.*, 1987; Blaschke & Blöschl, 1992). Firstly, user friendliness assists individuals who may have little experience in hydrological matters, in running a model and producing fine looking graphics. Secondly, models have grown more complex, implying a larger number of parameters and assumptions that are not always obvious to the user. Providing good estimates of parameter values is a problem for an experienced modeller (e.g., Bathurst, 1986) let alone inexperienced users.

Many approaches to addressing these problems have been suggested in the literature, ranging from licensing the use of models, project examination by independent engineers and the integration of expert systems (James & Robinson, 1981; Heatwole *et al.*, 1987; ASCE, 1988; Blaschke & Blöschl, 1992). All of these options may prove useful, however it is the authors' view that much of the responsibility should reside with the model developers. It is the modelers' responsibility to fully inform potential users of the capabilities, assumptions and constraints of his/her creation.

"Modelers often develop a vested interest in the success of their own creations and hence are in constant danger of losing their objectivity, like the mythic King Pygmalion who fell in love with his own Galatea" Hillel (1986).

ALTERNATIVE STRATEGIES

It is accepted that tools are needed to address the issues facing catchment managers but the problems discussed herein indicate that these should not be addressed by the combination of distributed parameter hydrological models with GIS. The development of distributed parameter models specifically designed to operate at the larger scale is an active area of research which may ultimately yield "hydrologically defensible" results but in the meantime, alternative strategies are required.

There is the need for a fundamental change in the expectations of model performance and in the desire for quantitative estimates, no matter how scientifically dishonest, from models of the type discussed herein. In some cases, these models can provide a qualitative understanding of distributed catchment response and the relative change in response resulting from altered spatial characteristics (i.e., the patterns of catchment behaviour) but not quantitative estimates because of the problems already discussed. Patterns of catchment behaviour, however, can often be provided by much simpler methods than distributed parameter hydrological models.

Decision making processes involve the syntheses of both qualitative and quantitative information. Traditionally, hydrological information has been provided in a quantitative form but as the type of information required becomes more complex it must be recognized that we rarely have models that can provide sound, quantitative estimates. Reliance on quantitative estimates should be replaced by a qualitative understanding of the pattern of hydrological response and this information should be combined with simple reasoning to assist in the decision making process. These methods may be no more accurate than complex models but they are much simpler and in the words of Hillel (1986) are *"modest and parsimonious"*.

We use the problem of identifying erosion features on the Wagga Wagga catchment as a simple illustration of this approach. A simple wetness index based on catchment topography is used to identify areas within the catchment that are prone to surface saturation (Barling, 1992). It is then reasoned that rainfall occuring on these areas will flow downslope creating the pattern of

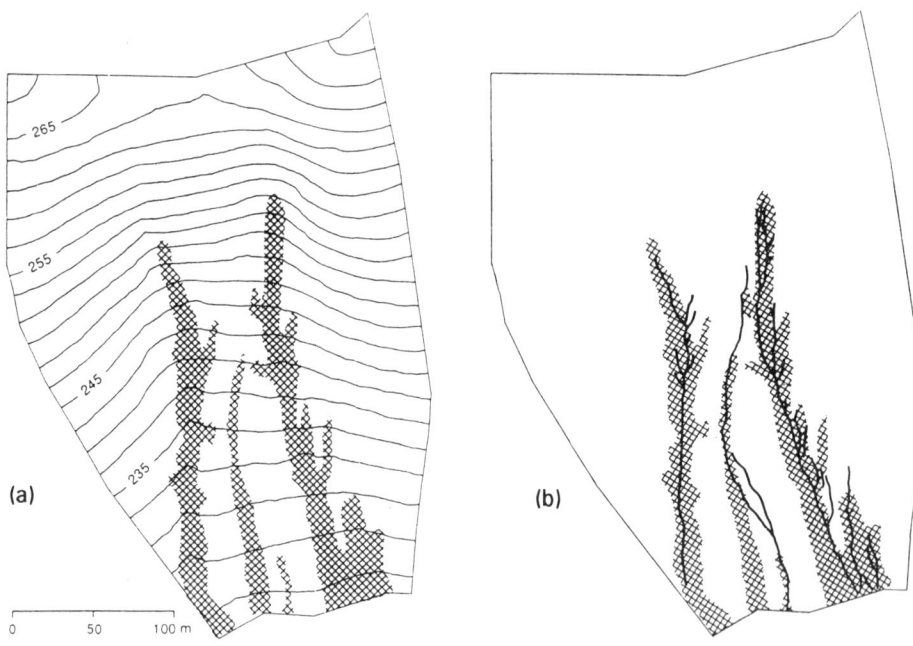

Fig. 2 Wagga Wagga catchment (7 ha), New South Wales, Australia. (a) Pattern of erosion risk for the saturated source area derived from a simple wetness index. Contour interval is 2 m. (b) Observed ephemeral gullies formed during storms in June 1987 overlayed on (a).

erosion risk shown in Fig. 2a. This is the area in need of protection (e.g., by a grassed waterway). Figure 2b shows the actual location of ephemeral gullies formed during a runoff event in June 1987 and illustrates that, for this example, the simple pattern/reasoning approach is able to capture the basic behaviour of the processes in question. Such approaches may be undertaken within or outside the GIS environment.

CONCLUSIONS

Despite the conceptual attraction of linking distributed parameter hydrological models with GIS a number of problems exist. These are due to the different scale between research catchments and management areas, the fundamental constraints of hydrological models and additional issues introduced by the GIS technology. There is a perception that GIS can generate information via interpolation but this is only possible under certain specific circumstances. In addition, the sophisticated graphics and data handling features of GIS can be used to seduce the user into an unrealistic sense of model accuracy. We therefore conclude that the spatial output should not be used for quantitative predictive purposes. This is particularly true when the models are applied to large scale management areas.

Alternative approaches to management issues are needed. These require a fundamental shift in thinking away from quantitative answers and toward a greater reliance on simple spatial modelling combined with qualitative reasoning. This approach is consistent with the availability of data and with our current ability to represent the system.

Acknowledgements The authors wish to thank the Australian Research Council, project no. SG 3911781, and the Fonds zur Förderung der wissenschaftlichen Forschung, Vienna, project no. J0699-PHY for financial support. Günther Blöschl is presently on leave at the Centre for Resource and Environmental Studies, Australian National University, Canberra, Australia.

REFERENCES

Abbott, M.B., Bathurst, J.C., Cunge, J.A., O'Connell, P.E. & Rasmussen, J. (1986) An introduction to the European Hydrological System - Systeme Hydrologique Europeen, "SHE" 1: History and philosophy of a physically-based, distributed modelling system. *J. Hydrol.* **87**, 45-59.
Anderson, M.G. & Rogers, C.C.M. (1987) Catchment scale distributed hydrological models - A discussion of research directions. *Progress in Phys. Geogr.* **11**, 28-51.
ASCE Task Committee on Turbulence Models in Hydraulic Computations (1988) Turbulence modeling of surface water flow and transport: part V. *J. Hydr. Engrg.* **114**(9), 1052-1073.
Barling, R.D. (1992) Saturation zones and ephemeral gullies on arable land in south-eastern Australia. PhD Thesis, The University of Melbourne, Melbourne, Victoria, Australia.
Bathurst, J.C. (1986) Physically-based distributed modelling of an upland catchment using the Systeme Hydrologique Europeen. *J. Hydrol.* **87**, 79-102.

Beasley, D.B., Huggins, L.F. & Monke, E.J. (1980) ANSWERS: A model for watershed planning. *Trans. Am. Soc. Agric. Engrs.* **23**(4), 938-944.

Beven, K. (1983) Surface water hydrology - Runoff generation and basin structure. *Rev. of Geophys. and Space Phys.* **21**(3), 721-729.

Beven, K. (1989) Changing ideas in hydrology - The case of physically-based models. *J. Hydrol.* **105**, 157-172.

Beven, K., Calver, A. & Morris, E.M. (1987) The Institute of Hydrology Distributed Model. *Inst. Hydrol. Rep. 98*, Wallingford.

Beven, K.J. & Kirkby, M.J. (1979) A physically based, variable contributing area model of basin hydrology. *Hydrol. Sci. Bull.* **24**(1), 43-69.

Blaschke, A.P. & Blöschl, G. (1992) Definig a pre and post-processing environment for groundwater modelling. In: *HYDROCOMP '92* (ed. by J. Gayer, Ö. Starosolszky & C. Maksimovic) (Proc. Budapest Symp., May 1992), 383-389. VITUKI, Budapest.

Blöschl, G., Kirnbauer, R. & Gutknecht, D. (1991) Distributed snowmelt simulations in an alpine catchment. 1. Model evaluation on the basis of snow cover patterns. *Wat. Resour. Res.* **27**(12), 3171-3179.

Freeze, R.A. & Harlan, R.L. (1969) Blueprint for a physically-based, digitally-simulated hydrologic response model *J. Hydrol.* **9**(3), 237-258.

Fürst, J. (1991) Entwicklung von Decision-Support-Systemen für die Grundwasserwirtschaft unter Verwendung geographischer Informationssysteme. *Österr. Wasserwirtschaft* **43** (11/12), 271-279.

Grayson, R.B., Moore, I.D. & McMahon, T. A. (1992a) Physically based hydrologic modelling: I. A terrain based model for investigative purposes. *Wat. Resour. Res.* **26**(10), 2639-2658.

Grayson, R.B., Moore, I.D. & McMahon, T. A. (1992b) Physically based hydrologic modelling: II. Is the concept realistic ? *Wat. Resour. Res.* **26**(10), 2659-2666.

Heatwole, C.D., Dillaha T.A. & Mostighimi, S. (1987) Integrating water research tools: process models, geographic information systems and expert systems, *Am. Soc. Agric. Engrs. Paper* 87-2043.

Hillel, D. (1986) Modelling in soil physics: A critical review. In: *Future Developments in Soil Science Research* A collection of Soil. Sci. Soc. Am. Golden Anniversary contribtutions presented at Annual Meeting, New Orleans, 35-42.

James, L.D. & Burges, S.J. (1982) Selection, calibration and testing of hydrologic models. In: *Hydrological Modeling of small watersheds* (ed. by C.T. Haan, H.P. Johnson & D.L. Brakenseik), Am. Soc. Agric. Eng., Monograph, no. 5, 435-472.

James, W. & Robinson, M.A. (1981) Standards for computer-based design studies. *J. Hydraul. Div. ASCE* **107**(HY7), 919-930.

Klemes, V. (1982) Empirical and causal models in hydrology. In: *Scientific Basis of Water Resource Management*, 95-104. Nat. Academy Press.

Klemes, V. (1983) Conceptualization and scale in hydrology. *J. Hydrol.* **65**, 1-23.

Klemes, V. (1986) Dilettantism in Hydrology: Transition or destiny? *Wat. Resour. Res.* **22**(9), 177S-188S.

Loague, K. & Gander, G.A. (1990) R-5 revisited, 1, Spatial variability on a small rangeland catchment. *Wat. Resour. Res.* **26**(5), 957-971.

Moore, I.D., Grayson, R.B. & Ladson, A.R. (1991) Digital terrain modelling: A review of hydrological, geomorphological and biological applications. *Hydrol. Processes* **5**, 3-30.

Moore, I.D. & Grayson, R.B. (1991) Terrain based prediction of runoff with vector elevation data. *Wat. Resour. Res.* **27**(6), 1177-1191.

Philip, J.R. (1972) Future problems in soil water research. *J. Soil Sci.* **116**, 328-335.

Philip, J.R. (1975) Some remarks on science and catchment prediction. In: *Prediction in Catchment Hydrology: a National Symposium on Hydrology* (ed. by T.G. Chapman & F.X. Dunin) (Proc. Canberra Symp., Nov. 1975) Australian Academy of Science, 23-30.

Vandenbroucke, D. & Orshoven, J.V. (1991) From soil map to land information system using GIS. In: *EGIS '91* (ed. by J. Harts, H.F.L. Ottens & H.J. Scholten) (Proc. Second European Conference on Geographical Information Systems, Brussels, April 1991), 1156-1164. EGIS Foundation, Utrecht, The Netherlands.

Wood, E.F., Sivapalan, M., Beven, K., & Band, L. (1988) Effects of spatial variability and scale with implications to hydrologic modelling. *J. Hydrol.* **102**, 29-47.

High resolution satellite imagery and GIS as a dynamic tool in groundwater exploration in a semi-arid area

PER GUSTAFSSON
Department of Geology, Chalmers University of Technology and University of Göteborg, 412 96 Göteborg, Sweden

Abstract High resolution satellite data together with thorough ground truthing, gives a powerful tool to aid groundwater exploration in arid and semi-arid areas. In bedrock areas with low primary porosity the exploration is directed towards finding fracture zones with locally increased permeability. Data from a semi-arid area in southeastern Botswana is used to show how interpretations of SPOT satellite data can be integrated with field data and geophysical investigations in a geographical information system to facilitate the identification of target areas in groundwater exploration. The conclusion of this study is that a digital approach implemented in the inception of a project, provides a time and cost effective exploration tool.

INTRODUCTION

The complexity of GIS and image processing software has previously prevented a full digital approach other than in very large groundwater projects. Training of operators, data entry and editing, etc. have been very time consuming and a digital approach has not been rewarding. During the last few years have, however, a large number of user-friendly and economically feasible image processing and geographical information systems emerged, which open up possibilities to use a digital approach in many groundwater exploration projects.

A common approach in groundwater exploration in arid and semi-arid areas is to combine existing data of an area, such as borehole information and major geophysical investigations, with either panchromatic aerial photographs or paper prints of satellite data to identify promising targets for groundwater abstraction. Based on the existing information a hydrogeological model is often delineated and used for prognostication.

Targets for detailed ground investigations are selected on criteria such as favourable aquifer bedrock, lineaments, geophysical anomalies, etc. The search is normally also narrowed by other factors, such as distance to population centres, existing groundwater abstraction, groundwater chemistry, etc. The established hydrogeological model is often used as a fairly rigorous platform for subsequent work in the project and results of detailed investigations are incorporated into *the existing model.*

Groundwater exploration in a semi-arid area can comprise several thousands of km^2 and include various types of valuable data. The amount of data often makes it impossible to do thorough analyses without entering the data into a computer database. Data entry and editing are however time consuming processes and few data are available in digital format of semi-arid areas, especially in developing countries. This has been preventive for a digital approach in many groundwater projects and only a part of the available data has been analyzed. The exploration process has thus been somewhat impaired.

STUDY AREA

The study area is situated in southeastern Botswana and comprise 400 km^2 of varied terrain, both with respect to topography and geology. The relief is moderate to high in the southern half and low in the northern half with the soil cover of the area being thin except around two major drainage systems in the northern half of the area. The vegetation is mainly of savanna type and there are pronounced differences between dry and wet season.

The bedrock geology is Archaean to lower Proterozoic in age and dominated by three major units; the Kanye Volcanic Group, the Gaborone Granite and the Waterberg Supergroup. The most important unit for groundwater exploration is the Gakobakwe Sandstone within the Waterberg formation. The sandstone has low primary porosity but is heavily intersected by fracture zones, i.e. secondary porosity, which could act as conduits for groundwater. The Gakobakwe formation has been subject to groundwater exploration for several decades and there are a large number of wells in this formation.

Available data of the area include a number of boreholes and geophysical traverses, both from previous projects and from an ongoing groundwater project in the area, the Molepolole/Mochudi Groundwater Exploitation Project. The area is also covered by a geological map in scale 1:125 000 and a topographic map in scale 1:50 000.

SATELLITE DATA AND INTERPRETATIONS

The satellite data used for this evaluation are SPOT XS and P from dry season, both standard processed, precision corrected data and specially enhanced images produced at the application development section at the Swedish Space Corporation (Österlund & Hansson, 1990). The former data have been tailor-made to match the topographical map sheets of Botswana. The data have been interpreted in two different image processing systems, EBBA GIS and ERMapper. The interpretations concentrated on lineaments and despite that the image processing systems are different and interpretations were made by two different operators on slightly different image sets, the results were very similar. A lineament map of the whole area is shown in Fig. 1.

The interpretation strategy was to map everything as accurate as possible on screen which *could* represent a true lineament and not to interpolate between or extrapolate any of the linear features mapped. The mapped lineaments were

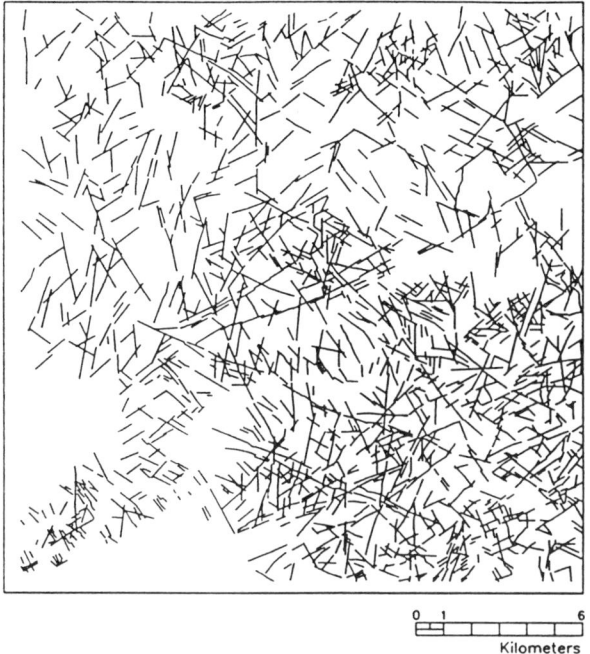

Fig. 1 Lineaments of the study area. Interpreted in the EBBA GIS image processing system at the Swedish Space Corporation.

complemented with a thorough field check of the area with the aim to try and verify or discard lineaments. No evidence of "false" lineaments, i.e. man-made lineaments, was found in the area. Interpreted lineaments were represented on the ground by either topographic lows, linear vegetation or both. Field checks were carried out both during wet and dry season. It was noted that during the wet season the fairly evenly distributed vegetation cover, especially in the southern part of the test area, obscure some of the interpreted lineaments, which are not accentuated by topography. During the dry season most of these difficulties are eliminated and vegetation differences stand out clearly. Choosing both the right registration date of the satellite data and the right time of year for field checking is thus very important.

Direction statistics have been calculated for interpreted lineaments, both the total distribution and the distribution in the major rock units of the area (Fig. 2). The latter has been performed using a GIS (Genamap), where the digitized geological map was merged with the lineament interpretations and a selection was made based on lineament centres. The GIS phase of the project is further described below.

GIS AS AN AID IN GROUNDWATER EXPLORATION

To handle the large amount of data of an exploration area, both data from previous investigations, such as boreholes, geophysical investigations,

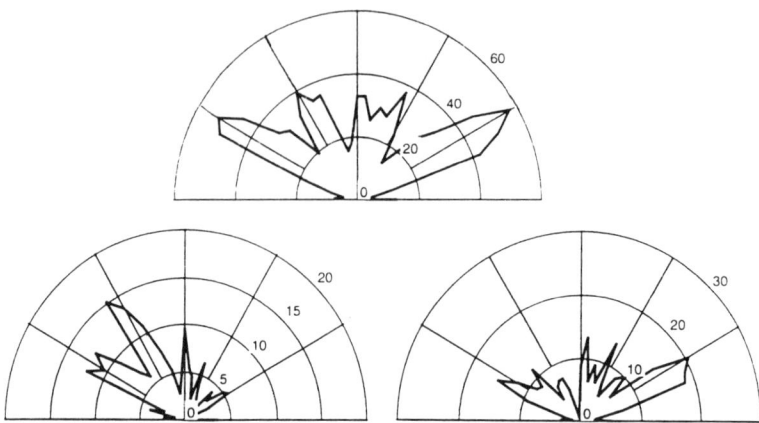

Fig. 2 Lineament distribution of the whole study area (top), in the Gakobakwe Sandstone (bottom left), and Kanye Volcanics (bottom right) with total lineament lengths (km) in five degree increments according to UTM.

geological and topographical maps, etc. and data from investigations in an ongoing project, it is necessary to use some sort of database system. Geographical Information Systems (GIS) have proved to be ideal for that purpose. A GIS can be used for data capture, storage, analyses, prognostication, presentation, and follow-up (Fig. 3) and constitutes an excellent tool for modelling work. It is important to start building the GIS database at an early stage of a groundwater project, enabling statistical analyses and prognostications as soon as enough data have been entered into the system.

Once the core of the database is established, it is easy to complement the data or add new types of data, with immediate recalculation of statistics and subsequent improvement of prognoses. It is important to regard the GIS as a dynamic or "live" database, reflecting the present knowledge of a project area, rather than just a tool for producing colourful maps and impressive data tables in the final stages of a project. In this study the following data have been incorporated in the Genamap GIS database:

a) SPOT lineaments, line data;
b) SPOT lineament centres, point data;
c) well locations, point data imported as ASCII text containing Easting and Northing coordinates;
d) well data, text data imported as columnar ASCII text;
e) magnetic anomalies, point data digitized with a z-value describing inferred type of anomaly;
f) electromagnetic (Slingram) anomalies, point data digitized with a z-value of estimated strength;
g) bedrock units, polygon data imported from Arc/Info in DXF format.

The quality of the borehole data varies, from very detailed descriptions, including lithological logs, water strikes, accumulative yield and pumping test data to old boreholes with no information other than a vague site description.

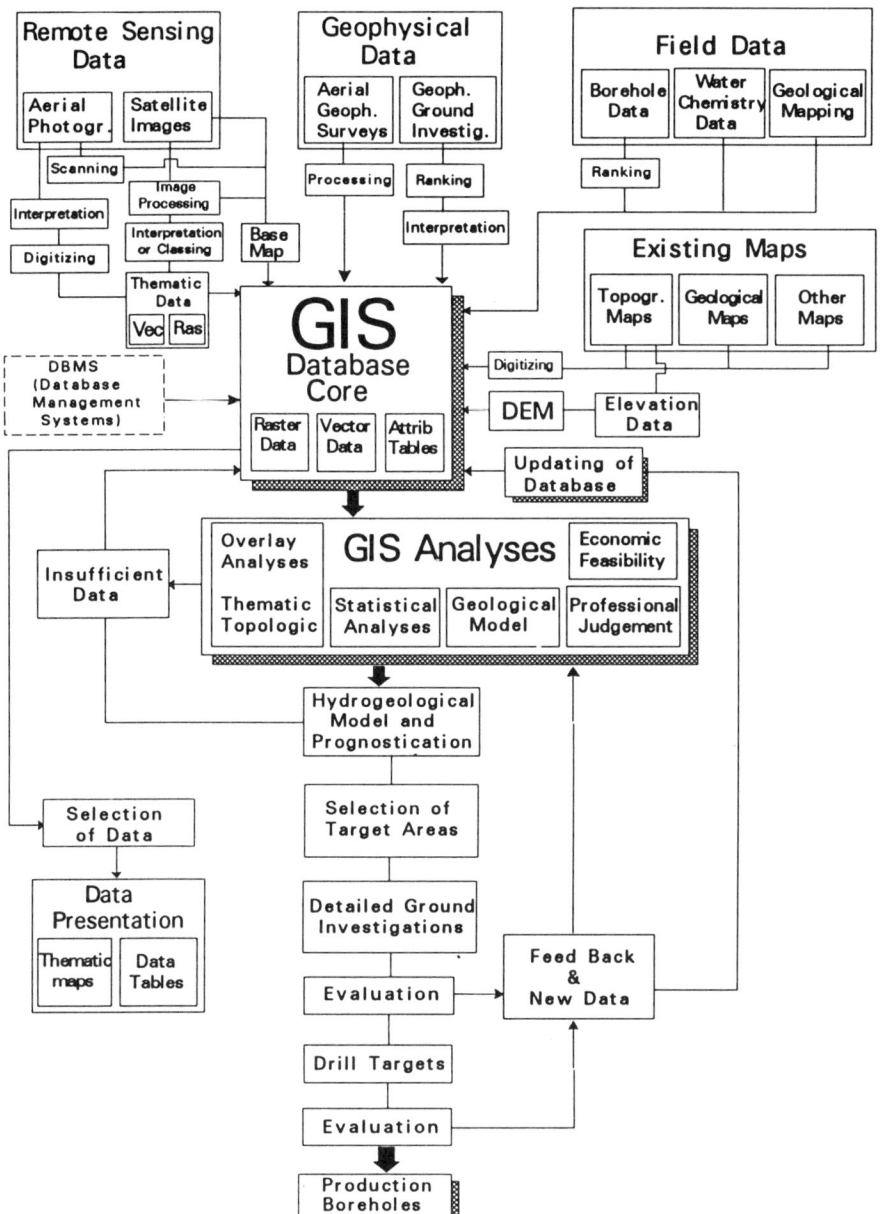

Fig. 3 The Geographical Information System as a tool in groundwater exploration.

This naturally restricts the analysis possibilities, which is shown below. The borehole data were, however, despite this put together in a small database for this methodology study.

GIS ANALYSES

An advanced GIS has a large number of analyzing tools which are applicable in a groundwater exploration project. It is often a question of deciding which analyses are relevant for a particular project and what limitations are set by the quality of the input data. GIS analyses can be separated into two main types; *selections* and *overlay* analyses, which both are applicable in groundwater projects.

In this study the correlations between interpreted lineaments and geophysical and borehole data have been studied. Buffer zones in 80 m increments from the lineaments were created and proximity analyses were performed to be able to compare geophysical ground investigations with lineaments interpreted from SPOT satellite data. Figure 4 show the correlation between electromagnetic anomalies (Slingram) and lineaments. A comparison was also made between borehole yield from around 50 boreholes in the northern half of the study area and lineaments (Fig. 5). It is important to take into account the area of the buffer zones, when making statistical analyses. In this study the buffer zone areas first increase up to 80 m from the lineament and then rapidly decrease as they meet buffer zones from other lineaments.

The comparison between borehole yield and distance to lineaments implies that satellite lineaments should be avoided for groundwater exploration, which is contrary to most opinions. This is probably explained by the fact that the borehole coordinates are very unreliable, with some believed to be several hundred meters off the location given in the borehole archives. This together with too few boreholes to motivate a statistical analysis of this kind, makes conclusions futile. The methodology is however promising for analyses of boreholes with reliable coordinates to e.g. identify lineaments or lineament directions with a possibly higher groundwater potential. A thorough well survey using e.g. GPS for positioning, can thus be a very valuable asset in a groundwater project.

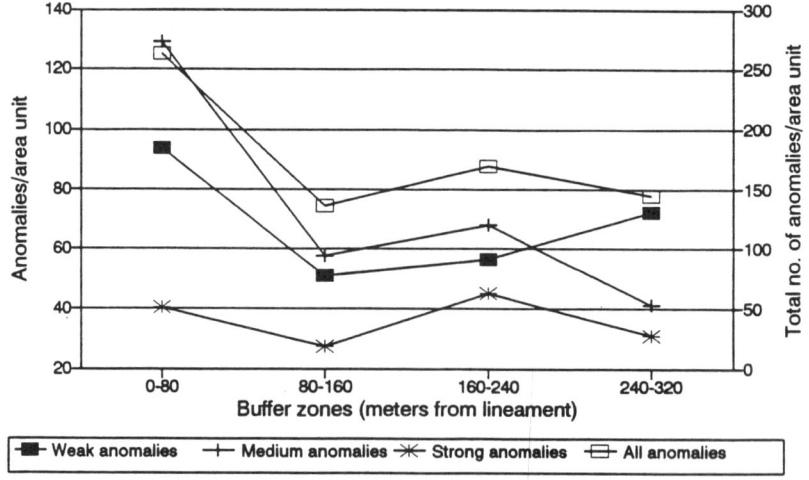

Fig. 4 Correlation between Slingram anomalies and lineaments. Buffer zones in 80 m increments from lineaments. Compensation made for decreasing area of buffer zones.

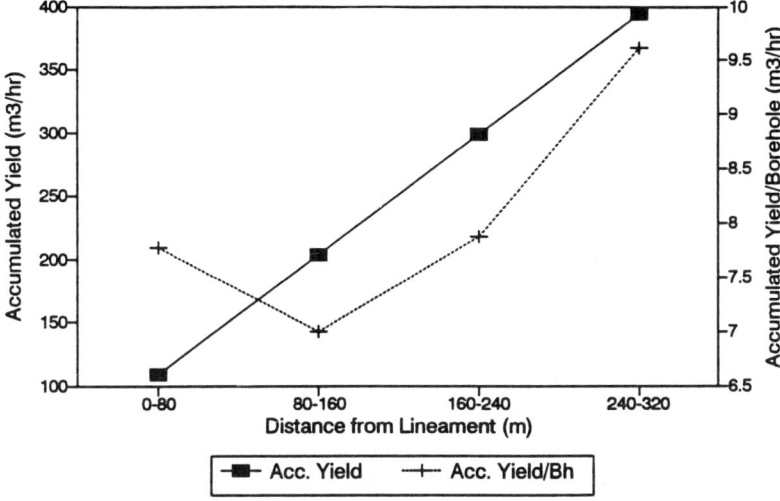

Fig. 5 Correlation between borehole yield (m^3 hour^{-1}) and distance to lineament. Buffer zones created in 80 m increments from each lineament. Continuous line representing accumulated total yield and dashed line representing accumulated total yield divided with the number of boreholes.

An advanced GIS should handle both raster and vector data and have import and export facilities for numerous formats. It is also valuable if the image processing is either integrated in the GIS or as with ERMapper, contain dynamic links to major geographical information systems. Genamap GIS which is used in this project handles both raster and vector data and have a large number of analyzing possibilities within the system as well as an open structure, enabling communication with e.g. advanced statistical packages.

CONTINUED WORK

This study has been a methodology study preceding an ongoing larger project, covering a 2000 km^2 area. The ongoing project is more in detail looking into GIS analyses and also comprise a more comprehensive interpretation of satellite data in the image processing system ERMapper. Interpreted features comprise lineaments, drainage patterns, and dry season green vegetation. A digital elevation model is incorporated into the GIS database and is being analyzed in the image processing system for lineament identification. The elevation data have been extracted from existing topographical maps.

The GIS phase of the ongoing project comprise correlation studies between lineaments and *GPS-positioned* borehole and geophysical data, which will enable a more concise evaluation of the value of satellite lineaments in groundwater exploration. The GIS will also be used to select promising target areas which can be used in future groundwater projects in the area. The compilation of a digital groundwater potential map of the whole project area is another important part of the project, which will continue until mid-1993.

CONCLUSIONS

The main conclusion from this study is that digital satellite data will increase in use as a very valuable source of data for groundwater projects, especially for planning of more detailed investigations. If areas for exploration can be narrowed by the use of image processing and interpretation, the more cost-effectively involved personnel can concentrate their talents (Drury, 1987). The large amount of interpreted data together with existing field data and geophysical data makes a GIS necessary to be able to in a satisfactory way analyze the data. It is, however, of great importance that the accuracy of the data entered into the GIS database is carefully evaluated, both the coordinates and attributes. GPS-technology is extremely valuable for acquiring accurate coordinates and will most certain increase in use in the future.

Entering and editing data is still very time consuming and will continue to be so until digital data are readily available of exploration areas. It is therefore important that methods for capturing analogue data in digital format are further developed and that geographical information systems have extensive import and export facilities. The GIS approach will greatly benefit from being implemented in the inception of a project, giving a powerful tool for prognostication and modelling. Once the core of the database is established it is easy to make updates, perform new analyses and recalculate statistics as well as producing maps and data tables to accompany for instance monthly reports.

Acknowledgements The Swedish National Space Board has provided funding for parts of this study. Permission to publish the material herein is granted by the Director, Department of Water Affairs, with the authority of the Permanent Secretary, MMRWA. Copyright of these data rests with the Government of the Republic of Botswana. The author would like to thank Paul Larkin of Aqua Tech and Jan Lindberg of VBBVIAK for supplying data from the groundwater project. In addition, the author would like to thank Mikael Gråsjö and Mats Öberg for the cooperation during the work with the report "Interpretation of tectonic structures on enhanced satellite images for groundwater exploration" (Gråsjö *et al.*, 1991), which has formed the basis for this paper and the continued work.

REFERENCES

Drury, S.A. (1987) Digital processing of images in the visible and near infrared. In: *Image Interpretation in Geology*, 118-148. Allen and Unwin (publishers) Ltd., London.
Gråsjö, M., Gustafsson, P. & Öberg, M. (1991) Interpretation of tectonic structures on enhanced satellite images for groundwater exploration. *Report FUX 830, Swedish Space Corporation, Solna, Sweden*.
Österlund, H. & Hansson, G. (1990) Tailor-made Satellite Images for Groundwater Assessment. Poster presented at International Symposium on Remote Sensing and Water Resources, Enschede, The Netherlands, 1990.

GIS application to water resources management in the land planning context: a methodological proposal

F. JEMMA
Consiglio Nazionale delle Ricerche (National Research Council), Institute of Land Planning, Via P.Castellino 111, 80131 Napoli, Italy

Abstract The water resources management is a very complex problem in several countries of the world. Any approach in solving this problem should take into account its wide range of physical, administrative and land planning components. In the present paper it is reported a new methodological approach to water resources management, at basin scale, achieved using GIS data processing scheme and characterized by a complexity increase with a contemporary synthesis level increment of the information.

INTRODUCTION

The water resources management as a whole, is made up by several distinct, but extremely correlated, components. Moreover, there are some others external components that are very important contributing factors in the management of the water resources system.

It does mean that problems related to this system present a very big complexity. Each action undertaken in one sector produces remarkable repercussions in other sectors, often unforeseeable if not analyzed as a whole system that in many circumstance can be represented by the environmental one.

The management of a so big system implies its knowledge at a very large spectrum scale, supported by adequately articulated and tested administrative and environmental planning procedures.

PHYSICAL APPROACH TO WATER RESOURCES

In Italy the available water reserves are about 2600 m^3 $year^{-1}$ for person (under the EEC average) (Berbenni, 1987). Rivers are widely used as discharge means of human and industrial wastes, groundwaters are under serious pollution risks, rain has becoming acid and water is still thought to be a renewable resources to squander.

The water demand for the growing residential and industrial development and for agricultural activity is increasing in contrast to a serious water management problem.

It is necessary the determination of both the real water demand (in terms of actual and prevailing uses) and the real water supply (in term of water quality and quantity) (Jemma, 1992).

An increasing water demand (classified in several water uses) needs a deep knowledge of the characteristics of water resources in order to offer a more rational water supply in terms of quality, quantity, continuity and costs.

In Italy only the water demand for civil average uses is known (from 150 to 450 l day^{-1} per person; Berbenni, 1987), without regarding the seasonal demand for tourist purposes; nothing (or very little) about the water demand for other uses (Bollini, 1990). The knowledge of characteristics and potentiality of italian water resources is quite little (Bollini, 1990) and fresh water is supplied without any criteria, allowing a lot of water squander.

Moreover, it is important that characteristics and problems about water resources are not simply related to a local scale but belong to an environmental complex system that should be analyzed at least at the hydrographic basins scale.

In Italy, all concerning water, from the single well to the water consumers basin, has mainly been thought as local problems (at different scales), unrelated to other environmental aspects. This has led to several geological and environmental emergencies.

A different approach to fresh water problems needs the pursuing of the following targets (Berbenni, 1987; Bollini, 1990; Jemma, 1992; Cardarelli et al., 1991):

(a) knowledge of hydrogeological parameters in terms of water flow/down flow ratio, surface/underground water changes, water table depth, groundwater supply and motion, chemical and physical quality of water, etc.;

(b) safeguard of the comprehensive biological water quality with insurance of suitability for several water uses;

(c) rationalization of water supply (locating alternative water sources for industrial and agricultural uses), water distribution (building a separate water supply network for drinkable use and with the) and water discharge (building a separate drain and sink networks);

(d) encouragement of used water recovering and re-utilizing together with water consume reduction in productive sectors;

(e) establishment of protection areas for wells supply and water re-loading zones together with the drawing up of groundwater vulnerability maps and water quality maps.

ADMINISTRATIVE WATER RESOURCES MANAGEMENT

The supporting action of the administrative procedures is the most important factor for an adequate water resources management.

The starting point is the application of laws about safeguard by water pollution risks, water reclamation, water supply regulation (according to a widespread meaning of water pollution not related to a particular water use) (Caravita, 1990).

Until the 70s, in Italy, the dominating thought was related to the inexhaustibility of the resources. The first law about water pollution safeguard was the no.319 of the 10/5/76. It introduces some fixed restrictions about companies outflow pollution, an authorizations system managed by the municipalities and a delegation to the districts for the compiling of a water restoration plan. This law failed completely, on the formal point of view, because the Public administration could not afford the task of management and control and, on the substantial point of view, because it did not take into account all environmental risk as a whole (Stile, 1977).

The second point deals with the establishment of an administrative management body that acts at a large scale (overcoming the administrative border limits) taking into account all environmental processes.

Only in the 1989 the above mentioned laws about water problems have been finally resolved in a breadth regulation (no.183/89) on soil safeguard creating the hydrographic management authorities. It is interesting to mention that this kind of law was already alive in France from the 1964 (Agences financiéres de bassin) and in England and Wales from 1973 (Water Act) (Cutrera & Nespor, 1990).

This law splits the national territory in several hydrographic basins classified according to the importance. The law is a technical and prescriptive tool necessary for the soil safeguard and acts as a Master Plan for the administrations. The weak side of this law is that competences are shattered between several Ministeries (Cutrera & Nespor, 1990).

The third point necessary to an adequate water resources management is the establishment of a national environmental information network able to manage all data regarding fresh water (water characteristics, water pollution monitoring, etc.).

In Italy there is a huge production of environmental data that are not linked yet to a national environmental network as it is supposed in the European Economic Community (EEC) requirements (Jemma, 1992; Nespoli, 1992). A little step toward the resolution of this problem has been the establishment of the Environmental Information System of Italy (SINA), not working yet that will create several data banks regarding the management of environmental data (i.e., scientific books, laws, alphanumeric data, imagery, etc.).

WATER RESOURCES PLANNING APPROACH

Land planning represents the field where environmental features (water resources included) must be managed in the context of all territorial problems. Unfortunately, in Italy, the main lines of land planning methodology do not mention any regulation about environmental problems apart those related to geological seismic risk.

In order to make land planning regulation coherent with all territorial and environmental features it is necessary to develop methodologies of analysis and research that leads "land planning to be thought as a continuous external input process directs to a dynamic system controlled by itself" (C.N.R., 1987), overcoming the traditional town-planning limits based only on the zoning approach.

The use of techniques and instruments (i.e. GIS) that allows a real time representation of territorial development processes represents (Cardarrelli *et al.*, 1991) the way to:
(a)   deal with the environmental problems in a systemic way;
(b)   solving the complexity through a multi-disciplinary work;
(c)   optimize the decision choices considering all the involved features.

GIS IN WATER RESOURCES MANAGEMENT

An example of environmental planning model regarding the water resources problems above mentioned has been achieved (within the C.N.R. Strategic Project "Clime, Environment and Territory in Southern Italy") making use of both the GIS characteristics and the classical relational database (RDB) architecture (Cardarrelli *et al.*, 1991; Atzeni *et al.*, 1985) throughout:
a)   analysis of the research problems;
b)   data identification in term of attributes, relative domains, availability, costs, and system implementation;
c)   definition of GIS relations necessary to the research methodological path;
d)   normalization of the RDB architecture in order to eliminate management anomalies and to simplify the GIS use.

The GIS data processing scheme (Fig. 1), inhere proposed, is characterized by a complexity increase contemporary with a synthesis level increment and by the possibility to process dynamic information (Cardarrelli *et al.*, 1991).

It is planned through five successive steps. In the first processing level, data integration is absent (there is only a data modification in function of assigned variables). At this step it is possible to assign weighted indexes to the data in order to allowed a subjective analysis. In the upper processing levels, data overlay is achieved in relation with the water resources management targets.

Each box of the GIS data processing scheme represent a layer of information, made up by one graphic map to which are associated one or more data tables.

The data set are split in the Resources, Uses and Management groups (Jemma, 1992; Besio, 1991; Janssens, 1991).

The Resources group, or natural system, represents the environmental picture in terms of potentiality and vulnerability. Data of this group are processed in order to draw the ground table susceptibility map.

Using several hydro-geological information, this map is produced overlaying maps regarding the vertical and horizontal permeability, the low permeability layer, the potential infiltration and the hydrographic basin contour.

The Uses group, or anthropogenic system, describes the kind of water resources use discharges (in terms of need and pollution hazard). Data belonging to this group are processed to draw the water supply map and the real and potential pollution map.

The water supply map is produced overlaying information about the hydrographic balance, the water quality monitoring, the aqueduct network status and the real water need for each land cover class.

In order to have the most true picture of the pollution hazard data are collected in two different way and then correlated to produce a measured map. In the first way a data monitoring program in wells and rivers produces the real pollution map. In the second way a deductive study of the polluting load of each land cover class gives the potential pollution map.

The Management group, or social, economic and political system, governs the previous. The drawing product of the processed data for this group is the management system map.

This map is produced overlaying all information about planning tools and laws.

The final map, achieved (Jemma, 1992) overlaying the above mentioned maps with an ARC/INFO PC system, give a synthetic picture of the real status. Ascribing some variable (i.e. time) or making some data modification to the data set of the first processing level the final representation begins a trend line or a project option representation.

CONCLUSION

The management of water resources is a worldwide complex problem. In Italy, the resolution of all these problems seems to be a long term target.

Apart from the physical and administrative components of these problems, inhere reported, it is my thought that the application of new land planning models together with the use of techniques and instruments, like GIS, could favor the resolution of several water resources management problems.

In the present paper it is reported a new methodological approach to water resources management, at basin scale, achieved using GIS data processing scheme and characterized by a complexity increase with a contemporary synthesis level increment of the information.

Starting from 27 layers of graphic/alphanumeric information, the final map, a synthetic picture of the real water resources status, is drawn throughout five overlay processing levels.

Intermediate products of the processing work are the ground water susceptibility map, the water supply map, the real and potential water polluting map and the management system map. The further possibility of this methodological GIS structure is the continuous test analysis and control in relation with the different land planning projects and the related environmental evolution processes.

REFERENCES

Atzeni, Batini & De Antonellis (1985) La teoria relazionale dei dati (Relational theory of data). Boringhieri Publ.

Berbenni, P. (1987) L'acqua potabile per il 2000 (Drinkable water for the '2000). Terra, 3.

Besio, M. (1991) Object oriented GIS improving environmental compatibility in italian rural landscape planning. In: *Proc. EGIS '91, II Europ. Conf. on GIS*, 2-5 April 1991, Brussels, Belgium.

Bollini, G. (1990) Il bacino idrografico: un approccio ecosistemico alla pianificazione territoriale (The hydrographic basin: an ecosystemic approach to land planning). Terra, 13.

Carravita, B. (1990) Diritto pubblico dell'ambiente (public rights of environment). Il Mulino Publ. Bologna

Cardarelli, U., Cipriano, F., Jemma, F. & Pedone, R. (1991) Remote sensing and GIS: Experimentation of two environmental technology tools applied to a new land planning proposal of the Sarno river valley (southern Campania - Italy) coherent with the hydrogeological reclamation of the area. *Int. Symp. Geoscience and Remote Sensing*, 3-6 June 1991, Helsinki.

Cardarelli, U., Cipriano, F., Greco, A., Jemma, F. & Pedone, R. (1991) Sperimentazione del telerilevamento da satellite e dei geographical information systems (GIS) per una nuova proposta di pianificazione territoriale della valle del fiume Sarno (Capania) (Experimentation of satellite remote sensing and geographical information systems (GIS) for a new land planning proposal of the Sarno river valley (campania). III Workshop "Informatica e Scienze della Terra" Sarnano (MC), Italy, in press.

C.N.R. (1987) Interazione e competizione dei sistemi urbani con l'agricoltura per l'uso della risorsa suolo: il quadro regionale in Emillia Romagna (Urban system interaction and competition with the agriculture in the soil resource use). C.N.R. P.F. I.P.R.A. ; Università degli studi di Bologna, Centro ricerche produzione animale Reggio Emilia. Pitagora Publ. Bologna.

Cutrera, A. & Nespor, S. (1990) La difesa del suolo e la politica delle acque (Soil safeguarde and water management). Quad. Riv. Giurid. Amb., 3. Giuffrè Publ. Milano.

Janssens, P. (1991) Land related information systems for research and planning. In: *Proc. EGIS '91, II Europ. Conf. on GIS*, 2-5 April 1991, Brussels, Belgium.

Jemma, F. (1992) Environmental planning model at basin scale achieved using satellite remote sensing data and geographical information systems. Int. Symp. of Photogram. and Remote Sensing, 2-14 August 1992, Washington, D.C, USA.

Nespoli, G. (1992) Una rete informatica per garantire l'efficienzy) (An informatic network to guarantee the efficiency). GEA, V, I. Maggioli Publ.

Stile, A.M. (1977) Prevenzione e repressione dell' inquinamento indrico nella legge 10/5/76 n.319 (Prevention and repression of water pollution in the law 10/5/76 n.319). Jovene Publ. Napoli.

Environmental modelling and GIS: dealing with spatial continuity

KAREN K. KEMP
National Center for Geographic Information and Analysis,
University of California, 3510 Phelps Hall, Santa Barbara, California 93106-4060, USA

Abstract Linking a GIS to a spatially distributed, physically-based environmental model offers many advantages. However, the relationship between the reality being represented by the mathematical model and the data models used to organize the spatial data in the GIS are generally incompatible. While many hydrological and other environmental models are based on theories that assume continuity, current GIS data models can only represent continuous phenomena in a variety of discrete data models. A strategy which enables modellers to work directly with the spatial data as spatially continuous phenomena is outlined. This strategy allows the manner in which the spatial data has been discretized and can be manipulated to be treated independently from the conceptual modelling of physical processes.

INTRODUCTION

Environmental issues are among the most important facing decision-makers today. The dynamics of the hydrologic and atmospheric systems of the earth imply that all environmental systems are tightly interrelated, dynamically and spatially. Impacts in one location usually have effects in others. Spatial data, systems for managing that data and analytical techniques for converting that data into information are now vital tools in the assessment and management of a healthy natural environment.

Considerable progress is being made in integrating spatial information systems and mathematical models of the environment (cf. Goodchild *et al.*, 1993). For most environmental modelling projects, GIS is seen as a convenient and well structured database for handling the large quantities of spatial data needed. Traditional GIS tools such as overlay and buffering are also important for developing derivative datasets that serve as proxies for unavailable variables. Many experts also expect that as better spatial analysis methods become incorporated into GIS, GIS will also become an important tool in all aspects of modelling, including model building, validation and operation. However, there are significant incompatibilities preventing true integration. GIS manages static and discrete data while environmental models deal with dynamic and continuous phenomena. GIS databases contain information on location, spatial distribution and spatial relationships while environmental models work on a basic currency of mass and energy transfer (Maidment, 1993). In order to fully integrate the two we need to add dynamics and continuity to our understanding of spatial data, and spatial interaction and functionality to the environmental models.

This paper addresses the first of these needs by considering the implications of working with continuous phenomena directly in the context of GIS. Following a consideration of how spatially distributed phenomena are characterized in environmental models, a strategy for representing these phenomena so that they can be efficiently and conveniently incorporated into the computer code of mathematical models is introduced.

CONTINUITY IN ENVIRONMENTAL MODELLING

Discrete representations for both equations about continuous processes and the continuity of space are widely available. Finite difference and finite element numerical solutions to differential equations discretize time and space into small units. Just as the equations are continuous, so are most of the phenomena being described. These continuous phenomena may be described by fields. A physical field is traditionally defined as an entity which is distributed over space and whose properties are functions of space coordinates and, in the case of dynamic fields, of time. Scalar fields are characterized by a function of position and, possibly, time whose value at each point is a scalar while the value at any location in a vector field is a vector (i.e. wind fields where the value at a location has both magnitude and direction). Since we cannot measure continuous phenomena everywhere, it is necessary to develop techniques for gathering information about fields at a finite number of points and representing it with a finite collection of data.

Physical fields are distinguished by their extremely high degree of spatial autocorrelation. Thus, while we cannot measure the value of a continuous phenomenon everywhere, we know that locations near those we can measure will have similar values. Knowledge of spatial autocorrelation, however, gives us little information about how rapidly and erratically the values change between locations at which we know the value. In order to represent and manipulate fields for mathematical modelling, we must have some way of linking the continuous variation of the field as it is observed in nature to the individual numbers or letters stored in the computer as representations of the value of the field at certain locations. The linkage between continuous reality and its representation is achieved by:
(a) dividing continuous space into discrete locations for which discrete values can be measured and recorded, and;
(b) establishing a rule for interpolating unknown values between these locations.

The first of these steps is known as *discretization*. The second step is accomplished through the use of *spatial data models*.

When considering the discretization of space for computer representation it is useful to recognize different levels of abstraction that are needed to transform spatial continuity into digital representations. *Geographic models* (a term proposed by Grelot, 1985) are those conceptual models used by environmental modellers and others as they evolve an understanding of the phenomenon being studied and extract its salient features from the background of infinite complexity in nature. *Spatial data models* are formally defined sets of entities and relationships used to discretize the complexity of geographic reality (Goodchild, 1992). The entities in these models can be measured and the models completely specified. *Data structures* describe details of specific implementations of spatial data models. Here we are concerned with the interface between geographic models and spatial data models.

SPATIAL DATA MODELS FOR FIELDS

In a literal sense, just as a hydrological model represents hydrology and a plant growth model represents plant growth, the term "data model" suggests the idea of a formal representation of *data*, not of reality. Unfortunately data modelling is often confused with issues of data structure (Goodchild, 1992) and becomes mired in questions of how points, lines and areas should be represented. In fact, this confusion of terms may be partially at fault for a lack of understanding about the fundamentally different ways these data models represent reality. Each one embodies one or more important assumptions about the form of the reality represented. These assumptions critically affect how the data model can be manipulated mathematically. For the representation of fields, there are six different spatial data models available: cell grids, polygons, TINs, contour models, point grids and irregular points.

A cell grid partitions the entire study area into rectangles which are evenly aligned in two perpendicular directions. Regular tesselations which are not rectangular (e.g. hexagonal tesselations) are considered polygon models. The most common example of a cell grid is a remotely sensed scene composed of pixels. The value of the phenomena over the entire area covered by each cell (the spatial element) is represented by a single value even though there may be considerable variation within the cell. As a result, values change abruptly at cell edges.

Polygons partition the entire study area into irregularly shaped contiguous regions. Like cell grids, the value of the phenomena within a single polygon is defined as a constant and changes abruptly at polygon edges. The boundaries of a set of polygons may be defined either by the phenomena (e.g. vegetation zones) or they may be independent of the phenomena (e.g. cut blocks, watersheds when used to partition soil characteristics). In environmental databases, polygon structured data is often categorical (e.g. soils, vegetation types, watershed). To be useful in mathematical models, these categorical datasets are usually linked to a relational table which describes various numerical and other properties of each class.

TINs (Triangulated Irregular Networks) partition the entire study area into triangular regions. The value of the phenomena is specified only at triangle nodes. However, since the sloping surface of each triangle is assumed to be flat, values anywhere on a triangle face can be calculated directly from the values at the nodes. While there is no abrupt change in value at triangle edges, there is an abrupt change in slope. The location of boundaries is defined by the location of the nodes. Thus the correspondence between the real surface and that represented by the surface of triangles is determined by the set of points (nodes) selected to define the critical points of the surface. Since TINs define continuously varying surfaces, TIN models can never be used to structure categorical, non-numerical data.

Point grids store the value of the phenomena at every intersection in a rectangular grid. These values represent the actual value of the phenomena at that location. The location of each sampling point is determined by the grid and is independent of the phenomenon.

Irregular point models store the value of the phenomenon at irregularly scattered point locations. The location of the points may be determined by the phenomenon, in which case, values may be assumed to be representative of neighboring locations (e.g. carefully selected representative locations for the collection of rainfall data). However, irregular point data may also be collected at locations determined by concerns other than the phenomenon under study (e.g. weather stations located at airports). In this case, the value at each point describes only the conditions at that particular location.

Contour models are unique amongst these spatial data models used for continuous phenomena. Unlike the other spatial data models, contour models are constructed by holding the value of the phenomena constant and determining the location. Lines are constructed to connect adjacent locations whose value matches that of the desired contour line value. This model explicitly identifies all places which exhibit a value expressed by one of the contour lines. However, the value of the surface is defined only along the contour lines. The location of the contour lines is determined by both the phenomenon and the selected values at which contour lines are drawn. Like TINs, contour models partition space into regions over which the value of the phenomenon varies. Between the lines the value of the phenomenon varies within the range defined by the values of the bounding contour lines. Finally, since contour lines must be measured on a continuous measurement system, contour models can never represent categorical data.

Goodchild has suggested that these six models represent two distinct ways of exploiting the spatial autocorrelation of fields (Goodchild, 1992). *Piecewise* models make use of the assumption that nearby locations are similar while *sampled* models exploit the fact that if we know the value at one location we can estimate the value at nearby locations. Piecewise models dissect the surface into contiguous regions. A value is defined at every location on the surface. The continuous variation of the value of the phenomenon within each region is described by a simple mathematical function. In two models, cell grids and polygons, this mathematical function is a constant while in the TIN model the function is linear. Thus if the values of the phenomena represented are drawn as a third dimension, grid cell and polygon models produce a stepped surface of horizontal regions, while the regions of the TIN model are sloping planes with the edges of each region coincident to those of its neighbors. The crucial assumption in all piecewise models is that the value or function assigned to each region is representative of the average value or general trend of the surface in the region. While each individual point may not be represented precisely, it is assumed that the integral of the values over this surface would produce the value or linear function assigned.

Sampled models use an entirely different approach. In these models, the phenomenon is sampled precisely at a number of different points. Sampling is done either at points as in point grids and irregular point models, or along lines as in contour models. No values are assigned to locations that have not been sampled and, except in the limited case of contour models, no information is provided about the variation in the value of the phenomenon between sample sites. In order to represent the continuous surface between these sample locations, an assumption must be made that variation between these points can be described by a mathematical function through the process of interpolation. The interpolation function chosen varies, even for a single dataset used in different applications.

Thus we have two groups of models with widely different basic assumptions. While piecewise models provide a generalized representation of the continuous phenomena, sampled models provide precise data at a limited number of locations. Sampling schemes may be unbiased (as in point grids) or biased (as in contours and some irregular point models). In terms of surface representation, it is useful to regard the six models in three distinct groups. Constant piecewise models depict a stepped horizontal surface with vertical breaks at cell or polygon borders. Surface models, TIN and contour models, depict a continuous surface with varying values within regions and continuity across borders (triangle edges or contour lines). Point models do not depict a continuous surface; interpolation must be used to construct one.

Data structures for field spatial data models

As Goodchild pointed out, data structures often become confused with data models (Goodchild, 1992). The reason for this is simple - there is a complex mapping between data models and data structures. If we consider only two large categories of data structures - raster and vector - the mapping between data models and data structures might look like this:

```
Cell grids         -> raster
Polygons           -> vector
TINs               -> vector
Contours           -> vector
Point grids        -> vector or raster
Irregular points   -> vector
```

Thus, a dataset may be stored in a vector format, but it may represent one of several different spatial data models. In order to use a spatial dataset appropriately, it is necessary to know which spatial data model has been used during the data modelling stage of database development.

Mathematics on fields

The computer is incapable of adding two continuous fields to produce a third continuous field. All fields must be reduced to simple finite numbers before mathematical manipulation can proceed. In order to manipulate two fields simultaneously, the locations for which there are simple finite numbers representing the value of the field must correspond. To add field A to field B, one must add the value of A to the value of B *at the same location*. Different spatial data models express location in ways which are generally incompatible. This implies that in order to perform mathematical operations on data in various spatial data models, we must first convert all models to spatially equivalent ones, or at least to extract estimates of values for locations in one field for which we have data in the other.

The method used to convert one spatial data model to another depends on the data model. Fortunately, the choice of the appropriate conversion procedure can be codified (Kemp, 1993). Similarly, when operations involve several different data models (e.g,. cell grid = contours x irregular_pts), it is necessary and possible to define a set of priority rules for conversion which determines which data model takes precedence and becomes the one in which mathematics is performed.

MODELLING WITH CONTINUOUS VARIABLES

For environmental modellers, designing and coding a mathematical model is an entirely different task than accessing and manipulating spatial data in a GIS. On the one hand modellers can use well-known and well-structured algebraic and computer languages, following widely accepted and proven rules for substitution and solution. On the other hand, when manipulating spatial data for use in the models, the modellers have only the idiosyncratic language of a specific GIS to work with. The procedures they must follow to get at the spatial data are not codified in any common language. There are no widely accepted common rules and defaults to guide how spatial data are used in environmental models. Thus, while modellers can use a common symbolic language to express the development of their mathematics and thus prove the validity of their approach, there is no simple way to express the transformations and manipulations that are necessary to incorporate the spatial data into the model. The consequence of this is that it is very difficult to assess the validity of the data incorporated into models which have been based on spatial data and, as a result, it is difficult to evaluate the validity of the model results.

A strategy for dealing with spatial continuity

What is needed are common strategies and techniques for handling spatial data about continuous phenomena in all its forms. A common strategy for handling data about fields in mathematical models provides a framework in which many issues related to the representation of continuous phenomena can be addressed. An awareness of the basic assumptions which are embodied in each field data model and a means for expressing exceptions to these assumptions can be provided for. Specifically this strategy should:

(a) allow expression and manipulation of variables and data about continuous phenomena in common symbolic languages. In other words, the strategy should be capable of being incorporated into computer language implementations of environmental models;

(b) eliminate the necessity to consider the form of the spatial discretization (the data model) whenever possible. While we believe it is desirable and possible to achieve this objective for most operations, it is necessary to provide for input of additional information for some operations;

(c) provide a syntax for incorporating primitive operations appropriate for environmental modelling with fields but which are not yet available in GIS or common programming languages. These include operations to perform discrete versions of "differentiation" and "integration" on variables representing fields and the incorporation of the concept of vector fields;

(d) guide and enable the rapid development of direct linkages between environmental models and any GIS.

The field data type and field variables

In order to manipulate data about spatially continuous phenomena, we suggest the definition of a *field data type* to be used in addition to the traditional data types (e.g. float, integer, character and so on). Variables declared as field data types are *field variables*. Field variables are the logical or functional representation of the concept of fields. These variables are spatially continuous and represent values of the field during a single slice or instant of time. Like other types of variables, fields are represented with symbols. For any field variable, it must be possible to determine a value at any location and these values may differ from location to location within the same field variable. Using a Cartesian coordinate system, we can refer to the temperature at a specific location in the field using the notation $T(x,y)$.

Declaring field variables

The essential elements of this strategy are the statements in which field variables are declared. Since the manner in which fields are represented within the computer is fundamental in determining how mathematical operations can be performed, properties associated with declared field variables describe the data model used and other critical characteristics related to information density, temporality and measurement system. These properties are critical in determining how the variable can be manipulated mathematically within the computer and help determine how variables represented by different data structures may be combined in a single mathematical statement. In most cases, sufficient information can be provided in the declaration of the field variables that no further input is required from the modeller when these variables are manipulated. In this manner, the objective of hiding specific details about how to perform operations on data from different spatial data models can be achieved. The issue is to determine what these critical properties are and how they can be clearly expressed. These topics will be discussed in subsequent papers.

Like declaration statements in standard programming languages, declarations establish certain functionality constraints and options particular to the specific data type. For example, in the language C, when a variable is declared as "char A[10,8]" memory is set aside to allow for a 10 by 8 array of characters. It is subsequently impossible to place a floating point or integer data value in this variable location, or to algebraically add two elements of this array. Similarly, the declaration of field variables establishes not only the type of data contained within the variable, but how it can be manipulated. Declaration of the field variable is not a difficult task. It simply provides a structure in which the essential properties can be unambiguously and completely described. These properties may used later by the compiler to determine the appropriate operations and conversions to perform.

WORKING WITH VECTOR FIELDS

If we accept the notion that fields can be handled like other variables, then the possibility of working with vector fields within GIS can be explored. Vector fields can be represented using the separable components of vectors: the length of unit vectors along each of the relevant axes, in two dimensions as (x,y) and in three as (x,y,z), or; distance and magnitude (a,r). Thus vector fields can be expressed using current data models by permitting multivalued attributes at each location included in the dataset. This approach is most suitable for the point based spatial data models since the vectors then describe the direction and magnitude of the flow of the phenomena at each point location. Cell grids and even polygons can be used to identify the location of vectors, though, like similar scalar fields, the vector is assumed to represent the average flow over the entire area included in the spatial element.

Although it may seem contrary, TIN and contour models, the two spatial models that do provide some information about the nature of the variation between recorded locations, cannot be used directly as models of vector fields. This is in spite of the fact that, unlike the other models, they do describe continuous surfaces. TINs are modeled by storing the value of the phenomena at the triangle nodes. Since these nodes join three or more triangles together, the direction and magnitude (slope) of the surface at these points is not be uniquely determined hence they do not provide convenient locations for the recording of vector values. While locations within the triangles are well described by vectors, if such vectors are attached to the triangles themselves, then the model becomes simply a triangular polygon model. In terms of contour models, while vector field lines, lines along which the tangent at any point gives the direction of the vector at that point, may create a representation which appears similar to contour lines, field lines give information only about direction, not about magnitude. Representations of field lines do not completely describe the field. Similarly, joining points of equal magnitude does not provide information about direction. Thus, only four of the six spatial data models are appropriate for representing vector fields.

How should vector fields be used in environmental models? If vectors are recorded in the (x,y) or (x,y,z) form, vector algebra can be readily performed on the vector elements as required by the mathematics. Analytical, finite difference and finite element solutions of complex equations may proceed in the normal scalar fashion with solutions being found for each spatial element and integrated for the whole. However, since vector fields have not been widely implemented in environmental models linked to spatial databases, vector operations such as divergence and gradient for GIS have yet to be adequately considered and thus true vector operations are not widely used in association with spatial databases. Further extension of this concept will only come with the incorporation of vector fields in conceptual data modelling and the implementation of vector operations for these fields.

Acknowledgements Support for this research by the US National Center for Geographic Information and Analysis, which is funded under the National Science Foundation grant SES-10917, is greatfully acknowledged.

REFERENCES

Goodchild, M.F. (1992) Geographical data modelling. *Computers and Geosciences* 18(4),401-408.
Goodchild, M.F., Parks, B. & Staeyert, L. (eds) (1993) *Environmental Modelling with GIS*. Oxford University Press, New York, in press.
Grelot, J.P. (1985) Archaic data models or hardware as a concept killer. In: *Proceedings AutoCarto London*, (ed. by M. Blakemore), 572-577.
Kemp, K.K. (1992) Environmental modeling with GIS: A strategy for dealing with spatial continuity. Unpublished PhD dissertation, Department of Geography, University of California, Santa Barbara, CA, USA.
Maidment, D.R.(1993) GIS and hydrologic modelling. In: *Environmental Modelling with GIS (ed. by* M.F. Goodchild, B. Parks & L. Staeyert), Oxford University Press, Oxford, in press.

Application of GIS for study of groundwater flow in arid regions using remote sensing data

Yu. L. OBYEDKOV & I.S. ZEKTSER
Water Problems Institute, Russian Academy of Sciences, 10 Novaya Basmannaya St., 107078 Moscow, Russia

Abstract Approaches to computer-aided remote sensing data processing are considered. Use of GIS programs is shown for constructing thematic hydrogeological maps. The groundwater flow of regions is qualitatively and quantitatively characterized by application of correlation factor analysis.

INTRODUCTION

Images obtained by remote sensing techniques contain such a quantity of information that powerful computers are needed for their processing. For instance, one image taken by the MKF-6 camera in all six spectral ranges contains about 170 MB of data (Sheremet *et al.*, 1982). Obviously most of this information is superfluous for solving hydrogeological problems. The use of methodologies for special visual interpretation and processing of remote sensing data is often of a subjective character and depends on the skill of the interpreter who, using a conceptual model, "filters" information useful for the solution of a geological and hydrogeological problem. The criteria for identification and classification of useful information are changes in phototone or colour hues, differences in the photo pattern and the character of topography, changes in vegetation and erosional ruggedness together with the spatial distribution and interrelation of the above objects.

An investigator usually works with available images. The interpretation of these images is significantly simplified if the objects or relief units are interpreted, contrasted, and accurately correlated to the topography. In recent years, to increase the objectivity and accuracy of remote sensing data processing and analysis, standardized methodologies have been introduced for interpretation, using methods of mathematical statistics and theory of information, mathematical logic, theory of algorithms and pattern recognition, methods of similarity evaluation using computer-aided systems for useful information selection. The development of these systems involves formalization of initial model interpretation and "teaching" initial elementary data filtration to computers. The following is achieved with the help of computers: geometric and tone correction of initial remote sensing data, the more contrast representation of image content, automated separation and classification of image structure and pattern, comparison of various remote sensing data, and presentation of different processing results in the form of images. The above initial data transformations pursue the main goal: to obtain, using different indices, the quantitative characteristics of the landscape pattern of the photographs of regions under study. The topography indices may include area of landscape types and features, their elevations, the dispersion and variation coefficient of distribution of these features, drainage network, coefficients of winding and branching of erosional topographic features, etc. The tectonic indices are orientation of faults (lineaments), their length, density, number of intersections, width of zones of crush, and the like. A set of similar indices may be obtained also for characterizing the lithological composition of deposits and hydrogeological features of the area.

In fact, transformation of initial data is made, using GIS programs (the ARC/INFO system in particular), which make it possible to pick up, in the same coordinate system, layer-by-layer information about different natural objects, their location, typology, and spatial characteristics and "saturation" of each layer with diagrams, tables, and surface investigation data for a specific physical object. This allows the geographic information system to meet different demands for a data base, involving the spatial distribution of objects, and to superimpose layers with different information, and to form buffer zones around objects under investigation. Finally, GIS provides computer-aided cartographic processing of data and makes it possible to obtain graphic and cartographic results, including thematic maps.

The use of quantitative methods for quantitative description of a photo image is of prime importance in automatization of the interpretation process. This complex process comprises successive and interrelated operations: (1) separation (filtration) of an object under study from all the remote sensing data; (2) automation of obtaining quantitative information about the object on the basis of photograph formalization; and (3) automation of pattern recognition on the basis of quantitative information obtained from remote sensing data. These operations are closely connected with general principles of identification of natural objects.

The filtration process in studying objects consists in transformation of remote sensing images, using photographic, optical, and electronic methods, into contour, structural, and different density images. The latter contain a larger body of information about object of interest to us as compared to original images.

The quantitative evaluation of geometric properties of objects is made from distributions of optical image density. Using special instruments, it is possible to obtain from the original image the number of objects, total and relative areas, lengths of projections, and regularity of their distribution.

At present, instruments with appropriate program packages and software for computer-aided processing of remote sensing information have been elaborated at the Institute for Space Studies of the Russian Academy of Sciences, Institute for Radiophysics and Electronics of the Ukrainian Academy of Sciences, Russian Research Institute for Remote Sensing Methods in Geology, Agroresursy Association, and Rosgiprovodkhoz Institute.

STUDY RESULTS

The main objects of geological and hydrogeological interpretation on remote sensing images are linear and areal landscape elements, which represent specific structural elements, hydrogeological phenomena and processes, geological and geomorphological setting, as well as deeper, concealed processes and elements. In thematic processing of images, it is necessary to proceed to schemes constructed on principles of quantitative analysis of interpretation results. This diminishes subjectivity in interpreting remote sensing information and makes it possible to obtain reliable parameters of natural objects (e.g., total density of lineaments or erosional downcuttings, specific density of faults of different directions).

Different parameters used for computations assume a great importance. Discrepancy commonly exists between variable geological and hydrogeological values taken from images and those obtained from on-site point data collection. To eliminate this discrepancy, it is necessary to average part of the parameters, though their reliability somewhat diminishes in this case. High data accuracy is required in studies, using GIS, of a specific object (e.g., a precipitation-surface runoff-pollution sources model for reservoir area). Average data are sufficient for characterizing spatially distributed hydrogeological models (e.g., for revealing regularity of distribution and quantitative assessment of surface runoff in a region in precipitation - surface runoff - tectonics - geological structure models). The latter is explained that the reliability of data on the subsurface structure is not very high in a number of cases. Soil data are highly unreliable, particularly outside agricultural areas. Geological data are usually taken from mapped exposures and from individual boreholes. The zone of aeration generally varies as to thickness and permeability. Aquifer transmissivity (in sedimentary rocks in particular) sharply varies horizontally and vertically.

However, indirect data are sufficient for characterizing water occurrence in a region, locating promising water points and potential groundwater pollution sources, as well as working out a strategy for water resources management and planning in the region.

The interpretation of space photographs of one of arid regions in Mongolia gave the following information: (1) the length, orientation, and distribution character of lineaments, width and depth of fracture zones resulted from faulting; (2) surface conditions of occurrence of rocks, their approximate composition, and types of groundwater flow media; (3) areas with hydrogeological phenomena and processes (bogs, solonchaks, takyrs, naturally moistened areas); (4) types of the river network and erosional topography ruggedness. Of practical interest is the evaluation of the relation between results of space photographs interpretation and land surface investigation data, characterizing water occurrence in the region (specific groundwater discharge values and water point yields).

To obtain an areal characteristic of space image interpretation schemes, the pickup of information with a definite set of elementary cell signs is required (this pickup of information should be continuous, using a sliding window, or discrete, using adjacent squares). Quantitative characteristics of signs are related to a point (cell centre), and a definite set of numerals, averaged for a given elementary cell, is assigned to this point. Correlation and factor analysis have been applied to the region under investigation.

Correlation analyses, using a large body of information, shows the degree of interdependence of parameters of interest to us. The same data are required for factor analysis. However, it describes the state of the environment or processes in it with factors whose number is much smaller than the number of parameters. Parameters of processes fall into certain number of uncorrelated groups. The number of these groups is equal to the number of factors. Data obtained from remote sensing and land surface geological and hydrogeological information are a matrix of initial signs whose lines are individual objects of investigation and columns are group of characteristic signs.

The analysis of all available remote sensing information and land surface data resulted in obtaining (1) a correlation matrix; (2) quantity of main factors whose number depends on the constant introduced, usually equal to 1; (3) a vector of factor values; (4) a matrix of factor loads A (M x K) where M is the number of signs and K is the number of factors.

We are usually pinning our hopes on factor analysis in regional subdivision by hydrogeological signs. When it is possible to distinguish several (2 to 3) main factors, one can compute factor values for each object (an object is understood to be either a water point or a part of the region under study selected by the net cell). To compute these values, the normalized initial information of signs for each object is multiplied by factor loads, then summed up and divided by proper numbers corresponding to main components:

$$Y_{i,j} = \sum_{r=i}^{m} A_{i,j} X_{y,i} / F_j$$

where X is matrix with dimensionality Mx (initial signs for all objects, N objects, M signs), A is matrix with dimensionality KxM (factor loads), F is vector with length K (factor values), Y is matrix with dimensionality N x K (factor projections on factor axes for each object). Having computed Y, it is possible to depict each object by a point in axes $\{F_j\}$. A compact cloud of points represents a class of objects. This type of dynamics may be easy to interpret for 2 or 3 factors only.

The analysis of the space photograph of Southwestern Mongolia shows that the heterogeneity of the structure of this region is due not only to the presence of Hangai and Gobi Altai mountains and Lake Valley located between them, but also to the different composition of water-enclosing rocks of these orographic elements. The interpretation, using GIS programs, resulted in compiling a series of thematic maps, including maps of erosional ruggedness,

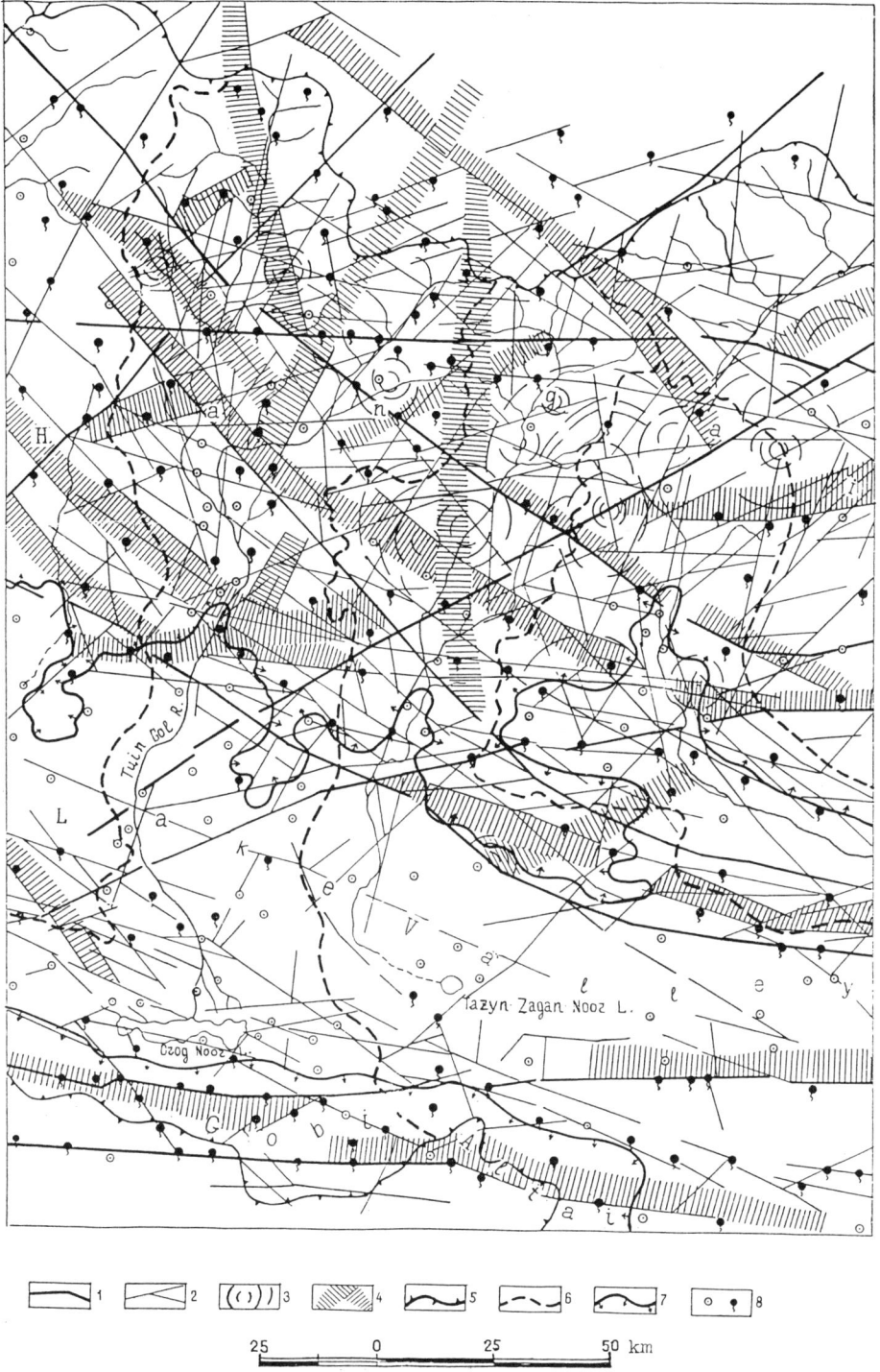

Fig. 1 Schematic structural and hydrogeological map of the Central Lake Valley artesian basin in Mongolia (from data of the interpreted synthetized space photograph taken from Salyut-6 on July 24, 1982): (1) regional deep faults; (2) other faults; (3) circular structures; (4) zones of fracture; (5) Lake Valley artesian basin boundary; (6) river basin boundary; (7) regional recharge area boundary; (8) water point (boreholes, wells, springs) boundary.

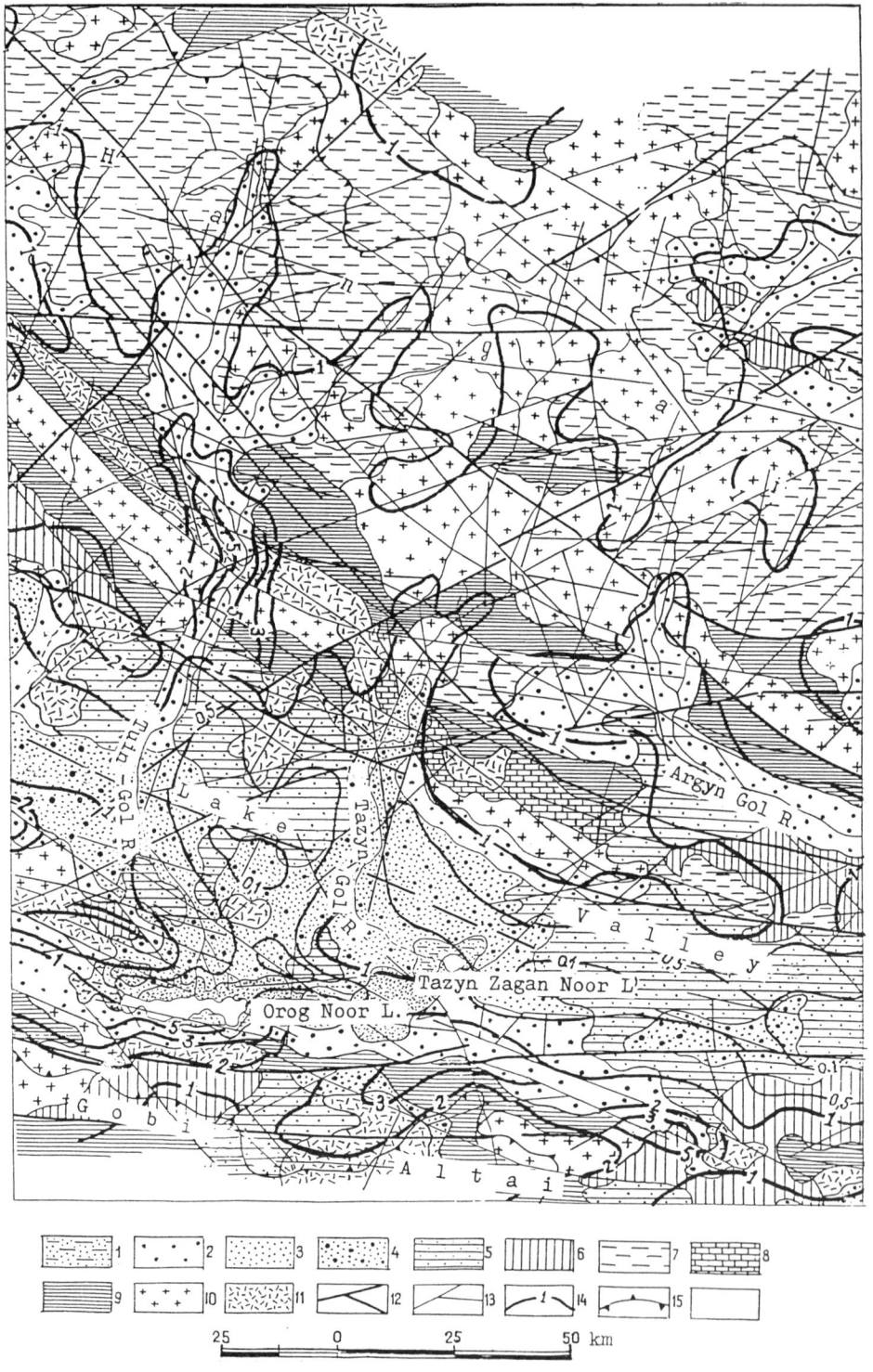

Fig. 2 *(Caption on opposite page)*

lineaments (Fig. 1), structural-tectonic, types of groundwater flow, geomorphological, and hydrogeological regional subdivision. The latter map is presented in (Fig. 2).

Based on the schematic maps constructed and land surface data, the correlation analysis of the dependence of groundwater flow in the region on a number of natural factors has been made. The work done was very labour consuming because it was necessary to pick up information, using the sliding window method, from almost 10 maps at different scales. The number (N) of cells was equal to 120. The results subject to interpretation were obtained after the region under investigation was divided into three parts, depending on its geomorphological structure: (1) the Lake Valley intermountain depression, (2) the Hangai old mountain system with smoothed topographic features, and (3) young pointed-top mountain structures of Gobi Altai.

Within the regions on the hydrogeological map, zones with large ($M \geq 1$ $1 s^{-1} km^{-2}$) and small ($M \leq 0.5 1 s^{-1} km^{-2}$) specific groundwater discharge values (moduli) were distinguished. Each of the zones was covered with a net (not necessarily rectangular, the net cell areas must be similar). Based on the correlation analysis of 51 signs, signs were excluded which had the lowest correlation with groundwater flow (areas of lakes, solonchaks, lineament orientation (except sublatitude-oriented faults). The groundwater discharge moduli with large ($M \geq 1 1 s^{-1} km^{-2}$) and small ($M \leq 0.5 1 s^{-1} km^{-2}$) values were retained among hydrogeological signs. Lithological signs were grouped so that the correlation between them and zones with a large groundwater discharge modules could be clearly expressed. For the Hangai zone, new signs were introduced, they are intermountain areas generally with river valleys. Large values of specific groundwater discharge were confined to these depressions. Climatic signs (mean annual precipitation) were also added.

Fig. 2 Map of groundwater flow in the Central Lake Valley artesian basin in Mongolia (from interpreted space photographs): (1) lacustrine and lacustrine-alluvial deposits (sands, sandy loams, clayey loams) with transmissivity (T) of 5-70 m^2 day^{-1}; (2) alluvial and alluvial-proluvial deposits (sands, coarse gravel, loams), T = 15 - 100 m^2 day^{-1}; (3) eolian deposits (sands), T = 80-100 m^2 day^{-1}; (4) indiscriminate marine coastal deposits (sands, gravel, coarse gravel), T = 20-120 m^2 day^{-1}; (5) slightly consolidated continental deposits (sands and clays), T = 5-80 m^2 day^{-1}; (6) sedimentary slightly consolidated deposits (sandy conglomerate formation), T = 180-250 m^2 day^{-1}; (7) sedimentary lithified deposits (sandstones, siltstones, claystones,conglomerates), T = 30-60 m^2 day^{-1} on water divides, T = 100-110 m^2 day^{-1} in river valleys, and T = 140-310 m^2 day^{-1} in fault zones; (8) karst formations (limestones, dolomites, marls) generally with T = 30-50 m^2 day^{-1} and T = 200-150 m^2 day^{-1} in zones of tectonic disruptions; (9) metamorphic formations (gneiss and shale strata) generally with T = 60-80 m^2 day^{-1} for fracture permeability and T = 15-200 m^2 day^{-1} in the near-fault zone; (10) intrusive formations (gabbro, sienites, granites), T = 15-30 m^2 day^{-1} in central parts of blocks and T = 350-500 m^2 day^{-1} near linear boundaries; (11) volcanogenic formations (basalts, andesites, tuffs) with linear-block permeability (T = 70-100 m^2 day^{-1} in Paleozoic volcanites and T = 100-15 m^2 day^{-1} in Mesozoic-Cenozoic volcanites); (12) regional deep faults; (13) other faults; (14) specific groundwater discharge values (moduli) in 1 s^{-1} km^{-2}; (15) Lake Valley artesian basin boundary.

Let us consider the results of the analysis performed for Hangai mountain structures. Signs were selected which consider the relation of hydrogeological conditions and tectonic and geomorphological features. That is, in the mathematical analysis of remote sensing and land surface information, the selection is important of concrete quantitative and qualitative signs, most fully characterizing the object under investigation. In particular, in the problem under consideration, based on a combined structural and quantitative analysis, the most informative signs of tectonic fracturing and their quantitative characteristics, as well as typical signs characterized by a priori land surface data have been selected. The number of these signs is 12 (Table 1).

Table 1 Group of natural signs for correlations analysis (Hangai region).

Sign type	Sign No	Signs
Tectonic	1	Zones of fracture
	2	Nodes of lineaments
Geomorphological	3	River valleys
	4	Water drainage divides
	5	Intermountain areas (superimposed structures)
Lithological	6	Eolian sands. Groundwater flow media of sedimentogenic type
	7	Groundwater flow media of lithogenic and fracture type
	8	Groundwater flow media of metamorphic, magmatic and volcanic types
Climatic	9	Precipitation 200-400 mm year^{-1}
	10	Precipitation 400 mm year^{-1}
Hydrogeological	11	Groundwater flow modulus 1 l s^{-1} km^{-2}
	12	Groundwater flow modulus 0.5 l s^{-1} km^{-2}

According to the methodology elaborated for the region under study, a regional information body, consisting of 70 objects (cells) has been formed. Zones with large and small groundwater moduli were placed in them; the number of the zones was equal and amount to 35.

The analysis of the correlation matrix obtained shows that the groundwater flow modules ≥ 1 l s^{-1} km^{-2} (sign 11) have a direct correlation relation with river valleys ($R = 0.72$), intermountain areas ($R = 0.92$), and eolian sands ($R = 0.75$). The reverse relation of sign 11 and drainage divides is $R = -0.55$ and groundwater flow media of the lithogenic-fracture type (sign 7) is $R = -0.51$. The low level of groundwater discharge ≤ 0.5 l s^{-1} km^{-2} has the same correlation relations, but with an opposite sign.

All this represents the fact that within the region under study intermountain areas enclose river valleys composed mainly of sand gravel, and water drainage divides are composed of fracture media, including carbonate, metamorphic, and magmatic rocks with an insignificant occurrence of fracture zones. A model of factor analysis has been used for the analysis of the internal structure of the correlation matrix. The essence of factor analysis consists successive determination of main components called general factors (Ariun *et al.*, 1986). These three factors are $F_1 = 0.72$, $F_2 = 0.61$, $F_3 = 0.43$.

The first factor F_1 may be called transmissivity factor. Values of factor loads on signs give the same picture that is given by correlation analysis. This factor is related to zones and nodes of intersections of the densest net of faults, occurrence of groundwater flow media of the fracture and fracture-vein types in the lower parts of slopes and river valleys. The second factor F_2 has large loads on climatic signs. It is called climatic factor. The third factor F_3 is geological, it has loads on the fracture sign and also on lithological signs. It may be interpreted as a factor characterizing the water content of rocks occurring along major zones of fractures and faults of sublatitudinal extension, identified on remote sensing images.

CONCLUSIONS

In summary the statistical interpretation of results of quantitative processing, using factor analysis, of remote sensing data and geological-hydrogeological information about Southwestern Mongolia shows that groundwater accumulation areas with large specific discharge values are confined to fracture zones and nodes of intersection of regional and deep faults, as well as to media of the fracture and fracture-vein types, occurring in lower parts of slopes and in intermountain depressions surrounding the Lake Valley trough, and to sand masses in valleys of large rivers within the trough itself. Over 80 promising areas, obtained from results of computer-aided processing of remote sensing and land surface information, coincide with known zones of large groundwater flow moduli, determined by analytical hydrogeological methods. This shows the efficiency of using mathematical methods of analysis with application of computers in determining the character of groundwater flow values in regions under investigations.

Evidently, the results obtained characterize the hydrogeological distinctive features of the region under investigation. However, significant correlation relation between signs distinguished in the region under investigation may not occur in an other region. In other regions, correlation relations between other signs may be of importance. By and large the methods of analysis of thematic maps, produced with the help of GIS programs, allow the researchers to obtain quantitative characteristics of objects under study, determine the regularity of their spatial distribution, reveal the extent of dependence and interrelation of different signs, characterizing the natural conditions of the region under investigation.

REFERENCES

Ariun, Z., Baasanzhav,L. & Tserendorzh, Z. (1986) K voprosu avtoomatizirovannogo prognozirovaniya mineral'nykh resursov na osnove kosmicheskoi informatsii (On the problem of computer-aided prediction of mineral resources on the basis of remote sensing information). In: *Nauchnye Experimenty v Kosmose. Soyuz - 39 - Salyut-6 - Soyuz 3-T4 (Scientific Experiments in Space. Soyuz-39 - Salyut 6 - Soyuz 3-T4)*. AN MNR Press, Ulan Bator, 121-136.

Sheremet, O.G., Moralev, V.M. & Gonikberg, V.E. (1982) O sposobe opredeleniya optimal'noi ploshchadi osredneniya geometricheskikh parametrov lineamentnykh setei (On the procedure of determining the optimal area of averaging geometric parameters of lineament nets). *Issledovanie Zemli iz Kosmosa* (4), 15-19.

Application of GIS and remote sensing in hydrology

G.A. SCHULTZ
Institute of Hydrology, Water Resources Management and Environmental Techniques, Ruhr University Bochum, P.O. Box 102148, 4630 Bochum 1, Germany

Abstract Remote sensing data from airplanes, satellites or groundbased radar represent a valuable source of data for hydrological purposes and are highly suitable to be used in GIS. After a brief description of the fields of RS applications in hydrology and some characteristics of typical RS data a few examples for the use of RS data in a GIS for hydrological purposes are presented. In the first example a technique is presented which allows to use data from polar orbiting satellites (Landsat, SPOT, NOAA) for the estimation of hydrological model parameters. In the second example it is shown, how a time series of historic runoff data in a river catchment can be computed with the aid of geostationary satellite data (Meteosat). In a third example the use of groundbased weather radar for the purpose of real-time flood forecasting is demonstrated. Along with all three examples the need and suitability of GIS is discussed. The following chapter discusses the necessity in hydrological modelling of merging data from remote sensing sources with other information. Two examples are given: (a) the merging of landuse classification raster data obtained from Landsat with soil maps in order to determine soil water storage maps and (b) the merging of landuse data with a digital elevation model in order to develop a map of so-called Hydrologically Similar Units. The paper terminates with a discussion of the use of GIS with increasing complexity (e.g., ILWIS, ARC/INFO, GRASS) and a perspective of future developments.

INTRODUCTION

Remote sensing (RS) data play a rapidly increasing role in the field of hydrology. Although only very few remotely sensed data can be directly applied in hydrology, such information is of great value since many hydrologically relevant data can be derived from remote sensing information. One of the great advantages of remote sensing data in hydrology consists in the fact that we obtain area information instead of the usual point data (e.g. at rain gages, evaporation pans, infiltrometers etc.). In hydrology, particularly in hydrological modelling we need, however, area information, which for certain parameters can be derived from RS information. This information is usually digitized in form of picture elements (pixels) the scale of which depends on the sensors which are used. Thus large quantities of raster data are collected permanently with the aid of various platforms and sensors all over the world. These large data amounts require large and well organized data banks as well as user friendly data processing hardware and software. Therefore GIS represent highly suitable opportunities for efficient handling of the large quantities of raster data. Furthermore they allow the necessary merging of remote sensing data with digital elevation models, digitized maps of various sorts as well as other hydrometeorological data.

Simpler GIS products to be generated with the aid of RS data for hydrological purposes comprise maps of e.g. snow cover, areas of inundation by floods, rainfall intensity maps etc. A more complex application of remote sensing data in hydrology consists in the use of such data either for the estimation of parameters of hydrological models or as input into such models. For both purposes the application of GIS and remote sensing improves the efficiency of such procedures significantly.

REMOTE SENSING IN HYDROLOGY

General information

The sensors acquiring RS data have to be mounted to platforms from which measurements are taken. Such platforms may be (1) ground observation platforms (masts, towers, vehicles); (2) balloons; (3) aircraft and remotely piloted vehicles; (4) rockets and (5) satellites. We distinguish between polar orbiting and geostationary satelllites. While polar orbiting satellites follow eliptical orbits (crossing near the poles) geostationary satellites are positioned always above the same point on the earth (e.g. Meteosat: 0° longitude, 0° latitude). There are different types of RS sensors, e.g. aereal photography, scanning radiometers, spectrometers, microwave radars etc. For hydrological investigations, electromagnetic energy is the most important medium through which observations are made. Only certain spectral bands are relevant which are discussed later. Further details have to be obtained from literature (Schultz, 1988; Schultz & Barrett, 1989).

Another important feature of RS is the resolution in time and space of the sensors under consideration. Some polar orbiting satellites produce a high resolution in space (e.g. Landsat pixels: 30 x 30 m; SPOT: 10 x 10 m, NOAA: 900 x 900 m). The resolution in time is not very satisfactory (16 days for Landsat, 12 hours for NOAA). Groundbased weather radar, however, produces a rather high resolution in time and space (e.g. 15 minutes in time, 1 km^2 in space). Geostationary satellites produce a very high resolution in time (30 minutes) but a rather coarse resolution in space (e.g. 5 x 5 km by the satellites Meteosat, GOES, GMS and INSAT). The resolution in time and space obtained from airplanes can be adapted to the problem under consideration.

A hydrologist who wishes to use RS data for the solution of his problems has to find out which available remote sensing data are suitable for his problem. Here the most important parameters are the resolution in time and space as well as the spectral bands available from the relevant platform and its sensors.

The main fields of remote sensing application in hydrology are: rainfall, evapotranspiration, soil moisture, groundwater, surface water, snow and ice, sediments and water quality as well as hydrological modelling. The use of RS data in hydrology requires efficient storage and retrieval of RS raster data in a data bank coupled to a GIS. Here the original RS data have to be stored for all spectral bands (e.g. Landsat: 7 channels) as well as the products which will be derived from the original RS data, e.g. landuse classification maps, snow cover maps, vegetation index maps etc. The GIS will aid the hydrologist to produce these derived maps from the original RS data and it will be again of assistance to him when he uses these derived data in hydrological modelling.

The satellite platforms which are presently of highest relevance to hydrologists are: Landsat, SPOT, NOAA, ERS-1, JERS-1, DMSP, Meteosat, GOES, INSAT, GMS. The most relevant spectral bands are visible (VIS) at various colours, infrared (IR), near infrared (NIR), thermal infrared, passive microwave and active microwave (Radar). Besides RS data collected on satellites also groundbased RS data are of importance to hydrologists, particularly those from groundbased weather radar which is suitable for area precipitation measurements.

In the following sections some examples will be given in order to show how remote sensing data are presently used in hydrology.

Polar orbiting satellite data for hydrological model parameter estimation

The condition of the land surface (landcover, landuse, vegetation status etc.) influences hydrological processes like interception, soil moisture, infiltration, runoff etc. In many cases such information is derived from digitized maps which are stored in GIS. This conventional technique has several disadvantages (maps are often outdated and of low accuracy). These problems may be overcome by landuse classification carried out with the aid of satellite imagery. The Landsat satellite is highly suitable for this purpose, since it has a very high resolution in space (30 x 30 m pixels) and 7 spectral bands.

The French SPOT satellite has a higher resolution in space, however, it has only 4 spectral bands. Therefore an analysis was carried out in order to investigate, whether landuse classifications generated by SPOT and Landsat differ significantly. Figure 1 shows for a subcatchment of the Mosel river in Germany landuse classifications by Landsat and SPOT and their differences. Although the SPOT image gives a more convincing impression of the landuse classification (e.g. the river Sauer can be seen very clearly), the result obtained with the aid of Landsat seems more trustworthy since it is based on information in 7 spectral bands. A slight difference in the geometric correction of both images gives rise to a large number of pixels with different landuse classifications.

Estimation of Vegetation Index and Leaf Area Index with the aid of Landsat imagery

The rainfall runoff process is influenced by the vegetation status. There are

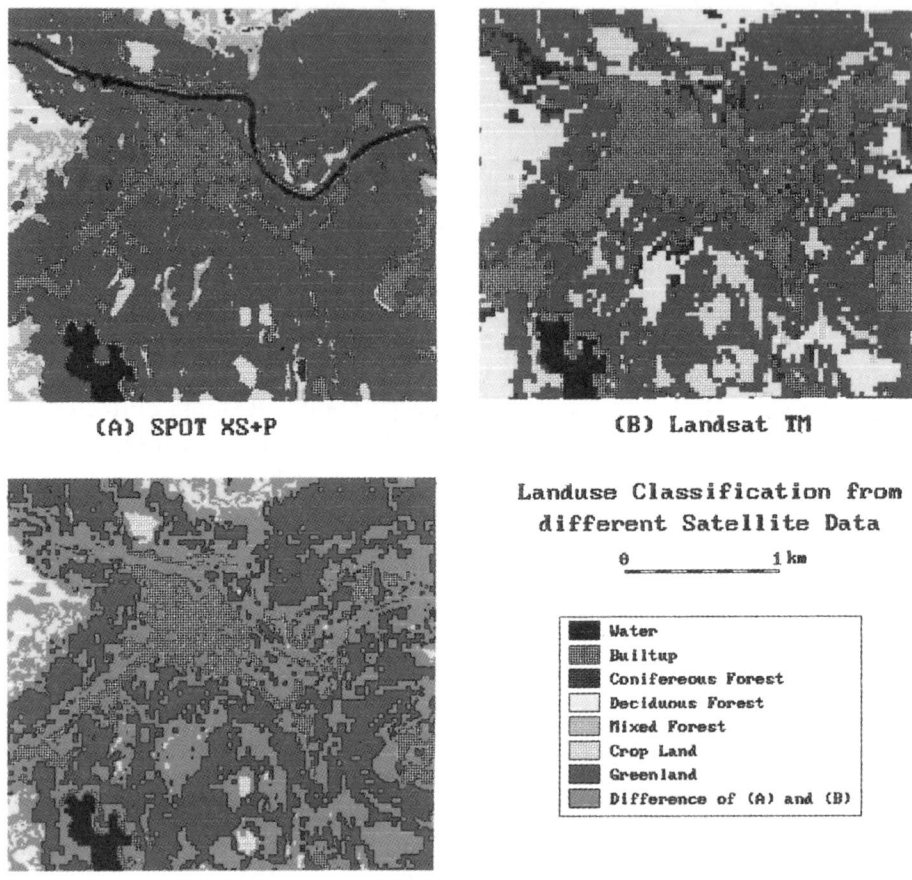

Fig. 1 Comparison of Landuse classification; Landsat vs. SPOT data. Section of Sauer catchment near Echternach (Mosel river tributary).

various indicators for this vegetation status, e.g. the "Leaf Area Index" (LAI) or the "Normalized Difference Vegetation Index" (NDVI). With the aid of two channels of Landsat or NOAA imagery it is possible to estimate the vegetation index. Figure 2 shows the Vegetation Index (NDVI) for the total Mosel river basin in two different seasons, Spring and Summer (based on NOAA data). The figure shows that in Spring the vegetation is highly developed, while in summer we observe already a less developed vegetation. For some hydrological processes this is relevant, e.g. for interception, evapotranspiration, infiltration etc. With the aid of seasonally varying vegetation indices it is possible to estimate different hydrological model parameters for different seasons.

Geostationary satellite data for generation of historic runoff time series

Besides data from polar orbiting satellites also data from geostationary satellites with their high resolution in time (but lower resolution in space) may

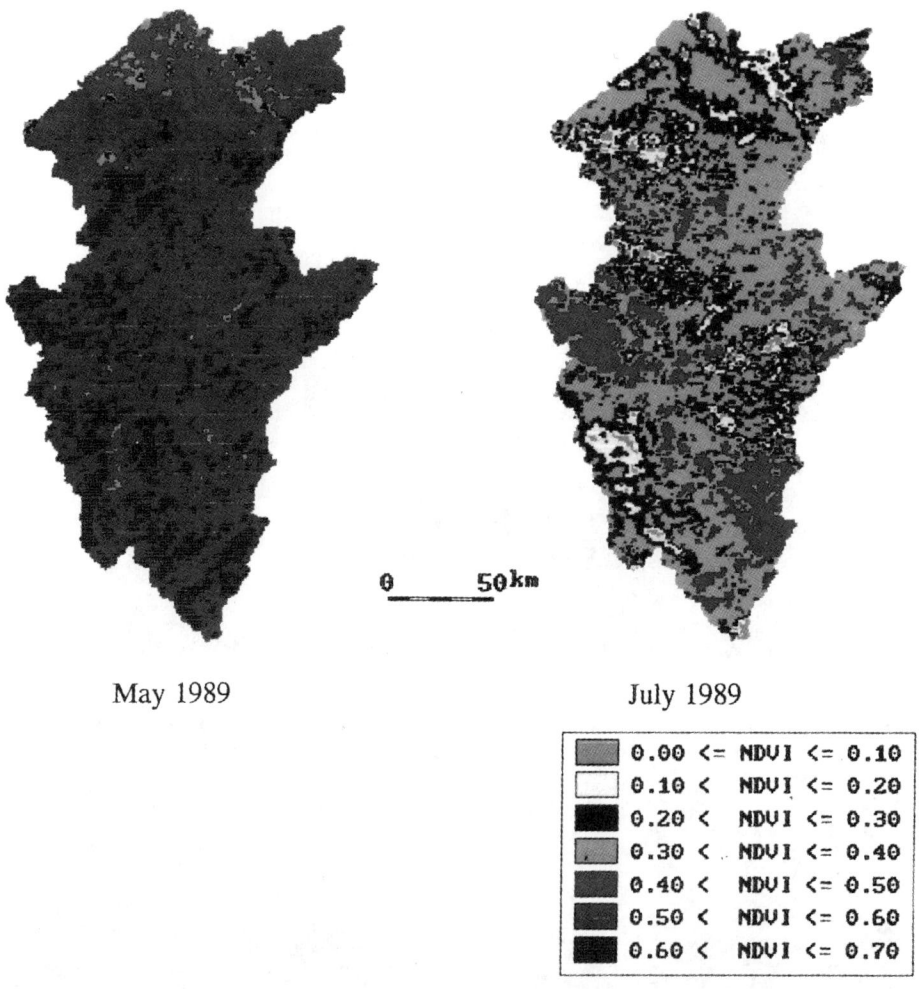

Fig. 2 Normalized Difference Vegetation Index (NDVI), Mosel river basin. Late spring vs summer. Basis: NOAA satellite data.

be used in hydrology. Here an example will be presented where such data are used in order to generate hydrological river runoff time series needed for the planning of water resources management projects. Frequently planning of water projects is hampered by non-existance of hydrological data. In such cases often rain gages and river gages are installed and the data obtained during the planning period (usually one to three years) are used for the design of a water resources system. Such short time series are not suitable for the evaluation of the reliability of the planned water project to fulfill its tasks. With the aid of data from geostationary satellites (e.g., Meteosat, GOES, GMS, INSAT) it is possible to reconstruct historical river flows. For this purpose it is necessary to have a hydrological model which relates the observed short term runoff data to simultaneous information obtained from the satellite with the aid of a mathematical model. Such a model was developed by Papadakis (Papadakis *et al.*, 1992) for a river catchment in West Africa. After calibration with

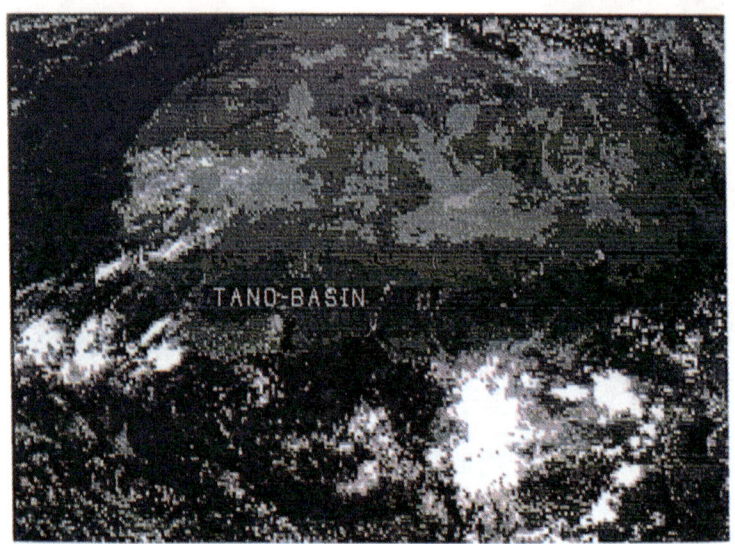

Fig. 3 Cloud development, Tano river basin, Ghana, West Africa. Two successive IR images of Meteosat.

groundtruth the model is able to reconstruct monthly runoff values for the Tano river in Ghana (17 000 km^2) on the basis of the Meteosat data alone. Figure 3 shows two Meteosat images in the infrared band for the catchment area from which rainfall estimates are derived. The rainfall values are used in a rainfall runoff model in order to reconstruct the historical monthly river flows in the Tano river. For the research project in which the technique was developed it was necessary to process about 20 000 Meteosat images (3 spectral bands). A GIS which is connected to a suitable data bank system (e.g., ORACLE) allows efficient handling and processing of this huge number of images.

Groundbased weather radar for flood forecasting

A typical case for the application of RS data as input to hydrological models is the use of measurements taken by a groundbased weather radar which after calibration can be converted into rainfall. Radar information is available at a high resolution in time (e.g., one hour or even 15 minutes) and a high resolution in space (e.g., 1 x 1 km). Since the rainfall intensities in a catchment area can be received by the radar in real time and transmitted to a hydrological center it is possible to use radar data in order to forecast expected flood hydrographs in real time (Klatt & Schultz, 1983). Figure 4 shows a sequence of three images taken by the radar of the German Weather Service in the city of Essen in which a severe storm with the danger of hail can be seen moving across the basin of the lower river Rhine from south-west to north-east. It is obvious that a large number of such images has to be processed in real time in order to compute a flood forecast for such a catchment which can be sped up by application of GIS. Figure 5 shows a comparison of a computed and observed flood hydrograph on the basis of radar rainfall measurements for a catchment in southern Germany.

MERGING OF REMOTE SENSING DATA AND OTHER INFORMATION WITHIN A GIS

General considerations

For many hydrological purposes RS data alone are not sufficient; they have to be merged with data from other sources. This is a field, where GIS's are of extreme value. Other data sources may be digitized maps, e.g. soil maps, geological maps etc. or digital elevation models (DEM) or digital terrain models (DTM).

The technique of merging information from different sources into one system has to be carried out with great care. It is necessary to use the same coordinate system, to work on the basis of equal pixel size etc. Since usually GIS's do not offer information on these problems in their manuals the merging of different informations often gives rise to severe errors.

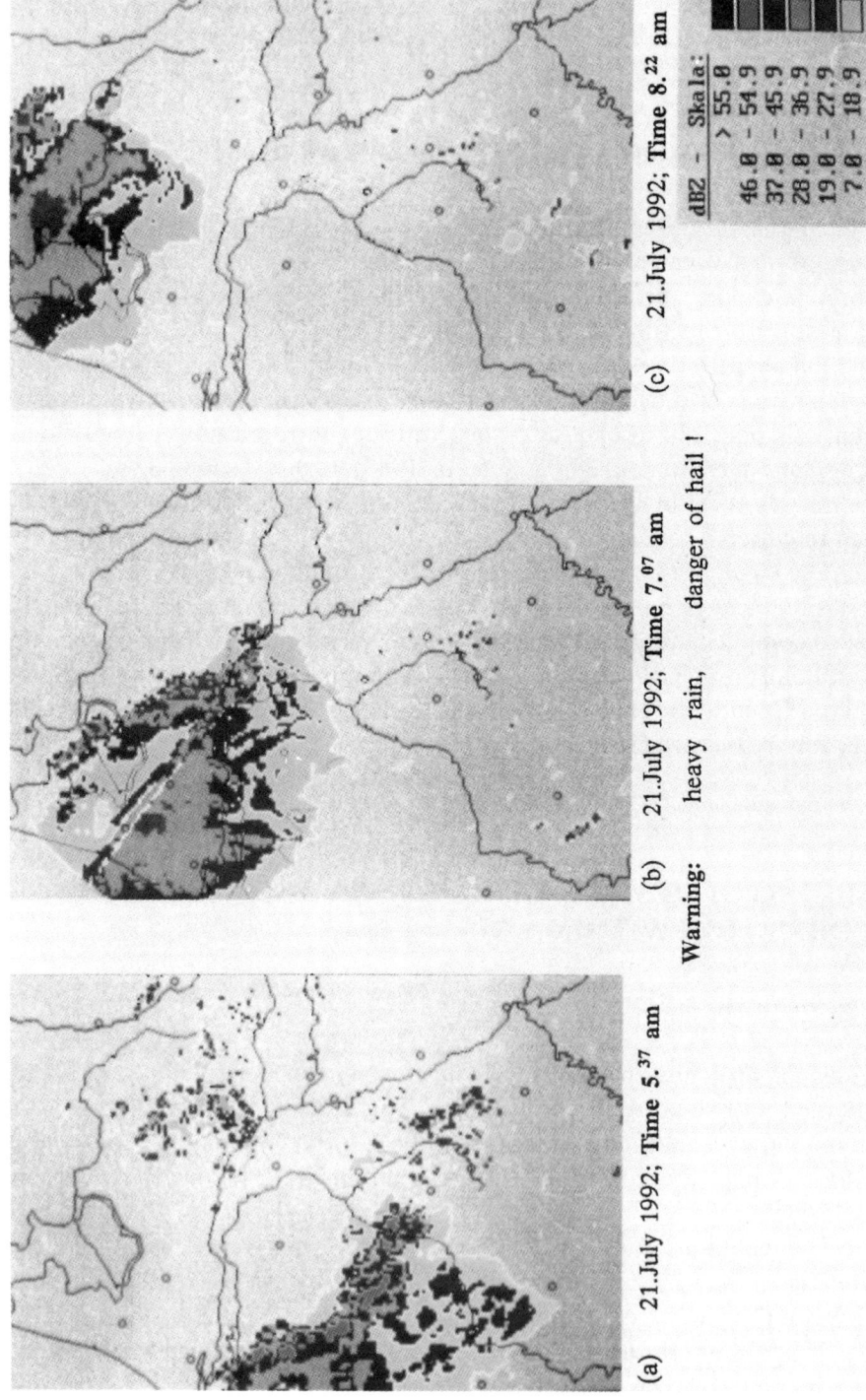

Fig. 4 Radar reflectivity of precipitation, river Rhein basin; Germany/The Netherlands (Data from DWD, Essen).

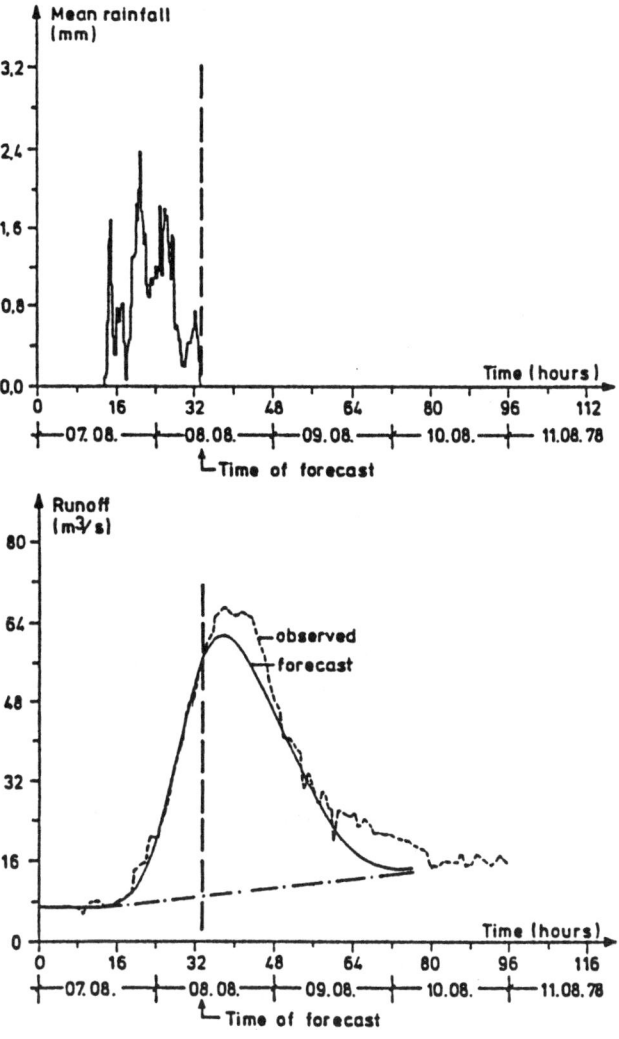

Fig. 5 Flood forecast based on radar rainfall measurement (river gauge Lauben, Günz river) after the end of rainfall.

Merging RS data and maps

For many hydrological purposes it is necessary to process information from different sources, and here the merging of RS data and maps is of particular importance, for which a brief example will be given. In hydrological rainfall-runoff modelling the water stored in the soil is of great relevance. This soil water storage depends on the landuse and on the soil types. Data on landuse and, derived from this, the plant root depth can be obtained with the aid of landuse classifications derived from Landsat imagery. Information on soil types and thus soil porosity can be derived from digitized soil maps. This way it becomes possible to estimate the maximum soil water storage for each pixel within a catchment area on the basis of merged RS and map information.

Figure 6 shows for the river Nims basin (Mosel river catchment) root depth (from RS), soil porosity (from a digitized soil map) and the product obtained by merging these two informations, thus indicating maximum soil water storage in the catchment which is needed for the hydrological rainfall-runoff model.

Merging RS data and Digital Elevation Model data

Another type of information which is frequently used in hydrological modeling is the subdivision of a drainage basin into so-called Hydrologically Similar Units (HSU's), which behave in the same way. The determination of the HSU's requires data on landuse, on elevation or slope and - possibly - on other data. Here an example is presented in which RS data are merged with Digital Elevation Model (DEM) data and a digitized soil map for the determination of HSU's. Figure 7 shows remote sensing data (landuse classification) and slope data obtained from a DEM as well as soil type data which are merged into HSU's, representing areas of equal hydrological behaviour due to their equal landuse, slope and soil type. A hydrological rainfall-runoff-model applying this HSU concept was presented in the literature (Ott *et al.*, 1991). Besides these two examples many other potential combinations of remote sensing with other raster data are applied in hydrology all of which require for the efficient use a GIS.

APPLICATION OF GIS WITH INCREASING COMPLEXITY TO HYDROLOGY

In hydrology various GIS of different complexity are in use. Examples of GIS with increasing level of complexity are (1) self-developed GIS for a special purpose, (2) GIS: ILWIS for PC, (3) ARC/INFO for workstation, (4) GRASS for workstations and mainframe computers.

Many years of experience with simple self-developed GIS at the author's institute are available. There can be no doubt that this type of GIS development is highly time consuming and not up to standard, anymore. Nevertheless this type of GIS has two advantages: (a) the developer knows exactly the theoretical background of procedures implemented in the GIS and their pros and cons; (b) it is rather easy to change formulae or procedures or components in such a GIS.

The experience with the second level, the ILWIS-GIS (developed by ITC in The Netherlands) which is implemented on a 486 PC has many advantages as far as speed and efficiency of the GIS is concerned. On the other hand some routines may not be to the taste of the user and cannot be changed easily, and the GIS is too "small" to be used for larger river catchments. Working with Landsat imagery for hydrological modelling, as discussed above, catchment areas above 1000 km^2 can hardly be handled.

The third level type of GIS (e.g., ARC/INFO) which is presently implemented at the author's institute show certain distinct advantages. The

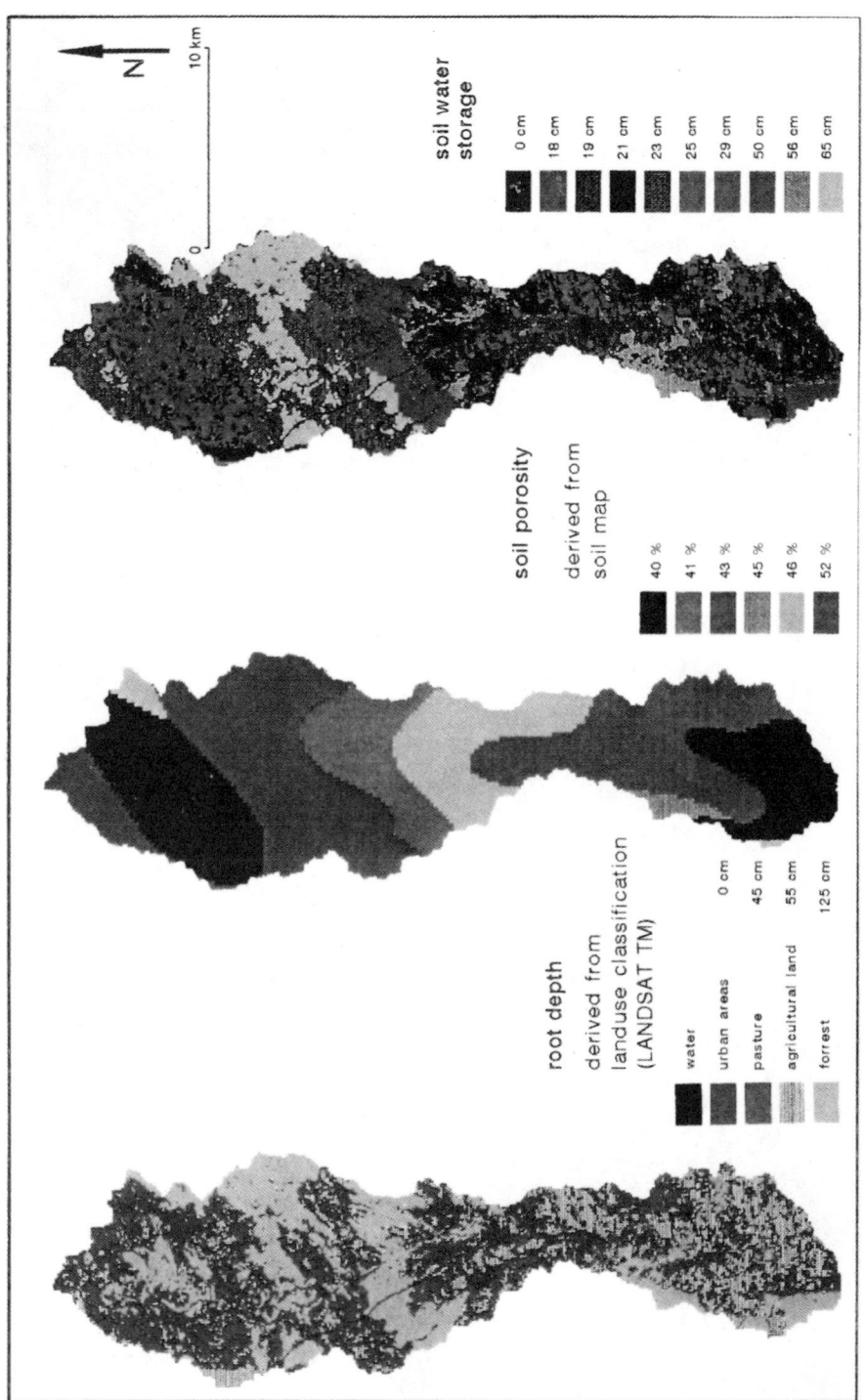

Fig. 6 Maximum soil water storage map (right) of the Nims river basin (Germany) as the product of root depth (left) and soil porosity (center).

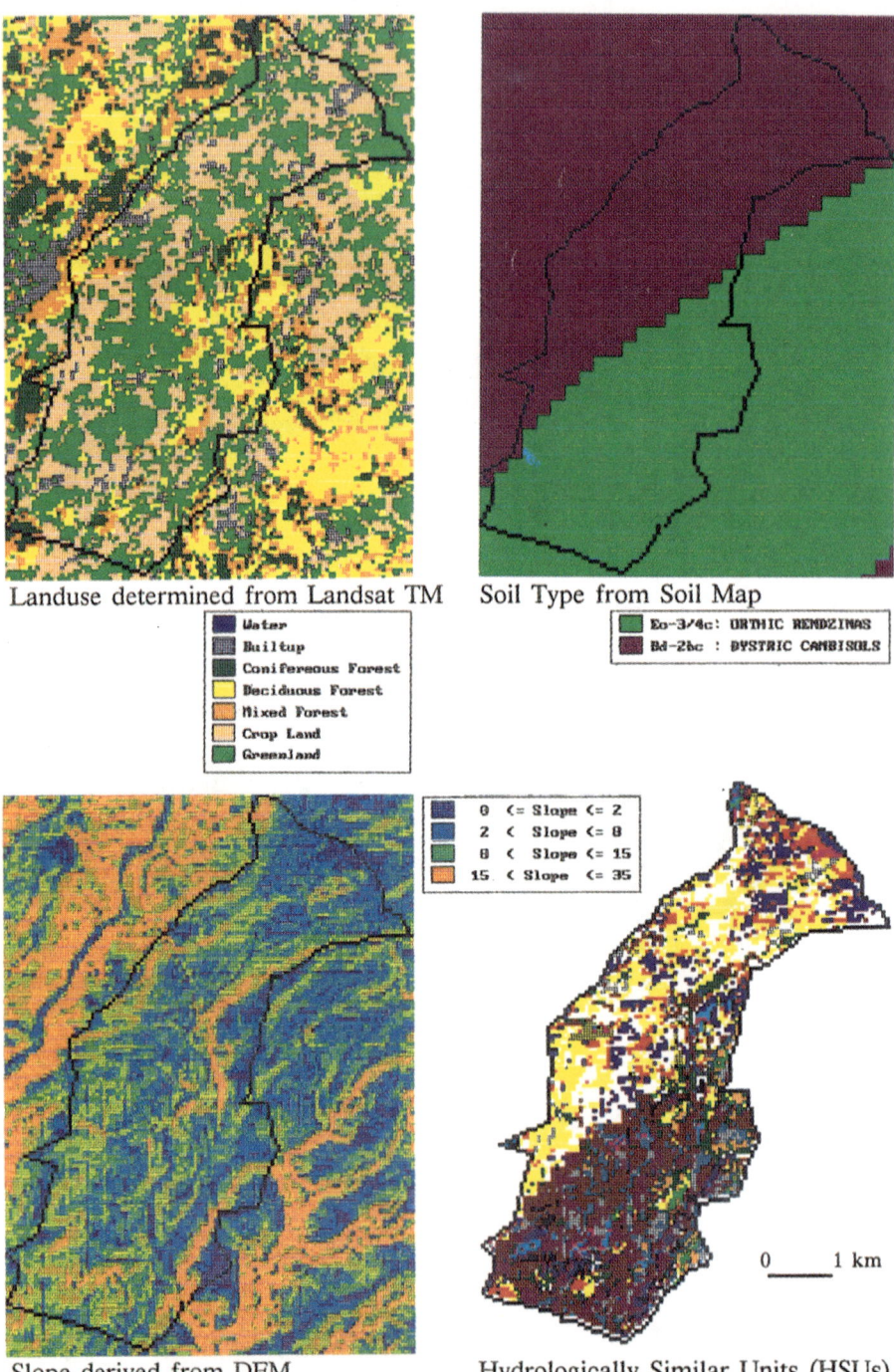

Fig. 7 Hydrologically Similar Units (HSU's) generated by merging data of landuse (from Landsat), soil type (from soil map) and land slope (from DEM). Nims river basin (Germany).

number of routines which can be handled is rather large and the handling of large data quantities for large hydrological drainage basins causes no problems. On the other hand, ARC/INFO - in contrast to ILWIS - was not primarily developed for hydrological purposes. Therefore many routines are unnecessary, while others do not exist. Furthermore, it is not possible to introduce new routines directly into the GIS. Required routines have to be programmed on the host computer and coupled with information which can be obtained from ARC/INFO.

The level 4 type GIS (e.g., GRASS) represents a GIS of high sophistication which allows treatment of large data quantities and containes many routines. The major advantage of such GIS consists, however, in the possibility to integrate routines which are required for hydrological purposes directly into the GIS. While the GIS of level 2 and 3 are rather rigid the fourth level GIS represents extreme flexibility. There can be not doubt that for the treatment of complex hydrological processes the tendency will move towards the level 4 type GIS.

It is necessary, that GIS for hydrological purposes have the capability of dealing with raster data as well as vector data. Particularly remote sensing requires large quantities of multi-temporal raster data, while in the field of hydrological modelling also vector data are needed: e.g. for the delineation of drainage basin boundaries, for the automatic generation of water courses within a catchment etc. The possibility of merging vector and raster data is therefore indispensable.

FUTURE DEVELOPMENTS

In this paper five sources of RS data were discussed in the context of GIS application for hydrological purposes: Landsat, SPOT, NOAA, Meteosat and groundbased weather radar. Considering the development of such applications of RS data in GIS for hydrological purposes it is possible to extrapolate certain trends. Before the availability of GIS RS data were used in form of simple self-developed GIS. At present we observe the use of available and suitable GIS on PCs (e.g., ILWIS) and on workstations (e.g., ARC/INFO). A comprehensive study in which about twenty GIS were compared which are available on the market (Büscher *et al.*, 1992), the conclusion was drawn that ARC/INFO is highly suitable for hydrological purposes if combined with other software, and the GIS GRASS for hydrological purposes in which the GIS shall be coupled to self-developed programs. It can be expected that future hydrological work will show a tendency towards the fourth level type GIS, like GRASS, in order to integrate software representing mathematical hydrological models into the GIS.

Such a GIS should contain not only one hydrological model for one purpose but it should rather contain a selection of hydrological models for each single purpose. Furthermore the GIS should contain - besides other relevant GIS information - RS data, i.e. the original data like Landsat, Meteosat or weather radar data etc., but also the processed data files containing information like soil water storage, Hydrologically Similar Units etc. This requires the coupling of the GIS with an efficient data bank (e.g., ORACLE). It can be expected, that computer "Expert Systems" will be developed, which help the user in the decision, which software (e.g., RS data processing, hydrological models etc.) should be used for which purpose. Such a complex "Hydro-GIS" does not yet exist and should be developed during the next decade or so.

In the future we will also have new types of RS data to be used for hydrological purposes at our disposal, since the space agencies are presently developing completely new space systems and networks of groundbased weather radar are under development in many countries.The new hydrological potential offered by this tremendous amount of new information will certainly require in the future more complex GIS hardware and software.

REFERENCES

Büscher, K., Kirchhoff, Chr., Streit, U. & Wismann, K. (1992) Analysis of the suitability of various GIS for regionalisation in hydrology. Study for the German National Science Foundation, Workshop report: Environmental- Agricultural-Geoinformation, volume 1, Institute of Geography, Westfalian Wilhelms University, Münster, Germany.

Klatt, P. & Schultz, G.A. (1983) Flood forecasting on the basis of radar rainfall measurement and rainfall forecasting. Proceedings, Symposium HS1, IAHS/IUGG General Assembly, Hamburg. IAHS publication No. 145.

Ott, M., Su, Z., Schumann, A.H. & Schultz, G.A. (1991) Development of a distributed hydrological model for flood forecasting and impact assessment of landuse change in the international Mosel basin. In: Proc. Int. Symp. on Hydrology for the Water Management of Large River Basins, Vienna. IAHS publication No. 201.

Papadakis, I., Napiórkowski, J. & Schultz, G.A. (1992) Monthly runoff generation by nonlinear models using multi-spectral and multi-temporal images. Proceedings, COSPAR symposium on Advances in the Use of Remote Sensing for Hydrological Science and Water Resources Management, Washington, DC.

Schultz, G.A. (1988) Remote sensing in hydrology. *J.Hydrol.* **100**.

Schultz, G.A. & Barrett, E.C. (1989) Advances in remote sensing for hydrology and water resources management. Unesco, *Technical Documents in Hydrology,* IHP-III, Project 5.1, Paris.

3 Digital Terrain Models and GIS

Integrating a physically based hydrological model with GRASS

S. CHAIRAT & J. W. DELLEUR
School of Civil Engineering, Civil Engineering Building,
Purdue University, West Lafayette, Indiana 47906-1284, USA

Abstract A procedure to integrate TOPMODEL with GRASS is described here. The Geographic Information System, GRASS, is used for hydrological parameter determination, hydrologic assessments and to link the hydrological model. A small agricultural catchment (3.38 km^2) is used for the preliminary simulations.

INTRODUCTION

The understanding and prediction of catchment response to rainfall through the several flow paths is an important part of water resources management and it is of practical interest in the estimation of flood events. A new trend of thought in hydrology has focused on the determination of the interactions of catchment geomorphology with spatial heterogeneity in watersheds. Thus, various researchers have attempted to explain the complexity of runoff production processes that arises from the spatial heterogeneity in topography, soil characteristics, soil type, vegetation covers and antecedent soil conditions.

Geographic Informations Systems (GIS) are computer based systems that provide very powerful data management facilities for handling spatial databases. With the increasing availability of these softwares, the task of manually dealing with large quantities of data is made easier. Further, GIS provide convenient analysis functions to maintain and analyze spatial and attribute data, integrate information, and display output in tabular and map format. Thus, the availability of GIS is an effective and convenient tool for hydrologists to study the spatially distributed basin characteristics and their influence on runoff generation.

This paper describes the procedure adopted to integrate a physically based hydrological model and a Geographic Information System. The topography based hydrology model is TOPMODEL, presented by Beven & Kirby (1979). The Geographic Information System used in this study is the Geographical Resources Analysis Support System (GRASS), developed by the Corps of Engineers Construction Engineering Research Laboratory. The Integration of TOPMODEL with GRASS provides a powerful tool for a better understanding of hillslope runoff production within the catchment and for investigating the influence of soil variability on runoff generation, as well as for incorporating the geomorphological characteristics and their impacts on catchment responses.

The product being developed uses GRASS as a platform to operate

TOPMODEL, using the GIS to create, manage, manipulate, analyze and display TOPMODEL's input and output. Several tools are created to assist the data preparation and the interface.

MODEL CONCEPT

A physically based topography model, referred to as TOPMODEL, is used to simulate the catchment response. This section presents a simplified description of the model based on the extended version of TOPMODEL theory given by Beven & Wood (1983) and Beven (1986).

TOPMODEL conceptualizes the soil water storage as a sequence of storages with different properties. First the interception store is filled to its maximum capacity before any water can infiltrate to the infiltration store. A leakage takes place at a constant rate from this store to the saturation store. This latter store is nonlinear and an exponential relationship between the subsurface outflow and the storage is assumed. A vertical drainage is allowed from the infiltration store to recharge the saturation store. Evaporation losses are allowed from these storages. Two types of dischargeable outflows are contributed to the stream runoff. A delayed flow component is supplied by the saturated zone as the baseflow, and a quick flow component is given by the contributing areas and by excess infiltration.

TOPMODEL is an event-oriented model that predicts the catchment responses following one or a series of rainfall events. Further, it maintains a continued accounting of the storage deficits allowing the identification of the saturated source areas within the basin. This is obtained by combining the spatial variability of topography and soil characteristics. A brief description of this relationship is given here.

The subsurface flow rate per unit width of contour length, q_i, at any point on the hillslope is approximated by (Beven et al., 1984; Beven & Wood 1983; Beven, 1986; Sivapalan et al., 1987):

$$q = T_i \tan \beta_i \exp(-\frac{S_i}{m}) \tag{1}$$

where T_i is the soil transmissivity, β_i is the slope angle, S_i is the local storage deficit, and m is used to describe the change in transmissivity with depth. Based on this exponential approximation the local deficit is derived as (Beven, 1986):

$$S_i = \bar{S} + m\gamma - m \ln \left(\frac{a}{\tan \beta} \right)_i + m \ln (T_i) \tag{2}$$

where $\overline{S}$ is the average storage deficit, $\ln(a/\tan\beta)$ is the topographic index, a is the cumulative area drained through a unit length of contour line, and γ is the area weighted topographic and soil index defined as:

$$\gamma = \frac{1}{A} \int_A \ln\left(\frac{a}{T\tan\beta}\right)_i dA \tag{3}$$

Equation 2 is used to predict the saturated contributing areas at each time step. A negative value of S_i indicates that the area is saturated and saturation overland flow is generated. During the simulation, the mean storage deficit $\overline{S}$ is updated at each time step, and the base flow is determined. Further, the distribution of the saturation zone depends on the topographic index and the value of T_i.

The model is applied to each of the subcatchments, obtained by dividing the catchment into smaller areas based on the channel network. The flow is then routed to the outlet based on the assumption of constant kinematic wave velocity.

WATERSHED DESCRIPTION

The study catchment is located in a field station for interdisciplinary studies near West Lafayette, Indiana, named the Indian Pine Natural Resources Field Station (Fig. 1). The field station encompasses two major watersheds and several small catchments. The larger watersheds are located along the two major creeks draining to the Wabash River. The Indian Creek watershed has a drainage area of 67 km^2 and the Little Pine watershed includes 139.6 km^2.

The study area used for the preliminary analysis is located at the head of the Little Pine watershed with a drainage area of 3.38 km^2. It is observed that the soil types are primarily silt loams and silty clay loams. The study area is predominately agricultural covered with pasture and corn. The area is flat in the upper part of the catchment and gets steeper along the main stream near the channel outlet. The slopes vary from 1% to 14% at the outlet.

DATA REQUIREMENTS

Two types of parameters are required for TOPMODEL: physical parameters and hydrological parameters. The physical parameters include the area of the subcatchments, channel network, stream lengths, slopes, elevation, aspect, and the upslope contributing areas from which the flow drains into a given point.

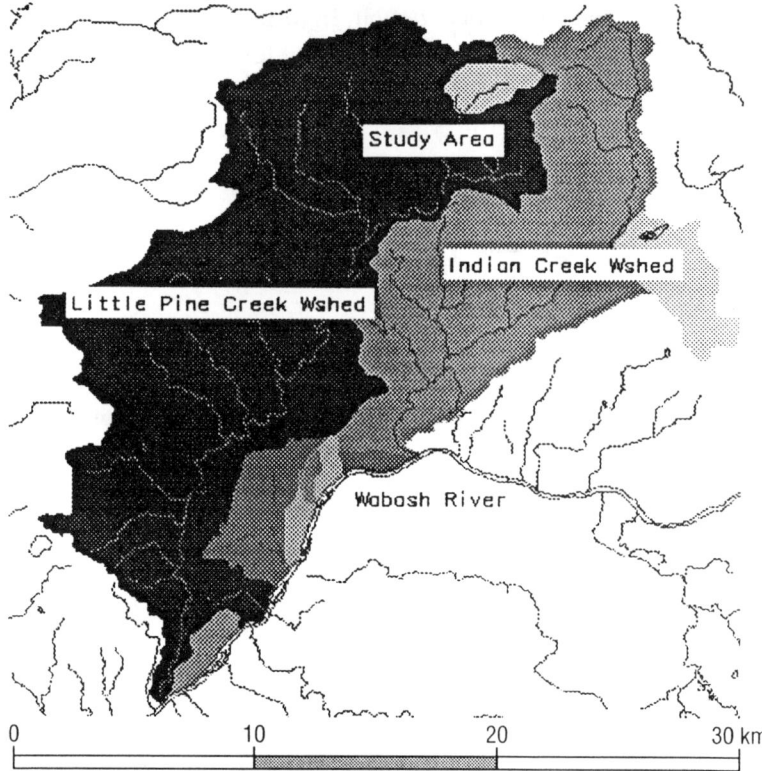

Fig. 1 Catchment location.

The hydrological parameters include the infiltration parameters, interception storage specifying the volume to be filled before flow enters the subsurface storages, rainfall events, evapotranspiration, hydrological properties of the soil and type of land cover.

In this study, the data required are divided into two types of input files. The first type of data input is provided by GRASS. These files provide the spatial and relational data describing the watershed characteristics. The second type of input provides the initial conditions and the temporal events (rainfall, discharges and evaporation). A detailed description of this process is given in the following section.

INTEGRATION PROCEDURE

The linkage procedure of TOPMODEL and GRASS is made at three levels. The first level corresponds to using GRASS tools for hydrological parameters determination. The second level consists of adapting the file formats for input and output to and from the hydrological model. The last step consists of TOPMODEL simulations. A schematic diagram of the integration procedure and the flow of information from one level to the other is shown in Fig. 2.

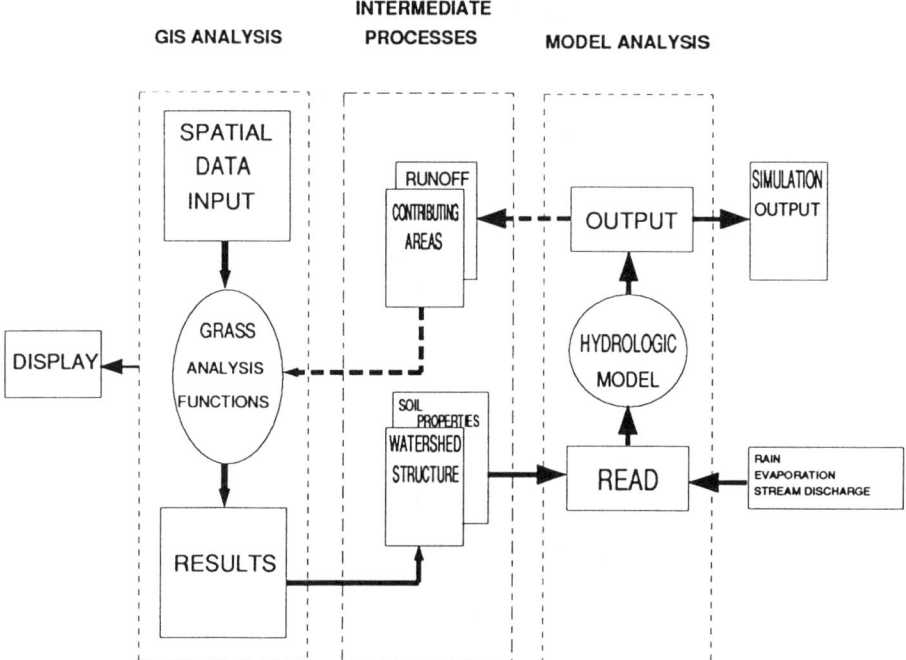

Fig. 2 Schematic diagram of the integration procedure of TOPMODEL on GRASS.

GIS analysis

A 30 m cell size was selected to store the elevation data, soils and land cover digitized from existing maps. GRASS provides powerful tools for the derivation of the topographic parameters. Specifically, GRASS provides a watershed basin analysis tool which is used for watershed and stream delineation. Thus, given the elevation layer (Fig. 3a), this tool is used to generate the subcatchment areas, the stream segments, the drainage direction within each cell and the accumulation map provides the upland area draining into each cell. Further, GRASS provides other tools for terrain analysis, and an inference engine that allows users to build their set of rules to create new maps. Also, GRASS allows users to perform a wide range of arithmetic calculations on existing map layers and to create new layers. These mathematical calculations are done on a cell by cell basis following the mathematical functions supplied by the user.

One of the fundamental parameters required by TOPMODEL is the topographic index which is obtained primarily from the spatial distribution of slopes and the drainage accumulation map layer. This can be easily obtained by using GRASS capabilities of map computations. This procedure computes the topographic index based on the single flow direction. This approach allows flow to drain into a given cell only in the direction of the gradient vector. Quinn *et al.* (1991) presented another approach which allows the accumulated upslope

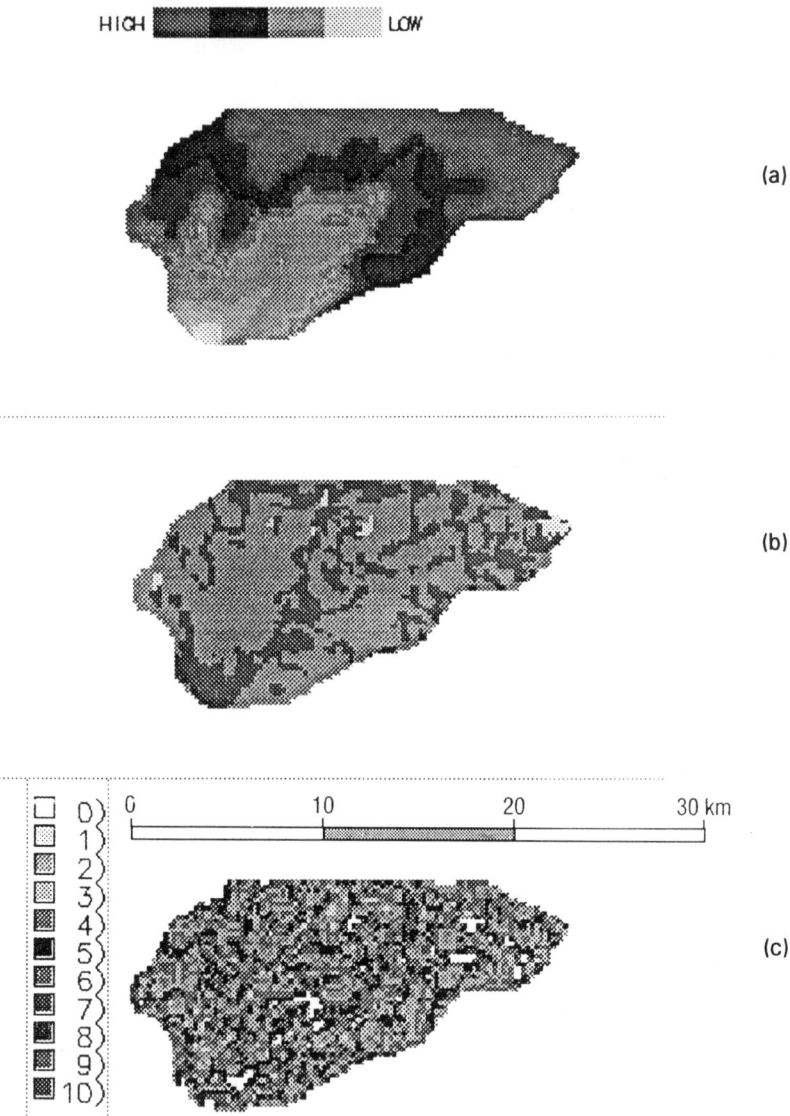

Fig. 3 GRASS map layers of (a) elevation, (b) hydraulic conductivity, and (c) topographic index.

flow from each cell to be distributed among eight flow directions by weighted proportions based on the slope angles and their directions. A GRASS tool is developed to determine the spatial distribution of the topographic index based on this multiple flow approach. Figure 3c shows the topographic index map layer obtained by applying this tool. Further details about this algorithm are given by Quinn *et al.* (1991).

The hydraulic conductivity for the digitized soil types (Fig. 3b) are extracted from the SCS Soil-5 interpretation data files. This database contains soil properties for the soil series in the United States.

Intermediate process

Tools for both analyzing and visualizing need to be developed to allow the flow of information from GRASS to TOPMODEL and for graphical display of the model results. The continuous arrows shown in Fig. 2 indicate that the link has been established, while the dashed arrows mean that work is in progress to develop those links. Thus, a tool has been developed to extract information from GRASS map layers into a file format readable by TOPMODEL. These files contain information about the topographic index distribution, the subcatchment area, the stream length and the soil properties.

The linkage of TOPMODEL's results to GRASS is still in progress. This tool will provide the display of the contributing areas for different time steps during the simulation. This provides a better understanding of the dynamic expansion and contraction of zones of surface saturation in the catchment, and identifies the potential contributing areas.

Model analysis

The last step of this integration procedure consists of simulating precipitation events with a time step of 15 minutes. The rainfall, evaporation and actual discharges information as well as the information provided by the intermediate step are input to the model simulation. The model generates the outflow hydrograph at the outlet, the baseflow separation, the subsurface flow component, the rainfall excess distribution and the model efficiency. A second output is directed to the intermediate process, which provides information to be displayed by GRASS about the saturated zones in the catchment for several time steps of the simulation.

Once the implementation of TOPMODEL on GRASS is completed, modification of the current routing procedure are considered. The utilities of GIS make the extraction of the detailed geomorphological characteristics easier to obtain. Thus, this information will be used to generate the geomorphological instantaneous unit hydrograph. Further, the model will be extended to simulate spatially variable rainfall intensities.

CONCLUSION

The Geographic Information System capabilities of putting together the large amount of spatial data required have been very efficient. Further, GIS has been very helpful for extracting the various basin characteristics such as the subcatchment areas, the stream network, slopes and the drainage direction. Although, certain parameters , i.e., the topographic index, had to be computed through new functions that are built into the software, this task was successful using the subroutines provided by GRASS.

The link between TOPMODEL and GRASS is an efficient way for data manipulation and updates and is a powerful tool to investigate the effects of soil heterogeneity on the catchment response. This link provides future capabilities of incorporating the effect of the catchment geomorphology and of implementing the model for larger watersheds. Further, the GIS utilities of surface fitting will be used to incorporate spatially variable rainfall intensities.

Acknowledgements The authors wish to thank Dr. K. J. Beven, University of Lancaster, UK, for his technical reports about TOPMODEL, Mr. Kurt Buehler, US Army Corps of Engineers Construction Engineering Research Laboratory, for providing several GRASS documentations, Dr. B. Engel and his graduate students of the School of Agricultural Engineering at Purdue University for providing the digital elevation model, and Dr. J.R. Wright, Director of Water Resources Research Center, for his helpful comments.

REFERENCES

Beven, K.J. & Kirkby, M.J. (1979) A Physically Based, Variable Contributing Area Model of Basin Hydrology. *Hydrol. Sci. J.* **24**(1), 43-69.
Beven, K.J. & Wood, E.F. (1983) Catchment Geomorphology and the Dynamics of Runoff Contributing Areas. *J. Hydrol.* **65**(1/3), 139-158.
Beven, K.J., Kirby, M.J., Schoffield, N. & Tagg, A.F. (1984) Testing a Physically-Based Flood Forecasting Model (TOPMODEL) for Three U.K. Catchments. *J. Hydrol.* **69**(1/4), 119-143.
Beven, K.J. (1986) Hillslope Runoff Processes and Flood Frequency Characteristics. In: *Hillslope Processes* (ed. by A. D. Abrahamms), 187-202, Allen and Unwin, Winchester, Mass.
Beven, K.J. (1987) Towards the Use of Catchment Geomorphology in Flood Frequency Predictions. *Earth Surf. Processes and Landforms* **12**(1), 69-82.
Quinn, P.F. (1990) Application of the Model TOPMODEL to the Catchment of Booro-Borotou, the Ivory Coast. Technical Report, Institute of Environmental and Biological Sciences University of Lancaster.
Quinn, P., Beven, K.J., Chevallier, P. & Planchon, O. (1991) The Prediction of Hillslope Flow Paths For Distributed Hydrological Modeling Using Digital Terrain Models. *Hydrol. Processes* **5**(1), 59-79.
Sivapalan, M., Beven, K.J. & Wood, E.F. (1990) On Hydrologic Similarity: 3. A Dimensionless Flood Frequency Model Using a Generalized Geomorphologic Unit Hydrograph and Partial Area Runoff Generation. *Wat. Resour. Res.* **26**(1), 43-58.
Wood, E.F., Sivapalan, M., Beven, K.J. & . Band, L (1988) Effects of Spatial Variability and Scale With Implications to Hydrologic Modeling. *J. Hydrol.* **102**(1-4), 29-47.

Matching standard GIS packages with urban storm drainage simulation software

J. ELGY
Department of Civil Engineering, Aston University, Birmingham, B4 7ET, UK

C. MAKSIMOVIC & D. PRODANOVIC
Department of Hydraulic Engineering, Faculty of Civil Engineering, University of Belgrade, Bulevar Revolucije 73, PO Box 895, 11 000 Belgrade, Yugoslavia

Abstract Existing urban drainage models are of sufficient detail and complexity to model accurately the effects of storms on an urban drainage basin. Their widespread use is curtailed mainly by problems of data acquisition, storage and manipulation. The combination of remote sensing and geographical information systems allows this to be done better. A series of experiments were carried out using different geographical information systems (GIS) to create the input files for two different urban drainage models. The grid based GIS IDRISI was the best for this use because of its simple file structure and ease of writing new modules. In general grid based GIS is best for this type of modelling method of calculating sub basins was developed and a program written to calculate areas of land cover in each basin to a file for direct input into an urban drainage model. The methodology is evaluated for an experimental catchment where the results were found to be compatible to a previous hand calculation but with major difference in calculated areas.

INTRODUCTION

Current urban drainage models are well established and generally adequate for the design of urban drainage networks. There is, however, a problem with the acquisition, manipulation and storage of data for the models. The combinations of the technologies of remote sensing and geographical information systems allow a solution to these problems. The development of general purpose GIS systems has urged and encouraged their application in various fields of engineering, particularly in the area of complex urban infrastructures. This application is spreading in two major directions:
(a) generation of maps and inventories;
(b) links with simulation models and the graphical presentation of results,

Although it is difficult to draw the dividing line between them it could be generally stated that the final product of the first type is a digital map and its accompanying database. The second group is more complex and it requires specific target oriented information processing (modelling). It is the second application that is of interest to the engineer.

Many details of urban drainage networks are already stored in GIS systems of the first type but the most effective use of that data relies on the second type of application. This paper examines how four commercial GIS packages were used to create input suitable for two standard urban drainage models. The packages concerned were: IDRISI, ARC/INFO, INTEGRAPH, and the CAD package AUTOCAD which with suitable macro programs offers a sort of GIS capability. The outputs from each of these models was than to be passed to two urban drainage models: WALRUS and BEMUS. IRTCUD in the University of Belgrade has a well-instrumented and documented catchment at Miljakovac in Belgrade and this was used as the study area to investigate the approach. The detailed objectives of the project were:

(a) to form the land cover and digital elevation models for a catchment;
(b) from the DEM map the contributing areas to each inlet to the sewer system and investigate how these can be modelled in each GIS;
(c) investigate the effect of flow diversion structures, for example, walls, road chamber etc.;
(d) calculate the statistics of each subcatchment required for WALRUS and BEMUS;
(e) compare the statistics from each method and traditional hand methods;
(f) prepare input files for the models from the GIS;
(g) investigate the different performance of the model for each method of deriving the input files;
(h) consider the sensitivity of output to each method of generating input files;
(i) asses the resources required for each approach in terms of manpower, training, computer power and software costs.

However due to time limitations many of these objectives have yet to be met. This paper reports on the use the IDRISI GIS system and how the outputs the data stored on this system was used to create input files for the BEMUS and WALRUS urban drainage models. Only the BEMUS model has produced output for analysis so far. The work continues.

CHOICE of GIS

The vector based GIS's proved difficult and slow for data aquisiton and modelling. Their complex data structures meant that it was also difficult to write suitable programs to extract the required data for the urban drainage models. Almost no progress was made using ARC/INFO and AUTOCAD proved useful only as a tool for digitization and to move data in the .DXF format to IDRISI. Digitizing the maps for INTEGRAPH proved very slow and though the accuracy of the maps produced was very high, failure to correctly brief the operators of the system meant that many features wanted for urban drainage modelling were not included in the datåbase. It was decided to temporally abandon this approach.

IDRISI is a low cost grid based GIS available from Clarke University in the USA. It is very popular as teaching system. Its simple file structure means that new modules can be readily written for the system. Two new modules were written for this project:

(a) to determine the contributing area to each inlet in the sewer system;

(b) to take these contributing areas and write the output files suitable for the urban drainage models.

In addition the IDRISI module to form a digital elevation model from point data was rewritten to make use of a transputer parallel processing board added to the micro computer system for faster execution. The only modules from the original IDRISI package used for this project were: the viewing module, COLOR; the slope and aspect modules, SURFACE; the reclassifying module, RECLASS; the area calculating command, AREA; and the geometric correction module, RESAMPLE. Other modules can be used for traditional GIS applications. The cross tabulation command CROSSTAB was used to verify the output files calculated from the post processing program generating direct inputs for the urban drainage models.

IMPLEMENTATION IN IDRISI

A coloured and photocopied map of the Miljakovac catchment were supplied to the University of Aston for digitization and initial processing. The maps

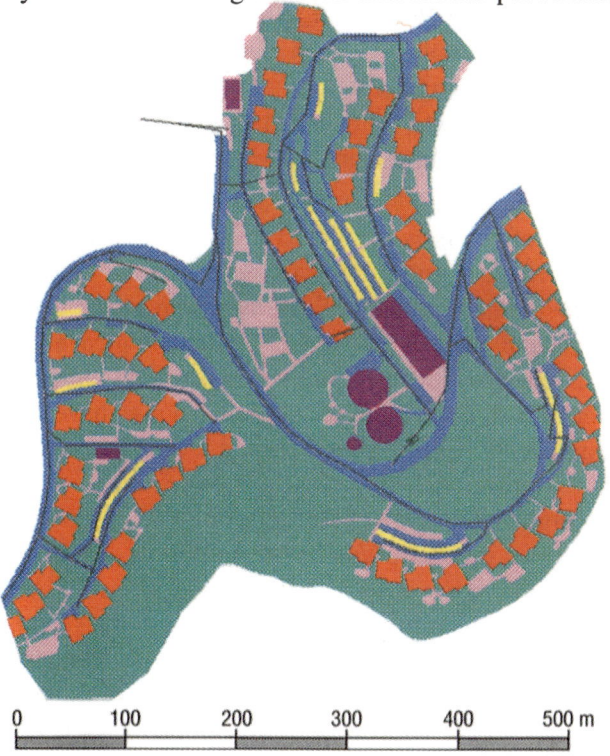

Fig. 1 Landcover image of the Miljakovac catchment. Each cell is 1x1 m (700 rows, 630 columns). Colour legend: red=domestic residencies, turqoise=grass, blue=roads, violet=paved areas not connected to the drainage systems, yellow=garages draining into rads, purple=industrial units with flat roofs directly connected to the drainage system, black=sewer system.

extended over several sheets of paper and distortions had been introduced by being photocopied at least twice. The colouring had been performed after field visits and identified permeable areas, roads and other impermeable areas. This coloured map was the basis of the land cover map (Fig. 1). It did not identify which of the roofs were connected directly with the sewer system, a fact which has made difficult later analysis. On digitization each building was given a unique identifier to enable them to be reclassified at a later date as connected or unconnected. The roofs were all reclassified as roof and the individual identifiers lost. control points were taken from an original map of the catchment and the digitized map resampled.

A cell size of 1x1 m was chosen giving an image of 630 columns (x - direction) by 700 rows (y - direction). Using the AREA command in IDRISI the following areas were calculated:

(a) grass and other pervious areas   141 231 m^2
(b) roads   44 606 m^2
(c) residential houses, with flat roofs   34 523 m^2
(d) garages, with flat roofs   4 259 m^2
(e) supermarkets and school   5 448 m^2
(f) footpaths and playgrounds   21 498 m^2
(g) boundary of meteorological station   29 m^2

The important part about these calculations was that they differed widely from the areas measured by hand. Firstly the roof areas measured by planimeter gave an total roof area of 26 770 m^2 as opposed to 44 230 m^2 by IDRISI, secondly the area of roads was 68 850 m^2 as compared to 44 606 m^2 by IDRISI. The difference in roof area was very large and in the absence of the original area measurement personnel and techniques used we were unable to arrive at a suitable explanation but conjecture that a typical area for a building had been found and the buildings counted to arrive at an total area. Garage roofs are not connected to the drainage system and may have been considered part of the roads area. Repeated checks failed to find any significant error in our digitization. Two separate digitizations were made of the catchment in IDRISI and though there were difference they were not of a magnitude sufficient to explain the difference. The area commands in AUTOCAD digitization were not used to verify the calculations.

The percentage of roofs directly connected to the sewer system is the most significant component in the peak flows in this catchment. Doubling the area of roofs has a significant effect on the magnitude of the peak flow, which was detuned by altering the percentage of roofs directly connected to the sewer system in the BEMUS model. Not surprisingly the percentage connected was reduced by almost half to 30%.

Two other IDRISI maps were produced: one containing the sewer pipe network and inlets to that network and one the digital elevation model. The sewer network was typed in as a vector file from source survey data, resampled to correspond to the road network. Though the errors produced by a small difference in the position of an inlet will have little effect on the performance of the drainage model. The inlet must, however, be on the roads or other paved areas to allow for the correct calculation of the contributing areas. The Miljakovac catchment is steeply sloping at the upper end and almost flat at the base. In addition there are some natural hollows or pits which do not contribute to the drainage areas. The maps contained spot heights, discontinuous contour lines (isolines) and shading of embankments. Many experiments were carried out to determine the best DEM:

(a) dividing the map into a 20 m grid and interpolating a value for the centre of each grid square from the contour (isoline) map;
(b) tracing contours where they can be found and using the contour to DEM module of IDRISI;
(c) taking point values from contours at a spacing along the contour equal to the spacing between contours;
(d) taking spot heights along roads and tops of embankments;
(e) taking heights from a stereo pair of aerial photographs.

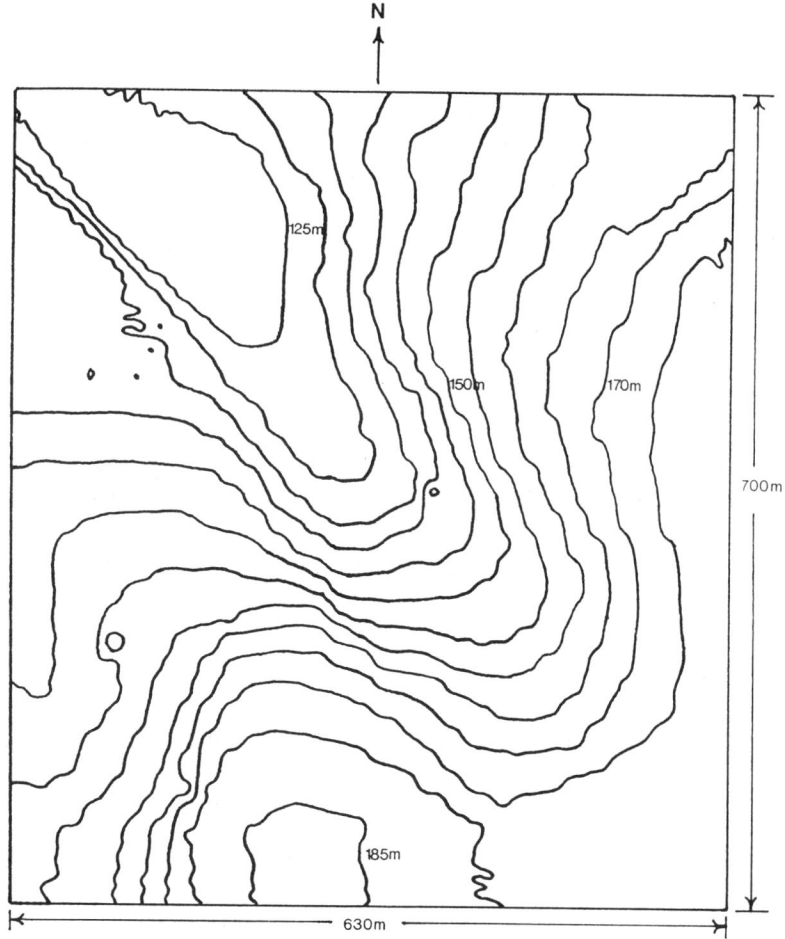

Fig. 2 DEM of Miljakovac catchment with 5 m contour intervals.

For some of these experiments it was discovered that even on a 33 MHz 486 computer the IDRISI interpolation routine was taking over 24 hours to arrive at a DEM. By writing a inverse square of distance weighted interpolation routine to use the 6 nearest points (with at least one in each quadrant) to run on a transputer based parallel processing computer this time was reduced to 20 minutes and allowed sensible experiments to be performed. It was found that experiment (e) produced the smoothest, best looking DEM. Whereas those making use of contours produced DEMs with steps in them. The modifying effect of roads to the contributing areas is significant. In the Miljakovac the road is often in small cuttings and this must be taken into account. At some parts of the catchment the interpolated heights varied by as much as 20 m! The effect of this is actually small, what was happening was that the horizontal position of a steep slope was modelled a few metres further south. The final DEM made use of almost all the data, over 8000 points in total:

(a) data spaced along contour lines every 20 m;
(b) data from the hand interpolated grids;
(c) data from aerial photography;
(d) all spot heights from the map and embankment data.

Unfortunately the aerial photgraphy was taken before much of the building work had taken place on the catchment and missed many embankments, otherwise we would use this method exclusively.

The few pits and hollows on this model were visible in all individual models and were thought to exist in the catchment. A 3 by 3 low pass filter was run over the resulting DEM many times to obtain a smooth surface. Figure 2 shows the resulting DEM. Notice that there are still a few hollows in the surface.

DELINEATION OF CONTRIBUTING AREAS

WATRSHED is a watershed or catchment delineation program supplied as part of the IDRISI package. From a number of points within the catchment it determines every cell which can flow into those cells. This is the catchment. It can not determine subcatchments of a major catchment or determine contributing areas to a series of inlets to the sewer system.

The conventional procedure for solving the problem of subcatchment boundary delineation can be described as:

(a) find the cell that represents the inlet;
(b) go around that cell and find out which cells can supply the water into the first cell. Those, whose aspect is towards the target cell within a given threshold (typically 90 degrees) called the "outflow angle" and are not already marked as part of a higher subcatchment are marked as part of the current subcatchment;
(c) repeat this procedure until no more cells can be added.

By multiple sweeps through the image, the subcatchment begins to grow.

If the slope of the cell is zero, or less than some threshold, called the slope threshold, then it is assumed to be able to flow in any direction.

This approach leads to very slow program execution. A program written in this way making direct file access to the hard disk of the computer (necessary to access the surrounding cells when the length of a row can exceed the buffer in the computer) ran continuously for several weeks on a 25Mhz 386SX computer with maths coprocessor at Aston University. It is unlikely that even with more powerful machines and holding the complete image in memory that this technique can be run in less than one day.

A new technique was developed. The general idea is that a temporary image is created in one pass that can hold all possible directions of inflow into each cell. Using this technique the slope and aspect models are used just once and just one comparison for each cell is performed. The input for the data can be easily buffered and the temporary image held in computer RAM for the fastest possible computation. This temporary file will be called the flow direction map. Exceptions from the land cover map can be easily incorporated into these rules.

Since each cell can have just eight neighbours a very compact method of data representation can be used using bit wise operations. If we assigning to each cell one byte then each neighbour can have an accompanying bit that indicate whether that cell can supply the current one with water or not. The first bit is used to designate whether the cell to the Northwest can supply the current cell with water; the second bit, west; and so on to the eighth bit which indicates the capability of the cell to the north of the current cell to supply it with water. A value of 3 means that the cell can be supplied with water from the west and north west; 131 from the west, north west and north. The value of the flow direction map for each cell caries sufficient information to build the next stage of the subcatchment image. The benefit of this approach is that the input images have to be read only once in a purely sequential manner. Using a 33 MHz 486 computer the forming of this map for the Miljakovac catchment took 10 minutes to create.

After making the flow direction map the subcatchments are calculated from the following procedure:

(a) find the highest inlet, mark it as current;
(b) search through the flow direction bits of the current cell, go to the first bit set and use that cell as the current cell;
(c) store the path used for the later return procedure;
(d) if all bits are zero, that cell is at the boundary of the current subcatchment, mark it as a member of the current subcatchment;
(e) step back one cell and mark it as the current one;
(f) if this is the last cell in the return path eend the search for the current subcatchment;
(g) if there are more cells in the return path, continue with the search procedure from step (b);
(h) move to the next highest inlet.

It is important to take the inlets from the highest to the lowest otherwise the lower catchment will overtake the higher ones.

Because of the bitwise operations the algorithm is very fast. Also it does not need to be changed for the introduction of any kind of exceptions introduced from the land cover file. The only major drawback is that for fastest possible execution the whole flow map must be in memory at one time and it is quite difficult to introduce any form of explicit buffering. The array is of the *unsigned char* type and for Miljakovac data this will be 440 kB long and only the results needed to be written to hard disk.

In preliminary runs of this program large areas were left unclassified, i.e. not in any subcatchment. The flow map model was modified to allow different outflow angles and slope thresholds for different land cover types. subjectively the best results were found with an outflow threshold of 90 degrees and variable slope thresholds of different land cover types of: grass, 2%; roads, 2%; houses, 0.1%; garages, 0.1%; supermarket roof (a very long roof behaving much like a road with separate inlets along the roof), 2%; other paved area, 2%. Figure 3 shows the resulting subcatchment boundaries superimposed onto the land cover map The shaded areas are unclassified by the subcatchment delineation program. Some of the unclassified areas clearly lie outside the catchment boundary, some are local pits and hollows.

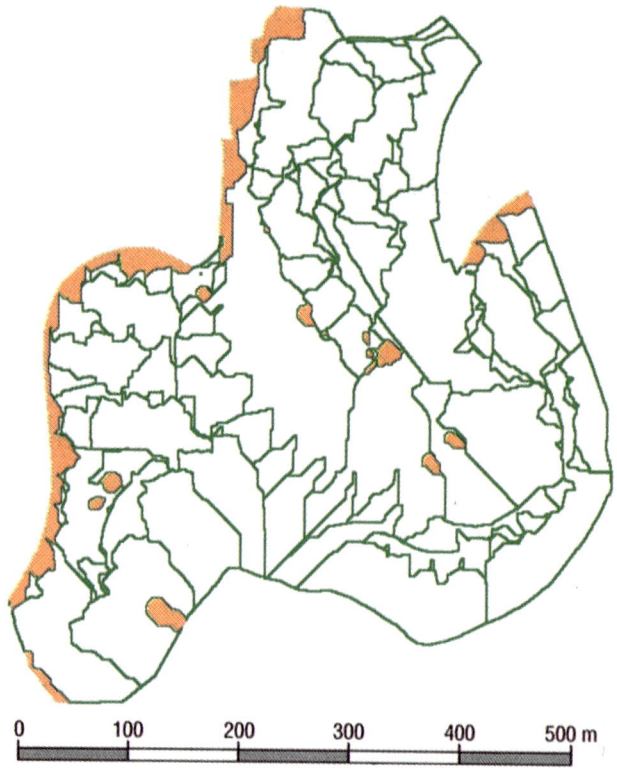

Fig. 3 Image of the Miljakovac catchment showing subcatchment boundaries.

To account for the flow diversion properties of buildings and roads, buildings were artificially raized by 5 m and roads dropped by 0.5 m. The "shadowing effect" of buildings caused considerable problems with large areas of unclassified land behind them. This effect was reduced by increasing the outflow threshold at building boundaries but the net result was not an improvement on the simply modified slope threshold.

RESULTS AND RECOMMENDATIONS

Much work still needs to be done to complete the objectives mention above and only one successful run of a simulation program has actually taken place. This run on the BEMUS storm water drainage package indicated that after tuning the percentage roofs connected to the sewer system the model accurately matched the measured outflow hydrograph. As did the traditional hand method. Provided that most of the catchment had been allocated to an inlet the exact boundaries seemed to matter very little. There very large discrepancies between the contributing areas to particular inlets from the hand and automatic methods. The lessons learnt from the GIS approach can be summarized as:

(a) initially the GIS approach is much more time consuming than drawing boundaries roughly by eye and measuring areas by planimeter, but this may be largely due to the steep learning curve of developing and using new technologies;

(b) it took 2 days to digitize the 4 sheets covering the Miljakovac catchment, though if this was only one sheet half a day would be sufficient;

(c) using aerial photography took only a few hours to create the DEM, map digitization took 2 days;

(d) formation of the DEM is easier from stereo pair aerial photography than by any other method;

(e) the GIS approach is more objective than traditional approaches;

(f) the availability of digital maps makes the GIS approach faster;

(g) existing urban drainage models no not need rewriting to take account of GIS technology;

(h) grid based GIS is easier to deal with than vector GIS for this purpose;

(i) once a GIS study has been carried out, changes to the catchment or to the sewer system can be readily investigated;

(j) more work is still needed on subcatchment boundary delineation, particularly with regard to flow diverting structures;

(k) levels of accuracy of the digitized maps and of the resulting output still need the judgement of an engineer, completely automatic techniques still do not exist;

(l) remote sensing offers an excellent possibility to obtain land cover data that is up to date and easily ported into a GIS;

(m) if the source of the information is an existing paper map then this is still the weakest part of the system;

(n) field visits are still essential, if only to discover which roofs are directly connected to the sewer system, this data could be directly entered onto a portable computer;
(o) more work needs to be done on examining other GIS approaches and urban drainage models;
(p) very flat catchments still prove a problem in determining contributing areas.

CONCLUSIONS

IDRISI proved a very flexible and useful tool for developing an objective approach to obtaining the input files for a traditional urban drainage model, BEMUS. The programs developed for calculating the contributing areas from the DEM and to create the input files for BEMUS and WALRUS are very useful tools. With further development the GIS approach could provide the standard method of gather inforamtion for urban drainage models.

Acknowledgements Much of this work was carried out on a workshop in Belgrade, Yugoslavia during May 1992. The worksop was funded by the EC under TEMPUS project JEP 2424-91/1. Special thanks should be given to Mr. Assen Dimov of the University of Architecture, Civil Engineering and Geodesy, Sofia, Bulgaria and Dr. Zhimin Chen of Aston University, Birmingham, United Kingdom who programed many of the algorithms used in this work.

Modelling the spatial variability of hydrological processes using GIS

I. D. MOORE, J. C. GALLANT & L. GUERRA
Centre for Resource and Environmental Studies, The Australian National University, Canberra, ACT 2601, Australia

J. D. KALMA
CSIRO Division of Water Resources, Canberra, ACT 2601, Australia

Abstract TAPES-G, a grid-based method of terrain analysis, SRAD, an approximate method of computing daily global short wave, net long wave and net radiation and extrapolating minimum, maximum and average temperature across a landscape, WET, an approximate method of computing spatially distributed soil water content and evaporation and catchment runoff and EROS, an approximate method for predicting erosion potential, are simple terrain-based models for analysing spatially distributed land surface processes. These models operate in a UNIX environment with X-Windows, have full graphics display facilities, and exhibit many of the features of a GIS. The spatial variability of the radiation, thermal, soil water, evaporation and erosion regimes on the 27 km^2 Lockyersleigh catchment in southeastern Australia is explored using these models.

INTRODUCTION

Models that include the key factors determining system behaviour, but which are based on simplified representations of the underlying physics of the processes, are effective tools for examining spatially variable hydrological processes using GISs. With this approach some physical sophistication is sacrificed to allow improved estimates of spatial patterns in the landscape. These models provide a means of relating "pattern to process" and can operate with "minimum data sets" - adequate spatially distributed data being a major constraint on the application of more sophisticated modelling strategies and GISs. TAPES (Terrain Analysis Programs for the Environmental Sciences) is a suite of integrated FORTRAN77 and C programs that have been developed at the Centre for Resource and Environmental Studies based on this philosophy. This paper describes and demonstrates the utility of some of the TAPES programs.

The TAPES programs have been specifically designed to analyse spatially distributed hydrological, geomorphological and ecological processes in topographically complex landscapes. They have been developed with full graphics interfaces using the UNIX operating system and X-Windows for displaying the results. In addition a number of tools have been developed for statistically analysing and fitting frequency distributions to the results. The TAPES programs are divided into three broad classes that correspond to the principal ways of structuring networks of elevation data: (a) Triangulated

Irregular Networks (TINs); (b) grid-based networks; and (c) contour-based networks (Moore *et al.*, 1991). Because remote sensing and many GIS systems are based on pixel or cellular structures, grid-based methods of analysis are particularly attractive for use in analysing spatially distributed environmental processes. Therefore, only the grid-based TAPES programs are described here.

TERRAIN ANALYSIS

TAPES-G is a grid-based method of terrain analysis that calculates slope, aspect, principal drainage direction, specific catchment area, profile and plan curvature and flow path length at each node in a grid-DEM. These attributes are useful in characterizing a wide range of hydrological, erosional and geomorphological processes occurring in complex landscapes (Moore *et al.*, 1991). The algorithm creates a depressionless DEM using the method of Jenson & Domingue (1988) if desired. Specific catchment areas are estimated using either the classical D8 algorithm, that allows drainage from one node to only one of eight nearest neighbours based on the direction of steepest descent, the quasi-random Rho8 algorithm, or the FRho8 algorithm that permits drainage from a node to multiple nearest neighbours on a slope weighted basis. The Rho8 algorithm produces more realistic flow networks than does the D8 algorithm, and the FRho8 algorithm permits modelling of flow dispersion in upland areas, which is important in convex topography (Moore *et al.*, 1992a). TAPES-G can also delineate sub-catchments draining to specified seed-cells (i.e., cells at the catchment outlets).

The first and second derivatives of the elevation surface in the x and y directions ($f_x, f_y, f_{xx}, f_{yy}, f_{xy}$) are estimated by applying a second-order, central finite-difference scheme centred on the interior node of a moving 3 x 3 square grid that is passed over the DEM. Forward and backward difference schemes are used to handle nodes on the edges of the DEM (Moore *et al.*, 1992b). Attributes such as slope, aspect and plan and profile curvature can be calculated directly from these first and second derivatives.

ENERGY AND THERMAL REGIMES

The surface radiation budget is a driving force for evaporation and photosynthesis processes occurring at the land surface and is highly dependent on topography. Vegetation diversity and biomass production have been shown to be related to radiation and temperature in many studies. Here, we develop an algorithm, SRAD (Moore *et al.*, 1993), for calculating net radiation and its components in topographically heterogeneous terrain. The algorithm also spatially extrapolates maximum, minimum and average temperature.

Radiation

The component terms of the net radiative flux density, R_n, include the direct, diffuse and reflected shortwave irradiance (the sum of which is the global shortwave irradiance) and the incoming or atmospheric longwave irradiance and the outgoing surface longwave irradiance (which make up the net longwave irradiance). The input parameters of SRAD include spatially invariant variables such as albedo, emissivity, sunshine fraction, and clear sky transmittance, as well as maximum and minimum temperatures (see below). The shortwave irradiance components are calculated based on Fleming's (1987) CLOUDY algorithm modified to account for the effects of shading from direct sunlight by surrounding terrain at enclosed sites using an improved version of the Dozier et al. (1981) solution.

Temperature

The spatial distribution of minimum, maximum and average air temperature is extrapolated from measurements made at a base-station using a modification of the simple approach proposed by Running et al. (1987). The method corrects for elevation via a lapes rate, slope-aspect via the ratio of short wave radiation on a sloping surface to that on an unobstructed horizontal surface, S, and vegetative effects via a leaf area index, LAI. In the southern hemisphere this approach increases temperatures on north facing slopes and decreases temperature on south facing slopes relative to a flat surface. This effect is greatest on poorly vegetated slopes (low leaf area index) and negligible in closed forests (high leaf area index). No leaf area index/radiation corrections are applied to estimates of minimum temperature as these occur during the night.

SOIL WATER CONTENT AND EVAPORATION

In complex terrain soil water distribution is controlled by vertical and horizontal water divergence and convergence, infiltration recharge and evaporation. The latter two terms are affected by solar insolation and vegetation canopy, and vary strongly with exposure. The divergence/convergence and solar insolation are dependent on hillslope position. Beven & Kirkby (1979) and O'Loughlin (1986) independently derived wetness indices for characterizing the spatial distribution and size of zones of saturation or variable source areas of runoff generation in a landscape. These indices have been used to model the distribution of soil water content (Moore et al., 1988). These wetness indices are derived from simple catchment drainage theory, and in their simplest form can be expressed in terms of terrain attributes and soil hydraulic properties as:

$$x_i = \ln\left[\frac{T_e}{bT\tan\beta}\int dA\right]_i = \ln\left[\frac{A_s}{bT\tan\beta}\right]_i + \ln[T_e] - \ln[T_i] \qquad (1)$$

where χ is the wetness index, dA_i is the element area (m²), b_i is the outflow width (m), β_i is the slope angle (degrees) and T_i is the transmissivity (m² day⁻¹) in the ith element, A_s is the specific catchment area (catchment area draining across a unit width of contour: m² m⁻¹) and $\ln(T_e)$ is the areal average value of $\ln(T_i)$. The integral term represents the upslope area draining across a contour segment of width b orthogonal to the flow. In equation (1) A_s is a measure of the steady-state subsurface drainage flux ($q \propto f A_s$, where f is the net recharge rate: mm day⁻¹), but assumes uniform infiltration over the entire catchment.

Equation (1) has been extended to allow estimation of the spatial distribution of the long-term (i.e., average annual) soil water content and evaporation using an equilibrium approach (Moore et al., 1993). Equation 1 can be rewritten as:

$$x_i = \ln\left[\frac{1}{b\tan\beta}\int \mu dA\right]_i + \ln[P] + \ln[T_e] - \ln[T_i] \qquad (2)$$

where P is the precipitation and μ is an area weighting coefficient that is dependent on the evaporation, deep drainage and precipitation in each element. This weighting coefficient can be written as:

$$\mu_i = 1 - \left(\frac{E+D}{P}\right)_i, \quad E = E_p[1 - (1-\theta)^{C/E_p}] \text{ and } \theta = \frac{\chi_i}{\chi_{cr}} \qquad (3)$$

where θ is the relative available soil water content (1=field capacity, 0=wilting point), D is the deep seepage, E is the actual evaporation, E_p is the evaporative demand, C is a constant and χ_{cr} is the critical wetness index at field capacity. The deep seepage, D, can be expressed as the product of the subsoil saturated hydraulic conductivity and a power function of θ.

The evaporative demand is estimated from the Priestley & Taylor (1972) equation for well watered vegetation under conditions of minimal advection:

$$E_p = \frac{\alpha_e R_n}{\rho\lambda\left(1 + \frac{\gamma}{\Delta}\right)} \qquad (4)$$

where α_e is an empirical constant (=1.26), Δ is the slope of the saturation specific humidity-temperature curve and γ is the psychometric constant (functions of air temperature and pressure), ρ is the density of water, λ is the latent heat of vapourization, and R_n is the net radiation estimated by SRAD.

Equations 2-4 are solved iteratively, beginning with the element of highest elevation and finishing with the element of lowest elevation at the catchment outlet. These equations are embodied in program WET.

Because of fluctuating rainfall intensities and the short duration of rainfall events compared to the travel time of subsurface throughflow, the subsurface flow regime in a catchment rarely reaches steady-state. Barling (1992) observed that during storm events subsurface flow is only affected by a small proportion of the contributing area directly upslope. He modified equation (1) by calculating an effective specific catchment area, A_e, using it instead of A_s. The method is based on the subsurface flow travel time-specific area curve, $a_s(t)$, where $a_s(t_e)=A_e$ and t_e is the time to equilibrium, proposed by Iida (1984). By specifying different drainage times, which might be the time since the last precipitation event, different values of A_e will be obtained. Program DYNWET performs these calculations using the terrain attributes calculated by TAPES-G as the primary input data. Using relationships similar to equations (3) and (4) and net radiation computed by SRAD, it is then possible to estimate spatially distributed daily soil water content and evaporation.

EROSION POTENTIAL

Moore & Wilson (1992) derived a dimensionless index of the sediment transport capacity of overland flow. They showed that this index was equivalent to the length-slope factor in RUSLE, the Revised Universal Soil Loss Equation (Renard et al., 1991), for a two-dimensional hillslope. This index can be easily extended to three-dimensional terrain to map the effect of hydrology, and hence topography on soil erosion. The erosion potential is a power function of specific catchment area (catchment area per unit width) and the sine of the slope angle. The exponents on these two terms range from 0.4 to 0.6 and 1.2 to 1.3, respectively.

STUDY SITE

The 27 km² Lockyersleigh catchment is located in the Goulburn-Marulan region of the Southern Tablelands of New South Wales, Australia (34°41'S, 149°56'E). Elevations range from 600-762 m and the terrain is undulating and largely cleared. The vegetation is a mixture of native and introduced grasses with open woodlands on the higher ground in the east and southeast. Tussock grasses occur in most creek depressions and native sedges are widespread, indicating impeded drainage. Soils are duplex with bleached sandy/silty A-horizons, changing abruptly to a yellow heavy clay B-horizon (Kalma et al., 1987).

Intensive field experiments, including soil water measurements, ground-based measurements of energy fluxes at four sites and tethered balloon and airsonde measurements of the boundary layer (temperature, humidity and wind speed) were carried out in February-March 1992.

RESULTS AND DISCUSSION

A 30 x 30 m digital elevation model (DEM) was developed for the catchment by line digitizing 10 m contour interval contour lines, streamlines, and spot

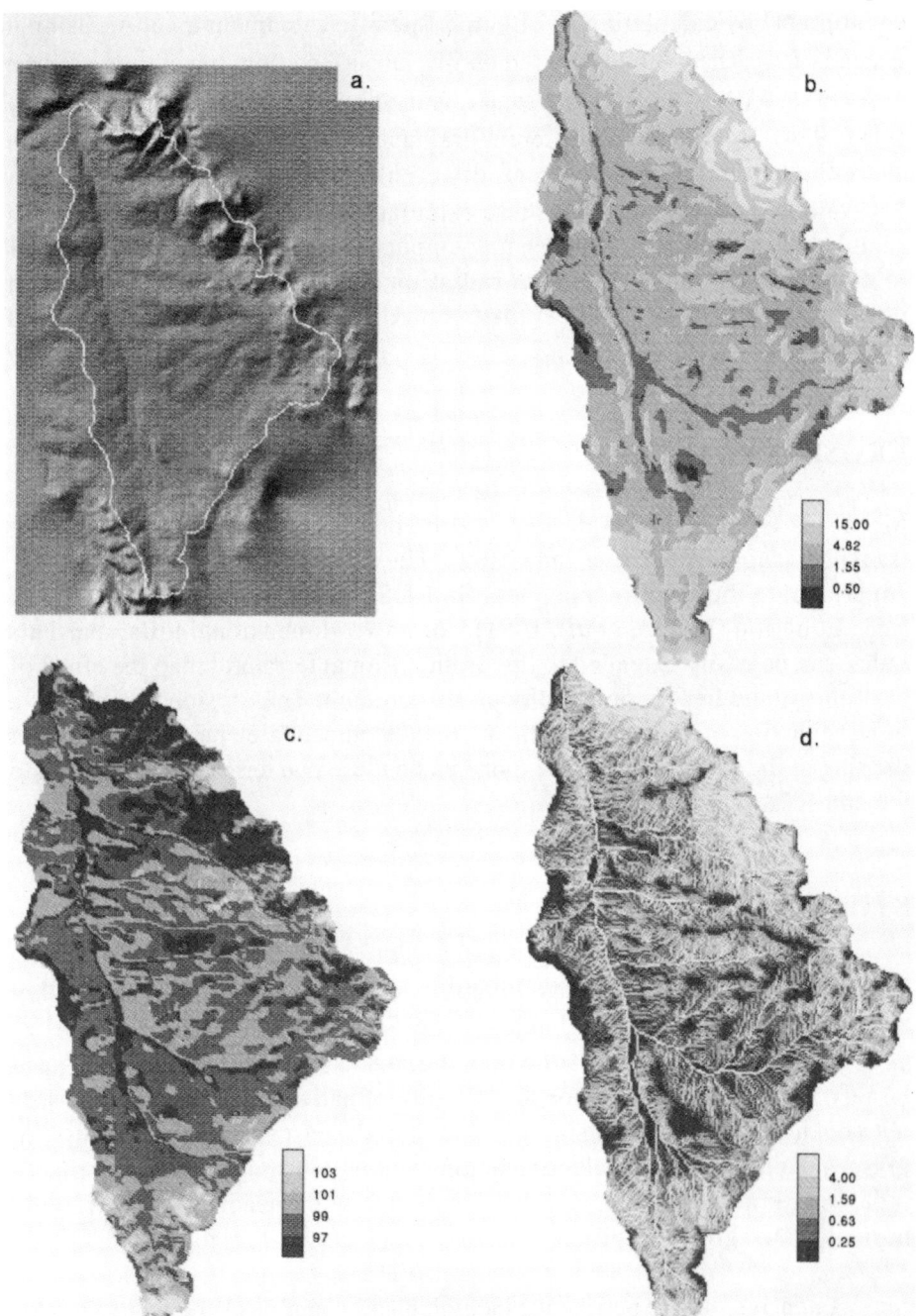

Fig. 1 (a) Shaded relief diagram of the study area, and predicted distributions of: (b) slope (%) within the Lockyersleigh catchment, (c) annual net radiation (W m^{-2}), and (d) erosion potential.

heights from a 1: 25 000 scale topographic map and then fitting an interpolating surface using ANUDEM (Hutchinson, 1989). ANUDEM is a finite difference interpolation method with a drainage enforcement algorithm that ensures fidelity with the drainage network and eliminates anomalous pits or sinks in the DEM. This DEM formed the primary data for the subsequent analysis.

A shaded relief diagram (x 5 vertical exaggeration) of the region, derived from the 30 x 30 m DEM using SHADEDEM, is presented in Fig. 1a. The

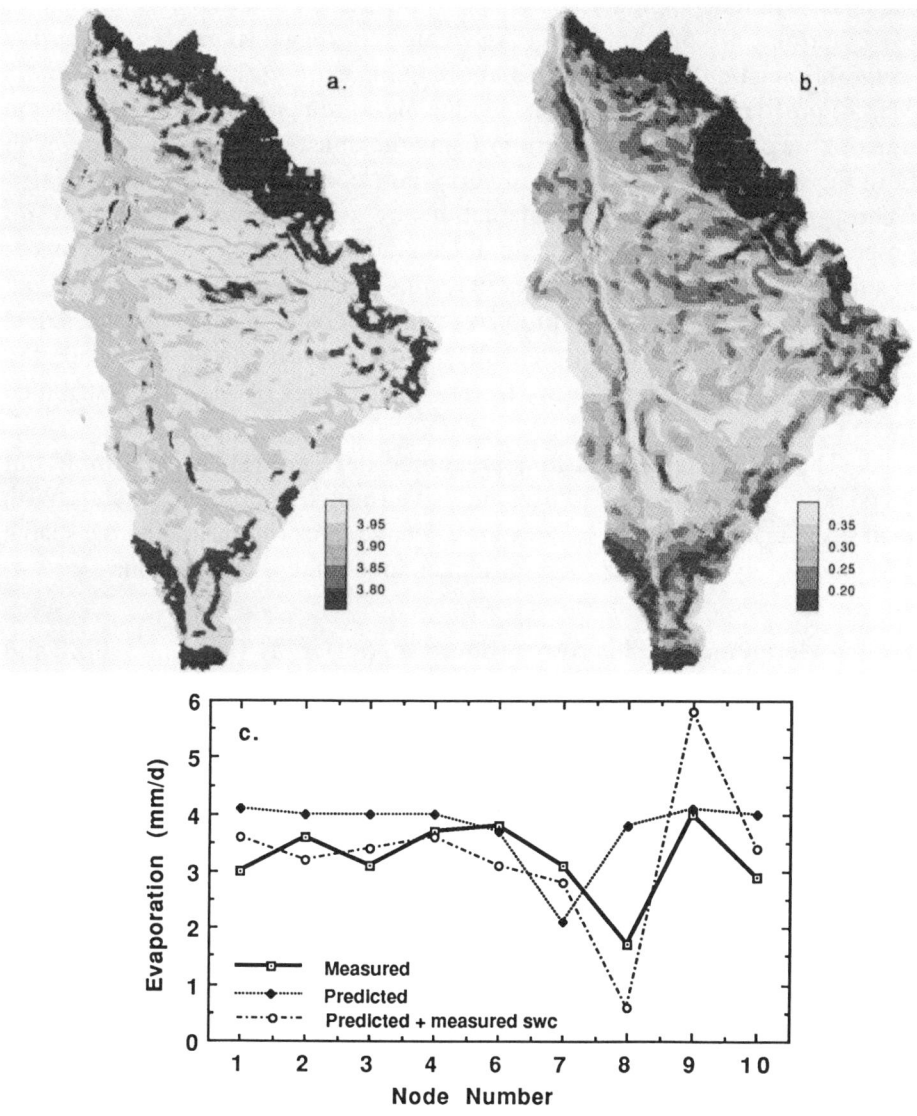

Fig. 2 (a) Predicted evaporation (mm day^{-1}) and (b) soil water content (m^3 m^{-3}) on 2 March 1992; (c) comparison of measured and predicted evapotranspiration on 2 March 1992 at nine nodes along a 0.54 km transect of the Lockyersleigh catchment.

spatial distribution of slope within the Lockyersleigh catchment computed by TAPES-G is presented in Fig. 1b. Figure 1c presents the spatial distribution of annual net radiation (W m^{-2}) calculated by SRAD. The computed global short wave radiation and net radiation on a horizontal surface are in close agreement with measured values (±5% on an annual basis). The spatial distribution of erosion potential is presented in Fig. 1d. Erosion potential is less than 1 and 5 over 51 and 90% of the catchment, respectively, and is highest in the steeper areas along the north-eastern perimeter of the catchment, as expected.

Evaporation and soil water content for 2 March 1992, calculated using the quasi-dynamic approach (both based on calculated values of net radiation for 2 March), are presented in Figs 2a and 2b, respectively. The effective evaporation calculated from differences in measured (on 28 February and 3 March) profile soil water contents at nine measurement sites along a 0.54 km transect are compared to predictions using the quasi-dynamic approach, with both the computed and measured soil water contents, in Fig. 2c (see also Guerra *et al.*, 1993). For these calculations we assumed spatially uniform soil properties, but the limited soil textural data indicates that there are differences in soil type over the catchment. These results show reasonable agreement, but suggest that there are potential problems with the spatial resolution of the DEM and the georeferencing of the field sites. Higher resolution DEMs (1:5000 to 1:10 000) appear to be necessary to drive these types of spatially distributed models.

The models and equations presented here are relatively simple but physically-based, are easy to use, and have limited data requirements. With their visualization capabilities, they are effective tools for examining the spatial variability of hydrological processes in real three-dimensional terrain.

Acknowledgements This research was supported by Grant no. 90/82 and Special Project 1991-92: ANU 3 from the Land and Water Resources Research and Development Corporation and by the Water Research Foundation of Australia.

REFERENCES

Barling, R.D. (1992) Saturation zones and ephemeral gullies on arable land in south-eastern Australia. Unpubl. Ph.D. thesis, University of Melbourne, Melbourne, Australia.

Beven, K.J. & Kirkby, M.J. (1979) A physically-based variable contributing area model of basin hydrology. *Hydrol. Sci. Bull.* **24**, 43-69.

Dozier, J., Bruno, J. & Downey, P. (1981) A faster solution to the horizon problem. *Computers & Geosci.* **7**, 145-151.

Fleming, P.M. (1987) Notes on a radiation index for use in studies of aspect effects on radiation climate. Unpubl. Manuscript, CSIRO Division of Water Resources, Canberra, Australia.

Guerra, L., Moore, I.D., Kalma, J.D. & Hofstee, C. (1993) Predicting spatially distributed evaporation using terrain, soil and land cover information. IAMAP-IAHS Joint International Meeting, Yokohama, Japan, 1993 (in press).

Hutchinson, M. F. (1989) A new procedure for gridding elevation and stream line data with automatic removal of spurious pits. *J. Hydrol.* **106**, 211-232.

Iida, T. (1984) A hydrological method of estimation of topographic effects on saturated throughflow. *Trans. Jap. Geomorph. Union* **5**, 1-12.

Jenson, S.K. & Domingue, J. O. (1988) Extracting topographic structure from digital elevation data for geographic information system analysis. *Photogram. Engr. & Remote Sens.* **54**, 1593-1600.

Kalma, J.D., Alksnis, H., Daniel, P. & Laughlin, G.P. (1987) The regional evaporation project: instrumentation and field measurements. Technical Memorandum 87/12, CSIRO Division of Water Resources, Canberra, Australia.

Moore, I.D., Burch, G. J. & Mackenzie, D. H. (1988) Topographic effects on the distribution of surface soil water and the location of ephemeral gullies. *Trans. Am. Soc. Agr. Engrs.* **31**, 1098-1107.

Moore, I.D., Gessler, P.G., Nielsen, G.A. & Peterson, G.A. (1992b) Soil attribute prediction using terrain analysis. *Soil Sci. Soc. Am. J.* (in press)

Moore, I.D., Grayson, R.B. & Ladson, A.R. (1991) Digital terrain modelling: a review of hydrological, geomorphological, and biological applications. *Hydrol. Proc.* **5**, 3-30.

Moore, I.D., Norton, T.W. & Williams J.E. (1993) Modelling environmental heterogeneity in forested landscapes. *J. Hydrol.* (in press).

Moore, I. D., Turner, A.K., Wilson, J.P., Jenson, S.K. & Band, L.E.(1992a) GIS and land surface-subsurface process modelling. In: *Environmental Modeling and GIS* (ed. by M.F. Goodchild, B. Parks & L.T. Steyaert), Oxford University Press (in press).

Moore, I.D. & Wilson, J.P. (1992) Length-slope factors for the revised universal soil loss equation: simplified method of estimation. *J. Soil & Water Conserv.* **47**, 423-428.

O'Loughlin, E.M. (1986) Prediction of surface saturation zones in natural catchments by topographic analysis. *Wat. Resour. Res.* **22**, 794-804.

Priestley, C.H.B. & Taylor, R.J. (1972) On the assessment of surface heat flux and evaporation using large-scale parameters. *Mon. Weath. Rev.* **100**, 81-92.

Renard, K.G., Foster, G.R., Weesies, G.A. & Porter, J. P. (1991) RUSLE Revised universal soil loss equation. *J. Soil & Water Conserv.* **46**, 30-33.

Running, S.W., Nemani, R.R. & Hungerford, R.D. (1987) Extrapolation of synoptic meteorological data in mountainous terrain and its use for simulating forest evapotranspiration and photosynthesis. *Can. J. For. Res.* **17**, 472-483.

Hydrologically oriented GIS and application to rainfall-runoff distributed modelling: case study of the Arno basin

P. LA BARBERA, L. LANZA & F. SICCARDI
Institute of Hydraulic, University of Genova, 1 Montallegro, 16145 Genova, Italy

Abstract When oriented to hydrological modelling Geographical Information Systems include automatic procedures implemented for drainage network delineation, hierarchical ordering and identification of geomorphological parameters. Such parameters are mainly related to the topographic description of the landscape through a distributed characterization of elevation, soil, and land use and cover. Cartographic sources are being currently replaced by remote sensing analysis based on the interpretation of satellite images. In this work a distributed rainfall-runoff model has been designed to include the spatial variability of the involved parameters and implemented in the framework of a hydrologically oriented GIS. The model is based on the distributed data handling capabilities of the GIS and relies on multisensor analysis for providing time dependent data for the input. The programs are implemented on personal computers with relatively small computational capabilities; an application of the model to the flood event of November 1987 on the Arno basin has been performed.

INTRODUCTION

The modelling and forecasting of hydrological processes have been traditionally developed through lumped models in order to include, in a limited number of global parameters, all the information needed when simulating a quite complex natural system (Todini, 1988). Such a system behaviour is obviously the outcome of several interacting contributions usually averaged in space and time over an appropriate multidimensional domain.

Nowadays the improvement of high performance artificial memories and computational resources makes the data handling capabilities of Geographical Information Systems (GIS) able to analyse large quantities of distributed geomorphological information. In recent years, a very big effort has been dedicated to the development of distributed hydrological models in order to overcome some of the traditional limitations of lumped modelling.

The knowledge of distributed landscape topography, as provided by Digital Elevation Models (DEM) (Ebner & Eder, 1992), is the basis of a series of automatic procedures able to derive drainage network structures at the appropriate scale of information (Brath *et al.*, 1989) and to support hydrological modelling. These procedures allows the derivation of traditional watershed parameters resulting into a set of useful informations for a detailed description of the landscape morphology (Quinn *et al.*, 1991).

The capabilities of GIS include the opportunity to overlay information provided by each thematic map according to a user-specified logic and to produce derivative map outputs (Vieux, 1991). Databases resulting from the joint use of several distributed information maps are able to support a series of hydrological models in the different fields of water resources management and environmental or civil protection strategies.

DRAINAGE STRUCTURES DELINEATION

A variety of specific algorithms has been developed in the last twenty years by several authors in order to provide automatic procedures able to derive hydrological basin network systems starting from the topographic information stored in a Digital Elevation Model.

The first procedure, implemented by Puecher & Douglas (1975), was based on the automatic investigation of the DEM matrix by means of a four mutually adjacent cells kernel identifying locally concave or convex configurations where river-course and ridge points respectively could be located. Though simple and rather efficient the algorithm showed its limitations when applied to complex morphologies and low resolution matrices. The derived networks present a large number of interruptions and isolated links not actually representing the natural configuration of the system. Some kind of hydrodynamic approach was later suggested by Moniod (1983) with the aim of simulating runoff concentration processes all over the basin through the delineation of paths of steepest slope; in particular every grid cell in the DEM matrix is modelled by receiving runoff contributions from the upstream draining cells and providing it, with an additional unit increment, to the downstream.

O'Callaghan & Mark (1984) and Mark (1984) implemented similar concentration procedures pointing out the relevance of the rising opportunity to derive drainage network systems based on specific morphological structures fixed by some threshold parameter expressing a minimum drainage area for each river-course element identification.

Quite different techniques were proposed by Palacios-Velez & Cuevas-Renaud (1986) on the basis of geometrical observations stating that, following paths of steepest slope in a downstream direction from a point to another and vice versa would lead back to the starting point then the path belongs to a "pure" streamline, otherwise it belongs to a river-course. The method seems to be intrinsically sensitive to low resolution matrices and DEM errors. Joint use of the Puecher & Douglas algorithm and some post processing procedure able to correct the drainage structures into a completely connected network was introduced by Band (1986); procedures are supported by pattern recognition methods leading to a hydrologically coherent drainage network system. Mark (1988) and Jenson & Domingue (1988) observed that pits, i.e. points surrounded by neighbours that have higher elevation, are frequently due to data errors in the DEM matrix and introduced preprocessing procedures in order to remove them before the application of automatic algorithms.

Carrara (1988) suggested a method based on the simulation of runoff concentration due to uniform rainfall occurrences over impermeable soil; flow paths are then simply controlled by landscape morphology. A preprocessing procedure deals with the problem of pits removal while a postprocessing one is able to derive a series of morphological parameters such as hierarchical ordering of network links and subbasin areas, slopes and exposures.

In a recent work Chorovicz *et al.* (1992) pointed out that pits may not just represent data errors or low resolution effects but also reproduce the actual nature of the landscape topography; a combined extraction process based on both geometrical and hydrological approaches is proposed for producing well connected networks and portraying the natural surface characteristics of the drainage system.

A set of quite efficient algorithms made up of two main procedures has recently been implemented (La Barbera *et al.*, 1992a) in order to derive drainage network systems and landscape morphological parameters as a basic structure for rainfall-runoff distributed modelling. Such procedures, for representing an essential feature of the proposed hydrologically oriented GIS structure, are herein presented in details.

The presence of peculiarities in the topographic matrix is first of all tested in the Lake preprocessing procedure by comparing each elevation value against the eight adjacent ones belonging to a 3x3 cell investigation kernel and once an isolated concavity is found the elevation at that point is modified until the examined cell is higher than the lowest neighbouring one. A fixed increment is used as previously established on the basis of the landscape topography. An iterative process is used to investigate the whole raster in this way resulting into a modified elevation matrix where no more pits exist and providing an updated DEM over which the delineation of drainage structures can easily be performed.

The Path procedure utilizes such a modified DEM identifying morphological flow paths on the hypothesis of space-filling drainage networks. The proposed automatic algorithm explores the surroundings from a given cell in order to find the direction which maximizes elevation gradients. Such a main direction, the existence of which is ensured by the preprocessing procedure, is assumed to be the actual flow path for the examined cell within the modified grid. It's important to stress the point that the resulting set of hydrological information does not represent the actual drainage system as laying all over the landscape up to the minimum scale corresponding to the original DEM discretization (space-filling network). The Path procedure is furthermore qualified to automatically derive a series of geomorphological parameters starting from the previously obtained space-filling network; the main parameters, widely used in hydrological analysis, rely on the dimension, shape, slope and elevation of the subbasins as well as on the Hortonian hierarchization of network links. Joint information from drainage structures and the main geomorphological parameters have been used in the development of an automatic procedure for network analysis able to evaluate the relative subbasin characteristics for each specific river section in the hydrological basin.

SCALE AND REPRESENTATION PROBLEMS

Space-filling drainage networks do not obviously represent the natural system of river courses within a hydrological basin. Every cell in the topographic matrix is indeed drained by some network link and this is not confirmed from the observed natural behaviour. Different kinds of filtering procedures have been suggested in order to derive drainage networks which are consistent with a specific scale of the basin representation.

Comparisons between automatically derived networks and the "blue lines" detected from the traditional cartography were initially performed, however the subjectivity due to cartographers interpretation makes such procedures unusable for automatic analysis. A morphological criterion widely implemented by several authors (Band, 1986; O'Callaghan & Mark, 1984) is based on the evaluation of contributing areas for every cell in the network. Cells are defined as channels when they exceed a threshold contributing area resulting in a drainage system that is actually suitable for hydrological interpretation.

The opportunity of linking such procedure to a fixed representation scale is obvious because the minimum value of contributing areas can be accurately chosen according to the cartographic scale which has to be considered the effective scale of the specific hydrological problem. Following the Lake and Path implementation a third procedure, named River, was developed in order to perform a river network filtering process according to a previously defined representation scale. Such a procedure allows for the selection of different threshold parameters among the minimum upstream contributing area, the drainage density, and the mean length of the first order Hortonian links.

A very interesting criterion for determining the appropriate drainage density at which to extract the network has been proposed by Tarboton *et al.* (1991) who basically suggest to derive the river structures at the highest drainage density that satisfy hortonian scaling laws which hold for channel networks.

Recent developments (La Barbera *et al.*, 1992b) allow to take into account the natural genesis of the system due to erosion and sediment transport processes within the hydrological basin. Filtering procedures are therefore developed on the basis of the cumulative probability distribution of the AS^k parameter, called hydrodynamic function and representing the combined effects of contributing areas A and local slopes S over the morphological shapening of the examined basin. The values of exponent k provide different shapes and interpretation keys to the hydrodynamic function:
(a) $k=0$ addresses a pure morphological distribution function and leads to the minimum contributing area criterion;
(b) $k=0.75$ produces the channel initiation function introduced by Willgoose

et al. (1991 a,b) in discussing the growth of drainage network;
(c) k=1 produces the energy dissipation function per unit length of a channel (Rodriguez-Iturbe *et al.*, 1992a);
(d) k=2 produces the energy dissipation function per unit area of a channel and was proposed by Rodriguez-Iturbe *et al.* (1992b) in the form of a principle of equal energy expenditure per unit area of a channel everywhere in the network.

The expression of the distribution function for the AS^k parameter has been derived by La Barbera & Roth (1992 a,b) in an analytical form and the selection of particular values leads to associated filtering criteria representative of a morphological, morphogenetical or purely energetical interpretation. The main feature of the energetic criterion is the high filtering effect occurring in the plain zones according to the natural behaviour of the system.

HYDROLOGICALLY ORIENTED GIS

A "hydrologically oriented" Geographical Information System should be able to store, manipulate and display geomorphological data related to the basin landscape domain resulting into an appropriate set of operational tools oriented to solve hydrological problems; databases are the fundamental skeleton over which information analysis can be performed.

The topographic information, which is referred to as the primary requirement for landscape knowledge, is the essential database to be included in the GIS structure. For hydrological purposes a series of further digital distributed maps such as soil type, land use and cover, soil moisture content, urbanization, etc. should be provided. Most of such parameters are currently suitable for remote sensing analysis based on the interpretation of satellite images. The development of algorithms for automated acquisition of remotely sensed informations is still a field of current research and seems to hold promising perspectives for wide applications in the near future.

Rainfall inputs are again suitable for a multisensor acquisition system that integrates satellite remote sensors with meteorological radars and ground based raingauge networks. However there is still an open question related to the implementation of automated procedures able to provide forecasted rainfall inputs, in a distributed form, on the basis of remotely observed meteorological systems (La Barbera *et al.*, 1992a).

As a GIS is not only a great amount of data stored in a distributed structure, it must be furthermore completed by specific algorithms in order to manipulate data and to obtain a variety of different interpretations of the system morphology and hydrological behaviour.

The Lake, Path and River procedures, like any other procedure based on the algorithms referred to for the automated detection of drainage networks at the correct scale of interpretation, play a fundamental role in hydrologically oriented GIS structures. Different procedures, further implemented to explain and evaluate other physical processes such as transpiration, infiltration, soil storage, etc., may give an important contribution to the analysis of the examined hydrological problem.

The high sensitivity of the hydrological basin behaviour to the spatial variability of landscape characteristics and rainfall inputs has been widely pointed out by several authors (Wood *et al.*, 1988). Thus the development of distributed approaches to rainfall-runoff modelling has received increasing impetus both in the research and technical field; GIS capabilities make the great number of model parameters, traditionally perceived as an indetermination index, to be now indicative of accuracy. On the other hand the current level of complexity inherent in distributed models leads to a variety of limitations in their practical suitability for technical applications. The simplified model proposed herein relies on a schematization of hillslope

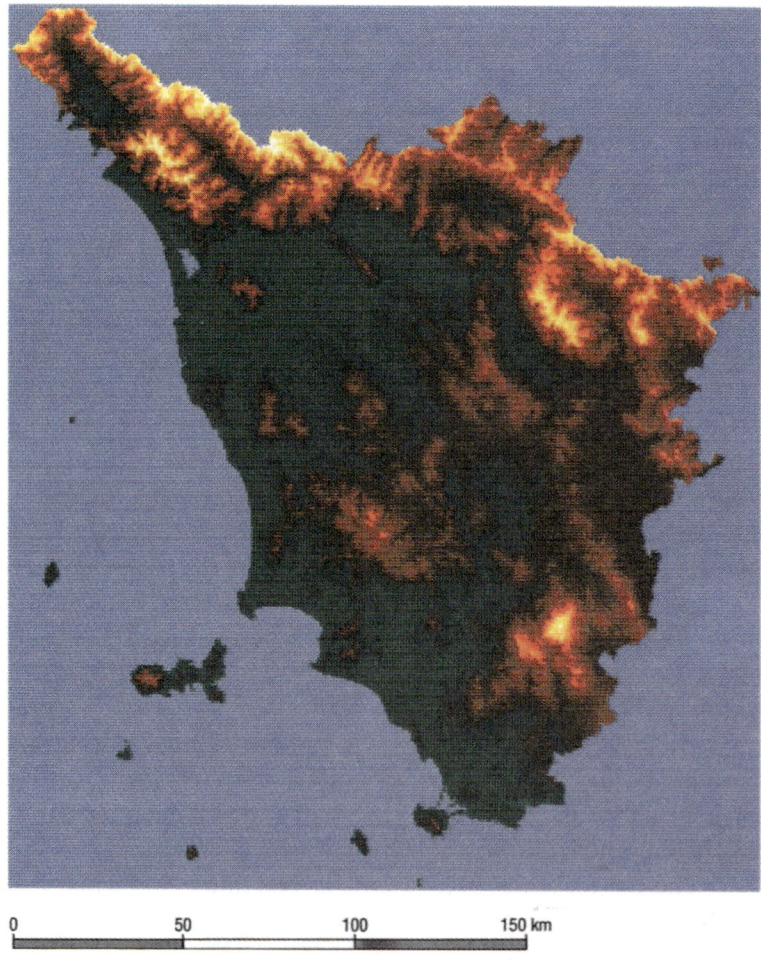

Fig. 1 Digital Elevation Map of the Tuscany region of Italy.

processes oriented to flood forecasting and is based on specific procedures using the morphological landscape characteristics, and particularly the slopes, to scale within the whole basin the hillslope parameters. Flood routing in the network is then modelled by the traditional Muskingum-Cunge procedure (Cunge, 1969).

THE ARNO BASIN: FLOOD SIMULATIONS

The Arno basin, geographically located in the northern Tuscany region of Italy, is about 4000 km² at the closure of Florence, which is one of the most important cities for Italian history and artistic heritage. Flood forecasting for this city is particularly necessary as relevant damages and losses to human lives have been experienced on several occasions in the last centuries up to the recent catastrophic flood of November 1966. Landscape description within the basin is based on classical databases collecting topographic elevation data in the form of a DEM with a cell size of 400×400 m (see Fig. 1) and eventually land use and cover characterization of coherent size.

Precipitation inputs all over the basin were derived from joint Meteosat estimates and raingauge network observations using the functional minimization procedure proposed by Filice *et al.* (1990). It must be pointed out that caution should be exercized in the application of such procedures to small

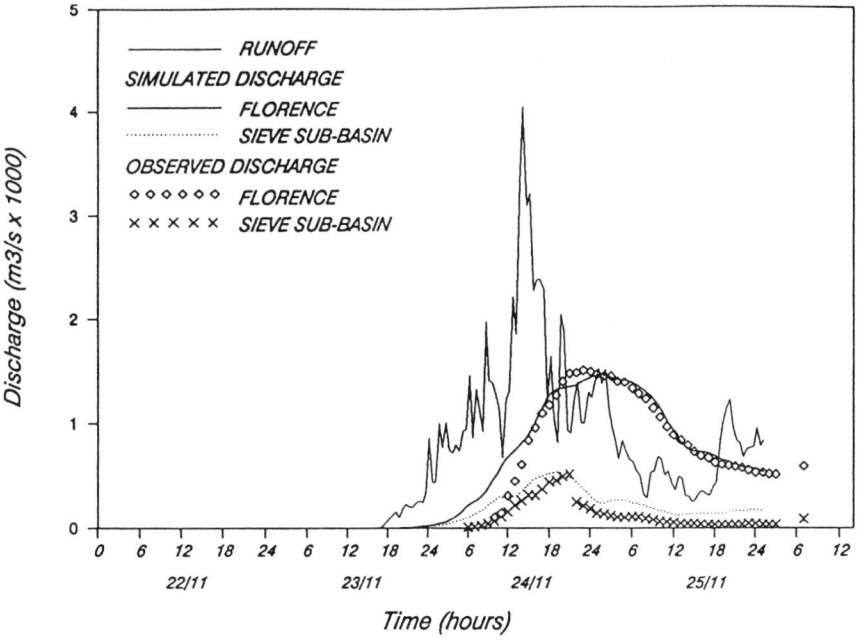

Fig 2. Flood simulations of the November 1987 event over the Arno basin compared to hydrometric observations.

sized hydrological basins because of the scale dependent reliability of geosynchronous satellite rain rate estimations (La Barbera *et al.*, 1992a).

Flood simulations were performed in the case of the November 1987 extreme event over the Arno basin and the proposed distributed model was applied to the whole basin domain and to the Sieve subbasin. The graphs in Fig. 2 show that, irrespective of the non-negligible geomorphological differences, the hydrological behaviour of the two basins investigated is well reproduced and the model result is validated when compared with hydrometric historical observations.

CONCLUSIONS

The distributed simulation of rainfall-runoff processes in order to perform acceptable flood forecasting is one of the main features of GIS application in hydrological modelling. Requirements in this case rely on specific procedures able to provide the practical implementation of a "hydrologically oriented" Geographical Information System and the appropriate filtering interpretation of the great amount of data stored in the database structure.

The application of GIS performances to rainfall-runoff modelling shows the managerial capability of the system in the face of the spatial variability of landscape characteristics and precipitation inputs. Relevant perspectives are opened in the field of real time flash flood forecasting oriented to early warning policies for civil protection. The main issue in this case relies on quantitative precipitation forecasting as well as on the development of automatic procedures able to derive remotely sensed rainfall data for the input which are consistent with the ground based rainfall observations.

Acknowledgements The present work was supported by an Italian National Research Council grant under the framework of the National Group for Prevention from Hydrogeological Disasters (GNDCI). Authors are especially grateful to Prof. G. Federici and Prof. I. Becchi of the University of Florence for providing the Arno basin DEM as developed in the framework of the "Progetto Strategico Arno".

REFERENCES

Band, L.E. (1986) Topographic partition of watersheads with digital elevation models. *Wat. Resour. Res.* **22**, 15-24.
Band, L.E. (1989) A terrain-based watershed information system. *Hydrol. Processes* **3**, 151-162.
Brath, A., La Barbera, P., Mancini, M. & Rosso, R. (1989) The use of distributed rainfall-runoff models based on GIS at different scales of information. ASCE Hydr. Div., 3rd Nat. Conf. on Hydraulic Engineering, New Orleans, August 14-18, 1989.
Carrara, A. (1988) Drainage and divide networks derived from high fidelity digital terrain models. In: *Quantitative analysis of mineral and energy resources* (ed. by C.F. Chung , A.G. Fabbri & R. Sinding-Larsen), D. Reidel Pub. Co., Dordrecht, 581-597.

Chorowicz, J., Ichoku, C., Riazanoff, S., Kim, Y.J. & Cervelle, B. (1992) A combined algorithm for automated drainage network extraction. *Wat. Resour. Res.* **28**(5), 1293-1302.

Cunge, J.A. (1969) On the subject of flood propagation method (Muskingum method). *J. Hydr. Res.* **7**, 205-230.

Ebner, H. & Eder, K. (1992) State-of-the-art in Digital Terrain Modelling. Proc. of EGIS'92, *Third European Conf. on GIS*, Munich, March 23-26, 1992.

Filice, E., La Barbera, P., Minciardi, R. & Siccardi, F. (1990) Predictive content of geosynchronuos satellite images in the evaluation of intensive rainfalls ground effects (in italian). *Proc. XXII Nat. Symp. on Hydraulics and Hydraulic Structures*. Cosenza, 4-7 October 1990.

Jenson, S.K. & Domingue, J.O. (1988) Extracting topographic structure from digital elevation data for geographic information system analysis. *Photogrammetric Engineering and Remote Sensing* **54**(11), 1593-1600.

La Barbera, P., Lanza, L. & Siccardi, F. (1992a) Flash flood forecasting based on multisensor informations. Subm. IAMAP-IAHS Joint Int. Symp., Yokohama, July 1993.

La Barbera, P., Roth, G. & Sguerso, D. (1992b) Scale problems in network identification from digital elevation maps. IAHS Workshop on Advances in Distributed Hydrology, Seriate (BG), Italy, June 25-26, 1992.

La Barbera, P. & Roth, G. (1992a) Scale properties in catchment morphology: probability distribution and their relations with the fractal characteristics of river networks. IAHS Workshop on Advances in Distributed Hydrology, Seriate (BG), Italy, June 25-26, 1992.

La Barbera, P. & Roth, G. (1992b) Invariance and scale property in the distribution of contributing area and energy in drainage basins. Submitted to {Hydrol. Processes}.

Mark, D.M. (1984) Automated detection of drainage networks from digital elevation models. *Cartographica* **21**, 168-178.

Mark, D.M. (1988) Network models in geomorphology. In: *Modelling in geomorphological systems* (ed. by M.G. Anderson), John Wiley, Chichester, UK, 73-97.

Moniod, F. (1983) Deux parametres pour characteriser le reseau hydrologique. Cha. Orstrom, ser. Hydrol. **20**(3-4), 191-203.

O'Callaghan, J.F. & Mark, D.M. (1984) The extraction of drainage network from digital elevation data. *Comput. Vision Graph. and Image Proc.* **28**, 323-344.

Palacios-Velez, O. L. & Cuevas-Renaud, B. (1986) Automated river-course, river and basin delineation from digital elevation data. *J. Hydrol.* **86**, 299-314.

Puecher, T.K. & Douglas, D.H. (1975) Detection of surface specific points by local parallel processing of discrete terrain elevation data. *Comp. Graph. Im. Proc.* **4**, 375-387.

Quinn, P., Beven, K., Chevallier, P. & Planchon, O. (1991) The prediction of hillslope flow paths for distributed hydrological modelling using digital terrain models. *Hydrol. Processes.* **5**, 59-79.

Rodriguez-Iturbe, I., Ijjasz-Vasquez, E.J., Bras, R.L. & Tarboton, D.G. (1992a) Power law distributions of discharge mass and energy in river basins. *Wat. Resour. Res.* **28**(4), 1089-1093.

Rodriguez-Iturbe, I., Rinaldo, A., Rigon, R., Bras, R.L., Marani, A. & Ijjasz-Vasquez, E.J. (1992b) Energy dissipation, runoff production, and the three-dimensional structure of river basins. *Wat. Resour. Res.* **28**(4), 1095-1103.

Tarboton, D.G., Bras, R.L. & Rodriguez-Iturbe, I. (1991) On the extraction of channel networks from digital elevation data. *Hydrol. Processes* **5**, 81-100.

Todini, E. (1988) Rainfall runoff modelling: past, present and future. *J. Hydrol.* **100**, 341-352.

Vieux, B.E. (1991) Geographic information systems and nonpoint source water quality and quantity modelling. *Hydrol. Processes.* **5**, 101-113.

Willgoose, G., Bras, R.L. & Rodriguez-Iturbe, I. (1991a) A coupled channel network growth and hillslope evolution model - 1: Theory. *Wat. Resour. Res.* **27**(7), 1671-1684.

Willgoose, G., Bras, R.L. & Rodriguez-Iturbe, I. (1991b) A coupled channel network growth and hillslope evolution model - 2: Nondimensionalization and applications. *Wat. Resour. Res.* **27**(7), 1685-1696.

Wood, E.F., Sivapalan, M., Beven, K. & Band, L. (1988) Effects of spatial variability and scale with implications to hydrological modeling. *J. Hydrol.* **102**, 29-47.

Developing a spatially distributed unit hydrograph by using GIS

DAVID R. MAIDMENT
Dept of Civil Engineering, University of Texas, Austin TX 78712, USA

Abstract Watersheds and stream networks can be delineated from a digital elevation model of land surface terrain by allowing water to flow from each grid cell to one of its eight neigbboring cells, thus creating a grid of flow direction. If a corresponding grid of flow velocity is added, the spatial velocity field for flow over a watershed is specified, from which its time-area diagram can be computed. If the watershed is divided into subareas A_i, $i = 1, 2,..., n$, by isochrones of time interval Δt, the ordinates U_i of its unit hydrograph of duration Δt are given by $U_i = A_i/\Delta t$. It appears that the unit hydrograph assumptions of constant time base and linearity of rainfall-runoff response are equivalent to assuming that regardless of the amount of rainfall, runoff always follows the same paths with the same average travel time, that is, that the runoff velocity field implied by the unit hydrograph is spatially variable but time- and discharge-invariant.

INTRODUCTION

A challenge for hydrology in the light of the independent emergence of GIS as a professional discipline is to determine how to use GIS in a useful fashion, either by linking hydrological models to GIS or by rethinking hydrological modelling in spatial terms so that better GIS-based hydrological models can be created. The unit hydrograph has been a basic tool of rainfall-runoff computation for many decades. It is a lumped model because it directly transforms rainfall into runoff without explicitly representing the internal distribution of flow within the watershed. The question arises as to how one could use GIS to construct a unit hydrograph which better reflects spatially distributed flow within the watershed.

Mathematical representation of the unit hydrograph also has a long history in hydrology. Clark (1945) formulated a unit hydrograph model by combining the time-area diagram of the watershed with a linear reservoir at the outlet. Nash (1957) proposed a cascade of linear reservoirs as a unit hydrograph model, and Dooge (1959) presented a unit hydrograph theory combining linear channels and linear reservoirs. Partial area flow was identified by Betson (1964) and the usual depiction of this concept involves source areas contributing to flow which expand and contract with time as the storm passes over the watershed, which is a partial area time-area diagram.

Many unit hydrograph models involving combinations of linear elements have been proposed, notably the theory of the geomorphic instantaneous unit hydrograph (Rodriguez-Iturbe & Valdes, 1979) in which Horton's stream laws are used to integrate over the watershed the delay effect of channel links characterized by a mean holding time to produce a unit hydrograph as the probability density function of travel time of water to the outlet. This approach implicitly assumes the runoff is produced by Hortonian overland flow throughout the watershed.

The spatially distributed unit hydrograph proposed here is similar in concept to the geomorphic instantaneous unit hydrograph except that GIS is used to describe the connectivity of the links in the watershed flow network, which eliminates the need for using probability arguments to combine the movement of water through the links. Moreover, the GIS-based approach permits the spatial pattern of excess rainfall to vary by isochrone zones within the watershed, thus relaxing the requirement for uniform excess rainfall over the whole watershed.

GRID-BASED FLOW PATTERNS

The current versions of some GIS systems, including the Arc/Info Grid system and the GRASS system, contain routines which determine flow direction over land surface terrain using the pour point model. As shown in Fig. 1a, water on grid cell is permitted to flow to one of its eight nearest neighbor cells. By taking a grid of terrain elevations (Fig. 1b), determining the slope of the line joining each cell with each of its neighboring cells, a grid of flow directions is created with one direction for each cell which represents the direction of steepest descent among the eight permitted choices. This grid is shown in Fig. 1c as a set of arrows but in fact is stored in GIS as a grid of numbers where each flow direction has a unique identifying number. The example grid used here is taken from ESRI (1992).

At first it seems that assigning water flow to one of only eight possible directions is an unnecessary restriction which does not accord with the reality that water can actually flow in any direction, but a very important simplification is accomplished by this device because it creates an equivalent flow network connecting the cell centers, as shown in Fig. 1d. There is thus a duality between the grid and the equivalent flow network and one might call the network so created a hybrid grid-network. Two-dimensional spatially distributed processes such as precipitation and infiltration can be modelled using the grid and the runoff from them can be routed to the watershed outlet through the associated one-dimensional flow network. Vieux & Westerfelt (1992) have constructed a cell to cell kinematic wave model operating in the GRASS grid cell system which performs flow routing over the grid-network in just this manner.

It can be added that there are other grid-networks which could be constructed using different criteria. For example, Sekulin et al. (1992) have delineated watersheds by taking an existing stream coverage, converting it to a grid, then assigning all the land surface cells to the nearest stream cell. Moreover, for the spatially distributed unit hydrograph approach proposed here, the land surface need not be subdivided into grid cells; it could instead be

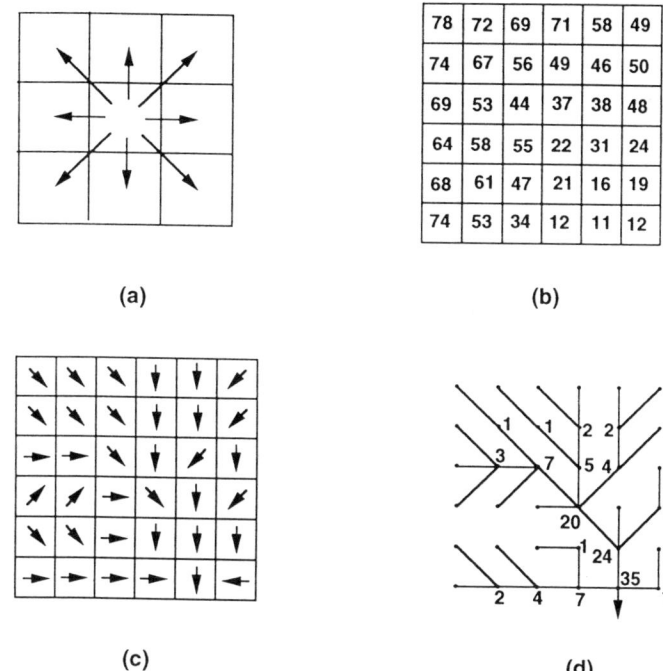

Fig. 1 Watershed terrain analysis using grid GIS methods: (a) the eight-direction pour-point model; (b) a grid of terrain elevations; (c) the corresponding grid of flow directions; (d) the equivalent network showing flow accumulation.

subdivided into polygons or into a triangulated irregular network. The first step in hydrological spatial analysis is to subdivide the total geographic space into subareas and which are connected to the watershed outlet by a one-dimensional flow network.

In GIS grid systems, watershed and stream delineation are accomplished by doing a flow accumulation on the equivalent network (Fig. 1(d)), where the number of upstream cells whose runoff passes through each downstream cell is noted; watershed divides are lines of cells having no other cells flowing through them and streams are lines of cells through which flows more than threshold number of upstream cells. In this manner, Lozar (1992) has delineated the drainage paths and watersheds of the whole earth based on the 5 arc-minute (~10 km) EOTOPE5 digital elevation model of the earth's land surface.

TIME-AREA DIAGRAM

Velocity is a vector quantity specified by magnitude and direction. The grid of flow direction shown in Fig. 1c is half of a velocity field; it specifies direction but not magnitude. Suppose a grid of velocity magnitudes was created based on land cover and slope. Sircar *et al.* (1991) have shown how this can be done using a velocity function of the form $V = aS^b$, where S is land surface slope, and a and b are coefficients related to land use taken from McCuen (1982) which are based on procedures of the USDA Soil Conservation Service. Since for each cell both flow direction and velocity are now known (Fig. 2a), and the pathway from each cell to the watershed outlet has been specified, it follows that one can create a grid of flow travel times where the value in each cell is the time taken for water from that cell to flow to the watershed outlet.

The cells may then be classified into zones i, i = 1, 2,....., whose travel time t falls into time intervals Δt, that is, the zone 1 has travel time $0 \leq t < \Delta t$, zone 2 has travel time $\Delta t \leq t < 2\Delta t$, and so on. The line bounding the outer limit of the cells in zone i is the isochrone of time of travel $t = i\Delta t$ to the watershed outlet. The total area of cells in zone i is A_i. In this way, the isochrone map of the watershed is created (Fig. 2b). The isochrone which has maximum time of

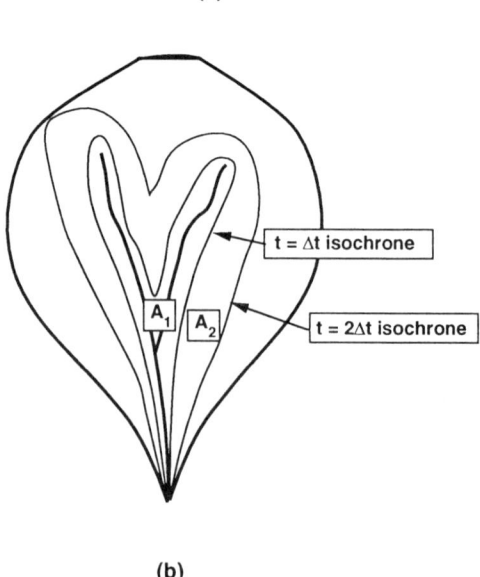

Fig. 2 Watershed time-area relationships: (a) a velocity field specified by the magnitude and direction of flow velocity; (b) a watershed isochrone map drawn by classifying a grid of time of flow to the outlet.

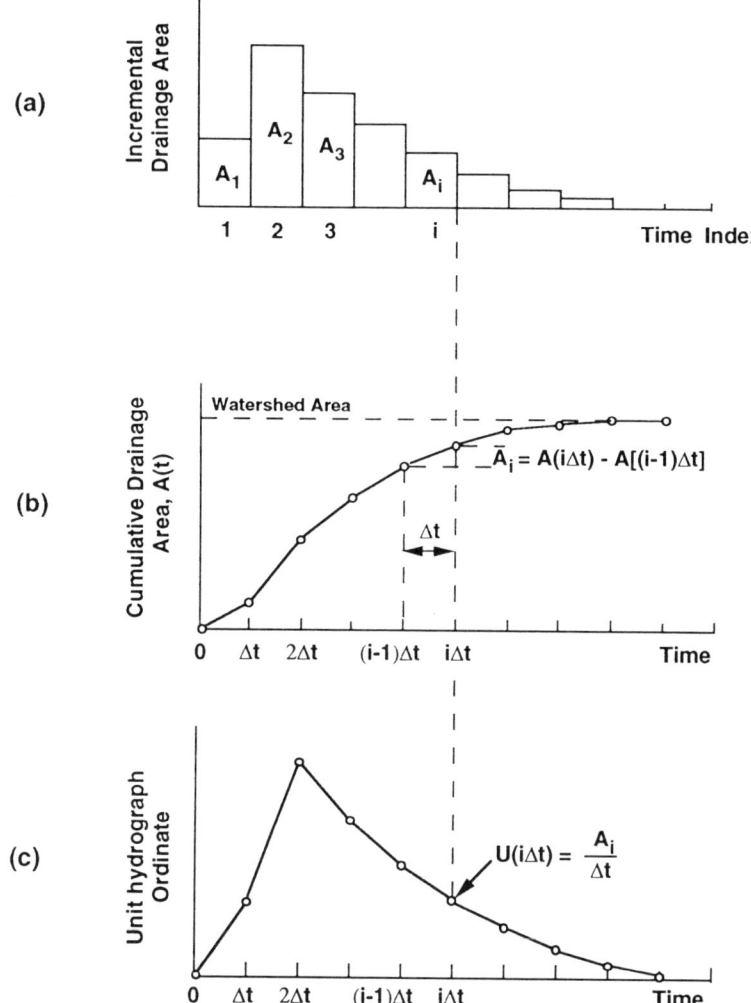

Fig. 3 The time-area diagram and the unit hydrograph: (a) incremental drainage area; (b) the cumulative time-area diagram ; (c) unit hydrograph found as the slope of the time-area diagram.

flow to the outlet is the time of concentration of the watershed, t_c, also sometimes called the time of equilibrium.

The time-area diagram is a graph of cumulative drainage area flowing to the outlet within a specified time of travel. It is constructed by summing the incremental areas Ai as shown in Fig. 3. Thus at time points t = 0, Δt, 2Δt, ..., iΔt,..., the cumulative area draining to the outlet A(iΔt) is given by:

$$A(i\Delta t) = \sum_{k=1}^{i} A_k \tag{1}$$

and conversely, the incremental areas are given by:

$$A_i = A[i\Delta t] - A[(i-1)\Delta t] \qquad (2)$$

It is important to note that the incremental time-area diagram (Fig. 3a) is a discrete time function having a single value over each time interval Δt, while the cumulative time-area diagram (Fig. 3b) is a continuous time function whose sampled value is given at regular time points by equation (1). Thus, if one classifies the time of flow grid into isochrones separated by a different time step $\Delta t'$, the appearance of the incremental time-area histogram will be altered but the cumulative time area diagram will simply be the same continuous curve sampled with a different time step.

UNIT HYDROGRAPH DERIVATION

If an excess rainfall occurs at rate I over the watershed, the runoff at the outlet is given by $Q(t) = IA(t)$, which is an S-hydrograph of runoff tending to an equilibrium discharge of IA where A is the total area of the watershed. The runoff response to a pulse of rainfall of intensity I and duration Δt is given by the unit hydrograph linearity principles as $Q(t) = IA(t) - IA(t-\Delta t)$, the difference between the S-hydrograph value and its value lagged by Δt time units (note that $A(t) = 0$ for $t < 0$). The amount of exess rainfall falling in this pulse is $P = I\Delta t$ inches or mm, so the corresponding response $U(t)$ to a unit pulse of rainfall (P = 1 inch or mm) is given by $Q(t)/I\Delta t$:

$$\dot{U}(t) = \frac{A(t) - A(t-\Delta t)}{\Delta t} \qquad (3)$$

Because the time-area diagram values are known only at time points, $t = 0, \Delta t, 2\Delta t,..,i\Delta t,...$, it follows that the unit hydrograph ordinates at the corresponding time points are given by:

$$U(i\Delta t) = \frac{A(i\Delta t) - A[(i-1)\Delta t]}{\Delta t} \qquad (4)$$

or by

$$U_i = U(i\Delta t) = \frac{A_i}{\Delta t} \qquad (5)$$

from equation (2). Equation (5) states that if one defines isochrones of time of travel to the watershed outlet at intervals Δt, the unit hydrograph ordinate at the end of the ith time interval is equal to the incremental area A_i whose drainage first reaches the outlet during that time interval divided by the duration of the interval, Δt. As shown in Fig. 3, this means that the unit hydrograph ordinate at time $i\Delta t$ is given by the slope of the time-area diagram over the interval [(i-1)Δt, iΔt]. In the limit as Δt tends to 0, this slope measure becomes simply the slope of the time-area diagram at time t so that the instantaneous unit hydrograph is equal to the slope of the time-area diagram.

The dimensions of U(t) in equations (4) and (5) are $[L^2 T^{-1}]$. These are consistent with units of discharge per unit of excess rainfall, which are the units which should be used for the unit hydrograph even though it is often stated just in units of discharge alone. By combining equations (1) and (5), the time-area diagram can be calculated from the unit hydrograph as:

$$A(i\Delta t) = \sum_{k=1}^{i} (U(k\Delta t)) \Delta t \tag{6}$$

DIRECT RUNOFF HYDROGRAPH

The portion of the rainfall which produces direct runoff is called the excess rainfall, and its values are symbolized by $P_1, P_2, ..., P_j, ...$, where P is the excess rainfall in inches or mm, and the corresponding direct runoff values are given by $Q_1, Q_2, ..., Q_n, ...$, where Q is a discharge rate at the watershed outlet measured in $ft^3 s^{-1}$ or $m^3 s^{-1}$. Given the excess rainfall hyetograph for a watershed, the direct runoff hydrograph is computed for the first time interval as:

$$Q_1 = P_1 U_1 \tag{7}$$

and by substitution from equation (5) it can be seen that:

$$Q_1 = \frac{P_1 A_1}{\Delta t} \tag{8}$$

Thus, if an excess rainfall of intensity $P_1/\Delta t$ begins falling on the watershed at time 0, after time Δt area A_1 is contributing to flow at the outlet so the direct runoff rate is $(P_1/\Delta t) A_1$ at that time. After time $t = 2\Delta t$, there are two rainfall pulses to contend with, P_1 and P_2, and direct runoff is computed by:

$$Q_2 = P_2 U_1 + P_1 U_2$$

$$= \frac{1}{\Delta t} [P_2 A_1 + P_1 A_2] \tag{9}$$

which includes the immediate impact at the outlet of P_2 flowing from area A_1 plus the delayed effect of P_1 flowing from area A_2. In normal unit hydrograph calculations, the excess rainfall is assumed uniformly distributed over the watershed in space so that any particular rainfall increment P_j refers to the average rainfall in time interval [(j-1)Δt, jΔt] on all areas A_1, A_2, ...,etc.

With GIS grid capabilities for rainfall mapping, this uniform spatial rainfall distribution is no longer necessary so that two subscripts are needed to characterize rainfall, P_{ij}, where P_{ij} is the average excess rainfall over all cells in isochrone zone i during time interval j. Direct runoff at time $t = n\Delta t$ is given by summing the runoff contributions from each of the applicable isochrone zones suitably lagged in time:

$$Q_n = \sum_{i=1}^{n} \frac{P_{ij} A_i}{\Delta t} \quad \text{where } j = n-i+1 \tag{10}$$

A concept similar to this was proposed by Tierstrep & Stall (1974) for urban drainage in the Illudas model in which they compute the runoff hydrograph from a time-lagged and area-weighted sum of "supply rates" of runoff from paved and grassed areas.

UNIT HYDROGRAPH INTERPRETATION

The theory of the unit hydrograph on a discrete time scale is rigorously derived in Chow *et al.* (1988) from the convolution integration of linear systems. The derivation presented here is consistent with that derivation but is much easier to understand. Once the time-area diagram is known, the unit hydrograph follows directly. Since the existence of the time-area diagram implies the existence of a velocity field over the watershed, it follows that if this velocity field is known, the unit hydrograph is completely specified without the need for any arbitrary mathematical functions or empirical formulas for time of concentration. Conversely, if the unit hydrograph is known, the time-area diagram follows from equation (6).

The unit hydrograph is based on the assumptions that the base time or duration of direct runoff is constant for excess rainfall of a specified duration, and that ordinates of direct runoff hydrographs of common base time are directly proportional to the total amount of direct runoff represented by those hydrographs (Chow *et al.*, 1988). If these assumptions are examined in the light of the spatial velocity field, it becomes clear that they are equivalent to saying that no matter how much excess rainfall occurs, runoff always follows the same paths with the same velocity. That is the only way that the constant time base and relative shape of the unit hydrograph can be maintained regardless of the amount of direct runoff which occurs. It appears, therefore, that the existence of a unit hydrograph of a watershed implies the existence of a velocity field for runoff which is spatially variable but invariant with respect to time and the magnitude of the discharge.

A time-area diagram derived from a spatial velocity field takes into account the time of transmission of water travelling across the land surface but does not account for the time delay caused by storage of water on the watershed. It is unclear at this time how the time of travel across a cell should properly be interpreted, whether as a pure transmission time determined from land characteristics or as an average holding time for water travelling through a storage system as is assumed in models employing linear reservoirs or other probability-based interpretations of the unit hydrograph.

EXAMPLE OF APPLICATION

The grid shown in Fig. 1 is analyzed as an example of the functions being discussed. A precursor to the time-area diagram is the distance-area diagram, which is the area draining to a watershed outlet within a specified flow distance of the outlet. If the cells in Fig. 1 have a size of 1 unit, flow between two adjacent cell centers in either the horizontal or vertical directions involves a distance of 1 unit while flow between diagonal cell centers involves a distance of $\sqrt{2} = 1.414$ units. Beginning in Fig. 1d with the cell adjacent to the watershed outlet (which has a flow distance of 0.5 units), calculations proceed cell by cell upstream along the flow network links, with the value stored in each cell being the flow distance to the outlet. A histogram of the cell flow distances is shown in Fig. 4a and a cumulative distance-area diagram in Fig. 4b.

These diagrams reveal an interesting shape: the distance-area histogram is negatively skewed which shows that the greatest concentration of cells are located far from the outlet rather than near to it. This seems counter-intuitive at first because unit hydrographs are usually positively skewed but upon reflection this distance-area histogram shape is reasonable when one thinks about watersheds being pear-shaped and narrowing down as the outlet is approached. The corresponding distance-area diagram shown in Fig. 4b has a concave upward shape, definitely not the S-curve shape that is customarily drawn for time-area curves! The simplest method of determining the velocity field, which is to assign the same velocity to each watershed cell, would create a time-area diagram of the same shape as that shown in Fig. 4 and would not lead to a unit hydrograph of the shape expected in Fig. 3c, at least for this simple example.

The flow accumulation values of Fig. 1d are presented as a frequency histogram in Fig. 5a. Of the 36 cells in the example grid, 19 have no cells upstream flowing through them, and the remaining 17 cells have from one to 35 upstream cells. If a velocity of 0.1 cell units per unit time occurs in the 19 cells with zero flow accumulation and 0.4 cell units per unit time in the remaining cells, the time-area diagram shown in Fig. 5b results. It is practically a straight line, which means that the unit hydrograph derived from it will be uniformly distributed over the duration of runoff. Several alternative velocity assignments were tried but they did not result in an S-hydrograph shape as expected for the time-area diagram.

The results of this example application are thus somewhat inconclusive.

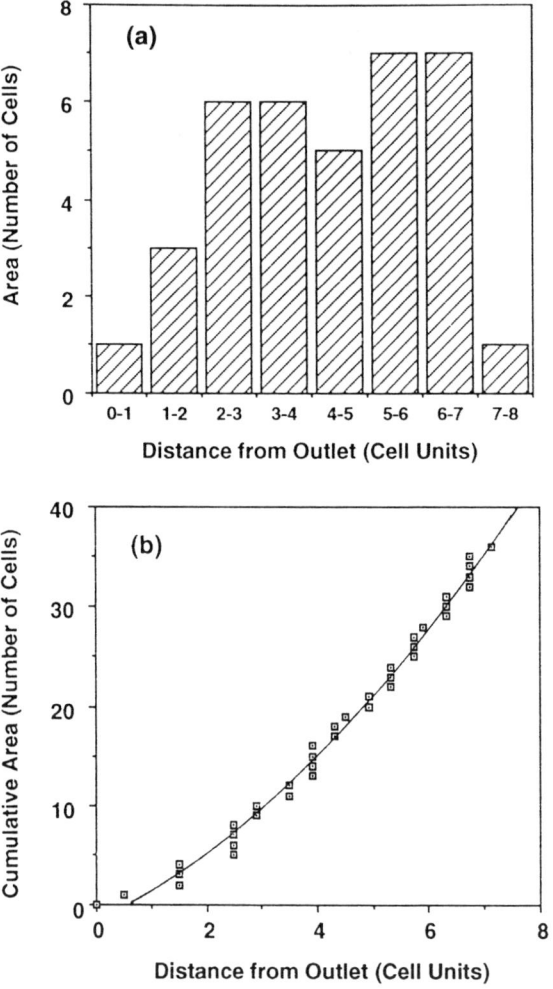

Fig. 4 Distance-area analysis of the grid shown in Fig. 1: (a) distance-area histogram; (b) cumulative distance-area diagram for distance of flow to the outlet.

It is possible that the example grid is just too small to give realistic results. The more realistically sized watershed grid studied by Sircar et al. (1991) did produce an S-shaped time-area diagram.

CONCLUSIONS

Given a digital elevation model for a watershed, a grid of flow direction is defined from each cell to one of its eight neighboring cells in the direction of maximum downhill slope. From this, a grid of flow distance can be compiled for the watershed by tracing from the lowest cell to all cells upstream and storing in each cell its flow distance to the outlet. By assigning a velocity of flow to each cell, a grid of flow times to the outlet can similarly be computed

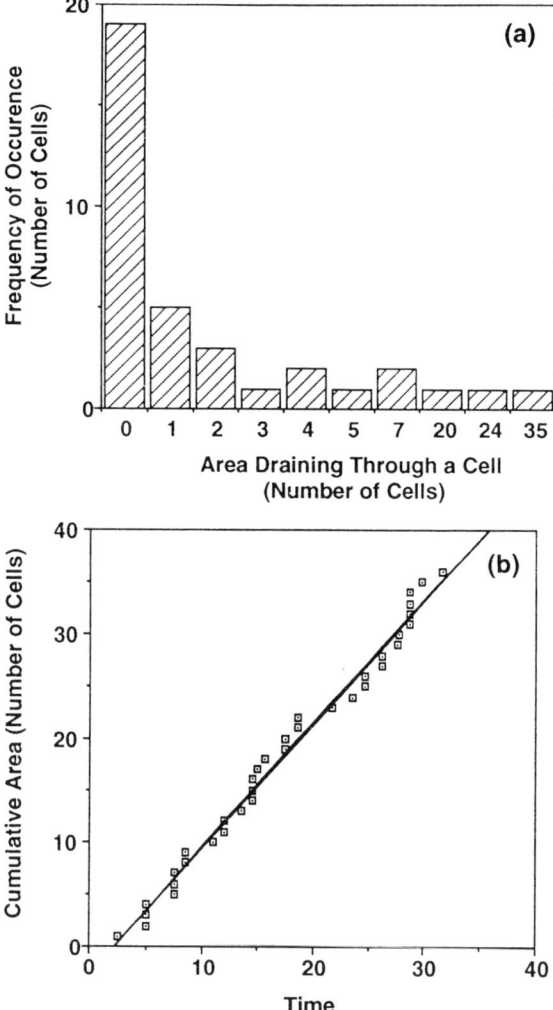

Fig. 5 For the grid shown in Fig. 1: (a) a frequency histogram of the flow accumulation in Fig. 1d; (b) the time-area diagram produced by assuming velocities of 0.1 and 0.4 cell units per unit time for cells having zero and more than zero flow accumulation, respectively.

from which isochrones of flow time at intervals Δt laid out on the grid. The incremental areas A_i, i = 1,2,..., of cells between isochrones can be determined and by accumulating these areas upstream from the watershed outlet, the time-area diagram for thewatershed is determined.

It is shown that the unit hydrograph ordinates U_i for excess rainfall of duration Δt are given by $U_i = A_i/\Delta t$. This leads to a conceptually simple way of computing direct runoff from a watershed where the runoff at any time point is the sum of the runoff contributions from each isochrone zone lagged by their time of flow to the outlet. In effect, each isochrone zone is performing as a mini-watershed in its own right and it can have a different excess rainfall rate than the other isochrone zones.

The relationship between the velocity field, the time-area diagram and the unit hydrograph derived here suggests that the normal unit hydrograph assumptions of constant time base and linearity of runoff response are equivalent to saying that regardless of the amount of rainfall, runoff always follows the same paths with the same average flow velocity. Thus the existence of a watershed unit hydrograph implies the existence of a velocity field which is spatially variable but time- and discharge-invariant. Study of a small example grid shows that the distance-area histogram is negatively skewed which must be offset by the spatial variability of the velocity field if a positively skewed unit hydrograph is to be produced. A simple assignment of the velocity field giving greater velocity to downstream cells carrying more flow and less to upstream cells produces an approximately straight line time-area diagram for the small example grid.

Further study of more realistic cases is needed before any definitive conclusions can be drawn about this method of deriving a spatially distributed unit hydrograph. A simple assignment of velocity to each cell neglects the delaying effects of storage on runoff. Indeed, one might question whether all of the watershed is really flowing to the outlet at all. The converse relationship, that the time-area diagram can be found from the unit hydrograph, may be a tool that could be combined with variable source area watershed studies to identify the area of the watershed that is actually contributing flow to the outlet during storms.

Acknowledgements This study was supported in part by the University of Texas at Austin and by the Hydrologic Engineering Center of the US Army Corps of Engineers.

REFERENCES

Betson, R.P. (1964) What is watershed runoff? *J. Geophys. Res.* 69 (8), 1541-1522.
Chow, V.T., Maidment, D.R. & Mays, L.W. (1988) *Applied Hydrology*. McGraw-Hill, New York, 201-236.
Clark, C.O. (1945) Storage and the unit hydrograph. *Trans. Am. Soc. Civ. Eng.* **110**, 1419-1446.
Dooge, J.C.I. (1959) A general theory of the unit hydrograph. *J. Geophys. Res.* **64** (1), 241-256.
ESRI (1992) *Cell-based modelling with Grid 6.1: supplement -- hydrologic and distance modelling tools.* Environmental Systems Research Institute, Redlands, CA, April.
Lozar, R.C. (1992) Global climatic management by watershed basin units. *Construction Engineering Research Laboratory, US Army Corps of Engineers, Champaign, IL.*
McCuen, R.H. (1982) *A guide to hydrologic analysis using SCS methods.* Prentice Hall, Englewood-Cliffs, NJ.
Nash, J.E. (1957) The form of the instantaneous unit hydrograph. IAHS Publication No. 45, 3-4, 114-121.
Rodriguez-Iturbe, I. & Valdes, J.B. (1979) The geomorphological structure of hydrologic response. *Wat. Resour. Res.* **15**(6), 1409-1420.
Sekulin, A.E., Bullock, A. & Gustard, A. (1992), Rapid calculation of catchment boundaries using an automated river network technique. *Wat. Resour. Res.* **28**(8), 2101-2109.
Sircar, J.K., Ragan, R.M, Engman, E.T. & Fink, R.A. (1991) A GIS based geomorphic approach for the digital computation of time-area curves. *Proc. ASCE Symposium on Remote Sensing Applications in Water Resources Engineering*, May.
Tierstrep, M.L. & Stall, J.B. (1974) The Illinois urban drainage area simulator. ILLUDAS, *Bulletin 58, Illinois State Water Survey*, Urbana, IL.
Vieux, B.E. & Westervelt, J. (1992) Finite element modelling of storm water runoff using GRASS GIS. *Proc. 8th ASCE Conference on Computing in Civil Engineering*, held in Dallas TX, Am. Soc. Civ. Eng., New York, 712-718.

Raster based modelling of watersheds and flow accumulation

BRODER MERKEL
WATEC Ingenieurgesellschaft für Hydrogeologie und Hydrochemie mbH, Im Wiegenfeld 4, 8015 Markt Schwaben, Germany

BARBARA SPERLING
AVENA GmbH, Marktplatz 15, 8015 Markt Schwaben, Germany

Abstract A tool is presented that enables hydrogeologists the computation of groundwater capture zones by means of a numerical groundwater flow model in combination with a raster based computation of watersheds, flow accumulation and residence time in the unsaturated zone. This tool might be helpful for hydrogeologists, groundwater engineers and authorities to save and/or improve groundwater quality.

INTRODUCTION

Groundwater protection becomes a task with increasing demand. Well head protection zones together with a catalogue of restrictions for industrial and agricultural land use with respect to certain zones have been proved to be the most evident protection strategy. Numerical flow and transport models are an important tool to define groundwater catchment areas and subareas by means of groundwater residence times due to variation of permeability, hydraulic gradient and the location of the production wells.

The residence time of seepage water in the unsaturated zone is not taken into account so far. According to a German guideline (DVGW Richtlinie W101) the hydraulic conductivity respectively the residence time within the unsaturated zone has to be considered for the calculation of well head protection zones if the aquifer is mostly fissured or karstified. With reference to new research results, the unsaturated zone should be taken into account in any case.

Moreover overland runoff may collect water from areas outside of the groundwater recharge area, but drain into the groundwater recharge area. Thus the surface catchment area should be considered as well.

Well head protection zones should therefore consider the residence time in the groundwater catchment area, the residence time in the unsaturated zone and the overland runoff if significant (Fig. 1).

GROUNDWATER FLOW MODEL

Two-dimensional steady-state groundwater flow in a heterogeneous, anisotropic and saturated aquifer is governed by:

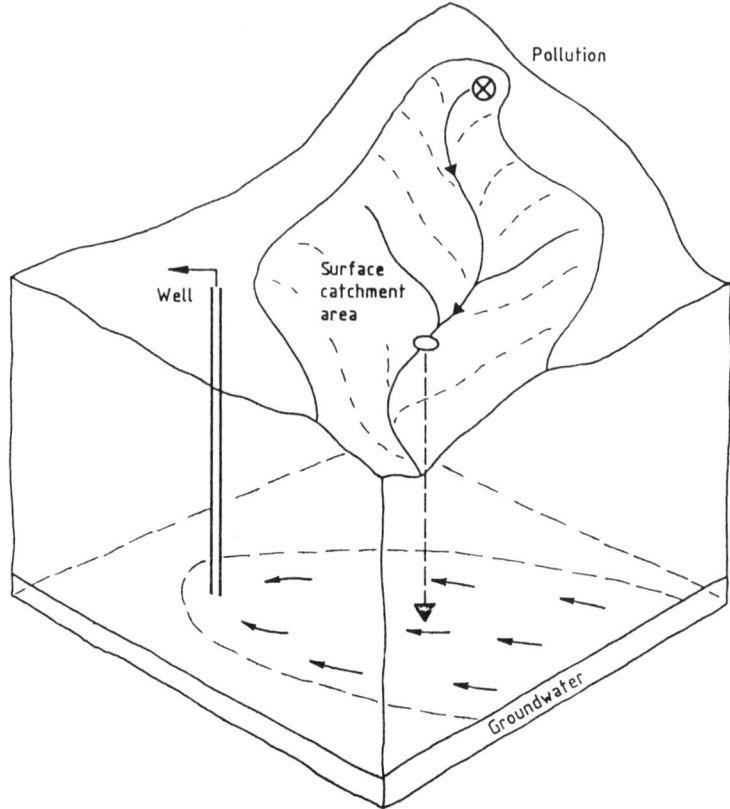

Fig. 1 Three-dimensional scheme of surface and groundwater catchment area.

$$\frac{\delta}{\delta x}\left(k_x \cdot b \frac{\delta h}{\delta x}\right) + \frac{\delta}{\delta y}\left(k_y \cdot b \frac{\delta h}{\delta y}\right) + Q_{xy} = 0 \qquad (1)$$

where: k_x, k_y = principal components of the hydraulic conductivity tensor;
  b = saturated thickness (unconfined aquifer);
     = aquifer thickness (confined aquifer);
  h = groundwater head;
  Q_{xy} = sinks and sources (volumetric fluxes);
  x,y = Cartesian coordinates.

Equation (1) is based on Darcy's law. Flow velocities can be derived from the groundwater heads easily and will provide the residence time. The partial differentials dx and dy may be approximated by finite lengths of x and y. This finite difference method is very common in groundwater modelling (Pinder & Bredehoeft,1968; Kinzelbach, 1986) and widely used in the groundwater industry. If an equidistant block-centered, rectangular grid would be used as difference scheme, input and output data would be compatible to raster based

Geographic Information Systems (GIS). If a solver for non-linear problems is available in the GIS used, the above described problem could be solved directly within the system. However since most problems need to be discretized on tracks with varying distances, standard GIS are not able to solve the problem directly.

Basically a couple of easy-to-use and powerful programs for numerical aquifer analysis are available (e.g., Frantz & Guiger, 1989), which may be used for this purpose. There is however some need for comfortable interfaces to swap data between numerical models and GIS.

It is important to stress the fact that groundwater flow in fissured or karstified aquifers does not follow Darcy's law. Thus either programs with more sophisticated algorithms have to be used or the solution has to be considered as a rough estimation of the reality only.

RESIDENCE TIME IN UNSATURATED ZONE

The residence time in the unsaturated zone may be calculated from the thickness and permeability of the strata. The strata might be soils but as well sediments (gravel, sand, clay) or hard rocks. The layers will usually show

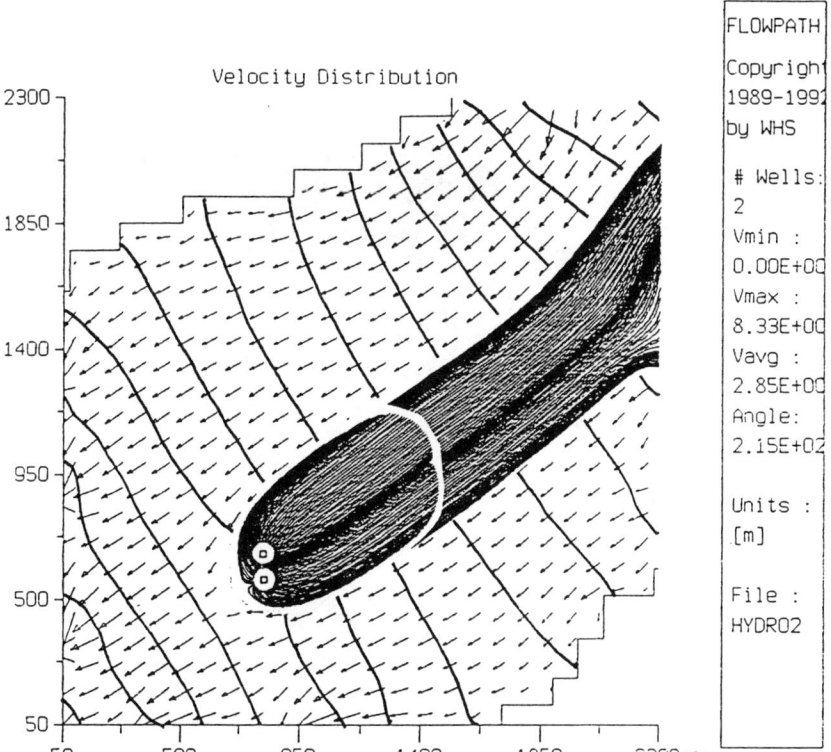

Fig. 2 Groundwater flow velocity (vmax = 8.33 m day^{-1}), contour lines and capture zones for 200 and 730 days without dispersion and diffusion.

considerable spatial variability. Furthermore the permeability in the unsaturated zone varies due to rainfall, evapotranspiration and temperature. A common approach is to use the permeability at field capacity to simulate the worst case conditions with respect to infiltration. Since thickness, pore volume and permeabilities varies intensively in space a vector based calculation is not recommended. Only a raster based GIS is able to solve the simulation which might be a multi-layer problem as well. Merkel & Tertilt (1992) have developed a GIS based tool to compute the groundwater recharge from surface infiltration and mean residence time considering the following parameters:

(a) rainfall taking into account elevation and spatial residues;
(b) evapotranspiration concerning slope and aspect as well as vegetation and surface runoff;
(c) infiltration due to rainfall and evapotranspiration and as a function of permeability, pore volume and flow accumulation.

The tool was developed to produce groundwater protection maps basically at a scale between 1:5000 and 1:50 000. However this tool may be used for other purposes as well.

WATERSHED COMPUTATION

Most of the algorithms used for watershed computation are based on a digital elevation model (DEM) and furthermore on a raster of the flow directions

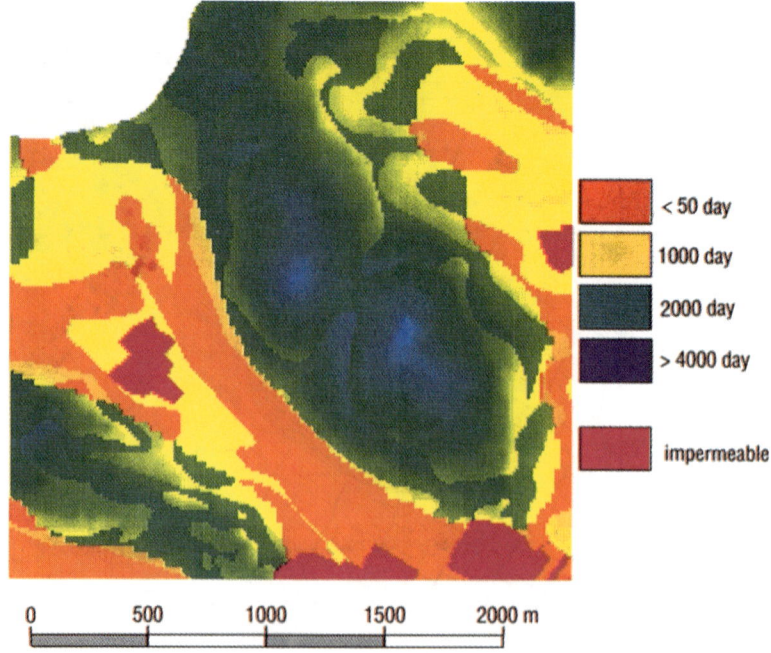

Fig. 3 Residence time (days) in the unsaturated zone, sealed areas are in magenta.

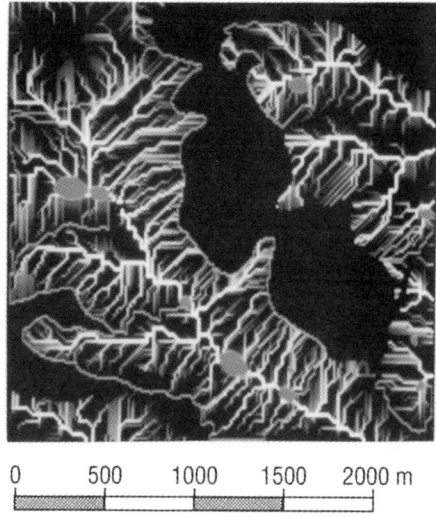

Fig. 4 Surface catchment area and flow accumulation.

derived from this. If a DEM is not available or has an insufficient precision, different algorithm might be used to compute a DEM from digitized contour lines respectively digital nodes. Triangulation seems to be a good technique but might run into time problems if the data base comes up to have 50 000 or even 100 000 digital points. Methods like piece-wise fitting, inverse distances and block-kriging will give faster answers but might be less precise.

There are generally two different approaches known to delineate the drainage network from raster based digital elevation models. Peucker & Douglas (1985) offered a method which works with a small window being moved along the raster. Cells at the bottom of the window are taken to be part of the network. The most serious problem which shortens the use of this algorithms is that it generates only discontinuous networks. Those flow systems requires an intensive postprocessing to connect the single parts to a complete network.

Another approach is the simulation of overland flow across a landscape. There are different approaches known which were developed by e.g., Jenson & Domingue (1988) and Martz & De Jong (1988).

Since depressions in the DEM cause severe problems for the calculation of flow directions - no matter whether they represent data errors or the reality - it is a common strategy to fill up in a first step all sinks to the level of the next lowest cell of the perimeter. Martz & De Jong (1988) did the depression filling after accumulating catchment areas along flow paths determined from the original DEM. After depression filling they modify the catchment areas to simulate the overflowing at the lowest point of the perimeter. Because only one overflow is allowed a decision between the potential pour points has to be made. Martz & De Jong (1988) follow the steepest flow path and add the total catchment of the flat area to the stream.

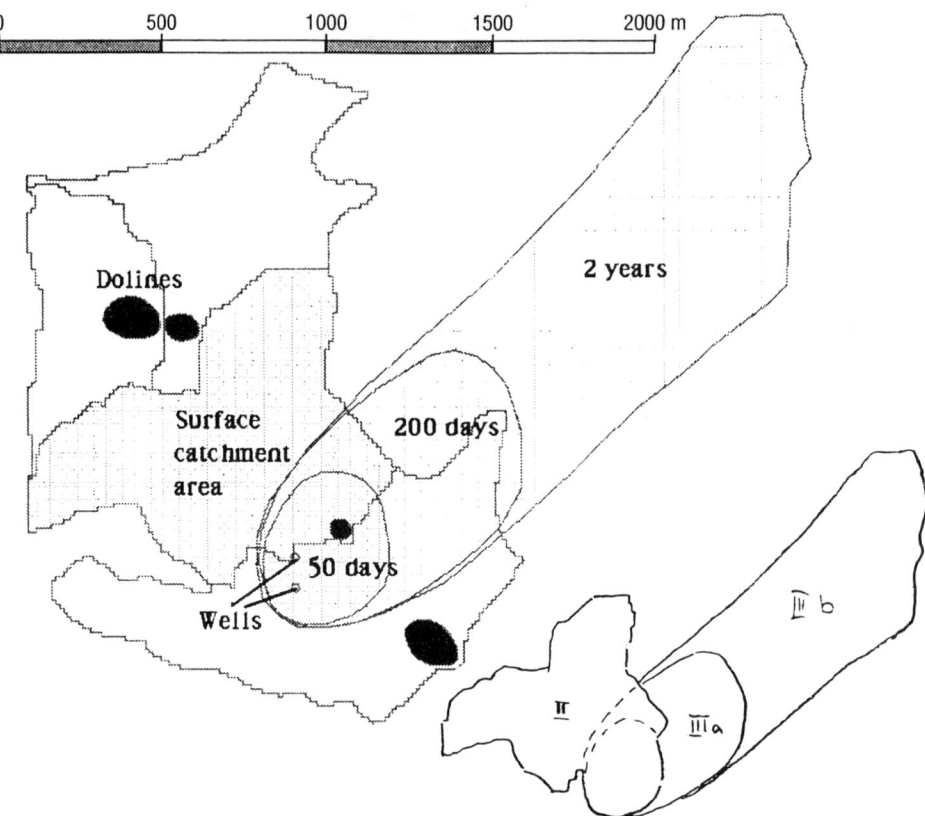

Fig. 5 Surface catchment area and groundwater residence time as well as scheme of proposed well head protection zones.

Jenson & Domingue (1988) fill the depressions in the first step as well and modify the flow directions in all flat areas so as to direct flow from each inflow cell to the nearest outflow cell on the perimeter of the sink. Multiple outflows do not require special treatment like the search for the steepest flow path. The total catchment of the area remains divided into several parts.

The difference of the two approaches lies in the way sinks are interpreted. Martz & De Jong (1988) assume that all sinks are real topographic features which should be treated hydrologically as ponds or reservoirs. The Jenson & Domingue (1988) approach assumes that the sinks are primarily data errors or artifacts.

These differences implies the applicability for DEM's according to their resolution. Martz & Garbrecht (1992) found that the strategy of Martz & De Jong (1988) works well for low relief landscapes with a horizontal resolution of 15 m and a vertical resolution of 0.1 m. They found the method not to be suitable to DEM's with lower resolution because flats could be artifacts of the limited elevation resolution as well.

APPLICATION

The well head protection zones for a small water work facility in Northern Bavaria was computed taking into account the above mentioned strategy. The groundwater modelling was done with FLOWPATH (Franz & Guiger, 1989) and MIPS was used for all other computations and drawings. MIPS (MicroImages) is a hybrid GIS on DOS, AMIGA and UNIX platforms. It comes along with a complete tool for computing watersheds and flow accumulations. With MIPS spatial manipulation language (SML) the raster based computation of the residence time can be done by the user. MIPS offers as well the possibility of raster to vector and vector to raster conversion, which is helpful for visualization and map layout. Figs 2 to 5 shows the way along the three steps to the final result.

REFERENCES

Franz, T. & Guiger, N (1989) FLOWPATH - Steady State Two-Dimensional Horizontal Aquifer Simulation Model. Waterloo Hydrogeologic Software (unpublished manual).
Jenson, S.K. & Domingue, J.O. (1988) Extracting Topographic Structure from Digital Elevation Data for Geographic Information System Analysis.- *Photogramm. Engng* **54**(11), 1593-1600.
Kinzelbach, W. (1986) Groundwater Modelling - An introduction with sample programs in BASIC.- Elsevier Science Publishers, Amsterdam.
Martz, L.W. & Garbrecht, J. (1992) Numerical Definition of Drainage Network and Subcatchment Areas from Digital Elevation Models. *Computer & Geosciences* **18**(6), 747-761.
Martz, L.W. & De Jong, E. (1988) Catch: A FORTRAN program for measuring catchment area from digital elevation models. *Computers & Geosciences* **14**(5), 627-640.
Merkel, B. & Tertilt, K. (1992) Verfahren zur Herstellung von Grundwassergefährdungspotentialkarten mit Hilfe eines hybriden GIS. WATEC GmbH, 8015 Markt Schwaben.
Peucker, T.K. & Douglas, D.H. (1975) Detection of surface specific points by local parallel processing of discrete terrain elevation data.- *Computer Graphics and Image Processing* **4**(3), 375-387.
Pinder, G.F. & Bredehoeft, J.D. (1968) Application of the digital computer for aquifer evaluation. *Wat. Resour. Res.* **4**(5), 1069-1093.

Hydrological terrain features derived from a pyramid raster structure

WOLFGANG RIEGER
Institute for Photogrammetry and Remote Sensing (I.P.F.), Vienna University of Technology, Gushausstr. 27-29, A-1040 Vienna, Austria

Abstract Hydrological terrain features such as drainage basins and channels are automatically derived from a DEM grid. A pyramid type data structure is proposed to overcome some disadvantages of the grid structure and to provide for efficiency. The channel network is obtained from a coarse simulation of water flow. Drainage basins are marked for each river section by use of a reversion of the summing algorithm. Applying successive stages of increasing grid resolution proves to be a promising tool for hydrological surface analyses.

INTRODUCTION

General purpose DEM-packages are usually optimized with regard to fast access by plane coordinates and fast local neighbourhood operations. There is nearly no structuring according to larger terrain features such as ridges, valleys, slopes, or drainage basins, as would be possible in "object oriented DEMs" with such features as objects. This is partly due to usual data capture methods (stereoscopic measurement, digitizing or scanning topographic maps, image matching) that do not yield any terrain features except for breaklines, formlines, highs, and lows. Besides, most applications do not use this information. And last but not least, this data structuring is very complex. Hydrological modelling on the other hand requires knowledge of such structures and particularly, information on river networks and drainage basins.

Mark (1979) demands, that the phenomenon should influence data representation, not computational considerations. According to this demand, Moore (1988) uses a contour-based DEM to construct the DEM-surface according to the regularities of water flow, but the structure is not easy to handle. An overview of hydrological and other applications of DEMs was given by Moore *et al.* (1991).

The idea of most approaches is to extract large features by local neighbourhood operations (especially in rectangular grid DEMs). Since the features to be extracted are large and spatially distributed, it is necessary to use iteration or to apply these operators in special sequences adapted to the terrain. O'Callaghan & Mark (1984), Jenson & Domingue (1988), and others use iteration of simple local operators to obtain river networks and drainage basins. Qian *et al.* (1990) use an expert system shell to check the results of local operator processing against rules concerning regularities in drainage basins. Here, a pyramid structure is proposed to obtain river networks and drainage basins in different generalization levels with an approximate knowledge of the feature from the respective coarser level (Rieger (1992b)). The structure can also be used for complex analyses of the terrain surface. The algorithm and details can be found in Rieger (1992a).

DATA STRUCTURE

The base structure is a rectangular grid stored as a matrix of grid cells. The x-, y-addressing is done implicitly by the matrix addresses. The data values are stored as the elements of the matrix and represent a value assigned to the corresponding grid cell. Vertically, this structure is expanded by a series of different layers (GIS concept) each of which has the same matrix size and structure, but different data information (e.g. elevation, slope, basin label, soil index). Horizontally, each layer is expanded by corresponding matrices with successively halved grid widths, thus obtaining different levels of a pyramid structure (Fig. 1). The pyramid may start at any base grid, but usually will start

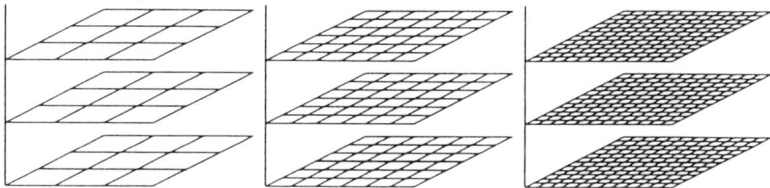

Fig. 1 Layers (vertical) and levels (horizontal). Each matrix represents the same region.

at convenient sizes, e.g. an area of several 1000 grid cells. Upon this structure, a set of functions is declared that work on a number of input and output matrices or scalars. Examples are element-by-element addition or multiplication of 2 matrices of the same level, filtering operations, or complex operations such as simulation of water flow with multiple of input matrices (e.g. elevation, soil conditions, steepness, ground water infiltration). The main advantages of this structure are:
(a) the simplicity of the data structure and the modular concept allow for fast implementation of the system as well as of new and sometimes very complex processes;
(b) common matrix functions can be applied for any matrices regardless of level and layer type, thus minimizing the programming effort and the number of functions necessary;
(c) the resulting modules may be used for a wide variety of applications as well as for very special tasks in geographic information systems;
(d) simpler operations may be easily combined to more complex ones, especially when using a macro-language as implemented in this study;
(e) algorithms and visualization techniques of digital image processing can easily be applied.
The total amount of storage for the pyramid structure is at maximum just about

30% higher than that used for the finest resolution. The maximum matrix size depends on the size of (virtual) main memory and on the number of matrices involved. The elements of matrices may be of type floating point and for some purposes of type byte.

There are two principal ways to obtain this structure. The first one is to calculate the grid iteratively at decreasing grid widths. Hutchinson (1989) proposed this method to derive a hydrologically qualitative grid, especially with removal of artificial depressions. The second way is to calculate the finest grid as a high quality DEM (in this paper the SCOP-DEM package is used (Molnar & Waldhaeusl, 1990)) and successively build coarser resolutions by using non-overlapping 2x2 blocks of grid cells to derive the elevation of one grid cell of the next coarser grid level. Simple arithmetic mean may be used.

DERIVING THE RIVER NETWORK

General

Some idealizations are used to accelerate the calculation of the drainage channels and basins: Rainfall shall be constant all over the region, which causes the total amount of water summed in one grid cell during the whole draining process to be proportional to the size of the drainage basin. Since no time considerations are necessary, water shall drain to all deeper neighbours at the same speed, i.e. all water of one cell shall drain to all its deeper neighbours during the time unit, independently of direction, slope, and soil conditions. Depressions are common in DEM's. Since they hinder water flow, they shall be removed. For the following considerations a depressionless DEM shall be used. Jenson & Domingue (1988), Hutchinson (1989), Rieger (1992a), and others show ways to obtain such a DEM.

Water always flows in the direction of the steepest slope. Normally, this direction is not restricted to neighboured grid cells (4 or 8 neighbours). In the grid DEM, however, one must decide, to which cell water shall drain. This can lead to gross errors along larger slopes, because water tends to float arbitrarily in grid direction. In reality all deeper neighbours get part of the outflow of a grid cell. A simple method to simulate this is the use of so-called drainage quota values: The drainage quota value (DQV) is defined as the quota of the accumulated water in a cell that will be sent to one of its neighbours. Let w_0 be the water accumulated in a grid cell; w_i, the amount of water that flows to one of the cell's neighbouring cells in 4-connectivity; h_i be the elevation of neighbour i, h_0 the elevation of the grid cell itself. The drainage accumulation value then can be calculated as (4-connectivity):

$$w_i = w_0 \cdot dqv_i \qquad (1)$$

The "drainage quota value" to neighbour i, dqv_i, is defined as:

$$dqv_i = \Delta h_i / \sum_{j=1}^{4} \Delta h_j \qquad \text{with} \quad \Delta h_i = \begin{cases} h_0 - h_i & \text{for } h_0 > h_i \\ 0 & \text{for } h_0 \leq h_i \end{cases} \qquad (2)$$

Calculation for one point: The starting cell is initialized with the water unit. Calculating the DQVs from up to low, a drainage band is obtained. Figure 2a shows the drainage band resulting from the drainage quota values, with white being quota values of (nearly or exactly) 1, gray of lower than 1, and black with quotas of (nearly or exactly) 0. Figure 2b shows the the path of maximum drainage quota values downwards in 4-connectivity. The drainage band is unrealistically wide-spread, which results from the simplicity of the assumption; in reality, not all water accumulated in one cell would drain to all deeper neighbours. Rather, there is only a channel in which water flows, and that channel often will drain to only one lower cell.

Fig. 2 Drainage in a slope. (a) Drainage quota values (DQV) highlighted; (b) Path of maximum DQV's on contour-bands (dark against light zones).

Calculation for a complete region

In a complete region, the drainage may be calculated as a piece. Since all this is done simultaneously, balance takes place between the outlet of neighbouring cells, therefore the wide spreading will cause only small errors. Furthermore, water tends to flow into channels, so even the spreading will be kept small. Therefore the method is well suited to yield the main drainage channels.

The calculation shall be done over the entire region in one process, the result being an array that shows to each grid cell the approximate (upper) drainage basin size of this cell. There are two possibilities to obtain this result: an iteration process, with each step consisting of a rainfall simulation and a following draining of each cell's accumulated water; and a one-step solution with one single rainfall simulation and draining from top to bottom. The iterative solution is best suited for parallel computing hardware and may be adapted to consider differing flow velocities and precipitation. The ordered calculation is much faster on conventional hardware and is preferred for deriving only the river network; yet it may not support the time component of water flow. Both concepts yield the same result.

The iteration process may proceed as this: at the beginning, two summing matrices, **S1** and **S2**, are filled with zeroes. One step of iteration now consists of a rainfall and a draining simulation: add the water unit to each single cell of matrix **S1** (rainfall). Drain the water from each single cell of **S1** to all deeper neighbours according to equation (1). These values are added in the corresponding cells in matrix **S2** (draining). Change matrices **S1** and **S2**, so that the result of one step of the iteration becomes the input of the next step.

The sorting process may be done by an index array. Each element of the index array points to one grid cell of the elevation matrix. The pointers are sorted according to descending elevations, so that the first element in the index array points to the highest grid cell in the elevation matrix, the second element points to the second highest cell, and so on. Sorting by quicksort, as implemented in the programming language C, has proved to be sufficiently fast (it takes the same time as 2 to 4 steps of the iterative calculation for models of up to over one million points, Rieger (1992b)).

Figure 3 shows a terrain in a combined representation of contours and hill-shading. The hill-shading value is calculated for each grid cell according to Tanaka's algorithm (Horn, 1982) as floating point value between 0.0 and 1.0. The elevation values are grouped in zones with varying values dependent on the zones (here, periodically changing values from 0.4 to 0.6 in 5 steps are applied, each step representing an elevation zone of 5 m). These two values for each cell are multiplied and the resulting data set is scaled from the range [0.0 .. ... 1.0] to [0 ... 255]. Figure 4a shows the summed water flow with white meaning much water. Figure 4b is a histogram equalization of Fig. 4a.

Fig. 3 Test terrain with hill-shading and contour lines (zones of elevation).

Deriving river networks

The drainage accumulation values as calculated by iteration or by the sorting process represent quite well the area of the catchment to each grid cell, expressed in numbers of grid cells. The data set is similar to the "flow accumulation" data set of O'Callaghan & Mark (1984) except that here each

 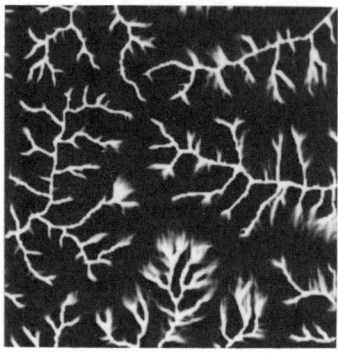

Fig. 4 (a) Drainage accumulation values over the whole region. White: much water; Black: little water. (b) Histogram equalization of (a).

cell may have more than one channel for water runoff (the term "drainage accumulation" is used here for distinction to "flow accumulation"). Along with the flow accumulation data set, a threshold value is used to mark all cells with a higher value as channel cells. This method is, in principle, also applied here. However, since water may spread across flatter areas when using the drainage accumulation values, channel lines may become wider than one single cell or may be interrupted. For these reasons and to obtain the hierarchical river network structure, channel lines are traced as follows:

Starting from outflow points along the edge of the terrain (detected as maxima in drainage accumulation and minima in elevation along the edge), channels are marked by tracing to all neighbours with higher elevation and drainage accumulation values higher than the specified threshold value. If there are more than one paths, a temporary node is installed, and all these paths are traced consecutively, starting with the one with the highest drainage accumulation value. Then a second threshold value is used that represents a minimum length for river channels to prevent parallel or very short stream channels in widening valleys. Channel pieces that are too short are rejected and the temporary node is deleted. If, however, two channels are long enough, the node is accepted and the two new channel segments are inserted. The channel with the larger catchment area (i.e., the higher drainage accumulation value) is expected to be part of the same river line as the lower course, the other one opens a new level in the hierarchy of the river network.

LABELLING THE DRAINAGE BASINS

The next step is to label all drainage basins to all river sections. This is done utilizing the existence of a sorted index array. Work takes place upwards. To each grid cell, all deeper neighbours (in 4-connectivity) are examined. The following conditions may occur:
(1)   no deeper neighbours do exist: A new basin starts;

(2) all cells belong to same basin: Test-cell assigned to this basin;
(3) cells belong to different basins: Test-cell assigned to basin with highest probability ("Catchment affiliation value", see later on).

The concept of the DQVs is adapted as the "catchment affiliation values" (CAVs) for eliminating dependencies on the grid: The base-cell of a basin is assigned a CAV of 1. The same is true for all cells along the marked river sections, which means, that all these cells belong to the basin. For processing reasons, the label is constructed as follows (*bas* be the river section's and basin's label number, *cav* the cell's CAV, *label* the value finally assigned):

$$label = \begin{cases} bas & \text{for } cav=1 \\ bas + cav & \text{for } cav<1 \end{cases} \quad (3)$$

The CAV is furthermore limited to a range of about 0.0001 through 0.9999 to avoid numeric zero for very small values (which would falsely be interpreted as a CAV of 1.0000) and numeric 1.0000 (which would falsely be interpreted as the basin with the next higher label-number, $bas+1$). Each cell i is assigned a CAV from all deeper neighbours j (with CAVs of cav_j) of the same basin k as follows:

$$_k cav_i = \sum_{j=1}^{4} {_k dqv_{ij}} \cdot {_k cav_j} \quad (4)$$

dqv is the drainage quota value defined in equation (2), now calculated only from those neighbours belonging to basin k. If there are two or more different basins, each basin gets its own CAV. Since only one value is provided, the cell is assigned to the basin with the highest CAV. The process can be finished in one step, if the affiliation values are obtained from bottom to top of the DEM, since each cell's CAV is calculated from only its deeper neighbours. Within a basin all cells have CAVs of exactly 1.0, while in the border regions the CAVs will be lower than 1.0. Figure 5a shows the CAVs for a part of the DEM of Fig. 3. Figure 5b shows the zones with CAVs lower than 1.0 in dark gray as an overlay

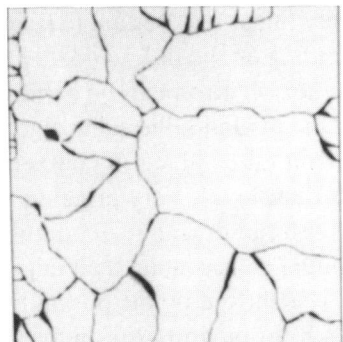

Fig. 5 (a) Basin affiliation values (CAVs) in the terrain of Fig. 3. White: CAV=1, black: CAV=0, gray: 0.5≤CAV<1. (b) Zones of CAV<1 as dark zones on the terrain.

on the terrain. The same algorithm is used for marking the outflow cells of depressions. In this case no CAVs need to be calculated, since all cells of the depression fully belong to the depression's basin.

To obtain the borderlines, all cells with neighbours belonging to different basins are examined, the cells with the highest elevation value and lowest CAV being marked as border cells. Figure 6 shows the final result, river-courses and basin borders on the terrain of Fig. 3.

Fig. 6 Final result: Potential drainage paths (black) and drainage basins (borderlines in white).

UTILIZING THE PYRAMID STRUCTURE

The algorithm shown above will work on matrices of any size. However, there are limitations by the hardware (memory sizes) and computational effort. With word-lengths of 32 bits for both integers and floating point values and a maximum of three matrices to be held in (virtual) memory simultaneously for the calculation of the drainage accumulation values, this means about 12 MB of storage for matrix sizes of about 1 000 000 grid cells. With modern workstations this is no real problem in terms of memory usage. But the computational overhead is out of proportion, since only linear features are to be extracted. The pyramid structure is now applied depending on the task to do.

Generally, scale-dependent generalization can be achieved by using suitable grid widths. On each level of the pyramid, the same algorithms can be used without changes. Calculation first takes place in coarser resolutions, yielding approximations for the next finer level. Very large matrices may be arbitrarily split into sub-matrices, with the drainage accumulation values obtained in lower levels impacted as initial values along their edges.

The main advantage, however, is achieved by the possibility to divide a region into drainage basins. Each single basin can now be used in finer resolutions. First the border will be refined, followed by an optional finer calculation of the river network and subdivision into subbasins. This process can be used to further divide a basin and work in even smaller subbasins or to

 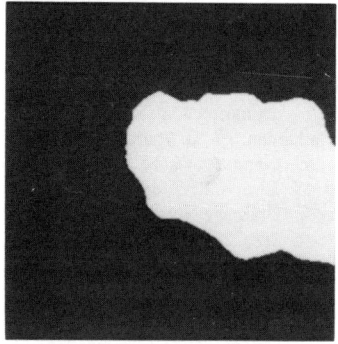

Fig. 7 (a) One single drainage basin highlighted in coarse resolution. (b) Enlarged mask of drainage basin for further work.

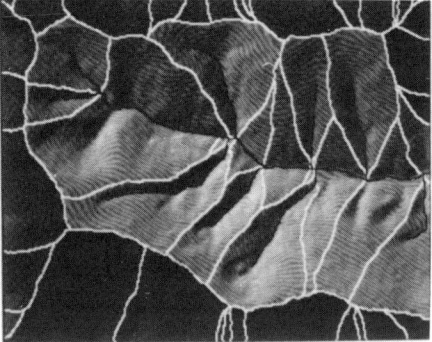

Fig. 8 River paths and subbasins in basin of Fig. 7 in next finer resolution.

work in the whole basin and its sub-divisions at once. Figure 7a shows the terrain of Fig. 3 with one basin highlighted. Figure 7b shows an enlarged mask of this basin, enlarged for more accurate deriving the border. Figure 8 shows the result of the described processes of channel detection and basin labelling in the next finer resolution.

REFERENCES

Horn, B.K.P. (1982) Hill shading and the reflectance map. *Geo-Processing* **2**, 65-144.
Hutchinson, M.F. (1989) A new procedure for gridding elevation and stream line data with automatic removal of spurious pits. *J. Hydrol.* **106**, 211-232.
Jenson, S.K. & Domingue, J.O. (1988) Extracting topographic structure from digital elevation data for GIS analysis. *Photogramm. Engng* **54**(11), 1593-1600.
Johnson, L.E. (1989) MAPHYD - A digital map-based hydrologic modeling system. *Photogramm. Engng* **55**(6), 911-917.
Mark, D.M. (1979) Phenomenon-based data-structuring and digital terrain modelling. *Geo-Processing* **1**, 27-36.
Molnar, L. & Waldhaeusl, P. (1991) Program System SCOP to create, maintain, and apply digital terrain models. *Product information of the Inst. for Photogrammetry and Remote Sensing, The Vienna*

University of Technology, Vienna, Austria.

Moore, I.D. (1988) A contour-based terrain analysis program for the environmental sciences (TAPES). *Trans. Am. Geophys. Union* **69**, 345.

Moore, I.D., Grayson, R.B. & Ladson, A.R. (1991) Digital terrain modelling: A review of hydrological, geomorphological and biological applications. *Hydrol. Processes* **5**(1), 3-30.

O'Callaghan, J.F. & Mark, D.M. (1984) The extraction of drainage networks from digital elevation data. *Computer Vision, Graphics and Image Processing* **28**, 323-344.

Qian, J., Ehrich, R.W. & Campbell, J.B. (1990) DNESYS - An Expert System for Automatic Extraction of Drainage Networks from Digital Elevation Data. *IEEE Transactions on Geoscience and Remote Sens.* **28**(1), 29-45.

Rieger, W. (1992a) Hydrologische Anwendungen des digitalen Geländemodelles (Hydrologic applications of the DEM). *Dissertation, The Vienna University of Technology, Vienna, Austria.*

Rieger, W. (1992b) Automated River Line and Catchment Area Extraction from DEM Data. *Proc. International Congress of the ISPRS*, Washington, D.C. 1992, Commission IV.

TOPMODEL as an application module within WIS

RENATA ROMANOWICZ, KEITH BEVEN & JIM FREER
Centre for Research on Environmental Systems and Statistics, Institute of Environmental and Biological Sciences, Lancaster University, Lancaster LA1 4YQ, UK

ROGER MOORE
Institute of Hydrology, Crowmarsh, Wallingford, OXON, OX10 8BB, UK

Abstract The Water Information System, WIS, has been developed at the Institute of Hydrology, Wallingford, as a four-dimensional database and GIS that includes procedures for manipulating digital terrain maps (DTM's) and hydrological time series. It is consequently an ideal basis for hydrological modelling studies. TOPMODEL is a set of concepts for distributed rainfall-runoff modelling, that makes use of detailed DTM data but is computationally efficient. Incorporation of TOPMODEL into WIS raises a number of problems, the most critical of which is the generalization of the model structure to allow for varying topographic and hydrological conditions with different levels of data availability. One approach to this problem is the use of "on-line" sensitivity analysis, parameter calibration and estimation of predictive uncertainty for different possible model structures, with the graphical presentation of the results within WIS for direct visual evaluation by the user. This allows qualitative as well as quantitative evaluation/validation of model simulations. MonteCarlo simulation is used for calibration and uncertainty estimation with a parallel processor linked to the dedicated graphic workstation running WIS to allow results to be presented within a reasonable time frame.

INTRODUCTION

Recent development of Geophysical Information Systems has allowed the storage and display of digitized land surface topography data, which is now becoming available in large quantities from a variety of sources, together with other features of the landsurface pertinent to hydrological modelling. Work towards the use of three-dimensional GIS in hydrogeological applications has been carried out to some extent (Turner, 1989). The Water Information System (WIS), which is being developed by the UK Institute of Hydrology in collaboration with ICL, is a four dimensional database which stores both spatial and temporal data on catchment characteristics and hydrological and water quality variables. Based on FORTRAN and embedded SQL data retrieval and storage, WIS is an ideal basis for hydrological modelling studies and, in particular, the application of distributed models. TOPMODEL has been implemented as one of a number of application modules within WIS. Other hydrology-oriented modules include river quality survey and the presentation of catchment and flow characteristics (see Bonvoisin & Moore, 1992).

Attempts to improve the description and understanding of hydrological processes have led to the introduction of distributed models, usually based on non-linear partial differential equations descriptive of the overland and subsurface flow processes. Because of the continuing computational problems of fully three dimensional descriptions (see Binley & Beven, 1992), most existing models use only quasi-three dimensional descriptions, with approximate solutions implemented on a variety of spatial discretizations. Regular finite difference (raster) grids, irregular finite element grids and triangular irregular networks have all been used (see for example, Abbott et al., 1986; Beven et al., 1987; Moore et al., 1986). Each type of grid creates different types of implementation problems within a GIS for the input and storage of catchment characteristics and distributed boundary conditions for the models.

TOPMODEL is a computationally efficient semi-distributed hydrological model, the operation of which is based on an analysis of the topographic form of the catchment (see for example Beven & Kirkby, 1979; Beven & Wood, 1983; Hornberger et al., 1985; Beven, 1986; Wood et al., 1990). Routines for the analysis of topographic data have been developed for raster elevation data to identify flow pathways and the river network (Quinn et al., 1991). Raster elevation data on a 50 m grid is being produced for the whole of the UK by the Institute of Hydrology based on the digital contour database of the Ordnance Survey. Other catchment characteristics such as different soil characteristics and vegetation characteristics can also be taken into account by the model. Model predictions of soil water status can also be mapped in space at each time step. Display of such predictions can provide a good qualitative impression of the spatial structure of the hydrology and the way in which the model is reproducing the flow processes. Thus, such a model is well suited to make use of the facilities of a GIS and, in particular, WIS.

There must also be provision for calibration of model parameters and, since any calibration process must result in some residual error, the estimation of the uncertainty associated with the predictions. The implementation of TOPMODEL within WIS includes automatic optimization routines and sensitivity analysis options to help in the analysis of model structures. In addition, the GLUE technique of Beven & Binley (1992) is used for uncertainty estimation. This Monte Carlo based technique is computationally intensive and a further novel aspect of the implementation is a direct link from WIS to the Lancaster University Meiko transputer parallel processor. The results are stored and may be displayed from within WIS.

GIS REQUIREMENTS FOR THE DISTRIBUTED HYDROLOGICAL MODELLING

Distributed hydrological models require tools to allow for quick retrieval, storage and modification of large quantities of spatial data in all three spatial dimensions and time. Until now, most GIS have been developed to handle time-invariant data. The novel feature of WIS is that it handles both spatial data and conventional time series data within a single system and it will ultimately allow the spatial data to vary with time. WIS provides facilities for user interactive preparation and display of atchment characteristics, of which the most demanding is usually high resolution topographic data, and initial soil moisture and river level conditions. For the future, there are plans to meet the increasing demcands for time variable spatial data, such as those arising from remote sensing images and those produced by the results of distributed models.

One of the most exciting possibilities in linking GIS with hydrological modelling is the possibility of dynamic spatial visualization of the simulation results, including user interaction for the display of both different simulated variables and choice of area or river cross-section to be displayed, depending on the kind of application. The real time, or near-real time three-dimensional visualization of simulated hydrological processes could greatly improve existing analysis of simulation results and provide a powerful tool for better understanding and solving real-world hydrological problems. The current implementation of TOPMODEL within WIS represents the first stage in the development of such a system built around a flexible and computationally efficient model and sophisticated SQL data structures to handle the large topographic and hydrological databases.

There is one very important aspect of the integration of GIS and distributed hydrological models that must be emphasized. Neither the definition nor the calibration of distributed hydrological models are well established (see Beven, 1989, 1993; Beven & Binley, 1992). The spatial predictions that can be produced by TOPMODEL are significant in that a comparison of dynamic maps of predictions with knowledge of runoff production processes in the field can be expected to lead to revision of the model structure. TOPMODEL has never been considered as a fixed model structure but rather a set of modelling concepts that can be adapted and modified by the user on the basis of such comparisons. Thus, a general implementation of TOPMODEL must allow for a user-interactive flexible adjustment of model structure depending on both predicted patterns of behaviour and the aims of the study.

There are at least three ways of integrating such a user-definition of model structure within a GIS. One is to provide a set of variant structures as compiled routines that may be selected from a menu presented to the user; a second is to allow the user to provide an executable program segment with standardized inputs and outputs; the third is to provide a meta-language within which the user may define and modify the model structure. The latter may be in either command line form (such as PV wave, see Clapp *et al.*, 1992) or GUI form (such as the SIMULINK interface for MATLAB). Within WIS, the first and second options are available.

TOPMODEL THEORY

TOPMODEL is a quasi-distributed rainfall-runoff model which allows the distribution of soil moisture deficits in the catchment to be predicted. It uses "physically-based" parameters and a spatially distributed topographic index computed from a map of digital elevation data (DTM). Instead of working on a spatial grid it evaluates soil moisture deficits for the discrete values of the frequency distribution of the soils/topographic index, $\ln(a/T\tan\beta)$, where a is the upslope drainage, β is the slope angle and T denotes local saturated transmissivity. The schematic structure of the model is presented on Fig. 1.

The dynamic of the catchment is evaluated for the catchment average soil moisture deficit, s_t, according to the state equation :

$$ds_t/dt = q_b - q_v \qquad (1)$$

where q_b (m hour^{-1}) is the total hillslope input to the stream channels and q_v (m hour^{-1}) is the total amount of water infiltrated to the saturated zone in the catchment.

It can be shown (see for example, Beven, 1986) that the steady-state solution of equation (1) under the assumption of an exponential dependence of subsurface flow on the soil moisture deficit gives the topography-dependent distribution of soil moisture in the catchment according to the equation:

$$S_i = S_t + m(\gamma + \ln(a_i/T_i \tan(\beta_i))) \qquad (2)$$

where $\quad \gamma = (m/A) \int \ln(a_i/\tan\beta_i) da - (m/A) \int \ln T_i da \qquad (3)$

and m is a recession parameter and A is the total catchment area.

Equation (2) describes a static relation between local storage deficit and the catchment average storage deficit that applies for any steady input rate. With its help we can obtain the maps of soil moisture deficits for each time step of a TOPMODEL run.

The objective function used for the evaluation of model performance can have the form of the sum of the differences between simulated and computed outflows from the catchment during the simulation period:

$$\min F = \sum_{t=1}^{T} \frac{(Q_{sim} - Q_{obs})^2}{(Q_{obs} - Q_{av})^2} \qquad (4)$$

where Q_{av} is the mean flow from the catchment.

WIS-TOPMODEL APPLICATION

Implementing the application of TOPMODEL within WIS requires fast, on-line data management : inserting, updating and retrieval of the database data in real time, while allowing for the visual effects of the choice of the model structure

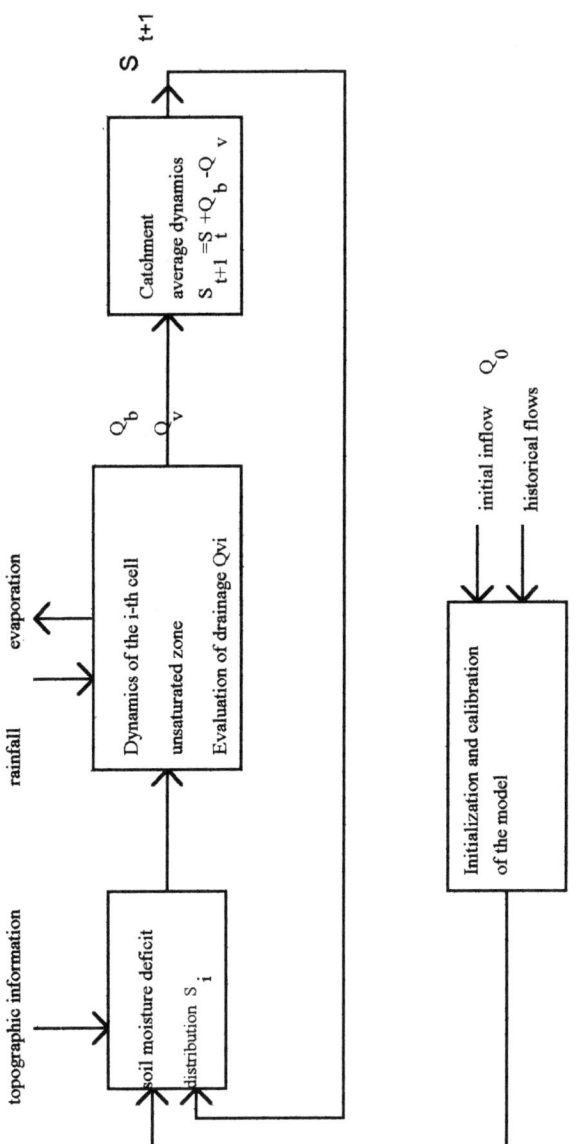

Fig. 1 Scheme of TOPMODEL structure.

and parameters to be compared. A TOPMODEL application consists of the five main options:
(i)   catchment characteristics;
(ii)   generalized likelihood uncertainty estimation (GLUE);
(iii) flow prediction;
(iv)   soil moisture-distribution;
(v)   map display.

The first option allows the choice of the catchment to be analysed and gives its description, which includes elevation and flow data. The second one switches WIS to the MEIKO transputer system to perform the sensitivity analysis of the catchment model. The third option is used to model the catchment runoff production. The fourth option evaluates the spatial distribution of soil moisture and evaporation fluxes in the catchment, and the fifth one is used to display the spatial predictions. The second option uses the GLUE procedure which employs Monte Carlo simulation, performed with the help of parallel computing (Beven & Binley, 1992). Due to the distributed nature of the catchment and the availability of only point measurements of flows, the objective function tends to be nonconvex with respect to the parameter space and often shows the existence of multiple optima. This poses a very difficult task for the optimization procedure. The GLUE procedure overcomes this problem by making a large number of runs of a model with different sets of parameter values chosen randomly from specified parameter distributions. Each set of parameter values is assigned a likelihood value. A distribution function of likelihood values for the parameter sets may be updated with a new series of measurements using the Bayes theorem (Beven & Binley, 1992). The WIS option allows switching onto the MEIKO transputer system, with up to 80 processors, where the Monte Carlo simulations are performed. The results are sent to the WIS directory, from which proper likelihood maps are formed and loaded onto the database. Graphic display of maps in WIS allows visual comparison of the results and the choice of optimal parameter sets.

The structure of the model and its parameters must be adjusted according to the catchment characteristics, available measurements and the goal of the modelling. For example, for flow prediction in a small catchment the results of the calculations may be accurate enough using the simplest version of TOPMODEL, with only two parameters, to be calibrated from continuous records of rainfall and the outflow from the catchment. That can be done using the third option. It involves sensitivity analysis, model calibration and validation. Different possible model structures may be chosen, including the choice of different goodness of fit functions for the evaluation of the model performance and different flow routing models. The sensitivity analysis shows which parameters of the chosen model influence a given objective function and what are their feasible ranges. Performance maps in the parameter space are created upon which the proper choice of parameters can be performed. From the point of view of WIS this requires the introduction and updating of the

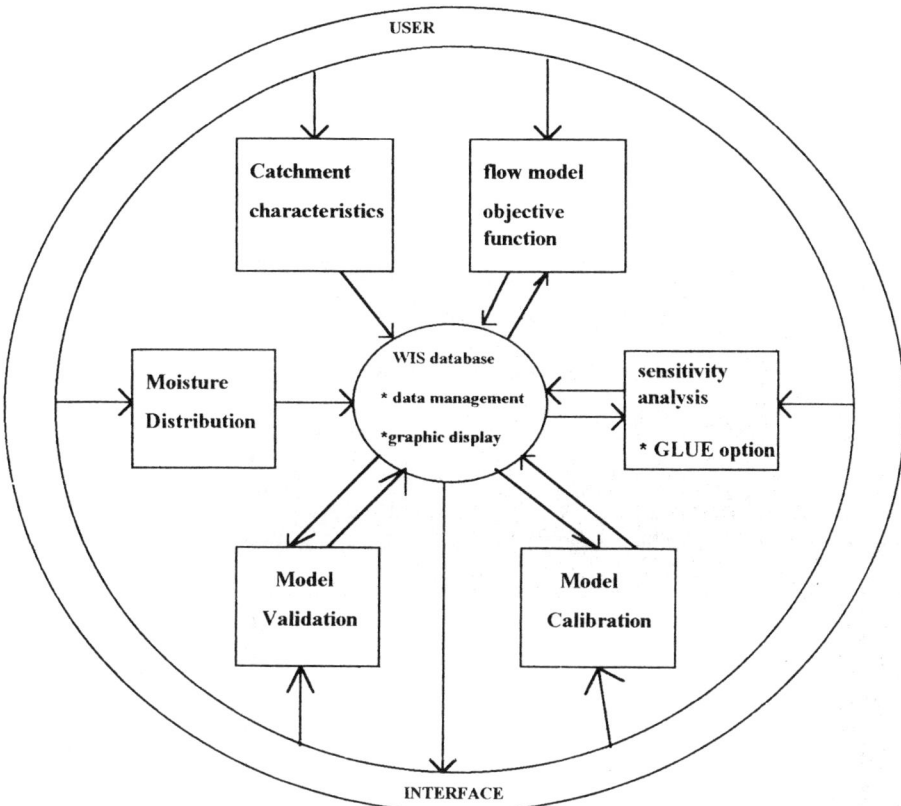

Fig. 2 Scheme of WIS/TOPMODEL application information flow.

sensitivity maps into the database and the subsequent visualization of the results using the graphic facilities available in WIS. The model structure may then be validated on a different time-period of the rainfall and discharge record (Klemes, 1986). Both calibration and validation need the visualization of rainfall and observed and simulated runoff time-series in graphical form.

In the soil moisture distribution option soil moisture maps are computed based on the TOPMODEL soil moisture deficit evaluations. The changes of subsurface soil moisture content and evaporation with time can be visualized in the map-display option in a quasi on-line mode, by the appropriate changes of the colour look-up table.

EXAMPLE APPLICATION

TOPMODEL within the WIS framework was applied on the river Severn at Plynlimon. It is a small forested catchment of an area of 8 km^2 with continuous hourly flow and rainfall data. One advantage of TOPMODEL is that the static storage equation (2) allows it to be relatively easily initialized for event simulation. The sensitivity analysis performed on the catchment showed that

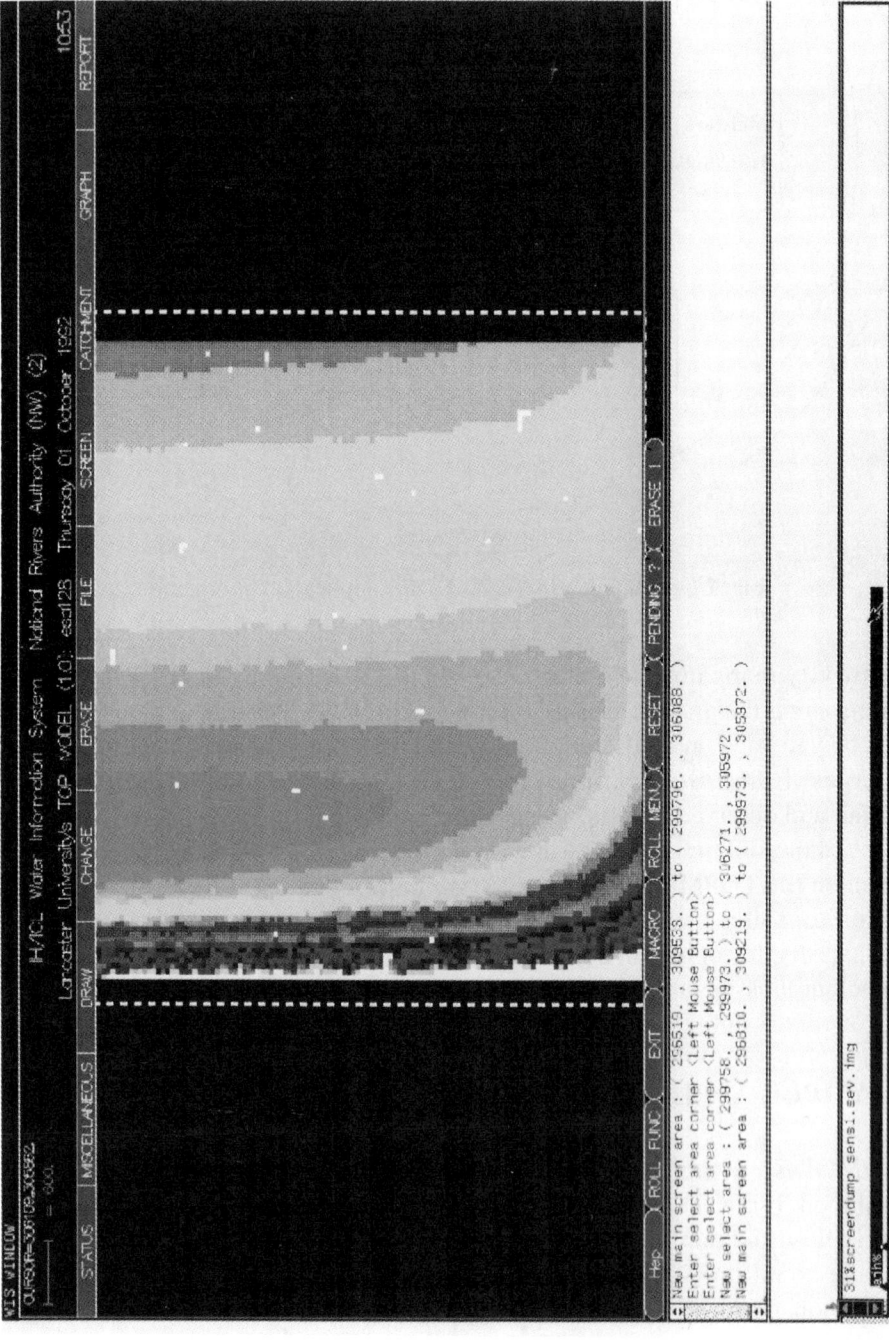

Fig.3 Two-parameter sensitivity map for the "maximum efficiency" objective function for Severn catchment.

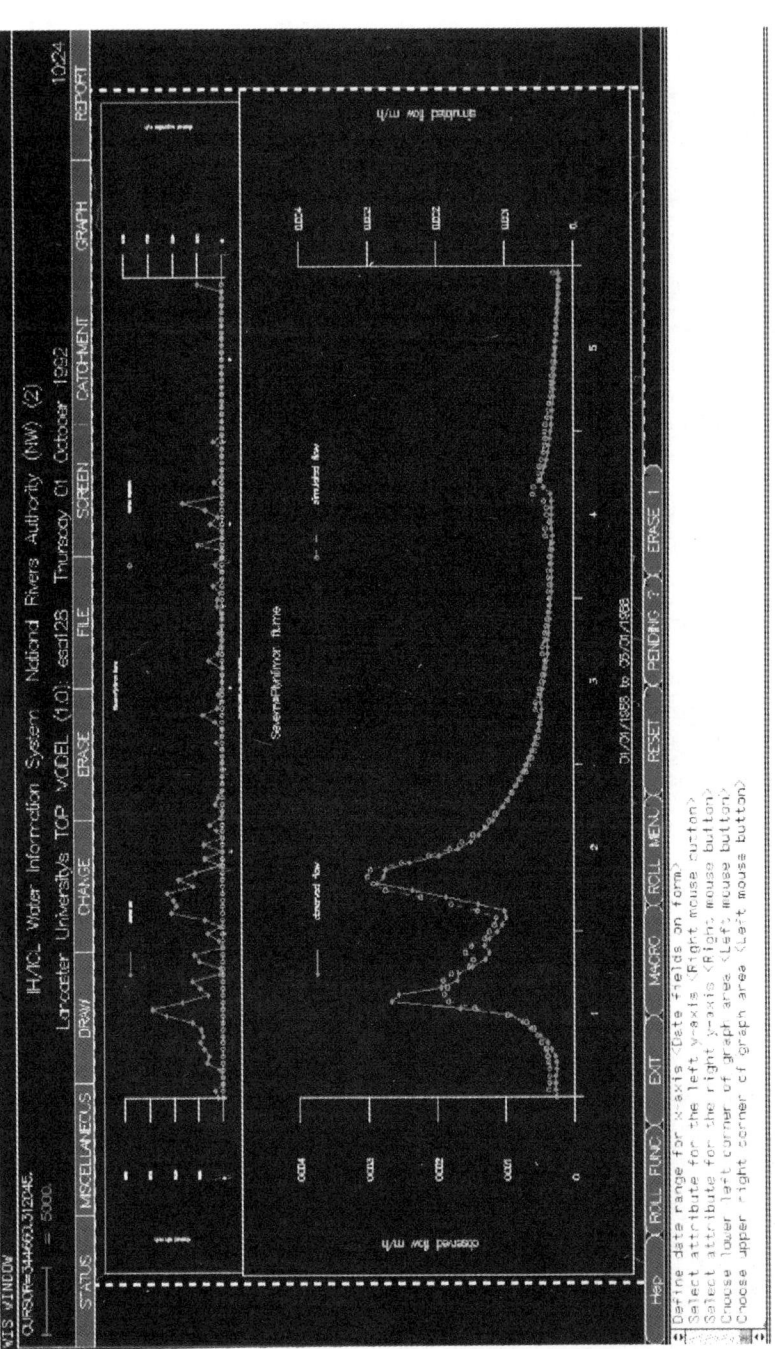

Fig. 4 Hydrographs of rainfall and simulated and observed flow for the validation period for Severn catchment.

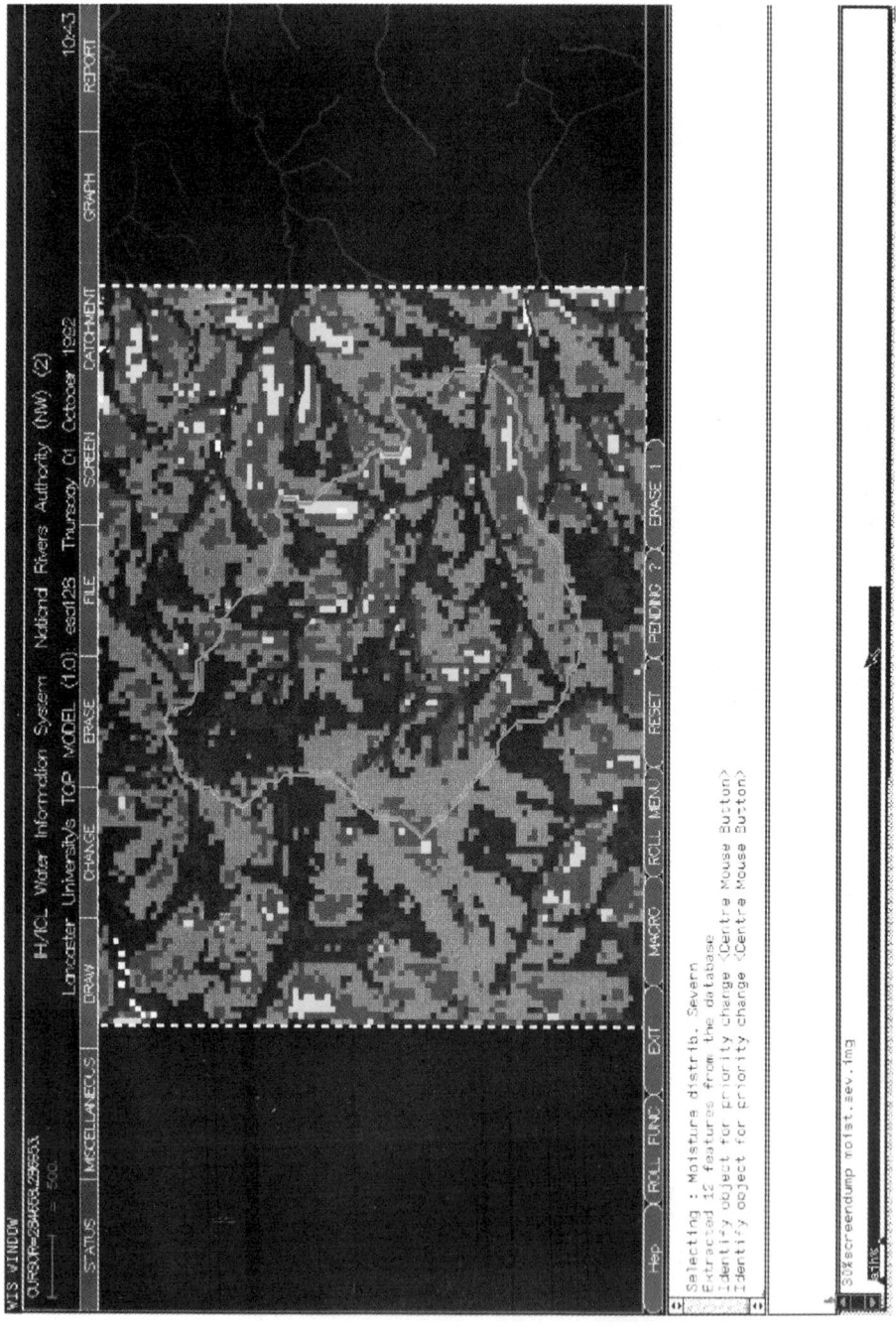

Fig. 5 Soil moisture distribution of the Severn catchment.

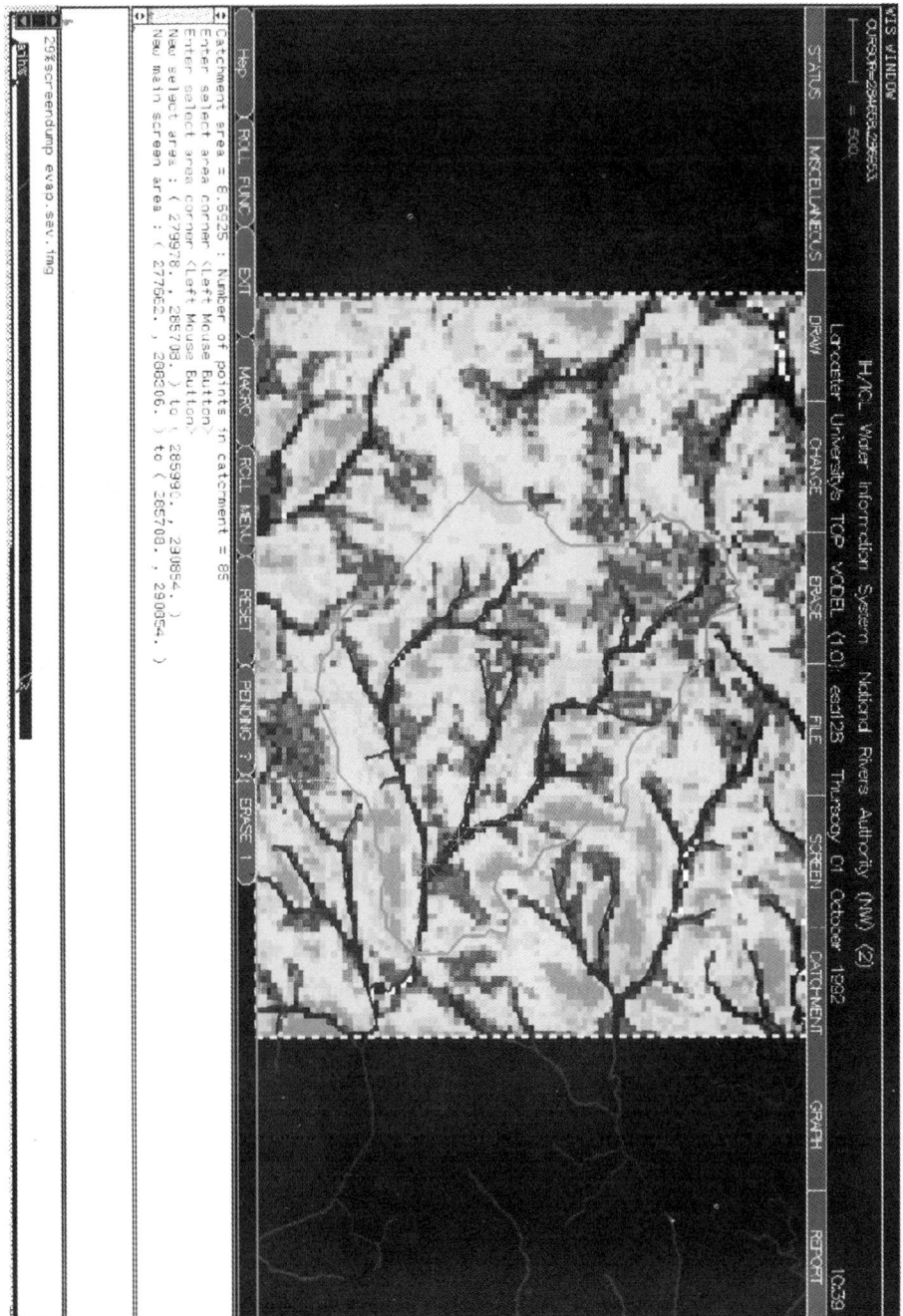

Fig. 6 Evaporation distribution of the Severn catchment.

only two parameters (the recession parameter m and the catchment average transmissivity, T) are influencing the objective function (4) significantly. The sensitivity map evaluated using the GLUE option is shown on Fig.3. The sensitivity analysis allowed the choice of suitable feasibility regions for the parameters. Later, the model was calibrated using the POWELL non-gradient optimization method with logarithmic transformation of variables to constrain the space domain. The hydrographs obtained for the validation period with parameters defined during the calibration procedure are shown on Fig. 4. Relative soil moisture content and evaporation rate distribution maps for the beginning of simulation period for the Plynlimon catchment are shown on Figs 5 and 6 respectively.

The work on the application of TOPMODEL as a module within the WIS framework has highlighted several specific problems connected with the character of hydrologic processes and their model representation. Until now the structure of WIS allowed for the transformation and visualization of data stored in the database, but there was no need of on-line interactive work with the database, as is required in the TOPMODEL application. This poses special requirements on the user-application and application-database interfaces. It has been shown that the dynamic display of hydrological processes is now available within the WIS framework. The initialization of the work on transputer system outside the WIS structure and subsequent visualization of the results of the computations on the WIS graphical facilities have also been demonstrated.

Acknowledgements This work has been supported by NERC grant GST/2 and the ENCORE program of the EEC. Nick Bonvoisin, Andrew Howes, David Morris and Robert Flavin from the Institute of Hydrology are thanked for their help and support in implementing WIS at Lancaster University.

REFERENCES

Abbott, M.B., Bathurst, J.C., Cunge, J.A., O'Connell, P.E. & Rasmussen, J. (1986) An introduction to the European Hydrological System - Systeme Hydrologique Europeen. *J. Hydrol.* **87**, 45-77.
Beven, K.J. & Kirkby, M.J. (1979) A physically based variable contributing area model of basin hydrology. *Hydrol. Sciences Bull.* **24**(1), 43-69.
Beven, K.J. & Wood, E. F. (1983) Catchment geomorphology and the dynamics of runoff contributing areas. *J. Hydrol.* **65**, 139-158.
Beven, K.J. (1986) Hillslope runoff processes and flood frequency characteristics. In: *Hillslope Processes* (ed. by A.D. Abrahams), Allen and Unwin, 187-202.
Beven, K.J., Calver, A. & Morris, E.M. (1987) The Institute of Hydrology distributed model. Institute of Hydrology, Report No. 98.
Beven, K.J. (1989) Changing ideas in hydrology - the case of physically-based models . *J. Hydrol.* **105**(1/2), 157-172.
Beven, K.J. (1993) Prophesy, reality and uncertaainty in distributed hydrological modellin. *Adv. Wat. Resour.* (in press).
Beven, K.J. & Binley, A.B. (1992) The future of distributed models: model calibration and uncertainty prediction. *Hydrol. Processes* **6**, 279-298.

Binley, A.B. & Beven, K.J. (1991) Physically based modelling of catchment hydrology : a likelihood approach to reducing predictive uncertainty. In: *Computer Modelling in Environmental Sciences* (ed. by D.G. Farmer & M.J. Rycroft),Oxford, 75-88.

Bonvoisin, N.J. & Moore, R.V. (1992) Use of GIS techniques to assess discharge consents and abstraction licences. In: *Application of GIS in Hydrology and Water Resources* (ed. by K.Kovar & H.P.Nachtnebel) (Proc. Vienna conf. April 1993), IAHS Publ. no. 211.

Clapp, R.B., Timmins, S.P. & Huston, M.A. (1992) Visualizing the surface hydrodynamics of a forested watershed. Proc. Conf. Computational Methods in Water Resources, Denver, Colorado, 765-772.

Hornberger, G.M., Beven, K.J., Cosby, B.J. & Sappington, D.E. (1985) Shenandoah Watershed Study: Calibration of a topography-based, variable contributing area hydrological model to a small forested catchment. *Wat. Resour. Res.* 21, 1841-1850.

Klemes, V. (1986) Operational testing of hydrological simulation models. *Hydrol. Sci. J.* 31, 13-24.

Moore, I.D., Mackay, S.M., Wallbrink, P.J., Burch, G.J. & O'Loughlin, E.M. (1986) Hydrologic characteristics and modelling of a small forested catchment in southeastern New South Wales. Pre-logging condition. *J. Hydrol.* 83 , 307-335.

Quinn, P.F., Beven, K.J., Chevallier, P. & Planchon, O. (1991) The prediction of hillslope flow paterns for distributed hydrological modelling using digital terrain models. *Hydrol. Processes* 5, 59-7.

Turner, A.K. (1989) The role of three-dimensional geographic information systems in subsurface characterization for hydrogeological applications. In: *Three dimensional applications in Geographical Information Systems* (ed. by J. Raper), 115-127.

Wood, E.F., Sivapalan, M. & Beven, K.J. (1990) Similarity and scale in catchment storm response. *Reviews Geophysics* 28(1), 1-18.

4 Application of GIS in Three- and Four-Dimensional Problems

Development of a three-dimensional hydrogeological framework model for the Death Valley region, southern Nevada and California, USA

CLAUDIA FAUNT, FRANK D'AGNESE & A. KEITH TURNER
US Geological Survey, Box 25046, Denver Federal Center, Denver, Colorado 80225, USA

Abstract Three-dimensional (3D) hydrogeological modelling of the complex geology of the Death Valley region requires the application of a number of Geoscientific Information System (GSIS) techniques. This study, funded by United States Department of Energy (DOE) as a part of the Yucca Mountain Project, focuses on an area of approximately 100 000 km^2 (three degrees of latitude by three degrees of longitude) and extends up to ten kilometers in depth. The geological al conditions are typical of the Basin and Range province; a variety of sedimentary and igneous intrusive and extrusive rocks have been subjected to both compressional and extensional deformation. GSIS techniques allow the synthesis of geological, hydrological and climate information gathered from many sources including satellite imagery and published maps and cross sections. Construction of a 3D hydrogeological model is possible with the combined use of software products available from several vendors, including traditional GIS products and sophisticated contouring, interpolation, visualization, and numerical modelling packages.

INTRODUCTION

Yucca Mountain at the Nevada Test Site in southwestern Nevada is being studied for a potential site to store high-level radioactive nuclear waste (Fig. 1). In cooperation with the Department of Energy (DOE), the United States Geological Survey (USGS) is evaluating the site as part of the Yucca Mountain Project. Because of the potential for transport of radionuclides from the repository to the accessible environment, studies are being conducted to characterize the Death Valley regional groundwater flow system which encompasses Yucca Mountain (Bedinger *et al.*, 1989).

This paper focuses on the initial phase of regional groundwater flow characterization, the development of a three-dimensional (3D) hydrogeological framework model. This framework model describes both the geometry and composition of the materials that form the natural hydrogeological system. The selection of numerical modelling parameters is facilitated by using attribute data stored in the data base that accompanies the framework model.

PHYSICAL SETTING

The study area, defined by the Death Valley regional groundwater flow system, is bounded by latitude 35°N and 38°N and longitude 115°W and 118°W (Fig. 1).

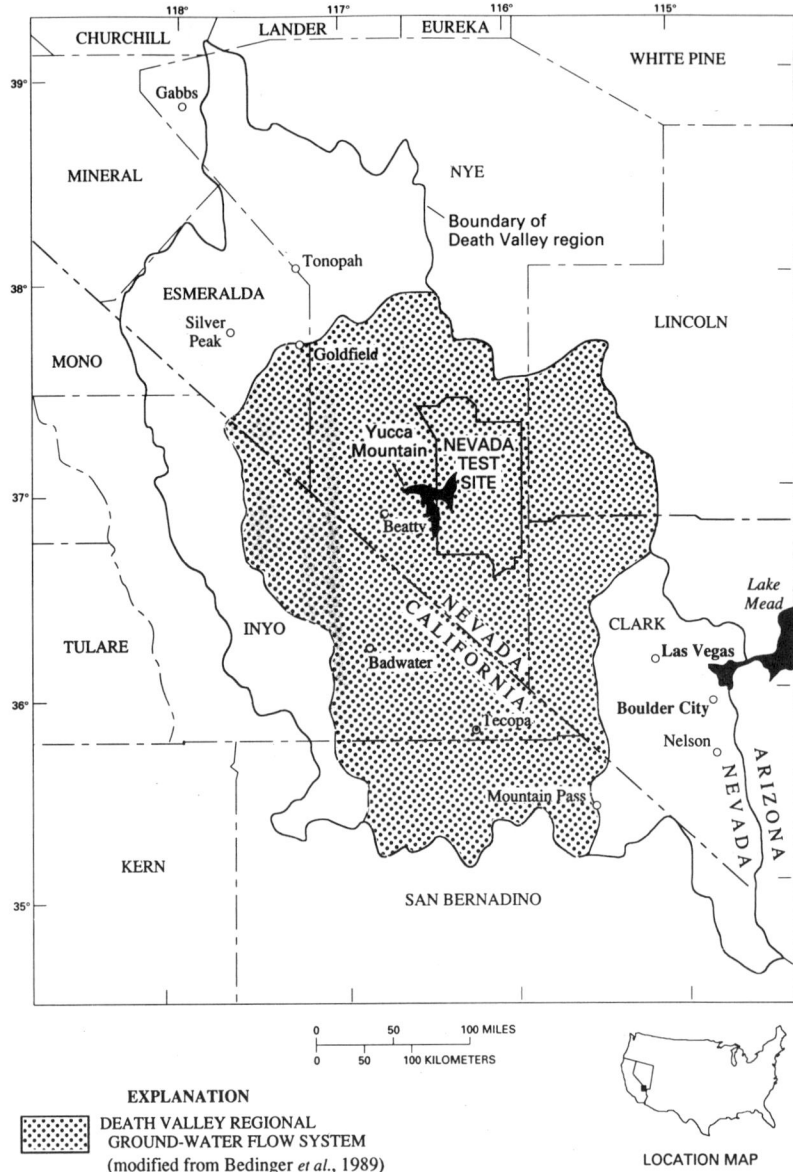

Fig. 1 The Death Valley Region, Nevada-California, in the southwestern United States.

The modelled area includes approximately 100 000 square kilometers and extends to depths of ten kilometers.

The study area has a semiarid to arid climate and is located within the southern Great Basin, a subprovince of the Basin and Range physiographic province. Topographic elevations range from 90 m below sea level to 3600 m above sea level; thus the region includes a diversity of climate regimes and associated recharge/discharge and infiltration conditions.

Geological conditions are typical of the Basin and Range geological province; a considerable variety of intrusive and extrusive igneous, sedimentary and metamorphic rocks have been subjected to several episodes of compressional and extensional deformation. Because of the complex geological and hydrological conditions, a number of GSIS techniques are needed to develop a 3D hydrogeological framework model.

ROLE OF GEOSCIENTIFIC INFORMATION SYSTEMS IN HYDROGEOLOGICAL MODELLING

An extension of the traditional two-dimensional (2D) geographic information system (GIS) method is required for 3D hydrogeological applications (Turner, 1989). These problems require representation of the depth dimension in addition to the areal extent of geological, hydrological, climatic, and ecological features, and linkages to various data manipulation procedures. As a result, the term "Geoscientific Information System", or GSIS, is used to differentiate these geologically oriented 3D systems from the more common 2D GIS products (Raper, 1989).

Some geological applications can be accomplished by representing the 3D subsurface volume as a quasi 3D representation through the use of surfaces. These surfaces, which can represent bedding planes, for example, can be contoured or displayed as isometric views. In these cases however, the elevation of the surface is not an independent variable, and so these systems are best defined as quasi 3D, or 2.5Dimensional (2.5D) systems. These 2.5D systems can accept only a single elevation (z) value for any surface at any given location. Accordingly, several important geological structures, such as folds or faults, which cause repetition of a single horizon at a given location, cannot be represented by these systems (Turner, 1991).

In contrast, true 3D systems, which contain three independent coordinate axes, can accept repeated occurrences of the same surface at any given location. The demand for detailed 3D subsurface data represented by a true 3D system, is critical when constructing a hydrogeological framework model that will be evaluated to provide the required parameters for groundwater flow simulation (Turner, 1991).

The use of GSIS for hydrogeological modelling involves six stages:
(a) development of a hydrogeological framework model that characterizes the 3D subsurface geological structures and materials;
(b) analysis of selected model components to define the physical boundaries and flow parameters of the system;
(c) evaluation of the mechanisms of groundwater recharge, discharge and flow to characterize hydraulic boundaries and flux conditions;
(d) development of a series of numerical model input arrays using an interface between the GSIS data base and the numerical model;

(e) numerical simulation of groundwater flow and evaluation of the model predictions through GSIS visualization; and
(f) repetition of the above stages to achieve numerical model calibration.

DATA COLLECTION AND PREPARATION PROCEDURES

Construction of a true 3D hydrogeological framework model is possible only with the combined use of many available software products, including traditional GIS, and sophisticated contouring, interpolation, and visualization packages (Fig. 2). Existing geological maps and cross sections, digital elevation models, satellite imagery, geophysical data, and hydrological data were converted into a consistent digital format utilizing the two-dimensional ARC/INFO[1] geographic information system (Fig. 2). Some data already exist in digital formats and only required extraction and transformation. Manuscript map data were scanned using a raster-to-vector Tektronix scanner and the resulting vector files were further processed to remove artifacts of the scanning process, transformed to a convenient geographic coordinate system, and edited

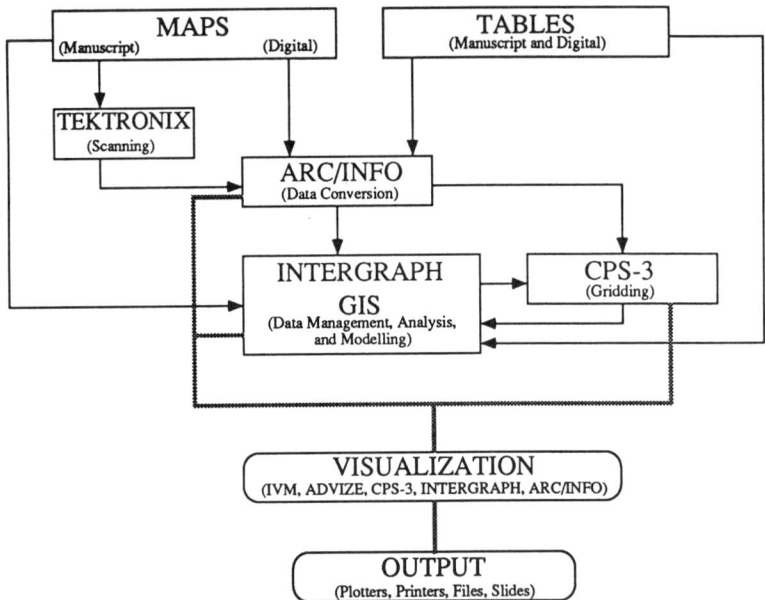

Fig. 2 Flow diagram showing the logical movement of data from raw format, through data conversion and analysis, and ultimately to model visualization.

---

[1] Any use of trade, product, or firm names is for descriptive purposes only and does not imply endorsement by the US Government.

to achieve accurate topology. The digital files were then moved to the Intergraph Corporation Modular GIS Environment (MGE) which allows the integration of raster and vector 3D data sets and associated data base files.

HYDROGEOLOGICAL FRAMEWORK MODEL DEVELOPMENT

Construction of the 3D hydrogeological framework model involves several steps. The digital geological maps and cross sections are accurately placed in their correct 3D spatial relationships and their features attributed using MGE (Fig. 2). The Radian Corporation's CPS-3 gridding system and fault handling package is then used to interpolate the fault planes and geological horizon surfaces between existing cross sections and boreholes. The hydrogeological system is then visualized (Fig. 3). After visualization, the data are re-entered

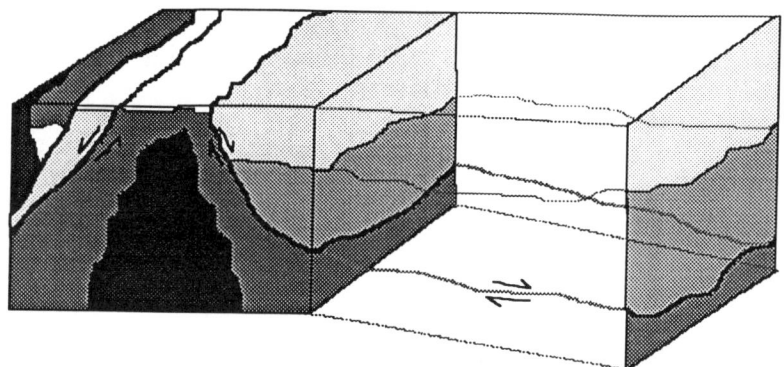

Fig. 3 Schematic block diagram showing the progressive construction of the 3D hydrogeological framework model.

into MGE for use in 3D groundwater flow modelling utilizing the MGE Environmental Resource Management Applications (ERMA) MODFLOW interface. Throughout the above steps, the geometrical model components are supported by a complex sequence of attribute information describing the hydrogeological properties of all components. This information is organized and stored in a relational data base in over one hundred tables organized into the data categories shown in Fig. 4.

ANALYSIS AND INTERPRETATION OF THE FRAMEWORK MODEL

The GSIS procedures allow the 3D hydrogeological framework model to be repeatedly analyzed and interpreted. Alternative strategies for evaluating the hydrogeological systems within the faulted terrains can be explored; for

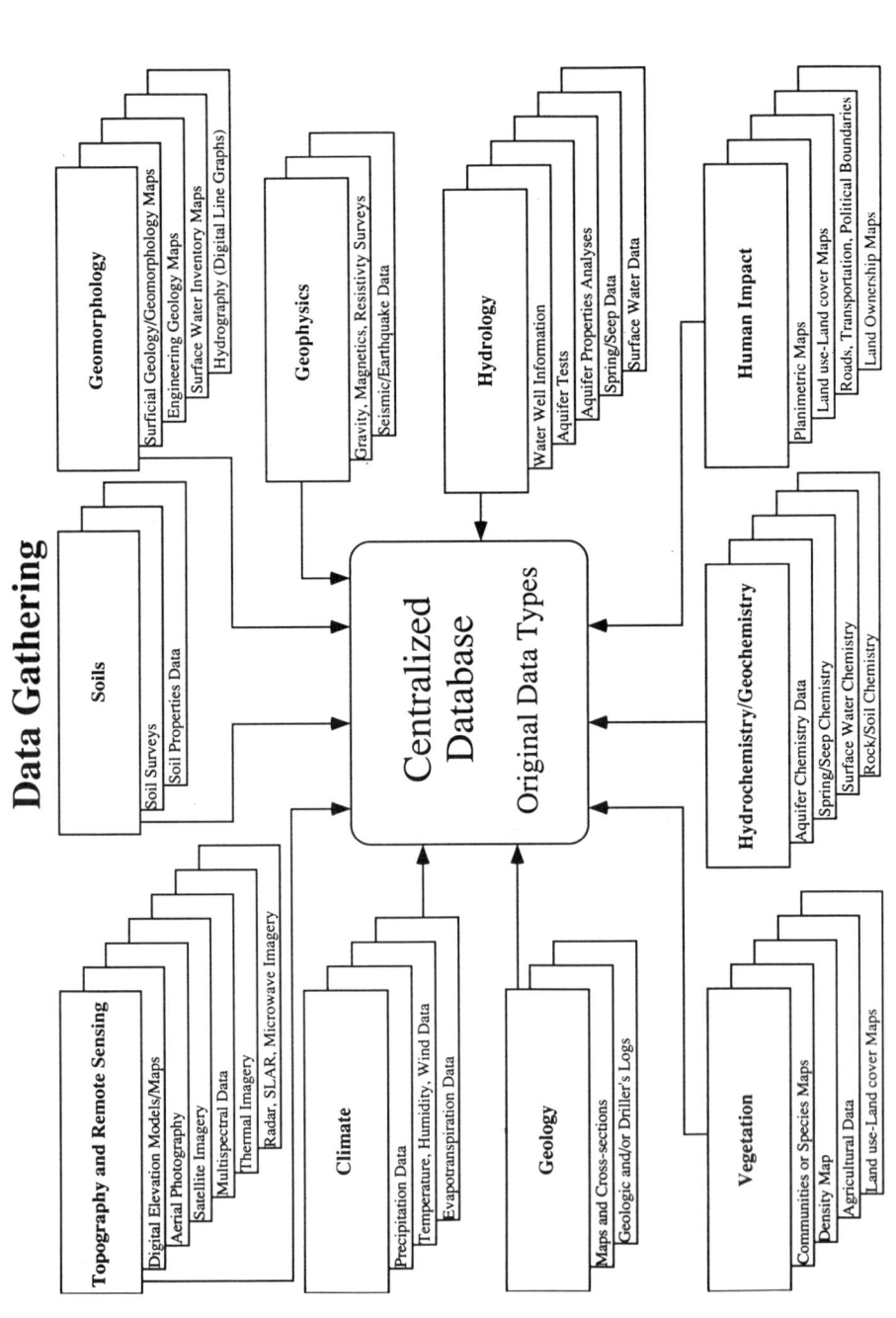

Fig. 4 Data categories and types stored in relational database that is linked to the graphical components of the 3D model.

example, they may be analyzed and interpreted as either porous media or fracture flow systems. For each case, the distribution and values of numerical flow model parameters are determined directly from the 3D hydrogeological data contained in the data base. These data include attributes gathered from aquifer tests, lithological and geological logs, and observations of degree of fracturing. The heterogeneous distribution of the flow parameters is estimated using stochastic procedures and probability distributions.

Groundwater recharge and discharge information gathered from satellite imagery, digital soils surveys, and climate data reports are compiled and interpreted using traditional 2D and 2.5D surface modelling techniques. Regional vegetation maps may be created from remotely sensed data using Intergraph's Imagestation Imager. These may be combined with maps describing surface hydrology, soils, and climate to determine potential regional recharge and discharge.

These data are then integrated to supply a series of layered input arrays suitable for processing by the numerical flow model. The model results are returned to the GSIS data base and visualized. The numerical model is then calibrated by making modifications to the input arrays on a node-by-node basis, or by revising components of the hydrogeological framework model, or the physical and hydraulic boundaries, fluxes, and flow parameters.

CONCLUSIONS AND POSSIBLE FUTURE WORK

Three-dimensional GSIS technologies are important tools for quantitative assessments of hydrogeological conditions. The role of GSIS is enhanced by interfacing with a variety of analytical techniques for subsurface characterization and numerical groundwater modelling. GSIS products support both spatial visualization and data management. Until now, the visualization of 3D subsurface features has been a major constraint to modelling. The ability to rapidly create and manipulate 3D images can materially accelerate the hydrogeologist's understanding of the subsurface.

A 3D approach to characterizing the regional groundwater flow system and estimating unknown boundary conditions for the Death Valley region should provide a reasonable simulation for the past and present regional hydrological regimes. This will provide a useful basis for studies of future groundwater flow systems that may result from changes in climate, geological framework, or water use.

REFERENCES

Bedinger, M.S., Sargent, K.A. & Langer, W.H. (1989) Studies of geology and hydrology in the Basin and Range Province, Southwestern United States, for isolation of high-level radioactive waste - characterization of the Death Valley region, Nevada and California: USGS Prof. Pap. 1370-F.

Raper, J.F. (1989) The 3-dimensional geoscientific mapping and modelling system: a conceptual design. In: *Three-dimensional Applications in Geographic Information Systems* (ed. by J.F. Raper), Taylor and Francis, London, 11-19.

Turner, A.K. (1989) The role of three-dimensional geographic information systems in subsurface characterization for hydrogeological applications. In: *Three-dimensional Applications in Geographic Information Systems* (ed. by J.F. Raper), Taylor and Francis, London, 115-127.

Turner, A.K. (1991) Applications of three-dimensional geoscientific mapping and modeling systems to hydrogeological studies. In: *Three-Dimensional Modeling with Geoscientific Information Systems* (ed. by A.K. Turner), NATO ASI Series C: Mathematical and Physical Sciences, vol. **354**, Kluwer Academic Publishers, Dordrecht, The Netherlands, 327-364.

Integrated three-dimensional geoscientific information system (GSIS) technologies for groundwater and contaminant modelling

THOMAS R. FISHER
Geographic Information Systems and Services Radian Corporation
8501 North Mopac Boulevard, P.O. Box 201088, Austin, Texas 78720-1088, USA

Abstract Overallocation of groundwater resources and the advent of groundwater quality regulation have turned attention to water quality and quantity analysis with particular emphasis on contaminant migration. Predictive modelling of contaminant plumes is a complex process requiring fusion of disparate data types into a meaningful and interpretable whole to create a geological framework. This geological framework may then be used to study subsurface processes and to supply support data on initial and boundary conditions to sophisticated groundwater and contaminant flow models. A 3D Geoscientific Information System (GSIS) is used for data fusion and data visualization. These systems provide continuous volumetric data structures which permit modelling of geological or other attributes in their true 3D spatial relationships. Case studies using commercially available 3D GSISs indicate the systems are as yet primitive in many ways, but are nonetheless highly useable when integrated with other computer technologies.

INTRODUCTION

With the full allocation or overallocation of groundwater resources and the advent of groundwater quality regulation the attention of hydrogeologists has turned to quantitative analysis of water quantity and quality with emphasis placed on contaminant migration (National Research Council, 1990). Predicting the movement of subsurface contaminants is a complex problem that begins with the integration and fusion of many disparate data types into a subsurface characterization. This means attempting to map geophysical, geochemical, hydrological and other site surveys into a multi-dimensional space where intersections of commonality may be found, i.e., a consistent solution to the problem (Olhoeft, personal communication, 1992). Once a suitable subsurface model has been achieved, analysis continues with the subsurface characterization data supporting appropriate groundwater or contaminant models (Moore *et al.*, 1992). The ultimate objectives are to determine the three-dimensional distribution, direction and rate of movement of contaminants in the subsurface.

This problem has no easy solution. The nature of the geological subsurface presents many obstacles to predictive modelling. Gross oversimplifications are often required to write manageable simulation programs. Not all physical processes or reactions that control contaminant migration in the subsurface at a site may be clearly known or understood and models of these processes are often insufficient for predictive purposes in the face of these complex heterogeneous and anisotropic environments (National Research Council, 1990). Additionally, the site's true characteristics may be only little understood, making selection of boundary and initial conditions difficult. Cost factors, the geography of a site, or concerns about disturbance of the hydrological system, may limit ability to obtain adequate sampling to quantify the natural spatial and temporal variability of a site.

The large number of parameters and data volumes to be dealt with in predictive hydrogeological modelling dictates the need for a 3D Geoscientific Information System or 3D GSIS (Raper, 1989; and Faunt & Kolm, 1991). Three-dimensional GSISs lend themselves to the iterative process of modelling (Turner, 1989) as well as the evolutionary nature of site characterization and remediation. They are a marriage of 3D computer graphic and visualization techniques with certain characteristics of more traditional 2D Geographic Information Systems (GISs). They provide a third dimension by accommodating subsurface data to afford solutions for efficient visualization, modelling, and interpretation of multiple geological or other attributes in their true 3D spatial relationships. The interpolated-imaged data can be extracted at selected locations as input arrays for traditional flow and transport models (Nichols *et al.*, 1992). Conversely, results from flow and transport models may be imported into the 3D GSIS for display and merge with the subsurface conceptualization (Fisher, 1993).

NATURE OF THE GEOLOGICAL SUBSURFACE

The science of hydrogeology involves conceptually understanding a site through the generation of models (Preslo & Stoner, 1991). The model itself is composed of site-specific data on initial and boundary conditions, material and fluid properties combined with numerical simulations of flow and transport. However, characterization of a site is always uncertain, for we are always sampling limited in some manner, such that the site's true spatial and temporal variability is never known completely. We must deal with incomplete and conflicting information over a range of scales; the parameters of importance changing as scales change (Van de Graaff & Ealey, 1989). However, many of the scale effects on rock properties are usually unknown.

We deal with two classes of "geo-objects" (Raper, 1989). Those that are "sampling-limited", i.e., the more samples the better defined the object (e.g., geological stratum); and those that are "definition-limited." Definition limited objects have no definitive natural "zero-edge" or boundaries that define the envelop of the object. An example is a contaminant plume. There is no definite "zero-edge" that can be defined because of the dispersion nature of the plume; a change of definition (e.g., change in the level of concentration of interest) may change the geometry. Subsurface geo-objects are both multi-dimensional and dynamic, their geometry, location and characteristics vary spatially and temporally. Obviously, they are heterogeneous. Scale effects may change their

PREDICTIVE MODELLING

Two of the main reasons for predictive groundwater modelling are optimization of placement of monitor wells at a site, and to predict the future behavior of contaminant plumes. The main obstacle in creating such models may be the disparity between simplifications required to write manageable computer codes and the great complexities that exist in real field situations (National Research Council, 1990). Most current models are two-dimensional and are inadequate for representation of complex groundwater and contaminant flow in heterogeneous media (Kolm et al., 1990).

The role of heterogeneity of an aquifer in simulation and predictive transport modelling has been discussed in Moore et al. (1992). They suggest that a systems analysis approach be taken. This allows for creation of coupled process models rather than single process approaches, and opportunity to incorporate the "truth of the physics" into the model. Turner (1989) suggests the process of hydrogeological analysis be considered in terms of four fundamental modules:
(a)   subsurface characterization;
(b)   three-dimensional GSIS;
(c)   statistical evaluation and sensitivity analysis; and
(d)   groundwater flow and contaminant transport modelling.

These modules are used in an iterative modelling process to integrate disparate data in a common spatial context for definition of a suitable geological framework model. This model can then be used to study geological and hydrological processes and for supplying parameters to sophisticated 3D groundwater flow and contaminant transport models.

THREE-DIMENSIONAL GEOSCIENTIFIC INFORMATION SYSTEMS (GSIS)

The nature of the subsurface, the evolutionary nature of site characterization, and the number of parameters required by complex groundwater and contaminant transport models demand a computerized system for integration of the many disparate and voluminous data sets created during site investigation. Kelk (personal communication, 1991) has outlined requirements for such a system, indicating what is required is a computer graphics system which allows interactive creation of spatial models of the physical nature of the subsurface. Such a system should provide capability to effectively model and visualize:
(a)   the geometry of rock and stratigraphic units;
(b)   three-dimensional spatial relationships between rock units;
(c)   variations in internal composition;

(d) displacement or distortion by faults, and other tectonic forces; and
(e) flow of fluids through rock and soil units.

These five points define the three key aspects required in the system (Fisher, 1993):
(a) geometry representation;
(b) geological framework representation; and
(c) process model incorporation.

Many of these requirements are already met in 3D GSISs, although they are yet somewhat primitive in nature and still evolving as commercial software applications. These systems provide continuous volumetric data structures and appropriate analytical functions without prematurely averaging the data. They may be able in some cases to provide data where adequate sampling density is too costly and allow conditions to be predicted in areas with no data or where non-intrusive survey techniques must be used (Fisher, 1993).

Most current 3D GSISs rely heavily on the power of computer graphics capabilities and stress geometry over interaction between geo-objects, but do attempt to provide solutions for creation, visualization, and analysis of complex geo-objects, spatial and entity relationships found in the real world (Fisher & Wales, 1991). Unfortunately, we are still at the stage of visualization of boundaries or visualization of internal attributes of an object rather than true modelling or interpretation. No one system yet meets all the needs outlined in the requirements, hence integration between multiple systems is required for complex problems. Most of these volume modelling or rendering methods rely on a semi-transparent depiction of both surface and internal features and have no manual equivalents (Van Driel, 1989). Solid volume techniques range from 3D representations using relatively simple polygon meshes and piecewise linear interpolation, complex 3D gridding and isosurface techniques, and voxels, to curve, surface, and solids modelling based on advanced mathematical functions. Most of these methods may be categorized as either volume or surface representations (Fig. 1.) Detailed discussions of these methods may be found in papers by Jones & Leonard (1990), Fried & Leonard (1990), and Fisher (1992).

CASE STUDY

Three-dimensional GSISs have been proven in actual site characterization studies. In one instance a 3D GSIS was utilized to determine volume and extent of several plumes of organic compounds which leaked from a plant site in the Industrial Corridor of the Northeastern United States. More than 100 monitor wells and soil borings were available for study. Almost all of the data were located on the plant site or near its periphery. The objective was to determine if the contaminants had moved from the plant site and effected populated areas near the plant.

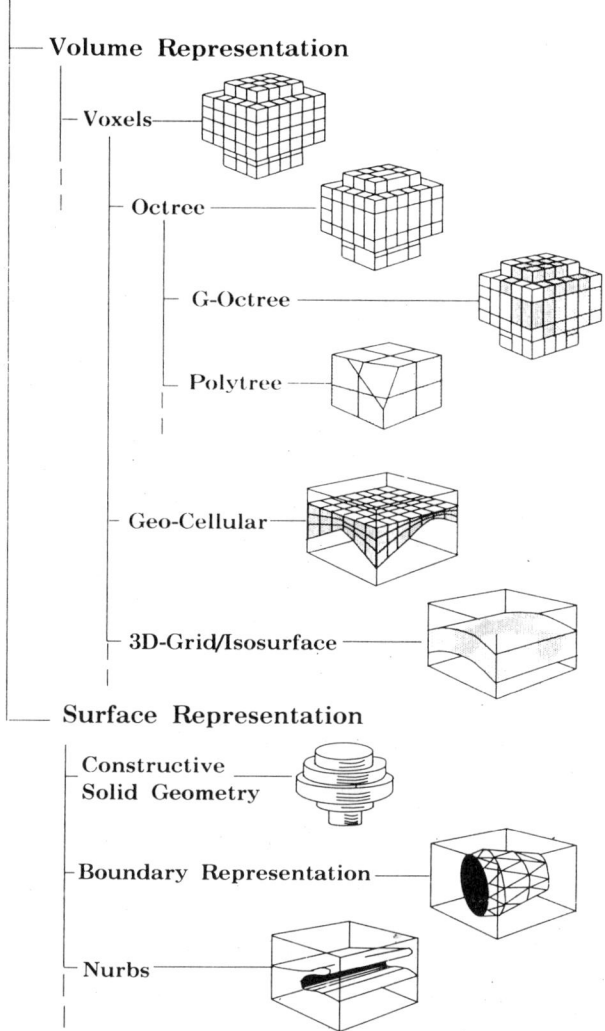

Fig. 1 Three-dimensional spatial representation methods (after Fisher, 1992).

An area of 1 by 0.5 km was studied to a depth of 75 m. A geological framework was first constructed in 3D based on existing regional and local geological data. The model was corrected and reconciled using monitoring well and soil borings data. Monitoring well and boring data were also used to provide information on the extent of contaminants in both the vadose and saturated zones. The scattered data set was used to interpolate values to a dense 3D grid lattice work representing the study area, using a 3D minimum tension algorithm. The resulting grids were "isosurfaced" using the "marching cubes" technique. Computed grids were constrained by geological horizons represented as 2D grid surfaces. The results were then brought to the screen for visualization and further study. New data created in the process of interpolation became available to supply groundwater and contaminant transport modelling systems. Results of numerical modelling was redisplayed within the original model framework. Some extrapolation was allowed in the process of populating the 3D grid lattice, with varying results.

The results of the modelling suggested a thin plume of contaminant existing in the vadose zone, resting atop the water table. Beneath this plume, in the saturated zone, contaminants were observed to form a multi-storied plume, which had spread along various horizontal permeability barriers (aquitards) before continuing their way downward to a zone of relative impermeability. Previous 2D contouring of the plume data, which forced averaging of the data over the entire geological interval sampled, had indicated a localized "semi-circular" area of contamination restricted to the plant site. Visualization of the data in 3D indicated that the contaminant plume was possibly moving off site and down geological dip, at a slight angle to regional groundwater flow.

The final models were used to determine placement of additional monitoring wells and recommend additional courses of action in remediation of the site. This model was also used to determine volumes of contaminant present in the subsurface and as a baseline model to monitor movement of plumes as remediation of the site begins. Accurate modelling of the site indicated possible sources of contamination based on relationships seen between water table elevations, engineered structures in the area, and contaminant concentration. Relationships between contaminant distributions in the subsurface and lithological variations in the strata were also observed, but not to the extent expected. Extrapolation routines tended to provide false data regarding contaminant levels in areas of sparse control.

REFERENCES

Faunt, C.C. & Kolm, K.E. (1991) GIS: a new tool for geohydrological characterization. *Association of Engineering Geologists 34th Annual Meeting Proceedings*, Chicago, 641-650.

Fisher, T.R. (1993) Concepts and approaches in multi-dimensional geologic modeling (to appear in *Die Geowissenschaften*, Essen, Germany).

Fisher, T.R. (1992) Use of 3D geographic information systems in hazardous waste site investigations In: *Environmental Modeling with GIS* (ed. by M.F. Goodchild, B. Parks & L.T. Steyaert), Oxford University Press, New York.

Fisher, T.R. & Wales, R.Q. (1991) Rational splines and multi-dimensional geologic modeling. In: *Three Dimensional Computer Graphics in Modelling Geologic Structures and Simulating Process* (ed. by R.Pflug & J.W. Harbaugh), Springer-Verlag, Heidelberg, 17-28.

Fried, C.C. & Leonard, J.E. (1990) Petroleum 3D models come in many flavors. *Geobyte* 5(1), 27-30.

Jones, T.A. & Leonard, J.E. (1990) Why 3D modeling? *Geobyte* 5(1), 25-26.

Kelk, B. (1991) *Personal Communication*.British Geological Survey

Kolm, K.E., Turner, A.K. & Downey, J.S. (1990) Design of a three-dimensional computer model for the regional ground-water flow system, Southern Nevada and Death Valley, California, USA. *Proceedings Sixth International Congress International Association of Engineering Geology* (ed. by D.G. Price), A.A. Balkema, Rotterdam, 55-64.

Moore, I.D., Turner, A.K., Wilson, J.P., Jenson, S.K. & Band, L.E. (1992) GIS and land surface-subsurface process modelling. In: *Environmental Modeling with GIS* (ed. by M.F. Goodchild, B. Parks & L.T. Steyaert), Oxford University Press, New York.

National Research Council (1990) *Ground Water Models: Scientific and Regulatory Applications*. National Academy Press, Washington, D.C.

Nichols, R.L., Looney, B.B. & Huddleston, J.E. (1992) 3-D digital imaging: revealing the location, depth and concentration of subsurface pollutants. *Environ. Sci. Technol.* 26(4), 642-648.

Olhoeft, G.R. (1992) *Personal Communication*.US Geological Survey.

Preslo, L.M. & Stoner, D.W. (1991) The overall philosophy and purpose of site investigations. In: *Practical Handbook of Ground- Water Monitoring.* (ed. by D.M. Neilsen), Lewis Publishers, Chelsea, Michigan, 69-95.

Raper, J.F. (1989) The 3-dimensional geoscientific mapping and modelling system: a conceptual design. In: *Three Dimensional Applications in Geographic Information Systems* (ed. by J.F. Raper) Taylor and Francis, London.

Turner, A.K. (1989) The role of three-dimensional geographic information systems in subsurface characterization for hydrogeologic applications. In: *Three Dimensional Applications in Geographic Information Systems* (ed. by J.F. Raper) Taylor and Francis, London, 115-127.

Van Driel, J.N. (1989) Three dimensional display of geological data. In: *Three Dimensional Applications in Geographic Information Systems* (ed. by J.F. Raper) Taylor and Francis, 1-9.

Van de Graaff, W.J.E. & Ealey, P.J. (1989) Geological modeling for simulation studies. *The Am. Assoc. of Petroleum Geologists Bulletin* **73**(11), 1436-1444.

Integration of three-dimensional groundwater modelling techniques with multi-dimensional GIS

JUDITH SCHENK & KEITH KIRK
Office of Surface Mining, 1020 15th Street, Denver, Colorado 80202, USA

EILEEN POETER
Colorado School of Mines, Golden, Colorado 80401, USA

Abstract The relationship between quality of overburden material at a mine site and the rise of water levels in the mine was explored using multi-dimensional GIS. Results of a groundwater model indicate steady-state water levels in the mine will be reached in approximately eight years. Three-dimensional analysis of geochemical data of the overburden material indicate acid-producing material in the overburden will mobilize sulfuric acid and iron in the groundwater during that time. A merged data display that includes transient water levels and three-dimensional analysis of the geochemistry of the overburden material indicate 1.39 million m^3 of acid-producing material will be exposed to oxidation during the eight years the mine is filling with water. Approximately 0.47 million m^3 of the acid-producing material will remain above the water table when steady-state water levels are reached. This portion of the acid-producing material will be exposed to oxidation as water moves through the unsaturated zone to the water table.

INTRODUCTION

We explored the relationship between the geochemistry of subsurface material overlying a coal deposit and rising water levels in a surface coal mine located in the southeast United States. Geochemistry of the subsurface material (commonly called "overburden" because it overlies the coal deposit) in the northern area of the mine was spatially analyzed in three dimensions. Analysis of the mine site required an integrated study of the hydrological and geochemical processes at the site.

This study has three components; hydrological analysis, geochemical analysis and a combined analysis of hydrological and geochemical information. Integration of temporal data with regard to water levels from the groundwater model and three-dimensional data of the quality of the overburden material were merged to provide a multidimensional analysis of conditions at the mine. Dynamic Graphics Earth Vision software was used in each part of this study.

GROUNDWATER MODEL

A two-dimensional finite-difference groundwater model was constructed of the mine using the United States Geological Survey groundwater modelling code, MODFLOW (McDonald & Harbaugh, 1988). The purposes of the groundwater

model are:
a) estimate the post-mining transient and steady-state groundwater levels in the mine;
b) estimate when steady-state water levels will occur;
c) determine the spatial relationship between acid-producing material in the overburden and water levels in the mine through time.

A finite difference grid of 38 rows and 19 columns overlays the area with active cells being those that overlay the mine area (Fig. 1a). The following parameters were examined to construct the groundwater model of the mine:
a) physical boundaries of the system;
b) recharge and discharge of water in the mine;
c) hydraulic parameters of the mine spoil.

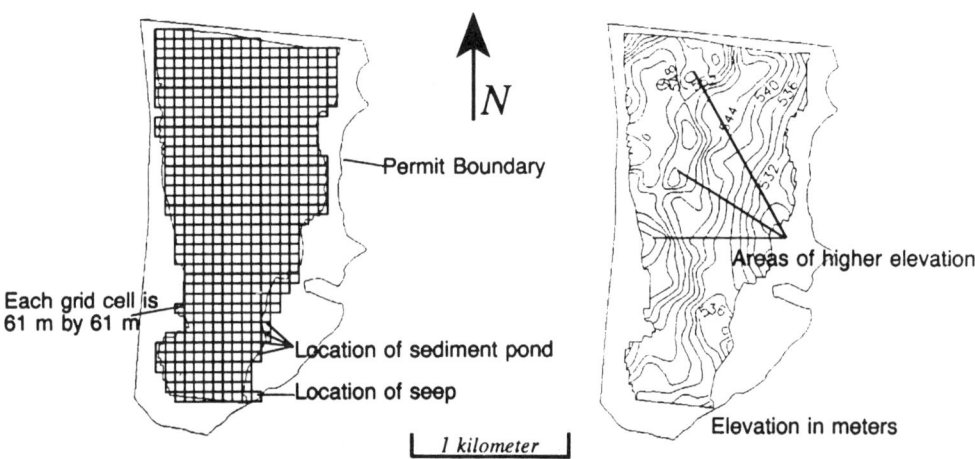

(a) Groundwater model grid   (b) Bottom of mined area

Fig. 1 a) Groundwater model grid and location of the sediment pond and groundwater seep. b) Bottom elevation of mine.

Physical boundaries

We assume the surrounding parent rock, composed of sandstone and shale, has a hydraulic conductivity much lower than the hydraulic conductivity of the overburden material backfilled in the mined area (mine spoil). Therefore, the highwall of the mine constitutes a no-flow boundary to the mined area because the amount of seepage out of the mine into the surrounding parent rock is much less than the seepage that will occur in other areas. The mine pit floor, composed of shale, constitutes a second no-flow boundary. Areas of higher elevation on the pit floor will obstruct the movement of groundwater as the mine begins to fill with water (Fig. 1b).

Recharge and discharge

A recharge rate of 30.5 cm year^{-1} (9.7x10^{-7} cm s^{-1}) was used in the groundwater model. We assume dewatering in the area adjacent to the mine has occurred and therefore, seepage into the mine from the surrounding bedrock is negligible. Groundwater exits the mine in the southeast area by seeping into a sediment pond (Fig. 1a). A seep discharges at the southeast corner of the mine into the adjacent river (Fig. 1a). We consider these the principal exit areas for groundwater.

Hydraulic parameters

No data were available to determine the hydraulic characteristics of the mine backfill material. Studies by Hawkins & Aljoe (1991, 1992) relate the hydraulic conductivity of mine backfill to the composition of the mine spoil. Hydraulic conductivity estimated from pumping tests ranging from 0.006 to 1.1 cm s^{-1} have been observed in mine spoil composed of 51% to 75% sandstone. Mine spoil at this site is approximately 50% sandstone and 50% shale. We used a hydraulic conductivity value of 0.018 cm s^{-1} and a specific yield of 0.2.

GROUNDWATER MODEL SIMULATIONS AND RESULTS

Because the groundwater regime is currently in a transient state, the groundwater model was run in reverse. A steady-state model run was completed to determine the water level at steady state. A transient simulation was done by using the steady-state water levels as the starting water levels and removing water at the recharge rate of -9.7x10^{-7} cm s^{-1}.

Present day water levels in wells located in the mine spoil indicate that we are currently at 8 years prior to steady-state conditions (Fig. 2). Therefore, steady-state water levels will be reached in the year 2001. Groundwater is impounded in the northwest area of the mine as the mine fills with water.

THREE-DIMENSIONAL ANALYSIS OF GEOCHEMICAL DATA OF OVERBURDEN

Multi-dimensional analysis was used to quantify the three-dimensional spatial analysis of geochemical data of the overburden material in the northern area of the mine (Fig. 3). The data include a three-dimensional location tied to a global coordinate system (easting, northing, elevation), and geochemical data or acid/base potential (ABP) measured in tonnes calcium carbonate equivalent per kilotonne of overburden (tCaCO$_3$ kt^{-1}). If the ABP is less than -5 tCaCO$_3$ kt^{-1}, then there is a likelihood of the subsurface material to generate sulfuric acid (H$_2$SO$_4$)

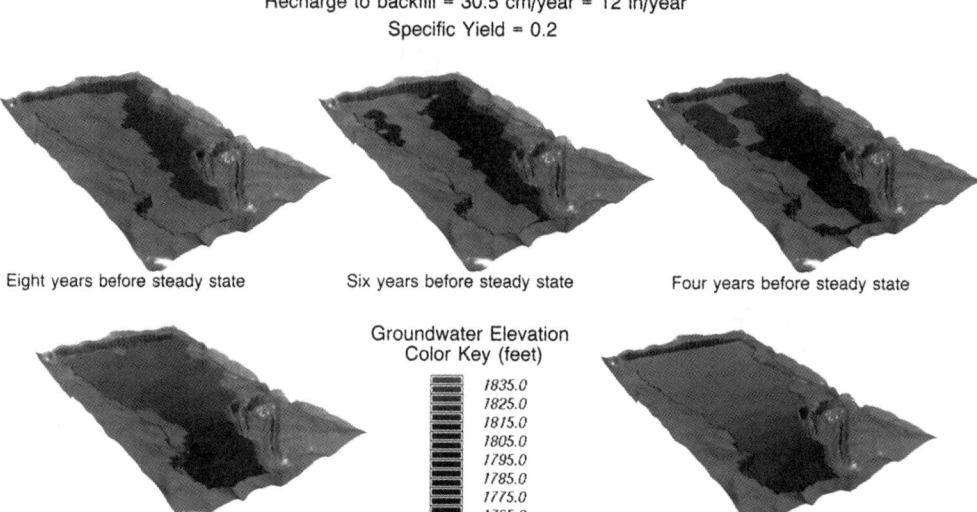

Fig. 2 Transient water levels in the mine.

and dissolved ferric and ferrous iron through the oxidation of pyrite (FeS_2) and the material is considered to be acid-producing.

The distribution of acid-producing material at the site is controlled both by the paleodepositional environment and local geological structure. To account for geological structure, elevations for the bottom of the coal seam obtained from bore hole data were used to create a two-dimensional structure contour grid (x, y, elevation) using minimum tension biharmonic smoothing gridding. The final grid was optimized by minimizing residuals through residual analysis to assure that the grid honored the scattered structural data. This two-dimensional grid was used to control the three-dimensional grid of toxic material so that it conformed to local geological structure.

The three-dimensional spatial distribution of toxic material was determined by creating a three-dimensional grid (x, y, elevation, ABP) representing the concentration of ABP in the subsurface above the coal (Fig. 3a). As with the two-dimensional grid the three-dimensional grid was optimized by minimizing residuals through residual analysis to assure that the three-dimensional grid honors the three-dimensional scattered data.

Because this is a true three-dimensional grid, the display can be manipulated so that the three-dimensional contours in space can be examined. The visual display shows qualitatively that a considerable proportion of the overburden contains acid-producing material as illustrated by exposing the -5 $tCaCO_3$ kt^{-1} contour (Fig. 3b). Total volume of the acid-producing material in this area of the mine is 2.06 million m^3 as calculated using Earth Vision. Calculation of this volume would be time-consuming using other methods.

Integration of three-dimensional groundwater modelling techniques 247

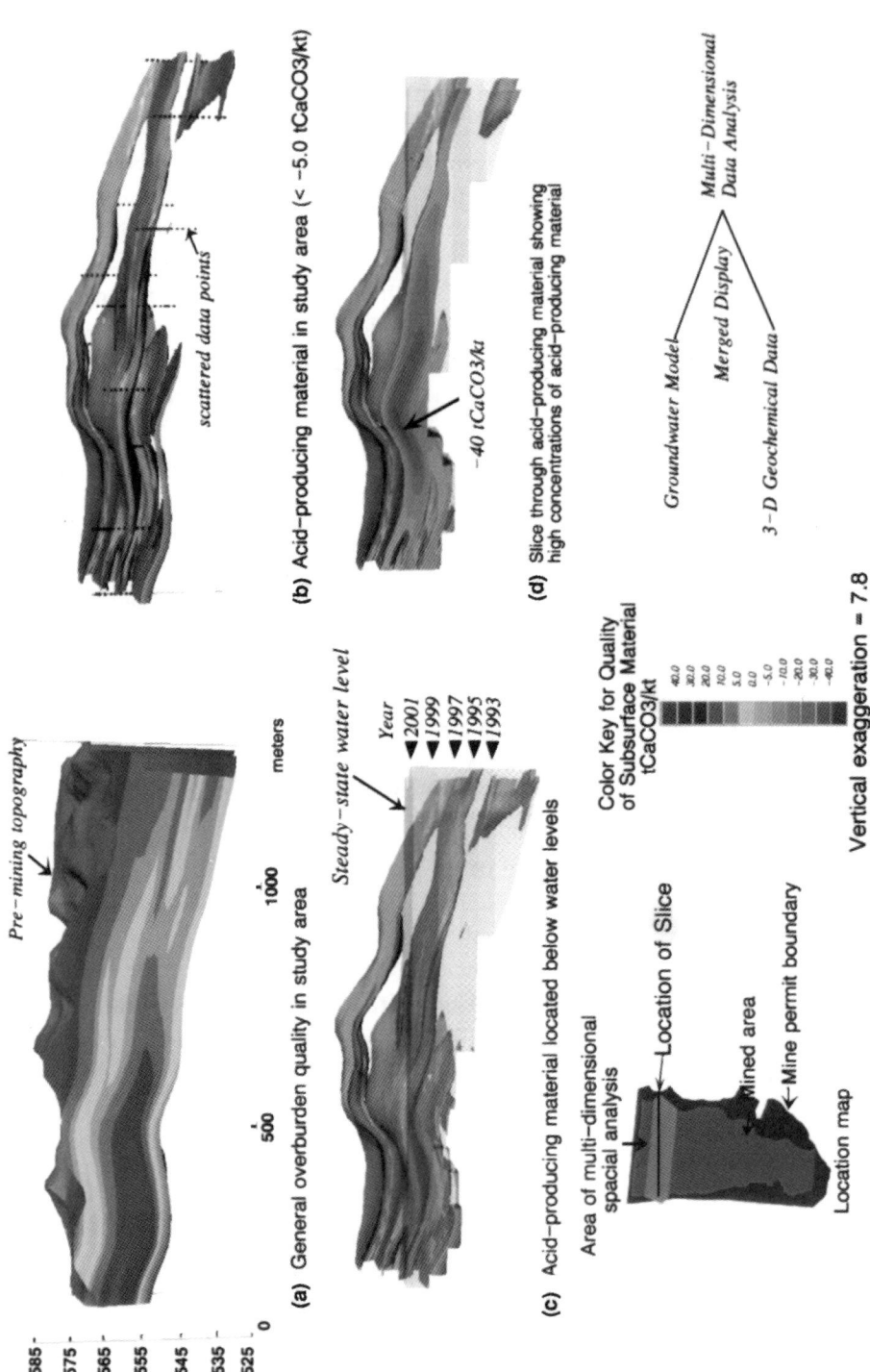

Fig. 3 Multi-dimensional analysis of groundwater model results and overburden quality.

MERGED DATA DISPLAY AND COMBINED ANALYSIS

The relationship between the temporal data of transient and steady-state groundwater levels, topography, geological structure and spatial distribution of toxic material is critical to the quantify the potential hydrological impacts. These interrelationships were quantified by creating a multi-dimensional model that combines topographic, geological structural, groundwater flow model and subsurface toxicity information. This merged model allows for the determination of volumes of various concentrations of toxic material between structural zones and within transient time periods to be calculated and simultaneously quantify the spatial relationships between all of these variables.

Oxidation of pyritic material as found in this acid-producing material in the northern area of the mine generally requires the presence of oxygen. Once oxygen is removed from the reaction process the oxidation of pyritic material is greatly reduced. Therefore the pyritic material is commonly placed below the water table to reduce contact with oxygen. Some oxygen will remain in solution within the groundwater through infiltration from precipitation. Therefore not all pyritic oxidation will be stopped. The spatial relationship between the pyritic material and steady-state and transient groundwater levels is required to determine how much pyritic material will remain above the groundwater table and for how long. This information was quantified using the merged display of the three-dimensional model of the geochemical data and the three-dimensional groundwater model (two-dimensional flow through time).

RESULTS OF MULTI-DIMENSIONAL ANALYSIS

The results of the multi-dimensional spatial quantification indicate approximately 0.47 million m^3 of pyritic material will remain above the steady-state groundwater table and 1.3 million m^3 of pyritic material will eventually be below the steady-state groundwater level (Fig. 3c). A transparency mode used in the display allows for better visualization of the relationship between transient water levels and acid-producing material (Figs 3c and 3d).

Steady-state water levels will occur in approximately 8 years from the present. Therefore, portions of the 1.39 million m^3 will be exposed to oxidation during that time period. Quantities of acid and dissolved iron will be added to the hydrological system and contaminate surface water and groundwater. In addition, acid-producing areas of up to -40 $tCaCO_3$ kt^{-1} are found in the overburden (Fig. 3d). Locating these areas will aid in formulating a plan to mitigate the hydrological impacts.

CONCLUSIONS

Without the use of multi-dimensional analysis of spatial and temporal data quantification of these variables at this site would not have been possible within a reasonable time period. This analysis enabled us to demonstrate convincingly that mining practices needed to be modified to reduce impacts on the hydrological system at this site. Use of multi-dimensional GIS tools have proved invaluable as a planning tool for predicting environmental consequences related to coal mining in the United States.

REFERENCES

McDonald, M.G. & Harbaugh, A.W. (1988) A modular three-dimensional finite-difference ground-water flow model. US Geological Survey. *Techniques of Water-Resources Investigations of the United States Geological Survey, Book 6, Chapter A1.*

Hawkins, J.W. & Aljoe, W.W. (1991) Hydrologic characteristics of a surface mine spoil aquifer. In: *Proceedings Second International Conference on the Abatement of Acidic Drainage, NEDEM - MEND.* Montreal, Quebec, Canada. 47-68.

Hawkins, J.W. & Aljoe, W.W. (1992) Application of aquifer testing in surface and underground coal mines, *Proceedings FOCUS conference on Eastern Regional Ground Water Issues.* National Ground Water Association, Dublin, Ohio. 541-555.

Integration of space and time in GIS applications: one step further to model reality?

BART KUSSE
National Institute of Public Health and Environmental Protection (RIVM), P.O.Box 1, 3720 BA Bilthoven, The Netherlands

COLIN GRAY
Intergraph European Headquarters, Siriusdreef 2, 2130 AH Hoofddorp, The Netherlands

HENK J. SCHOLTEN
Free University, Dept. of Regional Economics, De Boelelaan 1105, 1081 HV Amsterdam, The Netherlands

Abstract The use of a computer simulation model in the protection of the environment has become almost a day-to-day practice for researchers. Not only do they study the processes above and on the earth's surface, but also the subsurface is modelled. In addition, other models need to be considered in the protection of the environment: e.g. ecological modelling, modelling of air quality and pollution or modelling of the quality of the groundwater. Because there is a spatial component in each of these types for simulation models, Geographical Information Systems (GIS) are often used to store, integrate and analyze the data, and as a presentation platform for the modelling results. Different types of models which are used in environmental modelling are briefly described. In addition, because they all make use of the parameters z and t, we will discuss a more integrated approach to the use of time and space in modelling the environment. Finally a small project has been carried out to investigate the more practical side of the developed theoretical insight.

INTRODUCTION

One of the most challenging areas in which people are discovering the possible added value of GIS is public and environmental health. Measures to improve health and the quality of life are now receiving greater attention along with the need to protect and improve the environment. Environmental and health problems - which are diverse and complex - cross national boundaries and need to be dealt with internationally.

Accurate, adequate and accessible data and sophisticated tools are necessary for evaluating existing and future environmental and public health risks. One of the first challenges is the improvement of the quality, reliability, ease of access, and utility of currently held data on a local, national, regional, and international basis. One of the most important questions arising in public and environmental health therefore concerns the type of instruments that can be utilized to devise reliable and scientifically valid methods of rapid assessment, to be of help in health and environmental research and the planning, monitoring and evaluation of health and environmental programmes.

The information may be of various types, extensive in quantity, variable in quality, referring to areal units of different size, and be held by a large number of different data holders. All relevant information must be stored, managed, made available, analyzed and presented in a suitable form for use at different stages in the research and planning processes.

The continuous threat of environmental decline is the reason for major research for a considerable number of institutes and researchers. Almost every part of the environment is monitored by man and computers, and computer simulation models are developed to describe the processes that take place in each environmental compartment. These simulation models will be addressed to as *process-models* and the compartments that can be distinguished are: air, soil, and water. The quality of each of these compartments is a precondition for the quality of life of mankind. It is also clear that the understanding of the processes in one compartment is a first step to the overall approach which leads to an integration of the different processes through the different compartments and through different dimensions.

In this paper we start with a discussion of examples of processes in the different compartments: air, soil, and groundwater as a specific part of the soil compartment. Then we give examples of approaches in which an integration of the different processes and compartments takes place. In all sections the time and z-dimension play a role. We will give an overview of how these two dimensions are treated. In the last section, we try to take one more step in the direction of modelling the reality. However, the results of this exercise will be discussed at the conference itself. In the conclusions and recommendations, we give some possible tracks for future development in modelling the environment.

ENVIRONMENTAL COMPARTMENTS

Using his common visual approach, man is able to distinguish between the different compartments. In this way, he makes a difference between the air, his direct environment and the soil. Of all compartments, air is probably the most fascinating, because man is surrounded by it and dependent of it. As a result of this dependency, and as the visibility of air pollution (e.g. the smoke that comes from a steam engine or a waste incinerator) was directly linked to pollution in general, it was recognized at an early stage that forces had to be combined to try and avoid future health risks.

"Air pollution has existed for many centuries. However, man's activities now attack the natural fresh air so severely that a pollution control which aims at abatements in several areas of the globe is necessary to prevent a threat to human health, animals, plants and ecosystems......" (NATO/CCMS, 1980)

For the modelling of air pollution, a number of simulation models can be used which range from 1D to 3D. For the fourth dimension it is also possible to generate results for given timesteps. These models are used for a variety of purposes and a variety of chemical compounds. The basic concepts of air pollution consist of a sequence of interrelated processes (Fig. 1). Air pollutants are emitted from sources, then they are transported, diffused, and occasionally chemically transformed by means of various atmospheric mechanisms (Koussoulakou, 1990). The course of these process the pollutants will affect the concentration of certain chemical compounds and these, in turn, will have an effect on public health.

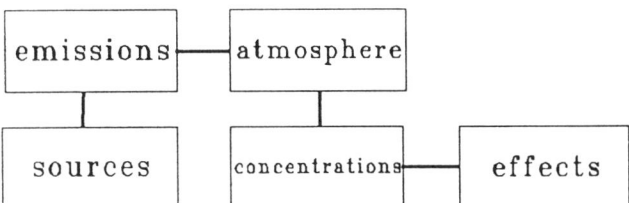

Fig. 1 Elements of Air Pollution (Koussoulakou,1990).

The purpose of these process-models is to establish a mathematical description of the processes which generate concentrations of pollutants in the atmosphere. They all take the z-dimension into account, and, depending on the type of model, this z-value can be divided into just one or more layers, ranging from 3 to 60 km. Also the size of the area for which the calculations can be done differs depending on the type of model. It can for example be either a city, or an area with a length of 1000 km. The amount of input is equal to the amount of detail needed by the model and the amount of detail in the result. A large amount of data is needed on actual meteorology (e.g., wind-speed, rainfall) and the type and source of emission that contributes to the amount of the chemical compound under consideration. Due to the rapid changes in quality, air process-models are very dynamic in character. This makes the time parameter t most important in air transport modelling. Only by correctly considering this parameter, can any air monitoring network accurately input data to the model so that it correctly provides air pollution forecasts for given times (e.g., at rush hours or smog episodes).

MODELLING OF GEOLOGICAL SURFACES AND PROPERTIES

The modelling of geological surfaces or properties in 3D has long been a part of the geologists' workflow but until recently computer techniques were not sophisticated enough to produce the type of 3D functionality required. Some geologists are concerned with a correct three-dimensional interpretation and representation for example in the modelling of a geological horizon. Once the 3D model is established, consideration can be given as to how to add extra dimensions such as the distribution of porosity within a reservoir or aquifer.

Other geologists, particularly structural geologists are concerned with time as a fourth dimension. In this case they wish to map the evolution of a sedimentary basin and its geological structures through time.

There are however various considerations. To begin with, the surfaces the geologist is trying to model are below the surface of the earth and so the model can never be fully compared with reality. Only in opencast mining where geological surfaces are shallow and are frequently exposed during extraction can the geologist have the chance to compare his map with the real surface.

A second problem is that much of the raw data of the geologist comes from boreholes and as a result it is sparsely and unevenly distributed in a given area. This is particularly so in petroleum geology where boreholes are primarily located on geological, geophysical, engineering and economic factors. This means that the data may not be ideally located on mathematical grounds adding extra uncertainty in the computer simulation of such data. The problem is most acute where large areas have to be modelled for example in regional maps of the North Sea. The problem is much less in areas of high data density such as individual oilfields, particularly those where modern 3D seismic has been acquired.

Despite the above considerations, geologists have been enthusiastic in trying to model their data. Sophisticated computer techniques are now available to geologists that firstly mimic established manual methods and secondly provide advanced methods of computation and visualization.

Originally 2D contour maps and 3D block (isometric) diagrams were constructed manually. These techniques were very useful but time consuming and so were amongst the first to be computerized. Today a number of commercially available programs are in routine use for this purpose.

This is achieved through the production of a 'digital terrain model' (DTM) either by various triangulation algorithms or by various gridding methods. The resulting output can be a planar map (2D) or a 3D wireframe model (Fig. 2). In the three-dimensional wireframe, the surface is generally displayed as contours, placed in their true 3D positions. This gives a good representation of the 3D nature of the model and can be rotated to display the structure to its best advantage. There are however several disadvantages to the 3D wireframe.

The first is that the model is not continuous for it has no values between the contours that constitute the wireframe. The geologist can only analyze the model on the contours and not in between.

The second disadvantage is that the DTM is generated only for a single surface thus modelling x, y and z. It cannot hold multiple values for the same x,y coordinates (i.e., model $x,y,z1,z2$). Thus, the DTM cannot easily model the intersection of two surfaces together, such as the intersection of a fault or an unconformity with a geological horizon or the spatial distribution of porosity throughout a reservoir for example. DTM's cannot model reverse faults or overhanging salt walls either. These limitations can be overcome but it requires careful and time-consuming work by the geologist. Finally, a third

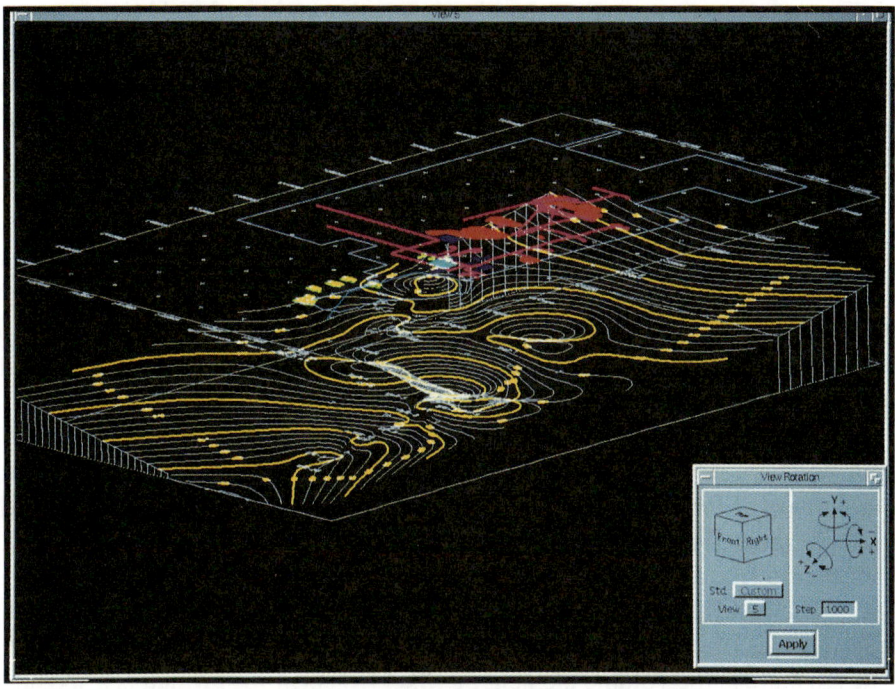

Fig. 2 Example of a 3D wireframe model, generated by Intergraph GIS.

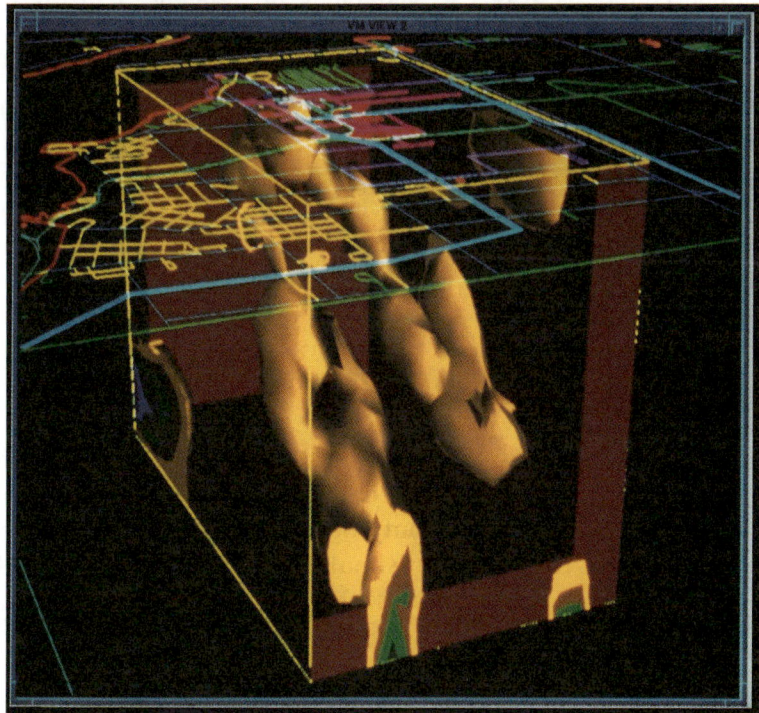

Fig. 3 Visualized distribution of chlorine concentration, generated by Intergraph Voxel Analyst (Intergraph, 1992).

disadvantage is that we cannot readily model x,y,z and t with DTM's.

Despite these problems, DTM's are in regular use by geologists in a variety of applications. With the establishment of DTM's the geological industry is now looking to three- and four-dimensional simulations and visualization as the next step forward in modelling techniques.

A logical extension of geological modelling is the inclusion of the z coordinate as an independent variable. In a four-dimensional trend series, for example, a dependent variable such as chlorine concentration in an aquifer is mathematically regressed along east-west, north-south and elevation coordinates. As a result, contour lines appear as envelopes containing points of equal value and the completed analysis is readily visualized as a 'solid' intrusion (Davis, 1973).

A variety of mathematical programs are now available to generate a 3D grid of a geological model. Discussion of these programs and their mathematical foundation is beyond the scope of this paper but the models include: water flow programs, kriging, reservoir simulation programs, finite element models, and many others discussed by other authors at this conference.

The resulting model is capable of storing many more attributes of the data $(x,y,z1,z2.....zn)$ and can use a variety of data structures. One example of a data structure is the volume element or voxel (Raper, 1989) which is a six-sided element (cube, cuboid etc.) The voxel can be irregular in shape so that it can better model a natural feature such as a fault. The term geo-voxel has been introduced for voxels that are geographically referenced, that is, they have true world coordinates not local arbitrary coordinates (Intergraph, 1992).

An important part of the workflow is the visualization of the 3D model. This allows the geologist to display and manipulate the model in a variety of ways so that he may better understand the relationships of the data under study. In Fig. 3 the distribution of chlorine concentration in an aquifer is visualized by the MGE Voxel Analyst software (Intergraph, 1992). In this example, $z1$ and $z2$ are the top and bottom surfaces of an aquifer and $z3$ is the chlorine concentration.

GROUNDWATER MODELLING

From the point of view of a hydrologist, the subsurface compartment is often referred to as the *soil* compartment. This compartment is considered to be the one most complicated, not only because we cannot "see" the geological formations that influence our simulation model or the detailed interaction between clay-particles, but because there is a variety of processes taking place.

The soil compartment is generally divided in two zones: the unsaturated and the saturated. In each of these we can distinguish hydrological-, chemical- and hydrochemical processes. Over the last decades the interest in groundwater flow has changed, from determining the amount of groundwater available as a resource for drinking water, to establishing methods and models to measure the quality of groundwater. For these purposes hydrological models have been developed to describe the flow of groundwater through permeable (sand) and less permeable layers (loam, clay). And, with the increasing threat of pollution to soil and groundwater, there is also a threat to the amount of groundwater that can be used for drinking water. Since this resource is far from infinite (as was found out in the Western part of The Netherlands when farmers were confronted with the problem of salt water intrusion from the sea), the last decade of groundwater modelling in the Netherlands has mainly been spent on modelling the quality of the groundwater. This was mainly due to the fact that "suddenly" a number of chemical contaminants were found in the drinking water which could not be extracted by the conventional treatment of drinking water. Up till then, there were no restrictions to the use of fertilizers, pesticides, and herbicides in agriculture so the source of this contamination could be readily explained.

To avoid this type of unforeseen circumstances in the future, considerable effort is spent in modelling the quality of groundwater. For this purpose a groundwater monitoring network has been set up by the Dutch National Institute of Public Health and Environmental Protection (RIVM, 1984). This monitoring network provides us with information on the quality of groundwater under a certain type of soil with a certain type of land-use at a certain moment in time.

Now we can benefit from years of research on groundwater flow. Most of the processes described are translated into groundwater simulation models which are developed and used to gain a better insight in the possible effects of all the human activities on the quality of groundwater. Depending on the geohydrological situation, either 1D, 2D or 3D models are applied. The 1D model is used for instance to trace the concentration of a contaminant along a pathline. The 2D model is often used in favour of the 3D model because the equations are less complex and the amount of data that has to be collected is acceptable.

If, however, the amount of data to be handled is no problem, the use of a true 3D process-model will provide the researcher with a better understanding of his problem at hand. For this purpose we can consider voxels. As with all groundwater modelling, the area of interest is changed into a grid-like structure, but also in the z-dimension the diversity of the layers is taken into account.

To obtain an insight in the results of the calculations and also of the actual processes taking place, at this moment the results are presented in concentration contours for each layer or cross-section.

However, how does the researcher interpret all these separate figures into a three-dimensional picture? This problem can be overcome by presentation software. But because of the linkage between the results of the process-model and its position on earth, the availability of these facilities within a GIS would be most appropriate.

As described earlier, also the parameter t is of importance in groundwater modelling. It can, for instance, give an indication of how the quality of groundwater will evolve in time, considering the current and future land-uses. In this way different policies can be compared with each other (Van Duijvenbooden et al., 1989; Kusse, 1990).

INTEGRATION OF THE DIFFERENT COMPARTMENTS

Since the publication of the Brundtland Report (World Commission on Environment and Development, 1987) a global concern has arisen on ecologically sustainable economic development. The crucial question here is whether economic progress is compatible with the ability of future generations to meet their own needs.

Clearly, the evaluation of the future against the present along two dimensions (economic progress and ecological sustainability) is far from easy. For instance, various hazardous waste materials do not only have an ad hoc effect on the environment, but may also have a long term non-reversible effect, with the result that future generations may be deprived of feasible desirable environmental options.

In order to answer the question four stages would have to be considered:
(a) environmental monitoring (inventory of hazardous materials in the external environment);
(b) biological monitoring (identification of causal chain mechanisms in living organisms);
(c) biological effect monitoring (assessment of impacts on living organisms); and
(d) health monitoring (evaluation of impacts on public health).

The notion "sustainable" has been chosen by several international and national institutes (e.g., the European Commission) as a concept of global development that is fair to the future and environmentally acceptable. They use the concept to develop a methodological framework on how to visualize, operationalize and promote even incremental moves in more environmentally benign directions.

It is clear that sustainable development (SD) has both a space and a time dimension, however, usually the models applied are non-spatial in nature. In support of sustainable development planning at a regional scale we have tried to integrate these non-spatial SD models with a GIS-approach (Scholten et al., 1991; Despotakis, 1991).

The SD-GIS link is based on the conceptual equivalence of economic stocks with geographical layers. In Murthy et al. (1990) stocks are also referred to as "reservoirs" or "levels" connected by flow paths. We may write e.g. for a spatial layer $A(x,y,z)$ and a stock S_i (which is a function of time), the following relationships:

$$A_1(x,y,z) \leftrightarrow S_1(t)$$
.
.
.
$$A_n(x,y,z) \leftrightarrow S_m(t) \quad (1)$$

We relate the idea of stocks with the idea of layers by realizing that the spatial contents of a specific layer "A" can be regarded to be stocks which dynamically change in time. In this manner we can generalize the concept of dynamic modelling of a specific phenomenon to include all necessary stocks S_i which are needed to describe the available spatial layers of information for a region.

From the above we further conclude that if there are m spatial layers available for a region, each with 1, m,...,q classes, then we need r=1+m+q stocks and flows definitions of the form as in equation (1). Our generalized spatial modelling of dynamic phenomena takes the form:

$$dS_1/dt = f_i(t,x,y,z; S_i; F_i; C_i), \quad i = 1, 2,..., r \quad (2)$$

with the initial conditions:

$$S_i(t_0,x_0,y_0,z_0) = \text{INIT}(S_i), \quad i = 1, 2,..., r \quad (3)$$

where F_i are the corresponding flows of the stocks S_i, and C_i are the values of the converters. Then the sustainable development considerations can be embedded in equation (2) and in each integration step dt by:
(a) imposing the necessary conditions that stocks S_i should fulfil;
(b) defining and examining the values of the indicators (but also the stocks, flows and converters themselves) as functions of the stocks, flows and converters; and
(c) defining and examining alternative scenarios which can be of any kind.

The proposed system is shown in Fig. 4. A first application of this approach was done for the Sporades Islands located at the central and western part of the Aegean sea in Greece (Despotakis, 1991). The layers are the land uses of the area, the digital terrain model (DTM), the terrain slopes and the aspects (SLOP and ASP) and the transportation (TRAN).

AN INTEGRATED MODEL

The IMAGE model (Integrated Model to Assess the Greenhouse Effect) is based on the principles discussed above. It is a parameterized simulation model developed for the calculation of historical and future emissions of greenhouse gases on global temperature and sea level rise and ecological and socio-economic interests in specific regions. The greenhouse problem is modelled as

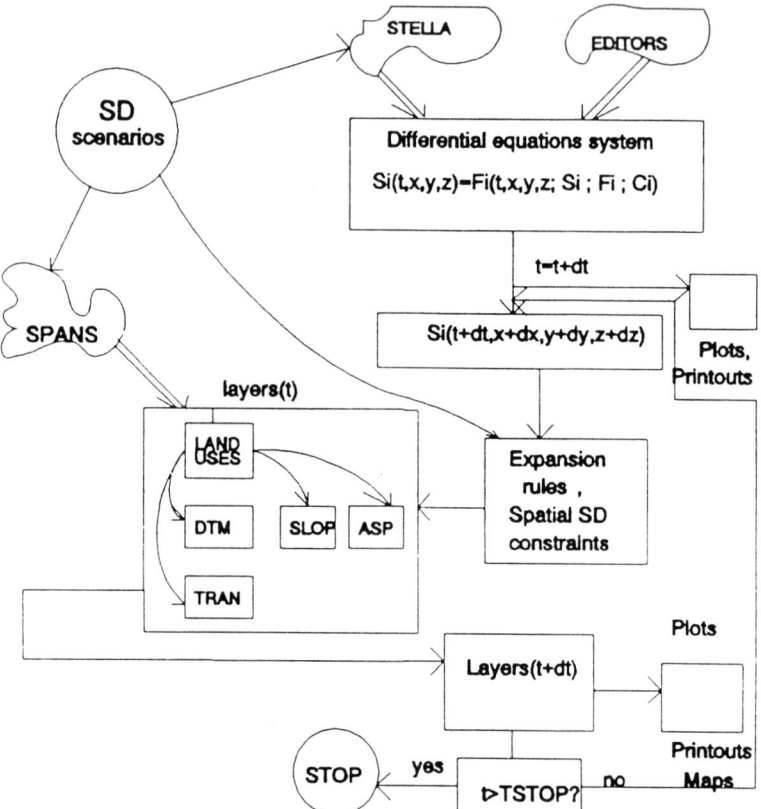

Fig. 4 The link between non-spatial and spatial dynamic modelling (Despotakis, 1992).

a dynamic system which evolves in time as a non-stationary Markov chain, with discrete time steps of half a year and a simulation time of 200 years, from 1900 to 2100. The system is split up into a number of different subsystems, which are modelled by linking sub-models. From a mathematical viewpoint, the model is a sequence of first order differential equations and algebraic equations. An excellent description of the model can be found in Rotmans (1990).

The similarity between the SD-model and the IMAGE model is the way they parameterize time. However, there are large differences when we look to the spatial component. In the IMAGE model there is a desegregation of the world into nine regions. The regional consequences of the model, e.g., coastal defense as a result of sea level rise, has been worked out for one country, The Netherlands, where the global output of the model has been used as input for a separate study which gives a general overview of possible consequences of sea-level rise.

Although the first applications of an integration of compartments with a time- and z-dimension exists, they are rather global. It is not easy at all to try to integrate the non-spatial causal chain mechanisms into the spatial environment and also to take into account the time dimension.

Therefore, it makes sense to go one step backwards. What has to be done and what are the state-of-the-art possibilities to integrate the time and space dimensions in the different processes before we try to integrate them. In Table 1, we have given an overview regarding the time- and z-dimension of the applications discussed in this paper so far.

Based on the overview in Table 1 it would be of interest to improve part of the modelling discussed in the air or groundwater compartment. The theory behind the Geo-voxels and the possibilities of including the time dimension have been applied in a case study to the modelling of the groundwater in the Netherlands. The results of this case study will be presented at the conference.

Table 1 An Overview of the t- and z-dimension.

Application	z-value	t - value	result
Air Modelling	x, y, layer z0	Diffusion model	z ++
	x, y, layer z1	t discrete	t +
	x, y, layer zn		z and t -
Groundwater Modelling	x, y, layer z0	t0, t5, tn	z +
	x, y, layer z1		t +
			z and t -
Geological Modelling	Geo-voxel	t0	z ++
			t ++
			z and t ++
Ecological Modelling	DTM-model and	Dynamic model	z +
	no z-dimension	t discrete	t +
			z and t +
General application	Geo-voxel/DTM	transport/	**Integration**
		t-continuous	**of z and t**

CONCLUSIONS

The conclusions and recommendations are based on the evaluation of the literature and our practical work in modelling the environmental processes. The ultimate goal of all our modelling work is to build decision support systems for environmental and public health. First of all research is necessary for the data architecture. By building a database, knowledge has to developed about the accuracy of spatial databases. It is clear that many institutes are collecting information. An important issue is the exchange of information and the harmonization of the definitions. The application of GIS in environment and public health requires specific analytical tools, and the building of a decision support environment makes it also necessary to develop evaluation techniques within a GIS-environment. We can expect that in the near future, from a technical point of view, many opportunities will be offered to improve the possibilities of visualization. It will be necessary to be critical to these possibilities, but also to benefit from them.

In this paper we have tried to evaluate the methods and techniques for the spatio-temporal aspects of environment and health. It is clear that in most cases the applications are two-dimensional. From a point of view of the existing software this does not come as a surprise, because the available software is in most of the cases two-dimensional. However, we see developments in the direction of the z-dimension and in the dimension of dynamic simulation. Based on both theory and state-of-the-art GIS-software, it becomes possible to enlarge the incorporation of time and z in spatial modelling. As discussed earlier, the data will again become the biggest barrier in modelling the reality.

REFERENCES

Davis, J.C. (1973) *Statistics and Data Analysis in Geology*. Wiley and Sons.
Despotakis, V.K. (1991) *Sustainable Development Planning Using Geographic Information Systems*. Ph.D. Dissertation. Free University, Amsterdam.
Intergraph Corp. (1992) *MGE Voxel Analyst Software*. Huntsville, Alabama, USA.
Koussoulakou, A. (1990) *Computer-Assisted Carthography for Monitoring Spatio-Temporal Aspects of Urban Air Pollution*. Ph.D. Dissertation. Delft University Press.
Kusse, A.A.M. (1990) The use of GIS in Combination with Hydrological and Hydrochemical Modelling. In: *Proceedings EGIS'90*, 10-13 April 1990, EGIS-foundation, Utrecht, The Netherlands, 640-664.
Murthy, D.N.P., Page, N.W & Rodin, E.Y. (1990) *Mathematical Modelling: A Tool for Problem Solving in Engineering, Physical, Biological and Social Sciences*. Pergamon Press.
NATO/CCMS (1980) *NATO/CCMS Pilot Study on Air Pollution: Assessment Methodology and Modelling*. Final report # 105.
Raper, J. (editor) (1989) *Three Dimensional Applications in Geographical Information Systems*. Taylor & Francis, London.
RIVM (1984) *Landelijk meetnet grondwaterkwaliteit: eindrapport van de inrichtingsfase* (The establishment of the National Groundwater Monitoring Network: Final report) (in Dutch). RIVM Report 840382001, Bilthoven, The Netherlands.
Rotmans, J. (1990) *IMAGE: An Integrated Model to Assess the Greenhouse Effect*. Kluwer Academic Publishers, Amsterdam.
Scholten, H.J., Despotakis, V.K. & Nijkamp, P. (1991) Workshop "Spatial Analysis". In: *Proceedings EGIS'91*, 2-6 April 1991, EGIS-foundation, Utrecht, The Netherlands.
Van Duijvenbooden, W.(editor) (1989) *De kwaliteit van het grondwater in Nederland* (The Quality of Groundwater in The Netherlands) (in Dutch). RIVM Report 728820001, Bilthoven, The Netherlands.
World Commission on Environment and Development (1987) *Our Common Future*.Oxford University Press, Oxford/New York.

Database development and GIS application in support of groundwater management: case study at the Austrian-Bohemian Border

M. STIBITZ
Institute of Geo-Data and System Analysis, Mining University Leoben, Peter Tunner Strasse 15, 8700 Leoben, Austria

Z. PATZELT
Czech Geological Survey, Malostranske nam. 19, 118 21 Praha 1, Czech Republic

J. WOLFBAUER
Institute of Geo-Data and System Analysis, Mining University Leoben, Peter Tunner Strasse 15, 8700 Leoben, Austria

Abstract A GIS study on the boundary between Lower Austria and Southern Bohemia in the vicinity of the towns of Gmünd and Ceske Velenice is presented. The study focuses primarily on the complex multilayer groundwater aquifer system. Recent activities with major environmental impact are sand exploitation and waste disposal in the abandoned pits adjacent to the border line. The aim of the study is to summarize the available environmental information pertinent to the mentioned human activities. The collected information is standardized, correlated and processed by means of the HADES environmental database system as well as by the ARC/INFO geographical information system. The framework for local decision-makers is subsequently being improved by groundwater simulations based on the MODFLOW and VAM3D three-dimensional groundwater models.

INTRODUCTION

One of the first cross-boundary GIS studies between Austria and Czechoslovakia is presented. It has been performed on the boundary between Lower Austria and Southern Bohemia in the vicinity of the towns of Gmünd and Ceske Velenice (Fig.1). In the vicinity of Gmünd and further to the northwest to Suchdol there are regionally important water resources in the Cretaceous basin sediments and quaternary aquifers of local significance. The groundwater system consists of quaternary and tertiary alluvial and eluvial sediments as well as cretaceous sediments of the Trebon/Wittingau Basin. The aquifer system thickness totals up to approx. 350 m. Major recipient is the river Lainsitz/Luznice. These water resources have been potentially endangered by the exploitation of feldspar sands and by landfill waste disposal.

The presented study has multiple objectives. The major one is the integration of environmental parameters as base information for the water resources management on both sides of the border in a geographic information system. The study has a few basic methodical steps: database and data processing, geographical information system and numerical groundwater

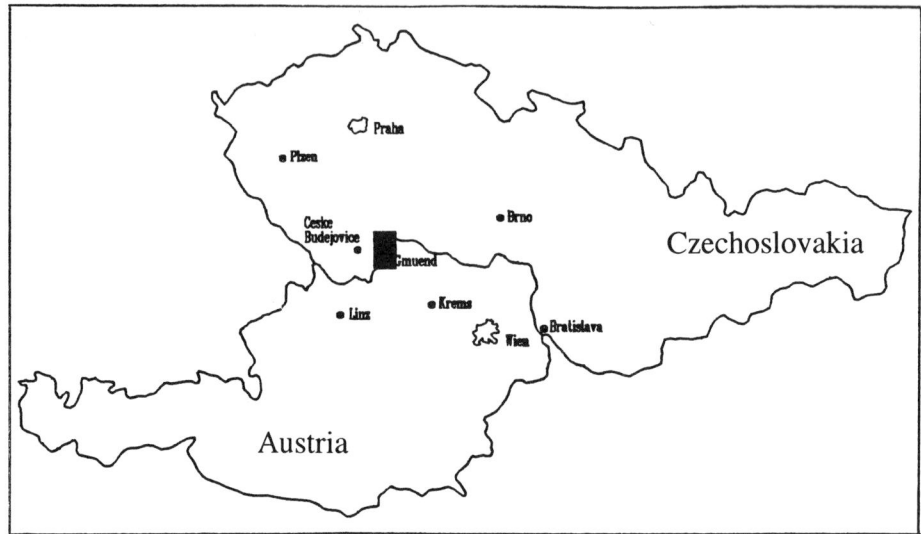

Fig. 1 Location of the study area at austrian-bohemian border.

simulation.

Large importance of the project activities lies in formal unification of data from the two countries, their basic geographic projection systems, as well as various institutions, and their correlation and interpretation for the cross-boundary region. There are three areal levels of information (Fig. 2) pertinent to the project (Stibitz et al., 1992) in the framework of which the presented paper has been prepared:

(a) area A of the regional groundwater model of the Southern Trebon Basin;
(b) area B of which the environmental data have been processed by means of the ARC/INFO geographic information system (studied area in the presented paper);
(c) area C of local groundwater model to simulate flow in the quaternary aquifer.

During the presented study, ARC/INFO is being used under VMS, UNIX and MS DOS; HADES and MODFLOW under MS DOS; and VAM3D under VMS, UNIX and MS DOS operating systems. The hardware used are VAX 4000/200, SUN SPARC workstations and IBM PC compatibles.

DESCRIPTION OF THE STUDIED AREA

The studied area for the GIS data processing and integration (Fig. 2 - B) has been chosen so that the impact of the sand exploitation as well as waste disposal on the groundwater system can be evaluated within natural boundary conditions. The quadrangular area of 286 km^2 is 22 km long in the north-south direction and 13 km wide in the east-west direction.

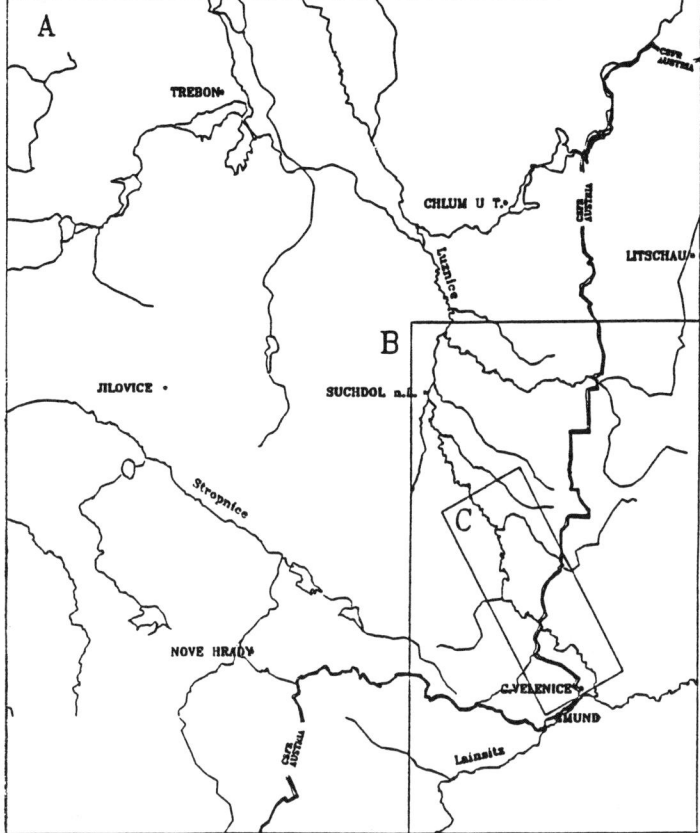

Fig. 2 Three areal levels of investigation: A - regional model; B - GIS study; C - local model.

Topography and morphology

The 15th meridian east of Greenwich and 48th northern parallel cross the studied area. The lower part of the studied area, the basin, has the average elevation of 460 to 480 m a.s.l., the crystalline hills reach heights of about 550 to 560 m a.s.l. with maxima of about 600 m a.s.l. at the south-eastern edge of the quadrangle.

The studied area lies on both banks of the river Lainsitz/Luznice between Altweitra in the south and Suchdol upon Luznice in the north. The northern boundary follows the creek of Reissbach/Dracice down to its mouth to Luznice at Suchdol. Major urban areas are the towns of Gmünd and Ceske Velenice.

The western part of the studied area is characterized by the flat landscape of the Trebon Basin. This landscape changes at the eastern and southern boundaries of the GIS area to the hills of the crystalline of the Bohemian Massiv. The crystalline and the basin are covered by significant quaternary layers which create again the flat relief or river terraces in the central part among Breitensee, Neu Nagelberg and Halamky.

The areas of sand exploitation at Breitensee and Halamky are characterized by an artificial relief with steep slopes, groundwater ponds and small forests. The still preserved tertiary areas are intensively used by agriculture.

Geology and hydrogeology

Two geological and hydrogeological units are of main interest at the presented study:
(a) late cretaceous and tertiary layers of the Trebon Basin including the Gmünder Bay; and
(b) quaternary sediments of the Lainsitz and its tributaries.

The base as well as the side boundaries of the sediments are formed by gneisses, granites or granodiorites of the crystalline. The basin sediment sequences resulted from more or less irregular accumulation of sands, silts and clays. Maximum thickness of the basin filling is about 350 m, the assumed thickness is approximately 100 m at the austrian-czech boundary. Thickness of quaternary Lainsitz sediments reaches the values of 10 - 20 m. Numerous faults have large significance for the geological and hydrogeological structural features.

The crystalline rocks are fractured and contribute to the groundwater flow in the basin. The cretaceous layers are characterized through regionally irregularly connected as well as local sandy confined aquifers. The basin aquifers may be discharged to the aquifer of the Lainsitzer gravels and sands in the occurrence of direct connections of sandy layers of the Cretaceous and Tertiary with the quaternary alluvium. The transmissivities of the aquifers are in the order of 1×10^{-4} to 1×10^{-2} $m^2 s^{-1}$.

The shallow groundwater flows towards local surface waters, primarily to Lainsitz. Deeper groundwater flows to the regional discharge tectonic zone which is assumed to be at the north-western boundary of the southern Trebon Basin.

DATABASE

HADES relational database (version 3.x, release March 1992) is used to store, retrieve and analyze all project data. The ARC/INFO layers are used as the background for graphic features which work with the areal information. HADES is a relational database developed as a software layer over dBASE III to store and manipulate geographical position and geological as well as other environmental attributes of objects such as boreholes, gauges, pits etc.

HADES is able to perform a full range of ordinary database functions (e.g., definition of new files, use of existing files, data entry, update, sort, and query. The project database is organized as a catalogue, while HADES maintains the relationship between corresponding files in the databank through the defined field (NUMBER in the case of this study) which is contained in every data file. The database uses an internal spatial indexing system to allow the user to easily associate and interrelate information from several files.

Special modules of HADES with powerful graphics are used for editing and leveling of the bore profiles as well as time series management and statistical evaluation.

Analog and digital data have been collected from seven institutions and companies in Austria and Czechoslovakia. All provided digital data have been overtaken and the database structure respected their existence although they may not have direct connection with the solved groundwater problems. The principal approach was to develop an environmental database.

The project database thus contains information on documented objects (file BASIC) as well as data on geology (GEOLOGY), hydrogeology (HYDROGEOLOGY), environmental quality (QUALITY), stream sediments (STREAM), groundwater quality (GW_CHEMISTRY), surface water quality (SW_CHEMISTRY), hydrometeorology (HYDROMETEOROLOGY), exploitation (EXPLOITATION), waste disposal (WASTE), and time series of water levels (WL_SERIES), precipitation (RAIN_SERIES), air temperature and sunshine (TSS_SERIES), humidity (HUM_SERIES), wind velocities and directions (WIND_SERIES), thermal radiation (RAD_SERIES), and the SO_x contents in the air (SOX_SERIES). The disk capacity required for the data is about 10 MB. The structure of the database is quite complex and is shown in Fig. 3.

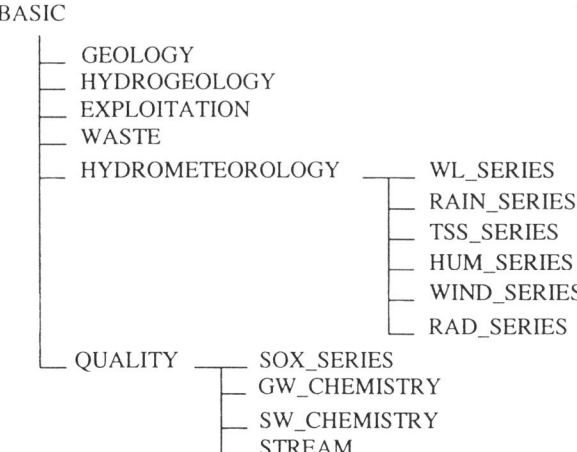

Fig. 3 Database structure.

GIS DATA INTEGRATION

ARC/INFO geographic information system has been applied to process and integrate information from various organizations, from Austria and Czechoslovakia (CSFR) One of the demanding tasks during the project has been the conversion of the czechoslovak Krovak system of conform conic projection to the austrian BMN system.

The following information layers have been collected for the study area of 286 km² (22 km N-S, 13 km E-W):
(a) topography;
(b) geology;
(c) hydrogeology;
(d) raw materials exploitation and waste disposal.

The following maps and plans have been used as analogue base information:
(a) topography: CSFR maps in the scale 1:50 000 (sheets 33-11 (Trebon), 33-13 (Ceske Velenice)); Austrian maps ÖK 15V in the scale 1:25 000 (sheets 05 Gmünd), 18 (Weitra));
(b) geology: Austrian geological map in the scale 1:50 000 (sheet 18 (Weitra)); Austrian geological special map in the scale 1:75 000 (sheet Litschau and Gmünd); CSFR geological maps in the scale 1:50 000 (sheets 33-11, 33-13);
(c) hydrogeology: CSFR hydrogeological maps in the scale 1:50 000 (sheets 33-11, 33-13);
(d) raw materials exploitation and waste disposal: plans in the scale 1:2000 of the exploitation of feldspar sands in the area of Halamky; and land-use plans in the scale 1:1000 for the area of Gmünd and Breitensee.

These data have been digitized and integrated into the ARC/INFO GIS. To standardize the topography on both sides of the border it was necessary to convert the different projection and coordinate systems used - BMN in Austria, Krovak in Czechoslovakia. ARC/INFO has been used as a primary tool in conjunction with the HADES database for the integration and synthesis of various data groups.

The resulting geological map has been thoroughly correlated and the interpretation of the geological structure, in particular tectonics, has been completed by additional field survey. The hydrogeological map has been finalized on the basis of the geological structure.

GROUNDWATER SIMULATION

The groundwater flow simulation of the southern Trebon Basin performed by Curda (1990) has been the primary hint for this study. Their model evaluated the groundwater flow in the basin with respect to the water supply potential. The model was however limited by the state boundary and did not reach the natural boundaries of the aquifer complex in the south on the austrian territory in the vicinity of Gmünd. The code used for this simulation was the three-dimensional USGS finite difference groundwater flow model MODFLOW (McDonald & Harbaugh, 1988).

In the framework of the recent study, the model has been extended to the whole area of the southern Trebon Basin (Fig. 2 - area A). The modelling has been performed in two phases. First, the regional system was simulated by MODFLOW. In the second step, after a stationary MODFLOW calibration on the mean water level data of the year 1989, the aquifer data were used as input to run the VAM3D (Huyakorn & Panday, 1991) finite-element code on the regional scale.

The MODFLOW cell-centered grid of the regional model has 28 rows, 40 columns in 4 layers in x,y-distance varying between 750 to 1500 m. This corresponds to 29 nodal rows, 41 columns and 5 layers in the VAM3D code. The division in 4 layers is purely geometrical and does not have to represent some physically existing layers in the basin.

PRELIMINARY RESULTS AND PLANNED ACTIVITIES

Results that may significantly improve the groundwater management in the border towns of Gmünd and Ceske Velenice in the context of other human activities have already been achieved in the areas of database design and development, GIS data visualization and analysis as well as groundwater modelling.

Significant improvement can be seen in the information transfer due to standardization and correlation of data from both neighboring countries - Austria and Czechoslovakia.

The VAM3D would be applied in the final part of the modelling part of the study for local simulation of the Quaternary in the zone of intensive human activities between Gmünd, Breitensse and Halamky (Fig. 2 - area C).

ARC/INFO should be used for graphical presentation of environmental relationships that have been analyzed by means of numerical simulation.

Acknowledgements The presented paper has been prepared in the framework of the research cooperation project System Study of Geogene and Hydrological Environmental Parameter for the Southern Trebon Basin/Gmünder Bay between Austria and Czechoslovakia. The project has been funded by the Austrian Ministry of Science and Research and by the Government of Lower Austria.

REFERENCES

Curda, S. (1990) Trebon Basin - Southern Part, Model Simulation of Ground Water Flow. Geoindustria. Prague.
Huyakorn, P.S. & Panday, S. (1991) VAM3D Version 1.5 - A Variably Saturated Analysis Model in 3 Dimensions with Capability to Handle Head-Dependent Flux Boundary Conditions. HydroGeoLogic. Herndon, VA.

McDonald, M. & Harbaugh, A.W. (1988) A Modular Three-Dimensional Finite-Difference Ground Water Flow Model. *Techniques of Water-Resources Investigations of the United States Geological Survey, Book 6, Chapter A1*, Reston, VA.

Stibitz, M., Wolfbauer, J. & Patzelt, Z. (1992) System Study of Geogene and Hydrological Environmental Parameter for the Southern Trebon Basin/Gmünder Bay. Joanneum Research. Leoben.

Deep waste injection, Western Canada: analysis of data and requirements for numerical simulation

J.R UNDERSCHULTZ & A.T. LYTVIAK

Alberta Geological Survey, Alberta Research Council, P.O. Box 8330, Postal Station F, Edmonton, Alberta, Canada T6H 5X2

Abstract The application of numerical models to deep waste injection problems requires stringent preparation of data from a variety of sources. These include physiography, topography, stratigraphic picks used to define the geological framework, and drill stem test, core analyses, and analyses of formation water chemistry used to characterize the hydrogeological regime. These data, together with modelling results, must be linked and displayed within a common cartographic system. Since no turn-key software is available which can characterize and integrate the wide range of data required for numerical simulations, the Alberta Geological Survey has assembled various purchased and in-house developed software for this purpose. This paper presents the techniques and associated software used for the spatial analysis and interpretation of various data types required for simulating deep injection of residual water.

INTRODUCTION

Huge reserves of hydrocarbons within the thick strata of the Western Canada Sedimentary Basin (Fig. 1) reside where Enhanced Oil Recovery (EOR) methods are required for their extraction. As a result, large volumes of residual

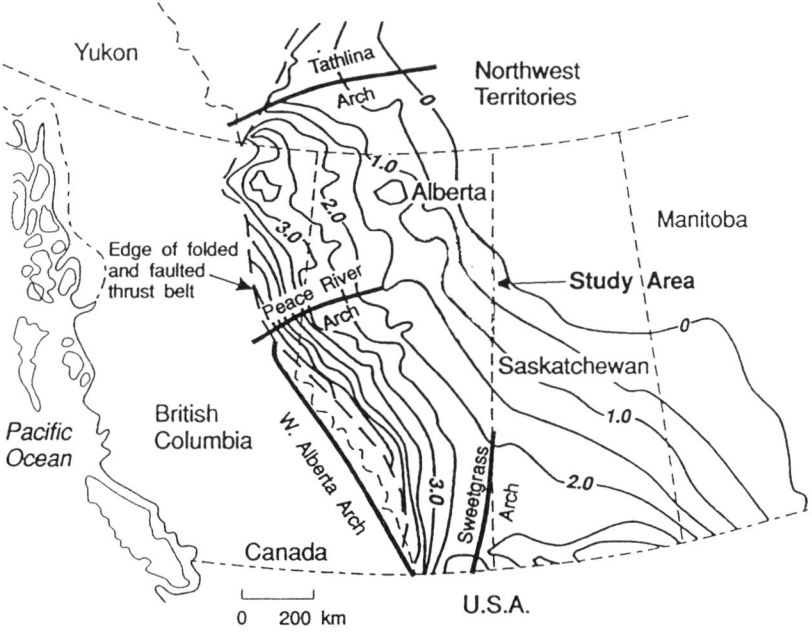

Fig. 1 Location and isopach of the Western Canada Sedimentary Basin.

water produced by EOR operations are disposed of by deep injection. Numerical models are commonly used to simulate the effect of injection on the resident hydrocarbon, mineral, and water resources. The confidence associated with simulation results is closely related to the quality of the input data. Thus, the application of numerical models to deep injection problems requires stringent preparation of data from a variety of sources. A computer based Dynamic Basin Analysis approach (Hitchon et al., 1987) as described by Bachu et al. (1987) is required to develop a comprehensive, qualitative and quantitative understanding of dynamic processes in the basin.

In order to use predictive modelling in the evaluation of the environmental effects of deep injection of liquid waste, a sequential approach to characterizing the rocks and the contained fluids is essential. Initially, the data is located and retrieved into an electronic data base. A gridding algorithm is selected and gridding constraints are chosen for the area of interest. The sedimentary rocks are described and characterized in terms of their lithology and variability of hydraulic properties. The pressure regime and the chemical composition of formation waters have to be analyzed and defined. Lastly, the numerical simulation is run.

DATA SOURCES AND DATA BASE

The Western Canada Sedimentary basin comprises two entities of interest in these types of study; (a) the rock framework, and (b) the contained fluids. Information on both entities reside with a variety of organizations both private and public, and in a variety of forms, both electronic and hard copy. Data of particular interest include stratigraphic picks and core analyses stored electronically and currated by various public organizations, and drillstem test and formation water analyses stored and currated by industry. In addition, an extensive amount of work has been conducted and documented in the literature which requires manual data entry. Most data are characterized by a geographic position, a depth, and various associated parameters, but the interrelations between the various types of data are very complex. Because of this complexity an appropriate Data Base Management System (DBMS) is essential.

The basic concept of a DBMS is to separate the description of the data from applications, and to provide basic tools for adding, deleting, and retrieving data. The current implementation of the Alberta Geological Survey Well Data Base uses the INGRES DBMS and contains stratigraphic, rock property and hydrodynamic data pertaining to more than 200 000 wells in western Canada. INGRES implements a relational data model and the SQL language. This allows the flexibility to support ad hoc queries characteristic of the data model.

AUTOMATED GRIDDING AND CONTOURING

Electronic grids are used to characterize parameters which represent a continuum, such as hydraulic head. Mathematical re-sampling (interpolation) of irregularly located parameter determinations into a uniformly distributed pattern (grid) is referred to as the gridding procedure. Maps are produced by contouring the grid node values and outputting them to a hard copy device. The CPS3 software developed by Radian Corporation is used for automated gridding and contouring. Alberta Research Council cartographic software is used to process point and vector data prior to gridding and integrate contour maps with geographic data in a common cartographic projection.

Integrated basin analysis requires a gridding algorithm that produces grids which meet essential criteria for all aspects of the project. The main criteria for algorithm selection are: (a) where data exist, the resultant grid must accurately represent the observed values; (b) for modelling purposes, the gridded parameters are required to be realistically characterized for areas in which there are little or no data; and (c) the gridding must accommodate the high variability in data density both areally and with depth, and various degrees of clustering. With these constraints in mind, a convergent gridding algorithm is often selected whereby grid-node values are converged upon through several iterations (Graf & Thomas, 1989). This iterative process produces a trend like solution in areas of sparse or no data, and an accurate representation where more control exists.

DATA PROCESSING AND GRID MANIPULATION

Each data type requires processing specific to the nature of the property they represent. For a sedimentary basin, data of concern can be divided into three categories, each with specific data processing requirements:
(a) the stratigraphic geometry of the rocks;
(b) their characteristic flow properties (porosity and permeability); and
(c) the hydrodynamic nature of the contained fluids characterize by pressure and ionic concentration distributions.

Stratigraphic geometry

In order to obtain a geologically realistic delineation of the stratigraphy, the following three-step approach is taken:
(a) interpreting and defining major geological events;
(b) data processing; and
(c) checking and correcting the resulting grids for internal consistency.

The method of building stratigraphic surfaces from "control surfaces" has been described by Jones & Johnson (1983) and Bachu *et al.* (1987) who define two categories of surfaces. The first category comprises "control surfaces", of which there is typically one for each package of conformable strata. The control surface is usually at the top of the stratigraphic package, has a large number of data points, and is either a structure top or base of a unit (elevation). The second category comprises "non-control surfaces", which make up the rest of the surfaces in a particular stratigraphic package. The approach starts with defining a grid for the control surface in a particular stratigraphic package. The non-control surfaces are then sequentially gridded starting from the one nearest to the control surface. If inconsistencies between surfaces arise (usually within zones of extrapolation), the "non-control" surfaces are modified by either adding or subtracting a geologically acceptable isopach from the control surface. In this way all inconsistencies are resolved in reference to the control surface.

The "control surface" approach can be further refined in that rather than grouping all "non-control" surfaces at equal status, a hierarchy can be used. The confidence which can be paced on the "non-control" surfaces depends on the amount and distribution of the data for any particular surface, and the quality of the data itself. Some geological formations have very distinct geophysical log signatures which result in extremely consistent picks. Other formations may be difficult to discern and therefore have picks of questionable consistency. Rather than gridding the surfaces in order from closest to furthest from the "control surface" they are gridded from highest to lowest confidence levels with inconsistencies resolved at each step. In this way the highest number of constraints are applied to areas where data are absent and extrapolation is necessary, and results in the best possible estimation of stratigraphic geometries which are geologically acceptable.

Porosity and permeability

Porosity data are obtained principly from core plug analyses but can be supplemented by geophysical logs. Permeability data are obtained from both core plug analyses and from drill stem tests. Because these data are extremely voluminous in the Western Canada Sedimentary Basin an automatic culling and data manipulation procedure has been developed in-house. Entire analyses and individual data are automatically rejected based on range and threshold criteria. The remainder of the data are partitioned into the stratigraphic framework described previously. A digitally based statistical approach is then used to characterize the spatially independent rock properties.

Porosity and permeability data obtained from core analyses represent volume averaged values corresponding to the plug scale (Baveye & Sposito, 1984). In order to characterize the strata at the formation scale, a scaling up procedure must be used. Because of the large difference in magnitude between the plug and formation scale, the scaling up is done sequentially (Cushman, 1984). Once the porosity and maximum permeability data are partitioned by hydrostratigraphic unit, they are scaled up first to the well scale and then to the formation scale using the procedure described by Bachu & Underschultz (1992).

Drillstem tests are interpreted using a Horner analysis (Timmerman & Van Poollen, 1972). The slope of the Horner plot together with Cl^- and temperature estimates are used to calculate the formation permeability. The Cl^- and temperature are needed to calculate the density and viscosity of formation waters according to relations published by Kestin et al. (1981) and Rowe & Chou (1970). Permeability values obtained from drillstem tests are already at the well scale. The flow toward the well during a drillstem test is three-dimensional in nature; thus, the measurement is not direction dependent and produces only a single value.

Pressure and formation water chemistry

Pressure data are obtained from drillstem tests stored in electronic form while formation water analyses are entered from hard copy. The drillstem test data require little data culling as the analyses are assigned a quality code at the stage of Horner interpolation, and only good quality analyses are reported from the data base for consideration. Formation water analyses require more stringent examination because contamination from drilling fluids and poor collection techniques result in a high proportion of poor quality analyses. Since the pressure and ionic concentration data represent a continuum, grid interpolation is used to characterize their distribution.

The formation pressure is obtained by extrapolating the Horner plot of pressure measurements. For constant density waters, a potential field driving the flow can be defined. Potentiometric surfaces are used in this case to represent the hydrodynamic field, identify horizontal flow directions, and calculate horizontal gradients (Fig. 2). The potentiometric surfaces are constructed on the basis of freshwater hydraulic head, thus representing the gravity driven flow component. In sedimentary basins the density is a function of salinity and temperature. Density driven flow enhances or retards the gravity driven flow of formation waters depending on the density distribution, hydraulic gradient and aquifer slope (Davies, 1987; Dorgarten & Tsang, 1991). The importance of the density driven flow component can be evaluated by calculating the driving force ratio as defined by Davies (1987) & Bear (1972).

The formation water chemistry data are culled electronically for unacceptable aspects of the analyses. Typical analyses can be contaminated, mixed with other samples, or be incomplete in some significant way. The automatic cull takes into account the presence and magnitude of OH, CO_3, Ca, Mg, SO_4, HCO_3 concentration, acceptable ranges of pH and density, and mixing of formation waters from several intervals. Subsequent to the automatic cull, the analyses are partitioned within the stratigraphic frame discussed previously. The distributions of individual ionic concentrations are then gridded and mapped for each flow unit.

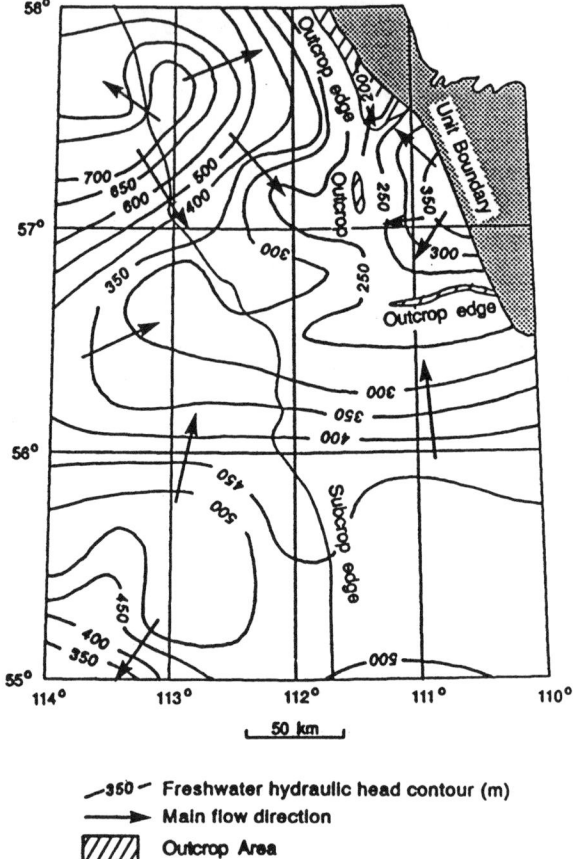

Fig. 2 Freshwater hydraulic-head distribution for the Beaverhill Lake-Cooking Lake aquifer system in northeast Alberta.

NUMERICAL SIMULATION

The objective of numerical simulation of formation water flow and fluid injection is to provide a predictive tool for determining the effects of deep waste injection on the natural hydrogeological regime in terms of hydraulic head buildup. After calibration, the system is perturbed by injecting fluids at specific rates. For a given stratigraphic configuration, the system response is indicative of the effects of injection in the various aquifers in the succession.

The modelling process is carried out in two stages. First the effect of injection is simulated on a regional scale using regional values for hydraulic parameters. The purpose of regional scale modelling is to check for interference between multiple injection sites and to establish the impact of injection on the natural flow regime (Fig. 3). Local values of hydraulic conductivity and specific storage could be different from the values used in regional scale simulations. Resolution at the regional scale grid is too coarse for a proper estimation of the pressure buildups at the injection wells

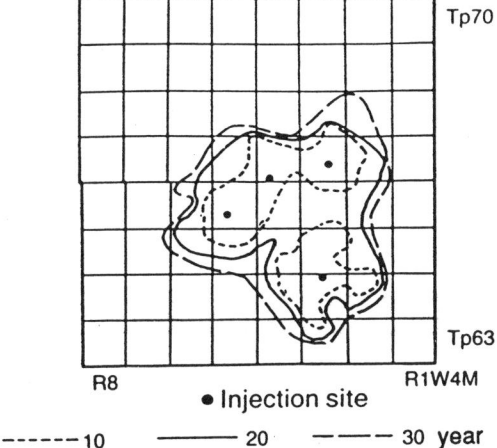

Fig. 3 Predicted spread of the pressure buildup for the basal Cambrian sandstone aquifer in the Cold Lake area, Alberta, using regional average hydraulic conductivity.

themselves (Bachu *et al.*, 1989). For these reasons the effects of injection are simulated on a local scale as a separate stage. At this scale, local values of hydraulic parameters are used in order to compare hydraulic head buildups with the fracturing threshold of the rocks (Fig. 4).

CARTOGRAPHIC SOFTWARE

Throughout the entire characterization and modelling procedure, the various data and output need to be tied to a common cartographic system for reference

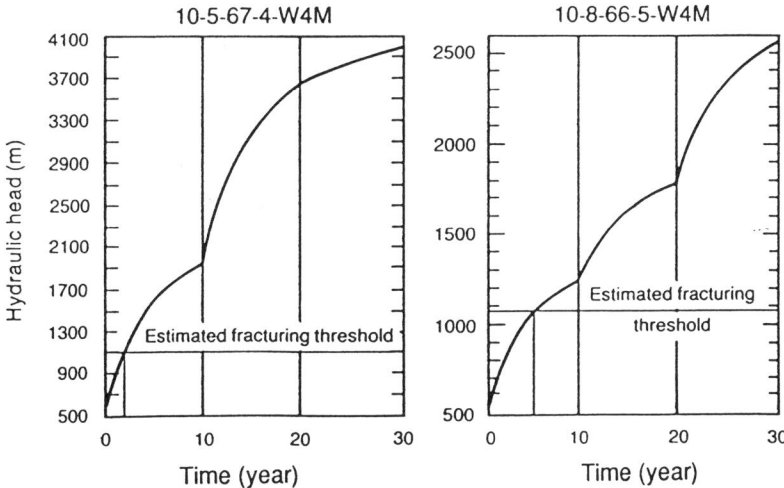

Fig. 4 Predicted effects of variable injection in the basal Cambrian sandstone aquifer at two sites.

and display. Since no turn-key software is available which can characterize and integrate the wide range of data required for numerical simulations, the Alberta Geological Survey has developed a cartographic package which serves to integrate digital output from surface manipulation and display systems, graphical support systems and simulation software. The central function of the cartographic package is to convert various geo-location systems (such as latitude and longitude) into a common engineering or cartesian coordinate system for a particular area of interest. It takes input of points, vectors, polygons, and surfaces and exports them in the format required by various in-house and purchased data manipulation and display software. The central cartographic package is the key to preparing registered input for numerical simulation of deep waste injection, and providing a means of outputting results in a uniform and integrated format.

REFERENCES

Bachu, S., Sauveplane, C.M. Lytviak, A.T. & Hitchon, B. (1987) Analysis of fluid and heat regimes in sedimentary basins: techniques for use with large data bases. *AAPG Bull.* **71**, 822-843.

Bachu, S., Perkins, E.H., Hitchon, B. Lytviak, A.T. & Underschultz, J.R. (1989) Evaluation of effects of deep waste injection in the Cold Lake area, Alberta. *Alberta Research Council Bull.* **60**.

Bachu, S. & Underschultz, J.R. (1992) Regional-scale porosity and permeability variations, Peace River Arch area, Alberta, Canada. *AAPG Bull.* **76**, 547-562.

Baveye, P. & Sposito, G. (1984) The operational significance of the continuum hypothesis in the theory of water movement through soils and aquifers. *Wat. Resour. Res.* **20**, 521-534.

Bear, J. (1972) *Dynamics of fluids in porous media*. Elsevier, New York, 654.

Cushman, J.H. (1984) On unifying the concepts of scale, instrumentation, and stochastics in the development of multiphase transport theory. *Wat. Resour. Res.* **20**, 1688-1676.

Davies, P.B. (1987) Modeling areal, variable density, ground-water flow using equivalent head-analysis of potentially significant errors. In: *Proc. of the NWWA/IGWWC Conf. - Solving ground water problems with models*. 888-903. NWWA, Dublin, Ohio.

Dorgarten, H.W. & Tsang, C.F. (1991) Modelling the density-driven movement of liquid waste in deep sloping aquifers. *Groundwater* **29**(5), 655-662.

Graf, K.E. & Thomas, M.D. (1989) *CPS subroutine library users manual, version 1.5*. Radian Corporation, Austin, Texas.

Hitchon, B., Bachu, S., Sauveplane, C.M. & Lytviak, A.T. (1987) Dynamic basin analysis: an integrated approach with large data bases. In: *Fluid flow in sedimentary basins and aquifers* (ed. by J.C. Goff & B.P.J. Williams) 31-44. Geol. Soc. Spec. Publ. no. 34.

Jones, T.A. & Johnson, C.R. (1983) Stratigraphic relationships and geologic history depicted by computer mapping. *AAPG Bull.* **67**, 1415-1421.

Kestin, J., Khalifa, H.E. & Correia, R.J. (1981) Tables of the dynamic and kinematic viscosity of aqueous NaCl solutions in the temperature range 20-150 C and the pressure range 0.1-35 MPa. *J. Phys. and Chem. Ref. Data* **10**(1), 71-87.

Rowe, A.M. & Chou, J.C.S. (1970) Pressure-volume-temperature-concentration relation of aqueous NaCl solutions. *J. Chem. Eng. Data* **15**(1), 61-66.

Timmerman, E.H. & Van Poollen, H.K. (1972) Practical use of drillstem tests. *J. Can. Petrol. Tech.* **14**, 31-41.

5 Coupling GIS with Hydrological Models

Multi-media: an integrated approach combining GIS within the framework of the Water Act 1989

CHRISTINE H. ASHTON
North West Water Ltd., New Town House, Warrington, WA1 2QG, UK

DAVID SIMMONS
GeoData, Southampton University, Highfield, UK

Abstract The Water Act, enacted under British law in September 1989, steered the way for the restructuring, and subsequent privatization of the water industry in England and Wales in a way which supported EEC pollution legislation. The act detailed the roles and responsibilities of private companies and government agencies with respect to water and environmental activities. It is current practice to review the operating criteria and performance of both these sectors in order to ensure the successful management of activities, improvement of standards and the effective communication of information and policy to all stakeholders. The purpose of this paper is twofold. Firstly, by focusing on one major activity - water resources management, we briefly demonstrate the requirements and current usage of GIS within the National Rivers Authority. Secondly, by way of a computer based prototype which seamlessly integrates GIS, Desktop Mapping and Multi-media we demonstrate the benefits and potential of such an approach, particularly for the purposes of communication and public access to environmental information. In conclusion, we describe the potential for the exchange and communication of environmental data and information using Multi-media on a european scale.

INTRODUCTION

The Water Act 1989 is one of the most detailed and, as a result, one of the lengthiest pieces of legislation ever passed in the United Kingdom (UK). The main thrust of this act is pollution-prevention rather than just regulation. Prior to 1989, the water industry in England and Wales consisted of public water authorities established by the Water Act 1973. These water authorities combined both utility and regulatory functions in order to facilitate an integrated approach to water management, but were often criticized for being both "poacher and game keeper". It had been the intention of the UK government to privatize the water industry some 2 years prior to the 1989 Act, during the 1986-87 legislative programme. However, plans intended to simply transfer all the functions of the water authorities to the private sector were abandoned as they were viewed as being only in the interests of the authorities themselves, and also contrary to the spirit of European Community water pollution legislation. In 1987, the government announced plans to establish a new public body, the National Rivers Authority (NRA), which would inherit the main regulatory and water-source management functions of the water authorities, leaving the utility roles of water supply and distribution and sewerage services for privatization.

THE CREATION OF THE NRA

The National Rivers Authority came into power on 1st September 1989. One of the strongest environmental protection agencies in Europe, its mission is to control pollution and improve the quality of the UK's river systems and coastal waters. The NRA's wide ranging responsibilities encompass water pollution, water resource management (including abstraction licensing), land drainage and flood protection, fishery and navigational functions, and also, a general brief to promote conservation and recreational interests. However, by far one of the most important of the NRA's functions, vital to the long term success of its policies, is the communication of environmental policy, and the raising of environmental awareness within government, industry, commerce, agriculture, the public and the European community. Furthermore, European Community (EC) legislation coming into force at the end of 1992, requires public bodies throughout all EC member states to provide the public with access to the environmental information which they hold. Clearly, providing access to raw data and ad-hoc information will not raise levels of awareness unless it is presented in an easily digestible form, ideally integrating a combination of media including geographic maps, video, sound, text, photographs, and satellite data all cross referenced and presented in an informative and easily accessible way.

THE ROLE OF GIS IN THE NRA

Long before the creation of the NRA, GIS had been recognized by the water authorities and environmental interest groups as the vehicle by which spatially related water and environmental data could be collected, stored, and managed. Many of the NRA's ten operating regions in England and Wales had begun, while still part of the former authorities to embark on GIS projects to establish information systems relating water resource, asset and operational data to large scale Ordnance Survey digital map data. Many of these systems were custodial in nature, often mainframe-based, and initially focused on the automation of traditional labour intensive activities such as map production.

The majority of potential GIS endusers in organizations like the NRA are hydrologists, land drainage engineers, catchment planners, resource and environmental planners. These people are professionals in fields other than computing, and their jobs revolve around making informed decisions in their own specialist area, rather than relying on a knowledge of Information Technology (IT). They need GIS systems which are easy to use, can perform spatial analysis and modelling, support common data formats, and can produce high quality hard copy output. It is these people who are responsible for deciding and publishing groundwater policy, catchment plans, areas worthy of capital investment, flood alleviation plans, conservation action plans, regional reports to NRA Head Office and so on. In addition, many of these people have an increasing amount of contact with third parties including Members of Parliament and the Public, Farmers, Secretary of State, Environmental Action groups such as Friends of the Earth and Green Peace. These "environmental stakeholders", the majority of which are not IT specialists, expect to be kept informed of project progress in key activity areas and strategic planning in a way which they can readily assimilate and understand.

Progress made using SPANS GIS

TYDAC Technology's SPANS GIS has been implemented by five of the ten NRA regions in England and Wales. The system was selected largely due to its ease of use, comprehensive functionality, analytical capability, and output facilities. The system has now been applied to a diverse range of projects, and as a result has delivered quantifiable benefits both in terms of cost savings and improved environmental management. Water Resources Management projects to date include:

a) groundwater resources management - the production of catchment action plans and vulnerability maps using multi-criteria models optimized to allow the amounts of water infiltrating into aquifers to be quantified;

b) assessment of nitrate sensitive areas - identification of the aquifers and other groundwater supplies at risk of exceeding EC nitrate standards and the remedial action required;

c) regional implications of land management activities - taking into account the use of fertilisers, pesticides, acid runoff and agricultural practices.

Projects in the area of groundwater resources management have enabled hydrologists and hydrogeologists in the NRA's Severn Trent region to utilize satellite data and GIS functionality to establish practical solutions for groundwater resources management. GIS has allowed professionals to use the technology to re-visit traditional problems and apply alternative approaches. However, the implementation of environmental plans and strategies cannot be done by the NRA alone but requires co-operation from a wide range of environmental stakeholders. Although GIS is an excellent tool for professionals, an alternative technique is required for demonstrating, communicating and sharing ideas on a regional and european scale with environmental stakeholders, european policy makers and funding committees.

MULTI-MEDIA - THE COMMUNICATION ALTERNATIVE?

GIS technology has developed and improved many times over in recent years as a direct result of research activity, user involvement and vendor cooperation. Such initiatives have increased users' awareness of spatial thinking and encouraged confidence in the widespread use and exploitation of GIS technology. It would appear, that given this appreciation of the power of GIS and spatial techniques in general, that the time is now right to progress to the next logical step - Multi-media.

What do we mean by the term Multi-media?

In managing water resources, as in other aspects of environmental management, we find that the majority of the material critical to the decision making process is not necessarily available in a convenient form.

Only on rare occasions is key data available in conventional formats such as columns in tables which can be easily imported into software packages such as spreadsheets and transformed into useful statistical information. Such data can be easily assimilated into a GIS to produce maps annotated to indicate sites and features of interest. It is more likely that the immediacy of an incident such as a flood will summon a helicopter equipped with video camera to record the extent of a flood event or the widespread wildlife mortalities caused by a pollution incident. Photographic sets of satellite imagery may be browsed to assess which images are pertinent to the analysis of a specific catchment area. The bare record of a species occurrence at a location will be given both relevance and context by an accompanying photograph or video sequence. Noise nuisance due to works, which can be difficult to describe can be recorded, digitally stored, and replayed.

Much legislation is based around the creation, update and interpretation of textual documents - discharge consents and abstraction licences are good examples in our context. Thick technical reports are the norm when compiling management recommendations. All the above are examples of "multi-media documents". Multi-media systems extend the information management capabilities of computers to this kind of material, and the term "hypermedia" has been applied to systems which provide structured, exploratory and explanatory navigation within these documents. Without a hypermedia system the user, whether water resources planner, district official or member of the public is left to make "links" as best he can. With such a system in place the user can answer such questions as: What is the precise location of a video sequence? Where was that sound recorded? What is the visual impact of a discharge on the environment? What will be the view from a location if a planning decision is approved?

A hypermedia system, entitled "Microcosm", developed for MS-Windows by Southampton University is used in our prototype. When coupled with real-time video compression and decompression as implemented in "ActionMedia" (a sophisticated product from IBM using Digital Video Interactive (DVI)), Microcosm can be used to build an information management system to handle a diverse range of multi-media data in order to facilitate answers to questions similar to the above. For example, dynamic links can be defined within a document to and from read only media such as CD-ROM and moving video. The continued rapid development in computer technology, making more and more processing power, speed and storage sizes available at the desktop, mean that such concepts can now be practically implemented.

The effective communication of environmental quality to the public is a growing requirement of legislation at an EC level. For example, the communication of up to date information to answer the question "What is the water quality in my river?" will not be achieved either by a general purpose and probably out of date leaflet, or at the other extreme, by obscure print-outs of aged water quality sample data. Using our approach an enquirer could go to a public terminal, zoom to his area of interest using a computer based map, select river locations of interest, call a brief video sequence to confirm his choice, and then ask for time series graphs of key water quality variables against relevant EC limits using the latest analytical data drawn directly from the on-line databases of the controlling authority. Using such a methodology, public awareness, understanding, and participation would be greatly increased as would the accountability of the private water companies and NRA. This technology is already within our grasp, and forms the basis of the prototyping activity which we go on to describe.

An example application - The Blackwater Catchment

Water resources management is a critical step towards the goal of effective catchment planning. Catchment planning involves managing the interactions and conflicts between a whole variety of users in the natural water environment. GIS has already been used to integrate a range of relevant data sets, e.g., Local Authority information, environmental quality data, flood defence standards of service, water quality and fishery data, topographic and map information, in order to produce catchment management plans. Currently, the methodology involves the production of a general text-based policy document for each catchment which includes the following information:

a) a description of the catchment;
b) a description of the current and future activities in the catchment such as water abstraction, angling, flood defence requirements;
c) a statement on the current status of the catchment in terms of water quality, water resources and physical features;
d) a detailed action plan to resolve issues and problems, together with estimated costs and time scales;
e) supplementary information to support and illustrate each section of the plan including: text, aerial and general photographs, site plans, maps and videos.

To date, Thames NRA has used GIS in the catchment management process. In the production of its Blackwater Catchment plan, Thames used SPANS GIS to store, manage and manipulate data sets and to produce supporting maps and statistics. Desktop mapping, a subset of GIS which allows non-expert users to easily visualize spatial data, has been used to facilitate communication of results and strategies amongst NRA staff and to stakeholders. However GIS and Desktop mapping can only support maps and data stored, and referenced in spatial databases, and cannot easily reference associated text, video sequences or photographs.

The Blackwater catchment Multi-media Project

The Blackwater catchment is being used as a prototyping environment to explore the potential for integrating GIS and multi-media and to investigate various technological solutions. The base for this work is SPANS GIS running in an OS/2 Version 2.0 environment, closely coupled to the OS/2 SQL database environment, with communication to the Microcosm hypermedia and comprizing ActionMedia based on DVI technology to allow windowed access to moving and still video and sound capture and playback facilities. Such a configuration enables full use to be made of the OS/2 Version 2.0 integrated desktop and help facility.

The possible uses of such a system are in three categories. First we are looking at multi-media integration to support operational needs. We are customizing the SPANS main menu so that hypermedia sequences can be invoked to describe and explain operational procedures. We are building a "point and follow link" facility. SPANS allows a location on a map to be selected so that map attributes and database contents can be viewed for that point. The integration of multi-media extends this to allow any multi-media document to be automatically retrieved and inspected, stored animations to be run, or indeed a sequence of documents illustrating some aspect of the location to be initiated, and other document links to be constructed. In fact the full range of hypermedia functions can be accessed from a user-selected map location. In the water management context, textual consent documents relating to discharge points could be selected and viewed on-screen from a geographic map and then linked to a video sequence of the discharge in operation together with on-going engineering works captured using a camcorder which in turn may provide links into related publicity information such as press reports and photographs.

Secondly, we are looking at the potential of multi-media to construct an environmental audit trail from data acquisition to map production. This would enable such items as instrument siting and weather conditions to be simply recorded using a camcorder, and then linked to tabular data.

Thirdly, we are looking at a whole range of educational possibilities. Clearly all material collected and linked can be compiled into instructional sequences to inform the public and train new and existing employees. Illustrated sequences might be embedded permanently into the GIS on-line help system as a customized super-help related directly to the specific uses in the organization to which the system is being put, and to enforce particular operational standards.

To prototype the possibilities of public access to water quality data we are co-operating in a Friends of the Earth project to use the Hampshire

Environmental Network (HEN). HEN also a pilot, is an online dial-up environment to enable local groups including libraries and community schools to access local environmental information. We intend to explore the use of GIS and multi-media in this project, with the aim of making local water quality information more accessible and comprehensible.

CONCLUSIONS

There is great potential for the adoption of multi-media as an environmental communication tool, both within the UK, and throughout Europe. The demand for increased public awareness and availability of environmental data on both a UK and European scale is increasing. Environmental action groups such as Friends of the Earth and Green Peace are actively lobbying governments, and also providing their own information services to facilitate progress. In the case of Europe-wide data and information, the planned European Environmental Agency (EEA) is likely to serve as a clearing house, however it will need to set standards for data interchange and access. By coincidence, European GIS data standards and transfer formats are currently under discussion. In this paper we have shown that GIS is an excellent vehicle for collecting, storing and managing environmental spatially-related data, and much progress is being made by the NRA's regions in the UK. We have also shown how the integration of GIS with multi-media is the next logical step, and an altogether more effective communication tool. In pursuance of european-wide environmental communication it might be sensible not only for the EEA to set standards for data quality and access, but also to sponsor a European Environmental Information Technology Architecture to encourage the collation of information in each member state on a truly national scale, and its subsequent communication between member states. Such an architecture should be able to support multi-media, GIS and online access from anywhere in Europe.

Acknowledgements This paper was written while Christine H. Ashton was a Marketing Consultant with TYDSC Technology Ltd., Henley-on-Thames, Oxfordshire, UK.

Groundwater modelling and GIS: integrating MICRO-FEM and ILWIS

A. BIESHEUVEL & C.J. HEMKER
Faculty of Earth Sciences, Vrije Universiteit, P.O. Box 7161, 1007 MC Amsterdam, The Netherlands

Abstract In groundwater modelling, the large number of geographical data necessary to create a model, such as transmissivities, hydraulic resistances, groundwater heads, recharges and discharges, often require laborious input by hand. By integrating MICRO-FEM (Hemker & Van Elburg, 1990), a finite element groundwater model, and ILWIS (ILWIS, 1992), a vector and raster GIS, most pre-and post-processing can be done by computer and it becomes possible to improve modelling methods. This paper focuses on pre- and post-processing by means of a GIS and on the development of two interface programs, ILFEN and FEMVAL (Biesheuvel, 1992).

INTRODUCTION

Nowadays it is necessary to increase our knowledge about the quality and quantity of groundwater and groundwater flow systems, since droughts and environmental pollution threaten our groundwater reserves. Groundwater modelling has proven to be an important tool in acquiring this knowledge. The groundwater models are used frequently by consultants, drinking-water companies and research institutes. In The Netherlands the MICRO-FEM groundwater modelling program (Hemker & Van Elburg, 1990) is widely used to simulate groundwater flow.

In spite of the user-friendly preprocessor of this package, setting up a complex simulation in a large area means a lot of editing work and preparations. In this case a GIS can be very useful (Fig. 1). This GIS should also run on IBM-compatible microcomputers and must be user-friendly. The ILWIS-program (Integrated Land and Water Information System) suits these requirements (ILWIS, 1992).

GROUNDWATER MODELLING WITH MICRO-FEM

The MICRO-FEM programs have been developed to create and analyze a wide range of multi-layer steady-state groundwater flow models with a maximum of four aquifers, 3000 nodes and 6000 elements. The computer is used to replace the laborious finite element administration and the file editing tasks by means of full graphic screen control. The presentation of model results includes water balance calculations, three-dimensional flowlines and travel times. The MICRO-FEM package includes:
(a)   FEMGRID : to generate a triangular network;

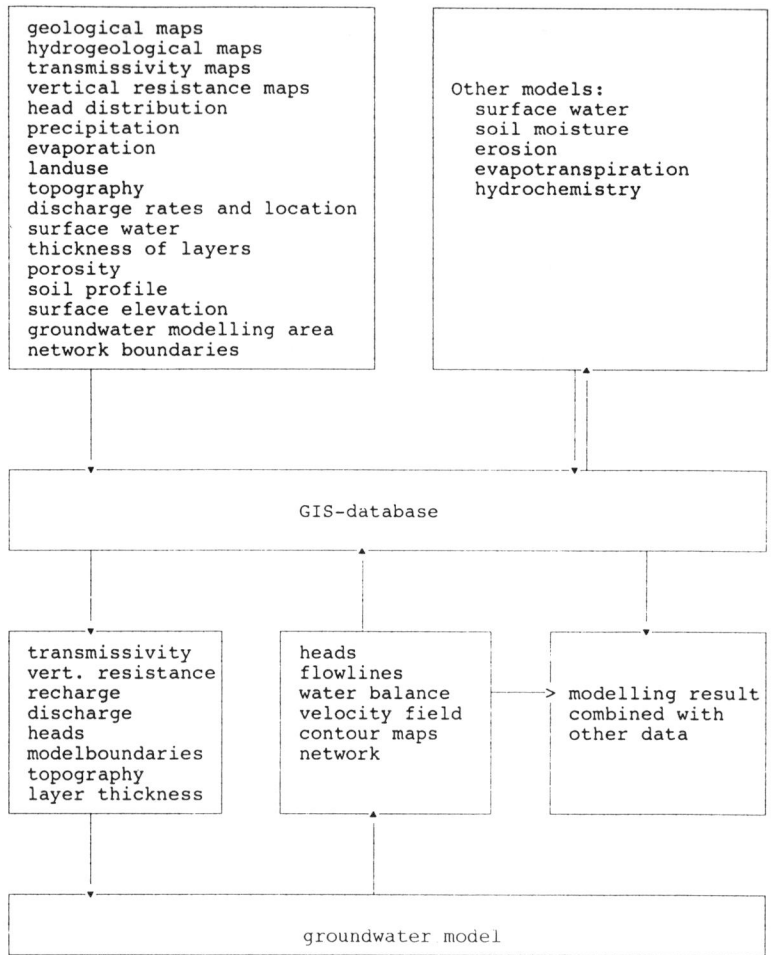

Fig. 1 The use of a GIS in groundwater modelling.

(b) FEMODEL : to modify the grid and to specify aquifer and aquitard parameters, discharges and boundary conditions, to provide a graphical representation of entered data (pre-processor); and to observe, analyze and interpret model results (post-processor);
(c) FEMCALC : to calculate nodal heads;
(d) FEMPLOT : to plot (or write to a file) a network, the distribution of heads, aquifer and aquitard model properties and flowlines;
(e) F3MODEL : to calculate and display three-dimensional flowlines;
(f) FEMPROF : to project flowlines on a profile.

OVERVIEW OF ILWIS

The ILWIS programs (Integrated Land and Water Information System) have been developed at the ITC (International Institute for Aerospace Survey and

Earth Sciences, Enschede, The Netherlands) and run on IBM-compatible microcomputers, with a coprocessor and an additional graphic screen. ILWIS is a GIS that can handle both raster and vector data. It integrates image processing capabilities, tabular databases and conventional GIS characteristics. All operations are performed through a user-friendly menu. Maps with point-, line- (isoline) or area-based information can be digitized and converted to raster maps.

INTERFACING BETWEEN ILWIS AND MICRO-FEM

The general ideas of the use of a GIS in groundwater modelling (Fig.1), are aplied to ILWIS and MICRO-FEM. Two interface programs, ILFEN and FEMVAL (Biesheuvel, 1992) have been developed to facilitate the integration and interaction of MICRO-FEM and ILWIS. ILWIS can support groundwater modelling with MICRO-FEM in four ways:
(a) building a network (grid);
(b) calculating and assigning values to the model;
(c) combining the results of groundwater modelling with other types of data;
(d) creating background maps which can be used during the modelling process.

Figure 2 displays the file and program structure of the interface between ILWIS and MICRO-FEM.

Building a network

Building a finite element network with ILWIS and MICRO-FEM requires four steps described below:
(a) define the modelling area by drawing all boundaries on a map, the outer boundaries to define the whole modelling area, the inner boundaries to define areas where a higher node density is required, for example because steep gradients in the head distribution are expected (e.g., near wells, rivers, etc.). To simplify the calibration process, it can be useful to incorporate the location of wells and measuring points in the triangular network. Put these points (with a fixed location) on the map together with the outer and inner boundaries;
(b) digitize all boundaries (segments) and points with ILWIS. Create closed regions (polygons) where a certain grid spacing is required;
(c) use ILFEN to check the digitized boundaries and convert them to a file that FEMGRID can use (an incomplete FEN-file). ILFEN calculates the spacing of nodes on each segment, depending on the node spacing of the regions on the right and left side of the segment;
(d) use FEMGRID to generate the grid from the incomplete FEN-file. All segment nodes, digitized or generated by FEMGRID and all digitized

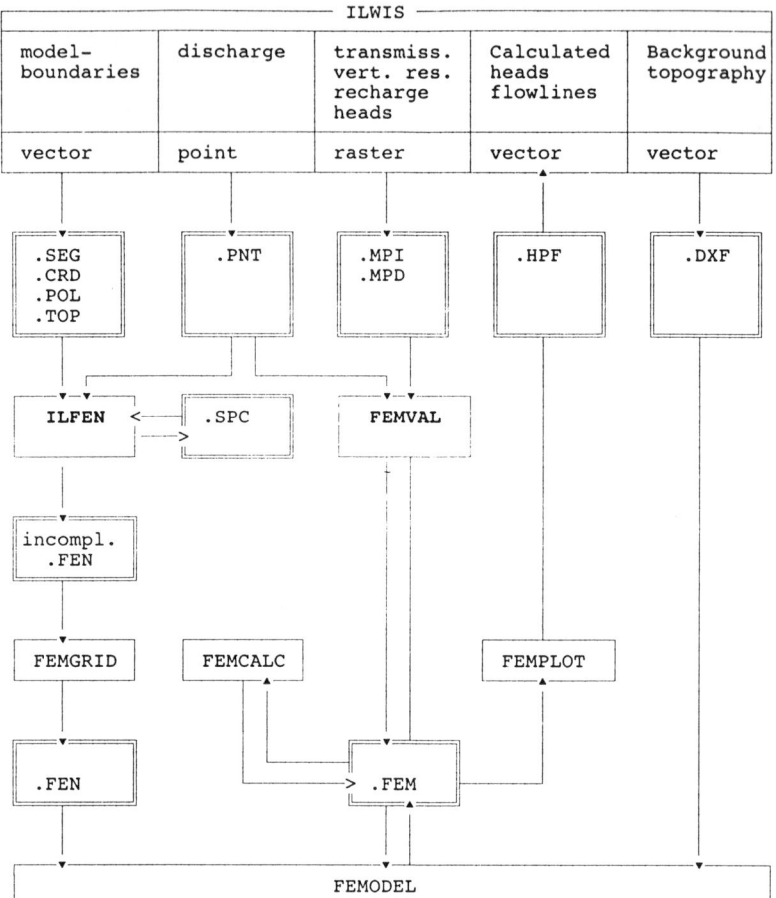

Fig. 2 File and program structure interface ILWIS - MICRO-FEM.

points will become nodes with fixed locations in the generated network. When the grid has been generated and the grid data have been saved in a FEN-file, start FEMODEL and enter the desired number of aquifers. Default values are assigned to all parameters and the resulting model can be saved in a FEM-file.

Calculating and assigning values to the model

Two steps are required to assign parameter values to the groundwater model :
(a) digitize or calculate all input maps, such as maps with transmissivities, vertical resistances, heads and recharges, with ILWIS and convert them to raster maps. Data derived from isoline-maps or point-data can be interpolated with ILWIS in various ways, including kriging. Discharges should be saved as "point-maps" (PNT-file);
(b) use FEMVAL to assign values from these maps to the model (FEM-file). Point-data (discharges) are assigned to the nearest node, if the distance to this node does not exceed a user definable threshold value.

Combining results of groundwater modelling with other types of data

The results of the modelling process, such as calculated heads and flowlines and the generated network, can be saved in HPGL-files (Hewlett Packard Graphics Language) using FEMPLOT. It is possible to import these files in ILWIS, to combine the results with other maps. In this way, it is very easy to integrate modelling results with other information, or to use the results as input for other models.

Creating of background maps

The use of background maps during the modelling process is supported by the latest versions of ILWIS (1.3) and MICRO-FEM (2.2). These background maps can contain, for instance, the location of roads, rivers, or areas of particular interest. Digitize these maps with ILWIS and save them as DXF-files (Drawing Interchange File, an ASCII file also used by AUTOCAD). The maps can be displayed with FEMODEL together with the nodes, the network or the results of the modelling process. There is a restriction : FEMODEL reads only the LINE information from a DXF-file.

CASE STUDY : THE VELUWE AREA

The developed interface has been tested in a case study in the central part of The Netherlands: the Veluwe area (Fig. 3). The central part of the modelling area is a relatively high area (maximum circa 100 m above m.s.l.) and in the northern part there are low lying polder areas (5 m below m.s.l.), separated from the "mainland" by lakes. Near the eastern boundary is the River IJssel, and near the southern boundary the River Rhine. The following points are important for the generation of a network for this area:

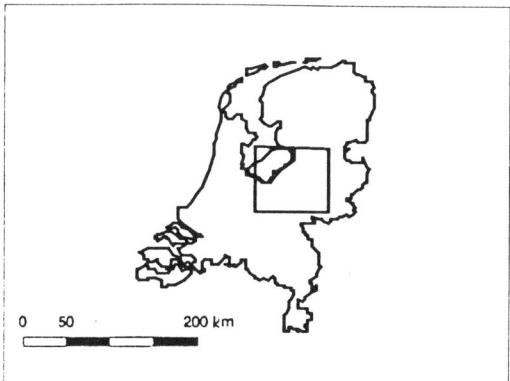

Fig. 3 The location of the modelling area in The Netherlands.

(a) a dense grid is required near rivers and lakes and near wells with large discharges;
(b) in the central part a less dense grid can be used;
(c) nodes with fixed locations are required for the rivers.

Figure 4 displays the digitized boundaries. Figure 5 displays the network generated using this input.

The three-dimensional model is represented by a layered system of three aquifers, separated by aquitards. Compiled maps of transmissivities and hydraulic resistance of aquitards are digitized and rastered in ILWIS and transferred to MICRO-FEM, using FEMVAL.

During calibration, special attention is paid to the anisotropic character of parts of the aquifers. It is assumed that average flow conditions are represented sufficiently accurately by steady-state models. Thus storativities are specified and transient flow during a 40 years period is simulated to investigate the fluctuations of the groundwater table on a regional scale.

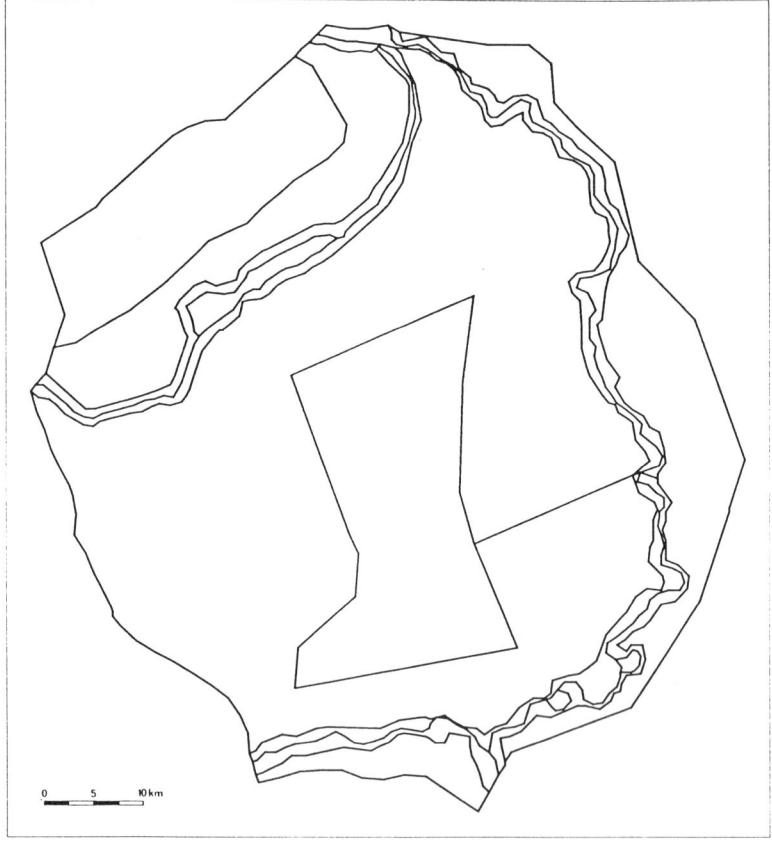

Fig. 4 Boundaries digitized with ILWIS. These boundaries define regions where a different grid size is required. At the eastern side River IJssel is visible, at the northern side the Flevopolders.

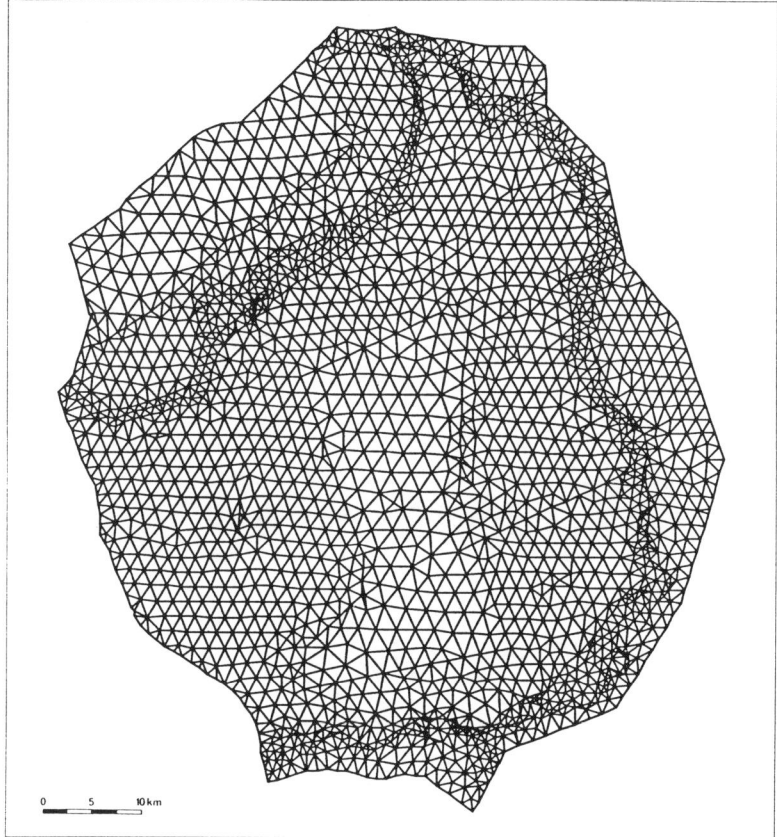

Fig. 5 Grid generated with FEMGRID. Notice the dense grid near the River IJssel and the less dense grid in the central area (Veluwe).

CONCLUSIONS

Groundwater modelling programs should have such extended pre- and post-processing possibilities, that they can be considered as GIS programs with a specific field of application. This has been achieved by developing interfacing programs which integrate the groundwater model MICRO-FEM with the ILWIS. The inerface is applied to the study of groundwater fluctuations in an area in the central parts of The Netherlands. It appears that both data entry and the presentation of model results are speeded up considerably by the easy exchange of vector and raster data with GIS programs. Nevertheless, the calibration of complex models is, and will remain the time-consuming part of groundwater modelling

REFERENCES

Biesheuvel, A. (1992) Groundwater modelling and GIS, Integrating Micro-Fem and ILWIS. Faculty of Earth Sciences, Vrije Universiteit, Amsterdam, The Netherlands.

Hemker, C.J. & Van Elburg, H. (1990) MICRO-FEM User's Manual, Version 2.0. *Microcomputer Multilayer Steady-State Finite Element Groundwater Modelling*. Amsterdam, The Netherlands.

ILWIS (1992) Integrated Land and Water Information System, User's Manual, Version 1.3. Computer Department ITC, May 1992. International Institute for Aerospace Survey and Earth Sciences, Enschede, The Netherlands.

Models, GIS, and expert systems: integrated water resources models

K. FEDRA
International Institute for Applied Systems Analysis, Advanced Computer Applications, 2361-Laxenburg, Austria

Abstract The use of computers for water resources planning and management and the modelling and design of components of the hydrological cycle is a well developed field with a substantial tradition. Most hydrological and water resources problems do have an obvious spatial dimension. Within the domain of modelling this is increasingly being addressed by more complex spatially distributed models, in part made possible by the rapid development of computer technology. Geographic information systems can be used as tools to capture, manipulate, process and display the spatial or geo-referenced data for and results from these distributed models. Consequently, the integration of these two fields of research, or sets of methods is an obvious and promising idea. More complex and integrated tools however, also require a better user interface to fully exploit their potential. Advanced information technology provides the tools to design and implement smart software where, in a broad sense, the emphasis is on this man-machine interface and a flexible problem representation. Symbolic and analog, graphical interaction, visual display and animation, integrated data sources and built-in domain knowledge can effectively support users of complex and complicated software systems. Integration, interaction, visualization, intelligence and customization are key concepts that are discussed in detail, using a number of operational water resources models as examples.

INTRODUCTION

Software and computer-based tools are designed to make things easier for the human user; and they should improve the efficiency and quality of information processing tasks. In practice, only very few programs do that. They make things possible that would not be possible without the computer, but they don't always make it easy on the user. Computer based tools, and models in particular, have to be seen as an integrated part in a much more complex information processing and, eventually, decision-making procedure. This involves not only just running the model, but certainly preparing its input, interpreting and communicating its results, and making them fit the framework of the existing institutional procedures or personal workstyles.

Basic concepts of easy-to-use models and information systems include integration, interaction, intelligence, visualization, and customization. They are necessary, but certainly not sufficient conditions for useful software systems.

Integration implies that in a software system for real-world applications, more than one problem representation, model, or tool is used conjunctively; that several sources of information or data bases, possibly distributed, are accessible; and finally, that a problem-oriented user interface combines these components in a common framework to provide a rich and useful information base.

Interaction is a central feature of any effective man-machine system: a real-time dialogue allows the user to define and explore a problem incrementally in response to immediate answers from the system; fast and powerful systems with modern processor technology can offer the possibility to simulate dynamic processes with animated output, and they can provide a high degree of responsiveness that is essential to maintain a successful dialogue and direct control over the software.

Visualization provides the band-width necessary to communicate and understand large amounts of highly structured information, and permits the development of an intuitive understanding of processes and interdependencies, of spatial and temporal patterns, and complex systems in general. Also, many of the problem components in a real-world planning or management situation, such as risk or reliability, are rather abstract: visual inspection of systems behavior and the effective presentation of information is supported by symbolic, graphical representation.

Intelligence requires software to be "knowledgeable" not only about its own possibilities and constraints, but also about the application domain and about the user, i.e., the context of its use. Defaults and predefined options in a menu system, sensitivity to context and history of use, built-in estimation methods, learning, or alternative ways of problem specification, can all be achieved by the integration of expert systems technology in the user interface and in the system itself.

Customization is based on the direct involvement of the end-user, and the consideration of institutional context and the specifics of the problem domain in systems design and development. It is the user's view of the problem and his experience in many aspects of the management and decision making process that the system is designed to support. This then must be central to a system's implementation to provide the basis for user acceptance and efficient use.

INTEGRATED MODEL AND INFORMATION SYSTEMS

Integrated information and decision support systems, built around one or more numerical simulation models or rule-driven inference models, and integrated with data bases and GIS, feature:
(a) an interactive, menu-driven user interface that guides the user with prompt and explain messages through the application. No command language or special format of interaction is necessary, the computer assists the user in its proper use; help and explain functions can be based on hypertext and possibly include multi-media methods to add video and audio technology to provide tutorial and background information;
(b) dynamic color graphics for the model output and a symbolic representation of major problem components, that allow easy and

immediate understanding of basic patterns and relationships. In parallel to the numerical results, symbolic representations and the visualization of complex patterns support an intuitive understanding of complex systems behavior; the goal is to translate a model's state variables and outputs into the information requirements of the decision making process, turning data into insight;

(c) the coupling to one or several data bases, including geographic information systems, and distributed or remote sources of information in local or wide area networks, that provide necessary input information to the models and the user. The user's choice or definition of a specific scenario can be expressed in an aggregated and symbolic, problem-oriented manner without concern for the technical details of the computer implementation;

(d) embedded AI components such as specific knowledge bases ensure user specifications in allowable ranges to be checked and constrained, and ensure the consistency of an interactively defined scenario;

(e) and they are, wherever feasible, built in direct collaboration with the users, who are, after all, experts in the problems areas these systems address.

In summary, integrated information systems are designed for easy and efficient use, managing potentially very large amounts of data but also usable in data-poor situations, and cater to the user's degree of computer expertise. Most importantly, they have to address the users specific information needs explicitly to be directly understandable and useful.

APPLICATION EXAMPLES

To better illustrate all the above concepts and ideas, a few operational software examples are described below. They are all drawn from customized development projects of the Advanced Computer Applications (ACA) group at the International Institute for Applied Systems Analysis in Austria. More detailed descriptions are given in Fedra (1991a,b; 1992a,b).

River water quality

A straightforward example of a single dynamic analytical model of water quality is implemented as a component of a Dutch national environmental information system, primarily designed for technological risk management. The river model is a complementary tool to a complex risk analysis system for process plants and the transportation system that focuses on atmospheric releases, fire, and explosion. However, chemicals might also end up in the river, and the model is designed to simulate and evaluate such spill scenarios.

The model simulates the propagation of an accidental spill of a chemical, represented by its initial mass, time pattern of the spill, and a first-order decay rate of the chemical substance. The underlying numerical model was developed by Delft Hydraulics (Bockholts & Heidebrink, 1988). The model uses the output from a complex hydrodynamic model to interpolate spatially distributed and, for longer model runs, dynamic flow fields for the rather complex Rhine-Maas system, including the operation of sluices and locks. Several control options allow the interactive definition of a spill scenario, and a number of model control options provide a set of display and analysis styles. There are several ways in which this problem specification is assisted by the system: for example, to obtain the necessary decay rates for a substance, the user can call up the chemical substances data base and select a substance for simulation. The required parameter, a lumped first order decay rate, is then automatically loaded from the data base into the simulation model.

Alternatively, the user can invoke the hypertext function from the respective editors and obtain information on typical decay rates for selected substances and substance groups. Similar information is available to help estimate the amount of the spill from the nature of the accident, or to obtain water quality standards for the substance. The model itself is then run interactively, with dynamic animation of the spill's propagation. The model can be run continuously or step by step, and allows to interactively query the display to read back concentration values for arbitrary locations at any point in time. The user can also set observation points and plot the time history of the spill for selected locations along the river (Fig. 1). The display also indicates a reference concentration so that violations of this water quality standard can be directly read from the graph.

River basin management

A prototypical example of an integrated information system for river basin management is currently being developed at ACA within the framework of the EUREKA project. Bringing together data bases and geographic information systems, and simulation models and expert systems, environmental considerations in water resources management are the central theme. Water quality in rivers, reservoirs and groundwater, primarily related to their use as drinking water, and various sources of pollution such as urban runoff and treatment plants, agricultural land use and land application of manure or treatment sludge, and land disposal of wastes are addressed by the system. Environmental impact assessment of major projects such as new reservoirs, irrigation systems, or industrial and urban development, and the environmental evaluation of water resources management policies in general, are another major focus of the system.

Rather than building one monolithic software system, the approach adopted in this project is modular and tool-kit oriented. Several alternative models, for example TOMCAT from Thames Water and QUAL2 of the US EPA, both simulating surface water quality, are being integrated. Within a common framework for information management, communication and model

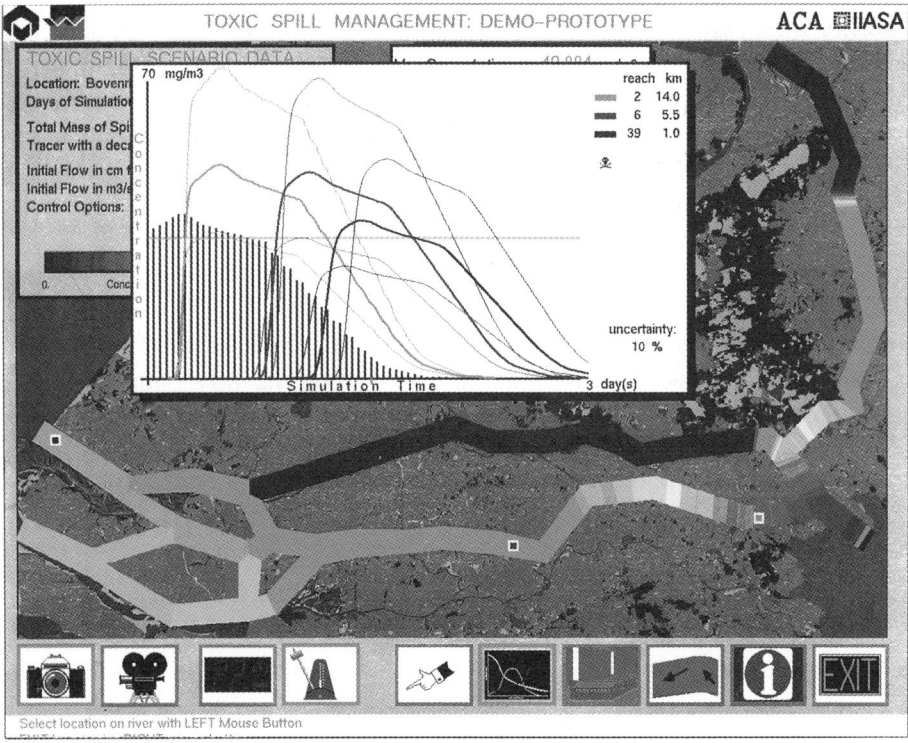

Fig. 1 Spatially distributed output from the spill model over the background map of The Netherlands. With the animation halted, diagrams for individual observation points can be drawn.

integration, different building blocks or "objects" can easily be assembled to work together. Exchange or addition of models or other building blocks should not require any major modifications in the overall system as long as the alternative or additional components comply with the generic interface definitions and are structurally equivalent.

The river basin is a natural and usually well defined unit. However, with users in the water industry or regulatory government agencies, administrative and political boundaries do not always coincide with the natural catchment borders. Embedded GIS technology provides the tools to manipulate and organize, analyze and display spatially referenced data with a flexible definition in a very efficient way.

A GIS is therefore in the core of the system, with the necessary linkages to the various data bases and models, and the multiple local study areas within one major river basin. And to obtain most recent and synoptic data with full coverage, satellite imagery such as Landsat TM or SPOT is used extensively (Fig. 2). Another important element are digital elevation models used for river basin definition. Together with the surface characteristics derived from the satellite imagery, soil maps, and hydro-meteorological station data, they provide the necessary information basis for surface runoff and routing models, and non-point source pollution or erosion modelling. The integration of necessary background data and the linking of tools to derive the necessary model input data makes sure that models are always ready to run, and that a set of default inputs is always available, that may only require small modifications to represent a new problem situation.

Within this framework of data bases and GIS, dynamic simulation models for scenario analysis are tightly coupled. Eventually, the system will comprise models describing rivers and reservoirs, the groundwater system, estuaries and the adjacent coastal marine areas. Water quantity, both for supply management and flood protection, and water quality must be considered, and different forms of water use, industrial, agricultural, and domestic as well as the regulatory framework of water use and environmental legislation provide the socio-political and economic context for a comprehensive approach to river basin management.

In the current prototype, the first model implemented is a groundwater model that simulates contamination problems due to landfills and possibly more wide spread sources of pollution such as agrochemicals and surface application of treatment sludges. The contamination of drinking water is the primary problem addressed. Alternative protection and clean-up strategies can be designed interactively and then simulated to evaluate their efficiency. Initial and boundary conditions for the 2D vertically integrated model are loaded from the GIS, where the respective data matrices are kept as topical maps or cell-grids (Fig. 3).

Groundwater management

For the management of hazardous waste, site selection and risk assessment of landfills, or the design and evaluation of groundwater remediation measures, a set of interactive groundwater models was developed at ACA (Fedra & Diersch, 1989; Fedra *et al.*, 1992). Using several basic 2D models, based on finite difference and finite element schemes respectively, these systems integrate a geographic information system for the manipulation of background maps such as geological, biotope, or land use maps, and input data sets for the model. The systems also incorporate a built-in expert system that assists in the definition of input parameters and decision variables such as source strength or pumping rates. One of these models is being integrated in the river basin management system described above.

The interactive model provides dynamic output, that can be viewed as a color coded overlay over various topical maps or a pseudo 3D display of concentration fields or groundwater head (Figs 4 and 5). The user can introduce new wells or contamination sources or modify existing ones, and then resume the simulation or "rewind" the model and start another run with the new scenario definition. To assist the specification of some of the input values

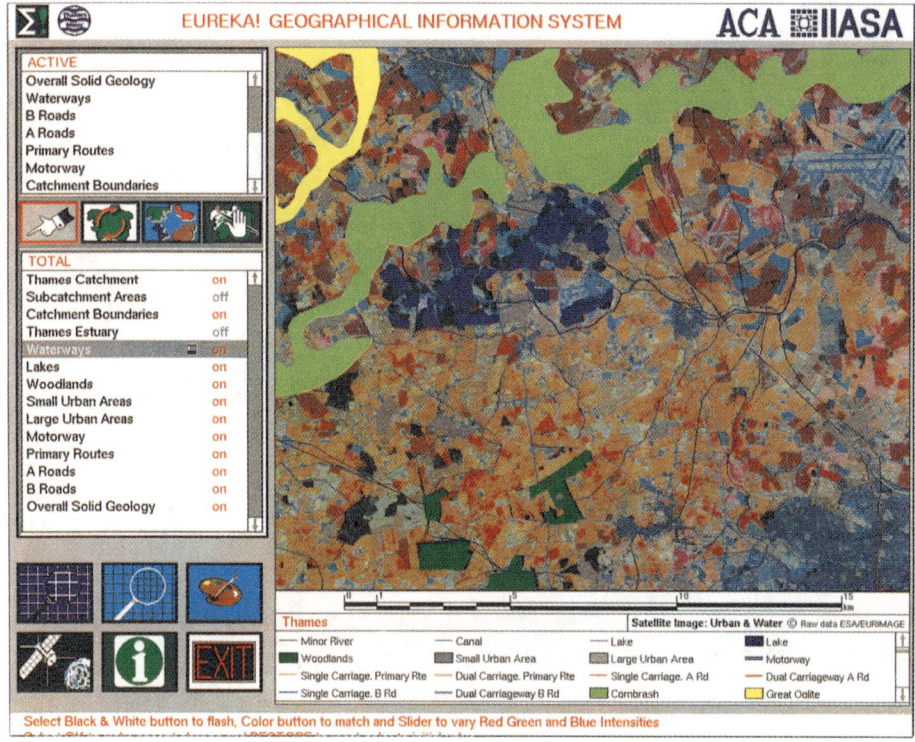

Fig. 2 The hybrid GIS of the River Basin Management System, combining polygon overlays from the land use and geology maps and line features over the basic Landsat TM scene.

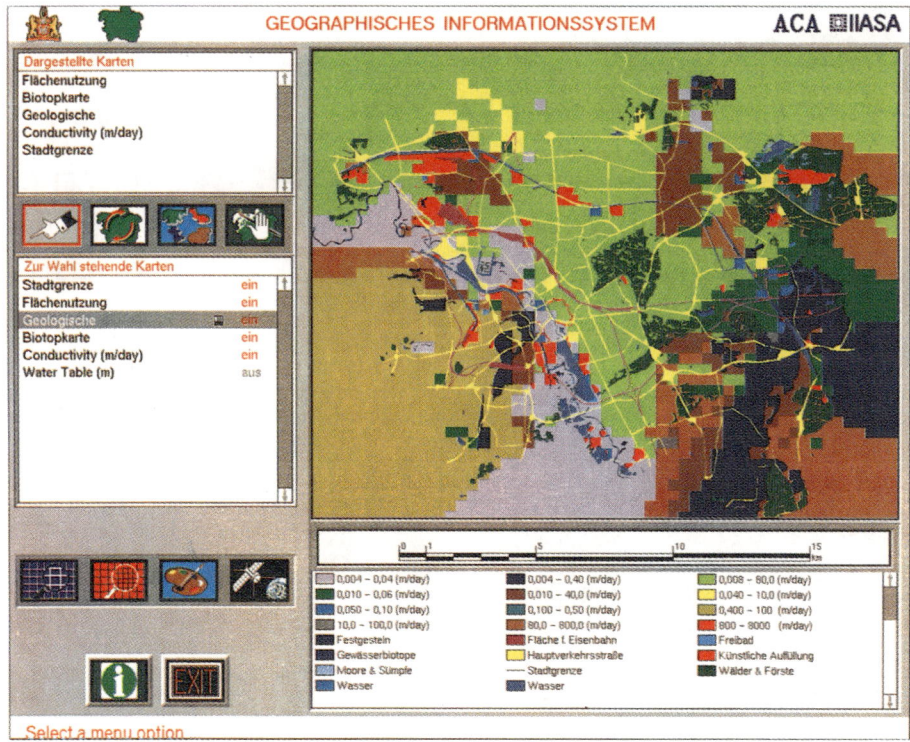

Fig. 3 Overlays of land use and geology over a cell grid file of conductivity. The conductivity map is directly used as an input data set for the 2D groundwater model.

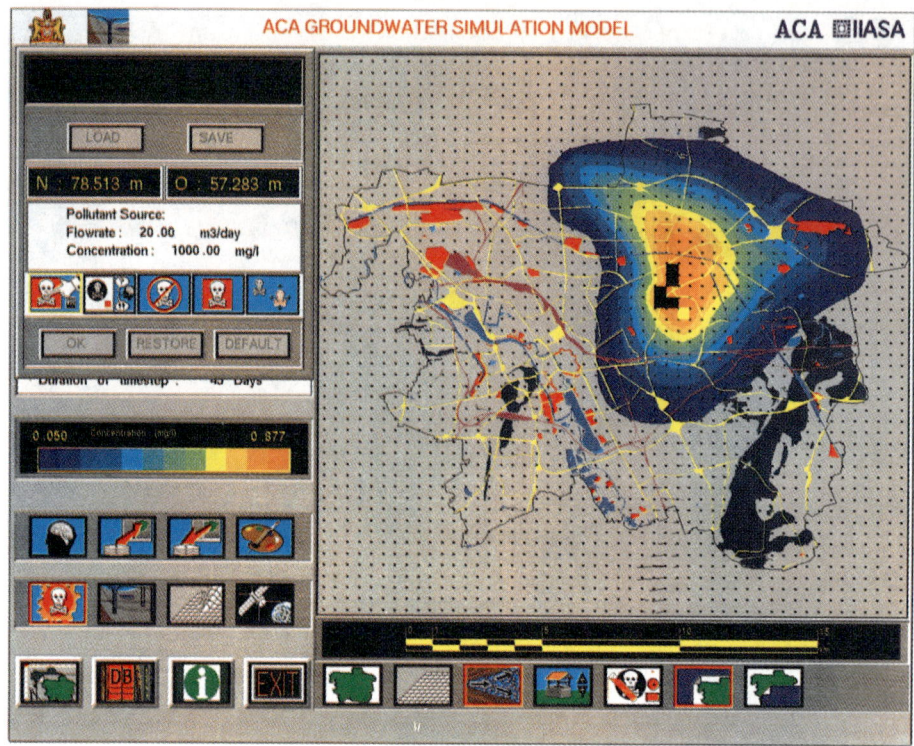

Fig. 4 Contaminant plume over a subset of map features. While the dynamic model is paused, wells or sources of contamination can be edited.

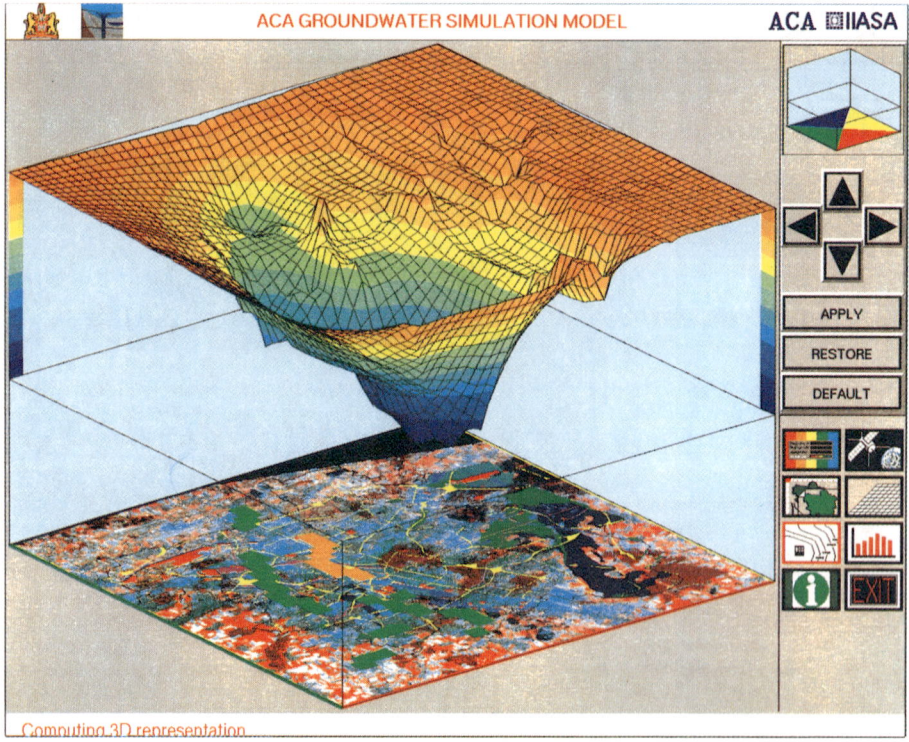

Fig. 5 A pseudo 3D view of the groundwater head over the background map. Different views can be generated by arbitrary rotation of the data set.

required, the system offers in addition to the basic editing function and hypertext support an embedded expert system. This will guide the user through a dialogue session, compiling information that will allow him to estimate the value in question. For example, when defining a pumping rate for a water supply well, the system will inquire about its purpose and then continue to compile information specific to this purpose. If, for example, an irrigation scheme is specified as the form of use, the system will inquire about the size of the irrigation area, irrigation technology and management, and crops irrigated. With reference to the GIS and its climate data base, crop water demand and irrigation water demand can be estimated by rules or simple models, and the estimated water requirement will be suggested to the user. The system will then check whether the expected drawdown from meeting these requirements is reasonable in relation to other, existing water uses in the area, and suggest a feasible pumping rate. During this dialogue, the user can display the rules used and, whenever the system reaches an intermediate or final conclusion, ask to have the underlying reasoning explained. The terms used in the rules are again linked to the hypertext system, so that further explanation can be provided for a better understanding.

The final value, or any intermediate one for that matter, can be modified by the user, overriding the expert systems suggestions. However, the system enforces some absolute bounds on the values a user can choose in order to ensure that the model is used within a reasonable range of parameters. The integration of expert systems functionality allows the user to draw on a larger information basis: often, exact parameter values or model inputs are not known, and have to be estimated. Many parameters are not directly measurable but require interpretation of usually scarce data, or, in the case of some of the control and decision variables, might have to be based on a lengthy engineering study. This requires considerable experience not only in the problem domain, but in many cases also in the use of a particular model. Both elements can be encoded in the knowledge base of the system, to provide an efficient but approximate answer for the experimental and explorative use of the software.

A common architecture

Other examples of integrated model and information systems include an expert system for environmental impact assessment of water resources development projects, Fedra *et al.* (1991), where data from the GIS are used both in the rule based deduction as well as in simulation models integrated into the inference trees; or models to simulate the effects of ocean outfalls on coastal water quality (Fedra, 1992c) (Fig. 6). They all share a common architecture designed to make them efficient to use and easy to understand: one or more simulation models are coupled with the necessary data bases, a GIS, and expert systems components, and integrated in an interactive, graphical user interface.

Linking the models to the GIS as a source of data describing initial and boundary conditions not only provides a convenient tool to download this

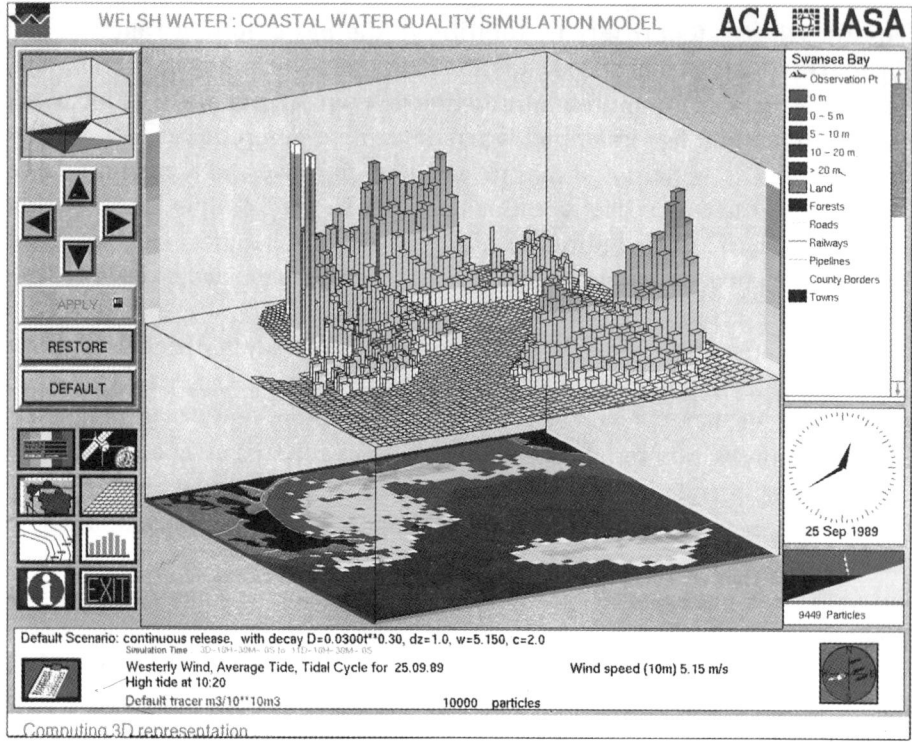

Fig. 6 Contaminant plumes from ocean outfalls in a combined 2D and pseudo 3D display. The concentration field and the models curvilinear grid are dynamic map overlays.

information. It is also a powerful tool to process some of the required information, for example, extracting catchment boundaries from a digital elevation model, or compiling transmissivity classes from geological maps (Fig. 3). However, most importantly the GIS provides the tools to efficiently display the data, also in combination with other related information for visual inspection. And if model output can be exported to the GIS, the same tools can be used to display and further analyze, e.g., by overlay analysis, model generated results as maps. If display functions and the dynamic models are tightly coupled, animated output can be generated (Figs 1, 4, 5 and 6)

Expert systems, in conjunction with simulation models, can serve two major roles: on the one hand, integrated with the user interface, they can assist in using a model or a data base by providing domain expertise and guidance. The estimation of model parameters, checking of completeness, consistency, and plausibility of scenario assumptions, the choice of appropriate tools, or paths through the information system, depending on context, are all possible examples (Fig. 6). The second major role is in modelling itself: expert systems components can be used as qualitative, logical models in combination with their algorithmic counterparts.

DISCUSSION

Integrated information and decision support systems bring together large volumes of background data, and interactively distill from it decision relevant information. Key technologies are workstations and networking, data bases and GIS, interactive graphics, modelling and optimization, and expert systems. The models, and their interfaces, are representations of the problems they address as much as of the planning and decision making processes they are designed to support. In the latter field, if not also in the former, their users are the real experts. Thus, their expertise and experience needs to be included in the systems. As a consequence, the user must be involved in the design and development, so that he can accept responsibility and ownership for the software system.

Institutional integration also must look at aspects such as user training, data entry, maintenance issues of keeping systems current and operational, providing adaptations and updates, etc. Any complex information system has more than one user at more than one level of technical competence and with different roles within an institution. Different users have different requirements that need to be supported: flexibility and adaptability are therefore important features. Systems must be able to grow with their users. Therefore, the institutional commitment and technical infrastructure to keep a system alive and evolving are as important as the scientific and technical quality of the original software system.

It is the easy-to-use "smart" interface, the fast and efficient operation, and the apparent intelligence of the programs that makes them attractive. For the specific model and GIS coupling, this means that their respective functions are fully and transparently integrated. The distinction between GIS and spatial model disappears. The system provides a coordinated set of functions or tools that cooperate in a common environment, within a single integrated system.

For the user, however, it is immaterial which method is used to generate the answer to his questions, to provide insights or arguments, help structure his thinking and communicate information within a group. And in fact, it will usually be the combination of several "methods" or tools that are required. This combination of methods of analysis, and the integration of data bases, geographical information systems, and hypertext, allows to efficiently exploit whatever information, data and expertise is available in a given problem situation. The integration of water resources management models and geographic information systems, expert systems, and interactive graphics, is an obvious, and a challenging and promising development in environmental research and applied informatics. The need for better tools to handle ever more critical environmental and resource management problems is obvious, and the rapidly developing field of information technology provides the necessary machinery.

REFERENCES

Bockholts, P. & Heidebrink, I. (1988) Modelling of accidental spills as a tool for river management. In: *Chemical Spills and Emergency Management at Sea*. Proc. of the First Int. Conf., Kluwer, Dordrecht.

Fedra, K. & Diersch, H.-J. (1989) Interactive groundwater modeling: color graphics, ICAD and AI. In: *Quantity and Quality* (ed. by A. Sahuquillo, J. Andreu & T. O'Donnell). Proc. of the Int. Symp. on Groundwater Management, October 2-5, 1989. Benidorm, Spain. IAHS. Publ. No. 188, 305-320.

Fedra, K. (1991a) Smart software for water resources planning and management. In: *Decision Support Systems* (ed. by D.P. Loucks & J.R. Da Costa), Wat. Res. Planning. NATO ASI Series. Series G: Ecological Sciences, Vol.26. Springer Verlag. 145-172.

Fedra, K. (1991b) A computer-based approach to environmental impact assessment. RR-91-13. Reprinted from Proc. of the Workshop on Indicators and Indices for Environmental Impact Assessment and Risk Analysis (ed. by A.G. Colombo & G. Premazzi), Joint Research Centre,Ispra, Italy. 15-16 May, 1990. Commission of the European Communities, Luxembourg. 11-40.

Fedra, K., Winkelbauer, L. & Pantulu, V.R. (1991) Expert systems for environmental screening. RR-91-19. International Institute for Applied Systems Analysis, A-2361 Laxenburg, Austria.

Fedra, K. (1992a) GIS and environmental modeling. In: *Geographic Information Systems and Environmental Modeling* (ed. by M.F. Goodchild, B.O. Parks & L.T. Steyaert), Oxford University Press. (in press.)

Fedra, K. (1992b) Intelligent environmental information systems. In: *Vorträge Wasserbau Symposium Wintersemester 1991/92 Ökologie und Umweltverträglichkeit*. Mitteilungen 85. (Lecture delivered at 22. Int. Hydrol. Engineering Conference, 3-4 January, 1992). Technical Univ., Aachen, Germany.

Fedra, K. (1992c) Marine systems analysis and modeling. Presented at Scientific Symp. "The Challenge to Marine Biology in a Changing World". 100 Years Biologische Anstalt Helgoland. September 13-18, 1992. Isle of Helgoland. Germany. Helgoländer Meeresuntersuchungen, Special Symp. Issue. (in press)

Fedra, K., Diersch, H.J. & Harig, F. (1992) Interactive modeling of groundwater contamination: visualization and intelligent user interfaces. *Advances in Environmental Sciences*. Springer, Heidelberg. (in press).

Application of a GIS for simulating hydrological responses in developing regions

STEFAN W. KIENZLE
Department of Agricultural Engineering, University of Natal, P.O. Box 375, Pietermaritzburg, 3201, Republic of South Africa

Abstract Geographical Information Systems (GIS) and Hydrological Modelling Systems (HMS) are separate entities that require coupling to enable communication between both systems for improving hydrological simulation by a more effective use of relevant geographical information. Present capturing, processing and manipulation of spatial data and information as well as coupling processes between the ARC/INFO GIS and the *ACRU* HMS (Agricultural Catchment Research Unit) is demonstrated for the Mgeni basin in Natal, South Africa. Decision support systems facilitate the "translation" of spatial features into meaningful hydrological variables for hydrological simulation. The link between GIS and HMSs will facilitate improved water resources planning and land management decisions to be made in regard to minimizing non-point source pollution originating from agricultural land and informal settlements. The loading of hydrological results back into the GIS for display, map production and the assessment of cause-effect relationships is considered an important aspect of coupling GISs and HMSs.

INTRODUCTION

The physical environment in southern Africa is characterized by a wide range of soils, vegetation types and a particularly variable rainfall pattern. In terms of water resources management, this high risk, natural environment is aggravated by a rapid population growth in the Pietermaritzburg and Durban regions of Natal, South Africa, which produce 20% of South Africa's Gross National Product (Breen *et al.*, 1985). Projected population increases in the region, from 3.6 million people in 1985 to between 9 and 12 million by the year 2025 (Horne Glasson Partners, 1989), and the associated rapidly accelerating water demand due to rural, urban and industrial development will exceed local raw water resources of the 4387 km^2 large Mgeni basin in the near future (Breen *et al.*, 1985). Concommitant with the exhaustive utilization of the Mgeni's water production is the expected deterioration of water quality with associated increased purification costs and health risks in areas where untreated water is widely used for domestic or recreational purposes.

In order to evaluate consequences of possible future scenarios of land utilization on the basin's water resources and to provide an aid to managers, regional planners and water boards, the *ACRU* (Agricultural Catchment Research Unit) agro-hydrological modelling system is being used and is under further development to simulate the quantity and quality of Mgeni's surface water resources. *ACRU*, which has been developed over the past decade in the Department of Agricultural Engineering at the University of Natal, is a physical-conceptual and multi-purpose modelling system (Fig. 1) revolving around multi-layer soil water budgeting (Schulze, 1989) and containing decision support systems (DSS). The model has been designed for small catchments up to approximately 50 km^2, uses a daily time step and is structured to be sensitive to land cover and drainage basin management changes. Output from *ACRU* has been verified successfully for a range of hydrological regimes (Schulze & George, 1987; Schmidt & Schulze, 1987; Smithers & Caldecott, 1991; Kienzle & Schulze, 1992), and streamflow and other variables can be simulated for a large basin such as the 4387 km^2 large Mgeni basin by using the distributed version of the model, whereby a large number of sub-basins, each relatively homogeneous i.t.o. hydrological response, can be interlinked (Tarboton & Schulze, 1991).

As is the case with most integrated Hydrological Modelling System (HMS) which are applied to large and heterogeneous basin areas, *ACRU* requires considerable spatial information on, *inter alia*, topography, climatic parameters, soils, land cover, reservoirs, population distribution and density, streams, weirs, sampling points and sub-basins. This information can be captured and stored in a Geographical Information System (GIS). In recent years, GISs have emerged as major spatial data handling tools and have been applied in the hydrological field worldwide (Jenson, 1991; Vieux, 1991) as well as in South Africa (e.g., Goulter & Forrest, 1987; Arnold *et al.*, 1989; Conley, 1989; Lynch, 1989; Herald, 1991; Myburgh, 1991). A GIS, which facilitates the capturing, storage, manipulation and display of geographical information, can be used to communicate more or less directly, via an interface, with a HMS (Tarboton, 1991). This paper:
(a)   introduces the more important input variables required by *ACRU*;
(b)   portrays the capturing of data and storing of spatial information on ARC/INFO, using the established Mgeni GIS as an example;
(c)   describes the preparation of hydrological input parameters on the GIS;
(d)   outlines the concept of coupling the HMS with the GIS i.t.o. information input;
(e)   discusses the coupling between the HMS and the GIS i.t.o. information output.

GEOGRAPHICAL INFORMATION REQUIRED BY *ACRU*

ACRU requires variables and estimates characterizing the physical features of the basin rather than using optimizing parameters. As is depicted in Fig. 1, a number of input variables are needed for the *ACRU* model. Many of these variables, such as soil type, are heterogeneous within a basin or sub-basin and variables such as land cover or population distribution vary both in space and in time. These variables need to be defined both spatially within a basin and i.t.o. their respective hydrological and water quality properties. The specialized database management system provided in a GIS offers an environment which

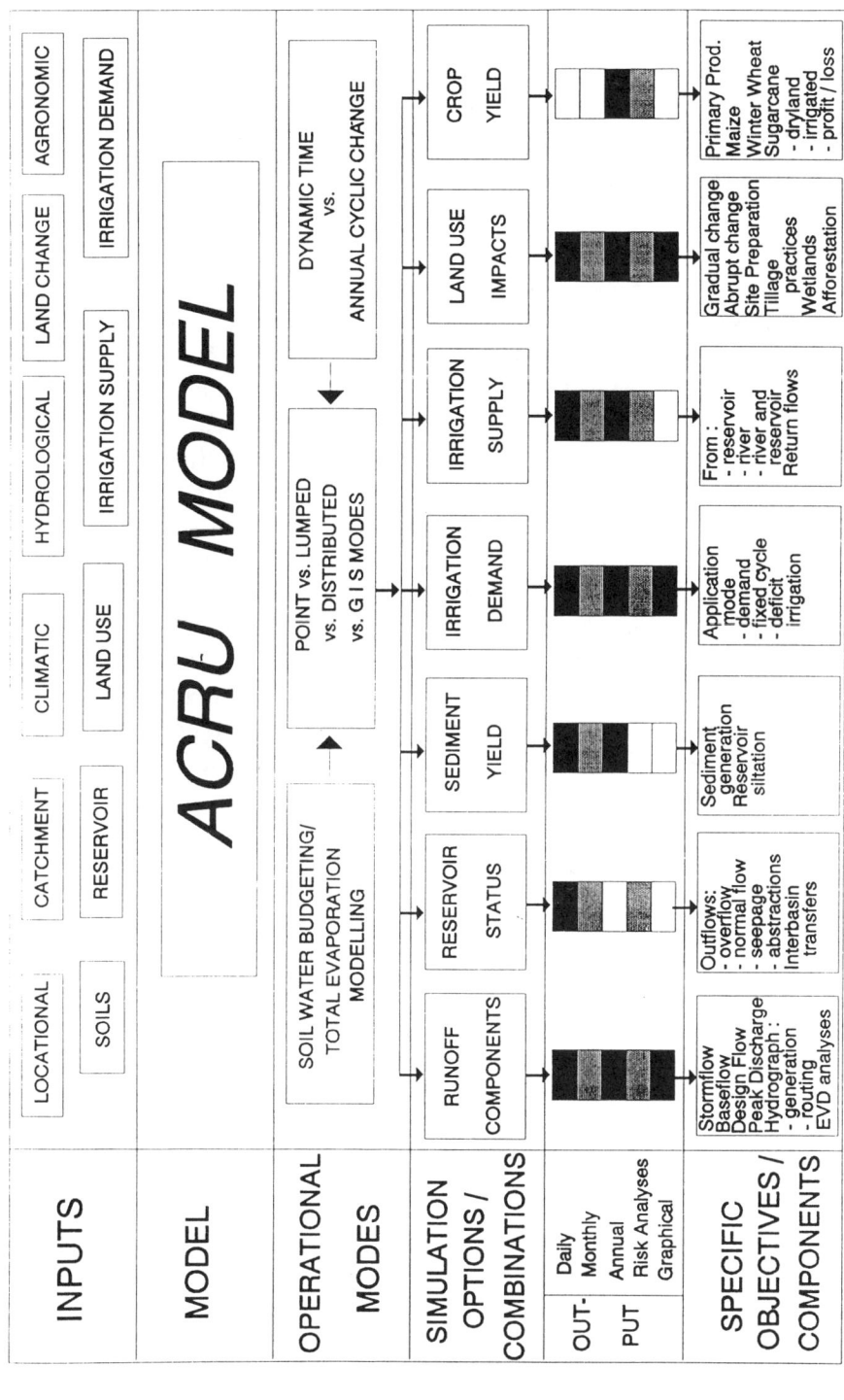

Fig. 1 Concepts of the *ACRU* agro-hydrological modelling system (after Schulze, 1989).

lends itself to not only capture, store and display spatial information, but also to attach non-spatial attributes to a spatial feature.

Therefore, a GIS is perceived as being an essential component, together with a series of DSS and the *ACRU* modelling system itself, to synthesize and manage spatial information and to integrate it into a unified system (Fig. 2).

PREPARATION OF THE GIS

Data and information were compiled and captured (digitized) in the form of point, arc and polygon coverages to form the foundations of the so-called Mgeni GIS. Maps at a scale of 1 : 50 000 formed the basis for most spatial information, and digitizing accuracy was kept to 1 mm (= 50 m in the field). Because of the size of the Mgeni basin and the unfortunate situation that certain areas in the Mgeni basin are inaccessible due to periodical unrest, information on land cover and in particular rural population distribution can only be acquired by remote sensing, which, in the case of satellite imagery, is in immediate GIS format. For land cover information, a multispectral SPOT image was geo-corrected and registered using control points digitized from 1:50 000 topographical sheets. Selected areas of the basin were displayed on the screen for visual classification and on-screen digitizing, with subsequent field verification of the land cover classification thus obtained (Kienzle *et al.*, 1992). Other GIS information, such as gridded values of rainfall and reference evapotranspiration, were generated using multiple regression functions (Dent *et al.*, 1987; Schulze & Maharaj, 1991). Additional GIS data were obtained from a number of co-operating organizations by directly importing already existing coverages. Table 1 gives a list of the different coverages which were obtained for the Mgeni basin. The attached information is either integrated directly into the GIS or can be accessed through a DSS.

A distinct advantage of having information stored in layers (i.e. separate coverages) is that, when coupling the GIS with *ACRU,* the information from each layer can be accessed individually or combined selectively with information from other layers to provide essential input to *ACRU*. For distributed hydrological modelling one needs, for example, to extract physiographic and climatic information from the GIS for each distributed sub-basin element. Hence, sub-basin delimitation and the creation of the sub-basin coverage is important, since this coverage interacts with all other layers in the GIS.

Initially, information extracted from the Mgeni GIS, through a combination of the sub-basin coverages with the physiographic and climatic layers, was not in a form that could be used directly in the HMS. Consequently, the coupling of the GIS and the HMS was required.

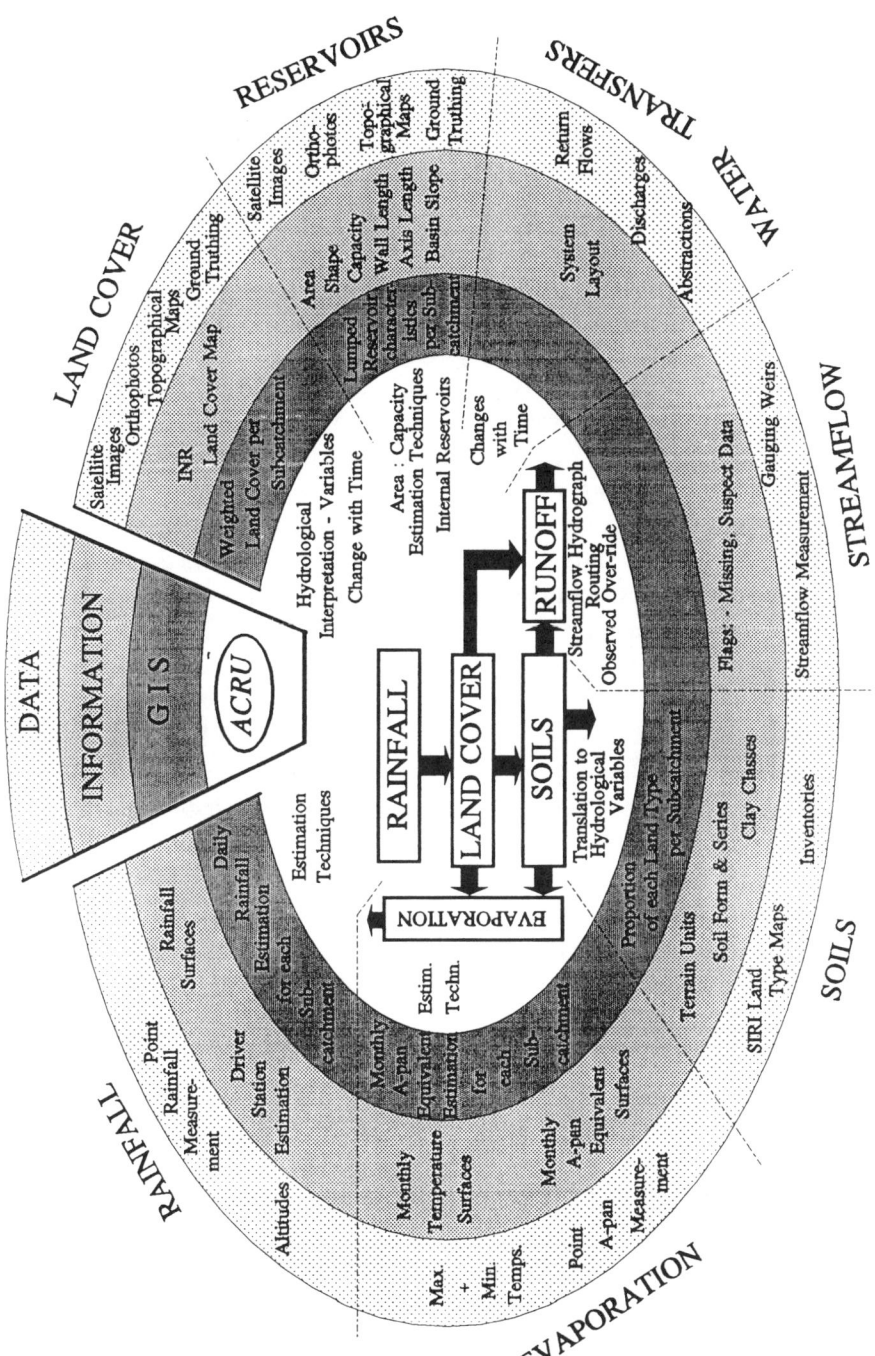

Fig. 2 Concept of integrated distributed *ACRU* hydrological modelling system (after Tarboton, 1991).

COUPLING THE GIS AND *ACRU*

GISs and HMSs are separate entities. Coupling of the Mgeni GIS and the *ACRU* model is the means by which the two systems can communicate and interact. Essentially, the GIS-*ACRU* coupling conceptualized in Fig. 3 consists of:

(a) output of spatial features from the GIS, such as land cover, soil type, reservoirs, settlements without provision of water supply and sewage, and their areal proportions and spatial distribution within a sub-basin;

(b) assignment of hydrological variables to these features via a set of DDSs, e.g., seasonally varying interception values, rooting depth and distributions, crop coefficients or fertilizer rate and application date, as well as non-transient attributes such as soil water retention characteristics, reservoir capacities and their surface areas, faeces production etc.;

(c) area weighting of relevant attributes within a delineated sub-basin or sub-basin in order to derive one representative set of hydrological properties for each sub-basin;

(d) semi-automatic input of the representative properties into the *ACRU* hydrological modelling system;

(e) operation of the model;

(f) loading the results into the GIS and displaying them in form of maps with the option of attaching graphs.

GIS PROCESSING

GIS processing includes combining the different coverages listed in Table 1 to obtain coverages with the desired input information required by *ACRU*. For distributed hydrological modelling in the Mgeni basin it is necessary to combine the sub-basin coverage with each of the coverages containing the input information required by the modelling system. This concept is illustrated in Fig. 4, which has two base coverages, the first containing a delimited sub-basin and the second a coverage with land cover information. In order to create a coverage containing land cover within the delimited sub-basin, the two coverages are "unioned", i.e. combined, resulting in a new coverage with the desired information. At the same time, attributes of the coverages are combined so that a summary of the information in the new coverage can be generated, as given in Table 2.

GIS processing of spatial information on soils and land cover involves the acquisition of the relative representation within each sub-basin by combining the soils or land cover coverage with the sub-basin coverage as described in Table 3. This table illustrates the result of the GIS's processing of land cover information for a selected sub-basin in the Mgeni basin. Similar processing of the soils coverage results in the percentage representation of each

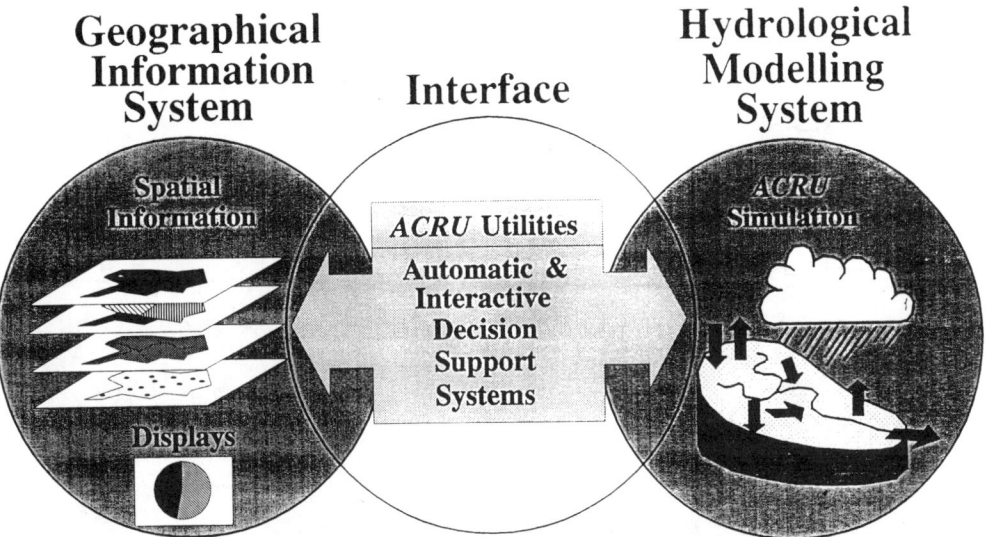

Fig. 3 Concept of the GIS-HMS-interface used for *ACRU*.

Table 1 GIS coverages and their attached attributes for the Mgeni GIS.

Coverage type	Geographical feature	Attributes attached to the feature
Point	Rainfall	S. Afr. Weather Bureau raingauge number, altitude, start year, end year, number of complete years, MAP, raingauge position
	Sampling station	current sampling frequency, first record, last record, median values and 90 percentiles for selected water quality variables
	Weirs	first record, last record, accuracy
Arc	Rivers	name, length, hydraulic properties, e.g. roughness
Polygon	Rainfall grid (1'x 1')	median monthly rainfall
	Temperature grid(1'x 1')	monthly means of daily max. and min. temperature
	Elevation grid(1'x 1')	altitude above sea level
	Evaporation grid(1'x 1')	mean monthly A-pan equivalent reference evaporation
	Soils	area, for A- and B-horizon: soil depth, texture class, wilting point, porosity, field capacity, drainage properties
	Land cover	area, crop coefficient, leaf area index, interception loss, root distribution
	Sub-basins	area
	Population	density, estimation of proportion without water supply or sewage
	Reservoirs	surface area, capacity, wall height, wall length, axis length, basin slope, dam shape

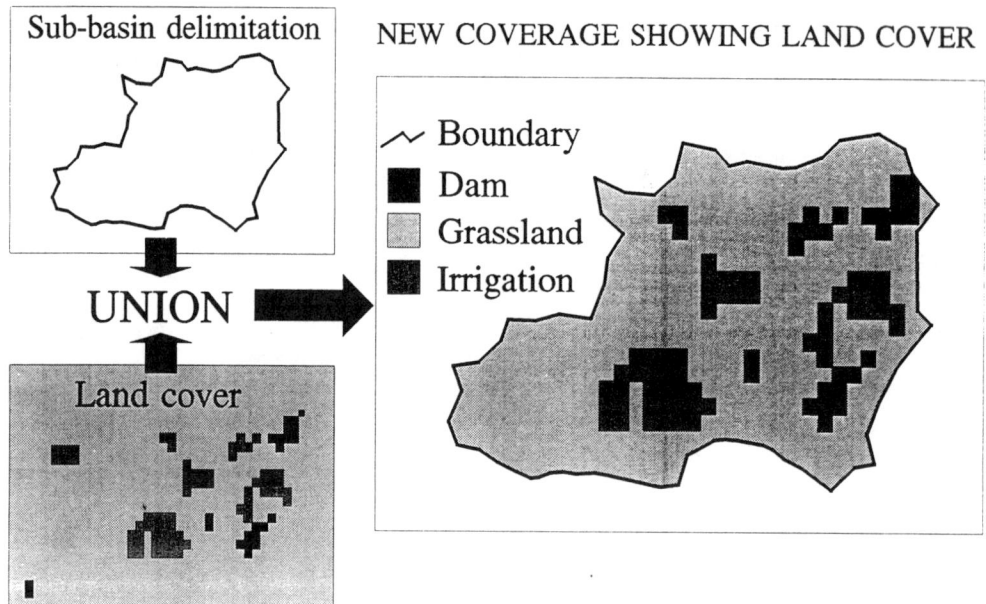

Fig. 4 Concept of combining coverages with a GIS (after Tarboton, 1991).

soil type within each of 123 sub-basins which make up the Mgeni basin. Further processing of the percentage representations of soils and land cover information to obtain hydrological variables for each sub-basin was performed by means of especially developed DSSs incorporated into *ACRU Utilities*, a computer program to guide the user through various applications.

The DSSs contain pre-programmed values of hydrological variables obtained from soils and land cover information. Land cover information from Table 3 serves as an example as to the manner in which spatial information obtained from the GIS is transformed into hydrologically meaningful variables. Table 3 displays the resultant area-weighted monthly values of crop coefficients, vegetation interception and proportions of roots in the topsoil. The table lists information that is obtained from DSSs and which can be attributed to the spatial features in a specific format automatically for direct input into *ACRU*.

Table 2 Example of land cover information within a sub-basin obtained by combining sub-basin and land cover coverages.

Land cover	Area (km^2)	Area (%)
Dams	1.25	4.16
Grassland	24.67	82.51
Irrigation	1.25	13.33
Sub-basin	33.01	100.00

Table 3 Hydrological variables obtained by entering land cover percentages into the *ACRU DSS* for the same sub-basin used in Table 2.

Variable	Jan	Feb	Mar	Apr	May	Jun	Jul	Aug	Sep	Oct	Nov	Dec
Crop coefficient	0.67	0.67	0.67	0.57	0.34	0.24	0.23	0.23	0.31	0.50	0.56	0.66
Interception loss (mm rainday^{-1})	1.12	1.12	1.12	1.03	0.94	0.94	0.94	0.94	1.03	1.12	1.12	1.12
Proportion of roots in topsoil horizon	0.91	0.91	0.91	0.95	0.98	1.00	1.00	1.00	1.00	0.96	0.91	0.91

CONCLUSIONS

With the increasing use of GISs worldwide and in South Africa, spatial information required by hydrologists and modellers become readily available, exchangeable and affordable. As soon as new or updated information on, e.g., land cover or population distribution, becomes available, it can easily be added to an existing GIS. By "translating" spatial information into hydrologically meaningful values as input into hydrological modelling systems, more meaningful results can be obtained.

The power - and the restriction - of integrated distributed modelling systems, such as *ACRU*, lies in the substantial amount of input required, in particular input on spatially distributed variables. If the modelling system has multi-level input options, as *ACRU* does, then with the availability of spatial input information, answers may be obtained more easily and can be portrayed more meaningfully for decision makers and hydrologists alike. The type of application presented in this paper is envisaged to be an initial step leading towards an era where GISs and environmental sciences will become more fully integrated entities.

Acknowledgements The author wishes to acknowledge the Water Research Commission for funding the project entitled "Development of a Distributed Hydrological Modelling System to Assist with Water Quantity and Water Quality Management in the Mgeni Catchment, Phase II", from which this study emanated. The Computing Centre for Water Research is also acknowledged gratefully for providing computer facilities. Messrs S.D. Lynch & K.C. Tarboton are thanked for their important contributions towards this project.

REFERENCES

Arnold, U., Datta, B. & Haenscheid, P. (1989) Intelligent Geographical Information Systems (IGIS) and surface water modeling. In: *New Directions for Surface Water Modeling* (ed. by M.L. Kavvas), IAHS Publ. No. 181, 407-416.

Breen, C.M., Akhurst, E.G.J. & Walmsley, R.D. (1985) Water quality management in the Umgeni catchment. *Natal Town and Regional Planning Supplementary Report*, 12, Pietermaritzburg, RSA.

Conley, A.H. (1989) Decision and information systems for South African water management - an overview. In: *Proc. Fourth S. Afr. Nation. Hydrol. Symp.* (ed. by S.W. Kienzle & H. Maaren), Water Research Commission, Pretoria, RSA, 417-431.

Dent, M.C., Lynch, S.D. & Schulze, R.E. (1987) Mapping mean annual and other rainfall statistics over southern Africa. Dept. of Agric. Engng., Univ. Natal, Pietermaritzburg, RSA, *ACRU Report*, 27.

Goulter, I.C. & Forrest, D. (1987) Use of Geographic Information Systems (GIS) in river basin management, *Water Science and Technology* 19, 81-86.

Herald, J.R. (1991) GIS: a tool for the hydrological modeller. In: *Proc. Fifth S. Afr. Nation. Hydrol. Symp.*, Univ. Stellenbosch, Stellenbosch, RSA, 9B-3-1 to 9B-3-6.

Horne Glasson Partners (1989) *Water Plan 2025*. Umgeni Water Board, Pietermaritzburg, RSA.

Jenson, S.K. (1991) Application of hydrologic information automatically extracted from digital information models. *Hydrol. Processes* 5, 31-44.

Kienzle, S.W. & Schulze, R.E. (1992) The simulation of the effect of afforestation on shallow ground water in deep sandy soils. *Water SA* 18 (4).

Kienzle, S.W., Weston, D.R. & Moolman, J. (1992) Image processing of land cover information from a SPOT image for application in a deterministic agro-hydrological model. In: *Proc. of the PICS Int.Conf.*, Pretoria, RSA, October 1-2, Microelectronics and Communications Technology, CSIR, Pretoria, RSA, (in press).

Lynch, S.D (1989) Geographical information systems in Hydrology. In: *Proc. SAGIS 89 Symp.*, Pietermaritzburg, RSA, Section 5, 1-13.

Myburgh, M.L. (1991) GIS: Present fad or future tool for the agricultural engineer? *Agric. Engng. S. Afr.* 23, 326-334.

Schmidt, E.J. & Schulze, R.E (1987) Flood volume and peak discharge from small catchments in southern Africa. Water Research Commission, *Technology Transfer Report*, 31/87, Pretoria, RSA.

Schulze, R.E. (1989) *ACRU*: Background, concepts and theory. Univ. Natal, Pietermaritzburg, Dept. Agric. Engng. *ACRU Report*, 35, *Water Research Commission Report* No. 154/1/89, Pretoria, RSA.

Schulze, R.E. & George, W.J. (1987) A dynamic process-based user-oriented model of forest effects on water yield. *Hydrol. Processes* 1, 293-307.

Schulze, R.E. & Maharaj, M. (1991) Mapping A-pan equivalent potential evaporation over southern Africa. *Proc. Fifth S. Afr. Nation. Hydrol. Symp.*, November 7-8, Univ. Stellenbosch, Stellenbosch, RSA, 4B-4-1 to 4B-4-8.

Smithers, J.C. & Caldecott, R.E. (1991) Development and verification of hydrograph routing in a daily simulation model. In: *Proc. Fifth S. Afr. Nation. Hydrol. Symp.*, November 7-8, Univ. Stellenbosch, Stellenbosch, RSA, 3B-4-1 to 3B-4-8.

Tarboton, K.C. (1991) Interfacing GIS and hydrological modelling : Mgeni case study. In: *Proc. Fifth S. Afr. Nation. Hydrol. Symp.*, November 7-8, Univ. Stellenbosch, Stellenbosch, RSA, 9B-2-1 to 9B-2-9.

Tarboton, K.C. & Schulze, R.E. (1991) The ACRU modelling system for large catchment water resources management. In: *Hydrology for the Water Management of Large River Basins*. (ed. by F.H.M. Van de Ven, D. Gutknecht, D.P. Loucks & K.A. Salewicz), IAHS Publ. No. 201, 219-232.

Tarboton, K.C. & Schulze, R.E. (1992) Distributed hydrological modeling system for the Mgeni catchment. Univ. Natal, Pietermaritzburg, Dept. Agric. Engng. *ACRU Report*, 39. *Water Research Commission Report*, No. 234/1/92, Pretoria, RSA.

Vieux, B.E. (1991) Geographic information systems and non-point source water quality and quantity modelling. *Hydrol. Processes* 5, 101-113.

Hydrological modelling within GIS: an integrated approach

NEIL STUART & CHRISTOPHER STOCKS
Department of Geography, University of Edinburgh, Drummond Street, Edinburgh, Scotland EH8 9XP, UK

Abstract Incorporating tools for physically based hydrological modelling within GIS offers mutual benefits. The presently favoured semi-distributed modelling approach seems particularly appropriate for closer linkage with GIS. Whilst specific hydrological models are best linked to GIS loosely through an interface, integrating a set of generic modelling tools within a GIS can create an advantageous environment for model development. To illustrate this, a set of tools are embedded within one commercial GIS and used to implement a semi-distributed hydrological model. Results show that this integrated approach has the advantage of producing a single environment in which users can conduct all their modelling work. This suggests that as GIS become more flexible, supporting a wider range of data models and more sophisticated customization languages, they may well become the preferred environment for hydrological modelling.

INTRODUCTION

Progress in hydrological simulation modelling and the development of geographical information systems (GIS) have largely occurred on parallel, but clearly separated tracks. This separation is surprising when one considers the common ground that the two research areas share and the scope for developments in each area to offer benefits to the other.

In the last ten years GIS have developed to the stage where large databases containing a diversity of geographic information in point, line, polygonal and cellular formats can be integrated, selectively manipulated and related to derive continuous estimates of many environmental parameters over a range of scales. Yet many GIS are still being used as advanced digital cartographic systems, oriented towards the maintenance of digital geographic data and relatively weak in respect of the higher analysis and modelling abilities that are required to make them effective decision support systems (Kehris, 1990; Densham, 1991).

Over a similar time period, researchers attempting to model hydrological processes at the scale of the drainage basin have found difficulty in synthesizing and aggregating data from a limited number of locations and in finding data of an appropriate quality and resolution for "fully distributed" models, such as the Institute of Hydrology Distributed Model (IHDM) or the Syteme Hydrologique European (SHE). It seems that the fine spatial resolution at which these models operate and their need for many parameter values which are difficult to measure in the field has limited their usefulness for practical purposes (Anderson & Rogers, 1987; Bathurst, 1988).

Partly because of these difficulties, the "semi-distributed" approach to hydrological modelling has gained ground as an attractive means of retaining the benefits of physically realistic results, but without these overheads of fully distributed models (Beven, 1989). The approach is more selective in the number of environmental variables employed and more sensitive in choosing an appropriate resolution for spatial partitioning. Some of the more attractive features of the semi-distributed approach include:

(a) adoption of a "minimalist" philosophy towards data selection. The importance of all available data is assessed and only the key variables are used as inputs to a model;
(b) simplicity is emphasized to retain understanding; only a subset of major processes are modelled by simple equations;
(c) the landscape is partitioned into units which reflect the spatial variability in the main driving processes and controls. The partitioning need not be into uniform cells, but into "response units" which have some hydrological significance or uniformity of process;
(d) an exploratory approach to model building is envisaged. Starting from a simple prototype, the effects of adding or removing variables can be explored to understand the sensitivity of the model. Also, the model can be run with different spatial partitionings and the results compared.

A CASE FOR CLOSER LINKAGE OF MODELS WITH GIS

The "semi-distributed" distributed approach has close parallels to methods used by workers developing GIS applications. A case for more closely linking model building to GIS can be made by identifying several convergent approaches in the modelling of data, in the partitioning of space and in exploring spatial data.

The concept of selecting key spatial variables such as the modelling substrate (e.g., elevation) and environmental controls on hydrological process (e.g., vegetation interception or soil porosity) is similar to the GIS concept of building a model of geographical space in which certain themes or layers are of interest.

Considering the spatial data models used in both fields, there is an obvious similarity between the "bottom up" approach of fully distributed models and GIS based on the raster or cellular data model; in both cases the area of study is partitioned at a fixed resolution by a fine mesh. The regular unit leads to the advantage of computing simplicity but the drawback of large volumes of data. On the other hand, the semi-distributed paradigm recognizes the idea of patches of the landscape within which conditions can be considered relatively uniform. This higher level identification of patches is similar to the notion of objects or features in some GIS. The higher level view has the advantage of simplicity in that we naturally classify a landscape into features and of efficiency in that grouping reduces the volume of data to be stored and processed.

A crucial aspect of semi-distributed modelling is the partitioning of a landscape based on a "response unit" which reflects local variations in the environmental conditions. The choice of variables and the appropriate spatial and temporal resolutions forms the structural basis of a model and can fundamentally affect the nature and reliability of results produced (Anderson & Burt, 1985). As Fig. 1 indicates, the concepts of classification and overlay can be used in a GIS to experiment with different layers and different spatial resolutions to derive a meaningful "response unit".

From an analysis of these convergent approaches it seems that a closer integration of GIS and hydrological modelling could be mutually beneficial. From a GIS perspective, the ability to characterize and model the spatial variations in hydrological processes would make GIS a more effective aid for planning and managing the use of land within a drainage basin. From a modelling perspective, GIS allow environmental data from disparate sources to be georeferenced and related; missing or discontinuous data values to be interpolated and many different combinations and resolutions of geographic data to be explored.

Whether or not these benefits lead one to conclude that GIS is only a complementary tool for model building, or in fact the possible future platform for model development, seems to depend on the human and technical obstacles to linking the two.

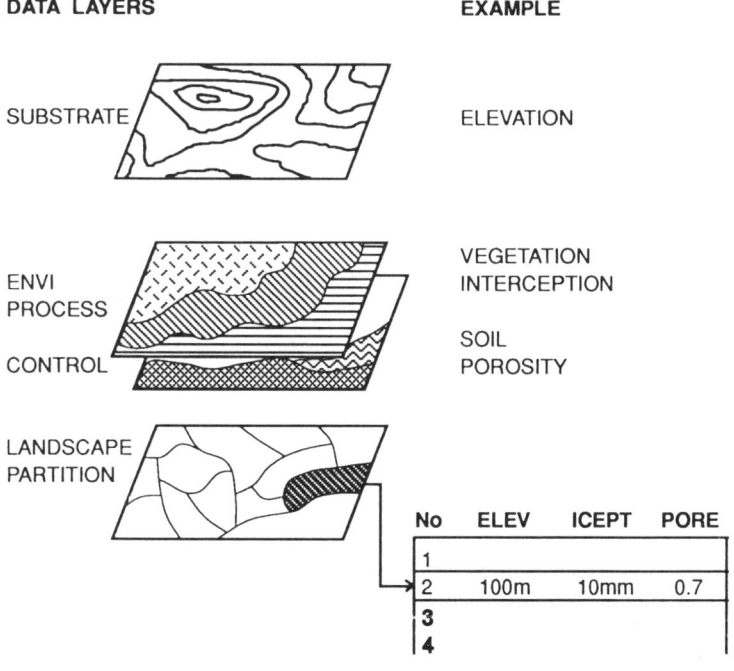

Fig. 1 Using GIS to overlay selected, classified data layers of hydrological significance, different spatial partitions of a catchment can be explored.

Opportunities and constraints on making the link

Until recently, GIS have been restricted in their complexity of computation to the mathematical operators offered by standard DBMS for performing calculations on attribute tables. This syntax limits the implementation of modelling equations to modest calculations of a few lines of code.

However, as GIS become more "open" and users demand greater ability to interact with them, more versatile customization languages are appearing, which allow users access to the underlying data structures and procedures. With these languages gaining more features of full programming languages, such as variables and looping constructs, it becomes possible to embed general purpose modelling tools within a customized GIS.

Some of the reasons for the separate development of GIS and hydrological models may therefore be understood as a previous inability of GIS

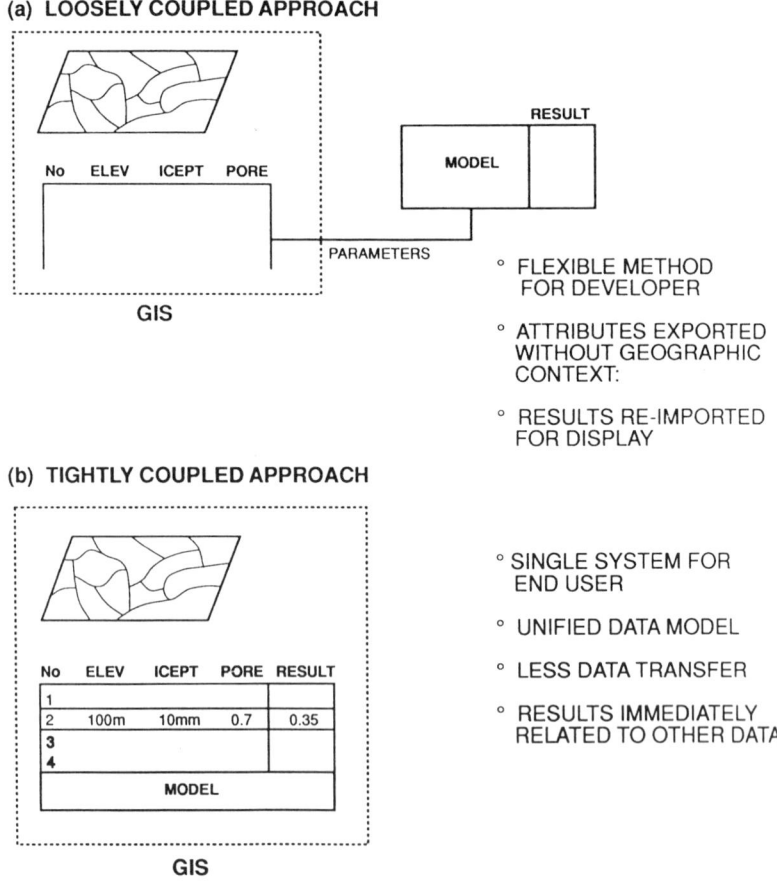

Fig. 2 Two alternative ways of linking a model to a GIS; in (a) the model is linked loosely through an interface; in (b) the model is encoded within the GIS and directly accesses the data structures and procedures of the GIS.

to meet the needs of the model builder for appropriate data models and flexible program development. In bringing modelling and GIS together at this time, two alternative strategies can be identified as shown in Fig. 2.

There are several well established hydrological models, implemented in standard programming languages. These may wish to use GIS for preparing and synthesizing data gathered in different forms and from different areal units and for experimenting with alternative spatial partitions. However, the model will still run externally, since the algorithms often exploit structures of the host programming language. Some recent prototypes are reported by Mallents & Badji, (1991) and Foster (1991). The investment in existing coding implies that a loose coupling to GIS is the best solution, at least in the short term. As Fig. 2a shows, the GIS supplies parameter estimates which are exported to the model; after performing the calculations, results that are locationally referenced may be reimported into the GIS for visualization.

In this case, each system is used for its designed purpose, but considerable engineering is required to bolt the two together; in most cases the link is not seamless as one system must be exited before the other can be run; Effort is needed to maintain the link, as there is no guarantee that the two packages will maintain their internal structures as software is upgraded. Often only the attribute data can be easily exported as ASCII tables, with the geographical context of the data sometimes being lost.

Whilst this method may suit researchers, end users with less technical expertise wish for a single environment and a unified data model as shown schematically in Fig 2b. Here, the user has more time available for developing and refining an appropriate model and wastes less time on the mechanics of data transfer. If a set of general purpose tools for model building can be provided within a system, then new models can be rapidly prototyped. To the user, this customized package seems more like a model development environment than a GIS.

A CASE STUDY ILLUSTRATION

To show how a set of general modelling tools can be embedded within a commercial GIS, using the functionality of an internal customization language, a set of tools have been written using the modelling language of the SPANS GIS. These tools are used to implement elements of the semi distributed TOPMODEL for predicting surface saturation. The test area selected is the Gwy catchment in west Wales. This is a small headwater catchment of some 4 km^2 and with a total stream length of 5.7 km. The catchment lies between 340 and 740 m and is distinctively upland; steep grassy slopes and brown earth soils mean that water is typically transmitted by lateral throughflow, often through an extensive network of natural soil pipes. The annual precipitation is around 2500 mm. The catchment was chosen for reasons of data availability and for continuity with previous studies. As the catchment has been instrumented by the UK Institute of Hydrology since 1968, there is a good supply of hydrological data. The land cover and soils within the catchment have also been mapped in some detail.

Model design

The hydrological algorithms used in this study were adapted from TOPMODEL (Beven & Wood, 1983; Beven 1989). This was selected for implementation because it uses a hydrologically proven, semi-distributed approach; its mathematics are relatively straightforward and well documented and its data demands modest, making it ideal for a modelling environment based on GIS. Also, TOPMODEL has previously been applied to the catchment, so key parameter values that require specialist calculation are available. Finally, as a result of previous work in Edinburgh by Stuart & Hartshorne (1992), digital datasets needed by TOPMODEL existed for the site in formats which were suitable for import into SPANS.

Using the basic TOPMODEL theory that explicitly links topographic form to subsurface water flow and the production of surface runoff, a set of physically based hydrological simulation tools have been implemented within SPANS. A catchment is partitioned on the basis of a topographic index describing hillslope form and further subdivided according to soil and vegetative conditions. The model is driven from precipitation records and the resulting patterns of surface saturation are modified by estimates of the spatial variations in evapo-transpiration, soil transmissivity and areas where pipe flow is known to occur (a factor frequently neglected in previous models).

Since the theoretical basis of TOPMODEL has been clearly reported by its authors (Beven & Kirkby, 1979; Beven & Wood, 1983), the example given below emphasizes how one of the principal equations, for deriving local subsurface flow, may be incorporated within a GIS and run upon a suitably constructed partitioning to visualize the spatial variations in subsurface flow and water deficit across a catchment.

Consider a segment of hill slope, draining through point (i) in a catchment. Assume that the subsurface flow rate (qi) can be related to soil moisture deficit (Si) by the relationship:

$$qi = To \tan(B) \exp(-Si/m) \qquad (1)$$

where $(tan (B))$ is the slope and is taken as an approximation to hydrological gradient at point (i), (To) is a soil conductivity or transmissivity parameter when the profile is just saturated (i.e. $Si = 0$) and (m) is a parameter relating to the rate of change of conductivity in the profile. For the case of a steady input rate r to the slope (precipitation intensity), then at any point the down slope flow must be given by:

$$qi = ar \qquad (2)$$

where (*a*) is the area of a topographic partial contributing area upon which precipitation is falling. The combination of equations (1) and (2):

$$Si = -m \ln(ar / To \tan(B)) \qquad (3)$$

provides the basis for the calculation of distributed local saturation deficit. Implementing this equation within SPANS is a two stage process; firstly distributed parameter values must be provided at the appropriate spatial resolution, using the GIS to derive and overlay multiple maps. Note that a map in the GIS is synonymous with a layer containing spatially distributed parameter values. Secondly the parameter values must be accessed by a modelling routine written in the internal Spans Modelling Language (SML).

A distributed hydrological database was established to provide common combinations of parameter values likely to be required for the modelling equations. Figure 3 summarizes the map data layers held as quadtree maps (.MAP) and the associated attribute tables (.TBB) which are linked to them. An attribute table is a look-up table, used to hold a range of hydrological data sets, such as rainfall records for different time periods.

The spatial partitioning of an area into units upon which a modelling equation can operate is achieved by overlaying simultaneously all maps containing parameter values required by the model. The subsequent map produced is partitioned into thousands of tiny areas; each area contains an individual combination of constituent variables from the overlain layers. Such a map is termed a "unique conditions map". Associated with it is a large attribute table, indicating for each unique area on the map, the class on each of the input maps from which it was derived. This map is used for modelling, rather than visualization, since it enables each unique area to be visited in turn and the value of each of the required parameters at that location to be extracted (or looked up). An overlay of the most important data sets for modelling saturation deficit involved combining eight maps to produce a "master" map and table of unique conditions. The function of this map and table as the common link between all other maps and their attribute tables is shown schematically in Fig. 3.

Hence, to evaluate equation (3) above across the catchment, minimally requires the overlay of the following five maps:
(a) partial contributing areas (part.map) for "a";
(b) local slope tangent (slopelut.map) for "tan(B)";
(c) soil transmissivity (soillut.map) for "To";
(d) rainfall intensity (rainfile.map) for "r";
(e) parameter m (veglut.map) for "m".

Executing the model can be represented simply by the following algorithm, where CALC implies a calculation on a raster map (in a style after Kwadijk & Van Deursen, 1990):
(a) TIME STEP;
(b) CALC Input.map = part.map x rainfile.map;
(c) CALC Substrate.map = soillut.map x slopelut.map;

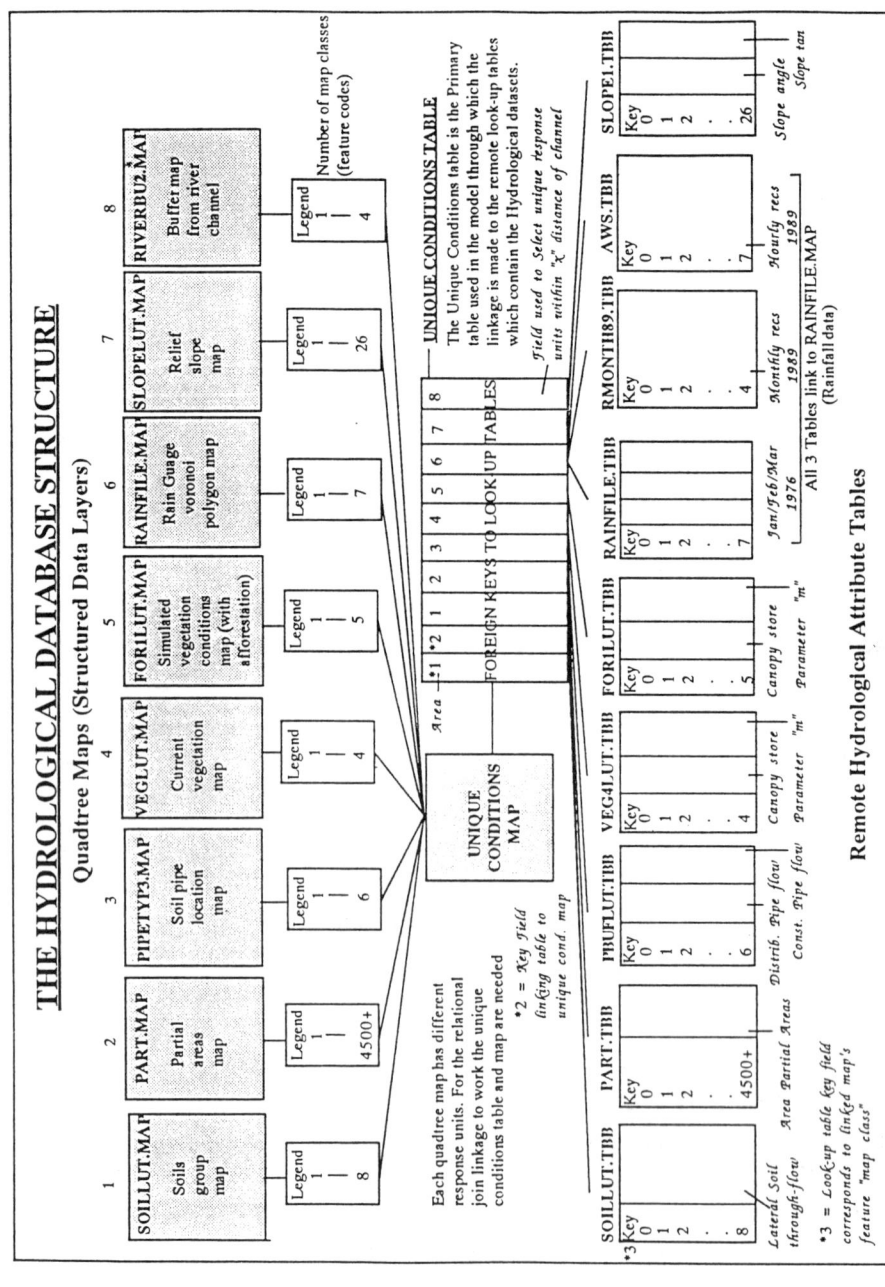

Fig. 3 The structure of the GIS database required for executing a range of hydrological modelling tools; Eight map layers (.MAP), each with one or more attribute tables (.TBB) are linked together through the composite unique conditions map.

(d) CALC Deficit.map = -(veglut.map) x Ln(input.map/substrate.map).

This expresses the idea of visiting each location, looking up values on a number of maps at that point and computing a modelling result. In fact it is actually the numeric values held in associated attribute tables that are being accessed and operated upon and a result is computed once for each unique area, not for each single cell.

Model results

As a result of this work, a toolbox of SML procedures has been produced. The tools presently implemented are those for retrieving and manipulating data and those for performing calculations according to a selected algorithm. For this example, whilst the algorithm is based on TOPMODEL, there is considerable flexibility in the selection and derivation of parameters. For example, facilities are included to derive a value for To at run time, representing either soil matrix flow alone, or including distributed pipe flow.

The results of running a modelling equation with and without the assumption of pipeflow can be assessed qualitatively by comparing the resulting outputs of Figs 4 and 5. In Fig. 4, maximum values for Si, approximating to areas of greatest throughflow are closely related to areas of steepest slope, which in this catchment are quite close to the channel. Figure 5 shows that by recalculating to account for the spatial distribution of soil pipes, many areas that could contribute to rapid throughflow expand and coalesce, which could markedly affect the hydrological response of the catchment.

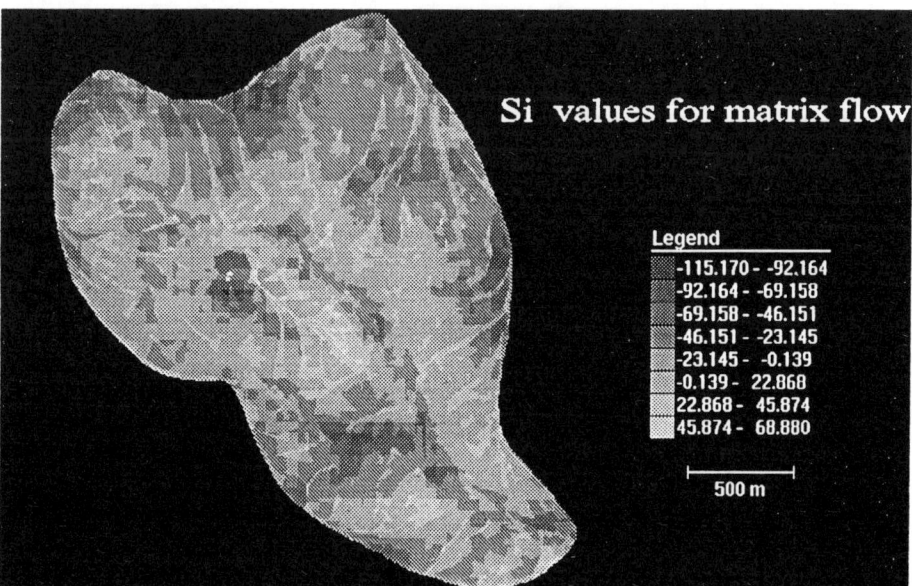

Fig. 4 Spatial distribution of saturation deficits in the Gwy catchment, estimated using TOPMODEL assuming no pipe flow.

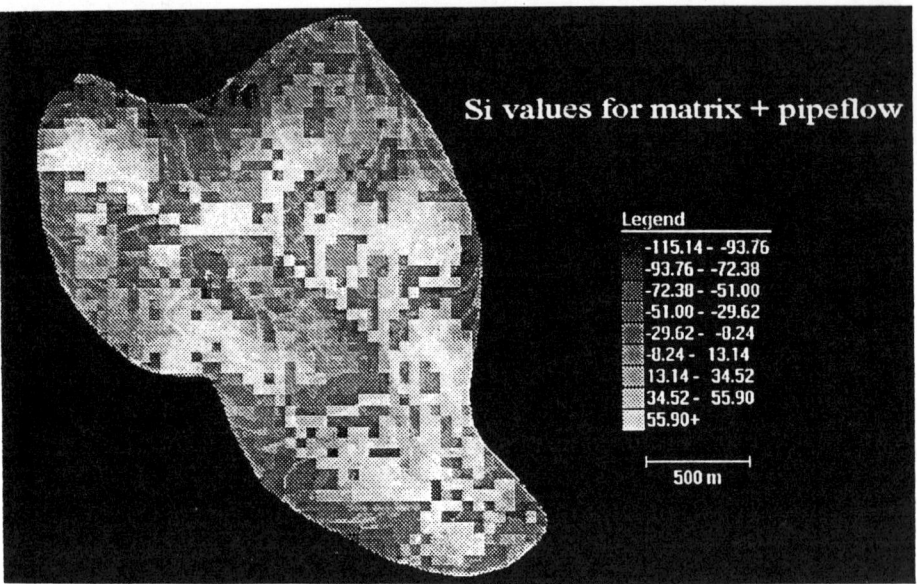

Fig. 5 Spatial distribution of saturation deficits in the Gwy catchment, assuming all soil pipes connected and flowing.

Individual tools in the form of blocks of code can be combined in different ways to produce variant models on a common theme. Other tools allow a user to define at run-time which pipe areas are to be simulated as ephemeral, which as perennial. Actual flow rates may be specified. There is a facility to simulate afforestation of the catchment and to explore how changes of land cover may affect the pattern of moisture deficits.

DISCUSSION AND CONCLUSIONS

The set of tools implemented are still simplistic in a number of respects. Most oversimplification results from the constraints of working within the limitations of the internal modelling language, which is still not as flexible as a high level programming language. Some examples of weaknesses are the lack of a looping construct, a limited number of variables and a limited precision for calculating with real arithmetic. As a result, the outputs produced may be interpreted as indicative, but without reliance on the absolute values being produced.

Whilst further work is required to validate and calibrate these models, this example seeks mainly to demonstrate some of the attractive concepts of integrating modelling tools within GIS. A single environment is created; within which all stages of model building can be undertaken from the initial selection and transformation of data sets, to the final visualization of results in their geographic context.

The concept of unique conditions modelling is similar to idea of the "response unit" in hydrological modelling except that being in a GIS the exact locations of the unit are known; the unique area contains all the parameter values required to evaluate a hydrological model for that location and the shape and size of the unit is also dependant on the layers of data selected. Models based on a few, generalized parameter maps will have few unique conditions, be quick to operate, but produce smoothed results. In comparison, using many parameters involves overlaying many maps and will produce a very complex mosaic that is very slow to run a model upon. This method therefore offers a flexible environment in which to experiment iteratively, seeking to find that optimal balance: a minimal set of maps which produces physically realistic results without the problems of "over parameterization" (Beven, 1989).

The investment in existing code means that many hydrological models will only take advantage of interaction with GIS through a loose form of linkage or interfacing. In the longer term, as GIS support a wider range of data structures and become open systems, bound together by more flexible internal languages, the challenge will be for hydrologists to customize GIS into environments for model development. The prototype set of general modelling tools illustrated here are hopefully a small step in this direction.

REFERENCES

Anderson, M.G. & Burt, T.P. (1985) Modelling Strategies. In: *Hydrological Forecasting* (ed. by M.G. Anderson & T.P. Burt), Wiley, chapter 1.

Anderson, M.G. & Rogers, C.C.M. (1987) Catchment scale distributed hydrological models: a discussion of research directions. *Progress in Physical Geography* **11**(1), 28-52.

Bathurst, J.C. (1988) Physically based distributed modelling of an upland catchment using the Systeme Hydrologique European. *J. Hydrol.* **87**, 79-102.

Beven, K.J. & Kirkby, M.J. (1979) A physically based variable contributing area model of basin hydrology. *Hydrol. Sci. Bulletin* **24**(1) 43-69.

Beven, K.J & Wood, E.F. (1983) Catchment geomorphology and the dynamics of runoff contributing areas. *J. Hydrol.* **65**, 139-158.

Beven, K.J. (1989) Changing ideas in hydrology - the case of physically based models. *J. Hydrol.* **105**, 157-172.

Burrough, P.A. (1989) Matching spatial databases and quantitative models in land resource assessment. *Soil Use and Management* **5**, 3-8.

Densham, P.J. (1991) Spatial decision support systems. In: *Geographical Information Sytems: Principles and Applications* (ed. by D.J. Maguire, M.F. Goodchild & D.W. Rhind) Longman, vol. 1, chapter 24, 403-412.

Foster, G.C. (1991) Integration of a semi-distributed model in a GIS environment. MSc. Dissertation, University of Edinburgh.

Kehris, E. (1990) A geographical modelling environment built around ARC/INFO. *Proc. EGIS 1990*, 556-564.

Kwadijk, J. & Van Deursen, W. (1990) Using the WATERSHED tools for modelling the Rhine catchment. *Proc. EGIS 1990*, 255-263.

Mallants, D. & Badji, M. (1991) Integrating GIS and deterministic hydrological models: a powerful tool for impact assessment. *Proc. EGIS 1990*, 672-679.

Mitchell, B. (1991) *Geography and Resource Analysis*. Longman, Chapter 6.

Stuart, N. & Hartshorne, J. (1992) Hydrological modelling with GIS - approaches and opportunities. *Proc. First European SPANS User Conference*, Amsterdam, 45-57.

6 Application of GIS in Water and Environmental Management

A national groundwater model combined with a GIS for water management in The Netherlands

W.J. DE LANGE
Institute for Inland Water Management and Waste Water Treatment (RIZA), P.O. Box 17, 8200 AA Lelystad, The Netherlands

J.L. VAN DER MEIJ
Institute of Applied Geoscience (IGG/TNO), P.O. Box 6012, 2600 JA Delft, The Netherlands

Abstract The Institute for Inland Water Management and Waste Water Treatment in cooperation with the TNO Institute of Applied Geoscience is developing NAGROM - a national groundwater model of The Netherlands (De Lange, 1990). NAGROM can be considered as an integrated part of the national geohydrological information system REGIS of TNO, which, in turn, is the national database for geohydrological data, such as borehole data, geophysical data and groundwater data. NAGROM can easily make use of the various original and interpreted data (sections and overlays of REGIS). As NAGROM is based on the analytic element technique (Strack, 1989), its integration with a GIS is different from models based on finite element or finite difference techniques. After brief descriptions of NAGROM and REGIS, some results will be presented. The results represent the present state of development.

INTRODUCTION

The Institute for Inland Water Management and Waste Water Treatment (RIZA) developes instruments for integral water management in The Netherlands. One of these instruments is NAGROM, the NAtional GROundwater Model of The Netherlands (De Lange, 1990). The model is used for scenarios on a national scale and serves also as a framework for detailed models. For national purposes NAGROM should be constantly updated and readily available for all types of groundwater (flow) problems. To that purpose, the concept of an easily adaptable and calibrated model has been chosen. The combination of the analytic element method (AEM) and a GIS allows for easy adaptation to problems and optimal use of the available geohydrological data.

A permanently adaptable model based on the AEM in combination with a GIS leads to a concept that is different from the combination of a GIS with finite element or finite difference methods (cf. Diersch *et al.*, 1991).
Elements with different geometry and parameter definitions and the absence of boundaries are the main differences between AEM and the other methods (FEM and FDM). With AEM it is also easy to adapt the existing element distribution to different problems in the same problem area.

The properties of the analytic element method with respect to a GIS are important for combining NAGROM and REGIS. Therefore, much attention will be paid to this subject.

DESCRIPTION OF NAGROM

NAGROM consists of about ten modular groundwater models that cover The Netherlands (Fig. 1); the model will be completed in 1993. It is used for analysis of national water management policies and serves at the same time as a framework for regional problems (De Lange, 1990). NAGROM simulates saturated groundwater flow with all its quantitative aspects, including density differences. The model is based on the analytic element method (AEM) developed by O.D.L. Strack (Strack,1989). The model code written in FORTRAN 77 mainly runs on 386 and 486 personal computers. Modelling with the AEM is based on superposition of analytic elements. The AEM method allows easy coupling of models as well as changes of type, geometry and parameters of individual elements.

In Fig. 2, the most important types of analytic elements are shown. The geometry of the elements can be: a point, a single line, a chain of straight or curved lines, or a quadrangular surface. The storage of input data is vector oriented.

DESCRIPTION OF REGIS

REGIS (Boswinkel, 1989) is an interactive geohydrological information system for the evaluation of geohydrological situations on a national and regional level. At present, REGIS is in a stage of rapid development. When completed, it will contain the following main components (Fig. 3):
(a) the ORACLE database, containing all relevant geohydrological data;
(b) functionality in ARC/INFO for (1) realization and visualization of data (see Figs 4 and 5), (2) transfer of data and information between the system and application programs (deterministic or stochastic models), and (3) presentation of results of applications in combination with data and other information.

Eventually, REGIS will be an information system containing all relevant information needed for groundwater management, for the exploitation of groundwater reserves, and for the protection of the subsoil.

THE INTEGRATION OF NAGROM IN REGIS

User requirements

The two NAGROM user groups mentioned in the introduction require different capabilities of the model. Performing large runs for national policy analysis requires automatic changes in selected areas of extraction rates, surface water levels and other variables. The large runs may be performed in background or batch mode and the presentation of results should be of a high quality. For the

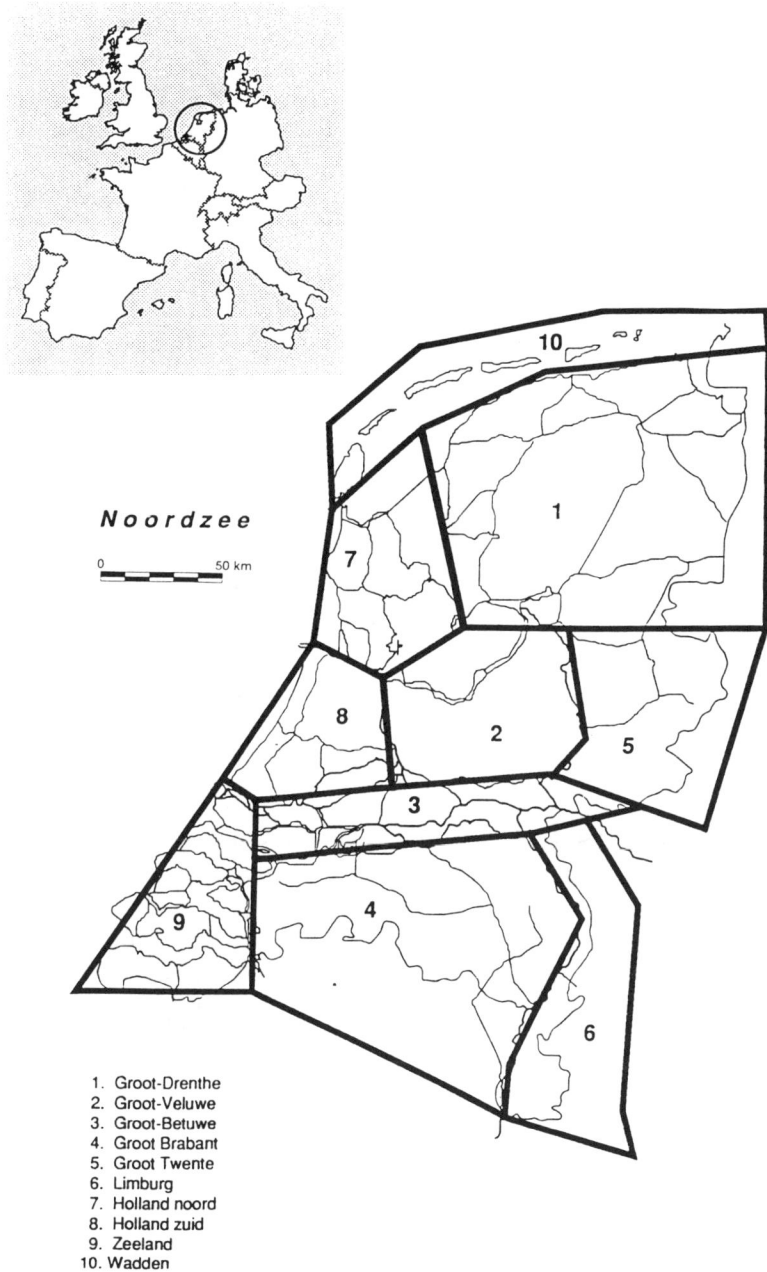

Fig. 1 Main models in The Netherlands (De Lange, 1990).

1. Groot-Drenthe
2. Groot-Veluwe
3. Groot-Betuwe
4. Groot Brabant
5. Groot Twente
6. Limburg
7. Holland noord
8. Holland zuid
9. Zeeland
10. Wadden

refinement of a part of NAGROM (a region), the presentation of basic or interpreted geohydrological information is needed in the form of overlays, sections and borehole data.

In the calibration stage the user requires interactive modelling. So runs alternate with changes of input data and presentation of preliminary results on screen. However, the final results need high quality output devices.

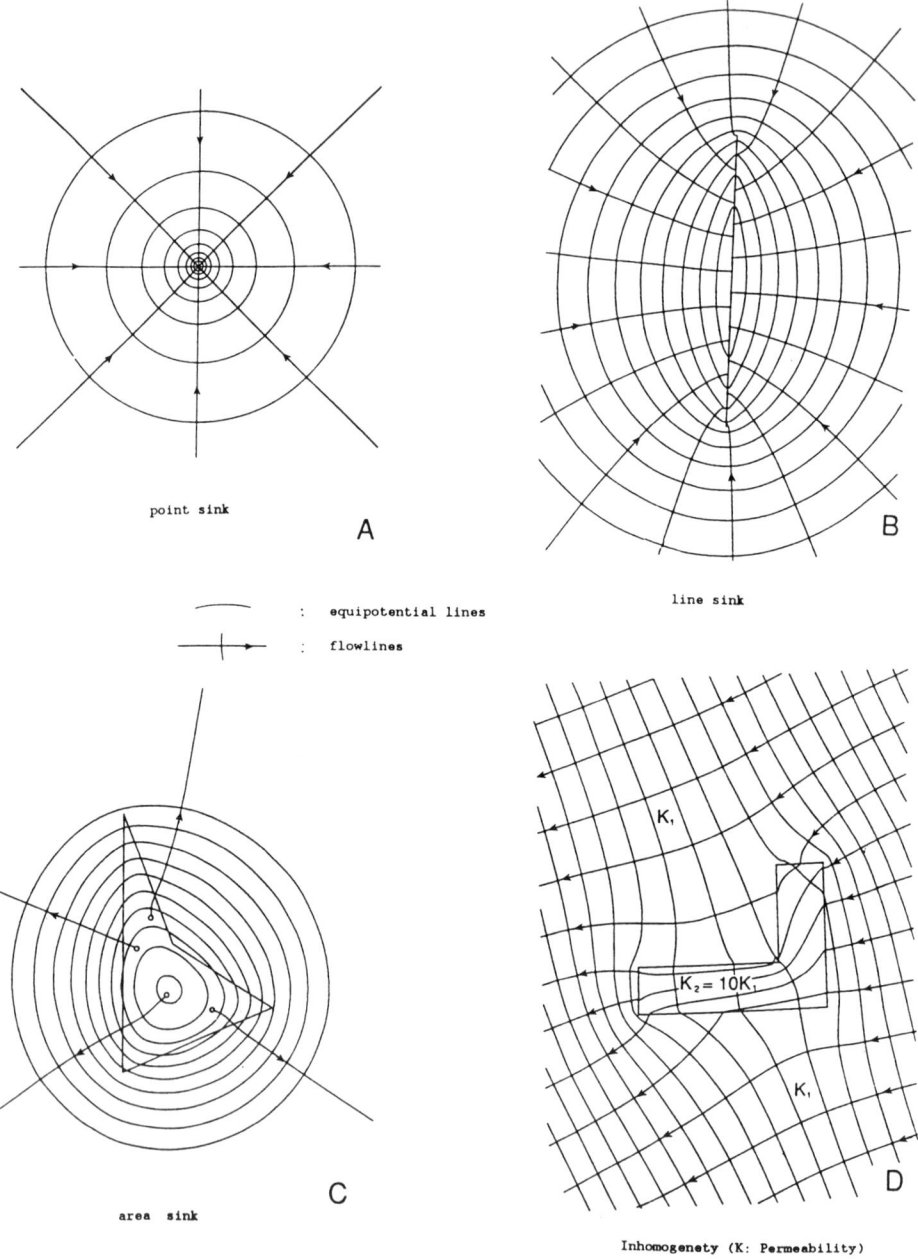

Fig. 2 Four basic analytic elements: (A) point sink, (B) line sink, (C) area sink, (D) inhomogeneity.

Technical description (Fig. 3)

On the input side, the analytic elements are prepared for the computation. Based on the analysis of the geohydrology, selection of analytic elements, refinement of the geometry of elements, and adjustment of the parameters are

A national groundwater model combined with a GIS for The Netherlands 337

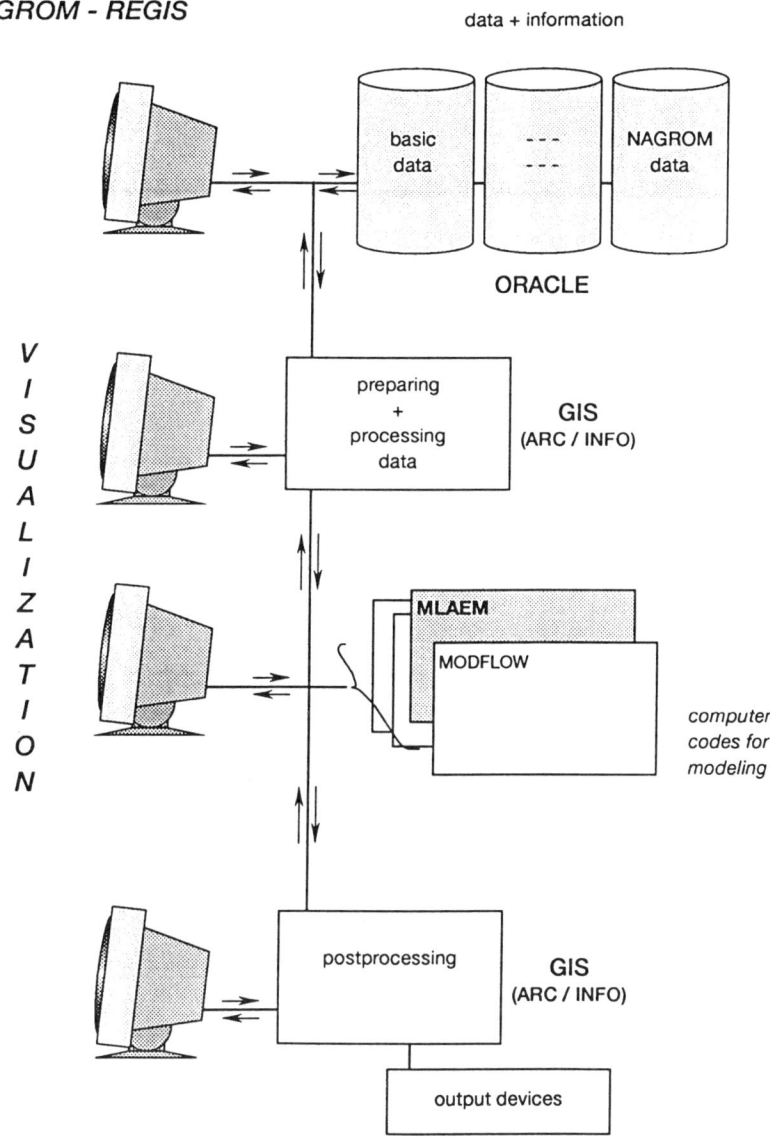

Fig. 3 Scheme NAGROM (REGIS) and MLAEM.

the usual actions in AEM modelling.

The geometry and properties of the analytic elements are stored in the database, organized according to element type, layer (aquifer or aquitard) and NAGROM sub-model. Before elements are included in NAGROM, basic checks are performed automatically by the program. A final check with respect to the elements is then left to the experienced NAGROM modeller. So, by selecting a domain of interest in NAGROM, a working model is generated more or less automatically.

The degree of detail of this model is as close as possible to the user-specified scale; however, elements grow larger towards the outside of the model as they reach the nearest hydrological boundaries defined in NAGROM.

After the model generation, the user can adjust the model while staying inside the GIS. The geohydrological information can be presented as underlays for the analytic element set-up, so that the parameter values of the elements and their geometry can be compared to the basic data. For the purpose of simulating national scenarios, tools for automatic adjustment of parameters, abstraction rates, surface water levels, etc. will be developed.

After the preparation phase, the input data are automatically transported to the analytic element program, for the computation of heads, fluxes and so on. Depending on the presentation required by the user, a grid of values, particular values, flowpaths, etc. is computed. The results are stored and automatically transported to the output part of the interface.

The results are presented by the GIS in combination with overlays. Interactive access to the computation method enables the user to carry out detailed analysis and particle tracking.

THE PROTOTYPE (1992)

A prototype interface is being developed for connecting NAGROM to REGIS. The different types of elements are imported from the existing NAGROM dataset in ASCII under pc-dos. The elements are stored in the REGIS database according to the scale of modelling and the type distinguished by the AEM (wells, line-sinks, area-sinks, doublets, etc.). In this way, the elements covering the entire country will be loaded in REGIS.

Using overlays of the geohydrological basic data in REGIS, the elements can be checked for geometry and parameter values (Fig. 4). The geometry and parameter values of individual elements can be changed interactively. This change can be a subdivision, a removal, or a creation of new elements (Fig. 5). By the end of the session, the adapted schematization is stored in a separate part (the user part) of the REGIS database. The data belonging to the schematization can be exported to input files for the AEM.

At present (1992), the output mainly depends on the output facilities of the AEM, which are sufficient for most modelling purposes. Combination with REGIS overlays is not yet possible. NAGROM output files combined with geophysical and subsoil maps have been used in a GIS (ARC/INFO, but not REGIS) for the preparation of input data and computation of subsoil flow and of its effects on agriculture and nature (Vermulst, 1992; Parmet & Mann, 1992).

APPLICATIONS

The prototype is being tested in several areas in The Netherlands with considerable differences in the surface, subsoil and deeper underground. Different types of analytic elements are used. In Fig. 4, an example of a

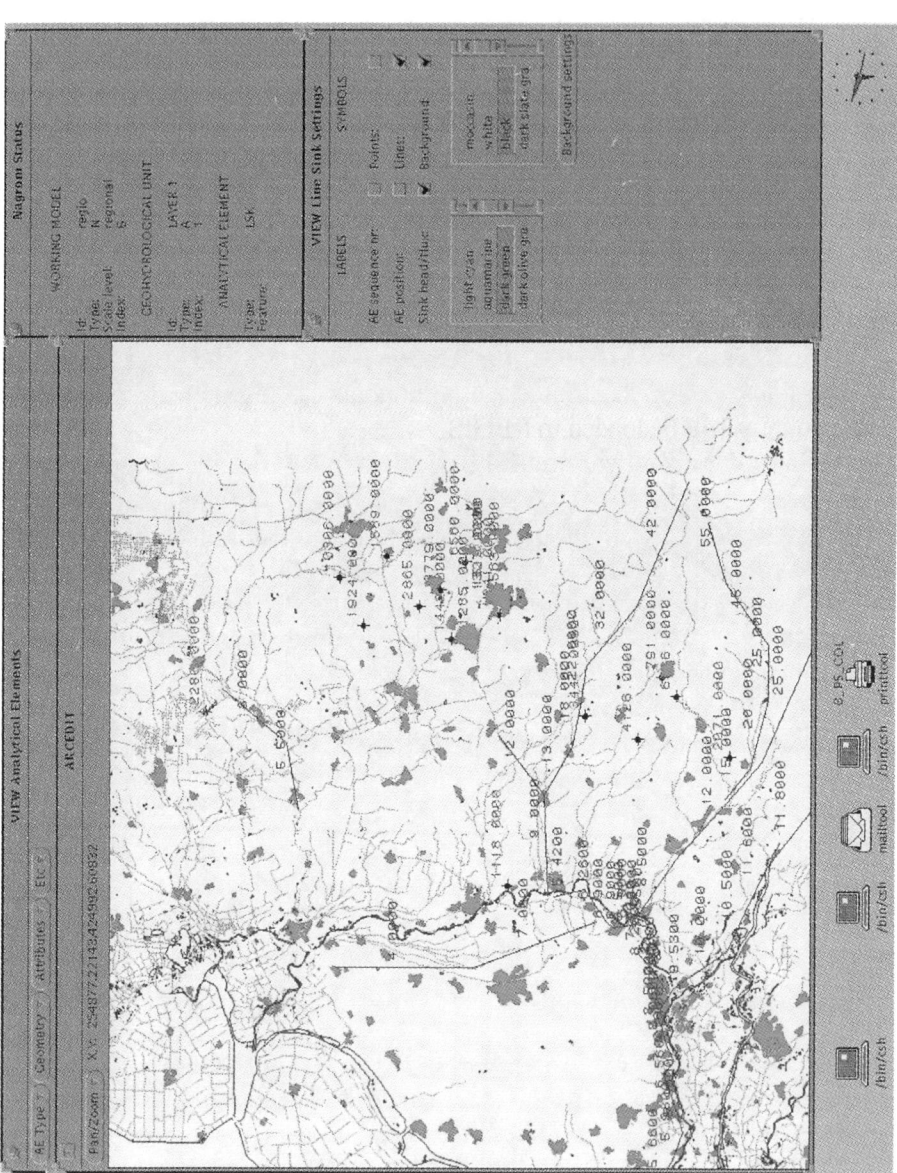

Fig. 4 Geohydrological overlay (rivers, urban areas) in combination with "point sink" and "line sink" type analytic elements (wells, rivers) in the "view"-window.

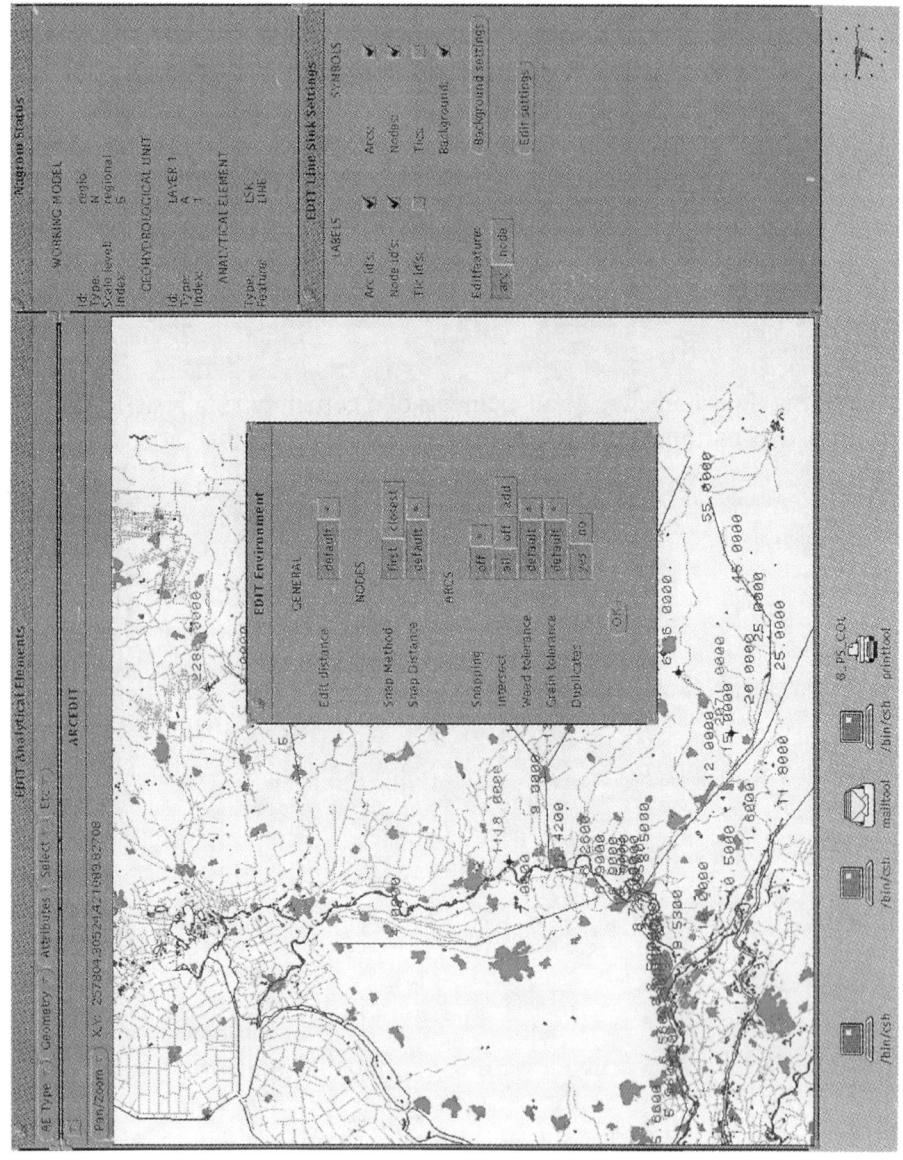

Fig. 5 Example of a menu in the "edit"-window for well-type analytic elements.

combination of geohydrological overlay with analytic elements of different types is presented. Point and line sinks are shown on an overlay of major rivers and urban areas in the east of The Netherlands (region 5, Fig. 1). In Fig. 5, an example of the "edit"-window is presented. In the "edit"-window, changes can be made in specified analytic elements in combination with different types of overlays. Changed elements can be stored directly in the NAGROM database.

THE FINAL SITUATION (1994)

From the viewpoint of a user, the interaction between the model in the GIS and the computation by the AEM should be flawless. Therefore, the interaction between the GIS and the numerical solution technique will be automatic. The solution technique will be implemented in a shell of REGIS, so the user stays with one interface only.

The input as well as the output is almost completely generated by the GIS. The input is always based on a schematization present in NAGROM. The GIS selects elements as close as possible to the specified scale in the user-defined window. Outside that window, up to the nearest geohydrological boundary defined in NAGROM, elements become coarser. These elements are selected from models on higher scales. The user can select individual elements or groups of elements at the same time for changes. These changes can vary from altering the parameters of all elements of a certain type in NAGROM, for the purpose of national scenario analysis, to changes in the geometry (e.g., repeated subdivision) needed for adapting the model to the local geohydrological situation. Presentation of results is in the form of maps and sections or output datafiles for postprocessing (e.g., statistics).

A COMPARISON WITH GIS & FINITE ELEMENT BASED MODELLING

The main differences between the combination of the analytic element technique and the finite element technique in combination with a GIS (see e.g., Diersch, 1991) are summarized in Table 1.

The GIS software, ARC/INFO, is vector oriented, so it should match naturally with analytic elements. Recently, its raster oriented capabilities have been advanced considerably. The different element types of the analytic element technique necessitate to distinguish different elements in the database. Different types of elements have different geometrical capabilities and different parameters. The need to store the aquifer or aquitard (property) per element is similar for both techniques. Although the analytic element technique uses different types of elements, the number of elements for a model is small

Table 1 Differences between analytic and finite element techniques with respect to GIS.

Finite element technique	Analytic element technique
• grid-point oriented	• polygon oriented
• large amount of input data	• small amount of input data
• triangular or quadrilateral elements	• straight or curved line elements and point-elements and multi-angular elements
• model in finite domain	• model in infinite domain
• connection of adjacent models not "natural"	• connection of adjacent models is "natural"
• refinement by combination of separate model	• refinement by changing elements in model

compared to that needed for finite element models (and so is amount of input data).

For finite element models, it is not common to store the input data permanently, because for each model a new interpolation of the basic data is needed. The input data is coupled with the elements and, in general, for each model a new mesh of elements is generated.

A model of analytic elements can easily be changed. Changes are made only in the area of interest; the rest of the model remains unchanged. So, input data of analytic element models can be simply re-used. In NAGROM, this advantage is used. The entire country is modelled at a certain scale. So, each user can start from a calibrated model which can be adjusted simply. The REGIS-NAGROM-shell is used to generate a proper model as close as possible to the needs of the user.

The boundaries of a model of analytic elements consist of a zone of coarse analytic elements around the area of interest, which simulate the groundwater flow to or from the surrounding area. In NAGROM, this can be generated automatically by the GIS-interface. So, the user can concentrate on his problem without being bothered by the definition of the model boundaries.

CONCLUSIONS

NAGROM, the combination of a GIS: ARC/INFO and a groundwater model, is different from those commonly used, because of the vector-oriented character of the analytic elements applied in NAGROM.

The concept of REGIS-NAGROM provides direct adaptions of (parts of) models for computation of nationwide effects as well as of local effects.

The combination of the REGIS-shell and NAGROM considerably facilitates the first steps of building a model for The Netherlands.

REFERENCES

Boswinkel, J.A. (1989) Regional Geohydrological Information System. In: *Proc. Int. Symp. on Hydrological Maps*, Hannover, Germany.

De Lange, W.J. (1990) A Groundwater Model of The Netherlands, RIZA note 90.066, RIZA, Lelystad, The Netherlands.

Diersch, H.J.G, Gruendler, R., Kaden, S. & Michels, I. (1991) Toward GIS-Based 3D/2D Groundwater Contamination Modeling Using FEM. In: *Numerical Methods in Water Resources* (ed. by T.F. Russel, R.E. Ewing, C.A. Brebbia, W.G. Gay & C.F. Pinder), Comp.Meth. in Water. Res. IX(1).

ESRI (1992) Environmental Systems Research Institute: ARC/INFO description, Redlands, CA.

Parmet, B.W.A.H. & Mann, M.A.M. (1992) Influence of Climate Change on the Discharge of the River Rhine - A Model for the Lowland Area, paper submitted to IAHS symposium J3, Yokohama.

Strack, O.D.L. (1989) *Groundwater Mechanics*. Prentice Hall, New York. USA.

Vermulst, J.A.P.H. (1992) Redesign DEMGEN: Toetsing van de geohydrologische schematisatie op afvoeren van de Schuitebeek (in Dutch) RIZA note 92.053X, RIZA, Lelystad, The Netherlands.

The use of GIS techniques to assess discharge consents and abstraction licences

N.J. BONVOISIN & R.V. MOORE
Institute of Hydrology, Crowmarsh Gifford, Wallingford, Oxon OX10 8BB, UK

Abstract The output of waste products into a water course, in the UK, is limited legally by an agreement termed a *discharge consent*. Similarly, the taking of water from a river or borehole is controlled by an *abstraction licence*. The *authorization* of a new consent or licence, or the review of an existing one, requires the quantification over space and time of water resources and their uses. This paper describes a data model used to store the information required to assess a new consent or licence and, subsequently, to monitor compliance with its conditions. It goes on to discuss how GIS techniques may be used to accelerate the estimation of river flows available to dilute discharges, or to meet abstraction needs. GIS methods also allow for the rapid analysis of existing uses which may interact with a new authorization, by looking both at limits already imposed and at time-series of water quality and quantity measurements.

INTRODUCTION

The discharge of waste into the UK river network is restricted by a number of laws, both UK and EC. In England and Wales, observance of these laws is principally monitored by the National Rivers Authority (NRA). Permission to discharge effluent to a water course or into groundwater is given in the form of a *discharge consent*; a set of conditions defining what may be discharged where, when and in what amounts. Before granting a consent, the NRA first assesses whether the requested discharge will have a significant detrimental effect on the river network downstream of the discharge point. The combined effect of existing and proposed discharges is evaluated so as to assess its likely impact on the ecology of the water course and downstream abstractions. Potential breaches of legal restrictions on water quantity and quality, and implications for achieving water quality objectives, are also checked.

Having decided on the level of discharge which it will allow, the NRA issues a consent. A single consent document may cover many discharges, and will specify the quantity and quality conditions with which each of those discharges must comply. These conditions may be expressed descriptively, or numerically in terms of a specific measured variable (or *attribute*), e.g. biological oxygen demand less than 20 mg l^{-1}. Relative conditions are also used, with the attribute value for the discharge at a downstream point (called the reference site) not being permitted to differ from the value measured at an upstream point (the comparison site) by more than a permitted amount, e.g. water temperature rise in a river. As well as placing limits on individual attribute concentrations or loads, limits may be placed on the total of a group of attributes, e.g. heavy metals. Restrictions may vary seasonally and may change over time as legislation and other factors change. Compliance may be required to be absolute, for example any value of biological oxygen demand found to be in excess of 30 mg l^{-1} is deemed a breach of the condition, or only required for a percentage of the time, usually 95%.

A similar process of *authorization* is followed when assessing a request to take water from a water course, or borehole. In this case, an *abstraction licence* is issued, this time taking account of the reduced quantity of water downstream (or within an aquifer) available for the dilution of pollutants, for fisheries and for other downstream users. Having granted a discharge consent or an abstraction licence, the NRA is subsequently responsible for testing compliance with the conditions.

The data involved in these analyses are diverse and large in volume. Therefore, the way in which the data are stored is crucial to the user's ability to manipulate them. This paper describes a data model used to store the information required to assess and test compliance of discharge consents and abstraction licences. The paper goes on to explain how this model can then be used to assist in quantifying impact, e.g. a change in flow regime, and to catalogue and appraise pre-existing uses. The model can also be used to hold the results of the monitoring process, for use in testing compliance. Finally, some conclusions are drawn as to how well the data model supports the application, and what changes could be made so as to make the impact assessments more accurate and their use more general. The functionality described in this paper is being developed within the Institute of Hydrology's Water Information System (WIS).

DATA MODEL

To allow assessment of consents and licences, a wide variety of information is held within a database. Before describing the particular way in which consents and licences are held, it is useful to examine the facilities in WIS for data storage at a more general level. This section is split into three parts: the basic data model, the use of the model to record consents, and the generalization of the model to record more types of authorization than just consents.

General data model cube

The basic data model is illustrated in Fig. 1; values are held within a three-dimensional space of *features*, *attributes* and time. A *feature* (or object) is something about which information is to be stored, while an *attribute* is a variable measured at a feature. The model allows the user to record the history of any feature. The design is totally general, the types of feature, and the data that describe them, being decided by the user. There are no limits on the size of cubes. Each cell within the cube contains the value of a variable describing the

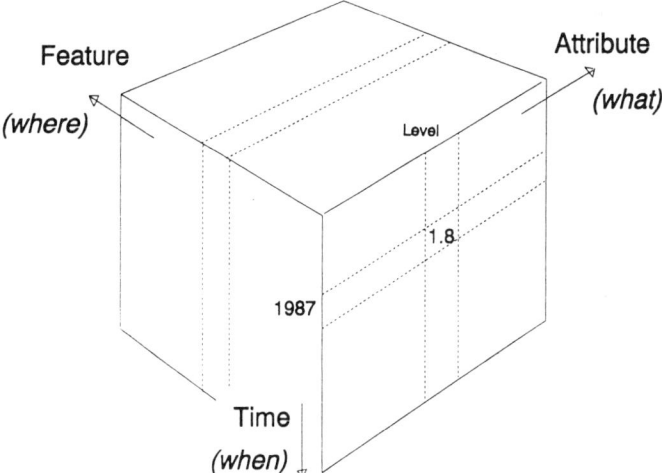

Fig. 1 The data model cube.

feature at a particular moment or period in time. For example, the feature may be a gauging station, the attribute water level in metres, the time 1145 h on 15 November 1987, and the value 1.8.

This model is applied whatever the feature, and whatever the attribute measured, and is similar to that proposed by Dangermond (1983). Dangermond identified a temporal axis, but had a split parameter axis which separated locational from other attributes. More common in texts on the time dimension of GIS, is to concentrate on spatial attributes. For example, Langran (1992) considered X and Y coordinates (and sometimes Z), together with time, as the axes of a three- (or four-) dimensional data space.

In the example above, the value was a floating point (or real) number. But many different forms of value may be stored, such as a name or a position. These different forms of attribute value are catered for within the data model and are termed *structures*.

One key structure is RELATE, which is used to identify a relationship between one feature and another. Such a relationship is often termed a parent-child relationship as it may imply a hierarchy. Within the data model, the structure records the type of relationship (e.g., "flows into", for a sewage discharge into a river stretch) and a pointer to the child feature; the structure is an attribute of the parent feature.

We thus have a single data model capable of holding all the water resource and environmental quality data necessary to assess a new discharge or abstraction. Most importantly it removes the need for separate structures for spatial and time-series information and, indeed, permits the storage of a spatial time-series.

Data model for conditions and regulations

The cube is useful for showing how many seemingly disparate data types can be conceived of as though they were held in a single storage structure. The faces of the cube represent the more frequently required ways of viewing the data. However, it is also useful to view the same data in the more conventional data model form; Fig. 2 shows this generically.

Let us consider a consent document, a piece of paper listing the conditions with which a discharger must comply. The top sheet of this document might contain the names and addresses of the consentee (discharger) and consenter (e.g., NRA), a textual description of the consent and a reference identifier (Fig. 3). A simple extension is to record a cross-reference between this consent and the one it superseded.

In this context, the discharge itself and its characteristics are not of interest. What is of interest is that each discharge is monitored by a sampling point and it is the data from this sampling point that are to be tested against the conditions that relate to it. This situation is represented in Fig. 4.

Each of the conditions describes the limits imposed on a particular

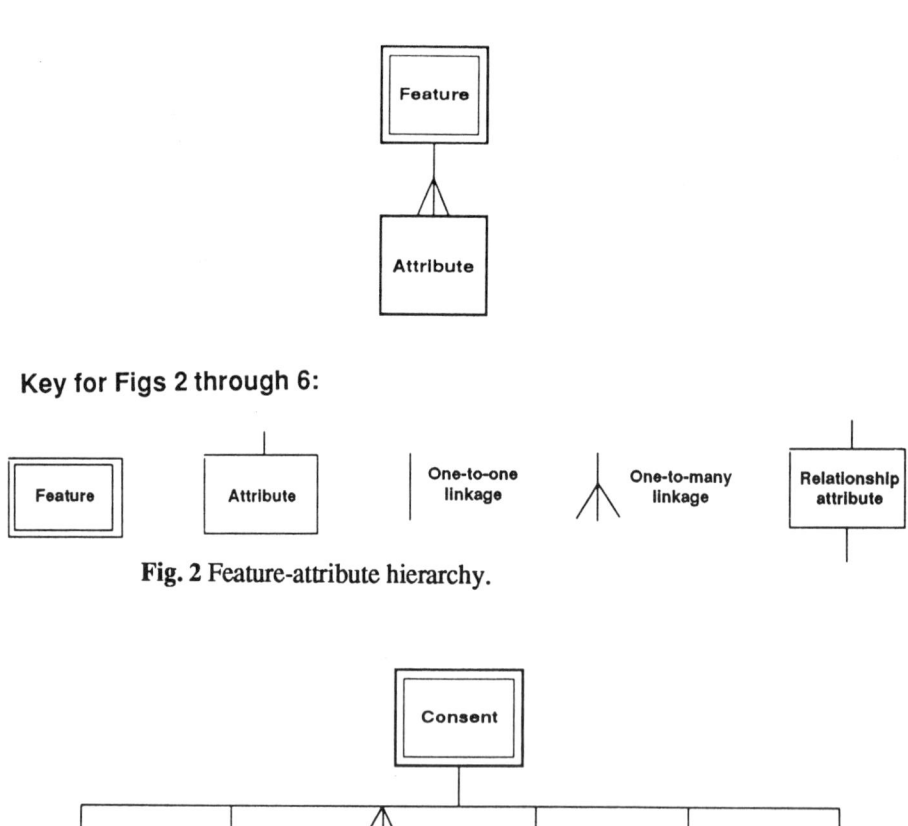

Key for Figs 2 through 6:

Fig. 2 Feature-attribute hierarchy.

Fig. 3 Example of a feature-attribute hierarchy for a consent (see Fig. 2 for key).

attribute, or group of attributes. By making the identity of the attribute which is being controlled, into an attribute of the condition, the situation becomes totally general. No changes to the application programs or the data model are required, should a new attribute become subject to controls. Thus the final model is as Fig. 5. With this model it is now relatively easy to link the monitoring data to the conditions and to establish the degree to which a discharge meets its conditions.

Generalizing the model

The model at this stage is entirely adequate for consents. However it is specific to consents. It could not be used simultaneously to hold conditions relating to both consents and, say, the "EC Directive on the Quality of Water for Freshwater Fish". However, if the term consent is replaced by the more general term authorization and the structure slightly rearranged (Fig. 6), two important questions "Which sites comply with this regulation?" and "Which regulations does this site comply with?" can be easily answered.

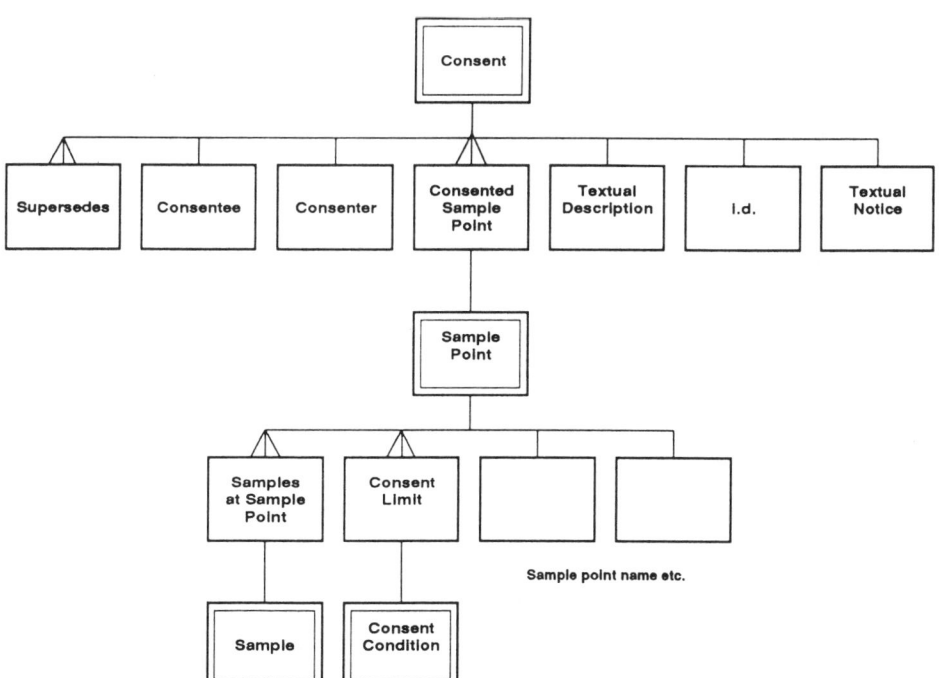

Fig. 4 Example of a feature-attribute hierarchy for a consent and sampling point (see Fig. 2 for key).

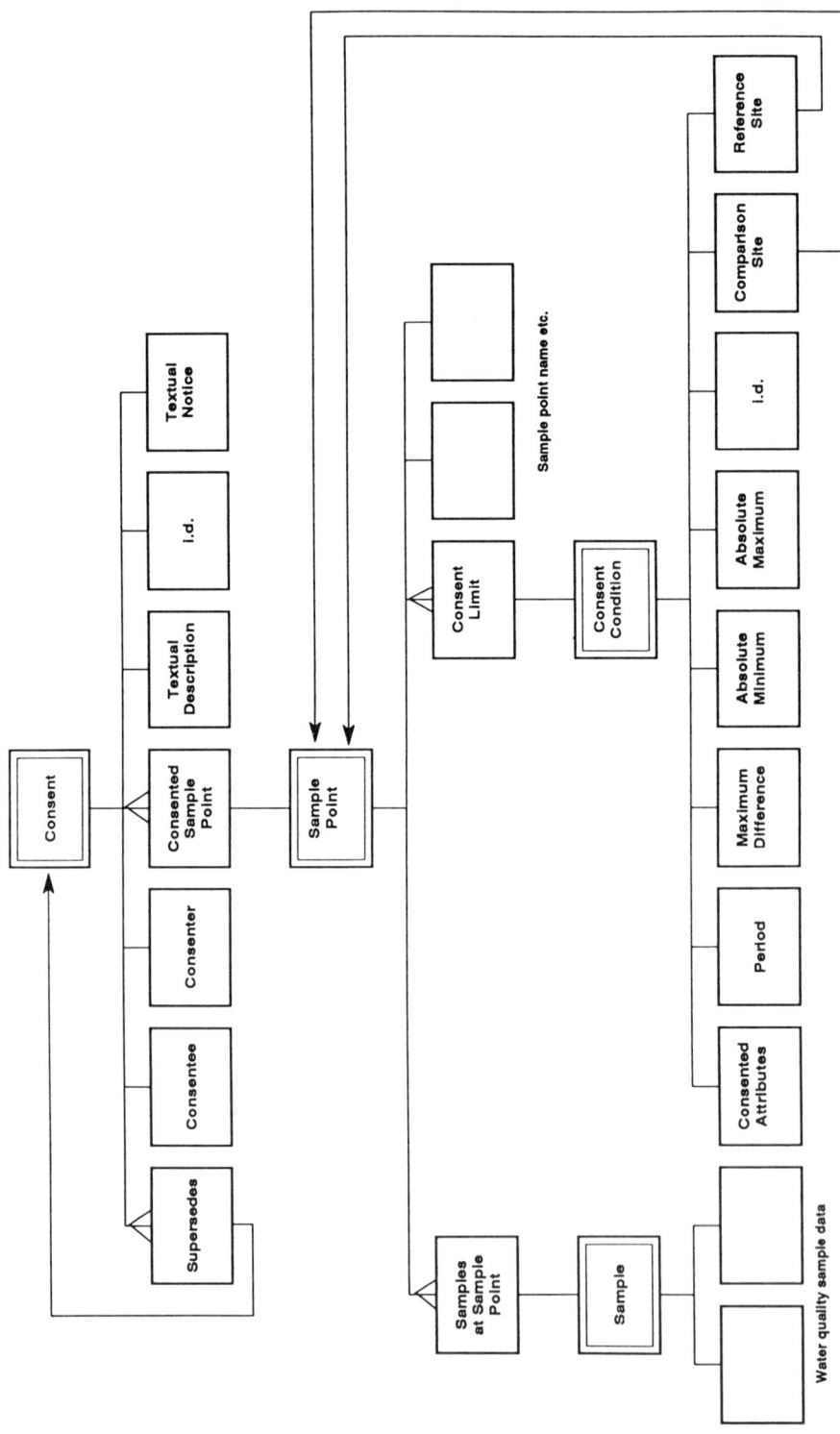

Fig. 5 Example of a full feature-attribute hierarchy for consents and conditions (see Fig. 2 for key).

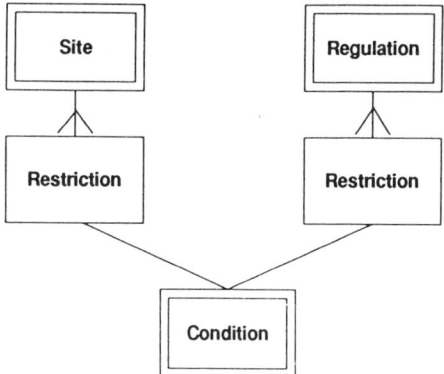

Fig. 6 General authorization hierarchy (see Fig. 2 for key).

GIS IN ASSESSING CONSENTS & LICENCES

The role of GIS

The application for a new consent or licence requires many issues to be considered before permission can be given; the more so as public awareness of the environment increases. A new discharge or abstraction has immediate implications for the river or aquifer in terms of flow, quality and biology. It also has much wider implications for the community which depends upon the river.

Plainly, it is not feasible to debate all the wider matters of principle for every new application. The licensing authorities therefore develop general statements of policy from which evolve a set of practical guidelines. To help assess applications, those charged with investigating them have developed tools for impact assessment, i.e. to see whether the guidelines will be broken. As understanding of the environment and the impact of abstractions and discharges upon it has increased, so these tools have become more sophisticated and have required ever increasing amounts of data. The role of information technology has been to enable the capture, storage and manipulation of these data. Particularly important has been the ability to hold data in a much more structured manner than previously, the data model for consents described above being but one example. The role of GIS has been to enable users to bring similar analytical techniques to bear on the spatial dimension of problems as has previously been brought to bear on the time dimension. Described below are some practical examples of the use of GIS techniques, and of how the data structure may be exploited.

Estimating basin watersheds

A starting point of many analyses is to establish the area of land draining to the point of interest. Traditionally, this is achieved by the use of a contour map of the area. However, the process is slow, prone to error, and two people given the same map seldom produce the same result. These problems are being addressed by the development of hydrologically appropriate Digital Elevation Models (DEMs) (Band, 1986; Morris & Flavin, 1990). Morris's DEM comprises several 50 by 50 m grids containing elevation, direction of steepest descent, cumulative basin area and other information, for the entire UK. The importance of data structure becomes immediately apparent when it is appreciated that each grid contains 100 000 000 values, and the objective is to be able to delineate any basin, whatever its size, at interactive speeds.

Other methods of automatic basin calculation which do not need a DEM are available, one such being the allocation of grid cells to the nearest river stretch (Sekulin et al., 1992).

The ability to delineate basins automatically and quickly, and the availability of many thematic maps in digital form, means that many basin characteristics can be computed automatically. Typical characteristics include long-term annual average rainfall and potential evaporation, and the proportion of the basin overlaying soils with different types of hydrological response, such as the Hydrology of Soil Types map (Boorman & Hollis, 1990). They can now be calculated with an ease and on a scale which could not previously have been contemplated.

Calculating flows

Central to the assessment of a new abstraction licence or discharge consent is the estimation of river flows. In the UK, widely used statistics are the mean daily flow, the 95 percentile one day flow and the mean annual minimum seven day flow (Gustard et al., 1992). More often than not, there are no historic hydrometric records at the site of interest. Therefore Gustard et al. employed regression to identify relationships between the statistics of time-series flow data and a variety of basin characteristics, obtainable from maps covering the entire UK. Hence, using the relationships, estimates of the flow statistics can be made on any UK river. Whereas originally these techniques were applied manually using paper maps, the advent of maps in digital form, and of spatial data handling techniques, has made it possible not only to automate the process, but also to gain access to a far greater proportion of the information that maps contain. An example of a hydrological model which can exploit this extra information is Topmodel, which is currently being enhanced to run directly out of WIS (Romanowicz et al., 1993).

Assessing other uses

In assessing the impact of a proposed use on existing users and the environment, the first problem is to establish the available data. The data to be searched may be large in volume and, more importantly, diverse in nature. The reason why universal data models such as the cube are so important is twofold. They provide a single structure which can handle the diversity. They also provide a framework upon which universal query mechanisms can be built, and hence allow this diverse and complex dataset to be interrogated. Languages such as the Structured Query Language (ISO, 1987) provide a great advance on their predecessors. However, they can only handle a limited set of value types (typically numbers and characters), and are aimed primarily at two-dimensional structures such as tables. WIS seeks to develop a new query mechanism capable of handling many more data types including those specifying geographical position, time and relationships to other features.

Frequent initial questions are "What is measured here?", "Where is this measured?" and "When is this measured?". Each of these questions results in a list and it was from this beginning that WIS's query facilities evolved. There are three types of list, corresponding to the three axes of the data model cube: *where-lists* (features), *what-lists* (attributes) and *when-lists* (time, occasions).

Lists can be created, combined, saved, retrieved, listed, deleted and used to drive data presentations or analyses, i.e. maps, reports, graphs and models. There are a variety of ways in which lists can be created, depending on the type of list. Two lists can be combined by the simple mathematical set operations of union, intersect and minus, to create a new list. Lists can also be referenced within a query to create a new list, for example: find all the abstractions downstream of a previously defined list of discharges; or find the occasions when river flows exceeded a threshold for a list of rainfall events.

Relating lists back to consent assessment and monitoring, this mechanism provides the means to find the gauging stations, quality monitoring sites, abstractions and discharges, upstream, downstream or in the vicinity of the site of interest. The data observed at these sites can be identified and produced as graphs, reports or maps, or the lists used by external models to locate the data upon which they should operate.

CONCLUSIONS AND FUTURE DEVELOPMENTS

In the preceding sections, only discharge consents and abstraction licences have been discussed. However, there are a wide variety of authorization issues of relevance to water, for example drinking water and fishing regulations. By providing a general model, and making the analyses less specific, a great range of authorization assessments should become possible. This would also allow questions of the nature "What regulations apply to this stretch of river, or to this basin?".

Developments in the relational database underlying this development will soon allow the introduction of some important new attribute structures, for example structures allowing storage of large documents, images and sound. Many other structures may be needed by users, such as one for the storage of a borehole log. Such a requirement could be met by allowing users to define their own structures within a data dictionary. The challenge is to find a genuine solution for the querying and manipulation of user-defined structures.

The current list functionality is very powerful, but could be developed further by considering what-where-, where-when-, when-what- and what-where-when-lists! For example, a where-when-list would identify as a set different occasions at different features. In contrast, a where-list combined with a when-list gives the same occasions at different features.

One final aspect, which needs to be considered, is the direct incorporation of models into the query facility. This would permit assessment of the effect of combining the existing chemicals in the river with a new discharge. By exploiting models such as Quasar (Piper *et al.*, 1992) it would become possible to answer "What if ..?" questions and to assess more complex interactions of substances in the watercourse.

Acknowledgements The work described in this paper is part of the development of the Water Information System, the result of a collaboration between the Institute of Hydrology and ICL (UK) Ltd. Many of the ideas presented in this paper have come out of this development with Andrew Howes and Jonathan Tilbury, of Tessella Support Services plc, providing much of the initiative. Our thanks to Isabella Tindall for preparing the figures.

REFERENCES

Band, L.E. (1986) Topographic partition of watersheds with digital elevation models. *Wat. Resour. Res.* 22(1), 15-24.

Boorman, D.B. & Hollis, J. M. (1990) A hydrologically-based classification of the soils of England and Wales. Proc. of Ministry of Agriculture Fisheries and Food Conf. of River and Coastal Engineers, Univ. of Loughborough, UK.

Dangermond, J. (1983) A classification of software components commonly used in geographic information systems. In: *Design and Implementation of Computer Based Geographic Information Systems* (ed. by D. J. Peuquet & J. O'Callaghan) (IGU Commission on Geographical Data Sensing and Data Processing, Amherst, New York).

Gustard, A., Bullock, A. & Dixon, J. M. (1992) Low flow estimation in the UK. Institute of Hydrology report no. 108, Wallingford, UK.

ISO (1987) Information processing systems - database language SQL. ISO standard 9075, Geneva.

Langran, G. (1992) Time in geographic information systems. Taylor & Francis, London.

Morris, D.G. & Flavin, R.W. (1990) A digital terrain model for hydrology. *Proc. of 4th Int. Symp. on Spatial Data Handling*, Univ. of Zürich-Irchel, Zürich.

Piper, B.S., Ishemo, C.A.L. & Whitehead, P.G. (1992) River quality models for environmental planning and management. *Proc. Ass. for Scientific and Technical International Cooperation Int. Conf. Environmental Protection: The Problems facing Eastern Europe*, Varna, Bulgaria.

Romanowicz, R., Freer, J. & Beven, K. (1993) TOPMODEL as an application module within WIS. In: *Application of Geographic Information Systems in Hydrology and Water Resources* (ed. by K.Kovar & H.P. Nachtnebel), Proc. Int. Conf. HydroGIS93, 19-22 April 1993, Vienna, IAHS Publ. No. 211.

Sekulin, A.E., Bullock, A. & Gustard, G. (1992) Rapid calculation of catchment boundaries using an automated river network overlay technique. *Wat. Resour. Res.* 28(8), 2101-2109.

Un système d'information géographique adapté à l'évaluation de la pollution agricole diffuse

D. CLUIS & E. QUENTIN
INRS-Eau (Institut National de la Recherche Scientifique, Université du Québec), C.P. 7500, G1V 4C7 Sainte-Foy, Québec, Canada

Résumé L'utilisation d'un système d'information géographique (SIG) pour l'évaluation environnementale des contaminations diffuses d'origine agricole est justifiée par la nature spatiale des données de base qui contrôlent les phénomènes physiques étudiés: utilisation du sol, pédologie, géomorphologie, structures d'entreposage, parcelles d'épandage, ... Avec les opérateurs classiques de modélisation cartographique contenus dans les SIG non spécialisés, on peut créer de nouvelles opérations spécifiques telles que la génération de données de précipitation et le cumul de contaminants des sous-bassins *amont* vers les sous-bassins *aval*. Toutefois, les fonctions de base restent encore trop limitées pour permettre d'intégrer dans leur totalité les fonctions de production et de transfert des modèles conceptuels autonomes. On peut donc espérer que les développements futurs de l'algèbre spatiale permettront de réaliser des outils d'évaluation "tout SIG".

INTRODUCTION

Durant la dernière décennie, la technologie des systèmes d'information géographique (SIG), devenue de plus en plus conviviale, n'est certes pas restée l'apanage de spécialistes en géomatique. En effet, les possibilités d'intégration de données multisources géoréférencées sont mises à profit dans une grande variété de domaines. Maguire *et al.* (1991) en ont dressé un panorama relativement complet. Or, si on recense les exemples d'utilisation des SIG pour des problèmes de qualité de l'eau, seule l'application de l'équation universelle de perte de sol (USLE) est abondamment documentée. Le potentiel de ce type de système comme outil d'aide à la gestion des ressources hydriques n'est pourtant plus à prouver (Wallis, 1988).

C'est justement pour faire face à la problématique de pollutions de source agricole diffuse qu'un SIG a été mis sur pied pour le Ministère de l'Environnement du Québec. Le logiciel commercial autour duquel s'organise le SIG, SPANS de Intera Tydac Technologies Inc., a été choisi pour sa structure matricielle avec compactage en *quadtree*, qui se révèle appropriée pour des analyses environnementales où de nombreux paramètres ont une variabilité spatiale prononcée. Les opérations disponibles restent de type général: il a donc fallu développer des procédures spécifiques aux processus hydrologiques et au transport terrestre des contaminants potentiels vers les cours d'eau.

APPROCHES DE MODÉLISATION INTÉGRÉE

Problématique considérée

Suite à la concentration de l'élevage intensif des 30 dernières années, la qualité des cours d'eau dans les bassins versants agricoles majeurs du Québec a connu une dégradation non négligeable. Une gestion adéquate des effluents produits s'est donc révélée prioritaire, puisque l'épandage de ces déjections animales s'effectue souvent à des doses dépassant les besoins agronomiques des plantes, entraînant alors des pertes vers l'environnement. Mais le contrôle de la pollution en milieu agricole reste particulièrement difficile du fait de la nature diffuse des sources de contamination. D'un autre côté, c'est cet aspect de variabilité spatiale du problème qui suggère l'emploi d'un outil permettant l'entrée, la gestion, l'analyse et la sortie de données à référence géographique.

Pour le développement d'un SIG adapté à l'évaluation régionale et locale de la contribution des activités agricoles en matières fertilisantes provenant des lisiers et fumiers, trois bassins agricoles pilotes de l'ordre de 5000 km^2 ont été sélectionnés (Quentin & Cluis, 1992). La structure des bases de données établies pour chaque bassin est résumée dans la Figure 1.

Un des problèmes typiques qu'on rencontre lors de la manipulation des données reliées au problème de pollution diffuse agricole concerne l'agrégation des données selon des unités administratives comme les municipalités, alors que les phénomènes de transport s'effectuent selon des unités de drainage reflêtant la géomorphologie des bassins versants. Lorque les critères de confidentialité des données empêchent d'avoir accès aux valeurs individuelles des attributs, on peut pallier à cet inconvénient à l'aide d'une procédure empirique de "ré-agrégation" des statistiques basée sur les surfaces (Cluis, 1992). On obtient ainsi la fonction de production pour les contaminants potentiels.

Il faut de plus disposer d'un module hydrologique permettant de "spatialiser" des données ponctuelles de précipitation mesurées aux stations météorologiques, afin d'évaluer ultérieurement des ruissellements et des débits.

Enfin, le cumul d'attributs de l'amont vers l'aval, selon un processus hydrique, requiert un module de drainage vers les cours d'eau avoisinants. Ce type d'algorithme fait l'objet d'études décrites par Depraetere (1991), Lammers & Band (1990) et Vieux & Kang (1990). De plus, dans les transports terrestre et aquatique, on doit tenir compte des atténuations qui diminuent la charge de contaminants transférés.

Modélisation dans les SIG

On peut distinguer actuellement quatre niveaux d'utilisation des SIG dans le cadre de modélisation hydrologique et de qualité de l'eau (Maidment, 1991). Ce sont par ordre de complexité croissante:

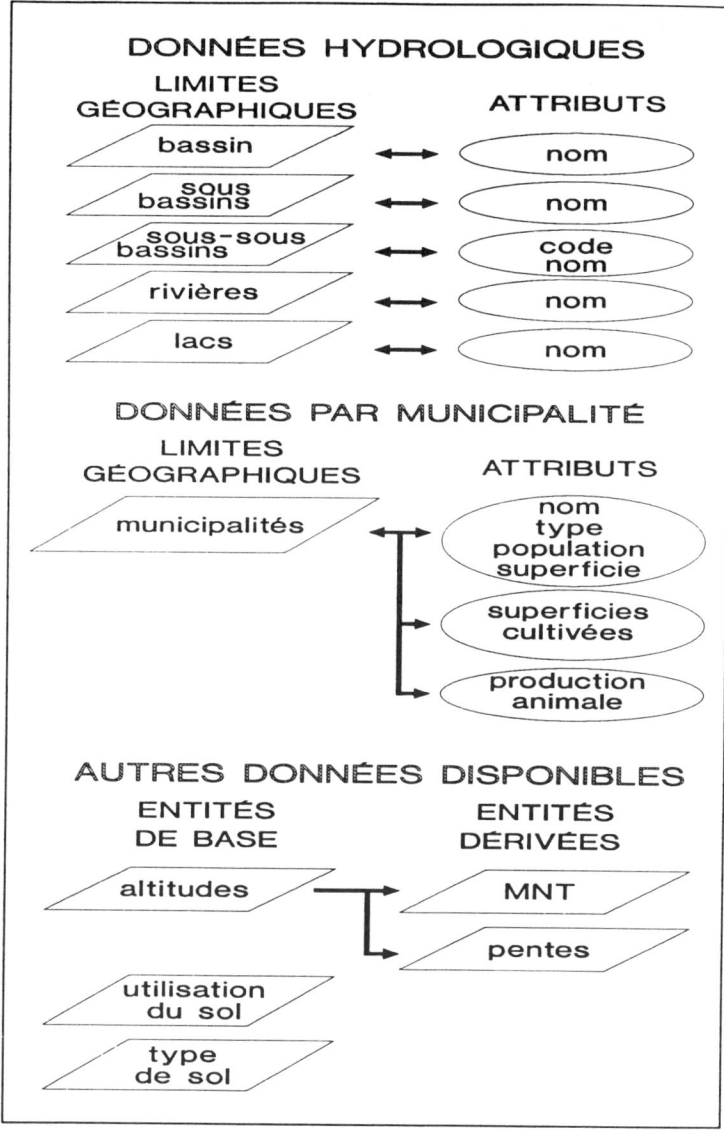

Fig. 1 Données intégrées au SIG.

(a) visualisation des résultats du modèle;
(b) évaluation de paramètres d'entrée au modèle;
(c) existence d'une interface entre le SIG et le modèle;
(d) modélisation intégrée au SIG.

Or ce dernier point est encore peu développé surtout si on considère les processus impliqués dans la pollution d'origine non-ponctuelle. Un module créé selon cette optique doit avoir les propriétés suivantes:

(a) spécifique au problème environnemental en question;
(b) général dans le sens qu'il peut être appliqué à toute zone d'étude pour laquelle les données de base existent;
(c) disponible au sein même du SIG.

EXEMPLES DE RÉSULTATS

Le schéma de la Figure 2 permet de situer la structure d'utilisation des opérations développées pour la problématique à l'étude.

Génération de données de précipitation

Lorsqu'on dispose de données ponctuelles sur un phénomène continu dans l'espace, il faut pouvoir attribuer une valeur à chaque cellule de la zone discrétisée.

Dans le cas des points d'altitudes numérisés à partir de cartes topographiques, on procède généralement par interpolation et on obtient ainsi le modèle numérique de terrain (MNT).

En ce qui concerne les précipitations, on sait qu'il existe une relation entre la hauteur de pluie et l'altitude. Dans un premier temps, on établit une

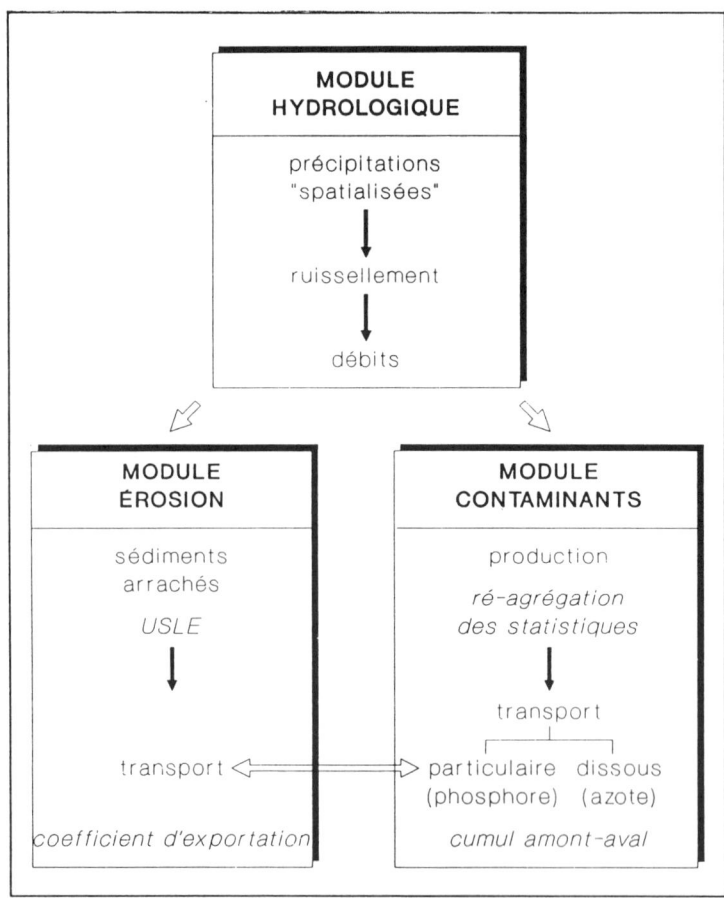

Fig. 2 Modélisation intégrée au SIG.

zone d'influence autour de chaque station météorologique sous forme de polygones de Thiessen, cette opération étant désormais classique dans les SIG. Puis, pour chaque cellule, la hauteur de pluie est calculée en fonction de l'altitude de la cellule et la précipitation reçue à la station la plus proche, selon les paramètres de la régression altitude-précipitation de la région considérée. La Figure 3 illustre le genre de résultat que l'on obtient.

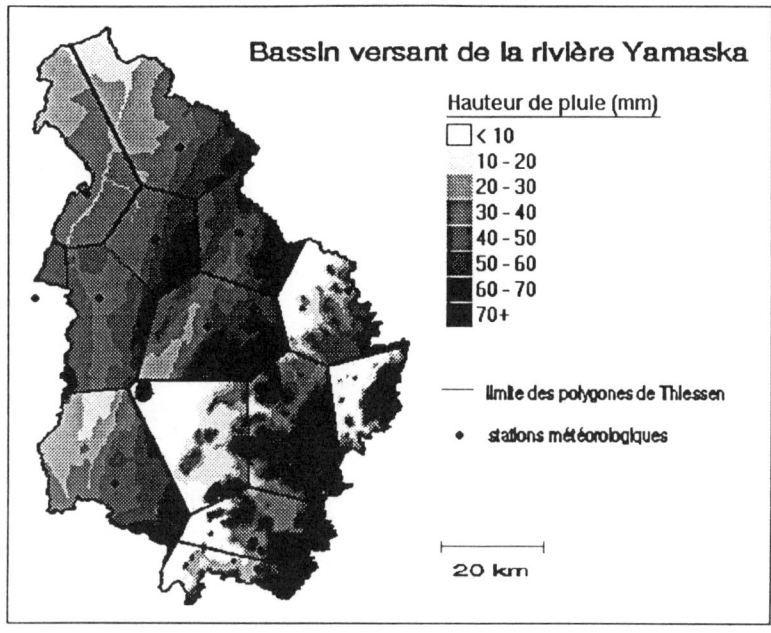

Fig. 3 Variabilité spatiale des précipitations.

Cumul des attributs des sous-bassins *amont* vers les sous-bassins *aval*

L'exemple fourni ici est une fonction de transfert terrestre sans facteur d'atténuation qu'on applique aux résultats de la fonction de production (macro-commande de ré-agrégation). On utilise un algorithme de drainage itératif, selon un arbre de déversement préétabli, mis sous forme d'une macro-commande qui requiert un fichier d'équations, selon le syntax du logiciel SPANS. Le résultat pour le phosphore est donné à la Figure 4.

Le langage de programmation offert par SPANS este limité, ce qui explique que la procédure implantée n'a pas une structure optimale. L'appel à un module externe en langage C ou Fortran offre une plus grande flexibilité de programmation. Par contre, la nécessité de transformer les données entre le format propre à SPANS et un format accessible au programme externe diminue l'intérêt de disposer d'un système d'intégration et augmente les risques d'accumulation d'erreurs et de perte de précision.

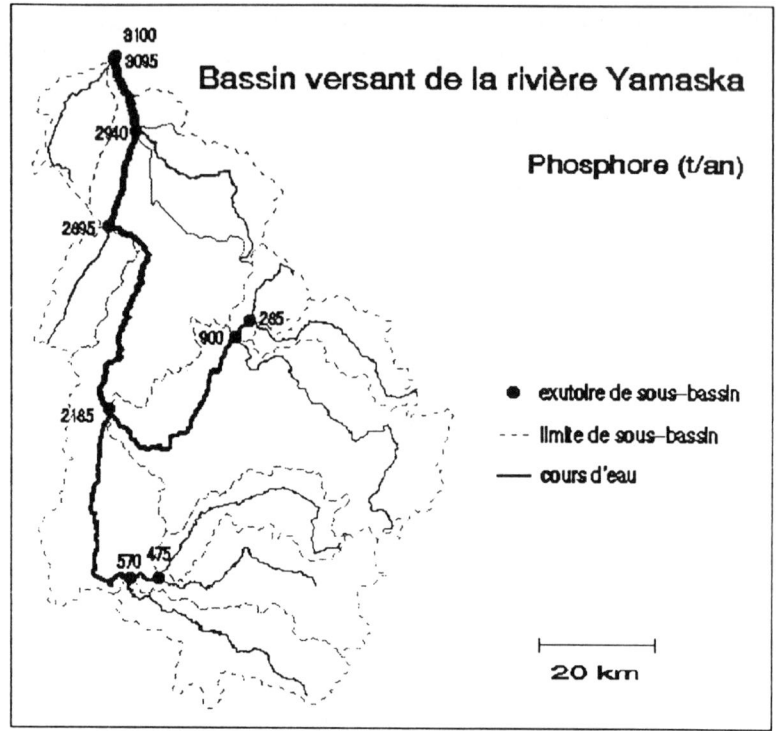

Fig. 4 Drainage par sous-bassin du phosphore des déjections animales.

CONCLUSION

La mise sur pied d'un SIG spécifique à une problématique de pollution diffuse en milieu agricole a su montrer l'intérêt de cette technologie non seulement en tant que base de données géographiques mais aussi pour les traitements internes qu'elle offre et les possibilités de modélisation intégrée à partir des opérations d'analyse de base. Pour l'application considérée, ces modèles ont généralement pour but d'identifier les zones à risque où prioriser des interventions et, suite à des simulations, évaluer l'efficacité des types de solutions possibles.

Toutefois, l'adaptation d'un outil qui tire parti de la discrétisation spatiale à une problématique pour laquelle le développement de modèles agrégés a été privilégié n'est pas encore pleinement complétée. En effet, intégrer à un SIG un modèle qui utilise la moyenne d'une variable sur la zone d'étude ne permet pas d'exploiter le potentiel novateur de ce type de système. D'autre part, peu de modèles ont été pensés en fonction de la structure spatiale offerte par un SIG. C'est donc vers une intégration complète des données et des modèles que l'avenir des SIG devrait tendre.

RÉFÉRENCES

Cluis, D. (1992) Des nouvelles technologies pour une gestion intégrée à l'échelle du bassin versant. *Association Québecoise des Techniques de l'Eau (AQTE)*, Assises annuelles, 8-10 avril 1992, Montréal.

Depraetere, D. (1991) *Démiurge: chaine de production et de traitement de Modèles Numériques de Terrain*. Orstom, Paris.

Lammers, R.B. & Band, L.E. (1990) Automating object representation of drainage basins. *Computers & Geosciences* 16(6), 787-810.

Maguire, D.J., Goodchild, M.F. & Rhind, D.W. (1991) *Geographical Information Systems: Principles and Applications*. Longman, London.

Maidment, D. (1991) GIS and hydrologic modelling. In: *First Conference/Workshop on the integration of GIS and Environmental Modelling*, 14-19 sept. 1991, Boulder, Colorado.

Quentin, E. & Cluis, D. (1992) Un système d'information géographique pour l'évaluation environnementale de la gestion des fumiers et lisiers. In: *Symposium sur la recherche et le développement en gestion environnementale des effluents d'élevage au Québec, 9-10 sept. 1992: textes des conférences*. Ministère de l'Environnement du Québec, Sainte-Foy, Canada. 425-236.

Vieux, V.E. & Kang, Y.T. (1990) GRASS waterworks: a GIS toolbox for watershed hydrologic modelling. In: *Proceedings of the Conference on Application of Geographic Information Systems, Simulation Models, and Knowledge-based Systems for Landuse Management*. Virginia Polytechnic Institute and State University, Blacksburg, Virginia. 309-317.

Wallis, J.R. (1988) The GIS/hydrology interface: the present and the future *Environmental Software* 3(4), 171-173.

GIS and multivariate ecological analysis: an input to integrated river channel management

A.M. GURNELL
GeoData Institute and Department of Geography, University of Southampton, Southampton SO9 5NH, UK

D. SIMMONS
GeoData Institute, University of Southampton, Southampton SO9 5NH, UK

P.J. EDWARDS
GeoData Institute and Department of Biology, University of Southampton, Southampton SO9 5NH, UK

J. BALL, J. FEAVER, A. McLELLAN & C. OGLE
GeoData Institute, University of Southampton, Southampton SO9 5NH, UK

Abstract This paper explores the potential for using multivariate ecological analysis of riverine plant species data to identify environmental gradients within British rivers. Plant species information is used to classify reaches along the River Blackwater and also to infer environmental characteristics of those reaches from the plant habitat requirements. A database management strategy based on SQL has been implemented. A GIS is closely coupled to the data management system and visualization, diffusion modelling and area analysis are combined to generate actual or surrogate environmental gradients which are found to match closely the inferred gradients derived from the plant species information.

INTRODUCTION

In the context of river survey and management, the application of multivariate ordination and classification techniques to information on macrophyte species within rivers in England and Wales by Holmes (1989) has clearly shown that at the national scale macrophyte communities reflect climate, rock type and water quality. Similar links between environmental controls and ecological characteristics underpin the development of systems such as RIVPACS (Moss *et al.*, 1987) and PHABSIM (Milhous *et al.*, 1989), which estimate the likely invertebrate and fish communities of rivers from measured site characteristics. This paper presents the results of a pilot project which explores the potential for identifying the general environmental character of river reaches from the plant species that are present. A linked series of analyses are presented. *First*, the spatial pattern of the results of multivariate ordination and classification of riparian and in-channel species are presented. *Second*, the known environmental tolerances of the species provide possible explanations for the observed spatial patterns. *Third*, the spatial distributions of potential environmental controls are analysed, to assess whether or not they match the inferences on controls derived from the species data. GIS coupled with a relational database manager has proved to be a key analytical tool in this work.

DATA ASSEMBLY AND PREPARATION

The River Blackwater (Fig. 1) was selected for this pilot study because it is a short river (approximately 30 km) within a small catchment (146 km^2) of limited relief (45-147 m OD) that is strongly influenced by man's activities. The small size and relief of the catchment limit the potential variability of climate, the underlying tertiary clays and sands provide a restricted geological variability, and the heavy disturbance of the river channel and predominance of urban and intensive agricultural land use are all likely to restrict the range of riverine plant species. This catchment, therefore, provides a severe test of the potential of riverine plant species information to aid in the identification of river environmental gradients.

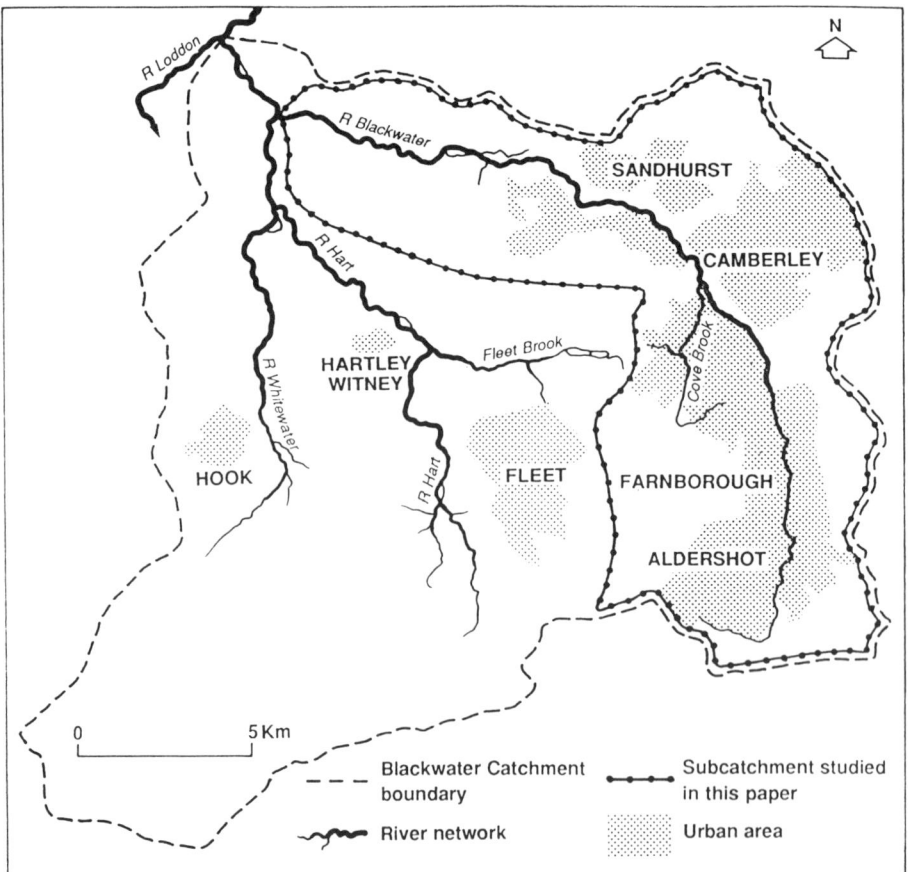

Fig. 1 The River Blackwater catchment.

A GIS for the River Blackwater catchment was assembled using the Intera Tydac Technologies SPANS product run under OS/2 version 2 extended edition on an IBM PS2 model 80 platform. A SPANS study area was constructed which included the watershed of the River Blackwater and its tributaries.

A River Corridor Survey (RCS) for the River Blackwater, commissioned by the National Rivers Authority, Thames Region, provided the basic data source for this study. Information in the RCS was relevant to two types of spatial entity: the River Blackwater itself, which was subdivided into fifty-nine 500 m reaches, and the river's floodplain, which was divided into polygons containing single classes of land use. For each river reach two plant species lists were provided, one for the riparian zone and one for in-channel vegetation. Field measurements of bank height and width (above the vegetation limit) and channel width and depth (below the vegetation limit) were also available for many of the reaches. Each flood plain polygon was allocated to one of 19 land use classes. Two additional sets of vegetation-related information were assembled for the tolerances of the riparian and in-channel species, respectively, to particular environmental characteristics. These data sets were developed from the habitat requirements of river plants listed in the appendix of Newbold *et al.* (1983), where a five-point scale (thrives in these conditions, common, occasionally colonizes, rarely found, absent) is used to relate each plant species to each habitat characteristic.

The vegetation data used in the study were managed as relational tables under control of the OS/2 Query Manager. A data table was thus constructed to hold the reach-based survey information (e.g. channel width) and the DECORANA axis scores and TWINSPAN classes (see (i) below) relating to in-channel and riparian vegetation. A code to name translation table was also available. A tolerance table contained rows for each species and columns identifying scores against different habitat requirements. A similar structure was implemented to manage floodplain habitat polygons and their species lists.

Two raster data sets provided contextual information on the catchment as a whole. Four tiles of the Ordnance Survey 50 m resolution DTM enabled detailed investigation and visualization of catchment topography and definition of watersheds for any location on the river network using the diffusion algorithm supplied within SPANS. Data for Landsat Thematic Mapper bands 3 (red) and 4 (infra-red) for August 9th 1984 were imported as a Normalized Difference Vegetation Image (NDVI = (IR-R)/(IR+R)) linearly stretched to the range 0-100, with image rectification and pre-processing carried out by IDRISI. The NDVI was evaluated and density sliced on the basis of training areas to estimate the spatial distribution of arable, grassland, woodland and heathland, and urban land use. Bare (harvested) arable fields and dense urban areas were difficult to distinguish using satellite imagery alone. To provide such differentiation, the 1981 census information of property density at Enumeration District level was used to generate a property density map based on voronoi boundaries. A SPANS matrix overlay procedure on the property density and

NDVI maps was then used to apply the logic that differentiated between urban and arable land uses.

Additional polygon and point datasets were supplied by the National Rivers Authority, Thames Region. Polygon-based information included the catchment and rock type boundaries, and the boundary of the maximum limit of known flooding. Point entities included the locations of sewage outfalls (for which the daily volume of the discharge consent was supplied) and water quality sampling points (for which 3 years of observations of a variety of water quality parameters were provided).

DATA ANALYSIS

The assembled data were used to explore three aspects of the river environment:

(i) Multivariate analysis of plant species information

The riparian and in-channel plant species lists associated with the 59 500m reaches of the River Blackwater were separately subjected to ordination and classification techniques. Ordination techniques of vegetation analysis produce an arrangement of samples (river reaches) to reflect their similarity. The ordination approach to vegetation analysis is usually used to compare species and environmental factors. Detrended Correspondence Analysis (DECORANA, Hill & Gauch, 1980) was the ordination technique used. The reach scores on the first

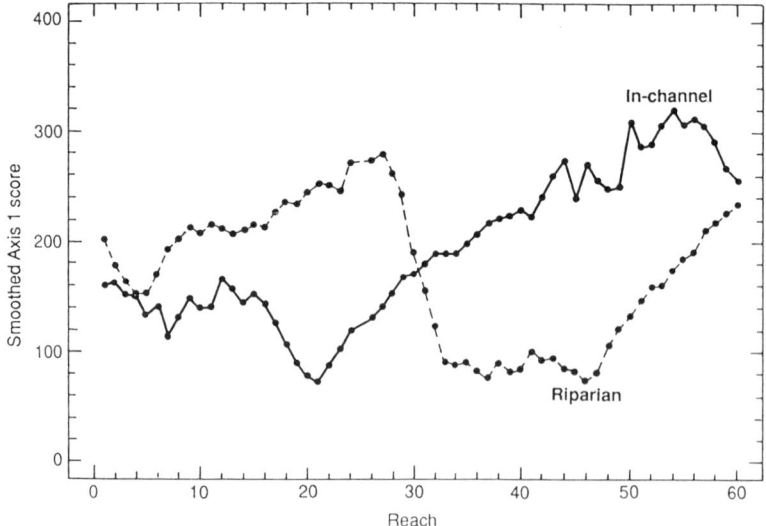

Fig. 2 The downstream pattern in reach DECORANA axis 1 scores for riparian and in-channel vegetation.

axis of both the riparian and in channel ordination generated a clear downstream pattern, which was not apparent in the lower order axes (Fig. 2), but in neither case was it the simple downstream trend that might be expected. Both riparian and in-channel axis 1 scores show clear changes in direction around reach 21 for the in-channel species and around reaches 36 and 46 for the riparian vegetation.

TWINSPAN (Hill, 1979) is a classification method which produces a classification of both plant species and sites or samples in the form of a two-way table. It proceeds by progressively splitting groups of samples into two, to give a hierarchical, dichotomizing classification. In much ecological research, TWINSPAN is used for identifying plant species that occur in association with one another, but in the present case the grouping of river reaches (i.e., the samples) was of more interest. In the present analysis this process of subdivision was allowed to proceed to the third level, so producing eight classes of river reach. Figure 3 plots these eight classes of reach for both in-channel (Fig. 3a) and riparian species (Fig. 3b) against their reach number. The TWINSPAN classes form a clear, if complex, spatial sequence. In the case of the in-channel vegetation, classes 000 and 111 contain only one and two reaches respectively and so were omitted from further consideration. Of the remainder, the downstream sequence of classes was 110, 101, 100, 001, 01 (i.e., 010 and 011 combined), with breaks between classes coinciding approximately with reaches 10, 22, 30 and 42. In the case of the riparian vegetation, the downstream sequence of classes was 00, 10/11, 01, with breaks coinciding approximately with reaches 30 and 50.

(ii) Habitat requirements of river plants and their apparent environmental gradients

Using the data on the habitat requirements of river plants given in Newbold *et al.* (1983), plant species were allocated scores of 1 to 5 for each environmental characteristic. A relational link between the habitat requirement score table for riparian or in-channel species and the appropriate species list for each reach, gave a series of scores for each environmental characteristic so that an average tolerance score could be calculated for each reach. The Twinspan class was used to group reaches into spatial units, and the average score was calculated for each environmental characteristic for each Twinspan class that had a distinct spatial location (Tables 1 and 2).

By identifying the highest scores for each characteristics, the in-channel plant species information indicates that the river has a clay substrate, with a tendency towards the coarser end of the clay range (higher scores on sand and lower scores on clay) in the middle reaches (10 to 42). Flow velocity is slow but decreasing scores on "slow" and increasing scores on "fast" downstream suggest some increase in velocity in a downstream direction. Flow depth is 0.1 to 0.5 m but with increasing scores on <0.1 - dry upstream and on 0.5-1 m and >1 m downstream suggesting a deepening of flow downstream. Flow width is

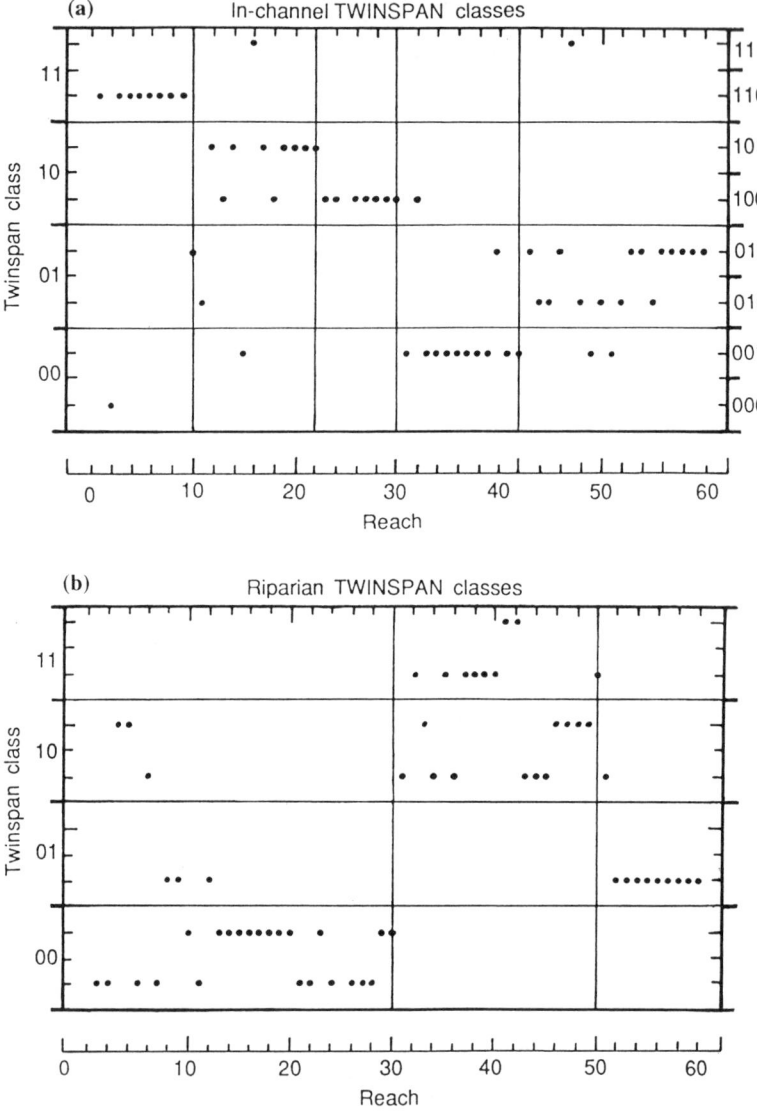

Fig. 3 TWINSPAN classes for river reaches: (a) in-channel, (b) riparian.

5-10 m but increasing scores on >10 m downstream and <5 m upstream suggest channel widening downstream. Base status of the water is neutral upstream but gradually changes towards rich downstream. Nutrient status is moderate upstream, increasing to rich downstream. There is relatively low shade tolerance which decreases slightly downstream, and moderate disturbance tolerance.

The riparian species indicate predominantly clay soil with a tendency towards the coarser end of the clay fraction (decreased clay scores and increased sand scores) in the middle reaches (30 to 50). Bank slopes are steep with the >60 degree class receiving the highest scores. Flow velocities are slow

Table 1 In-channel vegetation - average scores for each TWINSPAN class based upon plant species habitat requirements.

	TWINSPAN class ordered from upstream (left) to downstream (right)				
	110	101	100	001	01
SUBSTRATE					
Peat	1.57	2.04	2.01	1.98	1.68
Clay	**4.47**	**4.28**	**4.31**	4.18	**4.59**
Silt	3.73	3.82	3.82	3.81	4.10
Sand	1.97	2.06	2.03	2.27	1.97
Gravel	2.26	2.36	2.00	2.68	1.76
VELOCITY					
Negligible/slow	**4.65**	**4.58**	**4.53**	4.07	**4.49**
Moderate	3.28	3.28	3.01	3.21	3.13
Fast	1.47	1.67	1.72	2.36	1.61
DEPTH					
>1 m	1.71	1.71	2.23	2.45	2.20
0.5-1 m	2.73	3.08	3.67	4.13	3.41
0.1-0.5 m	**4.34**	**4.41**	**4.26**	**4.28**	**4.32**
<0.1 m - dry	3.97	3.10	2.76	2.65	2.72
WIDTH					
>10 m	3.39	3.63	3.70	4.04	3.85
5-10 m	**4.15**	**4.13**	**4.17**	**4.07**	**4.21**
<5 m	4.13	3.91	3.56	3.24	3.56
WATER CHEMISTRY					
Base poor	1.65	2.41	2.02	2.16	1.38
Base neutral	**4.26**	3.71	4.01	4.04	3.92
Base rich	4.02	**3.80**	4.06	4.19	**4.39**
Nutrient poor	1.65	2.15	1.81	2.00	1.32
Nutrient moderate	**4.07**	3.71	3.97	3.89	3.82
Nutrient rich	3.97	**3.78**	**4.11**	**4.19**	4.25
TOLERANCE					
Shade	2.50	2.56	2.40	2.37	2.32
Disturbance	3.68	3.56	3.65	3.53	3.58

Mean values are derived from allocating in-channel species within the reaches of a particular TWINSPAN class a score of 5 to 1 according to whether they thrive, are common, occasionally colonize, are rare or absent in these conditions (from Newbold et al., 1983). Highest scores for each characteristic are shown in boldtype.

and soils are neutral. Shade tolerance is moderate with the highest scores occurring in the middle reaches (30 to 50) of the river.

The data on habitat requirements presented by Newbold *et al.* (1983) were derived from a national survey. These identify the general character of the River Blackwater: a low velocity, narrow, shallow and disturbed river developed on alluvium derived from clay to sand based rock types. The clear, if not strong, spatial trends inferred from the vegetation species data are very surprising given the small size of the river and catchment and the limited range of apparent environmental gradients operating. The validity of these trends

Table 2 Riparian vegetation - average scores for each TWINSPAN class based upon plant species habitat requirements.

	TWINSPAN class ordered from upstream (left) to downstream (right)		
	00	10 and 11	01
SOIL			
Peat	2.28	2.49　2.16	2.19
Soft mud	2.57	2.51　2.61	2.57
Clay soils	4.21	4.10　4.19	**4.35**
Sandy soils	3.59	3.77　3.97	3.30
Loose shingle	2.20	2.46　2.39	1.89
Rock fissure	2.52	2.92　2.61	2.22
SLOPE			
>60°	3.71	**4.10**　3.90	4.03
30-60°	3.50	3.97　3.55	3.84
<30°	2.90	3.33　2.77	2.43
VELOCITY			
Negligible/slow	4.16	4.05　**4.26**	4.00
Moderate	3.78	3.90　3.94	3.89
Fast	2.79	2.95　2.77	3.03
CHEMISTRY			
Acid	1.62	1.87　1.55	1.78
Neutral	**4.07**	3.82　3.97	4.03
Alkaline	3.63	3.51　3.74	3.41
TOLERANCE			
Shade	3.21	3.51　3.39	2.92

Mean values are derived from allocating riparian species within the reaches of a particular TWINSPAN class a score of 5 to 1 according to whether they thrive, are common, occasionally colonize, are rare or absent in these conditions (from Newbold et al., 1983). Highest scores for each characteristic are shown in boldtype.

was, therefore, explored by estimating the actual environmental gradients or surrogates for them using the River Blackwater GIS.

(iii) The spatial distribution of potential environmental controls on plants species

In identifying the actual environmental gradients, it is the local (at-a-site and immediately adjacent) properties which are most likely to influence the riparian vegetation, whereas the in-channel vegetation is also likely to be influenced by the catchment area draining to that river channel reach. Therefore, in using the GIS to define actual or surrogate environmental gradients, three spatial units were used: the catchment area upstream from certain reaches; the area of flood plain adjacent to each reach as defined by the Maximum Flood Boundary and perpendiculars to the river channel generated from the reach end points; and the area defined by the river reach itself. The first two of these spatial units were generated within SPANS. The identification of subcatchments used a new SPANS module for diffusion analysis to ascertain from the DTM the areas flowing to, or "uphill" of specified reaches. The area of the floodplain adjacent to each reach was established from a voronoi map based on the reach centre points cut to the flood plain boundary.

The likely character of the river *substrate* and bank *soil type* was established from the distribution of rock types within the catchment. The proportion of different rock types within the catchment area was calculated from area cross tabulation between upstream catchment and rock type polygons. Figure 4 shows that the upper and lower reaches of the river are on London Clay, whereas the middle reaches are on the coarser Lower and Middle Bagshot Beds. The indication from the in-channel TWINSPAN classes of some coarsening of the substrate between reaches 10 and 42 corresponds with the zone where the river is underlain by Middle Bagshot Beds (Fig. 4a). The indication from the riparian TWINSPAN classes of some coarsening of the soil between reaches 30 and 50 appears to correspond with the pattern generated by rock type percentages within the catchment area (Fig. 4b), possibly reflecting

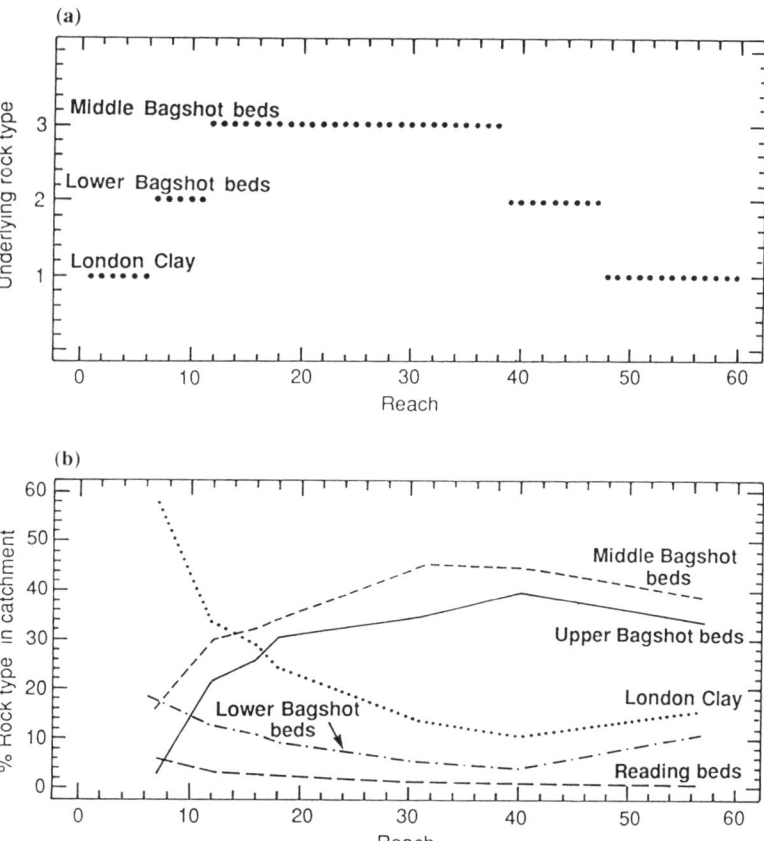

Fig. 4 Downstream pattern in (a) underlying rock type; (b) percentage of rock types within the catchment area.

the fact that the river banks are developed on alluvium (from the upstream catchment) rather than on the local underlying rock type.

The inferred sizes and trends in river *channel depth and width* and *bank slope* from the in-channel and riparian TWINSPAN classes are confirmed by the available surveyed cross section data (Fig. 5a). The tangent of the bank slope showed no clear spatial pattern, but the average tangent of the slope angle was 2.26 (66°), which falls within the predicted >60° class.

The likely pattern of *water chemistry* was estimated from three sets of information. The mix of rock types in the catchment (Fig. 4b) provides a control on background water quality, but this is moderated by land use, estimated through area analysis of the classified Landsat TM image for the subcatchment areas draining to seven reaches. A more direct indication of the urban impact on water quality can be gained from the discharge consent levels for sewage outfalls contributing to flow from varying subcatchment areas. Since river discharge, which increases with the catchment size, dilutes effluent discharged to the river, a Discharge Consent Index (DCI) was derived by dividing the accumulated discharge consents in each subcatchment by catchment area (Fig. 6a). Average determinations for selected water quality indices, particularly total oxidised nitrogen, over three water years (1989-90,

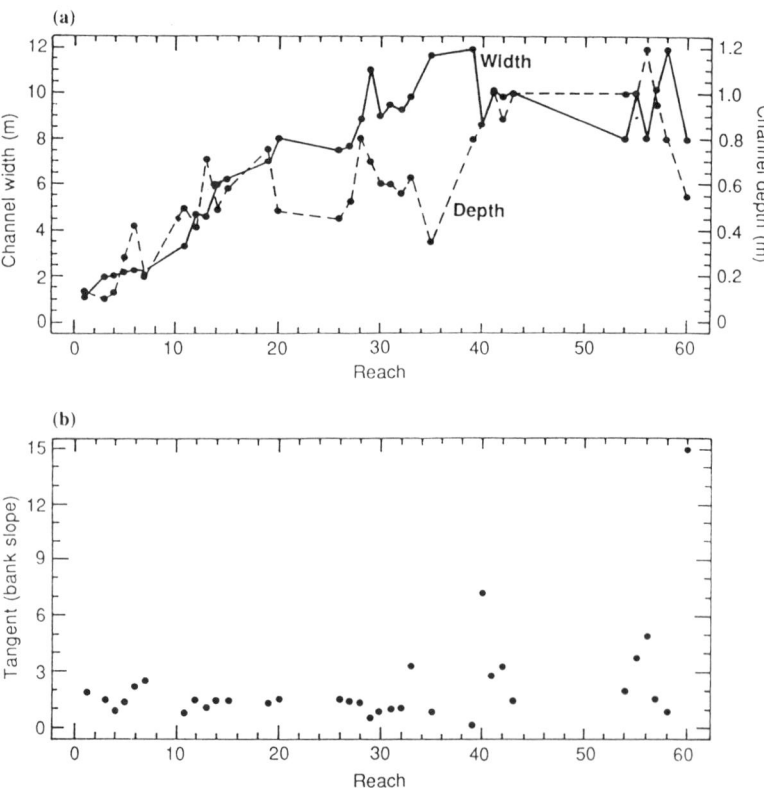

Fig. 5 (a) Channel width and depth; (b) tangent of bank slope, plotted against reach number.

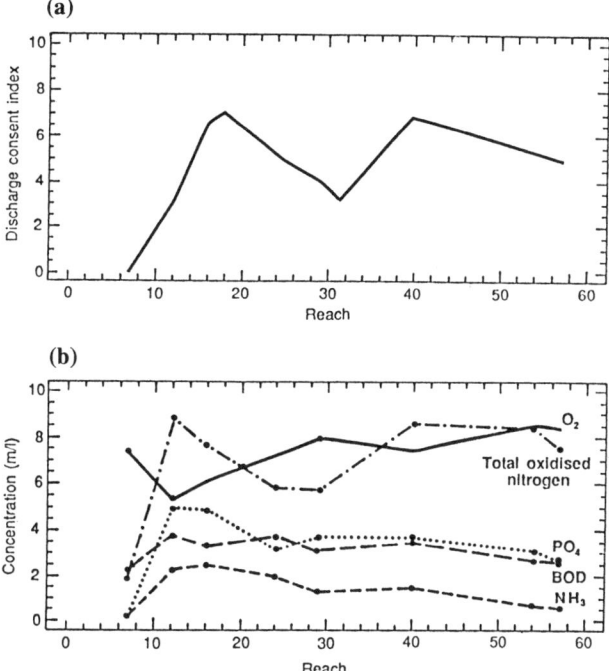

Fig. 6 (a) Discharge consent index (100 m^3 day^{-1} km^{-2}); (b) sampled water quality, plotted against reach number.

1990-91, 1991-92) at water quality sampling points (Fig. 6b) show a similar pattern to Fig. 6a. Analysis of flood plain land use adjacent to each reach shows a correspondence between woodland cover and shade tolerance scores for riparian and in-channel species.

CONCLUSIONS

This paper has reported on the combined use of ordination and classification of plant species information; inferences on environmental gradients from plant species habitat requirements; and the application of visualization, diffusion modelling and area analysis techniques within a GIS to illustrate and quantify actual environmental gradients or their surrogates. These techniques were applied to data from a pilot catchment which was specially chosen to minimize the potential for obtaining striking results. It is a testimony to the information content of plant species information that environmental gradients have been inferred from the analysis of these data, which have largely been confirmed from the observed environmental gradients or their surrogates. The power of some simple GIS functions in establishing catchment, flood plain and channel characteristics and their spatial distribution is also well illustrated by this pilot study. The Blackwater study shows the general power and value of combining a wide range of survey information in addition to integrating the more common types of map data within a GIS as a tool for understanding and, therefore, managing river channels in a more integrated manner.

Acknowledgements D. Simmons was seconded from IBM UK during the period of this research. The National Rivers Authority, Thames Region (particularly D. Mills, A. Driver, C. Woolhouse & J. Eastwood) is gratefully acknowledged for provision of data and for their encouragement throughout this project. Barbara Pack's cartographic and digitizing support is also very gratefully acknowledged. The channel dimensions were surveyed by Claire Hiscock.

REFERENCES

Holmes, N.H. (1989) British rivers - a working classification. *British Wildlife* 1(1), 20-36.

Hill, M.O. (1979) *TWINSPAN - A Fortran program for arranging multivariate data in an ordered two-way table by classification of the individuals and attributes.* Cornell University, Section of Ecology and Systematics, Ithaca, New York 14850, July 1979.

Hill, M.O. & Gauch (1980) Detrended correspondence analysis: an improved ordination technique. *Vegetatio* **42**, 47-58

Milhous, R.T., Updike, M.A. & Schneider, D.M. (1989) *Physical Habitat Simulation System Reference Manual-Version II.* Instream Flow Information Paper No. 26, US Fish and Wildlife Service, Biological Report 89(16), Washington, DC.

Moss, D., Furze, M.T., Wright, J.F. & Armitage, P.D. (1987) The prediction of the macroinvertebrate fauna of unpolluted running water sites in Great Britain using environmental data. *Freshwater Biology* **17**, 41-52.

Newbold, C., Purseglove, J. & Holmes, N. (1983) *Nature Conservation and River Engineering,* Nature Conservancy Council.

The importance of GIS in regional geohydrological studies

J.H. HOOGENDOORN, W. VAN DER LINDEN & C.B.M. TE STROET

Institute of Applied Geoscience TNO, P.O. Box 6012, 2600 JA Delft, The Netherlands

Abstract A methodology is presented for the execution of regional geohydrological studies, in which GIS enhances the opportunity for visualization and processing of data. This is illustrated with a description of geohydrological model input data and the assessment of the uncertainty in such input data. The methodology introduces the establishment of a regional geohydrological information system, which combines a geohydrological database and GIS functionalities.

INTRODUCTION

Regional geohydrological studies provide an analysis of the geohydrological conditions and often include a simulation of the groundwater flow system in a groundwater model. A major part of the work is to calibrate the model, to obtain a true and accurate representation of the geohydrological system. Once such a model is constructed, it may be used to simulate all kind of scenarios.

TNO uses a methodology for carrying out regional geohydrological studies in which a Geographical Information System (GIS) plays an important role. The methodology is not new nor unique, but it is important in relation to the development of so-called geohydrological information systems. Concurrently TNO is involved in the development of such systems.

GEOHYDROLOGICAL STUDIES

Geohydrological studies involve processing of large number of data of geological, hydrological, or other origins. In the past, such data was inventoried and collected at the start of the study and prepared according to the requirements of the selected model code. When another study was started in the same project area a number of years later, the same process of data handling was repeated, due to differences in the data requirements of the new and old model code. This meant that for every study a more or less new, model dependent database was constructed.

Nowadays it has become clear that the above described procedure is very inefficient and that it is more practical to construct one geohydrological database with an open structure, which is independent from the applications used in the geohydrological studies. Such a geohydrological database contains all data relevant to geohydrological studies and has an interface to process the data and for further analyses. The database combines data from different origins and from different scales and provides the facilities to keep the data consistent and up-to-date.

The obvious choice for processing and visualizing this data is a geographical information system (GIS). The GIS software contains all the tools to build a user-friendly menu-driven interface around such a database (together with the database software) to visualize and process data, to generate and present information and to transfer data and information to applications such as groundwater models.

The use of a database and GIS in the execution of geohydrological studies may be expressed as shown in Fig. 1. The process from defining a geohydrological problem to providing advice and suggestions for solution of the problem is carried out through five steps:
(a) data collection and retrieval of data from a database;
(b) processing of data and preparing model input using a GIS;
(c) running a geohydrological model: (i) calibration, and (ii) simulation of scenarios;
(d) interpretation of model output using GIS;
(e) visualization and translation of study results for discussion with involved parties and advice to principals.

The simulation of the groundwater flow is done using a groundwater model. The preparation of the input data, the calibration of the model parameters and the processing of the model output data is supported by a GIS.

The final step in the process is the translation of the (geo)hydrological effect to social, economical or environmental effects, which are increasingly becoming an integral and essential part of geohydrological studies.

The process described above may be incorporated in a so-called *geohydrological information system*. Such a system is currently in development at TNO in The Netherlands.

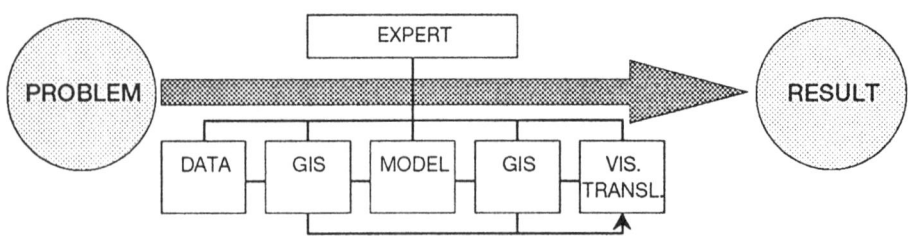

Fig. 1 Execution path of geohydrological studies.

GEOHYDROLOGICAL INFORMATION SYSTEM

The geohydrological information system REGIS (Regional Geohydrological Information System) is combined with a national database for The Netherlands, containing data relevant to geohydrological studies from all parts of the country. The data comprises the necessary information of the subsurface, but also information on surface water, meteorological data, information on land-use, topography, etc.

REGIS (Fig. 2) contains functionalities for realizing and processing geohydrological data and a large number of functionalities for visualizing data, generating and presenting information and transferring data and information to applications, such as groundwater models (Van Bracht *et al.*, 1993). The system is constructed with ARC/INFO as GIS and ORACLE as database management system.

The importance of a GIS-based system like REGIS is that it can be applied for specific projects, for which all available data are retrieved from the national database and transferred to the project environment. This means that the same basic data is available for any geohydrological project in the country. Furthermore the functionalities available in REGIS, are also available in the project environment.

DATA IN GEOHYDROLOGICAL STUDIES

The data used in geohydrological studies may be separated into basic and derived data. Both data types are stored in the geohydrological database. The main difference between the two types of data is that the derived data is subject to changes by the geohydrologist carrying out the project depending on the desired representation of the geohydrological conditions, while the basic data may be changed only by the database manager, e.g. because of reinterpretation of measurements. The derived data may be approximated more accurately using the results of calibration procedures.

Basic data comprises borehole information, geophysical information, pumping test data, measurements of the piezometric head, analyses of water quality, surface water data and rainfall/evaporation data. Derived data comprises extent of geohydrological layers, thickness and depth of layers, hydraulic conductivity of aquifers, resistance of clay layers, groundwater flow direction, groundwater flow quantities, groundwater recharge, and for transient conditions storage coefficient and variation of recharge.

The borehole and geophysical information provides the basic data from which the extent and the depth of the layers is derived. The borehole information and pumping test information provide the basic data from which the hydraulic conductivity of aquifers and the resistance of clay layers are derived.

Derived data specifying depth or thickness of a layer are contours. This line information may be generated directly from the basic point information (geohydrological columns) calculated by the GIS or may be determined indirectly from the basic point information by a (hydro)geologist (Fig. 3).

The point and line information specifying the position of the interface of a layer is stored in coverages. A TIN and consecutively a grid is created from each coverage. The grids are checked for consistency to trace incorrect spatial

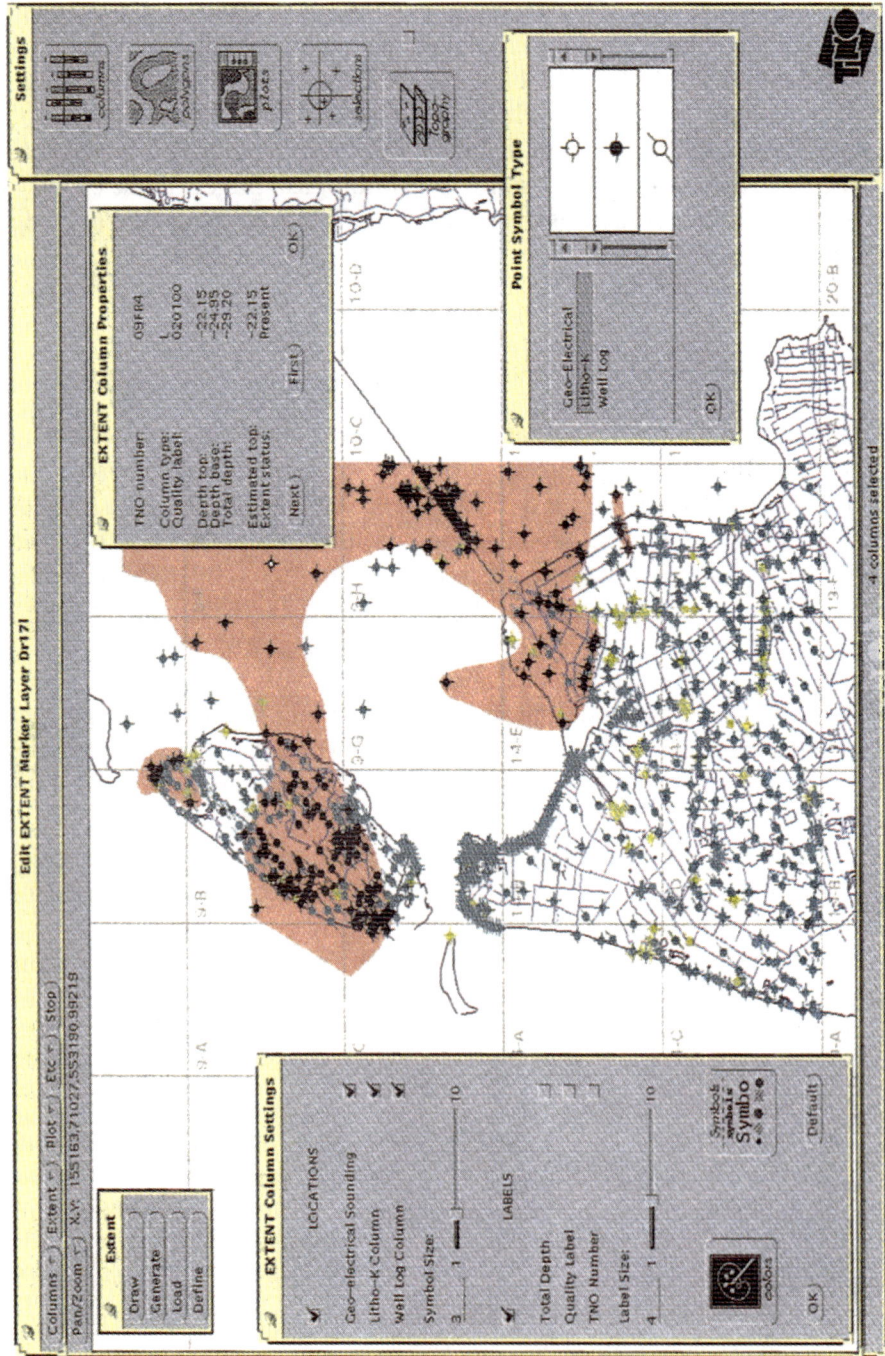

Fig. 2 Screen impression of geohydrological information system.

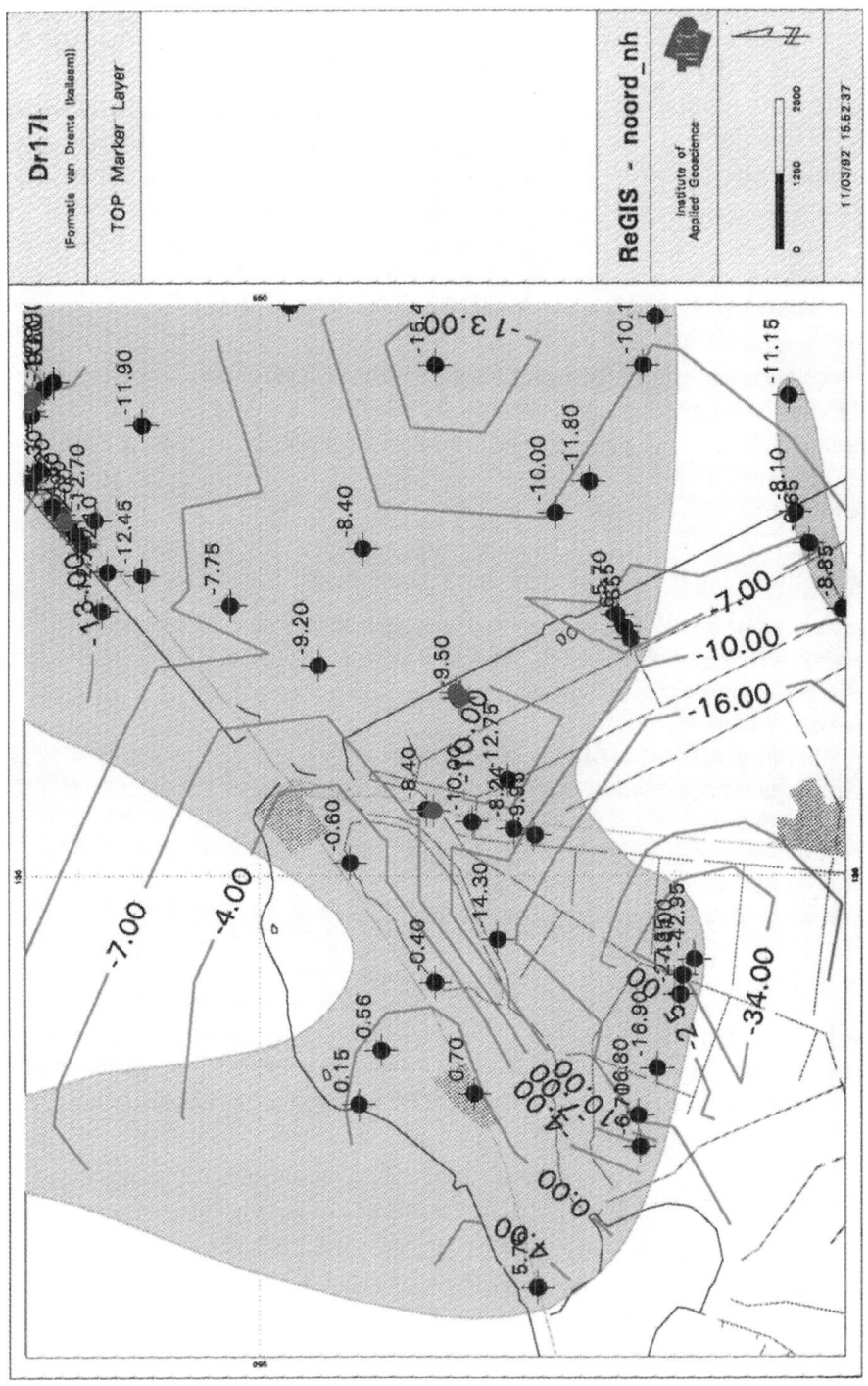

Fig. 3 Example of layer information.

information and correction of this information may be needed a number of times until the consistency of the layers is correct. Finally, the geohydrological model is constructed using the grids.

For the use of a groundwater model, the data stored in the GIS coverages or grids has to be converted to the model grid. This is done with the different functionalities available within the GIS environment. The discretized data is exported out of the GIS as ASCII files and reformatted with simple FORTRAN programs into groundwater model input files. The output of the groundwater model is transformed to input files for the GIS, also with simple FORTRAN programs. Visualization of the model results is done within the GIS environment.

DATA RELIABILITY AND MODEL CALIBRATION

The derived model input information given above has its own uncertainty. It can be classified according to (relatively) well known and less well known information. In general, geometry information -aquifer dimensions, location of water courses, location of abstractions, etc.- is relatively accurate. Also measured quantities like water levels, dimensions and bottom elevation of first order water courses, abstraction rates, etc., are relatively accurate. This well known information is referred to as "hard parameters".

The other information, referred to as "soft parameters", is adapted in the calibration procedure as described in this section. Soft parameters are for example:
(a) hydraulic conductivities;
(b) hydraulic resistance of clay or peat layers;
(c) conductance of water courses, which is equal to area/(resistance of the bottom);
(d) non measured surface water levels in second and third order water courses;
(e) bottom elevations of second and third order water courses;
(f) groundwater recharge.
   For the transient simulation the following soft parameters may be added:
(g) storage coefficients of the phreatic aquifers;
(h) the groundwater recharge, usually modelled as: $R = P - c_f E_p$; the crop-factor c_f is taken as a soft parameter (R = recharge, P = precipitation, E_p = potential evaporation);
(i) a damping and delay of the groundwater recharge with respect to the precipitation, which is related to the process in the unsaturated zone.

At TNO a procedure is used to calibrate the groundwater flow model with acceptance of a certain range of reliability of the soft parameters (*a priori* minimum-maximum values). It consists of the following steps:
(a) sensitivity analysis of the calibration criterion for variation of soft parameters. The effect on the calibration criterion (usually a function of

the squared differences between model results and measurements) over the total model domain and at measurement locations is calculated and visualized with a GIS. If the effect is small the parameter cannot be calibrated and the predefined range between maximum and minimum value is kept;

(b) sampling soft parameters and preparing input. With the Latin Hypercube sampling method (Stein, 1987) samples are taken of the soft parameters according to predefined distributions, correlations and min/max values. This stratified sampling technique results in better representation of the input space;

(c) Monte Carlo simulations. The input for the simulations is prepared automatically according to given soft parameters and their zonation and is calculated out of the samples. The pointers used for defining different zones are handled by the GIS;

(d) analysis of results and choice of accepted parameter sets according to the model criteria. The GIS is especially useful to visualize the results of the Monte Carlo simulations. Very often more than 10 000 model simulations are made, which are evaluated with specially developed FORTRAN programs and GIS. An example of the result of the Monte Carlo approach is presented in Fig. 4. This gives the variation of 20 accepted parameter sets (20 models) out of 16 000 simulations. These 20 models fulfilled a calibration criterion -squared residuals in 189 observation locations- within the range of 1 cm^2. The variation calculated per grid cell can be more than 150 cm. In this way the uncertainty of the model results -after calibration- is calculated/visualized efficiently.

TRANSLATION OF MODEL RESULTS

The model output results are usually piezometric heads and groundwater flows, which need to be translated into, e.g. social, economical or environmental variables, to obtain a result which can be understood by principals, politicians or other decision makers. Such a translation is straightforward in case a direct relation is existing, e.g. between calculated effects and economic damage. But very often such translations are complex and not easily represented by mathematics.

An example of a geohydrological effect is the lowering of the groundwater table through increased groundwater abstractions. This effect may cause subsidence of the ground surface, leading to damages to buildings or a decrease in agricultural crop yields or a reduction of nature values.

A GIS enables to visualize such effects and therefore plays an important role in the decision making. An example of a GIS presentation of the effects of a proposed pumping station (in this case on the groundwater head) is included in Fig. 5.

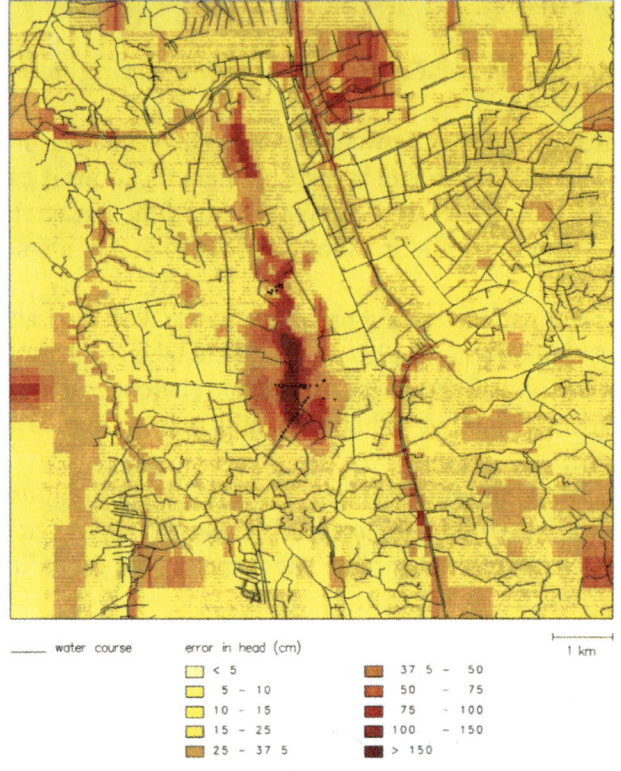

Fig. 4 Example of result of Monte Carlo approach: Maximum - minimum groundwater head at cells for accepted models (layer 1).

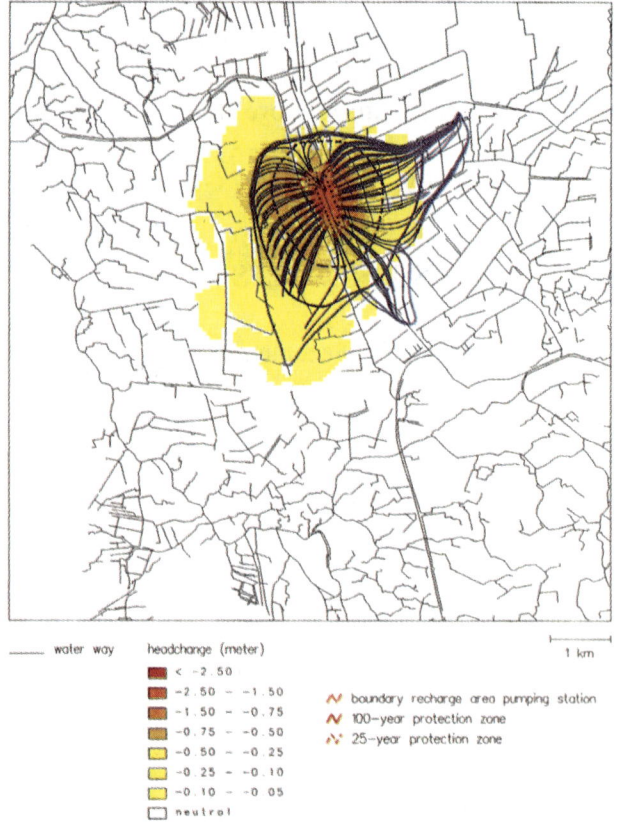

Fig. 5 Example of a GIS presentation of the effect caused by an alternative pumping station on groundwater head and flow.

LIMITATIONS OF THE GEOHYDROLOGICAL INFORMATION SYSTEM

The description of using geohydrological information systems has to be brief in this paper. However, it is possible to indicate a few limitations which may be encountered:
(a) the information used in the geohydrological information system has different origins and may therefore have different levels of accuracy and reliability. It is difficult to combine the different data types and at least correction of part of the data is necessary to prevent inconsistency;
(b) information out of geohydrological studies often involve temporal data which is not well covered by a GIS. Temporal information requires special software, which is usually a part of the database interface. It is processed by a GIS into maps of one specific date, e.g. represented in a contour map of the groundwater head.

CONCLUSION

In water management it is necessary to deal efficiently and adequately with occurring groundwater situations and problems. This requires the availability of a geohydrological information system based on a GIS, which provides up-to-date information which is quickly accessible and easy to process. The development of a geohydrological database with an open structure is an essential part of such a system.

REFERENCES

Stein, M. (1987) Large Sample Properties of Simulations Using Latin Hypercube Sampling. *Technometrics* 29(2),143-151.

Van Bracht, M.J., Broers, H.P., Hoogendoorn, J.H., Van der Linden, W. & Waardenburg, F. (1993) REGIS/Digital Groundwater Map; construction of a visualization and manipulation tool for groundwater related problems on a regional scale. Poster presented at Int. Conf. HydroGIS'93, 19-22 April 1993, Vienna.

GIS in water-related environmental planning and management: problems and solutions

STEFAN O. KADEN
WASY Institute for Water Resources Planning and Systems Research Ltd., Waltersdorfer Straße 105, D-1183 Berlin, Germany

Abstract In the last few years water sciences as well as water management and water-related environmental planning have been characterized by ongoing changes and new challenges in objects and objectives. Key words include global change and environmental impact assessment. Consequently, new tools are required to handle related information and decision support problems. As new tools, two interdependent developments will be considered - the application of geographic information systems (GIS) and databases for the management of spatial and temporal distributed data and information, and the development of simulation systems using this spatial (and temporal) distributed information within more or less physically based models. There is a large range of methods, tools and applications between the "simple" GIS applications and the highly sophisticated spatial distributed GIS-based model systems. This paper is an attempt to classify GIS applications in water-related environmental planning and management. A few GIS applications will be illustrated. Finally, experiences and problems of GIS-implementations in practice are discussed.

INTRODUCTION

Interactions between human society, biosphere, atmosphere, and hydrosphere have increased extensively, sometimes for the welfare of mankind and environment, but frequently for their harm. These interactions are characterized by increasing complexity, diversity, use, and misuse of natural resources, the latter permanently decreasing. And this holds true for any scale in space and time, i.e., from global to local and from long-term to short-term.

One of the most important water-related environmental problems concerns "climatic change". In the process of climate change, the classical hydrological cycle is or might be losing its stationary periodic character. The driving forces of hydrological processes and environmental conditions became subject to long-term man-made alterations and, vice versa, these driving forces and conditions are affected by large-scale hydrological processes. Consequently, periodic stochastic hydro-meteorological processes are superimposed on deterministic trends. The principal interrelations are depicted in Fig. 1. As this figure indicates, these global processes are in close relation to meso-scale and micro-scale problems.

On the regional and local scale the interactions between society, hydrosphere, and biosphere are relevant. Typical examples for water-related impacts are the construction of reservoirs, channels, drainage systems but also waste disposal sites, etc. In this case, the relations are basically unidirectional, i.e., impacts of water management on the environment, but drawbacks should not be neglected (e.g., impacts of deforestation). On the other hand man-made impacts on water resources, e.g. by waste disposal sites, have to be considered. In both cases environmental impact assessments are necessary.

Complex software systems are required to analyze the interrelationships depicted in Fig. 1 and to predict future developments. Such systems have to consider, as typical properties, the areal and temporal distribution of problems and parameters. Usually, they consist of a data and information system and a model system. In Fig. 2, the principal structure of such a system is depicted. The data and information system are inevitable. In recent developments, these systems consist of a GIS and an environmental (hydro-ecological) database. This distinction is being made because of the differences in managing predominantly geographical information, i.e., spatially distributed data of various surface properties, and time-dependent environmental data (station- or areal-related time series).

GIS is a valuable tool for overlaying and displaying large volumes of spatial and non-spatial data. In many disciplines of geosciences, the use of GIS is widespread for management, processing, and visualization of geo-referenced data. Goßmann (1989) has compared the role of GIS in modern geography with the role of the microscope in biology. This assessment may be transferred to hydrology and ecology, but in this case adding simulation as an important component. Consequently, there is a strong inter-relationship, between new GIS-technology and environmental modelling. In practice, this interrelationship but also the knowledge about the abilities of those tools, are weakly developed. As an example, a GIS definition from an IIASA-publication (IIASA, 1991) is quoted: "Databases may also combine information mathematically and display it on a grid. Such a database is called a GIS. A GIS can display environmental data such as height above sea level". GIS - that is much more!

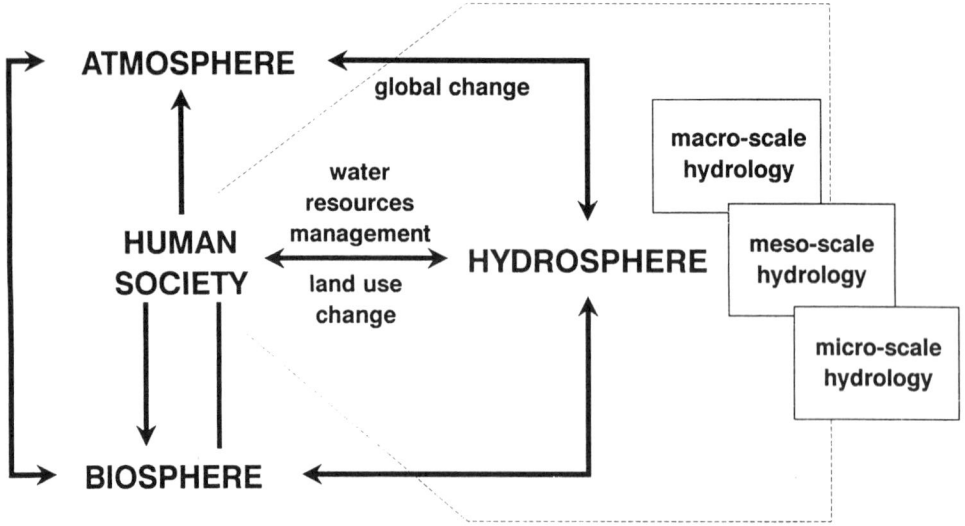

Fig. 1 Global change and hydrology.

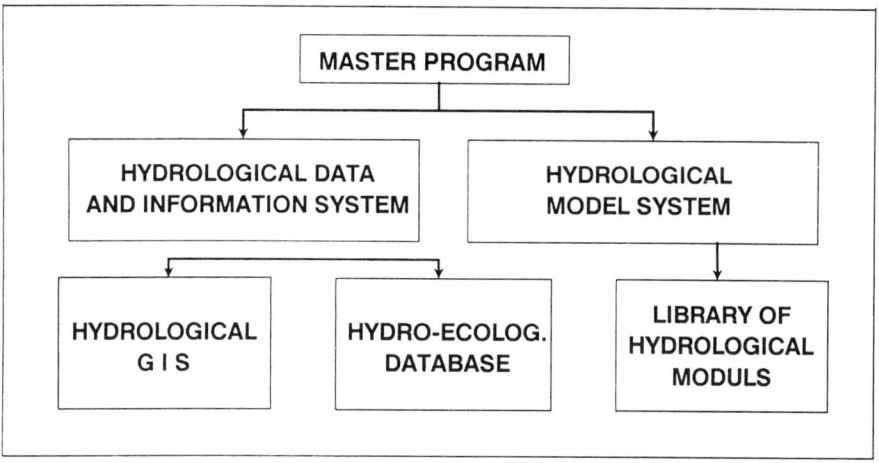

Fig. 2 Structure of an hydrological software system.

In the following section, the role of GIS will be discussed for water-related environmental planning and management, a role which is placed between databases and simulation models.

TASKS AND TOOLS

The terms water-related environmental planning and management summarize hydro-engineering activities affecting the environment, as well as other man-made activities affecting water resources as a component of the environment. This means, we consider "water" as the "perpetrator" as well as the "victim." Both types of activities are closely interrelated by nature. This may be illustrated considering environmental problems related to the construction of waterways. In Fig. 3, the so-called impact matrix is depicted. For example, the construction affects directly the surface water quality due to polluted sediment removal and changes of discharge. As an other example, groundwater lowering due to drainage of building pits affects the vegetation in the river valley.

In Table 1, an overview of the major areas and tasks of water-related environmental planning and management is given. The design of water-related facilities is not considered here, although GIS are very helpful tools in these cases. The tasks differ for each user according to characteristics such as:
(a) temporal scale (short-term, middle-term, long-term) and resolution (minutes, ..., years);
(b) spatial scale (local-, meso-, and large-scale) and variability (m^2, ... km^2).

In Table 1, the role of GIS is classified as "low" (not important), "medium", or "high" (very important). As expected, the role of GIS is especially high in case of middle-term meso-scale tasks.

The given tasks can be generalized into a number of basic activities. Keywords are for instance: assessment of problems and environment,

impacts	effected component of the environment	soil	ground-water	surface water	climate / air	vegeta-tion	fauna	human beiing	landscape recreation	cultural goods
claim of area	soil removal	+++	+	+		+++	+++		+++	+++
	soil disposal	+++		+		+++	+++		+++	+++
	sealing, compression	+++	+++			+++	+++		+++	+++
structural changes	cuting up of living space					+++	+++			
emission / immision	pollutants immison	+	+	+	+++	+	+	+++	+	+
	noice, shaking							+++	+++	
hydrological changes	flow changes			+++		++	++			
	infiltration changes		+++	+++		+++				
modifications of infrastructure	modification of traffic			++		++	++	++		
visual impacts	modification of structures as bridges								+++	

+   impacts of construction
++ impacts of management
+++ impacts of construction, structure and management

Fig. 3 Impact matrix for environmental impact assessment in traffic hydraulic engineering.

comparative analysis, monitoring of dynamic processes, evaluation of current conditions, detection of changes, forecast of future development, determination of the source of problems, planning of actions (e.g., mitigation activities), identification of regions that meet multiple criteria, and analysis of policy options.

GIS supplies tools to support these tasks, especially dealing with spatial scales and spatial variability, but only partially with temporal variablity. In Fig. 4 these features are related with the generalized tasks. According to that figure GIS are extremely helpful. But they do not solve any problem with regard to the given tasks. Related problems and requirements are discussed in a latter section of this paper.

EXAMPLES AND EXPERIENCES

In the following section, two examples are given, explaining the application of GIS in the discussed field. These examples should illustrate possibilites and experiences, rather than give a detailed description of the problems and solutions. Because of the restricted publication possibilies only one figure of GIS-applications will be depicted.

Nitrate transport in soils and groundwater in Germany

The cooperate project of the BMFT (Federal Ministry of Science and Technology), supervised by the Research Centre Jülich, involves the compilation of a regionally differentiated model on the transport of nitrate in the territory of former East Germany (Wendland, 1989). This model is planned to illustrate regionally differentiated hazard potentials and to serve as a decision

Table 1 Tasks in water-related environmental planning and management.

Area	Tasks of water-related environmental planning and management	Temporal characteristics scale	Temporal characteristics resolution	Spatial characteristics scale	Spatial characteristics resolution	Role of GIS
water resources protection and remediation	surface water pollution control	short	minutes-days	local meso	source-related	low
	groundwater protection against depletion and pollution	middle long	days-month	meso	$>>m^2$- km^2	high
	river basin sanitation	middle long	days-years	meso large	$>>m^2$- km^2	high
	groundwater sanitation	long	month-years	meso micro	m^2- $>>m^2$	high
	lake restoration	middle	month-years	micro	m^2- $>>m^2$	low
water resources management	assessment of groundwater resources	middle	month-years	meso	$100\,m^2$- km^2	high
	assessment of surface water resources (discharge, reservoirs)		days-month	meso large	$>>m^2$- km^2	high
	water resources monitoring: planning	-	-	meso	$>>m^2$- km^2	high
	operation	short	hours-days	large meso	point oriented	low
	management and efficient use of water resources available	short middle	minutes days	local meso	not of interest	low
	flood defense	short	minutes-hours	meso	not of interest	medium (flood plains)
hydraulic engineering and urban water planning and management	planning of reservoirs and of hydro-power systems	middle	days-years	local meso	m^2- $>>m^2$	low medium
	planning of waterways	middle	hours-days	meso	$>>m^2$- km^2	high
	water-related town planning	middle	not of interest	local meso	m^2- $<<m^2$	high
	planning of irrigation and drainage systems	middle	days	meso	$>>m^2$- km^2	high
	planning of sewer systems	short middle	hours-days	local meso	m^2- km^2	medium
	planning of storm water systems	short middle	hours-days	local meso	m^2- $>>m^2$	high
	planning of water supply and waste water treatment plants	middle	not of interest	local meso	m^2- $>>km^2$	medium

support tool for the design of water protection strategies against diffuse nitrate pollution of groundwater. The object of the study is to trace the nitrate transport in the groundwater from diffuse source input into the soil to the output into surface waters. Two basic methodological problems had to be solved:
(a) the regionally differentiated determination of nitrate inputs as a function of the agricultural, hydrological, and pedeological situation; and

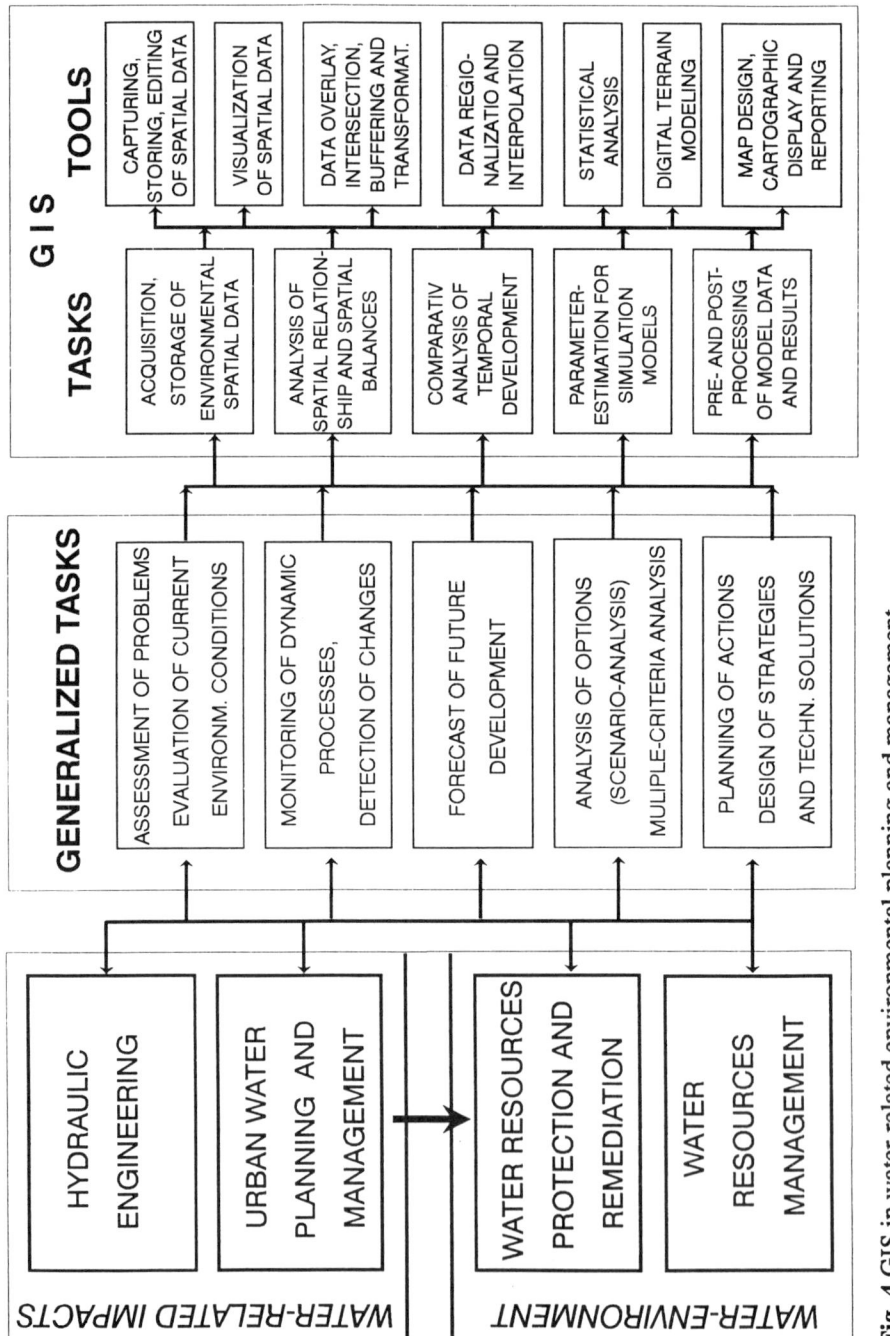

Fig. 4 GIS in water-related environmental planning and management.

(b) the transport and degradation of nitrate in the groundwater as a function of the geological, hydrogeological and geochemical conditions.

The model is based on a 3 by 3 km grid (in total 39 709 elements). WASY's task was the estimation of relevant hydrological and hydrogeological parameters for these grids. Among others, the following parameters have been estimated:

(a) precipitation;
(b) groundwater level below surface;
(c) permeability and effective porosity (of the top aquifer);
(d) groundwater catchment areas and their drainage network;
(e) hydraulic gradient and flow aspects;
(f) conditons for denitrification (in 5 classes).

In preparing the data, a number of difficulties had to be solved, mainly resulting from the heterogeneity of basic maps, i.e., varying scales, schematization and accuracy of those maps. Additional geometric distortions had to be considered in the territory of former East Germany (done for military security reasons). The necessary geometrical corrections have been done using the rubber sheeting GIS-function. In order to get a reliable and multiply applicable database, a combination of vector- and raster-based work has been organized. This will be explained for the map of the groundwater level below surface. The following steps have been taken:

(a) elaboration of a contour map of the groundwater level of the top aquifer (from the hydrogeological maps 1:50 000) at the scale of 1:500 000. This step was done manually with the help of scaling down and schematization;
(b) digitalization with ARC/INFO (vector-based);
(c) digitalization of the land surface from topographical maps 1:200 000;
(d) generation of digital elevation models for both surfaces (in ARC/INFO);
(e) transformation of the vector-based digital elevation models in equivalent 1 by 1 km raster (in ARC/INFO);
(f) calculation of the differences and arithmetic mean for the 3 by 3 km raster.

In this study, the GIS was a very helpful tool. It was used to work with maps, to rectify them, to overlay, to estimate digital terrain (elevation) models, to transform these models into raster maps and to visualize the results.

Study of non-point phosphorus input to surface waters

A joint research study has been performed by scientists of the Institute für Gewässerökologie und Binnenfischerei (Institute for Water Ecology and Inland Fishery in Berlin), the Zentrum für Agrarlandschafts- und Landnutzungsforschung (Centre for Agrolandscape and Land Use Research) and of the WASY Institute to define the low-scale limit for reasonable application of P-balancing and/or transport simulation models in pleistocene lowlands of North-East Germany.

In order to adjust and to improve conventional large-scale nutrient-balancing methods, results of local transport-path investigations should be used. For that, a reasonable spatial differentiation of balance factors has to be estimated. This differentiation depends on the degree of data availability and the modelling approaches.

For the pilot study, a meso-scale catchment (300 km^2) of a lake (Schwielochsee) was selected. It is large enough for a check of general data availability at different scales and for further regional differentiation (spatial resolution). For the catchment, long-term measurements of nutrient loads leaving the catchment with surface waters are available. These data have been used for model verifications.

For the study, a GIS database has been developed with the following themes at resolutions associated with scales between 1:50 000 to 1:200 000:
(a) elevation of the land surface and of the groundwater table in the top aquifer (1:50 000) (digital elevation models based on ARC/INFO-TIN);
(b) satellite imagery (raster-based) on land use (30 m raster);
(c) soil data from different sources (1:25 000, 1:100 000);
(d) stratigraphy of the top aquifer (1:50 000).

The phosphorous input was given for administrative and/or economic units (communities, agricultural cooperatives) in terms of supply of fertilizer, manure produced from livestock, etc. Based on that data, the balancing factors have been regionalized for those administrative and economic units.

One specific result of the study was the risk analysis of long-term non-point phosphorus input to the lake through groundwater outflow. Due to long-term phosphorous overloads in the groundwater, the eutrophication problems may continue even if point sources (detergents, fertilizers, etc.) are eliminated within this catchment area.

The given study could only be performed based on a consequent GIS application. The huge amount of spatially distributed data could not be handled without a GIS. Similar to the study on nitrate pollution, in this study most problems result from the difference in scales, resolution, and quality of basic maps and input data. In order to present the spatial database as well as the modelling results to non-experienced GIS-user the ESRI visualization system ArcView was used successfully.

PROBLEMS AND REQUIREMENTS

Based on a few years of experience with GIS application in hydrology and water related environmental planning, in consulting as well as in research and development, in the following several problems and requirements are discussed.

Systems integration

GIS applications in environmental planning and management usually require the integration of different software components. The basic components and their applications are depicted in Table 2.

Technically, this integration of (loosely) related components is more or less solved. Although most of the mentioned components are realized in stand-alone software systems, these systems usually do supply interfaces to other components. There is almost no system available offering the full complexity of applications (maybe one should not ask for one). One could agree to the opinion that "specialized technologies should be used for what they were designed for and linked to other tools that have similar special purpose missions, ESRI (1989)". The handling of different systems components and different GIS-related technologies requires specialists for each component, which is especially for smaller companies not easy to be done. To display maps on top of images is a simple but, under given technical and financial restricitons, frequently highly cumbersome job! Such problems are in close correlation to the problems of user-friendliness, etc.

The most realistic way for systems integration is the linkage between some of the components above for certain problem areas, i.e., linkages above the level of "simple" data interfaces. The integration of raster- and vector-based GIS is offered now by most GISs. ESRI's development of GRID, as an extension to the vector-based ARC/INFO, illustrates that finally there is now way outside of the hybrid GIS. But this was an internal company development. More complicated is the integration of different technologies (of different software houses). One of the promissing examples is the "mariage" between AUTODESK and ESRI with the product ArcCAD, offering the (almost) full spectrum of PC ARC/INFO under the AutoCAD surface. This enables the user to retrieve CAD drawings and raster images in association with spatial features

Table 2 Software components in water-related environmental planning and management.

Software component	Application	Importance
document scanning systems	background of GIS-maps, basis for vectorization	medium
image processing system	remote sensing	high
vector-based GIS	acquisition, storage, management and visualization of spatial distributed data	high
raster-based GIS	ditto	high
database management systems (DBMS)	acquisition, storage, management and visualization of temporal distributed data (time series)	high
computer aided drafting and designing (CAD)	report preparation	low - medium
video media systems	visualization, documentation of the state of the environment	low
surface modelling	basis for surface runoff and erosion models	medium
resources modelling and simulation	analysis of future development	high

on a map. And this opens a new, large market for GIS, but above all a new field of GIS applications especially to smaller companies in the field of environmental enginieering and consulting.

More complicated seems to be the disposal of open structures for the embedding of simulation models. GIS-based modelling is still in the initial stage with regard to practical applications. One example is the GIS-based modelling of macro-scale hydrological processes with areal distributed parameters and process-oriented models. Such systems should provide a description of the land-atmosphere interaction which is sufficiently realistic to simulate short-term behavior adequately but which is expressed in numerical parameters relevant over large areas and broad vegetation classes (Shuttleworth, 1988). Obviously for this task GIS are best suited as an integrated and integrating software component.

Dynamic problems - the 4th dimension in GIS

GIS are excellent tools to handle areal distributed data, but they are usually not designed to scope with data varying in time. In the same way as the third dimension in space (except digital terrain models implemented in GIS), the 4th dimension, time, is practically missing in GIS.

GIS can be used to handle restricted numbers of time levels as layers or covers, e.g. to compare them, but they fail practically for the analysis of real dynamic processes. They offer no usable formal mechanisms for representing the change of attribute values over time. At present for that purpose, multi-applicable solutions are practically not available. Frequently "hand-made" time series management systems (databases) are developed and used based on standard software (FORTRAN, file systems) - with the known advantages and disadvantages. In the future, the development of the integration between GIS and relational as well as object-oriented databases will and has to be continued for that purpose. The difficulties to develop such databases for practical applications result from the highly varying data structure in space and time (compare Table 1). In complex studies of environmental planning and management, such highly varying data structure has to be considered.

Availability of data

The availability of data is a technical problem (measurements in sufficient quantity and quality), but especially an administrative problem. The first aspect is solvable with the recent availability of modern technologies as remote sensing, global positioning system (GPS), etc. The technical development enables us to get improved data as long the required funds are available, which is obviously of special importance in developing countries and in the new states in Eastern Europe.

There is a number of problems related to data conversion and data accuracy, these are the different density of data in the case of large-scale applications, inaccuracies due to different scales (and related generalization), different age of data, out-of-date data sets, inhomogenieties due to changed sensors in remote sensing, changed algorithms, measurement and interpolation errors, and finally the use of data sets, being established for different applications. Consequently the verification of data is a very important task, because a GIS is only as good as its database! Some nicely coloured GIS applications present an accuracy which is not based upon a sufficient accuracy of data input systems.

Methods are available to scope with the nonuniformity of the space density of observation points, of the frequency and the quality of observations, the homogenization of measurement data and their regionalization (of point data). This is still a large open field for innovations, especially with regard to the integration of space and time.

Another aspect does not depend on the technical development, but on the tradition, the political / administrative system and others. For instance in Germany there are no public domain digital data sets available. Rules and regulations for the use and the sale of digital data sets are not clearly defined. This aspect restricts the application of modern GIS in many environmental related applications.

Technological changes

If we analyze the different lifetime of GIS-components, the aspects of technological changes become obvious. According to Bill & Fritsch (1991) we have to consider innovation rates with regard to GIS as follows:
(a) hardware   3- 5 years;
(b) GIS software   7-15 years;
(c) data   25-70 years.

These numbers are questionable, nevertheless, the drastic differences between lifetime of hardware, software and data seems to be realistic. Any permanent GIS application is affected by rapid changes in the field of hardware, but also GIS-software (forced by the market competition!). These rapid changes result in financial requirements, but, still more important, they result in strong requirements with regard to the qualification of GIS staff and GIS users. Due to the rapid development for both groups it is difficult to keep up with technologies and with technics. These rapid changes have to be compensated by easy implementation of the software and data on new platforms.

Further portability as far as possible, according to appropriate available and expected in the future hardware platforms, is needed. The data lifetime indicates that the database will usually be the most valuable part and (consequently) the most significant financial investment for a GIS!

User-friendliness

The increasing level of integration, complexity of problems and of software systems require special consideration to the user-friendliness of the software components, in the given case of GIS. This might be discussed from two points of view, from the viewpoint of specialized users, calling for sophisticated capabilities, and from the viewpoint of wide audiences, requiring the ease of use. In the past e.g. ARC/INFO was an excellent tool for specialists, but with high effort to produce easy-to-use applications for end-users. With the visualization software ArcView a new product was developed by ESRI, closing this gap remarkably. Now it is possible to group around a highly-sophisticated GIS (e.g. ARC/INFO, but also others) satellites of ArcView applications for non-experienced GIS users. This strategy is supposed to be successful especially in the field of environmental planning and management.

Project planning

In the case of GIS development, a well-designed planning is especially important because of the fact that failures in project managment in early project stages (design of GIS, digitizing, etc.) are hardly to recover in later stages! Project planning, proper data management for accuracy of data supply, etc. before production is a must! In environmental planning and management the interdisciplinarity has to be taken into the account also! Important aspects for project planning are the support of data exchange by supply of standardized interfaces and data formate. The usual stages of GIS-design and development are:
(a) conduct user needs assessement;
(b) evaluate existing data;
(c) define conceptual system design;
(d) define geographical database;
(e) plan quality assurance;
(f) develop implementation plan;
(g) realize pilot/prototype application.

For geographical database design, special consideration should be given to the accuracy of data and the sources of the data. Redundancy should be avoided as much as possible.

SUMMARY

For water-related environmental planning and management, a GIS is a valuable and frequently indispensible tool. It is used for documentation, management, storage, and visualization of spatial data and for parameterization of models. In hydrology, the application of GIS wil result in improved methods for regionalization as well as areal structuring of models, reducing the effort and the subjective components of these processes. Especially important is the easy consideration of changes in spatially distributed parameters such as land use, and the explicit linkages between spatial and descriptive data. GIS allow for a better sharing of data resources, less redundancy of data and the consistency of data content and format (standardization). GIS support different user views and development of user applications, a high flexibility of data retrieval, analysis, and reporting.

Without any doubts, the role of GIS in environmental planning and management will increase. But this process has to be accompanied by adequate improvements in both database systems (time-related data) and simulation models. User-friendliness, open interfaces, and portability are important features facilitating successful applications.

Acknowledgements This paper is based on project results of the WASY Institute and on valuable discussions with its members. Thanks!

REFERENCES

Bill, R. & Fritsch, D. (1991) Grundlagen der Geo-Informations-Systeme (Basics of Geo-Information-Systems). Bd. 1. Hardware, Software und Daten. Karlsruhe (Wiechmann).
ESRI (1989) Integration of Geographic Information Technologies. ARC News, Winter 1989 Issue, Vol. 11, No. 1.
Goßmann, H. (1989) GIS in der Geographie (GIS in geography). GEO-Informationssysteme (GIS), Jahrgang 2, Heft 3, 2-4.
IIASA (1991) *Environment, Development and Systems Analysis*. IIASA options, Dec. 1991, 4-10.
Shuttleworth, W.J. (1988) Macrohydrology - the new challange for process hydrology. *J. Hydrol.* **100**, 31-56.
Wendland, F. (1992) Nitrat im Grundwasser der "alten" Bundesländer (Nitrate in groundwater in the "old" Federal States). Berichte aus ökologischer Forschung, Band 8/1992, Forschungszentrum Jülich GmbH.

Interactive multi-media, GIS, and water resources simulation

DANIEL P. LOUCKS
School of Civil and Environmental Engineering, Cornell University, Ithaca, New York 14853, USA

Abstract Simulating the movement of water to, from, and within multiple watersheds and aquifers, and predicting the effects of various options taken for modifying the flows, storage volumes, and water quality within those watersheds and aquifers, is being made easier through the use of improved computer technology and modelling techniques. Increasingly, portions of Geographic Information Systems are being developed and linked to simulation models to provide a more effective means of managing spatial and temporal data. This paper explores the potential for incorporating video and sound within these GIS-based simulation models for increased efficiency in model use as well as for increased understanding of model output.

INTRODUCTION

Computer models are a well-established and accepted part of any hydrologist's or water resource manager's toolkit. The main purpose of these models is to create information useful for understanding, planning, designing and managing water resource systems. Computer simulation modelling, originating in the late 1940's and early 1950's, is arguably one of the most useful tools for estimating the hydrological, economical and environmental impacts associated with alternative water management plans or policies.

Any simplified simulation of a complex and uncertain hydrological, economical and environmental system requires numerous assumptions. These assumptions pertain to processes we can now measure and observe as well as to future conditions we can only guess might occur. Given any set of assumptions, e.g. future runoff resulting from changing land uses and possibly a changing climate as well, future demands, future costs or benefits, and future waste loads, computer simulations can give us an indication of the likely range of possible consequences of those assumptions. If we like or don't like what we see from the results of such simulations, we can try to manage our water resources accordingly. Simulation is a technique that is understandable, and becoming increasingly more comprehensive in scope and easier to use. Advances in computer simulation modelling have come mainly from advances in computer hardware and software technology.

Computer technology has changed considerably from the time Maas *et al.* (1962) introduced many of us to the idea and use of simulation on a digital computer. In case some of you were not around in what seems at my age to be a not too distant past, it is instructive to quote parts of what Manzer and Barnett (Maas *et al.*, 1962) wrote concerning river basin simulation:

"Because it is the essence of simulation to reproduce, in some cognate way, the behavior of a system in every important detail, simulation problems typically are large-scale and complex. Masses of input data, as well as the results of intermediate calculations, must be kept at hand.... Until digital computers of the magnitude of the (IBM 650 or Univac I) were constructed, the solution of large scale simulation problems was not feasible. Even the storage capacity of these computers was not commodious enough...(River basin simulation) had to await elaboration of the very large computers of the IBM 700 class, with an internal magnetic-core storage capacity of up to 32 768 words."

These words were published a mere 30 years ago! There were no graphic displays available then, neither interactive interfaces. And there were certainly no mice (just what are those?!), digitizers, or even color display monitors to help us input and edit data, run the simulation models, and display the results even in the form of graphs and pictures let alone over maps. There were no special programs for managing, displaying and analyzing spatial data, programs we now call Geographic Information Systems or GIS.

Today our computer technology is different. As I sit typing this paper with the aid of a word processor I am looking into a color display monitor attached to a small box that sits on top of my desk. This display monitor contains my text and a series of menu picking or keystroke options I can implement when I want to that will allow me to correct or modify what I have done. These options also allow me to call other programs, including perhaps an interactive hydrological simulation model coupled to a GIS. I can do this by just dragging a mouse that in turn moves a cursor on the display to one of numerous menu buttons and then "clicking" on that menu button. Would that now routine operation known today by elementary school children have been understood in 1962?

In addition to being easy to use, this little computer box contains over 100 times the 32k word storage capacity of the IBM 700 series machines referenced in Maas's book. The computing that occurs inside this box is substantially over 100 times faster than what was expected from that temperature-controlled room full of IBM 700 series equipment in 1962. Furthermore, it is done so inexpensively that users now don't even consider the cost of operating these computers any more than they consider the cost of operating their radios. Finally, the purchase price of this "microcomputer" and all its attachments and software is such that even individuals can afford to buy them. Thirty years ago, who would have ever predicted any of this? If someone had, who would have ever believed it?

Looking into the future is risky business. It is especially risky when considering a technology that is changing as fast as is computer technology. Nevertheless, in this paper I will attempt to look into the near future and discuss how we might begin to develop and use a relatively new part of computer technology, i.e., interactive multi-media and its links to GIS and simulation models. It's a technology that I think will be as common a decade from now as the microcomputer is today. This technology has just begun to be developed to the point that it could be applied to our area of interest, namely the study of hydrological and water resources systems. This technology should take us out of the largly silent and motionless world of interactive computer graphics into the world of sound, video, and animation. This paper speculates on how this interactive multi-media technology, coupled to GIS and simulation models, might be useful in the study and analysis of water resource systems.

Before discussing multi-media, it may be helpful to review briefly the use of GIS within simulation models. Following this discussion, the potential benefits of adding video and sound to a combined GIS-simulation modelling environment will be explored. Finally specific examples of some of the current work at Cornell University incorporating video and sound into GIS-based simulation models will be presented. The emphasis in this short paper, however, will be more on what we can do in the future, rather than on what little has been done, so far, up to the present.

COUPLING GIS TO SIMULATION MODELS

Simulation models of water resource systems, whether a small watershed or an entire region containing multiple river basins, require and produce spatial and temporal data. Geographic Information Systems are designed to manage spatial data. Increasingly simulation models and Geographic Information Systems are being combined to provide improved means of managing and displaying spatial data (Maidment, 1991; Moore, et al.,1991; ESRI, 1992).

The River Simulation System (RSS) model developed at the University of Colorado (Boulder), for example, includes a commercial GIS package. This GIS provides a vector map display of the river system being studied. The model has the capability of converting this vector representation of the river system into a node - link network representation of that system that is required for its object-oriented simulation (CADSWES, 1992). The Interactive River System Simulation (IRIS) model originally developed at the International Institute for Applied Systems Analysis and Cornell University (IIASA, 1990), permits the display of a digitized raster map background over which the required node - link network can be created for the simulation of any particular integrated surface and groundwater quantity and quality system. The MAPHYD hydrological modelling system developed at the University of Colorado (Denver) also uses digitized maps to access and display data (Johnson, 1989). These three models are among many that display maps having areas or sites that can be picked to access both input and output data pertaining to those specific areas or sites.

Photographs as well as plots of simulated time series data associated with specific sites in a region can also be accessed and displayed using the same pick-site operations on digitized maps of the region. Using the capabilities of GIS, particular user-selected features that may exist within a specified distance of a picked site can also be displayed in the form most appropriate. Picking sites on a map to access spatially dimensioned data increases the association of that data to its applicable location in the mind of the user. This approach to interactive data editing and display has been used in simulation modelling for at least a decade or more (e.g., Da Costa *et al.*, 1989; and French *et al.*, 1980). In most cases, the full power and capability of modern GIS programs are not being used, but this is not a prerequisite for the beneficial inclusion of appropriate portions of GIS within simulation programs.

In most of our previous applications linking GIS to simulation, we have tended, with few exceptions, to limit our displays to 2D static images. Simulated variable values that change over time are typically plotted on a graph where time is the independent variable - the x-axis of the plot. We don't actually see anything change over time; we just see in one display at one time perhaps one or more plots of those changes. These plots, digital maps and photographs are usually motionless. While portions of some vector displays may be animated, model users are not seeing what they would see, and not hearing what they would hear, if they were actually on the site in the region experiencing what the simulation indicates is happening. In fact they are not even experiencing what they would expect to experience if they were watching a televised view of the site. If there is both spatially and time varying information resulting from a simulation that can be better communicated or understood using full motion video or at least video-like images, and even sound, then we as developers of these simulation models should be thinking about how to make this happen, i.e. how to use video and sound effectively and creatively.

MULTI-MEDIA LINKS TO GIS AND SIMULATION

Water resource system simulations provide estimates of what might take place, over space and time, in a particular basin or region. Variables of interest might include flows, storage volumes, various concentrations of pollutants, and hydropower (both energy and power) or functions of any combination of these variables. Functions of those variables may define economical, environmental or social impacts that also will vary over space and time. The system being simulated can include numerous surface water river reaches, natural lakes and regulated reservoirs as well as multiple groundwater aquifers, all serving multiple purposes and demands. For a given set of input data and assumptions, the simulation results should tell us what will be happening at any particular site in the region at any particular time period, or what will be happening at a given site over a number of time periods. Can we visualize those simulated events as we would if we could actually see them happening? Perhaps we can if we begin to link multi-media capabilities to our model output.

Linking multi-media capabilities to GIS-based simulation is not a new idea. Narasimhan & Fox (1990) present a quick overview of some aspects of this subject. Just about every computer magazine currently being published contains articles on multi-media technology. Organizations working with land use data seem to be among the first to link multi-media to GIS database and display programs.

Interactive use of digital video and sound combined with hand-drawn renderings and a digitized zoning map of the area around Union Station in Washington, DC, give city planners at the National Capital Planning Commission an improved idea of how any proposed new development may fit into an existing urban environment. Their multi-media GIS offers an opportunity for them to fly low over the area to obtain a clear view of the area's buildings and parks and streets, and to hear the sounds of that environment, such as from traffic or industry. When they click on a city building, they see a video walkthrough and hear recorded digital sounds. Local planners can view their designs in a more comprehensive, dynamic and realistic way through the use of video and sound.

The Minnesota Department of Transportation is using multi-media linked to GIS in transportation studies and for real-time transportation planning. Real-time video images from cameras set up at intersections provide drivers of police, fire, and other emergency vehicles with views of existing traffic conditions at those intersections. Picking an intersection on a map provides a window view of what is currently happening at that location, as well as how noisy it is.

Property valuation in Jefferson County, MN, is facilitated through the use of a videotape and laser disc records of over 218 000 parcels of land. The video, showing the streets and front of each home and business in the county, is linked to the county's GIS database. These video files are used at hearings with the tax assessors or with the public works department. This video portion of the database apparently reduces the need for county officials to inspect each land parcel when considering reassessment or possible construction or maintenance of public infrastructure. The database is updated annually.

These and other applications of multi-media links to GIS have been reported by Lang (1992). It is clear this technology is just beginning to be used. Not every application will bring substantial benefits. But it seems there exists, for those who study water systems using GIS and simulation models, at least two potential advantages in using multi-media. The first is for helping the user apply or use a particular simulation model. The second is for providing the user with a more realistic view of the region and the water resource system being simulated. This view could be animated and stereoscopic. Until we discover how to effectively generate stereoscopic views of moving landscapes or rivers, for example, even moving video clips of landscapes, rivers and other objects overlayed with simulation results will provide a 3D realism we are not now used to seeing.

Consider, for example, the simulation of a regional water resource system that must include both surface and groundwater components, and the interaction between these surface and groundwater components, over a region the size of the State of New Jersey. This relatively small state bordering the Atlantic Ocean in the US contains 23 river basins and multiple layered aquifers. Surface water reservoirs and groundwater aquifers are used to store water for later use. Water can be transfered between surface water reservoirs and groundwater aquifers. If the water level of an unconfined aquifer drops, the base flow in the overlaying river reaches will decrease. This in turn could eventually result in unsatisfactory quality conditions from lack of adequate dilution. If this occurs, additional releases from upstream reservoirs may be needed for instream flow augmentation. In addition, low water levels in unconfined aquifers bordering on the Atlantic coast may cause saltwater intrusion into those aquifers. Furthermore, numerous demands are imposed on the entire surface and groundwater system by water supply utilities and industries. Given all these conditions and possibilities, how should this heavily populated and industralized State develop and manage their water resources? What policies should be defined for the integrated use of the State's surface and groundwater supplies, especially in times of drought, or in case of accidental (or purposeful) spills of toxic pollutants?

There are probably no obvious answers to questions concerning the management of water over a region the size and complexity of New Jersey. But to enable planners to assess what might happen if certain decisions are made, a simulation model is needed. This model must be able to simulate water quantity and quality in both surface and groundwater systems. It also must be compatable with the data that are available for model calibration. Model calibration is needed for flow routing, water quality prediction, and the transfer of water among multiple aquifers and between these aquifers and the various applicable river reaches, lakes, reservoirs and the ocean. The model must also be interactive, easy to use, and capable of readily accepting and simulating any changes in input data and assumptions. The model output must be presented in a form readily understandable by professionals and non-professionals alike. In short, it must be a very versatile model that will undoubtedly require considerable skill to calibrate and use correctly and effectively. On-line help will be essential.

How can the help provided by the "HELP" menu option of the model be most effective? How can the user of the model more fully understand just what is happening, or what happened, in the simulation? Attention to both these questions will be needed if the model is going to be of maximum value to its potential users. Research at Cornell University is currently examining the potential of video and sound as a response to, if not a complete answer to, these questions. Similar research is underway at the New University of Lisbon, Portugal, and at the Technical University at Aachen, Germany. Researchers at these and no doubt other institutions are working with a technology that is not yet fully developed. We are looking forward to the time when this technology will be standard equipment in most microcomputers and work stations.

A LOOK INTO THE FUTURE?

Currently, most interactive simulation models that have "HELP" menu options produce, when picked, some text display designed to guide the user in the operation of a particular portion of the model. Additional information, of course, can be found in model documents. But who reads documents, especially for specific information in the middle of a data entry and edit operation, a model simulation run, or when selecting from the data display options what data are to be seen, and how they are to be seen? Hence the need for on-line help files. But why limit these help files to text prompts? Now that we have the ability to open a window and display video clips (or snippets), why not supplement those text prompts with a video showing someone explaining just what the user needs to know? Users will be able to hear as well as see this information. And of course they can call up that help window again, or just back up the video, using a scroll bar, if they think they missed something.

In addition to using video to provide better "help" information, video displays of the region and its water resource system could also provide increased understanding of what is being simulated for those not familiar with its details. The capabilities of GIS could include the display of maps, aerial photographs (orthophotoquads), satellite images and video clips of selected sites in the region, and a flight over the river systems, at several different scales. Users of the model should be able to pick sites on any map or photograph or image of the region and zoom in to see what they want to see in whatever form they want to see it. They should be able to select two or more sites along a river and see a video of that river between those selected sites at any of several selected scales.

Perhaps the most challenging research will be in the integration of video with simulation output. Can we find a way of superimposing on our video of the river system or of the region the results of our simulations? Could we "see" on video the simulated river flows, the diversions, and the surface, and even groundwater, storage volumes changing over time? Can we figure out how to "see" the time varying concentration of various pollutants in those water bodies? Could we produce, on the fly, a Disney-like video similar to "Who Framed Roger Rabbit" combining reality with fiction - the results of a simulation run overlayed or superimposed on real video images? Probably not for a while, but if we think the results will be useful, it will happen, eventually.

To begin to build into our simulation models this interactive video and sound capability at any reasonable level of quality, improvements in hardware and software will be needed for capturing, compressing, storing and displaying video and sound. While the necessary hardware and software are not now routinely installed on all microcomputers operating under Microsoft DOS or Windows, they surely will be in the near future. Some multi-media capabilities are now available for microcomputers and workstations operating under these and other operating systems. But what is available today may not be what we will want to link to our simulation programs tomorrow. It is currently difficult to make choices among current alternative multi-media technologies simply because there does not yet exist a commonly accepted set of standards for multi-media (especially video) integration. Such standards are being developed, however, and should be in place soon (McManus, 1992).

Once sufficient standards and technology are available, we can begin to develop ways of accessing and displaying various video files from within our interactive simulation programs. Toolkits of interactive computer graphics functions useful for building the interfaces of our simulation models will need to include these functions for reading and displaying multi-media files. We will need software to edit video and sound. This software is already available for many computer platforms (Yager, 1992).

In the near future we should be able to provide sound as well as text whenever there is any need for a prompt to inform the user about just what is happening or just what is needed at the moment to continue the simulation exercise. In the more distant future we will be building into our simulation programs the ability to listen to what the user says, so that menu items, for example, need not be picked, but just called for - like "display map" or "pick color - red", or "set time" etc. More complex commands will follow. But even the widespread application of this simple command or menu calling capability, which has existed at some research facilities for at least the past decade, will have to wait until the required hardware and software improves and becomes inexpensive enough for popular use.

BACK TO THE PRESENT

The simulation model being developed for the study of New Jersey's water resources is a node-link network model. Nodes represent confluences, diversions, gage sites, monitoring sites, reservoirs, natural lakes, and groundwater aquifers. Links represent river reaches or the transfers of water among adjacent groundwater aquifers and the surface water system. Mass balances are calculated in each node and link. User-selected water quality constituent concentrations are also simulated, as is hydropower where applicable. This model is an updated version of the IRIS simulation model (IIASA, 1990). The network model is drawn on top of a digitized base map of the State. The geographic areas associated with each node and link, as applicable, are also identified. Users picking on any of these areas will be able to view either aerial photographs, satellite imagery, or video clips of the area. All on-line help text files will be supplemented with video files.

Building these multi-media links into our GIS-based simulation model has required us to make some preliminary decisions. Time will tell if those decisions were correct. One major decision was between analog video and digital video. We have decided on digital video. The main reason is flexibility. Computer users expect all computer controlled data files to be editable, shippable to other users, easily stored, capable of being embedded in a range of applications, and so on. Digital video permits all these functions.

Given a decision to use digital video, there is another decision to make between hardware-assisted compressors/decompressors and software-only "codecs". Intel's Digital Video Interactive (DVI) and the forthcoming Moving Pictures Experts Group (MPEG) standard for video use are examples of hardware-assisted codecs. These are now relatively expensive, but they enable full-screen, 30-frame-per-second video at CD-ROM data rates (i.e. 150 KBps). Use of this technology can provide good quality pictures. Software-only codecs are currently cheaper. We are experimenting with both options.

(Pictures showing some of the multi-media display capabilities being developed for users of the simulation model will be presented at the conference. Our more recent experiences in the use of multi-media will also be discussed.)

CONCLUSIONS

Advances in multi-media technology are making it possible for those of us using GIS and simulation models to break out of our largely silent and still world. While we have color and interaction, we are currently where the movie industry was in the 1920s, but without the music of the movie theater's piano player to go along with our graphics. This paper suggests it is time to start thinking about how the inclusion of video and sound might assist users not only in obtaining a better understanding of model use but also in understanding and interpreting the results of these GIS-based hydrological and water resources simulation models.

Applications linking multi-media to water resource simulation models and GIS are currently being explored at several universities in North America and Europe. At Cornell University we are examining how multi-media can enhance a regional GIS-based water resources simulation model. We are creating video files for added on-line help or assistance in model operation as well as for increased understanding of the region and water resource system being simulated. This involves the creation and editing of a considerable amount of video tape. Multi-media software (commonly called desktop video) is being used to edit this video and create the desired video clips that are callable by the model user. Audio files will allow model text prompts to be heard, as well as read. Calls for help on any particular aspect of the use of the model will result in a window opening up on a video clip showing someone explaining that particular aspect. Finally the user may view maps, aerial photographs, satellite images, or video clips, at various scales, of selected sites in the region.

In the future, we expect to be able to somehow overlay the results of our simulations onto these geographic images, including moving video images. We also forsee the possibility and advantages of having at least a limited ability to talk to the computer, as well as having it talk to us. Picking menu options may be done verbally, once we have the hardware and software that can hear and understand our voices. While this technology is not undiscovered or even unused today, it is not standard equipment in any of today's microcomputers or work stations.

Perhaps an appropriate ending of this paper is a recognition of the truth in what an engineering director of a multi-media firm has said regarding our use of this rapidly changing technology: "Whatever multi-media hardware and software we decide to use today, they will last as long as the computers into which we install them. Three years from now, all of it will be in the trash." (Yager, 1992) It is entirely possible even three to four months from now (November 1992), when this paper will be discussed at HydroGIS 93, that some of the ideas written herein will also be in the trash.

Everyone working in this field of interactive multi-media is on the beginning of their learning curve. Mixing sound and video with GIS-based simulation models offers considerable scope for innovation. Innovation is undoubtedly what is most needed today to take maximum advantage of this new technology for the study and analysis of hydrological and water resources management problems.

REFERENCES

CADSWES (1992) River Simulation System (RSS), User's Manual and Tutorial. University of Colorado, Boulder, CO.
Da Costa, J.R., Santos, M.A. & Loucks, D.P. (1989) Methodologies for Water Resources Policy Analyses. National Civil Engineering Laboratory, Lisbon, Portugal, report submitted to Scientific Affairs Division, NATO, Brussels, Belgium.
ESRI (1992) Hydrologic Modeling and ARC/INFO Rev. 6.1. ARC/INFO Software News, ESRI, 8-10.
French, P.N., Tayler, R.M. & Loucks, D.P. (1980) Water Resources Planning Using Computer Graphics. *J. Wat. Resour. Planning and Management* **106**, No. WR1, 21-42.
IIASA (1990) Interactive River System Simulation (IRIS): General Introduction and Description. Laxenburg, Austria.
Johnson, L.E. (1989) MAPHYD - A Digital Map-Based Hydrologic Modeling System. *Photogrammetric Engineering and Remote Sensing* **55**(6), 911-917.
Lang, L. (1992) GIS Comes to Life. Computer Graphics World. October, 27-36.
Maas, A., Hufschmidt, M., Dorfman, R., Thomas, H.A., Marglin, S. & Fair, G.M. (1962) The Design of Water Resource Systems. Harvard Univ. Press, Cambridge, MA, 324.
Maidment, D.R. (1991) GIS and Hydrologic Modelling. In: Proc. Int. Conf. on GIS and Environmental Modelling, Boulder, CO, September (notebook of collected papers).
McManus, M. (1992) Multimedia Development. Computer Graphics World, March, 69-74.
Moore, I.D., Turner, A.K., Wilson, J.P., Jensen, S.K. & Band, L.E. (1991) GIS and Land Surface-Subsurface Process Modelling. In: Proc. Int. Conf. on GIS and Environmental Modelling, Boulder, CO, September (notebook of collected papers).
Narasimhan, A. & Fox, E.A. (1990) Application of Multimedia in Simulation. memeo, Department of Computer Science, Virginia Polytechnic Institute and State University, Blacksburg. VA 24060.
Yager, T. (1992) Practical Desktop Video. Byte, April, 110.

A GIS-based hierarchical simulation model for assessing the impacts of large dam projects

R.A. McDONNELL & W.D. MACMILLAN
School of Geography, University of Oxford, Mansfield Road, Oxford OX1 3TB, UK

Abstract As reported in Biswas (1990) there is a growing awareness of the adverse environmental and social effects of water resource projects. Consequently, there is an increasing need to extend the analysis of such projects beyond the normal engineering and economical considerations. The work outlined in this paper is concerned with the impacts of big dam projects. The paper describes the development of a computer based system for assessing the environmental, economic and social impacts of large dam projects in an integrated way, using the SPANS Geographical Information System (GIS) coupled with a set of dynamic simulation models built with STELLA II. The application of the method to a recently completed dam scheme in the UK, the Roadford reservoir project, is explored. The eventual purpose of the system is to provide a tool that can be used at the design stage of dam projects, not just a means for evaluating completed designs.

INTRODUCTION

When a dam is built numerous impacts result from inundation, changes to the hydrological regime of the river, and alterations to the spatial structure and content of the local economy. Examples of these impacts include:

(a) alteration of the natural flow patterns of the river;
(b) changes to water quality through alterations to residence times and flow patterns;
(c) changes to the macro-benthic invertebrate and fish communities found throughout the river system;
(d) alterations to erosional and depositional processes, again throughout the whole river system;
(e) possible changes in microclimate and seismic stability;
(f) changes in natural habitats;
(g) disruptions to human habitation, land use and transportation systems both upstream and downstream of the dam;
(h) changes in settlement patterns, both local and in areas of resettlement;
(i) associated disruptions to employment patterns;
(j) changes in the nature and likelihood of occurrence of the hazards facing affected communities;
(k) introduction of new industries, often including tourism;
(l) changes to national water resource and energy portfolios, each of which has a variety of possible implications, well beyond the river basin;
(m) loss of cultural monuments.

For the first of these examples - the alteration to flow patterns - various possible effects are illustrated in Table 1. Similar tables could be constructed for all of the other examples but research work has tended to concentrate disproportionately on a few areas. In particular, there has been a focus on physical changes within the reservoir and the river immediately downstream of the dam (the literature on the effects on benthic invertebrate communities is particularly extensive). Petts (1984) gives a good general summary of the numerous impacts on the natural environment. Comparatively little systematic work has been done on the impacts on the social and economic environments, apart from the assessment of first-order economical costs and benefits. The other significant areas of research on human impacts have been the effects of resettlement and the spread of diseases like schistosomiasis in tropical dam projects.

ASSESSMENT OF DAM PROJECTS AND THEIR IMPACTS

The inadequacies of conventional cost-benefit analysis have started to lead to the inclusion of environmental factors in evaluations. The work of Pearce *et al.* (1988) characterizes this approach. Environmental Impact Assessment (EIA) has also begun to play a major role with many countries and organizations bringing into force legislation or formal guidelines requiring EIAs for major projects. Typically, these requirements have to be met to gain project approval or financial assistance. Recent changes in EIA policy, by organizations such as the World Bank, have included the need to assess both positive and negative impacts (World Bank, 1991). However, there has been little attempt at a truly holistic assessment of the environmental systems effected by major dam schemes.

A number of methods have been put forward for undertaking EIAs such as matrices and systems diagrams. Recently, engineering consultants have started to experiment in the use of GIS as a tool in dam project evaluations. In the work known to us (which is largely unpublished), this has involved using the GIS to hold environmental data, including remotely sensed information, and the generation of simple cartographic models from that data. Overlays, buffering and visualization techniques have been used to great effect. However the principal assessment and the mathematical modelling of impacts (if any is done) are undertaken outside the system.

DEVELOPMENT OF THE GIS-MODELLING SYSTEM

It is against this background that the current work was carried out. The research has been based on a particular dam project in the South West of England. This area regularly suffered water shortages during summer months in the 1970's and '80's as a result of falling supplies and increasing demands, particularly from the tourist industry. The development of water resources was hindered because of the outstanding natural beauty of the area and the associated

Table 1 Impacts of dams on the flow pattern of rivers (after Petts, 1984).

River/Reservoir, Country	Reported Hydrological Changes	Source
River Churchill, Canada	**Reduced Average Annual Runoff:** average discharge reduced from 1000 to 200 m^3 s^{-1}	Dickson (1975)
River Columbia, Dalles Dam, USA	**Reduced Seasonal Flow Variability:** maximum average monthly flow reduced by 50%	Trefethen (1972)
River Morava, Vir Dam, Czechoslovakia	**Altered Timing of Annual Extremes:** natural flow-regime characterized by two periods of high-flow in March and July, but regulated river has only one (in May)	Penaz et al. (1968)
TVA Scheme, USA	**Reduced Flood Magnitudes:** 1957 flood at Chattanooga; flood stage reduced from 7.3 to 0.7 m	Elliot & Engstrom (1959)
Upper Kennebec River, Maine, USA	**Imposition of Unnatural Pulses:** periodic instantaneous flow-fluctuations from 7.8 to 170 m^3 s^{-1}	Trotzky & Gregory (1974)
Missouri River, Fort Randall Dam, USA	**Imposition of Unnatural Pulses:** daily fluctuations from nearly zero to over 1000 m^3 s^{-1} in summer and 500 m^3 s^{-1} in winter, due to navigation and power peaking demand	Livesey (1963)

problems of environmental protection. One of the plans put forward to ease the water shortage was the building of the Roadford dam on the River Wolf, a tributary of the major River Tamar. The advantage of developing this new water source in the centre of the region was that it would support the urban areas both to the north and south of it. The dam was finally approved in 1985 and has now been functional for three years. It has a capacity of 37 000 Ml and a drainage basin of 31 km^2. Water is pumped directly from the reservoir to the north of the county and released into the river for abstraction lower down the system.

In order to represent accurately the environmental changes resulting from such a project, it is important to include both spatial and temporal elements. To this end, the SPANS GIS has been linked to the dynamic simulation modelling package STELLA II. The system was designed to be used in the context of design, forecasting and evaluation, and it covers environmental, social and economical impacts. Through the use of a variety of mathematical and cartographic models, the system allows both spatial and non-spatial representations of the current environment and facilitates predictions of reservoir and dam impacts throughout the river basin system. The models used vary in comprehensiveness. Some focus on specific aspects of the system such as the effects of the dam/reservoir on water quality whilst others synthesize the information generated by these first-order models to provide sectoral or aggregate views of the system. Thus, the system as a whole is hierarchical in structure.

TEMPORAL MODELLING

As noted above, the dynamic representation of the environment was effected through the STELLA II package. This software operates in the Apple Macintosh computer environment and is icon driven. One of the reasons it was chosen was its ease of use. The package represents entities in terms of flows, stocks and converters and describes the processes by a series of difference equations. The software has been used to model the changes in water flow, to analyze the dynamic and conservative determinands of water quality, and to represent the financial stocks and flows associated with dam construction and maintenance. An example of a model representation is shown in Fig. 1. The rectangles represent stocks, the circles represent converters and the pipe/arrows represent flows. As can be seen in the diagram, the rivers of the Wolf system are represented as a series of reaches. Inflows and outflows as well as instreamprocesses are simulated by using empirically derived time-series data or theoretically derived difference equations running "behind" the icons as shown in Figs 2 and 3.

The modelling of water quantity is based on strategies outlined by South West Water Plc, the managers of the water resources for the study area, and involves accounting for levels in another reservoir and river. These alternative sources are used to supply the southern region before Roadford water is used because of the expense of pumping the latter at the abstraction point.

The water quality modelling is based on conservation of mass theory. The work of Whitehead *et al.* (1981) & Warn (1990) has been adapted for the test site. The models are essentially empirical and have been calibrated and validated on existing field data taken before the dam was built. This field information, along with Monte Carlo simulated figures generated by STELLA II, have been used to produce predictions for different management scenarios. These flow sequences have been run for the dam and river system to give indications of resulting changes to the water quantity and quality regimes following impoundment. In turn, these model regimes have been compared with actual field measurements and correlate well with them. One of the purposes of this work in a design context is to make it possible to proceed from environmental constraints on water quality and quantity, through management scenarios designed to meet those constraints, to a dam and reservoir design capable of supporting those scenarios.

SPATIAL MODELLING

The spatial modelling in the system has involved the use of the SPANS GIS. Cartographic data have been captured in digital form on topography, rivers, geology, soils, agricultural land quality, human habitation and infrastructure, flora and fauna where available, and details of the dam and associated works. This data have been used in cartographic modelling of the river impoundment

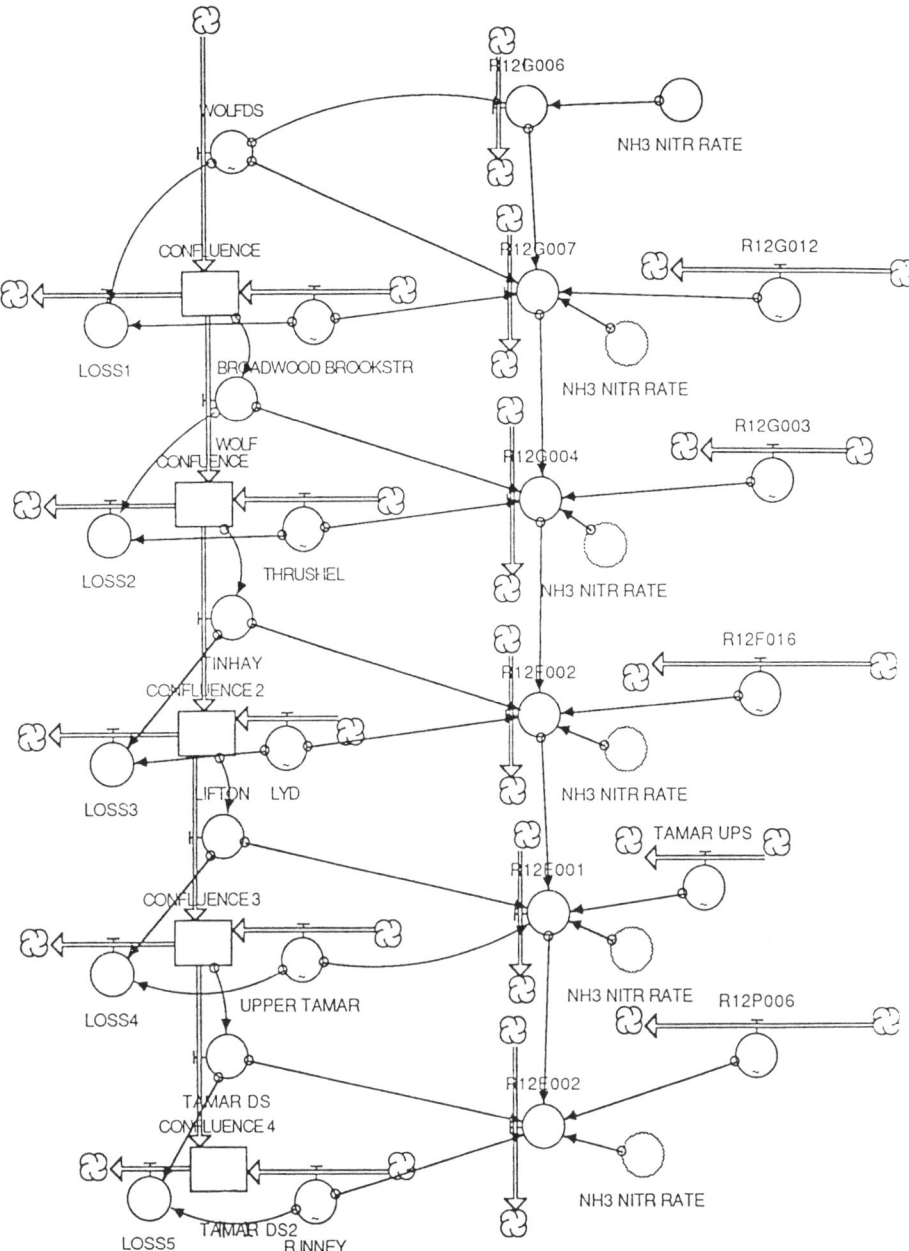

Fig. 1 STELLA II model representing the flow of ammonia downstream of the Roadford dam.

and the subsequent loss of land, ecosystems and cultural features. This has been effected through digital terrain modelling, overlaying, buffering and other inbuilt functions of the GIS. It is possible to display, for example, the areas of land loss by inundation for different dam heights and land most prone to waterlogging following a rise in water table.

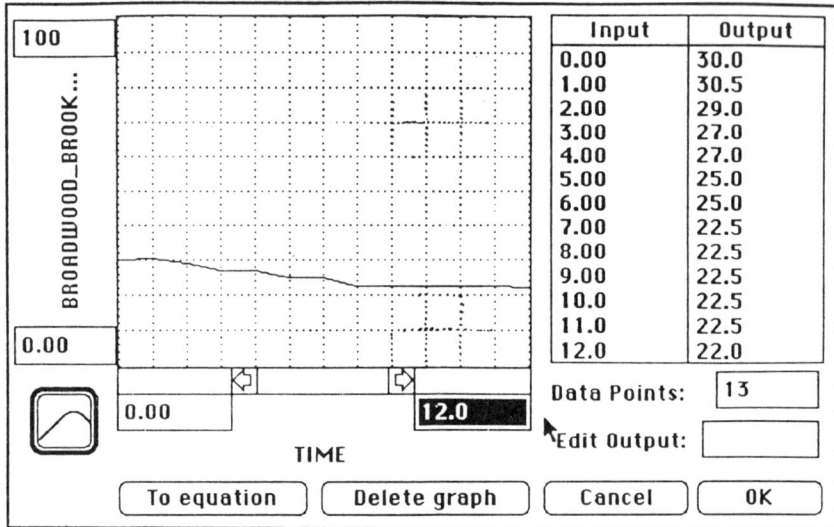

Fig. 2 Time Series data held within STELLA II.

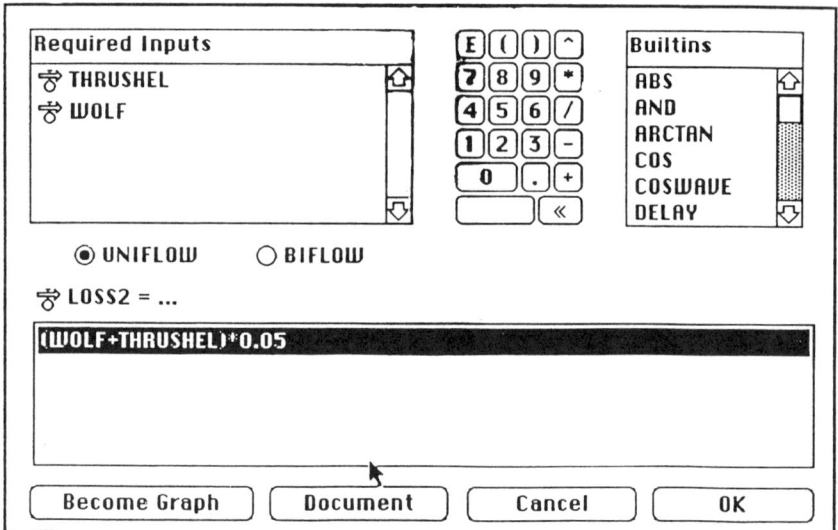

Fig. 3 Simple mathematical equations within STELLA II; as the figure shows many built-in functions are available for use.

By linking spatial economical models to the GIS it is possible to estimate economical impacts of projects in a fairly sophisticated way. Dams and reservoirs can be thought of as altering the structure and content of the spatial economy. They alter its structure in the sense that transportation and transmission networks are modified by the loss, alteration or addition of network links, including the possible use of the reservoir itself as an improved inland waterway. They alter its content by modifying the available qualities and quantities of land, water and power. Through the alteration of settlement patterns, the spatial distributions of labour and demand are modified. The complex interconnections between these processes can be captured with a tolerable degree of accuracy using computable equilibrium models. Since these models take the form of mathematical programming problems, they interface naturally with mathematical-programming based capital budgetting and project evaluation techniques. One of their notable advantages is their ability to generate shadow prices and their capacity for systematic sensitivity analysis. Both of these features are essential in the messy and frequently patchy data environments typically found in evaluation problems. As yet, progress on this part of the system has been limited to theoretical work on model structures and the technical problem of "plumbing" together optimization algorithms with GIS systems. The latter work has been successful in the context of GISPlus but has not been attempted yet with SPANS.

LINKING THE SPATIAL AND TEMPORAL MODELLING

Rather than develop each of the component spatial models, the project has focussed on the integration of the dynamic and spatial modelling elements. In particular, the STELLA II dynamic simulation models of river flow and water quality have been interfaced with the SPANS GIS system. The method that has been developed represents an advance on that employed by Despotakis *et al.* (1992). This link is important because it gives a spatial context to the temporal modelling. For example, the significance of a reach being identified through dynamic modelling as having a potentially high concentrations of nitrate levels depends on the intensity of agricultural landuse in the vacinity of that reach. This ability to couple temporal and spatial perspectives on a river basin suggests that the system could become a powerful management tool. To continue the example, the ability to identify nitrate-critical periods and river reaches could facilitate decisions about the release of water through the system to avoid negative impacts on river fauna.

PROBLEMS AND LIMITATIONS OF THE SYSTEM

With the development of system like this, a number of technical problems have been encountered, not the least has been the interfacing of STELLA II, an Apple Macintosh based package, with SPANS which runs in the Microsoft OS/2 operating system. This problem has been resolved but a more integrated system will be possible when a PC version of STELLA II is released in the Spring of 1993. The adaptation of models to a GIS environment has also proved problematic. Practical problems such as gaining access to and using many (often incompatable) forms of data have also been faced. The maintenance of database integrity and the minimization of compound errors have also been issues that have needed to be addressed.

It is important to acknowledge the limitations of such a system. Given the current state of knowledge of many of the impacts of a dam project there will obviously be areas where only crude predictions can be made. A computer system tends to lend credibility to any data that is generated by it so much caution is needed. Moreover, given the crudity of the data and of the modelling that is capable of being done in some areas, there is a need for economy of accuracy in those areas where more sophisticated work is possible. One of the attractive features of the STELLA II system is that it allows very simple models to be built of processes which are not well understood and it enables those models to be modified - played with, even - in a very straightforward way. In our view, the system has considerable potential and should be developed much further.

Acknowledgements The authors are most grateful to South West Water Plc, National Rivers Authority (SW Region) and the Ordnance Survey for generously supplying data for the work. The work was financially supported by the Mortimer May Research Scholarship.

REFERENCES

Biswas, A.K. (1990) Objectives and concepts of environmentally-sound water management. In: *Environmentally Sound Water Management* (ed. by N.C. Thanh & A.K. Biswas), Delhi, Oxford University Press.
Despotakis, V., Giaoutzi, M. & Nijkamp, P. (1992) GIS as a DSS tool for sustainable development strategies on Greek islands. *Proc. Third European Conf. and Exhibition on GIS*, 173-185.
Dickson, L.W. (1975) Hydroelectric development of the Nelson River system in Northern Manitoba. *J. Fisheries Research Board of Canada* 32, 10-16.
Elliott, R.A. & Engstrom, L.R. (1959) Controlling floods on the Tennessee. *Civil Engng* 29, 60-63.
Livesey, R.H. (1963) Channel Armouring below Fort Randall Dam. *US Department of Agriculture Miscellaneous Publications*, No 970, 461-70.
Pearce, D., Barbier, E. & Markandya, A. (1988) *Blueprint for a Green Economy*. London, Earthscan.
Penaz, M., Kubicek, F., Marvan, P. & Zelinka, M. (1968) Influence of the Vir River Valley Reservoir on the hydrobiological and ichthyological conditions in the River Svratka. *Acta Scientiarum Naturalium Academiae Scientiarum Bohemoslovacae-Brno* 2, 1-60.
Petts, G.E. (1984) *Impounded Rivers; Conservation and Management*. Chichester, John Wiley & Sons.
Trefethen, P (1972) Man's Impact on the Columbia River. In: *River Ecology and Man* (ed. by R.T. Oglesby, C.A. Carlson & J.A. McCann) New York, Academic Press.
Trotzky, H.M. & Gregory, R.W. (1974) The effects of water flow manipulation below a hydroelectric power dam on the bottom fauna of the Upper Kennebec River, Maine. *Trans. Am. Fisheries Society* 103, 289-300.
Trotzky, H.M. & Karpiscak, M.M. (1980) *Recent Vegetation Changes Along the Colorado River Between Glen Canyon Dam abd Kaje Mead, Arizona*. US Geological Survey Professional Paper, No 1132.
Warn, A.E. (1990) SIMCAT manual - Technical Background for Users. NRA Anglian Region.
Whitehead, P.G., Beck, M.B., & O'Connell, P.E. (1981) A systems model of flow and water quality in the Bedford Ouse River System - II. Water Quality modelling. *Wat. Res.* 15, 1157-1171.
World Bank (1991) *Environmental Assessment Sourcebook*. World Bank Technical Paper 139-145.

Application of GIS to water management in a developing region

J.J. OLIVIER & D.R. McPHERSON
Department of Water Affairs and Forestry, Private Bag X313, Pretoria, 0001 Republic of South Africa

Abstract Southern Africa is a water-poor region with a large, developing population. Many of the major rivers are located in international drainage basins, and the use of the water is governed by international agreements. The geopolitical and socio-political framework of the region add to the complexity of water management. Innovative and dynamic policies are required to achieve the best utilization of this scarce resource. In view of the shortage of skilled manpower, geographic information systems (GIS) is seen as an indispensable facilitating technology in the achievement of this aim. This paper discusses the underlying factors contributing to the complexity of water management, and looks at some GIS applications in this field.

INTRODUCTION

Background

The Republic of South Africa (RSA) is situated at the southern tip of Africa. It has a total land area of 1 221 000 km^2, with a population of approximately 40 million. The economically usable mean annual runoff is estimated at 34 000 million cubic metres (Conley, 1989). Sixty-five per cent of the RSA averages less than 500 mm year^{-1}, which is usually regarded as the minimum for successful dryland farming. The western regions receive less than 200 mm year^{-1}. Potential evaporation is significantly higher than the rainfall. Hydrological extremes, in the form of droughts and floods, are common.

Rainfall is poorly distributed relative to areas experiencing economic growth, decreasing rapidly and becoming increasingly variable from east to west. The runoff, which is even more erratic, is derived almost entirely from the rainfall.

The geological formations are predominantly composed of hard rock, and store insufficient ground water to stabilize the base flow of rivers. Over large areas the limited ground water is saline and slow to recharge, and cannot be relied on for bulk supply of water. Erosion is endemic and aggravated by poor landuse practices in developing areas. Transported sediment averages 120 million t year^{-1}, which reduces the cumulative reservoir storage by approximately the capacity of a medium-sized major dam each year (DWAF, 1986).

In the interior of the country the local resources are either fully utilized or overdrawn, and incremental investments in dam development show diminishing returns. Storage is subject to long carry-over periods (sometimes in excess of ten years) under conditions of high evaporation and increasing salination. To meet demands, it is necessary to import water from adjoining basins, resulting in higher unit costs of water supply. As an example, the RSA and Lesotho have embarked on the Lesotho Highlands Water Scheme, the first phase of which will cost in the order of US$ 2000 million.

Department of Water Affairs and Forestry: mission and objectives

The Department of Water Affairs and Forestry (DWAF) of the RSA is the central government department responsible for national water management. Its mission is to ensure the ongoing, equitable provision of water of adequate quantity and quality to all competing users at acceptable degrees of risk and cost under changing conditions (DWAF, 1986).

Since its inception as the Department of Irrigation in 1912, the function of the Department has evolved from providing water for irrigation, to constructing dams and supply schemes during a period of major economic growth in the 1960s. Its present function is perceived to be the holistic management of water as a scarce resource in the RSA through the application of appropriate high technology and multi-disciplinary expertise. By improving water management techniques and inculcating a greater awareness of the scarcity of water in the community, greater utilization can be obtained from existing resource development.

The complexity of water management in the RSA

Water management policy in the RSA is dictated by three major factors: supply and demand; geopolitical aspects; and socio-political aspects. These factors are considered in greater detail below.

As a central state department, the DWAF must take into account the influence political and social trends have on both water demand and its geographical distribution. The policies of the state, parastatal and non-governmental organizations all influence the formulation of water management policy.

The best utilization of water is not necessarily considered from the viewpoint of availability and the identified needs of user sectors only, but also from a more holistic viewpoint where econometric models play a significant role. For example, what is the benefit or disbenefit to the region as a whole in supplying water adequately, partially or not at all ?

Water supply schemes typically have lead times of 15-20 years, taking into account the planning, design and construction phases and the time taken for the reservoir to fill. To be effective, therefore, it is essential for the water management authority to adopt a proactive approach which will enable it to act dynamically to cater for anticipated changes in the macro-environment in which it operates.

Supply vs Demand The commercial, mining and industrial heartland of the RSA is situated on the Witwatersrand, which straddles the continental divide shedding water to the east and west coasts of the subcontinent. The problems of supplying water sufficient to sustain the growth of this region is an on-going challenge to the RSA's water engineers and hydrologists.

The discrepancy between where water is needed and where it is available typifies the problems facing water management in the region as a whole. In many areas, the demand for water is approaching the limits of what is economically exploitable locally. Repeated re-use of water by successive downstream users causes a progressive deterioration in water quality, evidenced by increasing salinity and heavy metal content.

Water supply management is complicated by the variety of users requiring water at different levels of assurance and quality. In the past, the difficulty in quantifying the requirements of forestry and the environment resulted in these sectors not formally being catered for in water resource planning. Because of the impact of rainfall interception by forests on runoff, however, and the growing public sensitivity to environmental issues, these sectors are now ranked with traditionally important users such as mining, industry and domestic consumers.

Geopolitical aspects Two of the RSA's major rivers, the Orange and the Limpopo, respectively form boundaries with the neighbouring states of Namibia, Botswana and Zimbabwe to the north. The headwaters of the Orange arise in the landlocked independent state of Lesotho. The waters of the Limpopo and other major eastward-flowing rivers pass through either Mozambique or Swaziland (or both) on their way to the sea. The areas drained by these rivers comprise almost two-thirds of the total land area of the RSA, and utilization of this water is subject to international agreements.

Socio-political aspects For several decades the migration of population in the RSA was subject to legal constraints. The relaxation of these constraints, combined with the effects of a deteriorating economy and the severe drought afflicting the entire subcontinent, have resulted in the large-scale influx of people from underdeveloped rural areas to the fringes of the major cities, and the consequent rapid development of informal or squatter communities. This development has brought with it the associated social, sanitation and infrastructural problems experienced throughout the world.

Changes in the socio-political environment have also resulted in increased pressures for social upliftment of underprivileged communities. This includes, *inter alia*, the provision or upgrading of housing, electrification, water and sewerage reticulation, etc. Meeting these needs, which could result in an order of magnitude increase in water demand, would have a major impact on the water supply infrastructure.

Social mores can also affect water demand. For example, in certain communities the number of cattle owned, or the number of children in the family, are an indication of wealth or status.

In southern Africa there is an overall lack of relevant digestible information and high-level manpower to properly address these complex issues.

GIS AS AN APPROPRIATE TECHNOLOGY

Why GIS?

Given the complex nature of water management in southern Africa, innovative solutions are essential. The water management community is small, but has an important responsibility to improve, through the equitable provision of water, the quality of life of a large, dependent population.

For these water managers to operate effectively, and to make well-informed decisions, it is essential for them to have detailed information on all factors influencing water demand. River systems and their drainage basins need to be assessed as integral units for development planning, and not as a patchwork of entities defined by political or administrative boundaries making autonomous decisions on development issues. The adoption of a consultative, holistic approach to water management is thus required to minimize the possibility of disputes between competing user sectors arising over water rights.

Improved methods of automated electronic data collection have resulted in prodigious quantities of data becoming available for analysis by planners and decision-makers. To make this data more readily digestible, sophisticated information-handling techniques are required. These allow information not available to conventional methods of analysis to be extracted and manipulated with relatively little effort by practitioners in a wide range of disciplines.

During the early 1980s the DWAF recognized that GIS was the most promising technology currently available capable of fully embracing these factors - particularly in view of the fact that almost all water management problems have a strong spatial component. Tedious, time-consuming data collection and validation searches could be automated and relegated to a lower level, allowing qualified personnel to dedicate more time to the conceptualization and analysis of alternatives to determine the most beneficial solutions to society as a whole.

GIS is considered of particular value as southern Africa is a developing region where high-level skills and expertise are in short supply. It is considered an ideal facilitating tool for integrating all information-based technologies, both existing and emergent. These include computer-aided design (CAD), global positioning systems (GPS), remote sensing (RS), image processing (IP), operations research (OR) and expert systems (ES), as well as traditional alphanumeric/tabular information systems (ISs).

When evaluating the value of technology in a developing country, the emphasis is often placed on the obvious economic advantages of labour-intensive practices. In many cases "appropriate technology" is equated with "low technology". In a field as important and complex as national water management, however, this approach would be inappropriate, and the use of high-tech systems such as GIS is considered essential.

The adoption of GIS as an appropriate technology does not, however, provide shrink-wrapped solutions. The implementation of a state-wide GIS is a major undertaking, which requires careful coordination and preliminary planning (Olivier *et al.*, 1989a). Although this holds true for implementation of GIS anywhere, certain aspects are particularly relevant in developing regions where access to funds is limited. These factors include overcoming a lack of awareness of information technology(IT) in general, and GIS in particular; maintaining management commitment in terms of funding and support; creating and retaining the necessary expertise to support the system; and developing and maintaining appropriate standards.

GIS IN THE DEPARTMENT OF WATER AFFAIRS AND FORESTRY

Having initially identified the need for, and the potential benefits of, GIS in its organization the DWAF has successfully installed what can arguably be considered the largest GIS installation in Africa. This has only been possible with enlightened management support. The existence of a Master Information Systems Plan, drawn up during the initial implementation period, greatly facilitated this process. The implementation of GIS has also been assisted by extensive cooperation with other state departments, both in the RSA and in neighbouring states. As one of the major regional players in the GIS field, the DWAF has promoted the use of this technology and the free exchange of information and applications throughout the southern African region. The DWAF has also been involved in efforts to establish national standards for data and data exchange.

One of the primary aims of the DWAF in implementing GIS and related technologies (CAD, GPS, etc) is the application of these to the modelling of alternative development scenarios, to derive better insights into the direction national water management policy should take. Implementing technologies such as these has had an impact on almost every division within the DWAF. The definitions and classification of everyday features such as rivers and dams were found to be deficient when viewed from an IT perspective. Data which had been acceptable for manual or limited computer processing was now perceived to be of too poor a quality for GIS processing.

The power of graphical interfaces for spatial data validation becomes quickly apparent when this data is displayed on a screen for the first time, and hitherto undetected errors in well-established data sets were soon revealed.

Some established functions (e.g., manual cartography, tracing, draughting, traditional surveying, manual planimetry, etc.), were perceived to be either inadequate or redundant.

The immediate implementation of these new technologies was impeded by the lack of accepted standards for, *inter alia*, data, data exchange and digital map symbology.

Hardware and software

The DWAF has established a sophisticated GIS platform which can support most intended applications with minimum reliance on outside organizations.

The present networked configuration comprises two Sun 690MP file servers, 25 Sun workstations, 50-odd PC workstations and two X-terminals distributed throughout the RSA. Output devices include an A0 colour electrostatic plotter, an A0 8-pen plotter, A3/A4 thermal wax transfer and colour Postscript devices, as well as laser printers, a film recorder for producing slides, and a high-quality video data projector. Input devices include A0/A1 digitising tablets and a 1 000-dpi colour A0 scanner.

Arc/Info software is used, the DWAF having some 40 Arc/Info workstation licences and 70 PC Arc/Info licences, together with ERDAS image processing software, VTRAK vectorising software, various relational database management systems and miscellaneous presentation graphics and application software.

People

An inherent difficulty in the large-scale implementation of a new technology like GIS in any large organization is the length of time it takes to implement changes to established post structures to cater for new personnel requirements. This makes it difficult to recruit and retain suitably skilled personnel to operate a system in which substantial capital investment has been made.

A further problem is that trained GIS personnel in the southern African region are scarce. To counter this, the DWAF has had to make provision for the initial and continuation training of personnel, and has successfully managed to create a thriving GIS environment which enjoys a remarkably low turnover of staff.

Finally, to successfully implement a GIS, sustained management support is essential. Experience has shown that a strict IT-orientated project-management approach should be followed, based upon a dynamic matrix organization structure. This ensures that the expertise of a few skilled practitioners can be applied to multiple projects running concurrently.

Data

The establishment and subsequent maintenance of a sound database is a *sine qua non* for any successful GIS. A problem experienced in many developing areas is the general lack of digital topographical, cadastral, geological and other thematic data conforming to acceptable standards.

The RSA's topographical mapping agency has started to make its data available digitally, at scales ranging from 1:50 000 to 1:500 000. The issue of whether this type of data should be copyrighted, however, and whether it should be made available free of charge to non-governmental organizations and individuals, is hotly debated world-wide - and no less so in the RSA. Because of the urgency of this requirement, several institutions have decided to undertake their own data capture programmes, adhering as far as possible to the national mapping standard.

In the mid-1980s it was realized that, in view of the effort involved in the capture of digital spatial data and the limited resources available in the RSA, cooperation between state departments in areas such as data capture and ensuring the maintenance and custodianship of data, was essential. This led to the implementation of a state-wide National Land Information System in the RSA.

Applications

GIS is used to capture, store, manipulate, analyze and depict spatially referenced information, as well as to disseminate the results of analysis using thematic maps.

At present a wide variety of GIS-related projects are being undertaken by the DWAF, some of which are discussed briefly below.

River basin studies Since the early 1980s the DWAF has been progressively carrying out river basin studies to assess the water resource potential and the historical development of water demand in major drainage basins. In these studies, civil engineering consultants experienced in water resource analysis are appointed to carry out thorough investigations of all aspects of water management in these basins, and to recommend alternatives for future development (Olivier *et al.*, 1989b).

By making GIS software available to the appointed consultants, the DWAF has ensured the establishment of a comprehensive database for each catchment in which these studies are carried out. This will facilitate the early identification of potential crisis or conflict situations, and allow alternative scenarios to be rapidly formulated and evaluated.

River system analyses The river systems supplying water to the major urban and industrial centres of the RSA are generally heavily developed. To ensure the best utilization of the available water, studies are carried out to optimize the operation of the reservoirs on these systems. These studies consist of detailed analyses and the application of sophisticated modelling techniques, which are considerably enhanced by integration with GIS.

Afforestation permit system Afforestation has a significant impact on the runoff-generating capacity of a basin. Timber is, however, a valuable commercial product, and the industry employs a sizeable workforce. To reconcile these factors, afforestation has to be carefully controlled, and this is done by a permit system. Before a permit is issued, a detailed analysis of the present state of afforestation and the impact on the water resources in the intended area is carried out. To expedite the issue of the permits, and to ensure that the interests of all parties are served, it is essential for an up-to-date spatial information system to be available.

Water Quality Permit System Owing to the widely varying hydrological regimes of the RSA, and the differing degrees of development in various areas, a dynamic, flexible water quality management policy has been adopted in which a receiving water quality objective is identified for rivers on a regional basis. Industries wanting to discharge effluent into a river are required to carry out an impact analysis, and to submit this with their application for the necessary permit. Owing to the high spatial component of this process, the assessment of the application and issue of permits (which are different for each site) are ideally suited to implementation on a GIS.

Hydrogeological mapping The southern African region has been experiencing its worst drought in many decades, and supply schemes based on surface water resources have failed in many areas. In regions where groundwater is not sufficient for extended bulk water supply, it may still be utilised for emergency supplies. A 1:500 000 national hydrogeological map series is being compiled on the DWAF's GIS.

Flood damage loss functions The RSA suffers hydrological extremes in the form of periodic droughts and floods. These can have significant socio-economic aftermaths. Owing to the long return periods of floods, and in spite of official discouragement, development often takes place in the flood plains of rivers. In urban areas this is manifested by the occupation of vacant land on the fringes by squatters, and by formal township development within the cities; and in rural areas by the cultivation of crops right up to the river banks.

When floods occur, the state is frequently called upon to carry a portion of the financial loss. The DWAF is at present formulating policy on the best way of managing this situation. The GIS provides an ideal tool for the determination of flood lines and the derivation of loss functions for various flood severities and land uses.

Hydrological and engineering modelling Most information used in hydrological and engineering models has a geographic or spatial orientation. At present, existing models making use of spatially referenced information are modified to accept this from, or introduce it to, the GIS database (McPherson, 1992).

Models undergoing integration at present include rainfall-runoff models, hydro-salinity models, dynamic feedback models and demographic models.

Future use of GIS

Proposed uses of the DWAF's GIS vary from simple plotting and mapping applications to spreadsheet-type, multimedia, and hypertext uses, as well as the further development of application-specific models, spatial decision-support systems and expert systems. Emphasis will be placed on the identification and development of critical management indicators to ensure that top management has convenient access to relevant information on a day-to-day basis. The use of software tools which make GIS data easily accessible to non-practitioners will be extensively investigated and combined with other decision support systems and executive information systems.

CONCLUSIONS

The DWAF is acutely aware of its responsibility for providing the necessary infrastructure and resource management strategies for an ever-growing, developing and dependent population. It firmly believes that the use of enabling technologies such as GIS will prove indispensable to decision-makers in the complex field of water management.

Acknowledgements The permission of the Director-General of the Department of Water Affairs and Forestry of the RSA to present this paper is gratefully acknowledged. The views expressed are those of the authors, and are not necessarily those of the Department.

REFERENCES

Conley, A.H. (1989) The value of high-technology for managing intersectoral and international sharing of inadequate water resources in a developing region of Africa. *Paper presented at Water'89: Water Decade and Beyond, 13-16 Dec 1989, Bangkok Thailand.*
Department of Water Affairs (1986) *Management of the water resources of the Republic of South Africa.*
McPherson, D.R. (1992) On the application of GIS to modelling in an engineering environment. *SAICE Young Water Engineers Conference, 5-6 March 1992, Midrand, RSA.*
Olivier, J.J., Greenwood, P.H., McPherson, D.R. & Engelbrecht, R.E. (1989a) Selecting a GIS. *ACSM/ASPRS Auto-Carto 9 Convention, 2-7 April 1989, Baltimore, Maryland, USA.*

Olivier, J.J., McPherson, D.R., Pullen, R.A., Conley, A.H., Van Zyl, F.C., Van Aswegen, J. & Langhout, C. (1989b) The use of GIS in integrated river basin management. *SAICE Computer Division 11th Annual Conference, 10 May 1989, Bloemfontein, RSA.*

GIS structure for the Nile River forecast project

JOHN C. SCHAAKE
Office of Hydrology, National Weather Service/NOAA, Silver Spring, Maryland 20910, USA

ROBERT M. RAGAN
Civil Engineering Department, University of Maryland, College Park, Maryland 20742, USA

EDWARD J. VANBLARGAN
Earth Information Services, 1240 North Mountain Road, Harrisburg, Pennsylvania 17112, USA

Abstract The Nile Forecast System is being developed to predict inflows into Aswan dam in Egypt. The system contains a set of hydrological models that are structured as a grid point, distributed modelling system. The system also contains a geographic information system (GIS) component in order to create map displays and to define the parametric hydrological input for each grid cell. A commercial GIS (ARC/INFO) was used to digitize data. However, software was custom written to provide the display and analysis capability. The terrain analysis software developed uses elevation and stream data to define important hydrological parameters such as the gridded flow connectivity, drainage areas, flow lengths, and slopes. The efficiency of using GIS technology allows using a grid point modelling strategy that would be prohibitive using traditional manual approaches.

INTRODUCTION

The Monitoring, Forecasting, and Simulation Project for the Nile River is currently underway. This joint project involves the Egyptian Ministry of Public Works, the United Nations Food and Agriculture Organization (FAO), the US Agency for International Development (AID), and the US National Oceanic and Atmospheric Administration (NOAA). The technical objective of the project is to develop an automated Nile Forecast System (NFS) to produce streamflow forecasts for the Nile River into the high Aswan dam in Egypt. Several contractors provide technical support which include the University of Bristol, the University of Maryland, and Earth Information Services.

The NFS is a sophisticated real time, hydrometeorological forecast system. It includes a stochastic model for the White Nile and a deterministic model for the Blue Nile. This paper discusses only the Blue Nile component. Real time rainfall gaging is not very dense or reliable over Ethiopia and Sudan. Therefore, the NFS uses satellite rainfall estimates plus any real-time rainfall reports to provide a gridded field of rainfall each day.

The NFS hydrological models use the rainfall estimates to produce short term flow forecasts and use historical rainfall to extend flow forecasts up to six months into the future. A distributed, grid point model is used to preserve the spatial information provided by the rainfall fields.

The NFS Blue Nile component includes four models for water balance, hillslope runoff, channel routing, and reservoir operations. The system is structured as a set of connected grid points. Each grid cell contains a stream channel and has an operation for all the models except the reservoir operations which are defined only at select locations. The water balance model produces runoff volume excess for each grid cell. The hillslope model provides a time distribution function for runoff to reach the stream channel. The channel routing then routes runoff from grid cell to grid cell. The modelling is continuous in time so there is a state variable for each model.

The entire Blue Nile is represented as a grid with cells roughly 5 km on a side. A separate channel routing scheme combines the Blue Nile with White Nile flows and routes them down to Aswan dam. Although the main points of interest are the reservoirs along the Blue Nile and at Aswan, the NFS can actually generate forecasts for any point in the gridded field.

Each grid cell has its own set of model parameters and state variables. There are approximately 10 000 grid cells for the Blue Nile. Defining the gridded flow sequence and parameters for each cell is not feasible by manual means. Thus a geographic information system (GIS) component was added to the NFS. The purpose of the GIS is for both analysis and display.

The main GIS objective is to produce parametric input for the hydrological models. This includes defining the gridded flow path connectivity and the model parameter values for each grid cell. The second GIS objective is to produce map background displays upon which the gridded rainfall and model outputs can be displayed.

The GIS component was designed specifically to serve the forecast component needs. As such, it had to perform in real-time and be integrated with the user interface of the NFS for display purposes. Because of these constraints, software for the GIS was custom written.

The purpose of this paper is to describe the structure of the GIS component of the Nile Forecast System. The requirements of the hydrological models are discussed first. Afterwards, discussion is presented on how GIS meets those needs.

HYDROLOGICAL MODELS

The Blue Nile is a grid point model. Internally, it is actually set up as a one dimensional sequence of cells. The sequence, or connectivity, array is defined in an upstream to downstream computational order and contains the flow path information for each cell in the basin. The flow path information includes the row and column in the gridded field for that point with a pointer to the next downstream sequence point. The row and column are used to extract the precipitation and parameters from gridded input files. The next downstream cell is not necessarily the next point in the array because of the branching nature of tributaries. Each grid cell has its own set of parameters and state variables for the hydrological models.

The water balance is a conceptual model with five parameters and two state variables. Parameters represent the soil moisture capacity and percolation rates for two soil layers. The upper zone provides a fast response while the lower zone has a slow response for simulating groundwater contribution. The water balance input is rainfall and evapotranspiration. The output is the runoff from the two soil layers which are combined into a total runoff volume. The hillslope model contains three parameters and one state variable which basically create a time lag for runoff to reach the stream. It is analogous to a unit hydrograph but it is nonlinear. The hillslope is a physical model with parameters for hillslope length, slope, and roughness.

The water balance and hillslope models operate only on the local flows within a grid cell and are independent of the flow path connectivity. The channel routing model uses the connectivity sequence and routes flow from grid cell to grid cell. At each cell it combines the upstream inflow with the local hillslope flow.

The channel routing model is a nonlinear physical model with six parameters and one state variable. The parameters are length, slope, roughness, two cross section shape parameters, and cell area. The area of each cell varies across the grid because of the projection of the satellite fields. The models were based on existing hydrological concepts but the software was custom written (Koren *et al.*, 1992).

GEOGRAPHICAL DATA

The GIS component was designed for both analysis and display capability. To be included in the geographical database, data had to be easily obtained and useful for either analysis or display. The end result was a database with the following types of data:
(a) stream vectors;
(b) elevation contour vectors;
(c) political boundaries vectors and names;
(d) city names and point locations;
(e) boundaries (vectors) for major hydrological basins.

All data types are used for displays, but only the streams, elevations and watersheds are used for analysis. A commercial GIS, ARC/INFO, was used to digitize the basic data. Digitizing used 1:2 000 000 maps of Africa published by the U.S. Defense Mapping Agency. The digitized data were transformed into an efficient binary format for use in the NFS. Attributes were added to the data which included contour intervals and names for streams and watersheds. The custom GIS software was then developed which works on both a DOS-based personal computer and an IBM RISC workstation.

A display program was developed to create map backgrounds for overlays of gridded hydrometeorological data. The GIS display program creates vector type maps in a Hewlett Packard Graphics Language (HPGL) format. A NFS user-interface was developed that can combine and display the GIS vectors with real time gridded data and hydrograph plots. The user has a menu system to zoom into any window, select colours, or only display select data by the attributes in the GIS database.

Analysis programs were developed to create new data sets from the basic geographical data. An interpolation program converts elevation contours into gridded elevations. Another program re-orders the stream vectors in a downstream to upstream sequence and adds attributes for stream length and stream order number.

GIS TERRAIN ANALYSIS

Specialized terrain analysis software was developed as part of the GIS component. The purpose of the terrain analysis is to combine gridded elevation and stream vectors to define certain hydrological parameters. The parameters which can be obtained directly from terrain analysis are the flow connectivity of the grid cells, flow lengths, and slopes.

The analysis also generates total upstream drainage area for each cell. In lieu of measurements, channel cross sections can be estimated for each cell using regional relationships of cross section and drainage area (Dunne & Leopold, 1978). These relationships have not yet been formulated for the Blue Nile. Some parameters cannot be derived currently by the GIS such as roughness and the water balance parameters. In concept, these parameters have a physical relationship to soils, land use, climatology, and/or elevation. Thus, future project work is planned to obtain and relate these physical data to model parameters.

The basis for the terrain analysis is a neighbour analysis of elevations where each grid cell flows to the lowest of its eight adjacent neighbours (Jenson & Domingue, 1988). The occurrence of pits, cells that are higher than all neighbours, complicates the analysis and must be resolved (Marks *et al.*, 1984). The algorithms developed in the NFS extend the neighbour analysis by adding stream vectors. The streams are used to constrain the elevation analysis such that any cell adjacent to a stream will flow to the stream regardless of the elevation. Also in flat pit areas, cells will drain to the nearest stream. It has been shown that use of stream data often results in faster and better terrain analysis than just using elevation data alone (VanBlargan *et al.*, 1990).

The terrain analysis is done only once. The end result is a gridded field of flow path directions with the same window as the elevation data. The flow direction for each cell is a number from 1 to 8 indicating which neighbour the cell flows to.

Another program was then written which defines the watershed boundary and flow sequence for any outlet point desired. The boundaries generated are vector files used for display. The flow sequence is a connectivity array used by the hydrological models. The sequence array was generated only once for the entire Blue Nile. Then, the program was run for selected points within the basin to create boundaries for display purposes.

BENEFITS OF INTEGRATING THE GIS AND HYDROLOGICAL MODELS

The GIS developed for the Nile forecast system is integrated with the operations and needs of the hydrological models. To estimate model parameters from GIS requires that parameters be physically based or related to measurable physical quantities. Currently the Nile GIS provides information on flow path connectivity, slopes, and flow lengths. Future work is needed to relate water balance parameters and channel cross sections relationships to physical data. The benefit of such relationships is that parameters can be defined without relying on sparse historical runoff data and subjective calibration. However, some parameters may always remain to be calibrated by conventional means.

The Nile Forecast System and GIS addresses important issues such as methods of combining raster and vector data, integrating hydrological models with hydrometeorological and geographical data, and estimating parameters with digital data. The efficiency of GIS technology allows using a grid point modelling strategy that would be prohibitive using traditional manual approaches.

REFERENCES

Dunne, T. & Leopold, L.B. (1978) *Water in Environmental Planning*. W.H. Freeman and Co., San Francisco, California, USA.
Jenson, S.K. & Domingue, J.O. (1988) Extracting topographic structure from digital elevation data for geographic information system analysis. *Photogrammetric Engng and Remote Sensing* 54(11), 1593-1600.
Koren, V., Barret, C. & Schaake, J.C. (1992) Nile River forecast system models. Nile Technical Note 54. Nile Project Office, National Weather Service/NOAA, Silver Spring, Maryland, USA.
Marks, D., Dozier, J. & Frew, J. (1984) Automated basin delineation from digital elevation data. *Geoprocessing* 2, 299-311.
VanBlargan, E.J., Ragan, R.M., & Schaake, J.C. (1990) Hydrologic geographic information system. *Transportation Research Record.* no. 1261, 44-51.

Groundwater system modelling and management using Geoscientific Information Systems

K.E. KOLM & J.S. DOWNEY
Dept. of Geology and Geological Engineering, Colorado School of Mines, Golden, Colorado 80401, USA

Abstract Hydrogeological problems require representation of subsurface conditions in addition to the areal extent of surface features. Geographical Information System (GIS) or the newer Geoscientific Information System (GSIS) may be used to assist digital modelling and management of complex aquifer systems. In many areas, the system complexity dictates the use of a three-dimensional groundwater modelling approach supported by true three-dimensional GSIS. The method presented here involves six steps using GIS or GSIS: 1) data gathering and preparation, 2) field (on-site) conceptualization, 3) surface characterization, 4) subsurface characterization, 5) geohydrological system characterization, and 6) numerical model simulation. The use of GIS or GSIS offers: 1) data management and data audit trails, 2) the integration of diverse data sources, 3) rapid development, visualization, and testing of alternative models, and 4) integration with the numerical modelling steps. The procedure presented combines the experience of the investigator with modern computer-based analysis tools, such as GSIS, to quickly develop digital models for the characterization of geohydrological systems. The results of using this method in the study of hydrological systems in the western United States are presented.

INTRODUCTION

The correlation and synthesis of a great variety of two- and three-dimensional data is a fundamental requirement in the development of digital groundwater flow models in areas of complex geology. This paper presents an integrated, step-wise approach for the conceptualization, characterization, and numerically simulating groundwater systems using either Geographical Information Systems (GIS) or Geoscientific Information Systems (GSIS). The multidisciplinary approach outlined here may be used for characterizing all hydrological systems. These systems may be characterized for groundwater exploration and development; aquifer simulation, evaluation, and management; and geotechnical or geohydrological site investigation, evaluation, and remediation.

Conceptual and numerical models simulating groundwater flow systems for management purposes may be developed using this hydrogeological, framework-based approach. This approach integrates and derives digital model input values, such as recharge and head, from the analysis of distributed hydrogeological parameters, such as type and quantity of vegetation, soils, and topography, as well as from hydrological data. This approach is to be distinguished from traditional modelling approaches that tend to rely mainly on limited point-source data, such as water levels from selected wells, precipitation, or stream discharge to characterize hydrological systems.

The precise representation of complex hydrological systems continues to be problematic, especially in three dimensions. Continuous and point data, collected in a variety of disparate formats and scales, must be integrated in order to formulate data sets for digital models. Therefore, either a Geographical Information System (GIS) or Geoscientific Information System (GSIS), a three-dimensional, scientifically oriented expansion of the traditional Geographical Information System (GIS), must be used to facilitate data management, analysis, interpretation, and visualization, and to develop input arrays for numerical groundwater models so that multiple working hypotheses may be easily investigated in a short period of time. GIS or GSIS also assist in supplying any necessary documentation for quality assurance, quality control, and management procedures.

GENERAL APPROACH

Hydrogeological problems require representation of the subsurface in addition to the representation of surface conditions, such as the areal extent of botanical, pedological, topographical, and geological features. Hydrogeological problems also require linkages to various geological data manipulation procedures (Turner, 1991; Turner & Kolm, 1991; Turner & Kolm, 1992). The ability to rapidly create and manipulate three-dimensional images, using a GIS or GSIS, can materially assist the hydrogeologist's understanding of the hydrogeological environment.

Calculations required in the solution of three-dimensional groundwater flow problems, using publicly available digital models, may be completed in only a few hours on a modern high-performance personal computer. However, the interpretation and visualization of the results from each model analysis is greatly hampered by a lack of 3D visualization tools. The model results are sensitive to the selection of input parameters, and traditional model calibration methods may fail to identify problems. In fact, as stated by Turner (1989), these models have outstripped our ability to supply the necessary data using traditional methods and a "Parameter Crisis" faces those who wish to use such models.

GIS or GSIS alone cannot solve the groundwater system analysis process, which involves six fundamental, interactive procedures: 1) data gathering and preparation, 2) field conceptualization, 3) surface characterization, 4) subsurface characterization, 5) geohydrological system characterization, and 6) numerical model simulation. The process starts when the investigator combines geological experience with limited field data to begin the surface and subsurface characterization. Surface data, such as vegetation cover, geological contacts, etc, are collected, mapped, and analyzed. Subsurface data, such as borehole information, are combined with the surface data for subsurface characterization. Given the sparseness of geological and hydrological data, geoscientists may develop one or more "most probable" or "equally likely" scenarios, or conceptual models, in order to expand the widely-spaced, known observations to create continuous geological and hydrological models of the entire subsurface (Turner & Kolm, 1991, 1992). During this process, the GIS or GSIS is designed to provide both data management and visualization support. A number of iterations are to be expected before the most probable subsurface conditions are defined. In some cases, a unique solution may not be achievable, and two, or more, alternative characterizations or working hypotheses concerning the character of the system must be used. A three-dimensional conceptual model of the hydrogeological system is developed, and hydrological system characterization proceeds.

Finally, a digital model of the geohydrological system is developed and model estimates may be compared with laboratory and field data. Using GIS or GSIS, the groundwater flow system can be visualized and adjusted during the modelling and calibration process.

HYDROLOGICAL SYSTEM CONCEPTUALIZATION AND CHARACTERIZATION PROCEDURE

The conceptualization and characterization process includes: 1) data gathering and preparation, 2) field conceptualization, 3) surface characterization, 4) subsurface characterization, 5) geohydrological system characterization, and 6) numerical model simulation (Kolm, 1993). During the data gathering and preparation step, the problem to be solved may be further defined. A literature search is conducted to locate and collect the basic data available for the study area. The basic data are organized and preprocessed into computer data files from which data can be accessed to aid conceptualization and numerical modelling requirements. At this time, several field trips to the study site should be made for it is in the field, on-site, where the ideas as to how the hydrological system operates are developed supported by the knowledge obtained during the data base development phase. This phase has been termed Field Conceptualization in this paper.

Surface characterization of the study area is conducted using the data bases developed during the data gathering phase. This step includes the following analyses: 1) man-made effects, 2) type and quantity of vegetation, 3) topography, 4) surface-water conditions, 5) climate, 6) pedogenic process and deposits, and 7) geomorphological process and deposits (Kolm, 1993).

A subsurface characterization of the study area is conducted using the results of the surface characterization and the geology and geophysics data bases. Interpreted information is produced during surface and subsurface characterization and used, along with the other data bases, to characterize the hydrogeological framework.

The conceptual model of the geohydrological system, developed during the field conceptualization step, consists of ideas regarding the type and distribution of recharge and discharge, possible groundwater flow paths, type and location of boundaries and configuration of the existing potentiometric surfaces. Recharge, discharge, boundary conditions, head, hydraulic conductivity and/or transmissivity, saturated thickness, and storativity values are estimated and/or assigned to characterize the geohydrological system conceptual model. These values are based on field and laboratory data, or are estimated based on the hydrogeological framework, distributed parameters, hydrological system conceptual model analysis, analytical calculations and/or various deterministic and statistical techniques. The hydrological system may be simulated, using a numerical model, by converting the quantified, conceptual model parameters into a series of input arrays either manually, or using either a geographical or geoscientific information system.

USE OF GEOGRAPHICAL OR GEOSCIENTIFIC INFORMATION SYSTEMS

The development of a hydrogeological conceptual model is the result of visits to the study area and the data collection phases of the study. Subsurface characterization, relying on knowledge of the geological framework, is difficult due to the variability of geological environments and the sparseness of geological data. In addition, a variety of data types must be combined or synthesized. This requires the use of GIS or GSIS that are capable of manipulating a great variety of data, and a system that allows the investigator to interact with the data base in order to visualize the subsurface in two- or three-dimensions (Turner, 1989).

The combined use of several sophisticated numerical procedures to support surface and subsurface characterization and the resulting hydrogeological and hydrological conceptual models demands careful data management and quality assurance/quality controls (Turner & Kolm, 1991). The data management procedures can be conceived of as having four stages: 1) data capture, 2) data-edit preprocessing, 3) data analysis for developing alternative conceptual models, and 4) data structuring and visualization (Turner & Kolm, 1991, 1992). The first stage, data capture, consists of both data gathering, the actual collection of new raw information, and data extraction, the selection of appropriate data items from existing data collections. The study purposes, goals, and objectives represent policy requirements and define the scope and type of data capture activities. Technological considerations may also constrain these activities.

The second stage in data management, data-edit preprocessing, involves data validation, data parsing, and data regionalization (Turner & Kolm, 1991, 1992). Data validation includes both the identification of errors and blunders in the database. Data parsing involves the review and conversion of descriptive data to consistent, standard terminology and formats. Data regionalization involves adjusting the information to represent appropriate levels of detail in order to accomplish the purposes of the study. These data-edit preprocessing procedures may produce multiple, alternative, standardized data bases from a single set of raw, original data.

The third stage involves the development of numerical models based upon conceptual models. Different interpretations of subsurface conditions from the same standardized data base may and usually do exist. By applying accepted geological concepts and knowledge, and by using data extraction, sampling and evaluation methods supported by GIS or GSIS, digital hydrogeological models may be developed and tested from each standardized data base.

The fourth and final stage involves data structuring and display. This allows the hydrogeologists using the system to evaluate, and then to accept, modify, or reject their conceptual and numerical models.

APPLICATION OF APPROACH

The hydrogeological framework approach using GIS or GSIS is being applied to several hydrological systems in the western United States to illustrate the techniques used for groundwater flow system simulation and to identify deficiencies and limitations. These systems include the Death Valley groundwater basin, California (Faunt et al.,1993), and the northern part of the San Luis Valley unconfined aquifer in southcentral Colorado (Williams, 1992).

These studies involve large areas, and are ideal for testing both GIS or GSIS, and the conceptualization, characterization, and numerical modelling procedures. The hydrogeology and the hydrology of these systems are complex, and well data are sparse. All of these areas are being impacted by human development, and groundwater management decisions are pending by various local, state, and federal agencies.

CONCLUSIONS

The process of groundwater system analysis recommended involves six fundamental procedures that are interactive using GIS or GSIS: 1) data gathering and preparation, 2) field conceptualization, 3) surface characterization, 4) subsurface characterization, 5) geohydrological system characterization, and 6) numerical model simulation. Conceptual and numerical models simulating groundwater flow systems can be constructed using a hydrogeological framework-based approach. Basic data, including topography, geomorphology, geology, geophysics, geobotany, hydrology, pedology, and geochemistry, are gathered, prepared, and analyzed. The geohydrological system conceptual model, including type and distribution of possible recharge and discharge, possible flow paths, and boundary conditions is developed. Recharge, discharge, boundary conditions, hydraulic conductivity and/or transmissivity, head, and storativity are quantified or estimated for use in the digital model based on the conceptual model. Finally, the geohydrological system is simulated using the digital model. During these procedures, a Geographical or Geoscientific Information System is used to facilitate data management, analysis, interpretation, visualization, and development of input arrays for numerical groundwater models.

REFERENCES

Faunt, C.C., D'Agnese, F.A. & Turner, A.K., (1993) Development of Three-Dimensional Hydrogeological Framework Model for the Death Valley Region, Southern Nevada and California, USA. In: *Application of Geographic Information Systems in Hydrology and Water Resources Management* (ed. by K. Kovar & H.P. Nachtnebel), Proc. Int. Conf. held in Vienna, 19-22 April 1993, IAHS Publ. No. 211.

Kolm, K.E. (1993) Conceptualization and Characterization of Hydrologic Systems: International Ground Water Modeling Center Publication, GWMI 93-01, Colorado School of Mines, Golden, CO, USA (in press).

Turner, A.K. (1989) The Role of Three-dimensional Geographic Information Systems in Subsurface Characterization for Hydrogeological Applications: In: *Three Dimensional Applications in Geographic Information Systems* (ed. by J.F. Raper), Taylor and Francis, London, 115-127.

Turner, A.K. (1991) *Three-Dimensional Modeling with Geoscientific Information Systems*. Kluwer Academic Publishers, Dordrecht, The Netherlands.

Turner, A.K. & Kolm, K.E. (1991) Three-Dimensional Geoscientific Information Systems for Ground-Water Modeling. Am. Congress of Surveying and Mapping-Am. Society of Photogrammetry and Remote Sensing Technical Papers Annual Convention, Baltimore, MD., Vol. 4 - GIS, 217-226.

Turner, A.K. & Kolm, K.E. (1992) New Approaches for Regional Ground-Water Modeling in Southern Nevada. *Proc. of 3rd Annual Int. High-Level Radioactive Waste Management Conf.*, April 12-16, Las Vegas, Nevada.

Williams, M.D. (1992) Characterizing the shallow ground-water flow system in the northern San Luis Valley, Colorado using a geographic information system and numerical model simulation. Master of Engineering Thesis, Colorado School of Mines, Golden, CO, USA.

National water resources management planning based on GIS

G. KESER
Department of Applied Informatics and Technical Graphics, Agrar- und Hydrotechnik Consulting Engineers Ltd., P.O. Box 100132, 4300 Essen 1, Germany

J.J. BOGARDI
Department of Water Resources, Wageningen Agricultural University, Nieuwe Kanaalweg 11, 6709 PA Wageningen, The Netherlands

Abstract GIS-based national water resources master plans are considered as flexible planning tools, able to overcome the shortcoming of the traditional large scale strategical plans. Based on the case study example of the ongoing master plan EAU 2000 of Tunisia, the GIS-based methodology is introduced. Potential extensions like creating a plan hierarchy of different scales and resolutions are discussed.

INTRODUCTION

The expected shortage in good quality fresh water has been identified as one of the global challenges of the next century (Biswas, 1991). Countries situated in the semi-arid or arid climatic zones are already confronted with temporary shortages. Thus, national development plans must be synchronized with the temporal availability and spatial distribution of water resources.

Long-term strategic water resources development planning is particularly relevant in countries with large spatial and temporal discrepancies between the availability of and the demand for water regarding not only quantitative aspects but also qualitative constraints.

By using statistical methods, future water availabilities and future water demands can be estimated for long-term planning. Water availabilities are deduced from water resources (like discharges, reservoirs, groundwater, water treatments, desalinization, etc.). On the other hand water demands are composed of rural, urban, industrial, and agricultural demands.

Water resources management policy must be derived on river basin scale and coordinated on national level, identifying surplus and shortage zones along with the technical and institutional options to improve the water balance of a country. Consequently, the importance of large scale strategical water resources plans has increased.

Computer technics (like relational databanks and geographic information systems (GIS)) have proved to be an excellent tool to support large scale resource planning and allocations allowing easy aggregation, overlaying and intersecting of resources and demands and components thereof according to physical or administrative subunits. Applicability of GIS within the framework of a national strategical water resources development plan will be demonstrated on the example of the project EAU 2000 in Tunisia. While the setup of this example is realistic, the paper focuses on the methodology rather than to draw conclusions as far as strategical options were concerned in the specific project.

EMBEDDING COMPUTER TECHNIQUES WITHIN A WATER RESOURCES MASTER PLAN

Frequently the basic question to be answered by a master plan can be reduced to a simple water balance comparing existing or projected water demands and their respective qualitative requirements with the available water resources. Out of the balance results regions of water deficits and surpluses can be identified for which water resources management measures have to be identified. Possible measures are: reduction of water consumption, e.g., rehabilitation of water supply systems, optimization of reservoir operation rules, activating of additional resources (like desalination of saline groundwater), increasing irrigation efficiencies, interbasin transfer, etc.

Consequently the first step for a master plan is a broad data collection, and analysis campaign. The checked data has been imported into a databank, structurized according to the needs of the responsible planning and management agency. The data can be brought up to date, modified and extended by an interactive databank program using menus and windows technics. Data import (of already existing data files) as well as data export (e.g., to the GIS) are managed also by menus. Simultaneously to the data collection the cartographic information of different maps will be digitized by using GIS technology. By adding the same codes to the geographic information such as for the data in the databank, an automatic transfer between databank and GIS can be set up in order to build up a "Nationwide Water Resources Geographic Information System".

Based on such an information system, water balances for different spatial aggregations can be derived using specific GIS-techniques like intersecting and overlaying. Figure1 displays the phasing of the project as a flow chart. Obviously the balance study can be repeated by considering not only the available data of resources and demands but also the impact of various water management options like incorporation of new reservoirs, transfer systems or the introduction of water saving technologies, etc. Consequently, a GIS-based water resources master plan is not simply a study, but an interactive tool for dynamic water resources decision making.

CASE STUDY

The execution of a GIS-based national water balance study as the core of a water resources master plan is demonstrated on the ongoing project EAU 2000, aiming to derive water resources management strategies for Tunisia up to the

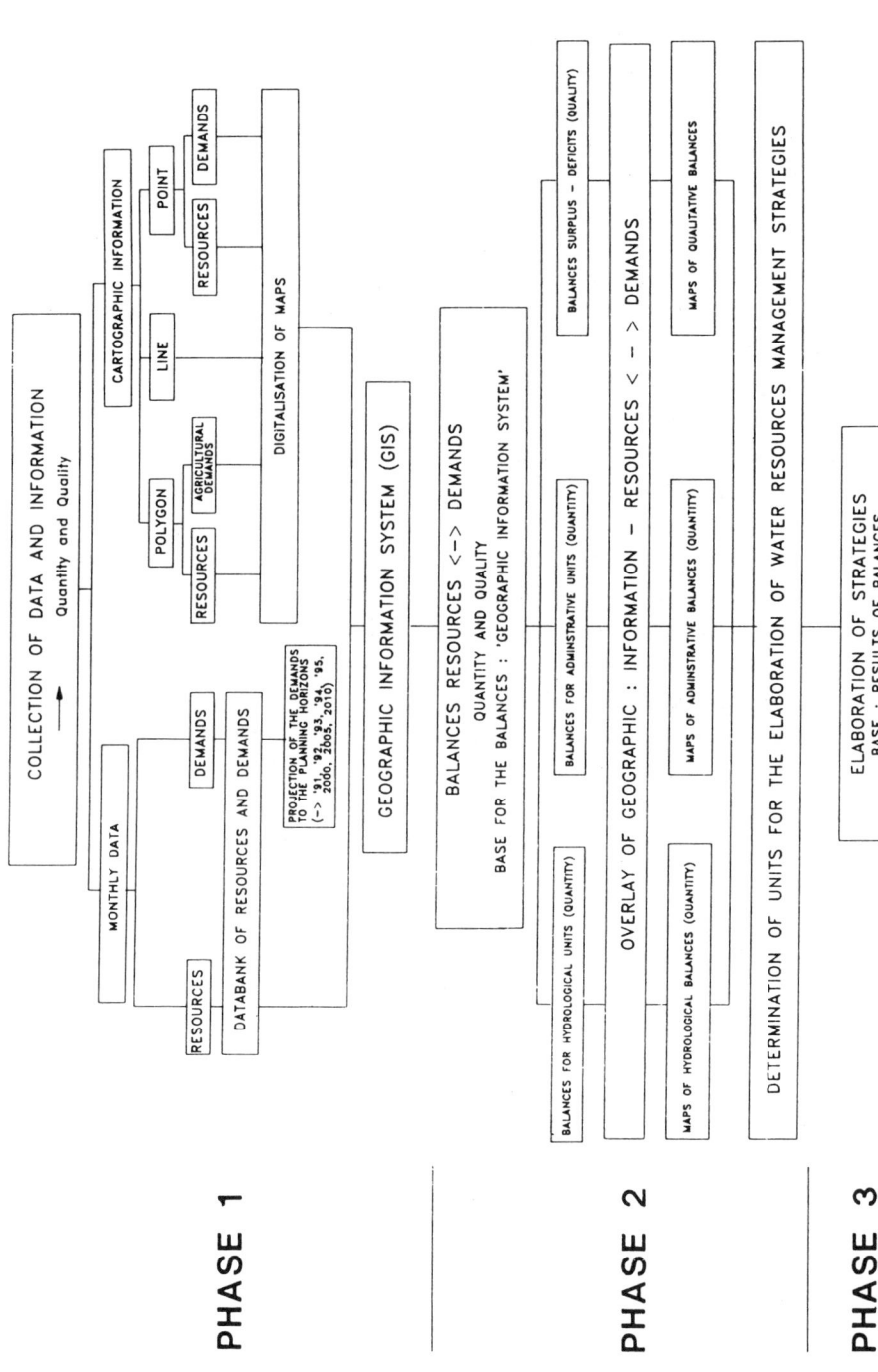

Fig. 1 Phasing of a GIS-based national water resources master plan.

year 2010. The country has been subdivided into 188 delegations (as administrative units, see Fig. 2a) and 40 hydrological units (catchments or groups of small catchments, see Fig. 2b). While the administrative allocation of surpluses or deficits in water balances is needed to reflect the decision making structure involved, the physical subdivision in hydrological units is needed to locate possible temporal surpluses to be harnessed and transferred to river basins with water balance deficits.

The results of Phase 1 of the project (see Fig. 1) are the following:
(a) implementation of a databank with consistent and plausible data for water resources and water demands at present and for the different planning horizons of the years 1995, 2000, 2005 and 2010;
(b) digitized cartographic information as point information (like location of reservoirs and water treatment plants), line information (like transfer system, topography, isohyets) and polygon information (urban water demand areas, irrigation perimeters, groundwater aquifers);
(c) elaboration of water resources and water demand coverages by appending stored data of the databank with the respective digitized geographic information;
(d) calculation of water resources and water demands concerning the delegations and hydrological units (water quantity and quality) by using GIS overlay procedures.

Phase 2 concentrates on the water balances and their display within the GIS. Based on the estimated water resources and water demands, 6480 monthly national balances have been calculated. These balances take into account:
(a) two balance units (hydrological and administrative units);
(b) three possible scenarios in the development of water demand upto the planning horizon year 2010;
(c) two initial water storages (empty, or half full) for the reservoirs at the beginning of their operation in the annual cycle (September), at the end of the dry season;
(d) three types of characteristic years (dry, median, wet) with respective monthly streamflows;
(e) no consideration of water quality and consideration of two water quality classes expressed by Cl^- concentration thresholds; and
(f) five different planning horizons (1992, 1995, 2000, 2005, 2010) with their respective demand, and projected water resources infrastructures.

Each national balance can be defined as a certain load on the existing water resource system in Tunisia. In order to obtain a critical load situation of the system as a base for defining future water resources management measures within a national strategy, the derived monthly balances have been analyzed statistically. For a preselected probability of 0.97 to meet future demands, the water resources management situation of the year 2010, month July, dry year, mean (most likely) development of water demand, no water storage in the reservoirs at the beginning of their annual operation has been identified. Figure 3 displays the delegations with balance deficits and surpluses.

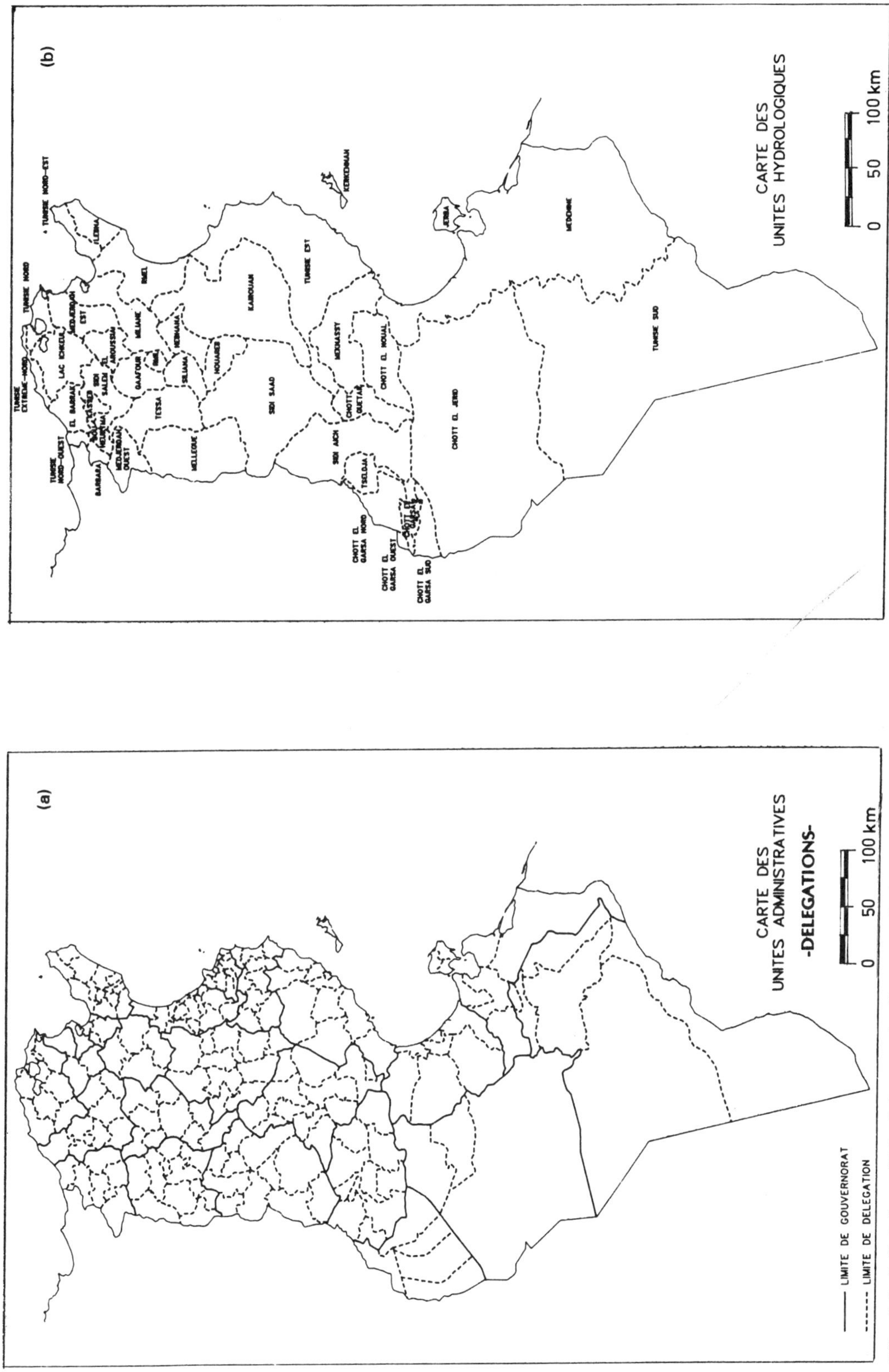

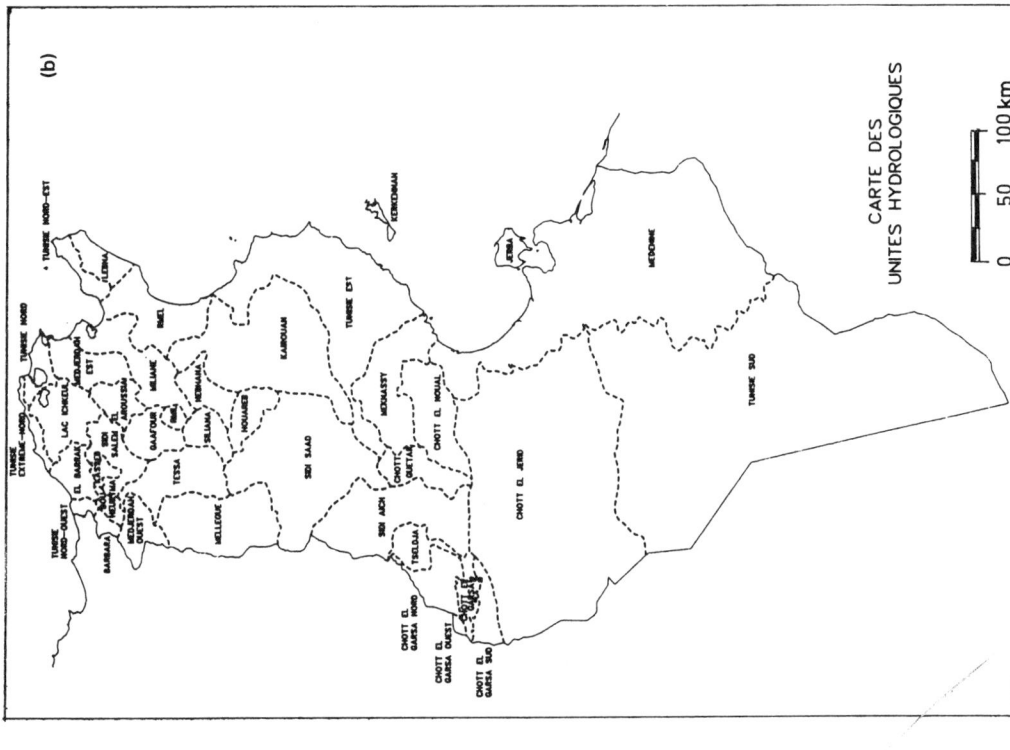

Fig. 2 Subdivision of Tunisia to units for water balances: (a) subdivision to delegations, (b) subdivision to hydrological units.

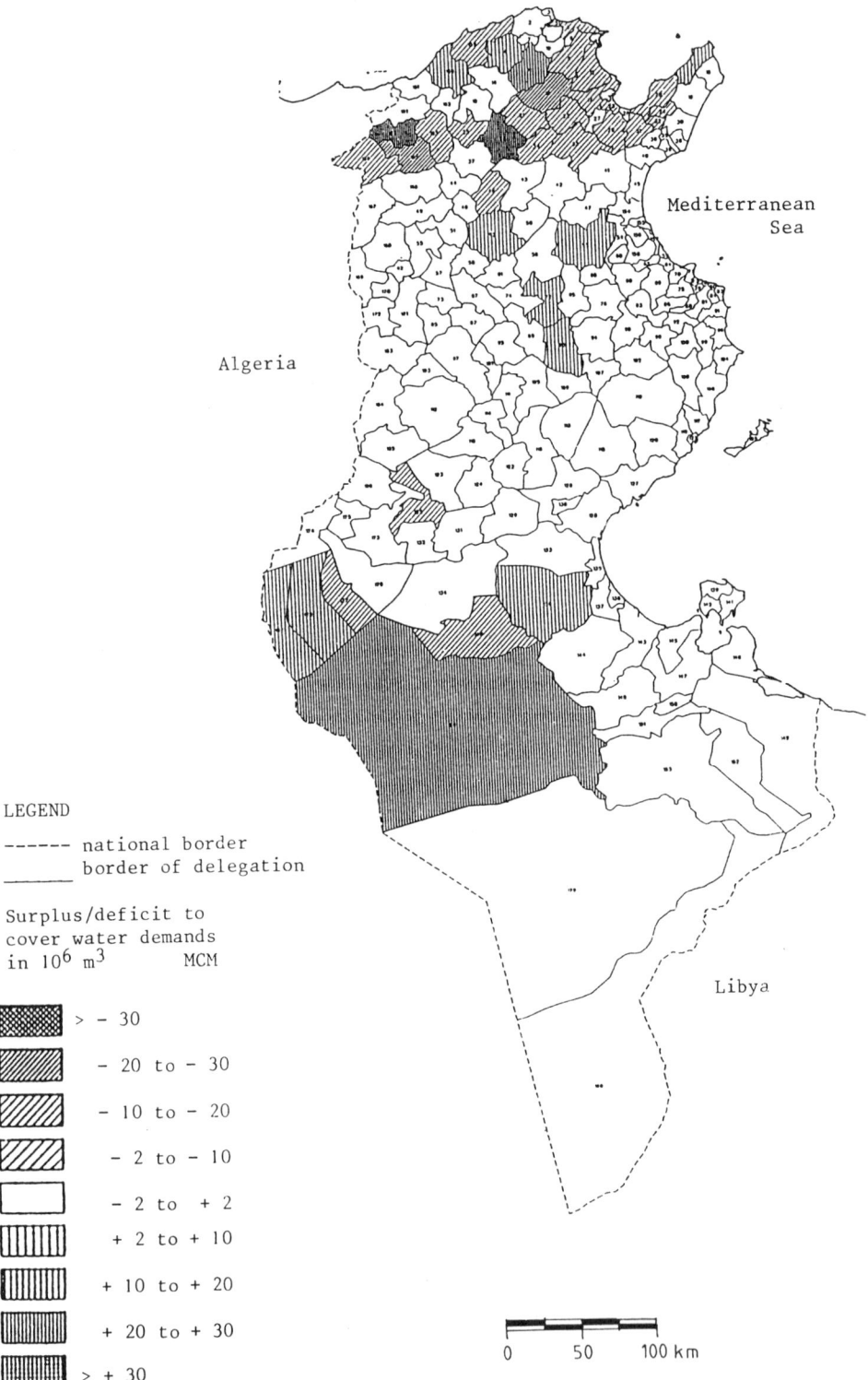

Fig. 3 Results of the water balances (resources minus demands) for delegations (dry year, first scenario for water demands, year 2010, initial water volume in reservoirs is 0 m³).

Phase 3 of a GIS-based national water resources master plan focuses on the elaboration of strategies in order to improve the projected water balances. Specific measures can be identified, their impact assessed and incorporated in the data base. Following the dual system of administrative and hydrological units these potential measures and remedies will be spatially allocated. Through the existing and envisaged transfer system, water balance of individual subunits can be changed. The present study concentrates on the following type of measures:

(a) reuse of treated sewage water;
(b) increase irrigation efficiencies;
(c) extend the existing water transfer system;
(d) limit municipal and industrial consumption;
(e) desalinization of the brackish water;
(f) extend the existing reservoir systems;
(g) increase the storage volume of certain large reservoirs;
(h) improved (stochastic dynamic programming-based) operation of existing and planned reservoirs;
(i) implement small-scale storage reservoir programmes;
(j) change envisaged water demand patterns.

The impact of selected combinations of the above outlined measures can be assessed. Consequently the water balance both on national and regional scale may change.

Together with the respective costs, change of performance reliability spatial water demand coverage etc. can be derived. These results form the base for the subsequent national policy decision. This step can be done either within a multicriterion decision support framework (Bogardi & Duckstein, 1992) or following more conventional decision making procedures.

DISCUSSION

"Traditional" water resources master plans have proved to be extremely time consuming, involving considerable costs (DVWK, 1984). Their impact remained very limited due to several reasons:

(a) results became outdated very soon;
(b) lack of updating;
(c) difficulty to overcome scale problems between strategical plans and specific project plans.

The availability of the coupled computerized data base - GIS technology offers the appropriate means to eliminate most of the shortcomings of the "traditional" master plans. As Fig. 4 displays, most of the steps involved in preparing the core water balances can be fully or partially automated. The technology can integrate or semi-integrate optimization techniques (SDP), GIS and data base management. Data and geographical information collection still remains a very tedious procedure. A well-structured and well-documented

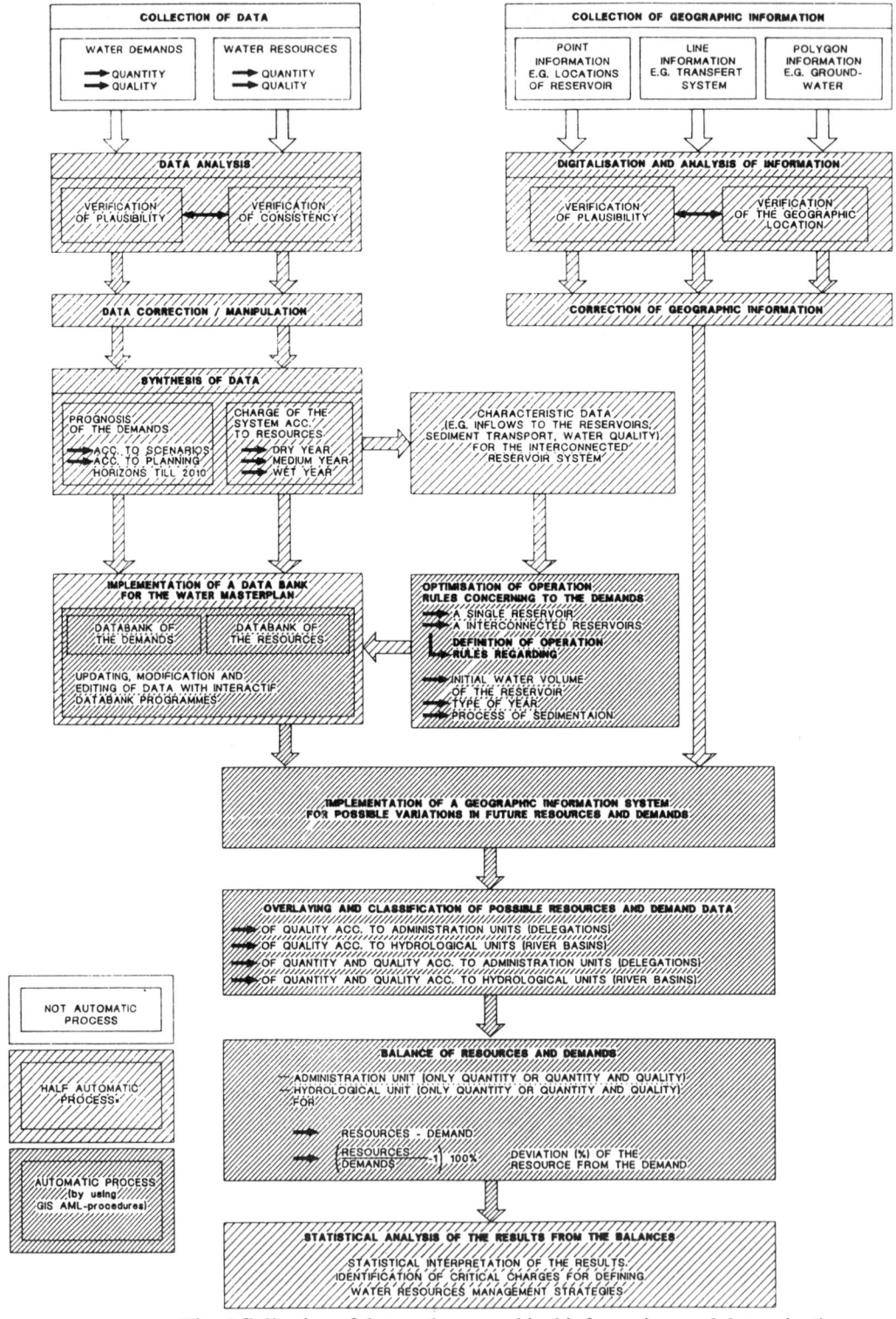

Fig. 4 Collection of data and geographical information, and determination of critical load situations of the water resource system in the framework of the national water masterplan.

databank facilities updating and the incorporation of new information.

As far as potential improvements are concerned the importance of the GIS-based master plan is even greater to overcome the scale problem between the different levels of the plan hierarchy. Figure 5 displays the possible linking of different planning levels using different resolution of information. The higher level plan does not only impose constraints on the lower level one, but may absorb changes as well, due to a "feedback" mechanism.

CONCLUSIONS

The application of integrated optimization-simulation-databank management and GIS technics, while still in its infancy, has already demonstrated considerable potential to revolutionize large scale strategical planning. The

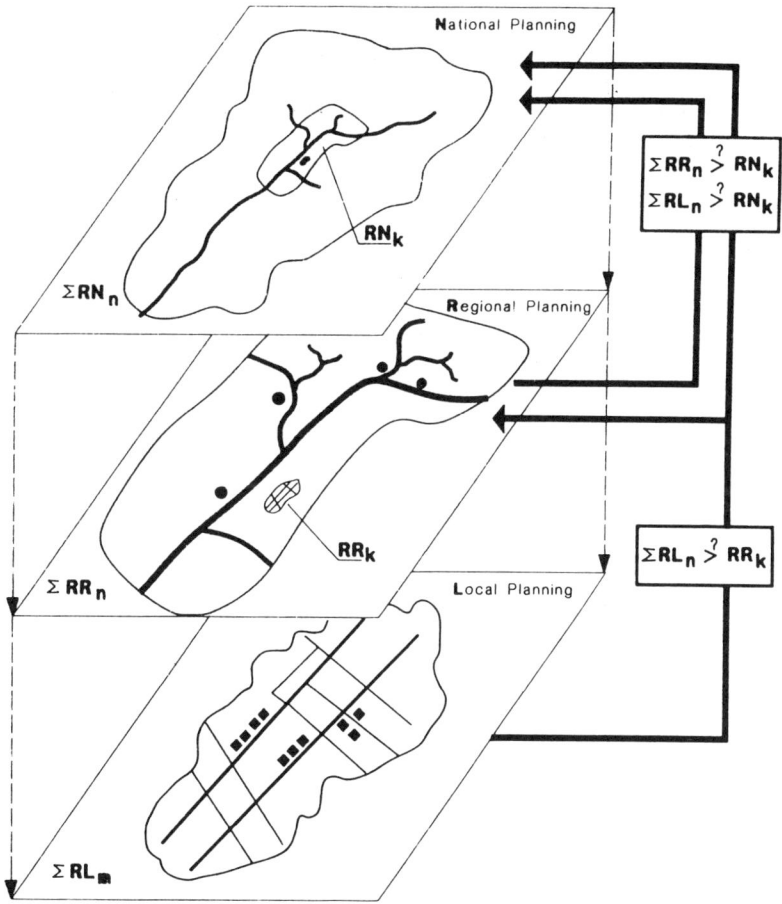

Fig. 5 Feedback from macro- to micro-planning concerning the available resources. The variable **R** indicates resources needed to meet the demands (n = 1,2, ... k, ...m).

applicability of the concept has been shown on the ongoing water resources master plan EAU 2000, covering whole Tunisia. While the available technology is impressive, some problems which may occur in course of its application should not be ignored. These concern mainly the human perception and ability to absorb information next to the need to acquire knowledge of managing and modifying databanks, GIS and specific incorporated algorithms. Further research is needed to specify the robustness of GIS-based water resources master plans, along with identifying what is needed by decision makers to derive sound, sustainable decisions. While these problems should not be neglected, it is believed that the available computer-based technology opens new avenues for well-founded and flexible water resources management.

REFERENCES

Biswas, A.K. (1991) Water for sustainable development in the 21th century: a global perspective. *Water International* 16(4), 219-224.

Bogardi, J.J. & Duckstein, L. (1992) Interactive multiobjective analysis embedding the decision maker's implicit preference function. *Wat. Resour. Bulletin* 28(1), 75-88.

DVWK (1984) Grossräumige wasserwirtschaftliche Planung in der Bundesrepublik Deutschland (Large scale water resources planning in the Federal Republic of Germany). Series Schriften Nr. 64. 1-115. Paul Parey Verlag, Hamburg.

7 Application of GIS in Surface Water Systems

Data constraints on GIS application development for water resource management

M. J. CLARK

GeoData Institute and Geography Department, University of Southampton, Highfield, Southampton S09 5NH, UK

Abstract GIS creates a dilemma for water resource managers. Few other techniques require such frequent access to data integration and modelling, and the scope for broad-based approaches to water resources is increasing steadily. Conversely, while GIS functions have developed rapidly alongside growing user needs, the same is not true of data acquisition and quality. With managers demanding more data, more robust data and more varied data, it is clear that data quality and quantity are now primary constraints on operational GIS. Data availability partly concerns data sources, but is also linked to data cost and quality, which often differentiate between data which are available in principle and those which can be acquired in practice. If technical and cost constraints on data acquisition are combined, then surrogate data may become the only realistic input. The discussion of water quality is similarly complicated by associated issues such as the relationship between data standards, data models and data error.

BACKGROUND CONSIDERATIONS

GIS is increasingly serving as an operational support for water resource management and planning, thus demonstrating that the introduction of this type of technology can now fully be rationalized in a business case analysis. While some aspects of the use of GIS focus on standard data access and integration functions which are readily available off-the-shelf, many of the more advanced operations have been developed primarily within a research environment. The migration of an IT application from a research to an operational setting involves a number of substantial implications for the application designer, and may require the designer or systems analyst to accept changes of both principle and practice in the way in which the application works. In this respect, data attributes and availability represent major constraints on the operational implementation of GIS applications in water resource management.

Within a research and development (R & D) environment, priority is often given to the system functionality that may well be seen as the main source of innovation in the development. Additionally, the fact that the GIS business case frequently emphasizes that functions add value to data means that again, functionality appears to be the key to both effectiveness and cost-efficiency. This is particularly true of cases in which complex models are developed or through which complex data structures are required. Within R & D, complexity may well be seen as an advantage, since it provides a path towards powerful functionality. On the other hand, in an operational environment simplicity is more likely to be welcomed as a means of ensuring cost-effective implementations, easy support and maintenance, and minimal disruption to existing work practices. Indeed, the managerial challenge of implementing substantial new software systems in an existing framework of well-established working practices will often take precedence over details of technical function.

The same contrasting approaches apply to questions of data acquisition, data management and data access. In an operational environment, 60-80% of GIS costs may well relate to data acquisition and management, with functional enhancement and even application development counting for relatively small costs. The contrasts with the R & D environment, which often performs satisfactorily with pilot data sets and with existing rather than newly acquired data, is sufficient to ensure that researchers and implementers often hold entirely different views on the nature and significance of data constraints. To the researcher/developer, data quality holds the key to effective GIS solutions. To the operational implementer, data quantity and acquisition are likely to be seen as at least equally important.

GIS thus creates a dilemma for water resource managers. On the one hand, few other techniques of operational management require such frequent access to data integration and modelling capability, and at the same time the potential scope for a broad-based approach to water resources is increasing steadily. On the other hand, while GIS functions have developed rapidly to keep pace with the growing and widening needs of the resource managers, the same cannot be said with such confidence of data acquisition and data quality. Despite the use of consumption conservation measures the residential, agricultural and industrial demand for water remains high, while the cost of water supply and sewage removal show no sign of falling. Across much of the world, the environmental standards required within channels and their drainage basins are becoming more rigorous, and the multiple use of waterways and water bodies impose great demands on both monitoring and management systems. With the consequent increasing requirement for data to support management decision-making and operations, it is apparent that the data sources employed by water resource managers are becoming more varied, and now include a full range of socio-economic variables to add to the more familiar engineering, physical, chemical and biological data.

With managers consistently demanding more data, more robust data and more varied data, it is clear that data quality and quantity now serve as primary constraints on operational GIS in water resource management. Data availability revolves partly around the consideration of data sources, but is also equally closely linked to a discussion of data cost, which often makes the difference between data which are available in principle and those which can be acquired in practice. If the hurdles of technical and cost constraint on data acquisition are combined, then it may well be that surrogate data become the only realistic input. The discussion of water quality is similarly complicated by associated issues such as the relationship between data standards, data models and data error.

DATA AVAILABILITY

The extent to which appropriate data are readily and cost-effectively available will in large measure condition the viability of most GIS-based water resource assessments. Of course, there are many cases in which a custodial GIS is implemented specifically to provide efficient storage, management and access control for existing water-related data sets. In such circumstances the data precede the GIS implementation, and data availability becomes a technical issue relating to the ease of conversion from pre-GIS to post-GIS formats and media. In other cases, particularly where a new water resource project is envisaged, the GIS design may take place to conform to a user specification rather than simply to absorb existing data. In such a context, data sources and attributes may well impose a strong constraint on GIS design and operation.

In as far as data acquisition represents a major cost and a major constraint on the pace of GIS implementation, it follows that data capture, entry and conversion technologies will play a critical role in the future. The potential for using GIS in water resource studies ranging from water quality assessment (Hirsch *et al.*, 1988) to project-related studies of the hydrological effects of groundwater extraction (Zelt, 1991) stress the extent to which the value brought by GIS to the project rests heavily on data integration. That such integration remains a technical challenge in many cases merely underlines how little has been achieved in bringing to water resource data the same kind of global homogeneity that one now expects of meteorological data. In terms of volume of data, both hydrology and water quality are relatively well served, though the proportion of data currently available in digital form is small. Where related water resource attributes such as land cover, surficial geology, topography, hydraulics and geomorphology are concerned, the position is very different and it must generally be expected that most of the required data will either have to be acquired *ab initio*, or that existing partial data sets will require substantial conversion to render them GIS compatible. Even in this data conversion process, however, GIS can often be of value in identifying erroneous or incorrectly assigned values.

Data commodification

Data represent the fuel which powers GIS. As with most other energy-related debates, however, it is economics rather than physics which dominates both the provision of alternative strategies and the selection of a preferred option. Do we use a positive business case to justify buying data in the quantities that are ideally required, or do we re-tune the engine to run effectively on existing (or converted) data? The answer to this dilemma appears to be at least in part regionally specific, in that data availability and national policies on data pricing vary massively from country to country. The paradox is starkly stated by Rhind (1992): most data (and almost all topographic data) are created in the public sector, but while system vendors and value added service providers favour subsequent data release on a cost-of-dissemination basis, governments with few exceptions regard public data sets as a commodity through which to recoup at least part of the acquisition cost that has been funded through public expenditure. Although the migration of data provision into a market-driven context will undoubtedly sharpen (and hopefully improve) the performance of data providers, system providers and service providers, the conflict between end user low-cost and high-cost strategies for data acquisition poses some uncomfortable challenges and can easily develop into a hurdle rather than a launch pad.

The basic topographic geography which provides both a locational framework and many of the derived drainage basin indices for water resource managers is in almost every case a market product. National mapping agencies are moving towards a fully digital product at varying speeds, but the outcome is output is a high value commodity which is often priced at a level which encourages GIS users to seek alternative data sources. In-house digitizing as a cost strategy is rarely viable since the base product usually embodies copyrights which themselves impose substantial charges. Early (pre-1984) Landsat imagery may offer a copyright-free source of small-scale geographic data, but requires georectification and considerable data extraction effort even if its date does not render it unacceptable. Air orthoimagery provides a useful source of large-scale information, but competes on cost only if much of the effort is in-house.

Initiatives to offer a path towards a quasi-standard topographic data base for European users are far from new. Work by the Comité Europeén des Responsables de la Cartographie Officielle on the European Territorial Database has been in hand since 1986, but a definitive outcome is still awaited. It is hardly surprizing that wherever possible, interest focuses on low-cost geographic data sources, including (for the European Community research environment) CORINE data, and for the rest of GIS users the Digital Chart of the World. Despite the scale of the latter (1:1 million), it will have immediate appeal to large-area analysts world wide, and the debate on its legality (Rhind, 1992) and commercial distribution will have considerable significance for drainage basin planners. The undoubted opportunities offered by low-cost data sources must be carefully balanced against the necessity to maintain (in a market-driven mode if this is necessary) high cost mapping agencies in order to enhance the overall and long-term interests of GIS users.

Water resource managers are increasingly finding themselves in the role of both purchaser and provider where data are concerned. The need to buy in topographic and other data to supplement the specifically water-based data of the agency concerned remains, but at the same time the agency's own data (hydrology, channel attributes, ecology etc) are becoming marketable products with many potential purchasers in both the private and public sector. At an informal level, data sharing agreements have proved useful, but these are difficult to devise and manage on a large scale. It may be inevitable that in a market-dominated economy, commodification of the agency data sets will offer a perceived simple route to regulating the flow of data, and to generating significant supplementary income in some cases. It is even the case that some water authorities are now seeing GIS services as a form of service diversification through which to achieve revenue generation and to maximize the return on investment. For example, with the split of the UK water industry into regional water companies and corresponding divisions of the National Rivers Authority, Southern Water plc was created in September 1989. At the same time, IT Southern Ltd was established within the Southern Water Group as an independent supplier of IT services (Elkins, 1990), with a strong interest in GIS. We are thus seeing commodification extending beyond data provision to include a range of value-added services.

Surrogate data

The notion of using surrogate data instead of direct measures of the variables of interest is commonplace in remote sensing, where in effect all raw data are surrogates for surface or near-surface properties. Remote techniques and their related surrogate data become most effective in regional scale projects, which renders them relevant to many water resources management projects. For example, the Metropolitan Water District of Southern California has both water distribution and treatment systems which are operating at near capacity during times of peak demand. Behind the concept of water use lie the land uses which demand water, so an appropriate water resources management response includes a major programme to delimit land use categories in relation to amounts of green vegetation using a combination of satellite remote sensing and GIS (Walklet & Hitchcock, 1991). Despite their success, however, examples like this exploit both GIS and the concept of surrogate data only marginally.

The arguments in favour of surrogates (which may be interpreted as being replacements or compromises) become more powerful when they are focused on a more positive-sounding term such as diagnostic index. Indices built from several contributing variables can offer increased rather than decreased power to a GIS, since if they are properly specified they can carry large amounts of information very efficiently (Edwards *et al.*, 1991). However, the fact that the index is generally synthetic in nature renders it somewhat project-specific, so that there is a strong argument in favour of retaining the raw data set for archival and recalculation purposes (not necessarily on line), as well as using the index for operational purposes. Indices nevertheless offer both data efficiency and data generalization, and in both contexts can provide clear benefit to managers in a multivariate field such as water resources.

DATA QUALITY

When so much management and research attention is devoted to quality it is somewhat surprising to find that the term does not have any really satisfactory definition that would help water resource managers set and monitor GIS data performance. In operational terms, the notion of quality implies a set of attributes that meet the reasonable expectations and requirements of the users, and a definition such as this clearly extends far beyond simple measures of "accuracy" or "error". In practice, however, quality issues continue to influence the viability of professional GIS implementation both in terms of distinctions between data sets and inconsistencies within single data sets (Briggs & Mounsey, 1989).

Professional GIS applications in the water resources management field will have a sufficiently high operational significance to necessitate associating each stage in the procedure with information to permit qualified authorities to evaluate the technical level and data quality involved. For example, a procedural design for assessing river maintenance requirements for the Thames Region of the UK National Rivers Authority (HR Wallingford, 1992) identifies data quality needs in terms of survey resolution (temporal and spatial), technical specification (including equipment and sampling design), and analytical/modelling specification. Such information would permit the designated authorities to judge whether the data and procedures were adequate to the task in hand, and to specify any additional data or technique refinement requirements. Both as a support for internal business standards and as a contribution to public accountability obligations, the establishment and monitoring of such quality benchmarks represents both operational enhancement and operational cost.

It is apparent that the concept of "data quality" encompasses a variety of attributes, all of which contribute to the overall efficiency and value of a GIS application (HR Wallingford, 1992). These include:

Accuracy specifies the extent to which the recorded attributes faithfully represent the variable that is of interest. In many cases this variable will be assessed indirectly through surrogate indices rather than being measured directly. For example, discharge may be approximated from stage; stage may be estimated through geomorphological characteristics; geomorphology may be evaluated from air photographs. Accuracy in this sense rests partly on the choice of the attribute to be recorded, and partly on the temporal and spatial sampling design.

Precision indicates the resolution (potential and achieved) of the measurement process, and is affected by instrument characteristics, operator characteristics and spatial/temporal sampling characteristics. Precision is usually easier to assess and record than accuracy, so that data quality assurance can become unacceptably dominated by precision.

Reliability subsumes a number of associated properties. In part it focuses on *objectivity*, since it is generally (but not necessarily correctly) assumed that an objective measure is more reliable than a subjective measure. Subjective records, even when they are produced by staff with great expertise and experience, will have less determinable reliability than objective (often instrumented) measures. The emphasis on properties which are "determinable" introduces the importance of the level of *documentation* associated with the data. Information for which there is no known source is regarded as inherently less reliable than information of established provenance. Full reliability requires documentation of instrumentation, sampling design and operator/observer characteristics. The achievement of consistency of observations or calculations is particularly important in cases where a procedure is applied across a range of operational areas by a range of operational staff.

Properties such as accuracy, precision and reliability can be determined/recorded at the time of observation, and can also in some cases be resurrected retrospectively from historical evidence. Given that the only useful data quality is *known* data quality, it follows that attention should also be paid to:

Data documentation determines the extent to which it is possible to identify data quality subsequent to the time of observation. Well documented data sets carry sufficient information for their quality to be assessed at any point, and for their suitability for new as well as existing tasks to be reviewed. Deficiencies in documentation substantially reduce the value of data sets (Briggs, 1991). It follows that the small additional cost of data documentation can fully be justified through the added value that comes with extended data use and multiple data use.

A data set that is documented with respect to its accuracy, precision and reliability has considerable operational value to the user, but in a large organization within which data move from user to user, or within which several users access the same data set (frequently the case in a multidisciplinary field such as water resource management), two other quality assurance issues are significant:

Information flow can impose some problems which neither the observer (surveyor) nor the end user envisage or recognize. General-purpose tabulation, for example, may introduce rounding, classification or generalization errors, some of which may be significant. This relates closely to the handling of error generation and propagation (see below).

Data version management is less often addressed in the research literature, but is of great importance in an operational environment. Version consistency is of critical importance with every type of data, and considerable inefficiency and error can be generated if different parts of a complex multiuser procedure such as water resource management use different versions of data. Without careful documentation and management of versions, all of the investment in data collection could be wasted if consultative or operation confusion was generated by the concurrent use of different data versions. As a documentation concern, version management can be approached through a networked system of version numbers and/or dates. In particularly sensitive areas, however, it may require some form of confirmatory procedure through a central index which identifies at any time the version that is current.

Data quality assessment is required by operational and decision-making staff as a basis for indicating how much reliance can be placed on the results of any analysis using the data, and also provides a foundation for public accountability. These two uses have rather different requirements. For in-house purposes, a clear technical set of data attributes can be tagged to the data set provided that the staff concerned are trained in the use of such evaluative metadata. In particular, this will involve instilling quality consciousness, whilst appreciating the practical value of data sets which do not yet meet the highest quality standards. It has already been established that ultimately, quality has to be defined in terms of a level which meets the reasonable requirements of the task at hand, and professional staff must be able to create and achieve such indicators of reasonableness. For public purposes, however, a different approach will be needed since the data quality record may become a part of the corporate operating procedure, and is therefore open to public review. However, for this review to be informed, it must be undertaken under circumstances in which the significance of the quality information can be properly explained and handled. In this context it must be repeated that quality indices tagged to any data set should be clearly purpose-specific. Thus, ecological data that may be of acceptable quality for assessing river maintenance engineering may be wholly inadequate for conservation review purposes. This linkage of quality assessment to the context of purpose must be included in procedures and in training/user awareness activities. It is also important that both the quality targets and the quality assurance procedures should be clearly specified prior to any data conversion contract, since deficiencies in this respect can significantly lengthen the data conversion process (Drake, 1989).

Data model and structure

It is self-evident that software, hardware, communication, operating system and data standards all contribute directly to enhancing the compatibility and interoperability of computer-based information handling systems. However, these fundamental components of "open systems" computing are so well covered in the It literature, and are so generic in nature, that they do not warrant extensive treatment in the context of water resources GIS. On the other hand, consideration of an appropriate data model for a GIS ranges across a spectrum of relevant issues from the mundane to the philosophical.

While graphic manipulation and display provide crucial power to a GIS, the fundamental control on system scope and efficiency is often closely related to the design of the database (Sinclair, 1990). In one of its more sophisticated modes, the debate focuses on the appropriateness of object-oriented data for use in GIS, and on the models that would be necessary to permit substantial movement towards a truly temporal GIS. Such matters are bound to generate controversy as new approaches struggle for recognition. Thus Dangermond (1989) argues cogently that an object-oriented structure is conceptually and technically complex and (more significantly in the longterm) that it constrains the user's flexibility in defining or dynamically creating feature relationships. In complete contradistinction, Piazza & Pessaro (1990) identify object-oriented structure as a powerful vehicle through which to bring a higher level of cognitive reasoning to GIS-based analyses, something which they regard as a necessary next step in GIS development. Hargis (1992) goes further in arguing that object-oriented data structures and modelling techniques provide a more logical path towards representing attributes and their geographies in a database, and suggests that the associated data structure visualizations are actually more intuitively interpretable by nonspecialists than traditional relational database designs. Whether or not their viewpoint is acceptable in any particular context, all of these papers indicate clearly that the data structure directly conditions what a GIS can do, and how effectively it can do it.

Over the past two years, object-oriented approaches (whether or not based strictly on an object-oriented programming language) have begun to emerge in major commercial GIS software, suggesting that they represent a competitive attribute - and there are signs that object orientation is becoming a preferred structure for parts of the water industry. Although users may be slow to determine the advantages and disadvantages, they are beginning to include object orientation in their statements of user requirement, so that water resource managers and the software developers who support them must move towards either satisfying this demand, or explaining convincingly why alternatives might be preferable or acceptable.

Error and uncertainty

Since it can be assumed that in a professional context GIS will generally be used to support either operational or planning decisions, it follows that possible error represents a significant problem. Surprisingly, although the research and development community has devoted considerable effort to conceptualizing the nature, generation and management of GIS errors, much of the professional community is far less demanding of error assessment in a GIS application than would be the case with most other areas of operation. In terms of source, we can assume that error divides between that which is inherent in the imported data (maps, statistics, images) and can thus in principle be tackled through quality assurance at source, and that which arises during data import, handling and interpretation, and is thus much more difficult to assess since there are no objective standards of correctness against which to measure an output. Furthermore, a moment's contemplation confirms that most source data sets in a field such as water resource management (whether spatial or not) are supposedly precise records of attributes which in reality lack sharp and robust attributes. Rivers change in size and position; rainfall events have indeterminate spatial and temporal boundaries; natural landcover types tend towards gradation or mosaic, not neatly delimited polygons. This inherent fuzziness of natural data sets implies that while error may be known and managed, it can never be eliminated.

The problem is even greater with attributes which are defined with a subjective element, such as soil type or vegetation dependence on water table. In such cases not only are the boundaries indeterminate, but the categories are uncertain. Far from undermining the value of GIS, this combination of error and uncertainty can (if properly approached) render GIS additionally useful. For example, the Thames Region of the UK National Rivers Authority is considering implementation of a data quality tag which would allow either a quantitative or qualitative quality assessment to be appended to all data sets, and would permit this quality tag to be displayed as part of any data visualization or analysis. Berry (1991) proposes using this type of concept to design a "shadow map" which incorporates an indication of both boundary and classification certainties, and allows this to be viewed alongside the attribute map based on the data concerned. The encouragement provided by such approaches lies in their demonstration that error can be confronted positively rather than either having to be ignored or having to be viewed as a threat to the GIS output.

This type of strategy then makes it possible to confront some of the implications of uncertainties in data model or data content. For example, Aspinall (1991) adopts a Bayesian approach to handle the uncertainty inherent in the task of integrating "high" resolution satellite image data and "low" resolution ground sample data (1 km^2 bird counts) to provide a composite map of habitat. While at first sight such an ornithological example appears far distant from the interests of water resource managers, the structural similarity between the applications is striking, with both needing to combine continuous representations of surface properties with statistically discrete area-based measures. The apparent ability to handle this task with a clear eye on the quality of the result is encouraging. Also of considerable potential interest is the use of various of the more sophisticated approaches to spatial (or network) interpolation. Particular interest has been shown over the last few years in the use of kriging in GIS to interpolate values at unsampled locations, to identify spatial patterns or to map from irregular to regular spatial sampling frameworks (Leenaers, 1990). Kriging is not in itself an error management technique, but it does clearly offer a robust route to minimizing the generation of error at the spatial modelling level.

As in the case of the term "quality", it is clear that error can in practice best be defined and managed specifically in the context of user requirements, since it is a relative rather than absolute property. Indeed, whether we are dealing with geographic (line/point) or attribute properties, there are some senses in which the "error" generation represented by processes such as generalization or classification offer positive benefits to professional GIS users: again, error is to be managed not eliminated. In the view of Maffini *et al.* (1989), error in digital geographic data is inevitable and cannot be removed simply by improving instruments and operators, though it can be controlled through such efforts. Instead of concentrating on error and uncertainty elimination, they suggest that we should improve the operational ability to cope with error, in part through better awareness of its significance and in part through the development of GIS tools to allow error specification to become a part of any working GIS application. Carver (1991) asks who cares about error modelling in GIS: even a brief consideration of the operational implications suggests that everyone should, but there is as yet disappointingly little indication that this challenge is being confronted in practice.

DATA MANAGEMENT

At first sight it appears self evident that the primary challenge of professional GIS for the rest of this Century will lie within data interchange and the many technical, legal and commercial issues which this engenders. Paradoxically, however, we are already beginning to become aware that interchange is merely the first of several hurdles to be surmounted, and that the more protracted challenge may relate to the management of the large data sets that are becoming increasingly accessible. Technical issues are by no means exclusively related to an individual system, since the move towards a distributed computing environment (including distributed relational database) places considerable demands on the management structures that will have to be put into place. Thus data standards and system standards have to be joined by metadata (information about data) standards if the desired aim of integration is to be achieved. Realization of these demands is beginning to emerge in professional GIS establishments, but is often a much lower priority within the R & D environment from which so many GIS innovations emerge. An early implementation requirement for any major water resource application should thus be to ensure that the standards and structures are open to migration towards the more complex data management contexts in which GIS will be used in the future.

Temporal data present a more substantial challenge, whether in the context of long-term archives or real-time (ie time-critical) management. Temporal GIS is still in its infancy, with many archives achieving little more than adding date as one their attribute fields so as to permit time-dependent query and display. For water resource managers derivatives such as change and pace of change are considerably more significant, and require much more creative conceptualization - including the notion of animated maps. Although aimed at the global scale, the work of Beller *et al.* (1991) introduces a number of fruitful directions for handling time-bound attributes of resource coverages. Once again, the challenge is as much based on data as on concept, since the best temporal data are often in the form of a time series which can easily be tied to a particular point, line or area for query purposes, but which is difficult to disaggregate within the system in order to prepare either time-slice maps or spatial maps of change. With global climatic change representing a policy-significant issue in water resources management, GIS will need to develop substantially if it is to support definitive analysis and modelling.

PERSPECTIVE

It would be a mistake to over emphasize the significance of data in constraining or steering the development of water resources GIS. Other external influences are also highly influential, including the widespread introduction of low-cost data viewers which disseminate GIS-like opportunities far more widely than has been possible hitherto. It is also important to recognize the influence of trends in the user interface, including the development of more intuitive and interactive systems, and the move towards multimedia formats. Nevertheless, the net result of all such trends is to continue the pressure for greater access to larger amounts of data by more and more people. The constraints on success are rarely functional, but most often are managerial or data-derived.

It should be stressed that data constraints represent a challenge rather than a barrier. As individual managers solve individual problems, it becomes apparent that this challenge can be met in principle even if it remains problematic in practice. The emphasis now must rest on attempts to introduce a generic function set that can offer water resources managers a broad-based structure within which to create specific applications for their own specific purposes.

REFERENCES

Aspinall, R.J. (1991) A landscape ecological approach to mapping distribution and abundance of birds with GIS. *Proc. AGI91, Third National Conf. and Exhibition, Birmingham, 20-22 November 1991*, 2.1.1-2.1.11.

Beller, A., Giblin, T., Khanh, Le V., Litz, S., Kittel, T. & Schimel, D. (1991) A temporal GIS prototype for global change research. *Proc. "What is Scientific GIS" Conf., Boulder, Colorado, September 19, 1991*. IBM, Boulder, Colorado, 1-16.

Berry, J.K. (1991) GIS facilitates error assessment. *GIS World* 4(8), 38-39.

Briggs, D.J. (1991) GIS development for broad-scale policy applications: the lessons from CORINE. In: *Geographic Information 1991* (ed. by J. Cadoux-Hudson & D. I. Heywood), The Yearbook of the Association for Geographic Information, Taylor & Francis, London, 113-120.

Briggs, D.J. & Mounsey, H.M. (1989) Integrating land resource data into a European geographical information system: practicalities and prospects. *Applied Geography* **9**(1), 5-21.

Carver, S. (1991) Error modelling in GIS: who cares? In: *Geographic Information 1991* (ed. by J. Cadoux-Hudson & D. I. Heywood), The Yearbook of the Association for Geographic Information, Taylor & Francis, London, 228-234.

Dangermond, J. (1989) GIS data structures: objects vs. layers. *The 1989 GIS Yearbook*. GIS World Inc.

Drake, P. (1989) The use of different data conversion options for a GIS application in the water industry. *"GIS - a corporate resource", Proc. AGI First National Conf. and Exhibition, Birmingham 11-12 October 1989*. Association for Geographic Information, 4.2.1-4.2.8.

Edwards, P.J., Clark, M.J. & Gurnell, A.M. (1991) Ecological data and geographic information system for planning and industry. In: *Terrestrial and Aquatic Ecosystems: Perturbation and Recovery* (ed. by O. Ravera), Ellis Horwood Ltd., 62-68.

Elkins, P. (1990) Water under the bridge - the background to Southern Water's experience with GIS. *Mapping Awareness* **4**(8), 38-40.

Hargis, J.E. (1992) Object Modeling Techniques ease relational database design. *GIS World* **5**(6), 80-83.

Hirsch, R.M., Alley, W.M. & Wilber, W.G. (1988) *Concepts for a National Water Quality Assessment Program*. US Geological Survey Circular, no. 1021.

HR Wallingford (1992) *Standards of Service Reach Specification Methodology*. Report no. EX 2652 to National Rivers Authority, Thames Region. Hydraulics Research, Wallingford, UK.

Leenaers, H. (1990) Multiscale spatial modelling of metal pollution in the fluvial environment. *Proc. EGIS '90 - First European Conf. on Geographic Information Systems, Amsterdam April 10-13 1990*. Vol. **2**, 664-674.

Maffini, G., Arno, M. & Bitterlich, W. (1989) Observations and comments on the generation and treatment of error in digital GIS data. In: *The accuracy of spatial databases* (ed. by M. Goodchild & S. Gopal), Taylor and Francis, London, 55-69.

Piazza, P.A. & Pessaro, F. (1990) A cognitive model for a "smart" GIS. *The 1990 GIS Yearbook*. GIS World Inc., 273-279.

Rhind, D. (1992) War and peace: GIS data as a commodity. *GIS Europe* **1**(8), 24-26.

Sinclair, N. (1990) Practical approaches to GIS database development. *Proc. EGIS '90 - First European Conf. on Geographic Information Systems, Amsterdam April 10-13*. Vol. **2**, 1035-1043.

Walklet, D.C. & Hitchcock, P.E. (1991) The California drought: GIS and remote sensing provide vital data. *GIS World* **4**(5), 52-55.

Zelt, R.B. (1991) GIS technology used to manage and analyse hydrologic information. *GIS World* **4**(5), 70-73.

Validation of the ANSWERS catchment model for runoff and soil erosion simulation in catchments in The Netherlands and the United Kingdom

A.P.J. DE ROO
Department of Physical Geography, University of Utrecht,
P.O. Box 80115, 3508 TC Utrecht, The Netherlands

Abstract Field and laboratory experiments are carried out in two small agricultural catchments in the loess area of the Netherlands, and one catchment in Devon (UK), to evaluate and validate the ANSWERS deterministic distributed catchment model for runoff and soil erosion simulation. The ANSWERS model is linked to a raster Geographical Information System to achieve a high accuracy and flexibility. The model is validated by comparing measured and simulated hydrographs runoff and sediment yield at the catchment outlet. Also, to validate the model in a distributed way, the spatial soil loss estimates, measured in two catchments using the ^{137}Cs technique, have been compared statistically with the simulated soil loss maps. Correlations of simulated soil loss with observed ^{137}Cs values in the Yendacott and Etzenrade catchment are low, but statistically significant. It is concluded that quantitative spatial modelling of soil erosion still involves large percentage errors, especially when the model is used in areas other than those for which it was originally developed and tested.

INTRODUCTION

Soil erosion and runoff are serious problems in South Limburg (The Netherlands) because of the unique combination of susceptible loess soils, sloping land, intensive agriculture, human settlement and ecologically important areas (De Roo, 1993). In order to understand how to reduce the magnitude of this problem it is necessary to make and test quantitative models of runoff and erosion which can be used to evaluate alternative strategies for improved land management. The ANSWERS model has been evaluated in two catchments in the loess area of South Limburg (The Netherlands): the Etzenrade catchment (225 ha) and the Catsop catchment (46 ha). To test the model under different conditions, it has been used in the Yendacott catchment, a small catchment in Devon, UK (147 ha). In these catchments, surface runoff and soil erosion have been measured and simulated. The three catchments are described in De Roo & Riezebos (1992), De Roo & Walling (1993), and De Roo (1993).

THE 'ANSWERS' MODEL

The distributed model ANSWERS (Areal Nonpoint Source Watershed Environment Response Simulation) (Beasley *et al.*, 1980; Beasley *et al.*, 1982), was used in this study for modelling surface runoff and soil erosion. The

ANSWERS model is designed to simulate the hydrological behaviour of catchments having agriculture as their primary land use, during and immediately following a rainfall event. Its primary application is in planning and evaluating various strategies for controlling surface runoff and sediment transport from intensively cropped areas. The original version of the model has been fully integrated within a Geographical Information System (GIS) (Burrough, 1986) so that data can be entered easily and the results can be displayed as maps and tables (De Roo et al., 1989). The GENAMAP GIS and the PC-RASTER package, developed at the University of Utrecht, have been used.

A catchment to be modelled is assumed to be composed of square cells. Channel flow is analyzed by a separate pattern of channel elements. Values of variables are defined for each element, e.g. slope, aspect, soil variables (porosity, moisture content, field capacity, infiltration capacity, USLE erodibility factor K), crop variables (coverage, interception capacity, USLE crop and management factors C and P) (Wischmeier & Smith, 1978), surface variables (roughness and surface retention) and channel variables (width and roughness). These values may be derived from digital maps or from interpolation of point data. From the above it is clear that several detailed maps of land attributes are needed to run the model. These maps can be stored and edited in a GIS.

A rainfall event can be simulated with increments of one second up to several minutes, depending on the grid size, thus taking into account spatial and temporal variability of rainfall. The continuity equation is used to establish the composite response of the single elements. The output of up-slope elements becomes the input of down-slope elements. Several physically-based mathematical relationships are used to describe interception, infiltration, surface retention, drainage, overland flow, channel flow, subsurface flow, detachment by rainfall and/or overland flow and sediment transport by overland flow (inter rill erosion) (Beasley et al., 1980).

Thus, the most important processes related to soil erosion are incorporated in ANSWERS. More detailed descriptions of the model can be found in Beasley et al. (1980) and De Roo et al. (1989).

MODIFICATIONS TO THE 'ANSWERS' MODEL

The ANSWERS model has been chosen to simulate surface runoff and soil erosion in South-Limburg, and to test the model under different conditions, in the Yendacott catchment (Devon, UK). However, it is inevitable that the model does not describe all processes relevant in these catchments: a 'complete' soil erosion model does not exist at this time, and probably will or can never be developed. Therefore, using a model in areas other than those for which it was originally developed and tested may necessitate model modifications. Field observation and hydrograph analysis confirmed that saturation overland flow is important in one of the research catchments, the Yendacott catchment. Therefore, the capability to simulate saturation overland flow, using the concept of variable contributing areas, has been incorporated.

Furthermore, even with the relatively small grid size used here (10 and 20 m) for the simulations, not all landscape details can be described and taken into account. But, some of those small landscape elements are important, such as the effects of asphalted roads, smaller than the pixel-size, in South-Limburg, where the increasing area of built-up areas and asphalted roads is a major cause of increased soil erosion and flooding problems. Thus, the model has been modified to incorporate the effects of roads. The location and width of roads (in dm) were used as additional input. Infiltration on roads was set to zero. Thus, a 20 m pixel with a 5 m road produces a minimum of 25% overland flow, regardless the infiltration capacity of the other 75% of the pixel.

Another important factor is the lack of some important input data needed for the model. It is impossible, at least for a research with limited staff, to measure spatially and temporaly varying data such as the soil moisture content, infiltration capacity, soil erodibility, soil cover by crops, and surface roughness continuously in space and time. Most of the time, those variables are measured once or a few times at a limited number of locations. Thus, solutions have to be found to deal with this situation, sometimes referred to as the "parameter crisis" (Burrough, 1989). One possibility is to develop sub-models that estimate parameters based on readily available information. In this study, the soil moisture content has been estimated based on daily rainfall, daily evaporation, and crop type (De Roo & Walling, 1993). Furthermore, infiltration rates as influenced by crusting have been estimated using cumulative kinetic energy of rainfall and soil tillage information (De Roo & Riezebos, 1992). Another possibility to deal with the parameter crisis is to use the information of parameter uncertainty for model simulations using e.g. Monte Carlo procedures (De Roo et al., 1992).

VALIDATING DISTRIBUTED HYDROLOGICAL AND SOIL EROSION MODELS

Distributed hydrological and soil erosion models can be validated by comparing measured and simulated values of discharge and sediment concentration at the catchment outlet. Also, overland flow patterns can be observed in the field and compared to the simulated patterns. Furthermore, soil erosion patterns can be evaluated using ^{137}Cs (Walling et al., 1986; De Roo, 1991; Quine & Walling, 1991). Thus, the distributed model can be evaluated in a distributed way. Other data to validate the model, such as comparing measured and simulated soil moisture content, were not available.

COMPARING MEASURED AND SIMULATED HYDROGRAPHS

Comparison of measured and simulated hydrographs is a necessary test but cannot be considered a sufficient test of models that are meant to simulate the internal responses of a catchment (Beven, 1989). However, for these catchments, no other hydrological data were available. Furthermore, additional testing of the model has been achieved by comparing measured and simulated soil loss patterns (see below).

The two catchments in South-Limburg were simulated twice, with and without using the crusting sub-model, which was developed based on field and laboratory measurements of loess soils from the catchment (De Roo & Riezebos, 1992). The sub-model simulates crusting, which decreases infiltration rates, as a function of cumulative kinetic energy of rain since last tillage. The model validation results are indicated best by McCuen's coefficient, which has been developed for hydrographs (McCuen & Snyder, 1975). In the Etzenrade catchment, the introduction of the crusting model does not improve the model predictions (Table 1). In the Catsop catchment (Table 2), the crusting sub-model does improve the results.

Table 1 Summary of the goodness of fit indices for hydrographs for the Etzenrade catchment and the most important correlation coefficients between observed and simulated data, for simulation runs with and without the crusting model.

23 rainstorms	not crusted		crusted	
	median	st.dev.	median	st.dev.
Pearson's correlation	0.717	(0.194)	0.732	(0.169)
McCuen's coefficient	0.230	(0.148)	0.153	(0.220)
Coefficient of Efficiency	2.448	(40.077)	-9.333	(109.81)
Total discharge	0.035		-0.045	
Peak discharge	0.799		0.751	
Time to peak discharge	0.983		0.977	

Table 2 Summary of the goodness of fit indices for hydrographs for the Catsop catchment and the most important correlation coefficients between observed and simulated data, for simulation runs with and without the crusting model.

14 rainstorms	not crusted		crusted	
	median	st.dev.	median	st.dev.
Pearson's correlation	0.548	(0.283)	0.562	(0.284)
McCuen's coefficient	0.109	(0.174)	0.140	(0.207)
Coefficient of Efficiency	-0.004	(0.576)	-0.087	(0.549)
Total discharge	0.558		0.578	
Peak discharge	0.817		0.919	
Time to peak discharge	0.969		0.969	

Table 3 Summary of the goodness of fit indices for hydrographs for the Yendacott catchment and the most important correlation coefficients between observed and simulated data, without and with calibration on lambda, and for the validation set.

32 rainstorms	before calibration		after calibration		validation	
	median	st.dev.	median	st.dev.	median	st.dev.
Pearson's correlation	0.637	(0.269)	0.617	(0.268)	0.616	(0.276)
McCuen's coefficient	0.246	(0.198)	0.504	(0.236)	0.287	(0.227)
Coefficient of Efficiency	-0.499	(12.898)	-0.221	(1.606)	-0.918	(10.964)
Total discharge	0.616		0.864		0.752	
Peak discharge	0.378		0.973		0.502	
Soil loss	0.632		0.809		0.610	
Time to peak discharge	0.865		0.926		0.870	

Although the crusting sub-model is a useful attempt to simulate soil crusting in the Holtan infiltration model used in ANSWERS, the effect of cracks in the crust arising in dry periods and the effects of soil animals breaking the crust are not incorporated, which may explain the results in the Etzenrade catchment. Also, spatial variability of rainfall in the larger Etzenrade catchment may be the cause of the less satisfactory results.

The concept of variable contributing areas has been introduced in ANSWERS to simulate saturation overland flow in the Yendacott catchment (Devon, UK) (De Roo & Walling, 1993). The catchment has been simulated three times: using the original method to estimate lambda (Beven & Kirkby, 1979), calibrating all individual storms on lambda, and using an independent validation set of 24 storms, using an equation to estimate lambda derived from the first eight storms (Table 3).

The modified model results, with lambda estimated using an equation derived from calibrating the first eight storms, are improved, but even with the improvement provided by incorporating the "variable contributing area" concept, the hydrograph still cannot be simulated closely. This may be caused by the fact that other important sources of water such as subsurface lateral flow caused by perched water tables or macro pores, are not represented, and groundwater flow is poorly simulated in the modified ANSWERS model.

COMPARING MEASURED AND SIMULATED SOIL LOSS MAPS

To validate the distributed soil erosion model in a distributed way, the spatial soil loss maps, produced by the model, have been compared to soil loss maps, obtained using the ^{137}Cs method. The ^{137}Cs data used for the Yendacott catchment are described Walling & Bradley (1988), and the data for the Etzenrade catchment are described in De Roo (1991). The ^{137}Cs data were used in two ways to validate the erosion model. The first approach is to

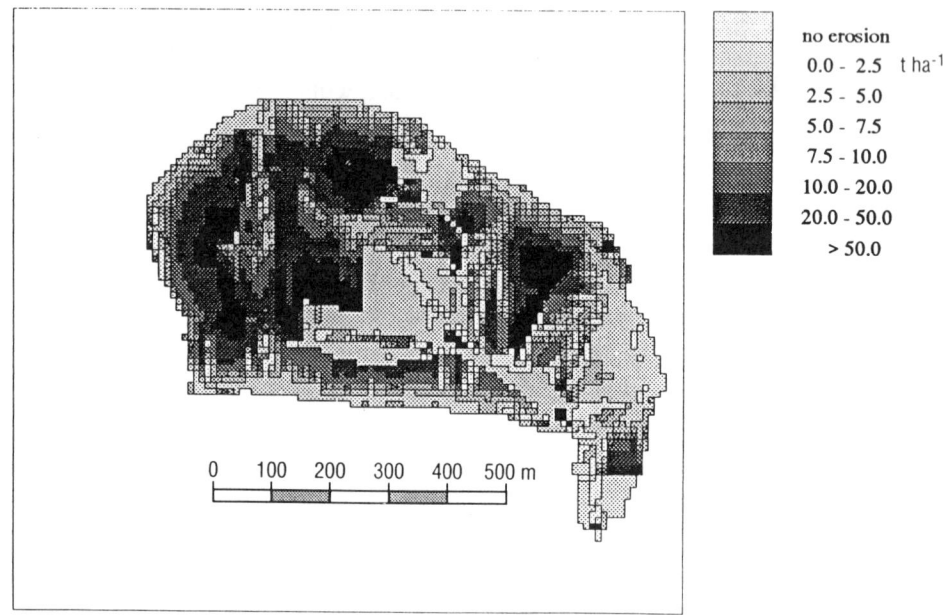

Fig. 1 Total soil loss between autumn 1986 and spring 1991, Yendacott catchment, Devon, UK, simulated with the modified ANSWERS model.

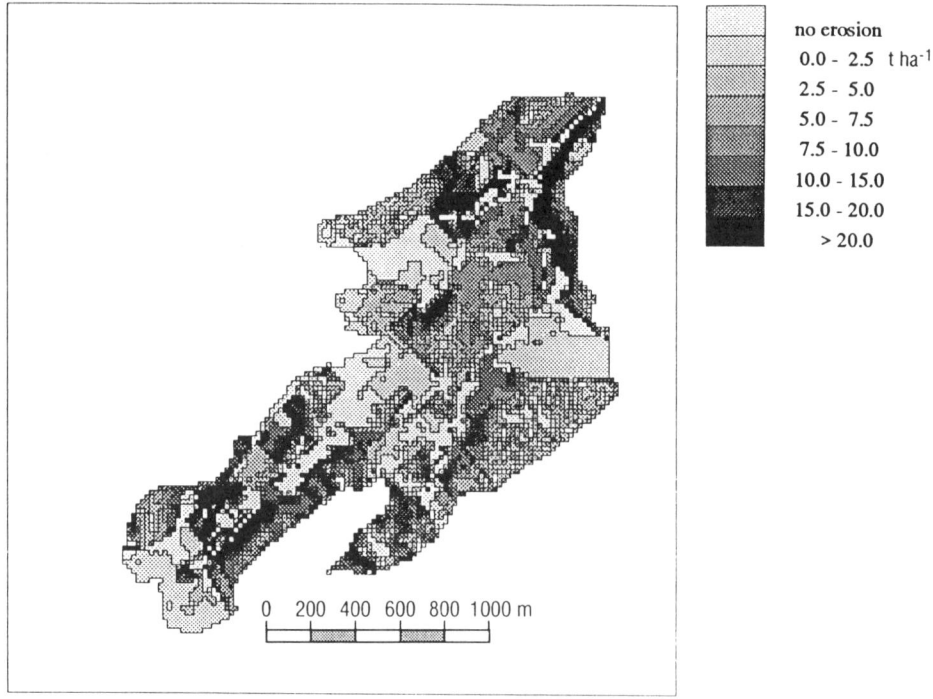

Fig. 2 Total soil loss between autumn 1987 and summer 1990, Etzenrade catchment, South Limburg (The Netherlands), simulated with the modified ANSWERS model.

Validation of the ANSWERS catchment model for runoff and soil erosion 471

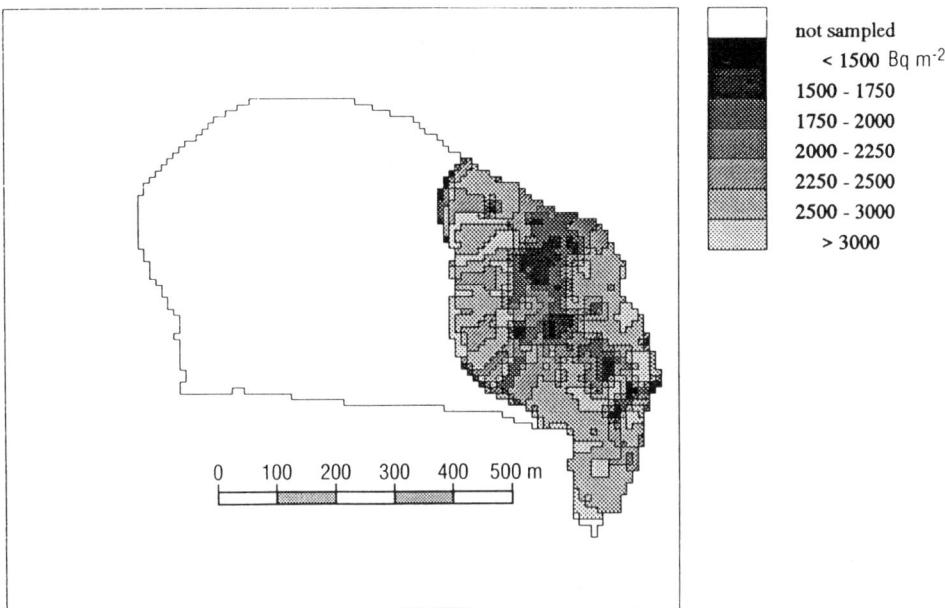

Fig. 3 ^{137}Cs distribution in the Yendacott catchment (based on Walling & Bradley, 1988) (low values correspond with high soil erosion rates).

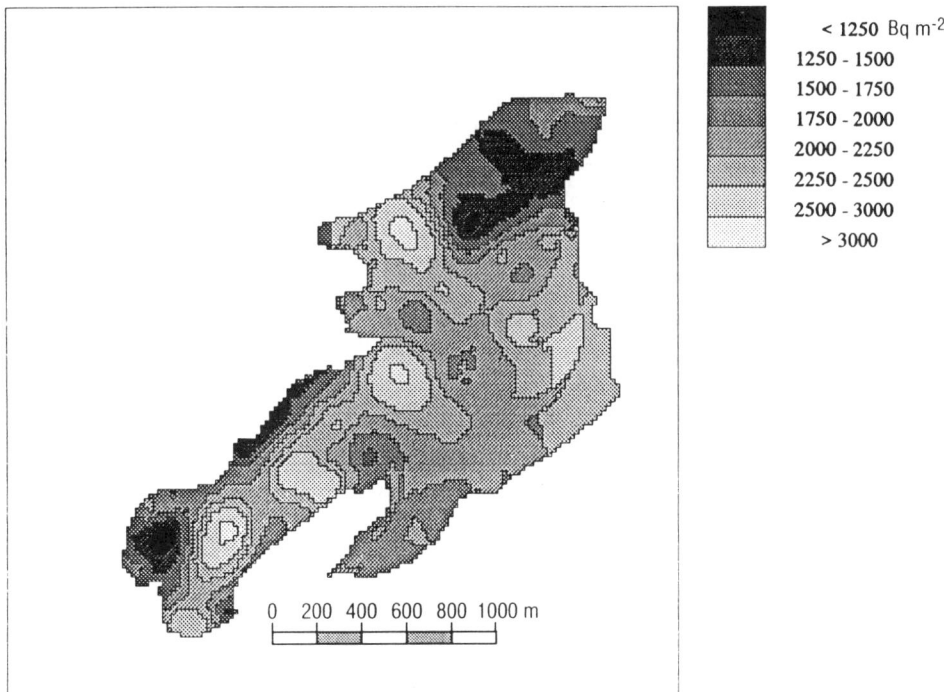

Fig. 4 ^{137}Cs distribution in the Etzenrade catchment (low values correspond with high soil erosion rates).

compare the caesium inventory for each individual sampling point with the soil erosion estimate provided by the model. The second approach is to interpolate the point caesium data to a map using the geostatistical block kriging method, and compare this map with the soil erosion maps simulated by the model. Maps showing total soil erosion between autumn 1986 and spring 1991 in the Yendacott catchment, and total soil erosion between autumn 1987 and spring 1991 in the Etzenrade catchment, produced by adding all the simulations with the modified ANSWERS model, are shown in Figs 1 and 2. The ^{137}Cs maps for both catchments are shown in Figs 3 and 4.

Both approaches were used here to compare measured ^{137}Cs data with simulated soil erosion rates. Correction of the ^{137}Cs content for ploughing and variable atmospheric input were not made. The results for the Etzenrade catchment are shown in Table 4.

In general, the correlations between simulated soil loss and observed soil loss are low for both catchments. The correlations between the individual point values of ^{137}Cs and the simulated soil loss at that location are significant for all Yendacott simulations, except the 1986 results, and significant for all Etzenrade simulations, except the 1988 results. This might be explained by the extreme rainstorm that occurred in May 1988 in Etzenrade, which influences the simulations results drastically. However, due to the large number of observations, the correlations between the individual raster cells of the ^{137}Cs maps of both catchments, interpolated using block kriging, and the raster cells of the simulated soil loss maps of the catchments are significant, although they are low.

Table 4 Spearman rank correlation of the measured ^{137}Cs in the soil and the simulated soil erosion rates in the Etzenrade catchment: total annual values.

	^{137}Cs (block)		^{137}Cs (point)	
	not crusted	crusted	not crusted	crusted
number of observations	5397	5397	134	134
Soil loss 1987-1990	0.230	0.244	0.201	0.251
Total erosion 1987	0.193	0.216	0.252	0.207
Total erosion 1988	0.200	0.207	0.140	0.150
Total erosion 1989	0.278	0.282	0.304	0.333
Total erosion 1990	0.224	0.233	0.307	0.291

0.289 = significant at 95% confidence level
0.150 = significant at 90% confidence level

CONCLUSIONS

Field and laboratory experiments are carried out in two small agricultural catchments in the loess area of the Netherlands, and one catchment in Devon (UK), to evaluate and validate the ANSWERS deterministic distributed catchment model for runoff and soil erosion simulation. The ANSWERS model, which is linked to a raster GIS to achieve a high accuracy and flexibility, is validated by comparing measured and simulated catchment runoff and sediment yield. Also, to validate the distributed model in a distributed way, the spatial soil loss estimates, measured in two catchments using the ^{137}Cs technique, have been compared statistically with the simulated soil loss maps. The ANSWERS model was modified to incorporate the variable contributing area concept to simulate saturated overland flow, using only the information of the Digital Elevation Model by combining slope gradient and upstream area. The validation results, obtained using non-calibrated data sets, indicate that the hydrographs still cannot be simulated closely in some cases, due to the difficulties to simulate soil crusting properly (Etzenrade & Catsop), the importance of subsurface lateral flow (Yendacott), and spatial variability of rainfall (Etzenrade). Correlations of simulated soil loss with observed ^{137}Cs values in the Yendacott and Etzenrade catchment are low, but statistically significant. This is not unexpected since it is difficult to simulate the hydrograph properly. It is concluded that quantitative spatial modelling of soil erosion still involves large percentage errors, especially when the model is used in areas other than those for which it was originally developed and tested. Although this type of modelling is useful, e.g. for the estimation of spatial effects of soil conservation scenarios, there is a substantial risk of gratuitous application.

Acknowledgements The author wishes to thank Prof. D.E. Walling and his co-workers of the University of Exeter for the cooperation in the Yendacott project, the Waterboard "Roer en Overmaas" for their cooperation in the Etzenrade project, and Ir. L. Eppink and his coleagues from the Agricultural University of Wageningen for their cooperation in the Catsop project. The Dutch Ministry of Agriculture, the Waterboard "Roer en Overmaas" and the Province of Limburg are thanked for financing the Catsop research. Prof. P.A. Burrough is thanked for his constructive comments on the research and the manuscript.

REFERENCES

Beasley, D.B., Huggins, L.F. & Monke, E.J. (1980) ANSWERS: A Model for Watershed Planning. *Transactions ASAE* **23**(4), 938-944.
Beasley, D.B., Huggins, L.F. & Monke, E.J. (1982) Modeling sediment yields from agricultural watersheds *J. Soil and Water Conservation* **37**(2), 113-117.
Beven, K. (1989) Changing ideas in hydrology - the case of physically-based models. *J. Hydrol.* **105**, 157-172.
Beven, K.J. & Kirkby, M.J. (1979) A physically based, variable contributing area model of basin hydrology *Hydrol. Sciences Bulletin* **24**(1), 43-69.
Burrough, P.A. (1986) Principles of Geographical Information Systems for Land Resources Assessment. Monographs on soil and resources survey no. 12, Clarendon Press, Oxford.
Burrough, P.A. (1989) Matching spatial databases and quantitative models in land resource assessment. *Soil Use and Management* **5**(1), 3-8.
De Roo, A.P.J (1991) The use of ^{137}Cs as a tracer in an erosion study in South-Limburg (The Netherlands) and the influence of Chernobyl fallout. *Hydrol. Processes* **5**, 215-227.

De Roo, A.P.J. (1993) Modelling surface runoff and soil erosion in catchments using Geographical Information Systems; Validity and applicability of the ANSWERS model in two catchments in the loess area of South Limburg (The Netherlands) and one in Devon (UK). Dissertation, University of Utrecht.

De Roo, A.P.J., Hazelhoff, L. & Burrough, P.A. (1989) Soil erosion modelling using ANSWERS and Geographical Information Systems. *Earth Surface Processes and Landforms* **14**, 517-532.

De Roo, A.P.J. & Riezebos, H.Th. (1992) Infiltration experiments on loess soils and their implications for modelling surface runoff and soil erosion. *CATENA* **19**., 221-239.

De Roo, A.P.J., Hazelhoff, L. & Heuvelink, G. (1992) The use of Monte Carlo simulations to estimate the effects of spatial variability of infiltration on the output of a distributed hydrological and erosion model. *Hydrol. Processes* **6**(2), 127-143.

De Roo, A.P.J. & Walling, D.E. (1993) Validating the ANSWERS soil erosion model using ^{137}Cs in the Yendacott catchment, Devon (UK). *Soil Technology* **6** (in press).

McCuen, R.H. & Snyder, W.M. (1975), A proposed index for comparing hydrographs. *Wat. Resour. Res.* **11**(6), 1021-1024.

Quine, T.A. & Walling, D.E. (1991) Rates of soil erosion on arable fields in Britain: quantitative data from caesium^{-137} measurements. *Soil use and management* **7**(4), 169-176.

Walling, D.E., Bradley, S.B. & Wilkinson, C.J. (1986) A Caesium-^{137} budget approach to the investigation of sediment delivery from a small agricultural drainage basin in Devon, UK. IAHS Publ. No.159, 423-435.

Walling, D.E. & Bradley, S.B. (1988) The use of caesium-^{137} measurements to investigate sediment delivery from cultivated areas in Devon, UK. In: *Sediment budgets* (Proc. of the Porto Alegre Symposium). IAHS Publ. No. 174, 325-335.

Wischmeier, W.H. & Smith, D.D. (1978) Predicting rainfall erosion losses - a guide to conservation planning. US Department of Agriculture, Agricultural Handbook No. 537, Science and Education Administration, US Department of Agriculture, Washington D.C.

Exemple d'utilisation d'un SIG pour la gestion des données d'un modèle hydrologique à mailles carrées

F. DELCLAUX & G. BOYER
Laboratoire d'Hydrologie, Institut Français de Recherche Scientifique pour le Développement en Coopération (ORSTOM), B.P. 5045, 34032 Montpellier, France

Résumé Le couplage SIG-modèle hydrologique est d'un intérêt particulier pour la gestion de données spatialisées dans la modélisation hydrologique. Nous examinons ce problème sous le double aspect pratique et méthodologique à partir d'outils existants, le modèle distribué à mailles carrées MODLAC (ORSTOM) et le SIG ILWIS (ITC). Après une description de leurs fonctionnalités, nous définissons les opérations à réaliser dans le SIG pour traiter les informations nécessaires au modèle, puis nous évaluons les capacités d'ILWIS à réaliser ces tâches. Nous définissons enfin les caractéristiques minimales d'un SIG que l'on désire interfacer avec un modèle hydrologique distribué ou global.

INTRODUCTION

Le cadre de cette étude est la connexion entre logiciels hydrologiques et géographiques existants et non conçus l'un en fonction de l'autre. Nous avons utilisé le modèle hydrologique distribué MODLAC (Girard, 1992) ce modèle, basé sur une discrétisation de l'espace en mailles carrées, nécessite une quantité importante d'informations géographiques. En ce qui concerne le SIG, nous avons utilisé le logiciel ILWIS (ITC, 1992) mis au point par ITC (Pays Bas).

Cependant, au delà du travail particulier du couplage d'un SIG avec un modèle hydrologique, nous avons essayé de répondre à une question plus générale et plus pragmatique : quel SIG pour quel modèle ?

DESCRIPTION DU MODELE HYDROLOGIQUE MODLAC

MODLAC est un modèle hydrologique conceptuel basé sur une discrétisation du bassin versant en mailles carrées et fonctionnant au pas de temps journalier. Ses objectifs sont la mise à disposition d'un outil mathématique permettant de simuler divers scénarios d'aménagement.

Le conceptuel discrétisé

Le mode conceptuel repose sur deux fonctions qui simulent le fonctionnement du bassin versant :
(a) la fonction de production qui évalue la partie de la pluie qui contribue à l'écoulement et qui répartit la lame d'eau écoulée entre les différentes formes de l'écoulement;
(b) la fonction de transfert qui transforme les lames d'eau produites en

débits écoulés à l'exutoire du bassin.

La discrétisation permet de diviser le bassin en surfaces élémentaires homogènes sur chacune desquelles ce schéma conceptuel est appliqué.

Les données géographiques

Le maillage Le bassin est inscrit dans un premier maillage régulier (quadrillage primaire) qui permet une première approche de la discrétisation et un calage aisé des photographies aériennes ou images satellite. Ce quadrillage initial étant trop grossier pour représenter correctement la complexité spatiale du bassin, il est découpé en mailles plus fines, les mailles secondaires, qui permettent de respecter les caractéristiques de la zone étudiée : superficie, localisation des postes pluviomètriques et des aménagements hydrauliques, informations physiographiques.

Caractéristiques et attributs des mailles Des attributs sont définis pour chaque maille : physiographie, altitude minimale, sens de drainage, type de fonction de production, nature de la maille (une maille est déclarée "rivière" si elle contient un poste hydrométrique ou un aménagement hydraulique, ou si elle est comprise entre deux mailles "rivière". Sinon elle est dite maille "bassin").

Les données hydrologiques

Les données hydrologiques sont classées en deux catégories, les séries chronologiques (pluie, débit, hauteur d'eau, ETP) et les données des aménagements.

Les séries chronologiques Les données de pluie et d'ETP sont les seules données chronologiques fournies en entrée. On attribue à chaque station sélectionnée une aire d'influence. Les séries de hauteurs d'eau dans les retenues et de débits aux stations hydrométriques sont utilisées pour le calage du modèle. Le pas de temps est journalier.

Les données des aménagements Chaque aménagement hydraulique est localisé sur une maille et possède des caractéristiques propres. Pour une retenue, nous aurons par exemple : le niveau d'eau, la relation hauteur/surface, la date de mise en service, les débits dérivés, etc.

Les paramètres du modèle

Ce sont les paramètres des fonctions de production qui optimisent la restitution des données. L'opération de calage nécessite une bonne connaissance du fonctionnement mathématique et hydrologique du modèle car la procédure est manuelle, de type essai/erreur. Ces paramètres sont classés en deux groupes :

Tableau 1 Les paramètres du modèle MODLAC - 1er groupe.

Stockage dans le sol par classe physiographique
CRT : niveau correspondant à la capacité moyenne de rétention dans le sol. Une diminution de sa valeur conduit à une augmentation de la lame d'eau écoulée;
DCRT : capacité minimale de rétention dans le sol. Une augmentation de sa valeur conduit à une augmentation des forts débits (pics) sans trop faire varier le volume global;
RI : quantité d'eau initiale dans le sol : une augmentation de sa valeur conduit à une légère augmentation du volume écoulé total et à une forte augmentation en début de période.

Tableau 2 Les paramètres du modèle MODLAC - 2ème groupe.

Réservoir conceptuel par classe physiographique
FN : valeur maximale possible de l'infiltration : sa variation doit aller en parallèle avec celle de CQI; une augmentation de sa valeur implique une répartition plus importante de l'écoulement souterrain dans l'écoulement global;

Ecoulement superficiel par classe physiographique
QRMAX : niveau maximal du réservoir superficiel : une augmentation de sa valeur conduit à une diminution du nombre de grandes crues;
CQR : coefficient de vidange du réservoir superficiel : une diminution de sa valeur conduit à une diminution de l'écoulement dans la période de crue;

Infiltration par classe physiographique
QIMAX : niveau maximal du réservoir souterrain : une augmentation de sa valeur conduit à une augmentation du volume écoulé dans la période d'étiage;
CQI : coefficient de vidange du réservoir souterrain : une augmentation de sa valeur conduit à une augmentation rapide des débits de base dans la période d'étiage.

les paramètres intervenant dans l'ajustement du volume écoulé global et les paramètres intervenant dans la forme de l'hydrogramme. (cf. Tableaux 1 et 2.)

DESCRIPTION DU SIG ILWIS

Principes, fonctionnalités

ILWIS, mis au point par ITC (Pays Bas), est un SIG qui fonctionne dans un environnement informatique de type compatible PC.

La base ILWIS gère les données vecteur, raster et table. On peut à tout moment importer de nouvelles cartes, modifier ou détruire une couche. ILWIS permet à chaque type d'information d'être géré indépendamment, de mettre directement en relation les données géographiques avec les attributs. ILWIS dissocie pour une même couche les informations descriptives, les informations géographiques (stockées dans des tables) et les données purement graphiques. Les attributs sont gérés par un gestionnaire de base de données relationnelle (pseudo SQL).

Les traitements ILWIS inclut les traitements nécessaires à l'analyse spatiale. On peut appliquer toute forme de combinaison sur les couches d'information : arithmétique, logique, relationnelle et mathématique. Il possède également des fonctions de traitement d'image (histogramme, fonctions de transfert, filtrage, classification). Sont également inclus des modules de géoréférençage et de calculs statistiques. Enfin l'utilisation d'ILWIS peut être entièrement automatisée par un langage de commande (fichiers "batch").

Les entrées/sorties ILWIS posséde un module de digitalisation de cartes et d'import/export pour les données vectorielles, raster et table : format .DXF, .ARC pour les couches vectorielles; TIFF, GIS, BMP en raster; .DBF, .DIF pour les tables.

L'INTERFACAGE MODLAC/ILWIS

Méthodologie

L'utilisation d'un SIG a pour but de faciliter la mise en oeuvre du modéle dans l'acquisition et la préparation des données géographiques. Le problème se pose de savoir comment va s'organiser le flux des données entre les deux outils et quelle sera la nature des traitements à effectuer. Notre travail s'effectuera en deux temps :
(a) description des fonctions nécessaires au SIG pour opérer sur les informations MODLAC;
(b) évaluation du potentiel d'ILWIS à résoudre les problémes posés.

Définition des liaisons MODLAC/SIG

Caractérisation des mailles La complexité et le savoir-faire hydrologique nécessaires à la génération du maillage ne permettent pas d'imaginer un SIG possédant cette fonction. Nous supposerons donc que le maillage a été généré à l'extérieur du SIG et mis sous forme d'un fichier vecteur polygone où chaque maille est référencée par un numéro d'ordre. De manière classique, le modélisateur superpose manuellement la représentation du maillage à la carte thématique et en déduit les valeurs à attribuer à chaque maille. Un SIG effectuera cette discrétisation automatiquement par des opérations de croisement de couches: le plan de base est constitué par le maillage, les autres plans contenant les informations thématiques. La couche résultante sera constituée par les mailles caractérisées par les données projetées. Le maillage ainsi défini est exporté sous forme de tables pour mettre à jour le fichier géométrie du modèle (voir Figure 1):
(a) attribut "type de fonction de production": on combine les plans "occupation du sol" et "géologie". La couche résultante est classifiée et

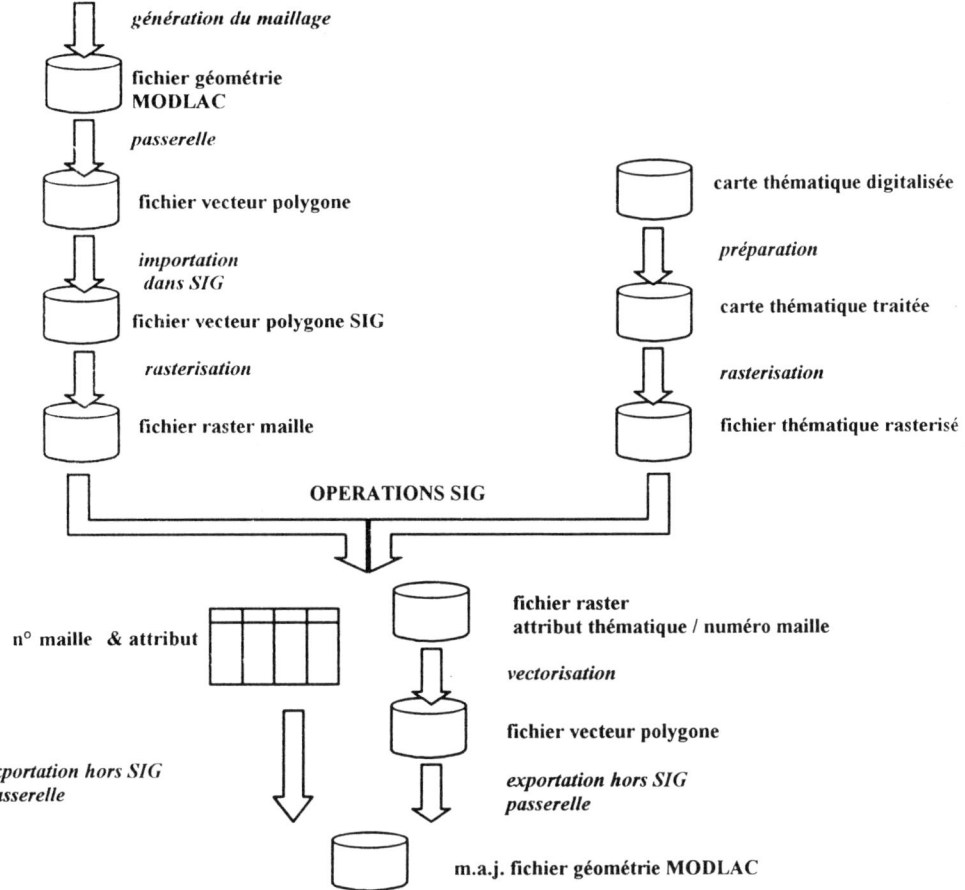

Fig. 1 Caractérisation du maillage - Procedure génerale.

projetée sur la couche maillage. La couche finale est alors constituée de mailles dont l'attribut est l'index correspondant au type de fonction de production La détermination des paramétres des fonctions de production sera examinée ultérieurement;

(b) attribut "altitude minimale": le SIG utilisera un MNT, soit généré à partir des courbes hypsométriques importées, soit calculé par un logiciel externe. Le MNT est projeté sur le maillage, puis, à l'intérieur de chaque maille, on retient l'altitude minimale;

(c) attribut "sens de drainage": l'utilisation d'un MNT est ici évidente. Néanmoins, la détermination du sens de drainage est complexe car MODLAC gére un sens de drainage par maille et non par pixel. Une premiére solution consiste à utiliser un MNT importé. Il faut regrouper les pixels appartenant à une même maille, puis rechercher la maille contigue d'altitude minimale. Le résultat final sera une table des indices de mailles associés aux mailles drainantes. Cette méthode n'est pas satisfaisante car il n'y a pas de procédure de contrôle par l'utilisateur sur les incohèrences topographiques. Une autre solution consiste à préparer une couche "réseau de drainage" à l'aide d'un logiciel adapté et possédant ses propres outils de vérification et de correction, puis à faire subir à cette couche les opérations décrites plus haut;

(d) attribut "type de maille": la recherche des mailles "rivière" s'effectue en parcourant le réseau hydrographique de l'exutoire jusqu'à la dernière retenue. C'est cette portion de réseau définissant les mailles "rivière" qui sera projetée sur le maillage: les mailles correspondantes seront déclarées et codées en "rivière".

Les données de pluie et d'ETP La gestion des données de pluie et ETP a deux aspects : les séries chronologiques par station (tables) et les aires d'influence des stations (données géographiques).

(a) La gestion des séries chronologiques par le SIG nous semble peu réaliste : (1) la simulation du fonctionnement d'un bassin versant sur une dizaine d'années conduirait le SIG à gérer des tables de plusieurs milliers de lignes par station, et (2) ces données sont généralement gérées par d'autres logiciels, ce qui multiplierait le nombre de fichiers correspondants. Nous choisirons donc une gestion externe des données

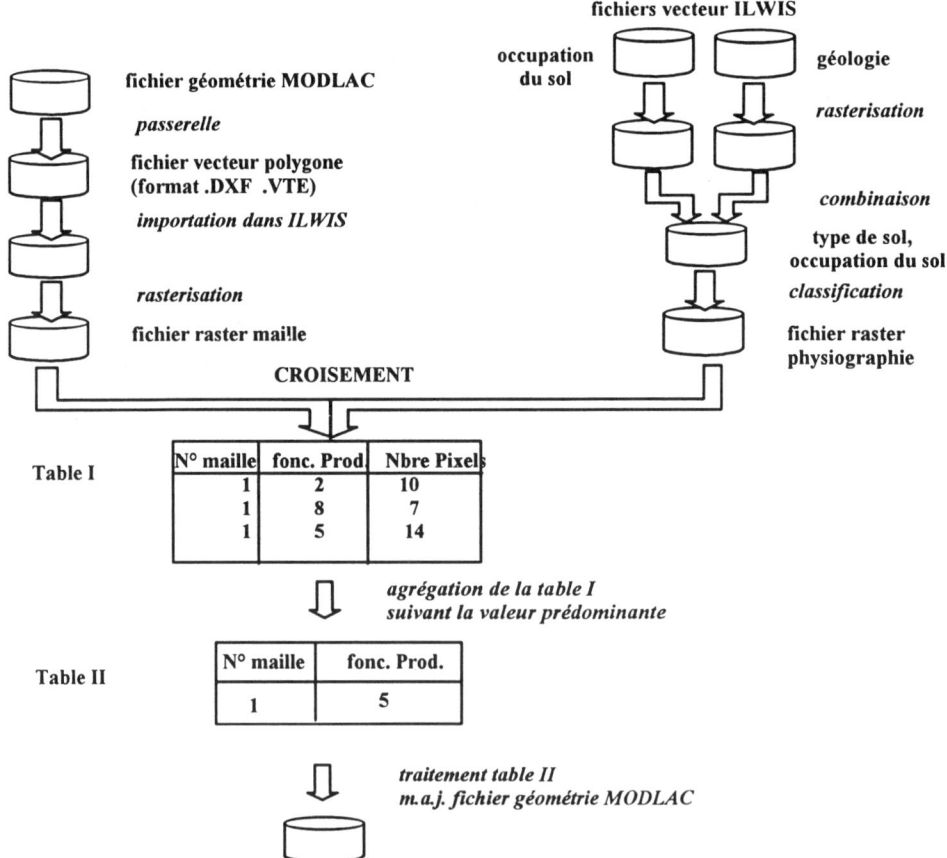

Fig. 2 Détermination indice fonction de production par maille.

de pluie par des outils spécialisés, tel le logiciel PLUVIOM (ORSTOM).
(b) Le calcul des aires d'influence et le croisement avec le maillage est par contre typiquement une opération SIG. Celui-ci devra être équipé d'une fonction automatique de délimitation d'aires (méthode de Thiessen par exemple). Sinon, il sera nécessaire de calculer ou de digitaliser cette information et de l'importer dans le SIG. La projection de la couche "aires d'influence" sur la couche maillage attribuera à chaque maille le descriptif de la station qui la caractérise.

Les données de débit et hauteur d'eau Les données de débit n'ont pas besoin de spatialisation. La gestion des données de débit est laissée, elle aussi, à un outil spécialisé (HYDROM ORSTOM).

Les données des aménagements Le SIG peut être utilisé pour déterminer plus facilement la localisation des retenues, le pourcentage d'occupation de la maille par la retenue, les surfaces irriguées, etc.

Les paramètres du modèle Les paramètres dépendent des classes physiographiques précédemment définies. Ces paramètres peuvent être estimés par une relation entre les thèmes "occupation du sol", "géologie" ou d'une autre variable caractéristique de l'état du bassin versant : l'utilisation du SIG sous cet aspect facilitera la détermination systématique de ces paramètres et leur optimisation.

ILWIS, le potentiel

Nous présentons les procédures à utiliser sous forme d'organigrammes. La majorité des opérations décrites dans ces organigrammes ont été réalisées dans ILWIS. Par contre, les manipulations externes à ILWIS n'ont pas été programmées et sont à l'état de projet. La Figure 2 permet de générer la table n° maille <=> indice fonction de production. Le problème principal est hydrologique : comment combiner les informations occupation du sol et géologie pour définir une valeur de la physiographie. La Figure 3 montre le cheminement effectué pour déterminer une altitude minimale par maille. Le résultat final est une table n° maille <=> altitude minimale. La Figure 4 indique comment générer la table n° maille <=> n° station. Nous avons ici utilisé un module d'ILWIS de calcul d'aire suivant la méthode de Thiessen.

Aucun diagramme satisfaisant n'a pu être réalisé pour l'affectation d'un sens de drainage aux mailles. Cette question pourrait être résolue à l'échelle du pixel d'un MNT, entre autres grâce au logiciel DEMIURGE (ORSTOM). Mais MODLAC utilise des mailles de taille variable (1 à 32 pixels) : il faut donc travailler sur des ensembles de pixels. De plus, nous ne disposons dans ILWIS d'aucun moyen de contrôle interactif lors de la constitution de ce réseau : nous ne sommes donc pas à l'abri des incohérences hydro-topographiques parfois contenues dans un MNT.

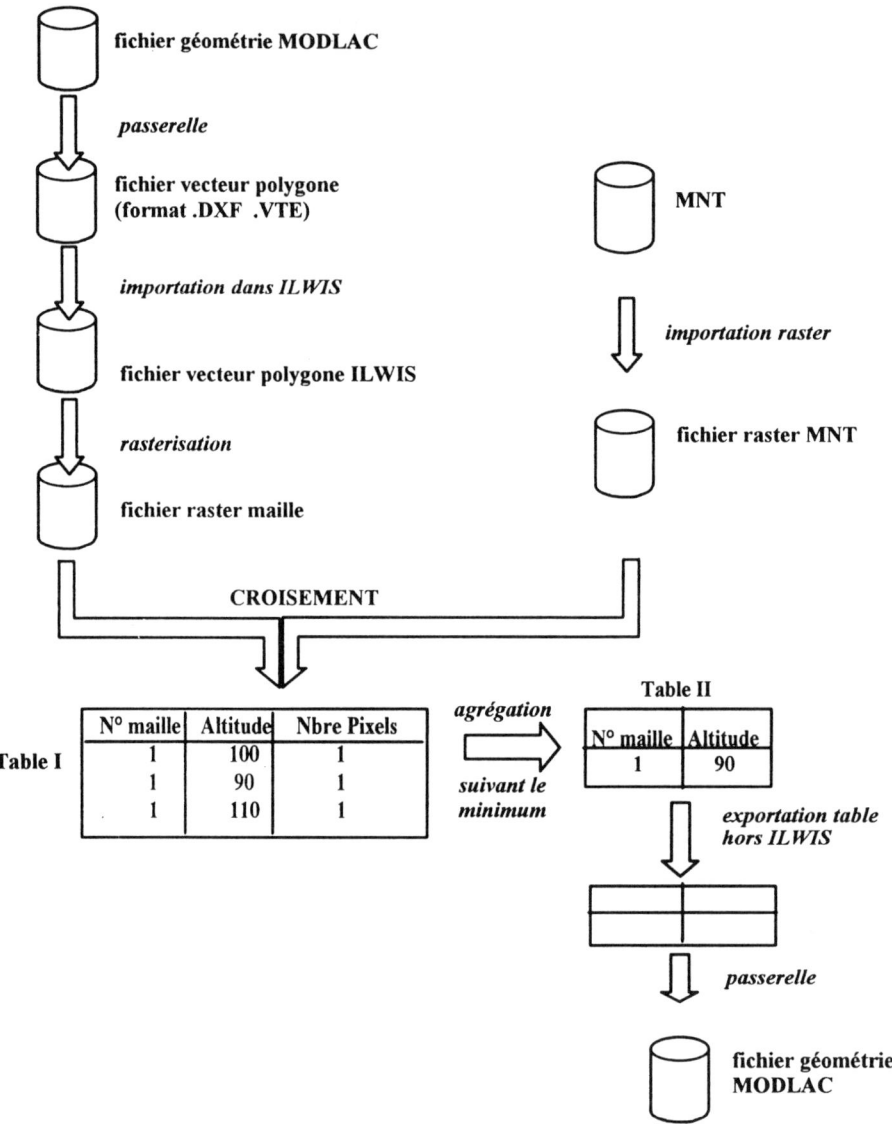

Fig. 3 Détermination de l'altitude minimale par maille.

CONCLUSIONS

Adéquation MODLAC/ILWIS

Certaines contraintes imposées par MODLAC dépassent les possibilités d'ILWIS (génération du maillage, sens de drainage). Ainsi, il sera difficile de mettre en oeuvre ce genre de procédure. En ce qui concerne les autres opérations, ILWIS répond aux problèmes de discrétisation des données géographiques sur un maillage de type MODLAC. Le travail est à compléter

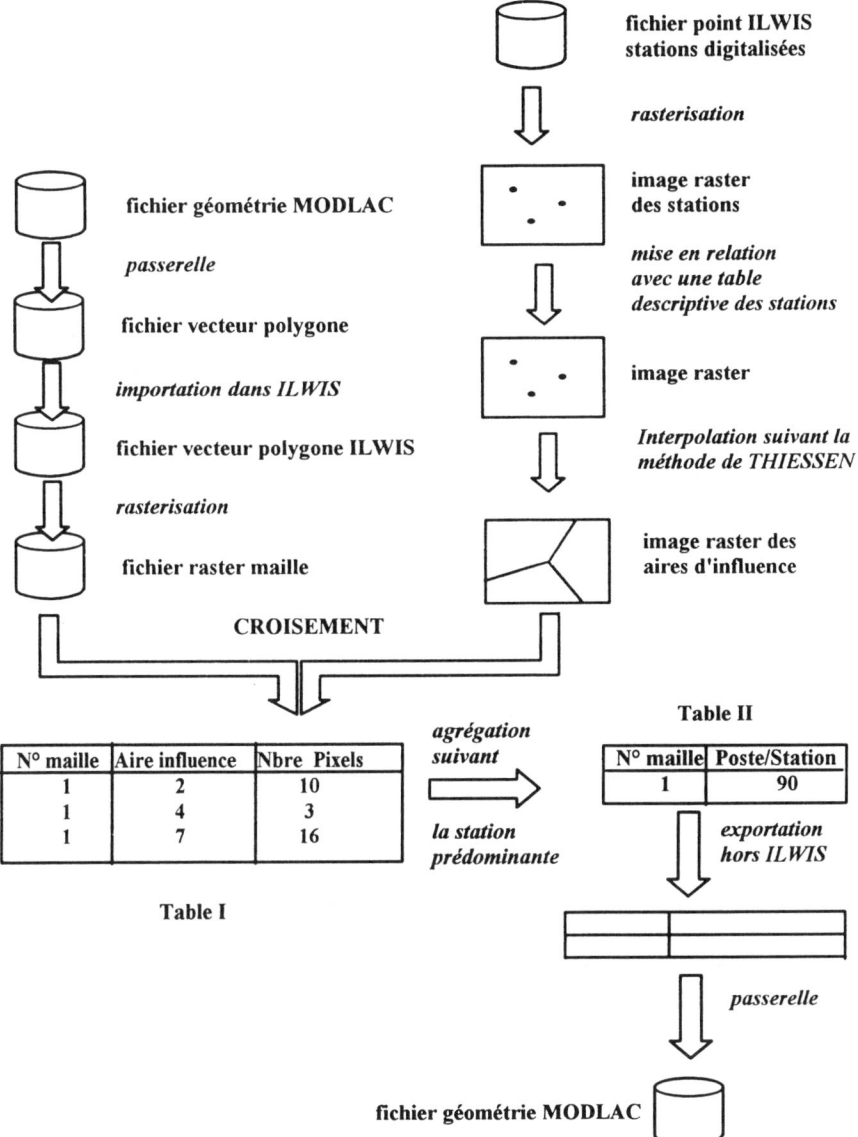

Fig. 4 Détermination des aires d'influence de stations.

par le développement des outils nécessaires pour assurer les passerelles entre formats de fichiers.

Quel SIG...

Un SIG utilisé dans une optique de couplage avec un modèle hydrologique doit pouvoir offrir les fonctionnalités suivantes :
(a)   capacité à travailler en mode raster et vecteur et à pouvoir passer d'un mode à l'autre pour faciliter l'intégration d'informations pixel (image

(b) existence de fonctions d'analyse spatiale élargies Il faut pouvoir élaborer de nouvelles informations par des opérations mathématiques, logiques ou relationnelles sur les divers plans thématiques disponibles;
(c) ouverture aux structures de données standard L'utilisation de données d'origine très diverse, la liaison avec les logiciels existants (tableurs, outils cartographiques) et avec les périphériques facilitent le travail d'interfaçage entre produits existants;
(d) possibilité de travailler en mode "batch" L'automatisation de tâches de calcul et du déroulement de procédures permet : (1) de multiplier les traitements et de faciliter des opération de tests et de calage; (2) de répéter les mêmes opérations sur des sites différents, pour définir des liens entre bassins versants dits représentatifs ou dans la perspective de tâches opérationnelles;
(e) existence de fonctions dédiées à l'environnement Des procédures telles le traitement de MNT, la gestion de réseau ou l'interpolation spatiale élargissent le potentiel du modèle hydrologique associé.

... pour quel modèle ?

Le choix du modèle dépend principalement de la problématique hydrologique, de la nature du site étudié et du type et du volume de données à traiter.

L'intérêt d'un SIG dans le cas de modèles distribués est évident : la saisie et la gestion de données géographiques, l'élaboration de nouvelles informations spatiales, les possibilités de cartographie automatique et de restitution sont autant de facteurs permettant une optimisation dans la mise en oeuvre de ce type de modèle.

Le cas des modèles globaux est différent. Dans cette approche le bassin versant est considéré comme une entité unique, il y a peu d'informations spatiales à traiter, les paramètres du modèle ont un sens physique difficile à définir, et il n'est pas certain que le travail d'acquisition des données représente réellement un gain. On peut néanmoins envisager le couplage SIG-modèle global dans l'étude de la représentativité de bassins versants : la recherche systématique de corrélations entre le comportement des bassins versants et leurs caractéristiques géographiques peut être grandement facilitée par le SIG.

REFERENCES

Girard, G. (1992) *Manuel d'utilisation et d'exploitation du modelè MODLAC.* Institut Français de Recherche Scientifique pour le Développement en Coopération (ORSTOM), Rapport interne

ITC (1992) *Integrated Land and Water Information System, ILWIS, User's Manual, Version 1.3.* Computer Department ITC, May 1992. International Institute for Aerospace and Earth Sciences, Enschede, The Netherlands.

Mapping and management of flood plains

P. FARISSIER & P. GIVONE
CEMAGREF (French Institute of Agricultural and Environmental Engineering Research), Hydrology and Hydraulics Division, 3 bis, quai Chauveau, CP 220, 69336 Lyon Cedex 09, France

Abstract The different topographical representations needed to map the flooded areas, make complex the use of a regular GIS This representation can be unified thanks to the identification of common topographical objects of hydraulic type. This analysis allows us to integrate in a simple and direct way the functions of the GIS needed by our hydraulic and cartographic models.

INTRODUCTION

Hydrology and hydraulics division proposes a new flood plain management method based on the comparison of hydraulic information (the area flooded by a flow with a T1 return period) with land use information (land use information are transformed into protection needs expressed by a T2 return period) (Gautier *et al.*, 1992). The use and crossing of distinct information layers - of hydraulic, topographical and geographical type - require the use of a GIS (Geographical Information System).

However, in order to avoid the addition of a GIS, which is a complex and heavy tool, dealing with a large amount of data, to other essential software such as the topographical and hydraulic numerical models, we consider that a global model having all the needed functions including those typical of the GIS, should be developed.

USING A GIS

A GIS models the real objects through points, arcs, polygons and their combinations. The hydraulic or topographical objects we are dealing with (flood plains, plots of land with specific land use, minor and major river beds, etc.) seem to be well suited to this type of polygonal modelling. But the GIS basic objects data acquisition is, in fact, "not natural" and not well embedded into the general processing used in hydraulics and cartography.

On the contrary, we have to deal with several different independent types of representation, and therefore different geometric objects (cross sections, irregular network, polygons, etc.) for river topography processing, hydraulic modelling and flood plains as well as land use mapping. So, the development of complex interfaces between these different models and an increase of the number of the numerical and topographical databases is required, although it is possible to choose a single representation and modelling method for all the processed objects.

We did not develop several interfaces between a "classical" GIS (ARC/INFO, for example) and our topographical and hydraulics tools, but we tried to rebuild each one of these processing steps "in the way" of a GIS So we used a single common basic object, with all the needed attributes. This basic object conveys the constraints and needs of all the types of processing. This will be translated into the building of a single topographical database, which will be used by all the software, including the hydraulic models.

TOPOGRAPHICAL MODELLING

The minor and major river beds or thalweg modelling is based on the acquisition of different basic objects depending on the fact that we want to display a topographical map (a set of irregular points is sufficient) to run a 1D hydraulic model (cross sections are required), to run a 2D hydraulic model (a network is needed) or to map the land use (from polygons identifying the homogeneous areas, lines modelling the different networks: roads, electricity, water, etc.).

We want to retain only one representation method for each kind of basic geometric object. We propose to use a network based on pseudo-rectangular patches of variable size, computed from the cross sections of the river, or from these cross sections (Farissier, 1992) and a set of scattered points (in the process of tests).

This generalized network is built in three stages:
(a) operating the cross sections of the valley, which remain the preference acquisition method of a natural geometry to run an hydraulic model. Noteworthy points and directions, such as the minor and major bed axes, the limits of these two beds (bank of the minor bed, limit of the major bed, Fig. 1a) and the bottom of the minor bed, are automatically identified. A set of scattered points, structured by guiding lines, translating the topographical and hydraulic constraints, is obtained;
(b) processing of this set of points in order to build the pseudo-rectangular patches, leaned, on the one hand, on a spatial generalization of the guiding lines (according to the axis of the river, Fig. 1b), and on the other hand, on the cross sections (original and interpolated sections, Fig. 1c). The result is a generalized 3D network (Fig. 1d);
(c) assignment of several attributes to each mesh of the obtained network, in order to constitute the necessary additional information layers. According to the type of information to process, it is possible to modify the size of the meshes and their number, in order to really adapt the network to all the constraints.

This network becomes our single Topographical DataBase from which we can equally select cross sections for the 1D models, as well as a network for the 2D or 3D models. A cross section at a given abscissa, is obtained as the intersection of the network with the corresponding vertical plane.

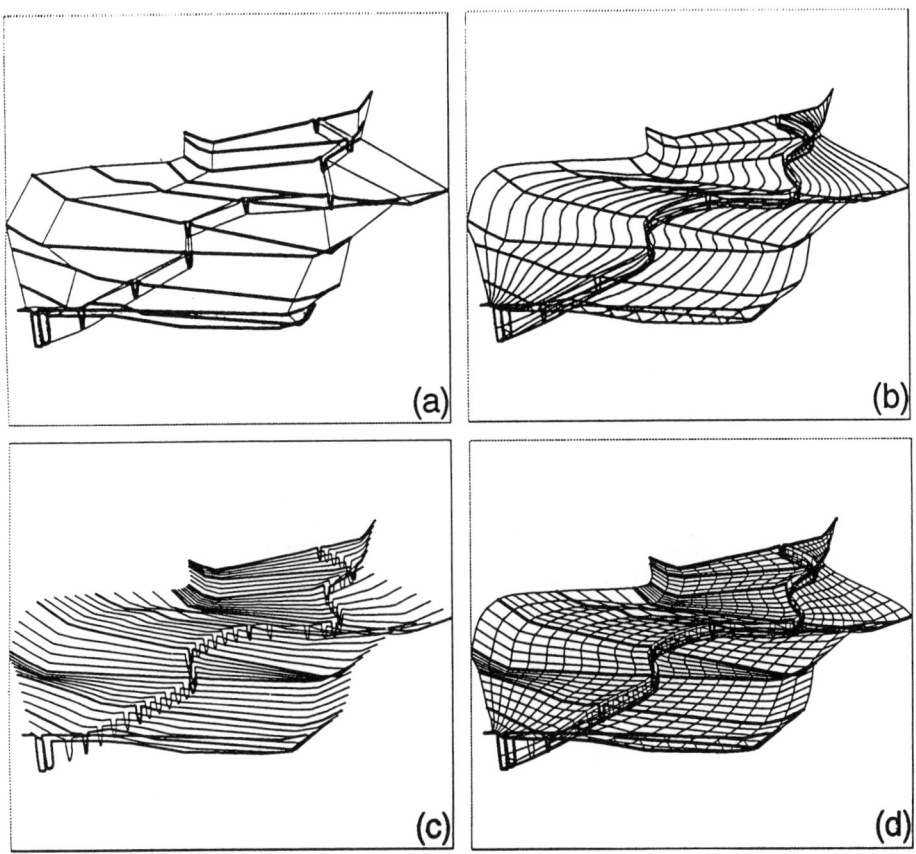

Fig. 1 (**a**) Original cross sections and guiding lines, (**b**) spatial generalization of the guiding lines, (**c**) interpolated cross sections, (**d**) obtained network.

HYDRAULIC MODELLING

The 1D and 2D hydraulic models computes water depths using different geometrical representations, but all based on the same patches. The hydraulic constraints determine the number and the size of the patches to have a good representation of the river geometry:

(a) case of 2D models: The maximum water level, directly obtained from the hydraulic model, is assigned to each patch. The limits of the flooded area is determined as the intersection of the surface representing the bottom of the valley (the network) with the free surface (the set of the water levels, Fig. 2a);

(b) case of 1D models: The model computes a single water level in each cross section. The correspondance between the cross sections and the patches enables us to assign a water level to each patch. Now, it comes down to the previous case.

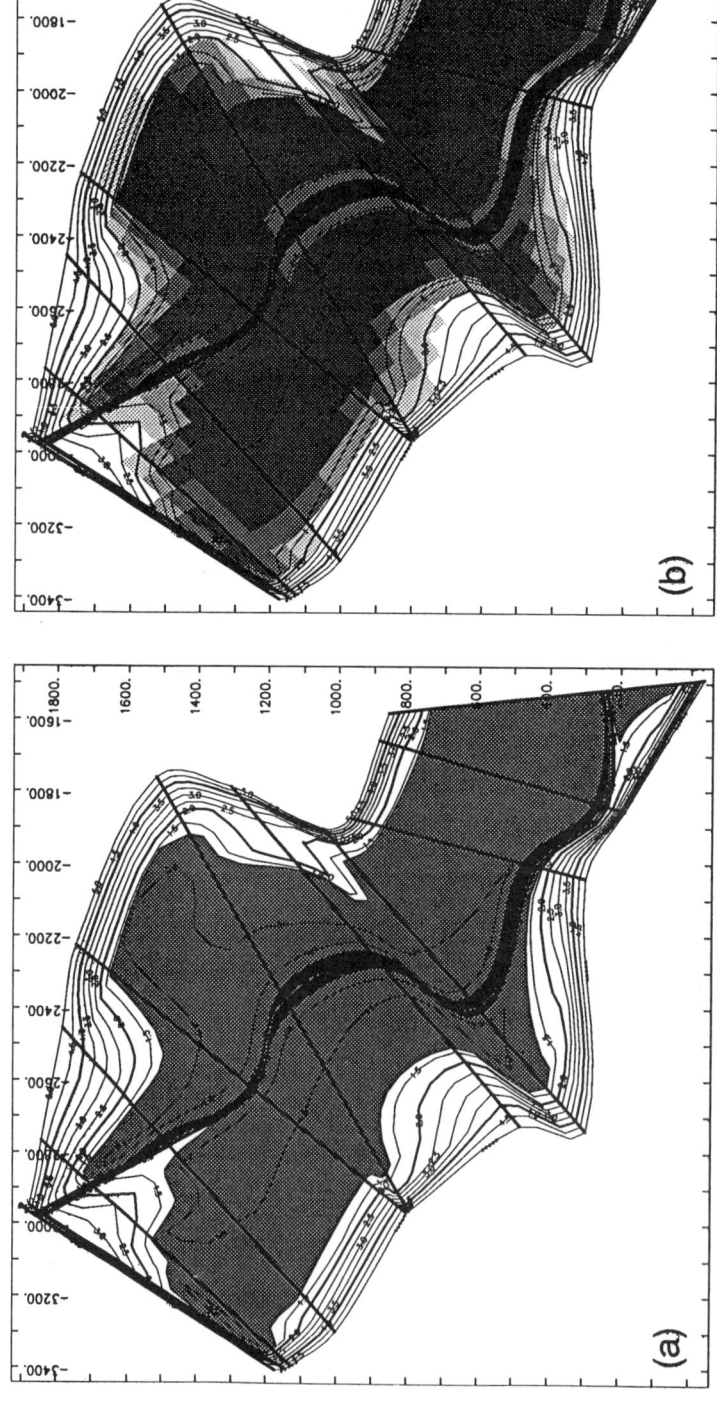

Fig. 2 (a) Flooded area map, (b) flooding risk map.

This method enables a direct comparison, for each mesh, of the water level (maximum flooding) and an information like the land use (the vulnerability toward the flooding risk).

GEOGRAPHIC INFORMATION

The land use is finally modelled as one or several attributes assigned to each patch. To achieve it, for each considered criterion (the flooding risk vulnerability criterion in this case) homogeneous areas are identified from usual topographical or geographical maps. Then these areas are mapped as the corresponding part of the network, and the criteria translated into the attributes assigned to each mesh.

The comparison between the vulnerability toward the flooding risk (or any other criterion related to the land use) and the hydraulic information is automatically processed by the comparison of the corresponding attributes at each mesh (Fig 2b). Displaying the results means to colour some patches in accordance to a pre-defined convention (Chastan et al., 1992). In this way, we are supposed to separate the areas whose protection needs are, or are not, fulfilled.

The usual computation of flooded surfaces, water volumes, etc. ,which are classically reserved to the GIS, are made easier thanks to our network.

CONCLUSION

The GIS concepts and functions are well enough in agreement with the map-making process of the flood plain management, but the current GIS implementations requires the development of complex interfaces which have to take into account the topographical and hydraulic tools used in this field. A better solution is to identify and to define a single type of basic geometrical object from the set of geometrical, geographical and hydraulic constraints. In this way, a single topographical database is built, on witch all main functions of GIS are developed. These functions are integrated into the different cartographic and hydraulic modules, used to simulate the river flows.

REFERENCES

Chastan, B., Farissier, P., Gautier, J.N. & Givone, P. (1992) INONDABILITE: Une méthode globale pour la gestion rationnelle des zones inondables. *Proc. of INTERPRAEVENT 1992, 29 June-3 July*, Berne, Suisse, Vol. 3, 179-190.

Farissier, P. (1992) Définition d'un modèle cartographique spécifique pour les lits de rivière et les modèles hydrauliques. *Ph.D thesis, to be published, University of Lyon, France*.

Gautier, J.N., Farissier, P. & Gilard, O. (1992) INONDABILITE: A tool for the management of flood plains by fitting the flood risks to needs of protections. *Proc. Conf. HYDROCOMP'92*, 25-29 May, Budapest, Hungary, 375-382.

Application of GIS in modelling winter orographic precipitation, Gunnison River Basin, Colorado, USA

L. E. HAY
US Geological Survey, Water Resources Division, MS 412, Denver Federal Center, Denver, CO 80225, USA

W. A. BATTAGLIN
US Geological Survey, Water Resources Division, MS 406, Denver Federal Center, Denver, CO 80225, USA

M. D. BRANSON
Atmospheric Science Department, Colorado State University, Fort Collins, CO 80523, USA

G. H. LEAVESLEY
US Geological Survey, Water Resources Division, MS 412, Denver Federal Center, Denver, CO 80225, USA

Abstract Accurate prediction of the spatial and temporal distribution of winter precipitation is important in the assessment of the effects of climate change on water resources within the Gunnison River basin of Colorado. The Gunnison River basin is a mountainous basin where reliable precipitation data are sparse and restricted to lower elevations. In this study, the Rhea-Colorado State University (RHEA-CSU) orographic precipitation model is linked with a geographic information system (GIS) to produce a flexible model that predicts winter precipitation in mountainous regions over a range of spatial scales. The RHEA-CSU model was used to simulate 30 years of winter precipitation in the Gunnison River basin using 10-, 5-, and 2.5-km elevation grids (grid cells of 100, 25 and 6.25 km^2, respectively) generated using a GIS, to determine the effects of changes in grid-cell size on model results. The RHEA-CSU model output at the three spatial resolutions was visualized and compared through use of a GIS to assess the effects of changes in grid-cell size on model results.

INTRODUCTION

The Gunnison River basin, located in southwestern Colorado (Fig. 1), is an important source of water for the Colorado River system, providing over 40% of the Colorado River streamflow at the Colorado-Utah state line (Ugland *et al.*, 1990). The basin has a drainage area of 20 770 km^2, and elevations that range from 1400 to 4400 m. The effects of climatic change on water resources in the Gunnison River basin are currently being studied by using hydrological models. Simulating the spatial and temporal distributions of precipitation is an important component of this effort. The overlapping data requirements for precipitation and hydrological modelling applications in the Gunnison River basin is managed through the use of geographic information system (GIS) technology (Hay *et al.*, 1992).

The spatial distribution of precipitation in mountainous areas, such as the

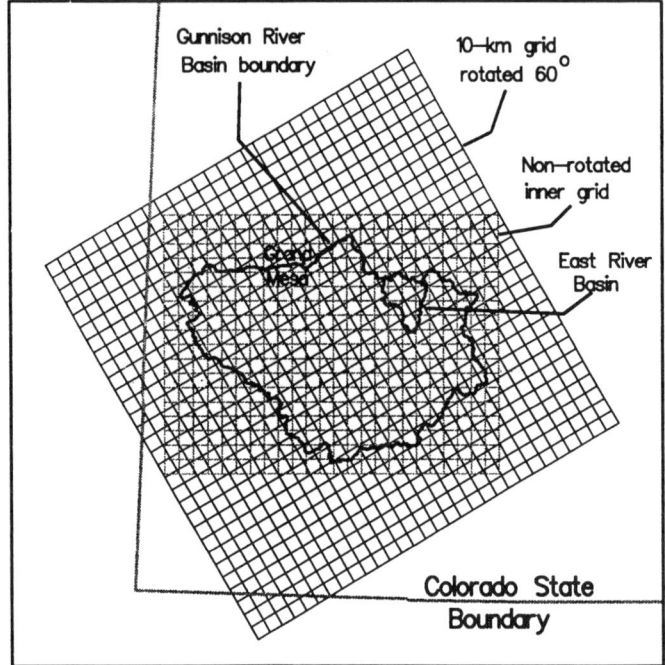

Fig. 1 RHEA-CSU rotated and non-rotated model grid.

Gunnison River basin, is difficult to determine because orographic effects make the precipitation distribution highly variable, and the recording of precipitation occurs primarily in the lower elevations near population centers. Snow is the major source of available water in the Gunnison River basin, with seasonal accumulation and storage located above 2800 m. However, no long-term precipitation station exists above this elevation. Defining the distribution of precipitation is a critical element in calibrating a watershed model within the basin (Kuhn & Parker, 1992). Evaluation of hydrological response to climate variability and change depends on accurate estimates of winter snowpack at these higher elevations.

An important question regarding the use of the RHEA-CSU precipitation model is what grid-cell size should be used to generate reliable precipitation values. In this paper, the Rhea-Colorado State University (RHEA-CSU) orographic precipitation model was used to simulate 30 years of daily winter precipitation (October-April, 1961-1991), at three spatial resolutions: 2.5-, 5-, and 10-km grids (grid cells of 6.25, 25 and 100 km^2 respectively). Precipitation output at the three spatial resolutions were analyzed using GIS technology to assess the effects of a change in grid-cell size on model results.

RHEA-CSU OROGRAPHIC-PRECIPITATION MODEL

The RHEA-CSU orographic-precipitation model was developed by Owen Rhea in the late 1970's (Rhea, 1977). The model is steady-state, multi-layer, and 2-dimensional; one dimension is along the prevailing 700 millibar wind direction and the other is vertical. The model simulates the interaction of air layers with the underlying topography by allowing vertical displacement of the air column while keeping track of resulting condensate or evaporation. The model requires as input (1) twice daily soundings from nearby or surrounding upper-air stations and (2) gridded elevation data. The soundings provide measurements of atmospheric pressure, temperature, relative humidity, wind direction, and wind speed at various altitudes. Nine elevation grids are generated using the topographic data, one for every ten degrees of rotation from 0-80 degrees; grids at complementary and supplementary angles are derived from these nine. An elevation grid is selected for each 12 hour sounding that corresponds to the 700 millibar wind direction at the center of the study area, rounded to the nearest 10 degrees. Upper-air data for each 50 millibar interval between 850-350 millibars are interpolated to the upwind edge of the selected elevation grid. Precipitation estimates are calculated at each point of the selected rotated grid and interpolated to the non-rotated inner grid using the inverse-distance squared from the four nearest grid points. The interpolated inner grid is for the model output and covers an area common within all of the rotated grids. Figure 1 shows the 60° rotated grid (wind directions of 60°, 150°, 240°, and 330°) and the non-rotated inner grid.

The RHEA-CSU model was linked with a GIS. Linking the model with a GIS automates development of elevation grids from Digital Elevation Model (DEM) point values making it possible to simulate precipitation over a range of spatial scales, and enables the user to choose the method of topography characterization (i.e., mean, maximum, or minium elevation (Hay *et al.* 1992)). In Rhea's original version of the model, mean elevations were characterized manually for each grid cell for every 10 degree change in wind direction using 2.5-km grid cells. From each rotated elevation grid, average elevation was computed for both a 5-km and 10-km grid interval by averaging the 9 or 25 surrounding 2.5-km points, respectively. However, only the 10-km grids were used in the original study, since it was found that the use of the 5-km grids quadrupled the running time of the model (Rhea, 1977).

In this paper, 30 years of winter season precipitation (October -April, 1961-1991) were simulated using the RHEA-CSU model using 10-, 5-, and 2.5-km mean-elevation grids. A GIS was used to calculate mean elevation values for each grid cell at each grid scale, by overlaying the nine rotated grids on DEM point elevation values spaced at every 900 m (Hay *et al.*, 1992). The parameterizations of the physical process within the RHEA-CSU model were kept independent of grid-cell size in order to compare the output from the three spatial resolutions. Output from the RHEA-CSU model was visualized and compared through use of a GIS to assess the effects of changes in grid-cell size on model results.

PREVIOUS INVESTIGATIONS

In past studies, the RHEA-CSU model was tested by comparing seasonal sums from the model versus three measured values: snowpack water equivalent at selected points, and spring and summer streamflow. (Rhea, 1977; Branson, 1991). Rhea (1977) tested the model in Colorado for 13 winter seasons using 10-km mean-elevation grids; the results were well correlated to snowpack and streamflow measures. Rhea noted that large discrepancies often existed between the model and measurements on a daily basis, but that the model frequency distributions of daily precipitation totals were realistic. Branson's study (1991) extended the study period to 27 winter seasons and the correlation coefficients for model-simulated precipitation values and the measured snowpack and streamflow maintained good agreement throughout the 27 year historical period. Both of these studies analyzed the accuracy of the model by comparing seasonal sums of model precipitation simulations versus snowpack and streamflow measurements using correlation analysis. Average correlation of model precipitation simulations to measured snowcourse water equivalent and spring and summer streamflow were on the order of 0.7 (Branson, 1991).

Hay *et al.* (1992) compared RHEA-CSU model estimates with measured precipitation using two spatial resolutions (5 and 10-km elevation grids) and three methods of elevation characterization (grid cells characterized using either maximum, mean, or minimum elevation). Problems with interpolating precipitation from a grid cell to a point location for comparison of measured versus simulated precipitation were found. When model results were interpolated to a station location, mean elevation gave the most realistic estimates except when the precipitation stations were located above or below all four surrounding grid cells that were used to interpolate to the station. In these cases, error resulting from not including elevation in the model's interpolation scheme to the inner grids was evident. Comparison of spatial resolutions showed 5-km precipitation estimates produced less precipitation overall (more precipitation at mountain peaks and less on mountain slopes) than 10-km estimates.

Another method of verification, described in more detail in Leavesley *et al.* (1992) and Parker *et al.* (1992), is to use gridded precipitation estimates as input to a precipitation- runoff model and compare predicted runoff with measured runoff in sub-basins within the Gunnison River basin. This method avoids the problems reported by Hay *et al.* (1992) of interpolating from a grid cell to a station location. Leavesley *et al.* (1992) and Parker *et al.* (1992) tested this method in the East River sub-basin (see Fig. 1 for location) of the Gunnison River basin and showed that simulated runoff volume using precipitation (1) from a single weather station at Crested Butte and (2) from the RHEA-CSU model output are similar to measured runoff volume of the East River sub-basin. However, improvement in simulated streamflow timing for the simulations using the RHEA-CSU model precipitation indicate that the spatial distribution of precipitation and the inter-annual variability of this distribution are better defined by the RHEA-CSU model than by the current extrapolation methods for distributing measured precipitation at a point.

RESULTS

In the previous investigations described above, the RHEA-CSU model output was examined on a seasonal basis using correlation analysis by Rhea (1977) and Branson (1991). This type of analysis will not show magnitude errors; that is, measured and simulated precipitation could increase or decrease at the same rate, but be of different orders of magnitude. Correlation analysis also will not identify model accuracy in predicting dry days or daily precipitation intensities. Hay et al. (1992) examined the RHEA-CSU model output by comparing simulated output with measured precipitation on a daily basis. Problems transferring precipitation from a grid cell to a point were found. Precipitation volumes produced by the RHEA-CSU model on a daily basis were examined by Leavesley et al. (1992) and Parker et al. (1992). This type of study examined the accuracy of the model in predicting precipitation volumes, but comparison of runoff integrates the spatial errors that may be present in the model output. In this paper, the effects of changing the grid-cell size on precipitation output was examined using GIS technology. This type of study examines the RHEA-CSU model output on a grid-cell basis. Sensitivity tests of all model physical processes including grid-size dependence will be tested in the future.

GIS technology facilitated the comparison between output from the three spatial resolutions by allowing the modeler to easily create visual (both interactive and hardcopy) displays of the results from the model runs. Arithmetic features of the GIS were used to compare the output at the three spatial resolution using grid subtraction and accumulation. Visualizations of output using a GIS helps the modeler gain insight into model results at the three spatial resolutions, verify that the model is functioning correctly, and communicate model results to other scientists and non-scientists.

Figs 2a to 2c show the 10-, 5-, and 2.5-km mean-elevation grids for the non-rotated inner grid and Figs 2d-f show the corresponding precipitation produced using the RHEA-CSU model accumulated for an average winter season: October 1, 1968 - April 30, 1969. Examination of elevation (Figs 2a to 2c) for the three grid-cell sizes shows an increasing amount of detail and higher maximum elevations with a decrease in cell size. The corresponding cumulative precipitation maps for the 1969 winter season (Figs 2d to 2f) display the same general pattern in precipitation over the basin with higher maximum precipitation values occurring as grid-cell size decreases.

A GIS was used to sum the 5- and 2.5-km precipitation output to the 10-km grid cells for comparison purposes. Figs 3a to 3c show precipitation-elevation plots for the 10-km precipitation grids (Fig. 3a), 5-km precipitation summed to 10-km grids (Fig. 3b), and 2.5-km precipitation summed to 10-km

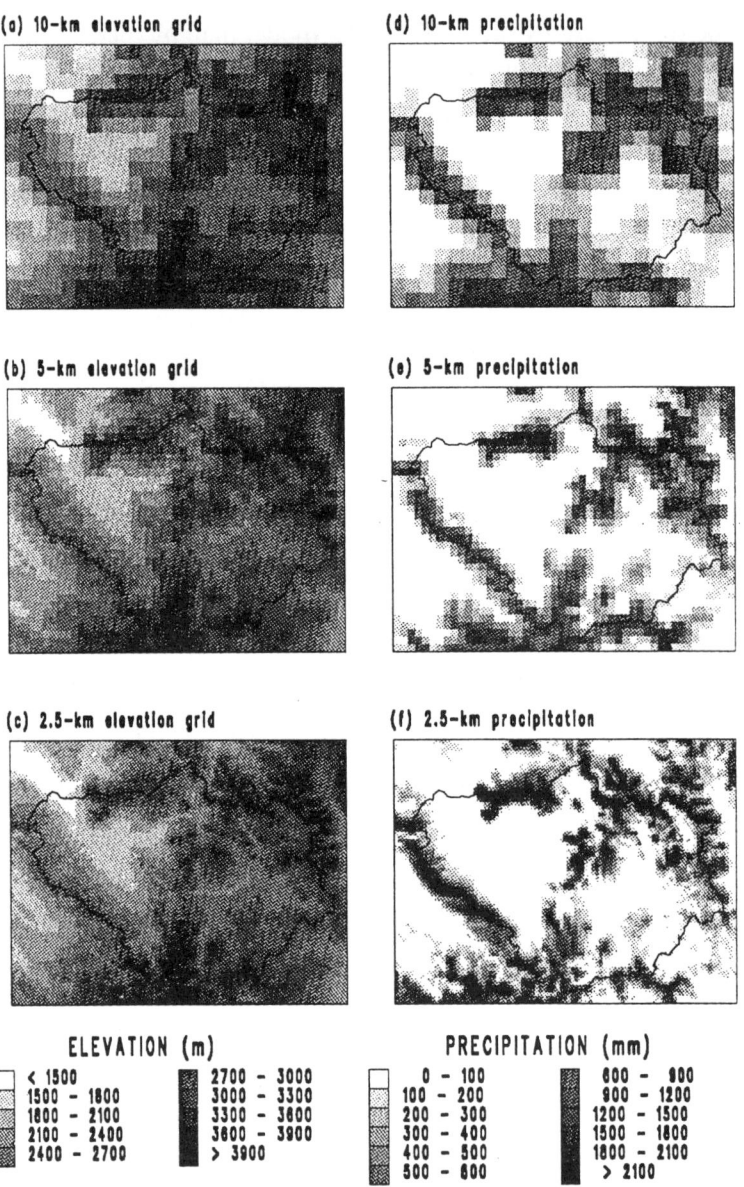

Fig. 2 Mean elevation characterized using (a) 10-km, (b) 5-km, and (c) 2.5-km grids and cumulative precipitation (10/68 -4/69) simulated using (d) 10-km, (e) 5-km, and (f) 2.5-km grids.

grids (Fig. 3c). The precipitation values are cumulative over the 1969 winter season (October 1968 - April, 1969). It appears that below approximately 2000 m in elevation, the RHEA-CSU model produces very little precipitation at any of the grid scales. Above 2000 m, larger scatter and higher maximums in precipitation are associated with a decrease in grid-cell size. Highest precipitation sums do not occur at the highest elevations in the basin for any grid size.

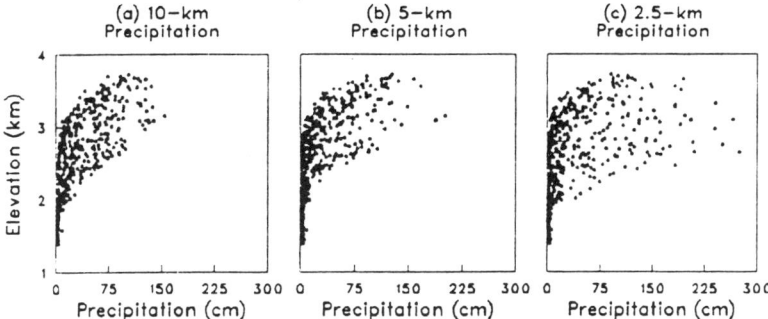

Fig. 3 Cumulative precipitation (10/68-4/69) versus 10-km mean elevation for (a) 10- km grids, and 9b) 5- and (c) 2.5-km grids summed to 10-km grids.

To further examine these differences produced by changes in grid-cell size, a GIS was used to sum the cumulative precipitation, for the 1969 winter season, over the non-rotated inner grid by elevation band. The 5- and 2.5-km cumulative precipitation output was summed to the 10-km grid cells for comparison purposes and the elevation bands were chosen to have an equal number of grid cells in each. Table 1 shows the results for the 10-, 5-, and 2.5-km grids. An examination of sums of the entire amount of precipitation produced over the basin indicates that the 2.5-km grids produce more precipitation than the 10-km grids, but the 10-km grids produce more precipitation than the 5-km grids. This behavior is consistent for all 30 winter seasons simulated. Divided by elevation band, this ordering is seen for every band except the lowest band (less than 2200 m) in which the 5- km grids produce almost the same amount of precipitation as the 2.5-km grids (16 mm more over the entire basin) and 742 mm more precipitation over the entire basin than the 10-km grids. The largest discrepancies between grid-cell size are seen in the middle elevation bands; 2200-2700 m and 2700-3100 m. Output from the 2.5-km grids falling within the 2200-2700 m elevation band produces double the amount produced using 5-km grids. In the highest elevation band (greater than 3100 m) differences in precipitation are less exaggerated.

Table 1 Grid-cell precipitation summed by elevation band, October, 1968 - April, 1969.

Elevation band	precipitation (mm)		
	10 km	5 km	2.5 km
< 2200 m	5 523	6 265	6 249
2200 - 2700 m	33 349	22 986	45 926
2700 - 3100 m	43 914	36 682	54 976
>3100 m	78 111	77 085	81 023
Precipitation sum	160 897	143 018	188 174

To analyze the spatial distribution of changes in precipitation resulting from changes in grid scale, a GIS was used to examine the differences in precipitation over the basin. Figs 4a to 4b show the 5- and the 2.5-km cumulative precipitation for the 1969 winter season, both summed to 10-km grid cells. Figs 4c to 4d shows the result of subtracting the precipitation grids shown in Figs 4a and 4b, from the 10-km precipitation grid (Fig. 2a), respectively. Grid cells with negative precipitation differences are shown filled with stripes and outlined in black (cells in which the 10-km grids have less precipitation than the grid being subtracted), positive differences in precipitation are shown with levels of gray shades (cells in which the 10-km grids show more precipitation), and grid cells with no change are shown in white. No change is seen at the lowest elevations since the current version of the RHEA-CSU model produces very little precipitation at the lower elevations at any grid scale. Figure 4c shows positive precipitation anomalies on the slopes (10-km grids have more precipitation on slopes) and also some negative precipitation anomalies at some of the higher elevations (more precipitation at higher elevation from the 5-km grids). In Fig. 4d, the ridge in the southwestern

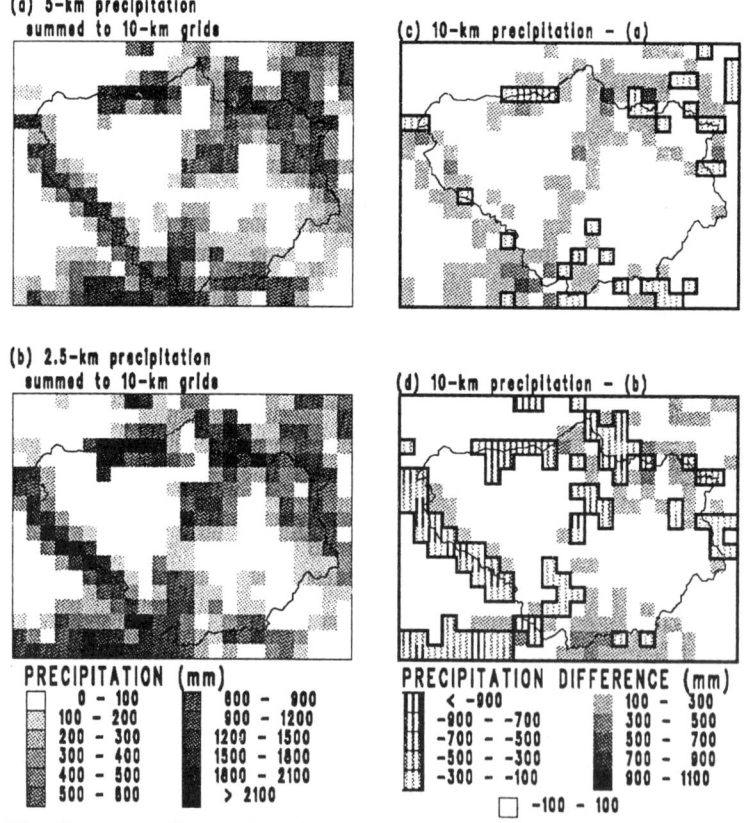

Fig. 4 Precipitation grids (10/68 -4/69) summed from (a) 5- to 10-km and (b) 2.5- to 10-km, and the difference in cumulative precipitation between 10-km grids and grids summed from (c) 5- to 10-km and (d) 2.5- to 10-km.

corner of the basin shows high negative precipitation anomalies at the highest elevations as well as on the southwestern slope of the ridge. The northeastern slope of the ridge shows positive precipitation anomalies. The higher maximum precipitation values at the higher elevations, produced using 2.5-km mean-elevation grids are easily visualized in Fig. 4d. Grand Mesa (see Fig. 1 for location) stands out in Figs 4c and 4d as an area that receives more precipitation as the grid-cell size decreases.

CONCLUSIONS

The RHEA-CSU model output using 10-, 5-, and 2.5-km mean-elevation grids were compared through use of a GIS's visualization and arithmetic features to assess the effects of a change in grid-cell size on model results. Comparison of output from the RHEA-CSU model, for every winter season over the 30-year simulation period, showed the 2.5-km mean-elevation grids produced more precipitation over the entire inner grid area than that produced using the 10- or 5-km grids, but 10-km grids produced more precipitation than the 5-km grids. This was contrary to what was expected; with a decrease in grid-cell size accompanied by no changes in any of the physical process parameters, it was expected that an increase in precipitation would be seen with a decrease in grid-cell size. These results indicate a need for a more detailed evaluation of the process algorithms in the model in order to asses the effects of a change in grid-cell size on RHEA-CSU model results.

REFERENCES

Branson, M.D. (1991) An historical evaluation of a winter orographic precipitation model. Master's Thesis, Department of Atmospheric Sciences, Ft. Collins, Colorado.

Hay, L.E., Battaglin, W.A., Parker, R.S. & Leavesley, G.H. (1992) Modeling the effects of climate change on water resources in the Gunnison River basin, Colorado, using GIS technology. *First Int. Conf./Workshop on Integrating Geographic Information Systems and Environmental Modeling*, September 15-19, 1991, Boulder, Colorado.

Kuhn, G. & Parker, R.R. (1992) Transfer of hydrologic model parameters from calibrated to non-calibrated basins in the Gunnison River basin, Colorado. *Proc. from AWRA 28th Annual Conf. and Symposia*, Reno, Nevada, Nov 1-5, 1992.

Parker, R.R., Kuhn, G., Hay, L.E. & Elliot, J.G. (1992) Effects of potential climate change on the hydrology and the maintenance of channel morphology in the Gunnison River basin, Colorado. *Proc. of the Effects of Global Climate Change on the Hydrology and Water Resources at a Catchment Scale*, Tsukuba-shi, Japan, Feb 3-6, 1992.

Leavesley, G.H, Branson, M.D. & Hay, L.E. (1992) Investigation of the effects of climate change in mountainous regions using coupled atmospheric and hydrologic models. *Proc. from AWRA 28th Annual Conf. and Symposia*, Reno, Nevada, Nov 1-5, 1992.

Rhea, J.O.(1977) Orographic precipitation model for hydrometeorological use. Ph.D. Dissertation, Colorado State University, Department of Atmospheric Science, Ft. Collins, Colorado.

Ugland, R.C., Cochran, B.J., Hiner, M.M., Kretschman, R.G., Wilson, E.A. & Bennett, J.D. (1990) Water resources data for Colorado, water year 1990, Volume 2, Colorado River basin, US Geological Survey Water-Data Report CO-90-2.

Linking sediment and nutrient export models with a geographic information system

E. KLAGHOFER & W. BIRNBAUM
Federal Institute for Land and Water Management Research, A-3252 Petzenkirchen, Austria

W. SUMMER
Vienna Technical University, Institute for Hydraulics, Hydrology and Water Resources Management, Karlsplatz 13/223, A-1040 Vienna, Austria

Abstract A grid based GIS was used in combination with different erosion models to estimate the sediment and nutrient export from a small lower alpine drainage basin. These hydrological processes are affected by the spatial variability of soils, topography, landuse and cover, climate and human induced changes and management. The linkage of the spatial data handling capabilities of the GIS IDRISI with the applied hydrological models EPIC and AGNPS show the advantages associated with utilizing the full information content of the spatially distributed data to analyse the hydrological processes.

INTRODUCTION

One of the major problems in modern agricultural management is the loss of soil due to surface runoff as well as the pollution in runoff, seepage or perculation from these management activities. Besides agricultural deficiencies, they have major impacts on the quality of surface as well as subsurface water. But also the increase in ecological concerns forces landscape professionals to analyze the characteristics of a drainage basin and its behaviour by computer models.

The hydrological system is predominantly derived from mapped attributes of the landscape. Topography, soils and landuse are the primary data sets which are then used to derive additional model variables such as slope, aspect, permeability or other factors related to soil erosion simulation. The spatial resolution of a computer based basin and its manageable size of the data base is often compromised by the reduction of the number of components describing the overall hydrological system. Should it be important to maintain the spatial variability this procedure limits an ecological approach.

APPLIED GIS-TECHNOLOGY

Geographic Information Systems (GIS) provide a practical solution for handling the detailed spatial variability of a drainage basin. By using IDRISI (Eastman, 1990) its technology was selected for creating, storing, analyzing and displaying information. The basic strategy required creating spatial and non-spatial databases for elevation, soil mapping units, agricultural landuse, hydrography, drainage basin, hydrology and field boundaries from which factors related to the USLE were spatially derived.

IDRISI was used to represent a lower alpine drainage basin of an area of 65 ha with intensive agriculture. The available Digital Terrain Model DTM was based on a rectangular altitude grid which covers almost all of Austria. The grid size varied between 30 to 50 m. The GIS then performed the determination as well as the assembling of catchment parameters (Fig. 1) which are required in estimating the soil erosion risk in each grid element of the basin.

In general a GIS is a highly specialized database management system for spatially distributed information. They are classified in vector and raster based systems, characterized by the way in which the analysis is performed on the spatial data. By converting points, vectors and polygons into digital form, areal information contained in maps are digitized in the first analyzing step. Once in digital form the analysis is performed on a raster of pixels (picture elements) replacing the digitized polygons. IDRISI allows the overlaying of the digitized lines on top of the raster image. As a raster based GIS and linked to the soil erosion model EPIC, IDRISI performed all necessary analysing and overlaying functions as well as displays in a fast manner, even on a PC.

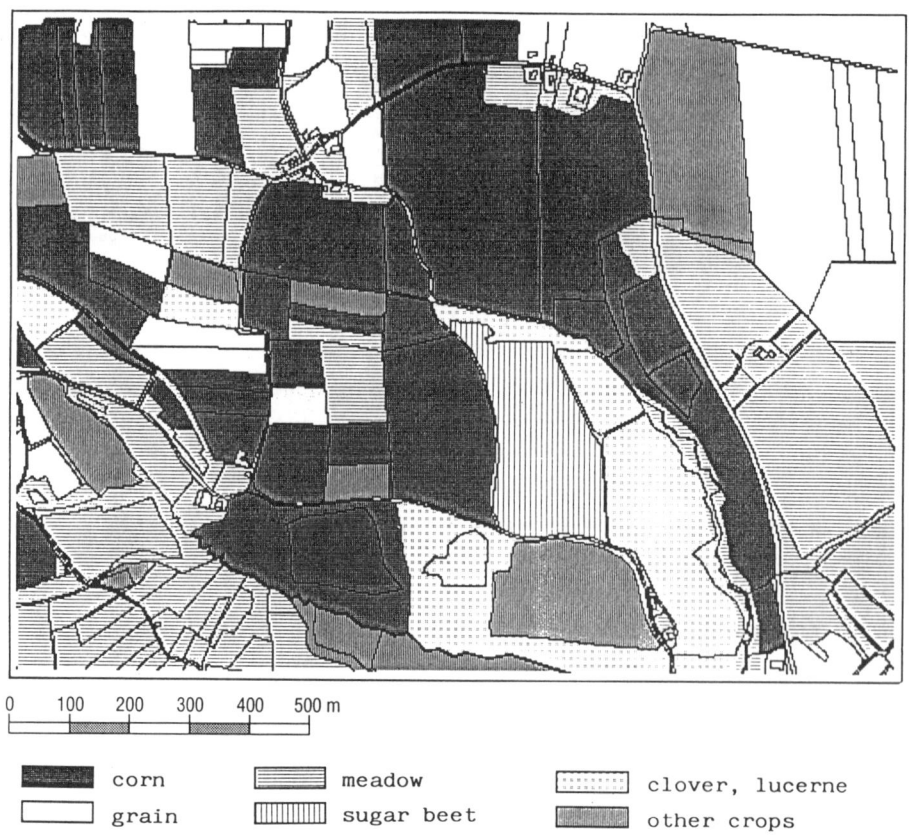

Fig. 1 Spatial distribution of the landuse in the drainage basin.

LINKING IDRISI WITH THE SOIL EROSION MODEL EPIC

The soil erosion model EPIC (Erosion Productivity Impact Calculator) (Williams *et al.*, 1983) - a field scale model - is based on a deterministic approach to estimate soil losses at a specific site. To run EPIC, information in the EPIC's major components are needed. This data including topography, soil types, climate data, cropping pattern and tillage were assembled in a spatial manner by IDRISI. With the provision of all parameters as well as a soil erosion matrix, considering several possibilities of different slope gradients and soil types, it became possible to run EPIC. Overlaying four maps with the information on landuse, topography or gradient, erosive slope length and soil type, the resulting map distinguishes areas of different soil erosion conditions. Assigning the data of the soil erosion matrix to it (Fig. 2), a detailed map showing the spatial distribution of the erosion rate in the basin could be produced (Fig. 3). Applying the concept of a Sediment Delivery Ratio (SDR) to each erosion unit leads to the amount of sediment entering the river network.

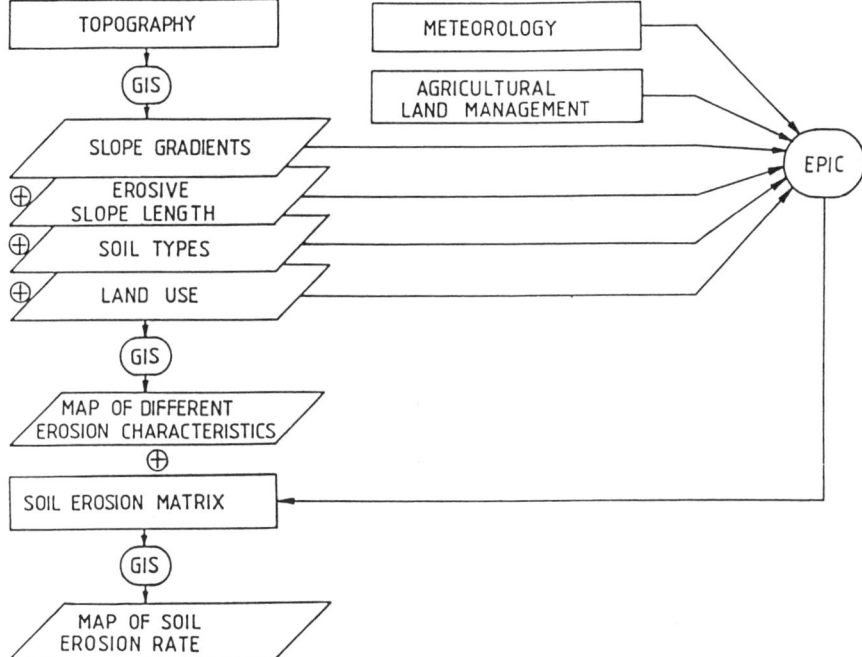

Fig. 2 GIS overlay producing EPIC input information as well as output display.

AGNPS - MODEL DESCRIPTION, PARAMETERS AND ASSUMPTIONS

AGNPS (Agricultural Non point Source Pollution) (Young *et al.*, 1987) was

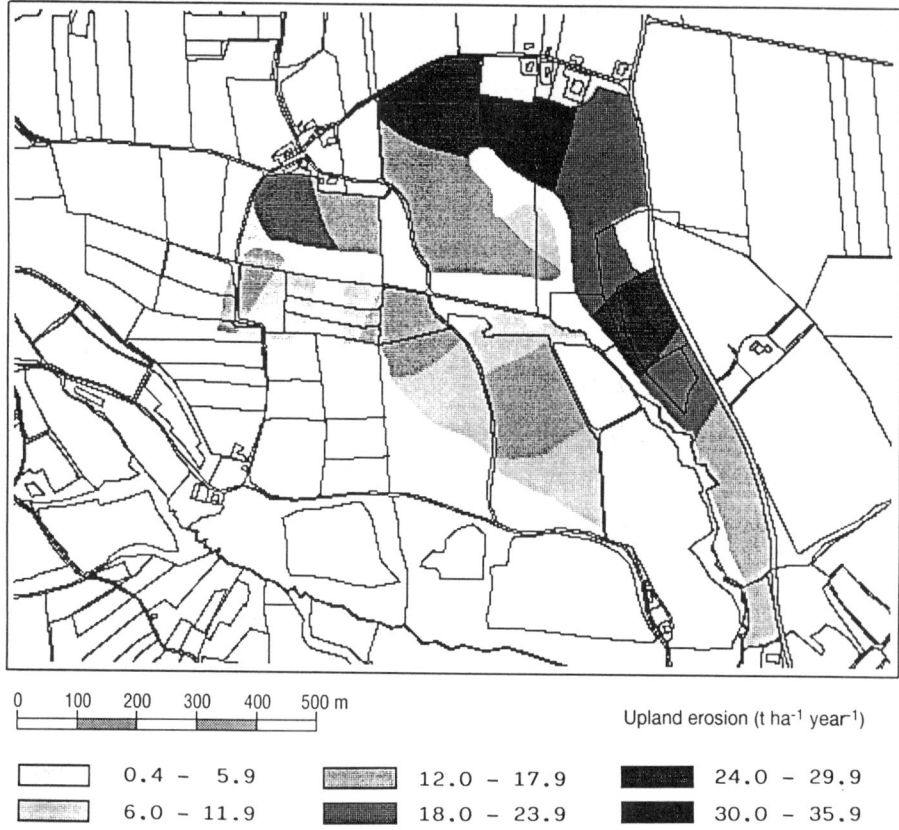

Fig. 3 Spatial distribution of the potential soil erosion rates in the drainage basin.

designed by the U.S. Agricultural Research Service, Morris, Minnesota and the University of Minnesota for the Pollution Control Agency as a tool for assessing the impacts for agriculture on surface water quality.

To analyse the water quality impacts of nonpoint source pollution, it predicts the runoff volume, the peak rates, soil erosion as well as sediment, nitrogen, phosphorus and chemical oxygen demand concentration in the runoff. Hence, such hydrological and physical domains that the model attempt to represent are approached by varying degrees of empiricism, functional representation and deterministic description of water quality and quantity processes. Though AGNPS models distributed processes, all erosion processes are modelled by simplistic and empirical relations. At the drainage basin scale, AGNPS simulates all processes on a distributed level, because the model evaluates information at grid cells ranging between 0.4 and 16 ha interior to the basin (Fig. 4). But AGNPS is also well suited to simulate hydrologically induced processes at a large drainage basin scale. Considering a simple way to standardize the preparation of input data, a linkage between IDRISI and AGNPS can be done by a "Query File". It contains only pixels, covering the

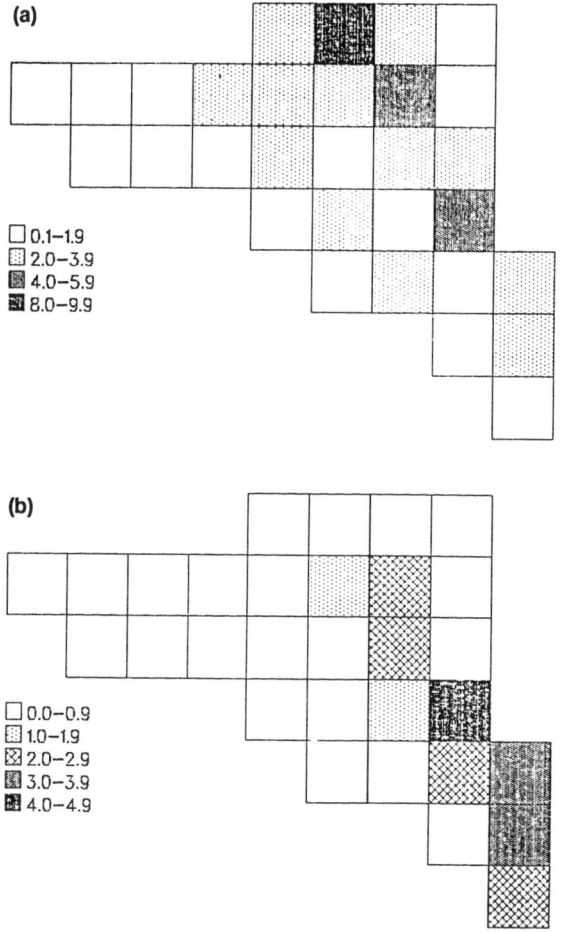

Fig. 4 AGNPS - Grid based data for 6 June 1991: (a) erosion rate (t ha^{-1}), (b) sediment yield of the cell (t).

complete basin. Both tools are grid based systems, hence, spatial information can be easily exchanged between both analysis systems.

RESULTS AND CONCLUSION

Analysing the erosion units evaluated by EPIC as well as the sediment yield determined by AGNPS showed that during periods of single rainstorm events the SDR in the drainage basin of a size of 65 ha has a value of approximately 1%. Applying such approaches to larger drainage basins by using these tools - IDRISI, EPIC, AGNPS - could lead to a general relation between the SDR and the catchment size for lower alpine areas as well as single rainstorms for the first time.

REFERENCES

Eastman, J.R. (1990) IDRISI - *A Grid Based Geographic Analysis System*. Clark University, Graduate School of Geography, Worcester, MA 01610, USA.

Williams, J.R., Dyke, P.T. & Jones, C.A. (1983) EPIC - A Model for Assessing the Effects of Erosion on Soil Productivity. Analysis of Ecological Systems: State-of-the-Art in Ecological Modeling. *Developments in Environmental Modelling* 5, Elsevier Scientific Publishing Company, Amsterdam-Oxford-New York, 553-572.

Young, R.A., Onstad, C.A., Bosch, D.D. & Anderson, W.P. (1987) AGNPS: Agricultural Nonpoint Source Pollution Model: A Watershed Analysis Tool. USDA-ARS, *Conservation Research Report,* 35.

RHINEFLOW: an integrated GIS water balance model for the river Rhine

W.P.A. VAN DEURSEN
Resource Analysis, Zuiderstraat 110, 2611 SJ Delft, The Netherlands

J.C.J. KWADIJK
Department of Physical Geography, University of Utrecht, P.O. Box 80.115, 3508 TC Utrecht, The Netherlands

Abstract This paper presents the GIS-based RHINEFLOW model. This model describes the changes in the water balance compartments of the river Rhine on a monthly time basis. The model is build with the PC RASTER package as an integrated part of a GIS. PC RASTER is a set of utilities for hydrological and geomorphological modelling that can be linked to a raster GIS. With these utilities water balances can be modelled on each of the cells of the raster system. A geomorphologically based routing model allows the surface water computed from the water balance to drain down to the outlet of the catchment. The utilities allow for evaluation of model results including their spatial and temporal distribution. The RHINEFLOW model performs well in calculating the monthly discharge of the river Rhine and its tributaries.

INTRODUCTION

Geographic Information Systems are very suitable for storing, analysing and retrieving the information needed for running hydrologic models. GIS can hold many data on the, albeit static, distribution of land attributes which form the control parameters, boundary conditions and input data for the model (Burrough, 1989). However, with most GIS it is usually necessary to export the maps from the GIS and convert them into input data sets for the model, which is run seperately. The results can be brought back into the GIS for further analysis. Especially when running several models for a certain analysis, or using different sets of input maps when analysing different scenarios, this ad hoc approach is cumbersome and for routine research the models should be an integral part of the GIS. A useful general approach would be to extend the GIS with a (limited) set of general tools that can be used to build hydrologic models to meet the users requirements. Instead of implementing a limited number of specific solutions for geographic problems, the GIS is a toolbox, with which an experienced user can build solutions for numerous spatial problems. This is basically the approach used by Tomlin (1983) when designing the Map Analysis Package.

CONCEPTUAL MODELLING AND GIS

A GIS toolbox to be used for hydrological modelling must at least contain

procedures to:
(a) *procedure A:* read input data for a certain location (x) and time step (t);
(b) *procedure B:* carry out the necessary calculations to determine the water balance at (x,t). For spatially distributed water balance modelling, the flow direction at all locations must be known since the water balance at location x may also depend on the water production in the neighbourhood;
(c) *procedure C:* perform some output operations to enable others (users or models) to use the results of the model.

The GIS toolbox PC RASTER PACKAGE, contains the following programs for hydrological modelling (procedures A, B and C).

Procedure A

Procedure A is represented by an input module (TIMEINP). This module reads input data from time series. Input comes from ASCII data files that store values for a certain input (e.g. precipitation, temperature) for the time interval used in the model. TIMEINP reads the values for these inputs sequentially from these datafiles, and produces raster maps with the (distributed) value for the inputs for a time t.

Procedure B

Procedure B is represented by:
(a) a raster map calculator CALC. This module can evaluate almost any arithmetical point-operation on gridded map-data. It has a list of built in functions, such as LN (natural logarithm), SQRT (square root), which can be applied to the raster maps. Logical functions can be implemented with MIF (IF Map ... THEN .... ELSE....) using operators LT (less than), GT (greater than), EQ (equals);
(b) the program WATERSHED. This program enables the user to extract drainage patterns, a Local Drain Direction matrix (LDD), from elevation data and use these patterns for the geomorphologically based routing modules. The basic concepts for drainage pattern analysis were taken from published descriptions of recursive basin delineation given by Marks *et al.* (1984) and publications by Jenson & Domingue (1988) and Morris & Heerdegen (1988). The approach followed is discussed extensively in Van Deursen & Kwadijk (1990) and Van Deursen (1993);
(c) the module ACCU. This is a tool to route water through the local drain direction matrix (LDD). In a water balance model such as RHINEFLOW the inputmap for this module contains surface water calculated for each cell. The outputmap will contain the surface water accumulated through the Local Drain Direction Matrix.

Procedure C

Procedure C is represented by:
(a) the module DISPLAY, which is used for display of any of the maps through the model timeinterval;
(b) the module TIMEOUT which is used to create timeseries of the results of the model.

MODELLING THE RIVER RHINE DRAINAGE BASIN

In this part we will examine the steps needed to build a spatially distributed water balance model, called RHINEFLOW, for the river Rhine drainage basin

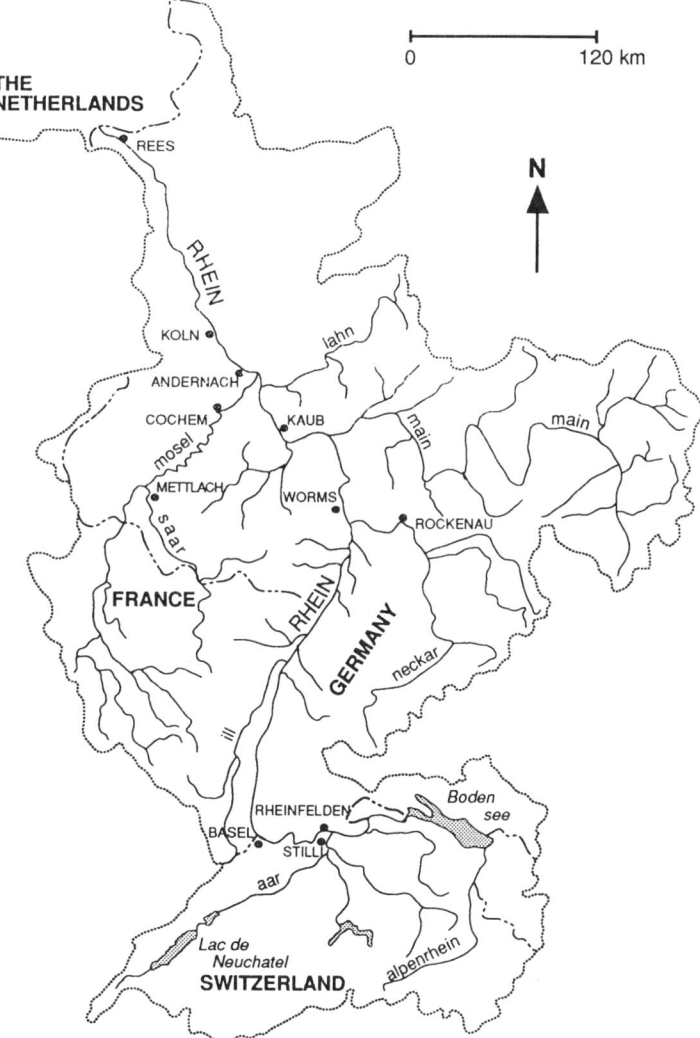

Fig. 1 Catchment of the river Rhine.

(Fig. 1). The model has been designed to investigate possible changes in the water balance compartments in the river Rhine basin, due to climate change. RHINEFLOW is written with the syntax developed for the PC-RASTER PACKAGE. Figure 2 shows a flow diagram of the model. The entire model takes up only 40 lines of code. The model runs on IBM PC-AT or compatible computers.

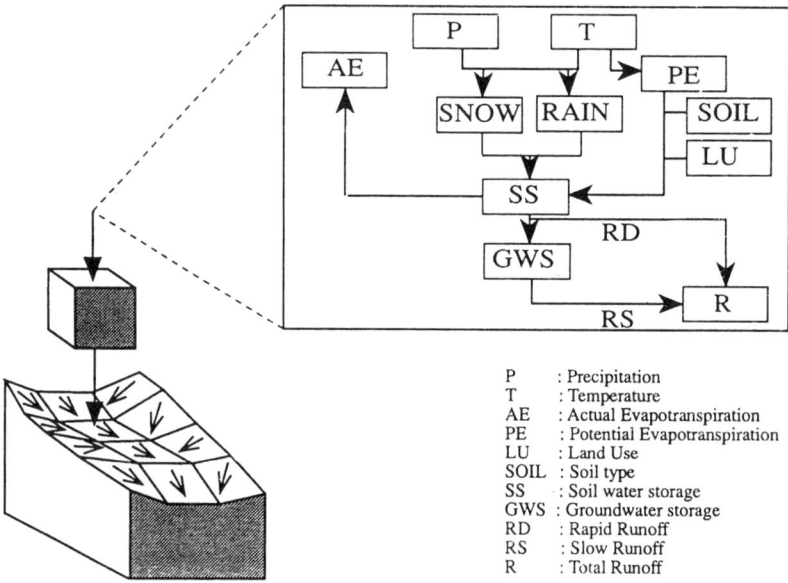

P	: Precipitation
T	: Temperature
AE	: Actual Evapotranspiration
PE	: Potential Evapotranspiration
LU	: Land Use
SOIL	: Soil type
SS	: Soil water storage
GWS	: Groundwater storage
RD	: Rapid Runoff
RS	: Slow Runoff
R	: Total Runoff

Fig. 2 Flow diagram of the RHINEFLOW model.

Water balance modelling

The discharge regime of the Rhine is governed by five major natural storages and controls:
(a) the amount of precipitation and its spatial distribution;
(b) the temperature distribution in the catchment, which mainly depends on the topography. This distribution forms the main control for the amount of snow storage, snow melt and potential evapotranspiration in the catchment;
(c) soil moisture storage which gains water from the surplus of precipitation and looses water to evapotranspiration, to seepage to the groundwater and to the direct runoff;
(d) groundwater storage which gains water from the soil water seepage and looses water to the river baseflow;
(e) the distribution of the land use and soil water storage capacity in the catchment, which control the actual evapotranspiration.

Finally, the discharge regime is not only influenced by the local conditions but is also by the water production in the upstream area:

(f) the spatial connectivity of the geographical sub-elements in the catchment.

Data requirement and data sources

For the RHINEFLOW model a raster GIS dataset was created. The different maps in this dataset were digitized and rasterized to a grid size of about 3x3 square kilometer. Each of these raster cells in the database represents one calculation element in the RHINEFLOW model and is the smallest element in the spatial connectivity analysis. Figure 3 shows a diagram of the data flow.

The following set of data is used as input variables and parameters for the model:

(a) monthly areal precipitation and temperature data for a large number of stations in the Rhine catchment. Temperature data was interpolated to cover the complete grid with the use of elevation data from the digitized Digital Elevation Model (DEM);

(b) a DEM. This DEM was digitized from an elevation map. The digitized elevation map (Fig. 4a) was interpolated using an inverse distance method;

(c) a soil map. The soil type map was relabeled into a soil storage capacity map (SSmax.map) using published data (Groenendijk, 1989);

(d) a land use map. The land use map was relabeled into a cropfactor map. This map was used to adjust the calculated potential evapotranspiration for different crop types;

(e) from the DEM a local drain direction map (LDD) was created using the program WATERSHED. This LDD map represents the drainage pattern of the Rhine catchment. This map is used to route the water produced in each cell (see next section) to neighbouring cells and eventually to the outlet of the basin, thus representing the spatial connectivity of the geographic sub-elements. From this drainage pattern map, WATERSHED can determine the upstream area map, which for each cell represents the number of grid cells draining through that cell. This upstream area map is given in Fig. 4b.

The original maps are all published by the Commission Hydrologique du Basin du Rhin (CHR/KHR, 1977).

The following data set is used for calibration and validation:

(a) monthly discharges for the main tributaries of the river Rhine and 7 stations along the main river. This time series covers periods ranging between 1870-1980 and 1956-1980. These data were derived from the German Hydrological Office (BfG);

(b) average monthly changes in snow water equivalents (SWE) for 30 mountain weather stations in the Alps. These data were derived from the geographical institute at the ETH (Switzerland) (Martinec *et al.*, 1992);

(c) average monthly evapotranspiration for different crop types for several stations in Germany, published data on the average monthly

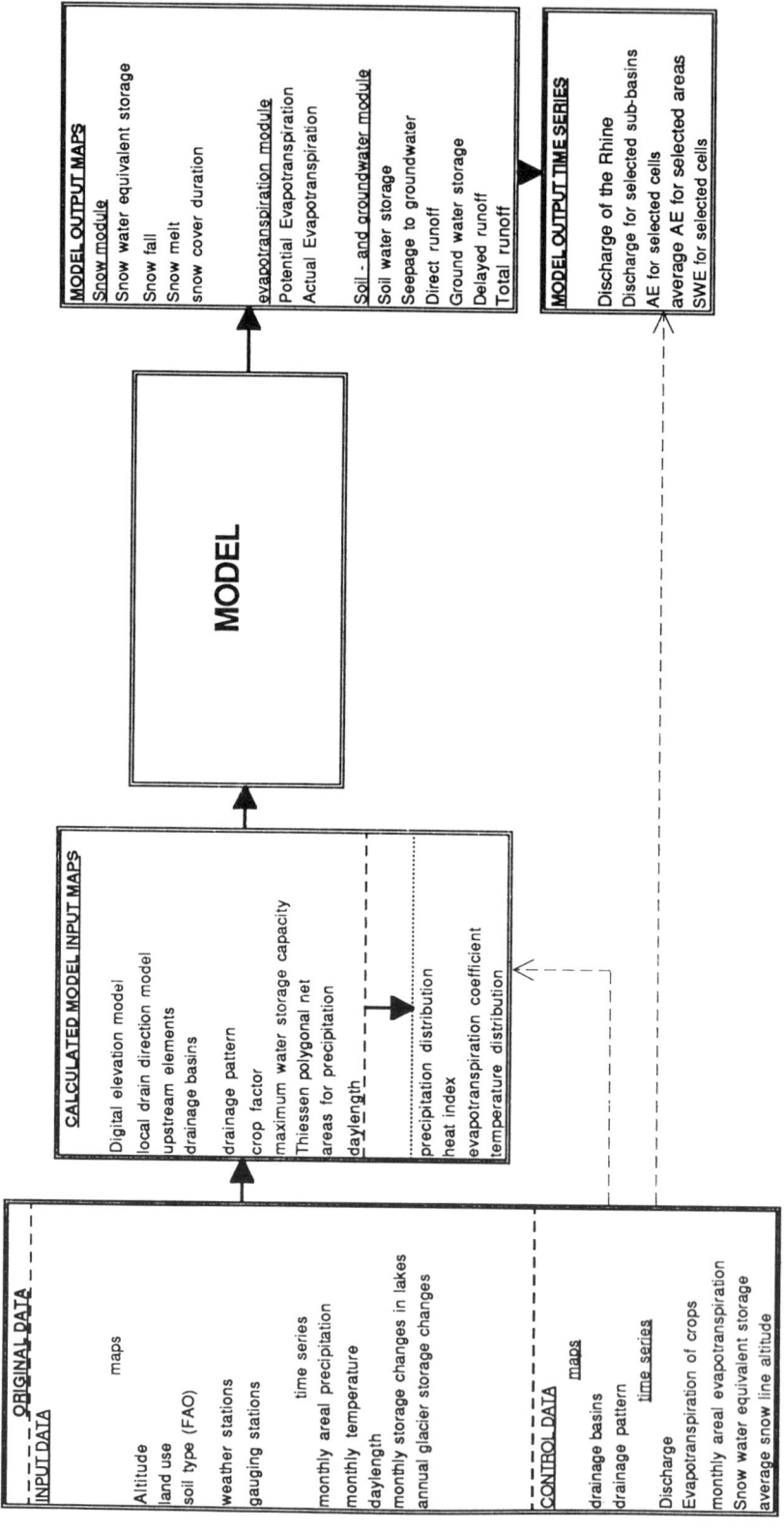

Fig. 3 Data flow diagram of the RHINEFLOW model.

RHINEFLOW: an integrated GIS water balance model for the Rhine 513

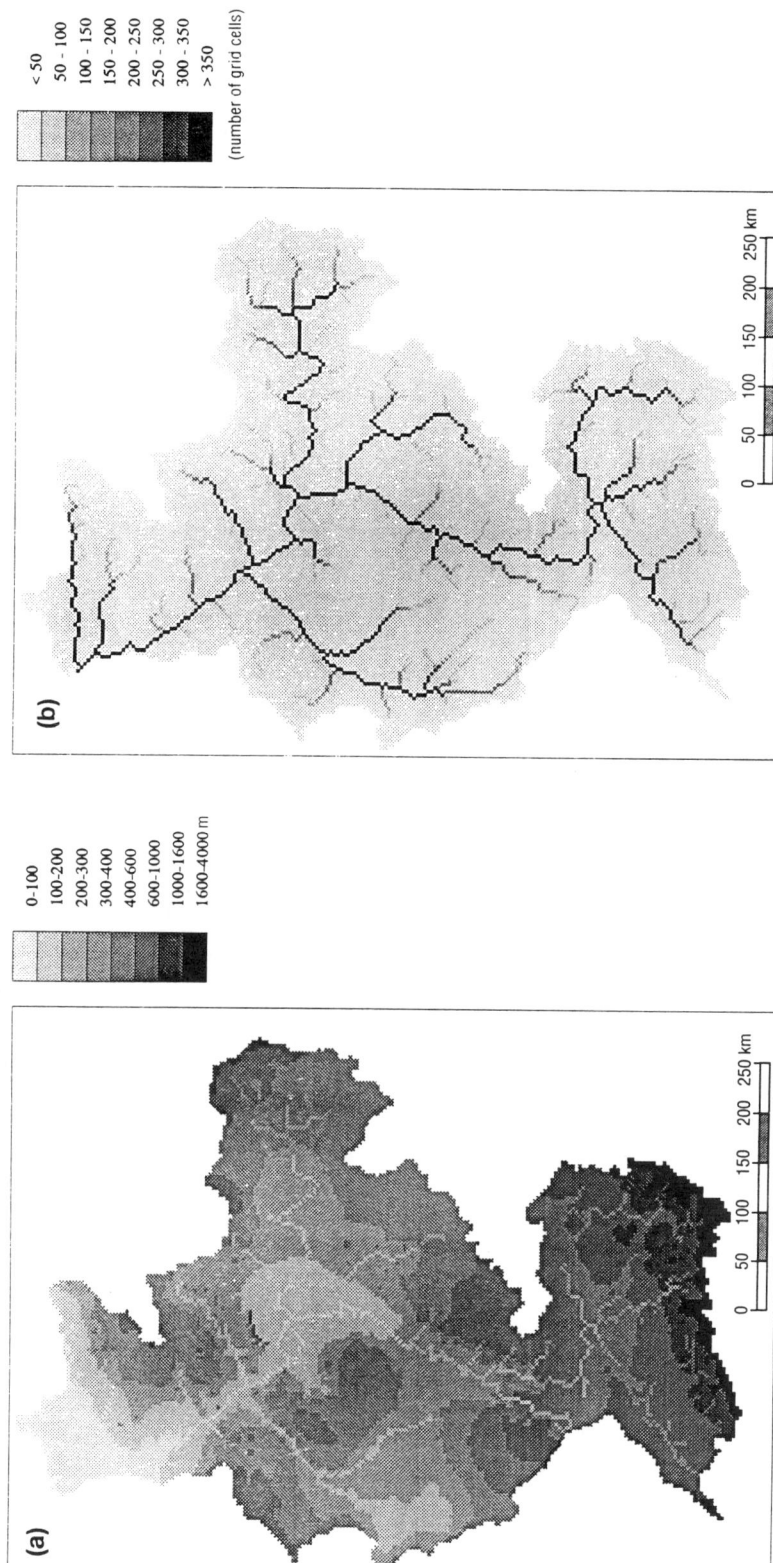

Fig. 4 (a) digital elevation model (m); (b) calculated upstream area map (number of grid cells).

evapotranspiration in the middle and lowland area (CHR/KHR, 1977) and published data on the evapotranspiration in the Alpine area (Schädler, 1985).

All meteorological data were obtained from German and Swiss meteorological Offices (DWD and SHA, respectively).

Model calculations

Evapotranspiration and soil moisture

The Potential Evapotranspiration (PE) is estimated on a monthly basis using a Thornthwaite approach, corrected for different types of land use (Thornthwaite & Mather, 1957; Kwadijk & Van Deursen, 1993). From the PE the actual evapotranspiration (AE) and soil storage (SS) were calculated as follows:

(a) if P>PE then AE=PE and SS is filled until it reaches field capacity (SSmax). The water surplus above the field capacity storage (SPW) is assumed to drain to the groundwater;

(b) if P<PE then SS is estimated with a Thornthwaite-Mather approach:

$$SS = SSmax \times EXP(-APWL/SSmax) \qquad (6)$$

$$AE = P + dSS \qquad (7)$$

in which APWL is the Accumulated Potential Water Loss, this is the sum of monthly water shortages during a dry period.

The results are formed by maps in which all raster cells contain a value that is an estimate for the local potential and actual evapotranspiration and the soil water storage. Consequently these maps show the spatial distribution of AE and soil water storage for the drainage basin in the month of concern.

Runoff production

From the above calculations for individual cells the seepage from the soil is known (SPW). Runoff is seperated into rapid and delayed runoff using a runoff coefficient of 0.2. This is an average value when using the rational method (Ven Te Chow et al., 1988) for runoff seperation. The direct runoff (RS) is assumed to come to discharge within the month of concern. The delayed runoff (RD) is stored as groundwater. The flow from this water balance compartment to the river is described with a linear recession equation:

$$RS = GWSTOR/T \qquad (8)$$

in which T is a recession parameter which is calibrated for all tributaries containing a gauging station near their outlet (t), GWSTOR is the volume water stored as groundwater (mm).

Now the water balance for all individual cells of the raster map is known. The geomorphological routing routine (ACCU) routes the water produced through the LDD map (which connects the cells within the watershed with the outlet of the basin). A cell in the resulting DIS.MAP contains the average runoff production for that cell and the cells upstream of (draining to) that cell. With the tool TIMEOUT the calculated runoff series of the cells representing a gauging station can be produced. Figure 5 shows the syntax used to calculate the evapotranspiration, soil moisture and the runoff production.

Snowfall and snow-melt RHINEFLOW calculates snowfall and snow melt with a temperature - index method. It is assumed that all precipitation falls in form of snow if the monthly temperature is below zero degrees Celsius. Snow starts to melt if the temperature rises above zero. Maps showing the spatial distribution of snowfall, water equivalent snow storage (SWE) and the melt water production per cell form the output of this module.

Model results

The RHINEFLOW model produces maps and tables at all calculated timesteps, not only for the runoff at a certain location, but also for the other calculated hydrological variables including their spatial and temporal distribution. As an example, Fig. 6 shows the spatial distribution of the calculated potential evapotranspiration. The RHINEFLOW model was calibrated for the period 1965-1970. The calibrated model was tested for the period 1956-1980. The runoff results have been tested on the goodness of fit using the coefficient of efficiency (Nash & Suttcliff, 1970). This coefficient ranges between -1 and 1, a value of 1 describes a perfectly fitting curve while zero means that the average is an equally good estimate for the runoff as the calculated curve. This test provided results between 0.7 and 0.85. An extensive discussion on the model results is published elsewhere (Kwadijk & Van Deursen, 1993).

CONCLUSIONS

Conclusions drawn from the river Rhine example show that even with these simple approaches the RHINEFLOW model is able to describe monthly changes in the water balance compartments for both the entire basin and its tributaries quite accurately. Since the topographical and climatological characteristics of the tributaries are different, the model seems accurate over a wide range of both climate and other environmental conditions. This allows the conclusion that the model can be used to estimate possible changes of the river Rhine discharge to a change in temperature and precipitation.

The Rhine study shows that the PC RASTER PACKAGE offers a general approach for modelling different kind of processes, and includes the possibility to combine detailed spatial resolution with simple (conceptual) models. The toolbox allows validation of water balance models not on

```
INPUT MAPS                                      INPUT TIMESERIES
Thiessen net temperature (TSTAT.id)             Monthly temperature (T.dat)
Precipitation areas (PAREA.id)                  Monthly areal precipitation (P.dat)
Elevation map (DEM)                             Daylength coefficient (d)
Heat-index (I.map)
Thornthwaite coefficient (A.map)
Maximum soil water storage (SSmax.map)
Initial soil water storage (SS.map)
Initial APWL (APWL.map)
initial ground water storage (GWS.map)
Upstream elements (UPS.map)
Crop constant (LU.map)
Basin recession coefficient (T.map)
Output id-files (OUTP.id, AE_OUT.id, DIS_OUT.id)
```

```
MODEL CALCULATIONS
loop
  timer = t
  1  TIMEINP TSL.map   = T.dat,TSTAT.id,timer
  2  TIMEINP P.map     = P.dat,PAREA.id,timer
  3  CALC T_CELL.map   = TSL.map - 0.0065 * DEM
  4  CALC PE.map       = MIF(T_CELL.map GT 0
                        THEN LU.map*d*1.6*{T_CELL.map/I.map}^A.map ELSE 0)
  5  CALC PEFF.map     = P.map + SM.map - PE.map
  6  CALC SS1.map      = MIF (Peff.map GT 0 THEN SSmap+Peff.map
                        ELSE SSmax.map*EXP(-APWL.map/SSmax.map))
  7  CALC AE.map       = MIF (Peff.map GT 0 THEN PE.map ELSE P.map + SS.map-SS1.map)
  8  CALC SPW.map      = MIF (SS1.map GT SSmax.map THEN SS1.map-SSmax.map ELSE 0)
  9  CALC SS.map       = MIN (SS1.map, SSmax.map)
  10 CALC RS.map       = SPW.map*0.35
  11 CALC GWS.map      = GWS.map+(SPW.map-QR.map)
  12 CALC RD.map       = GWS.map/T.map
  13 CALC GWS.map      = GWS.map-RD.map
  14 CALC R.map        = RD.map+RS.map
  15 ACCU DIS.map      = R.map, ALT.ldd
  16 CALC DISBAS.map   = DIS.map/UPS.map
  17 ACCU AETOT.map    = ALT.LDD, AE.map
  18 CALC AEbas.map    = AETOT.map/(ups.map)
  19 TIMEOUT AEBAS.dat = AEbas.map, AEOUT.id, timer
  20 TIMEOUT AE.dat    = AE.map, OUTP.id, timer
  21 TIMEOUT DIS.dat   = DISBAS.map, DIS_OUT.id, timer
  timer = t+1
End loop
```

```
OUTPUT MAPS                                      OUTPUT TIMESERIES
Potential Evapotranspiration(PE.map)             Actual evapotranspiration for selected cells(AE.dat)
Actual Evapotranspiration(AE.map)                Average areal actual evapotranspiration(AEBAS.dat)
Actual Soil water storage(SS.map)                Montly discharge for selected cells (stations) (DIS.dat)
Seepage to groundwater(SPW.map)
```

```
CONTROL TIMESERIES
Observed evapotranspiration for different crops
Observed areal evapotranspiration
Observed discharge for selected gauging stations
```

1 Temperature (sea level) data file is converted to a temperature map
2 Areal precipitation data file is converted into a precipitation map
3 Cell temperature=temperature at sealevel- (altitude_temperature coefficient * altitude)
4 Potential evapotranspiration = if temperature > 0 celsius then: F(Temp, landuse,daylength) Else: 0
5 Available water= precipitation + meltwater - potential evapotranspiration
6 Soil water= if available water more than 0 then initial soil water + available water
       Else soil water storage F(maximum soil water storage, APWL)
7 Actual evapotranspiration = if available water > 0 then AE=PE Else precipitation + soil water storage loss
8 Seepage to groundwater = if actual soil water storage > maximum soil water storage then
              actual soil water storage - maximum soil water storage, else 0
9 Final soil water storage = minimum between calculated and maximum soil water storage
10 Direct runoff = seepage * seperation coefficient (0.35)
11 Groundwater storage = initial groundwater storage + (seepage - direct runoff)
12 Delayed runoff = groundwater storage / delay factor
13 Final groundwater storage = groundwater storage - delayed runoff
14 Total runoff = delayed runoff + direct runoff
15 Total basin runoff = accumulation of produced runoff for individual cells through the drainage network
16 Average basin runoff = total basin runoff / basin size
17 Total areal actual evapotranspiration = accumulated AE through the drainage network
18 Average basin AE = total basin AE / basin size
19 Generating timeseries of average AE for selected areas
20 Generating timeseries of AE for selected cells
21 Generating timeseries of discharge for selected basins

Fig. 5 Syntax used to calculate the evapotranspiration, soil moisture and runoff.

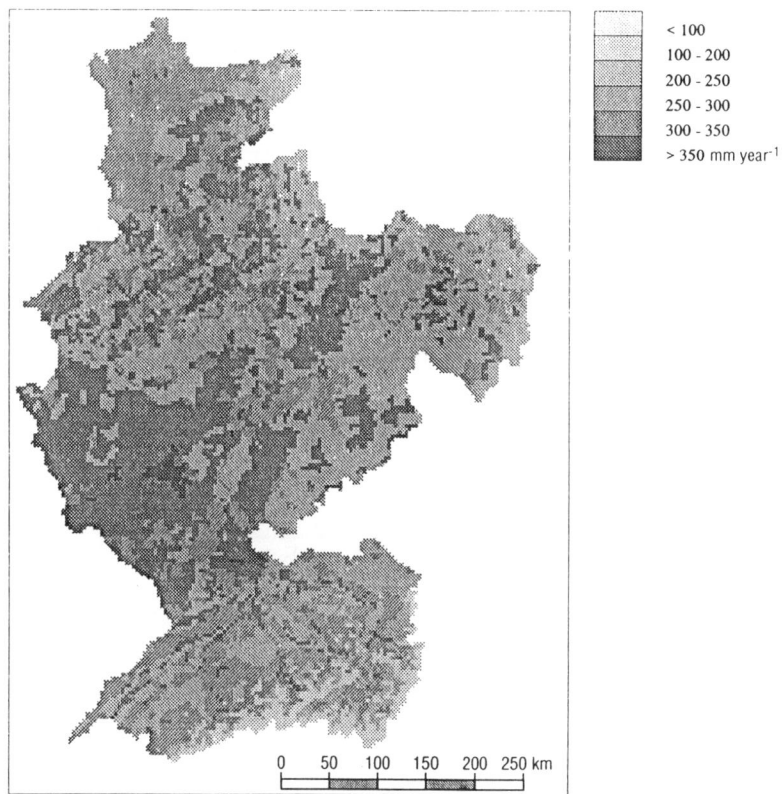

Fig. 6 Map of average calculated potential evapotranspiration (mm).

discharge only but also on the spatial and temporal distribution of other hydrological variables. Since the discharge output can be obtained for any cell along the flow line, it forms a very flexible tool to test the model for the calculated runoff.

REFERENCES

Burrough, P. (1989) Matching spatial databases and quantitative models in land resource assessment. *Soil Use and Management* **5**, 3-8.
CHR/KHR (1977) Le Basin du Rhin / Das Rheingebiet. Monographie hydrologique / Hydrologischer Monographie. Part A: texts, part B: tables, part C: maps and Diagrams. CHR/KHR.
Groenendijk, H. (1989) Estimation of waterholding-capacity of soils in Europe. *Proc. Conf. on Climate and Water,* Helsinki, Finland 11-15 September, 1989. 1. Publications of the academy of Finland, Helsinki. 293-300.
Jenson, S. & Domingue, J. (1988) Extracting topographic structure from digital elevation data for geographic information system analysis. *Photogrammic Engng and Remote Sensing*, 1593-1600
Kwadijk, J.C.J. & Van Deursen, W.P.A. (1993) *Development of a GIS-based water balance model for the river Rhine.* To be published in the CHR/KHR report serie (in prep).
Marks, D., Dozier, J & Frew, J (1984) Automated basin delineation from digital elevation data. *Geo-Processing* **2**, 299-311.
Martinec, J., Rohrer, M. & Lang, H. (1992) Schneehöhen und Schneedecke, Hydrologische Atlas der Schweiz.

Morris, D. & Heerdegen, R. (1988) Automatically derived catchment boundaries and channel networks and their hydrological applications. *Geomorphology* **1**, 131-141.

Nash, J.E. & Sutcliff, J.V. (1970) River flow forecasting through conceptual models. Part 1-a discussion of principles. *J. Hydrol.* **10**, 282-290.

Schädler, B. (1985) Der Wasserhaushalt der Schweiz (Water Management of Switzerland). Mitteilungen no.6. Bundesamt für Umweltschutz. Abteilung: Landeshydrologie, Bern/Berne, Schweiz.

Thornthwaite, C.W. & Mather, J.R. (1957) Instructions and tables for computing potential evapotranspiration and the water balance. *Publications in climatology* **X**, 183-243.

Tomlin, C.D. (1983) Digital cartographic modelling techniques in environmental planning. Unpublished Ph.D. thesis, Yale University, Connecticut.

Van Deursen, W.P.A. (1993) Watershed: general purpose utilities for hydrological and geomorphological processing of raster GIS data. *Geomorphology* (in press).

Van Deursen, W.P.A. & Kwadijk, J.C.J. (1990) Using the Watershed tools for modelling the Rhine catchment. *Proc. First European Conf. on GIS,* Amsterdam, April 1990, 254-263.

Ven Te Chow, Maidment, D.R. & Mays, L.W. (1988) *Applied Hydrology.* McGraw-Hill, Civil Engineering Series.

Application of GIS in determining sources and loads of pollutants transported by regional rivers into The Netherlands

A. MOLENAAR, W. BLEUTEN, M.J. ZEYLMANS & N.F.M. VAN LEEUWEN
Department of Environmental Studies, University of Utrecht, P.O. Box 80.115, 3508 TC Utrecht, The Netherlands

Abstract Surface water quality and factors that influence have been investigated in transboundary rivers flowing into The Netherlands. Besides the three main rivers (Rhine, Meuse and Scheldt), many small rivers and brooklets, here referred to as regional rivers, flow into this country. Sources and quantities of pollutants transported by regional rivers are not well known. GIS has been applied on two different scales. On national scale, surface areas of all catchments have been defined and used for calculating missing discharges and loads. On regional scale, in the framework of catchment orientated case studies, a land use GIS has been created in order to study the effects of agriculture as a non-point source of water pollution.

INTRODUCTION

Especially in The Netherlands, much attention has been payed to transboundary water pollution. Besides three large rivers (Rhine, Meuse and Scheldt), approximately 100 relatively small, the so-called regional transboundary rivers and streams, flow into The Netherlands (Molenaar & Bleuten, 1992). From ecological point of view, the regional lowland rivers are very valuable as they are characterized by a high diversity in habitats. In general, streams often function as a backbone of the so-called ecological infrastructure. Transboundary systems are especially important from international point of view, connecting nature reserves in different countries. Unfortunately, these regional rivers and streams are often heavily polluted but the sources and cumulative quantities of pollutants are not well known. To a large extent, this is because upstream parts of the catchments fall outside the Dutch Water Board Districts. Besides, water management of regional rivers is separated into 17 different water boards and provincial governments, using different data management systems, resulting in fragmented and incomplete time series which are not compatible. Consequently, the possible effects on water quality and wet ecosystems in The Netherlands are not quite clear.

The purpose of this paper is to give a first determination of the contribution of regional rivers and the spatial variation in total transboundary river pollution to The Netherlands, supported by a GIS of all watersheds. Also the relation between variations in river water quality along the Dutch border and the agricultural land use per region will be illustrated. Finally, relationships between land use and water quality by using a catchment approach and a land use GIS, are discussed.

In the framework of the first two subjects, GIS has been applied to determinating load quantities and land use/water quality relationships on a national scale. A second application has been carried out on regional scale.

TRANSBOUNDARY WATER POLLUTION ON NATIONAL SCALE

The Netherlands receive drainage water from five upstream countries. The total foreign drainage area equals approximately 5 times The Netherlands itself, which is including regional systems nearly 210 000 km^2. The catchment of the river Rhine accounts for 75% of the total surface area. The remaining 25% is drained by the rivers Meuse, Scheldt and 100 relatively small lowland rivers and brooklets which cross the Dutch border, consequently causing a quite diffused source of water pollution.

Data management and the role of GIS

In order to quantify the cumulative transboundary solute load transported by regional systems, time series of discharge and water quality have been collected. This information was stored in an Integrated River Information System, so-called IRIS (Fig. 1).

Loads of individual rivers are based on concentration levels of solutes and discharge. Often no time series were available, only 15% of all regional systems were monitored on discharge and 60% on water quality. A survey of rivers and brooklets whose water quality has been monitored is given in Fig. 2. Determination of missing discharges was based on first order relationships between catchment surface area and net precipitation. This was accomplished by digitizing watersheds of individual catchments, using 50 topographical and hydrological maps (scale 1:50 000). Where possible, catchment boundaries were defined as groundwater watersheds, which often differ from surface water watersheds. A GIS of catchment areas of all regional transboundary river basins related to The Netherlands (Fig. 2) has been composed and fitted into IRIS. This data system serves as a frame to the model for determination of transboundary water pollution on a national scale. The final model for determining solute transport per catchment and cumulative loads of pollutants on national scale is also shown in (Fig. 1).

Dimensions, discharge, water quality and loads

Together, all foreign parts of regional catchments which drain towards The Netherlands measure 9600 km^2. This equals nearly 25% of area of the country (CBS, 1990) and 50% of the total watershed area of the second largest transboundary river, the river Meuse (Van der Made, 1972). The surface area located in The Netherlands and directly influenced by these regional

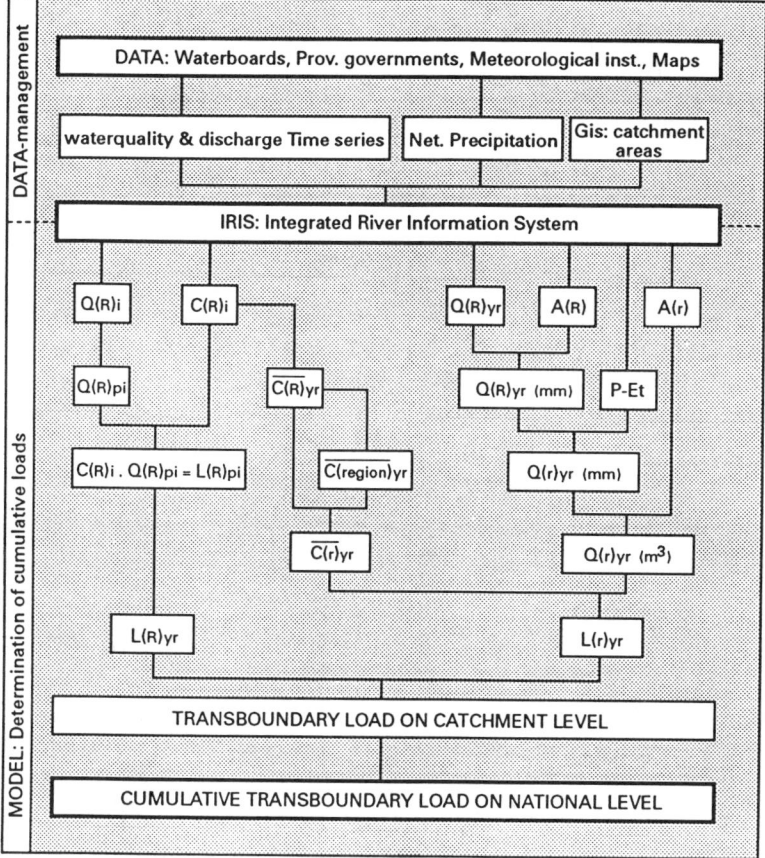

Fig. 1 Determination of cumulative transboundary load using an Integrated River Information System.

transboundary systems (including transboundary groundwater) approximately equals the foreign drainage area. The catchment area frequency distribution is skewed and shows that 75% of all basins do not exceed 5000 ha. With 30 m^3 s^{-1} as average discharge, the river Roer is the largest regional system (number 32 in Fig. 2).

Regional river basins in Germany and Belgium yearly discharge between 2.8×10^9 and 3.7×10^9 m^3 water, which corresponds with 4.3% of the Rhine discharge. From year to year both individual and cumulative contribution of regional systems into the transboundary discharge can be very different.

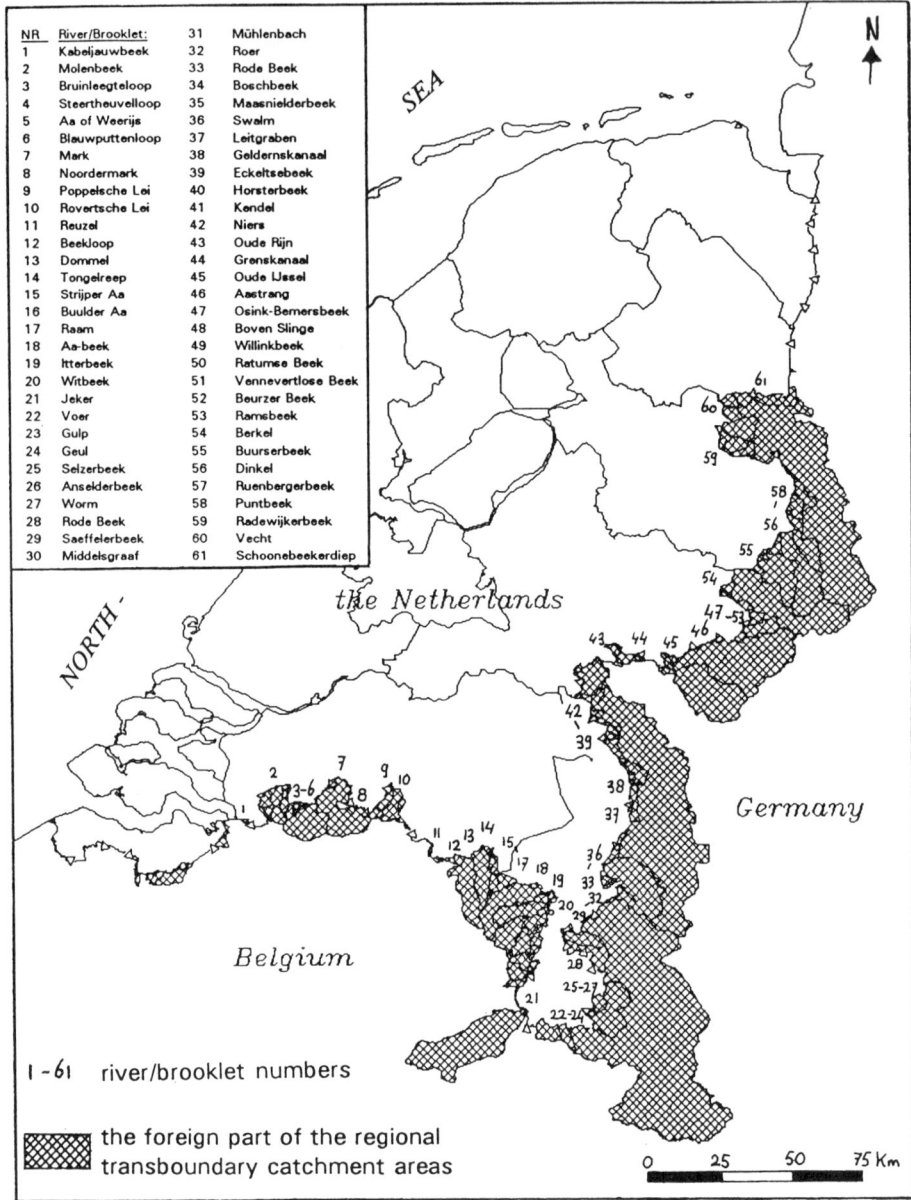

Fig. 2 Location of main regional transboundary rivers and brooklets and foreign parts of the catchments.

Catchment response depends strongly on characteristics of geology and soil within the drainage basin.

Water quality of the transboundary rivers and brooklets also shows relatively high variation, both in time and in space. Average solute levels of some nutrients and heavy metals (Table 1) are relatively high and exceed national standard levels (Ministry of Transport and Public Works, 1989). Maximum annual levels of heavy metals, like cadmium and zinc, and the

Table 1 Regional rivers: water quality and cumulative transboundary load (t year^{-1}) into The Netherlands (in 1987).

	C_{aver}	s	C_{max}	n	Standard level	Load (t year^{-1})	% of Rhine load
Cl	49	26	173	61	-	306 800	2.5
SO$_4$	77	33	173	42	-	323 500	4.8
NH$_4$-N	1.8	1.9	9.7	61	-	6 015	11.1
(NO$_3$+NO$_2$)-N	7.0	4.3	19.8	61	2.2	22 700	6.1
TPO$_4$-P	0.7	0.66	2.75	57	0.15	2 110	6.0
Zn-tot	105	165	1056	46	30	320	10.1
Pb-tot	7.0	7.2	32.3	38	25	36	10.0
Cd-tot	1.1	3.9	26.8	46	0.2	3.5	28
Cu-tot	6.1	2.9	12	36	3	26	5.1

C_{aver}= average of all mean annual concentrations of regional transboundary rivers (in mg l^{-1}), except metals (in µg l^{-1}); s= standard deviation; C_{max}=maximum of all annual concentrations; n= number of rivers used for calculation.

nutrients nitrogen and phosphorus measured in individual rivers and brooklets even exceed national standards by a factor of 100 and 10, respectively.

Individual discharges of the regional systems are relatively small especially in view of the national scale. However, it is often not realized that, except for lead and chloride, mean annual solute levels of the average regional transboundary river also exceed levels measured in the river Rhine. It has also been ascertained that solute values for the river Rhine, especially heavy metals, have declined between 1985 and 1990, whereas levels in most smaller rivers have showed a significant increase during the same period. Consequently, until now the contribution of regional rivers to the total transboundary load of pollutants, especially of heavy metals, has been underestimated in most National environmental surveys. It is calculated that the transboundary load of regional rivers, in particular heavy metals, can exceed 20% of the Rhine load (Table 1).

In order to quantify the cumulative pollution transported into The Netherlands, an estimation of natural concentrations on national scale has been made for some solutes (Table 2). The natural solute levels are based on both

Table 2 Actual versus natural transboundary load of pollutants transported into The Netherlands by regional rivers (1987).

	natural level µg l^{-1}	excessive load[1] t year^{-1}	relatively[2]
(NO$_3$+NO$_2$)-N	500 - 2000	21 156 - 26 825	74-93%
TPO$_4$-P	50 - 150	1530 - 1917	73-91%
Zn-tot	10 - 20	252 - 286	79-89%
Cu-tot	1 - 2	18.7 - 22.4	72-86%

[1] actual cumulative transboundary load minus load based on hypothetical natural levels;
[2] excessive load as percentage of actual total load.

literature (Geochem-research, 1992; Molenaar, 1992) and on median values found in reference areas which are located in northwestern Europe. For this study, calculations were limited to natural values on national scale. However, in future, regional natural values should be used where-ever possible, because without human activities regional differences with respect to geology, geohydrology and soil would reflect in the spatial variation in streamwater chemistry. First attempts to build up a GIS of geological and lithological characteristics per river basin have already been made.

Water quality/land use relationships

On base of spatial variation of water quality (Fig. 3), it seems possible to distinguish several geographical districts in which river basins have partly similar water quality characteristics. It has been found that generally these water quality regions are stable from year to year. For example basins in districts with Pleistocene sandy deposits, located in the eastern and southern parts of The Netherlands, contend with relatively high nitrogen and phosphorus levels in both groundwater and surface water. Rivers crossing the German-Dutch border mainly are influenced by agricultural activities, this is reflected in relatively high nitrate and phosphate concentrations. A group of rivers and brooklets (numbers 12-19 in Fig. 1) crossing the Belgium-Dutch border has also been contaminated with heavy metals (Fig. 3).

In this part of Europe, river basins are almost completely occupied by anthropogenic activities. Industrial and agricultural activities in foreign parts of the transboundary rivers can change streamwater chemistry or a group of streams in a negative way. In order to analyze possible relationships between upstream agricultural land use and water quality near the Dutch national border, more insight in surface areas of different land use types was necessary. German and Belgian bureaus of statistics have supplied land use data on municipality scale. In total 267 boundaries of municipalities were digitized. This GIS has been used as overlay on the watershed map, to determinate land use percentages per basin. Roughly, the districts with high levels of phosphorus and nitrogen correspond with catchment groups which have relatively high land use percentages of pasture and maize (Fig. 3). Like in Dutch part of the sandy areas, in neighbour countries the increased use of manure fertilizer and organic manure has also caused water quality problems. Maize and nearby pasture lands are used to dump the over-production of slurry which is produced by intensive animal husbandry (De Wit *et al.*, 1989). On heavily grazed grasslands, grazing itself can be a significant contributor to high nitrogen and phosphorus loads (Heathwaite *et al.*, 1990). In German parts of the catchments, the area of maize and pasture lands has doubled in the past 4 years.

High nutrient levels do not always correspond with large areas of maize and pasture in upstream parts of the basins, but sometimes have been caused by inflow of unpurified sewage water. The latter often can be observed in rivers which have source areas in Belgium.

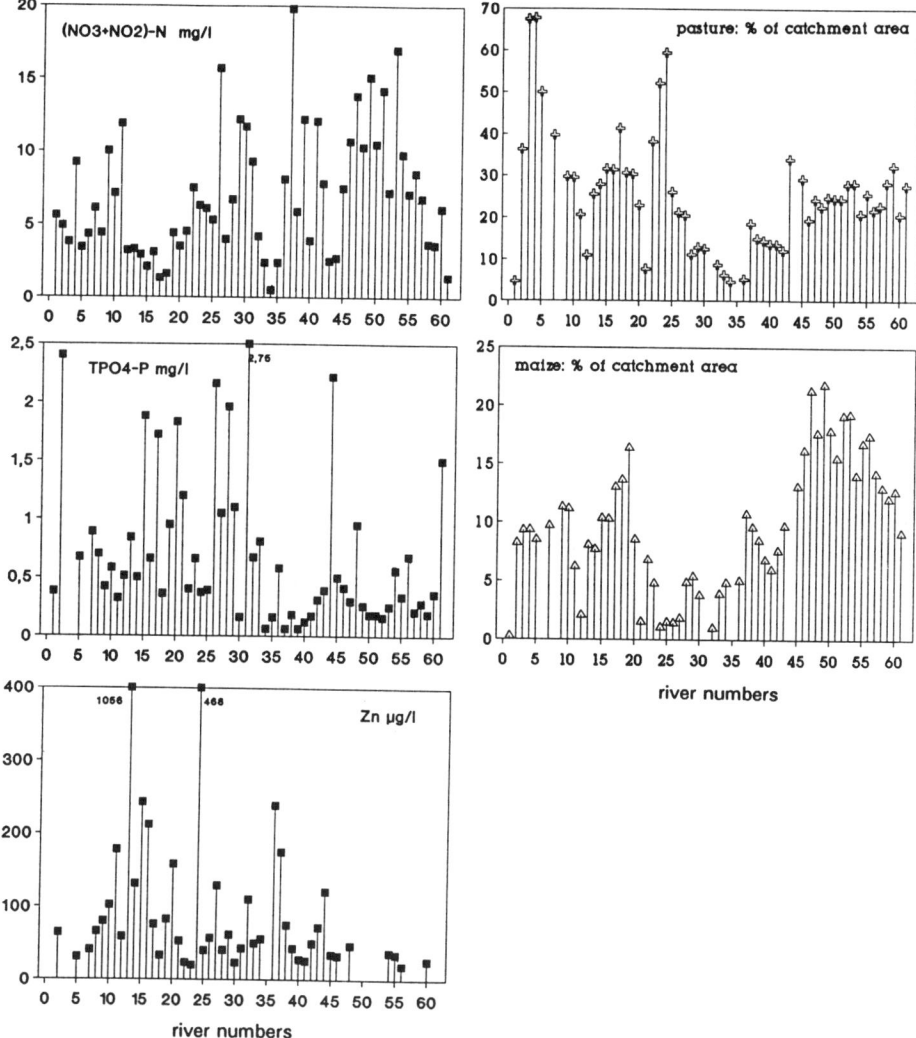

Fig. 3 Annual concentrations of nitrogen, phosphate and zinc and land use percentages per basin (river numbers as used in Fig. 2).

The load of pollutants on top of the natural load is caused by both point and diffuse sources. For trace elements such as zinc and copper it is more complicated to designate sources of pollution. Except for the northern region of Belgium, increased levels of both elements are mainly caused by point sources in upstream parts of the catchments. Here atmospheric deposition related to old fashioned metallurgical industries is an important diffuse source of zinc and cadmium, which is illustrated by relatively high background concentrations in the upper reaches and source areas of transboundary rivers. Copper is brought into the aquatic environment both by agricultural and industrial activities.

TRANSBOUNDARY WATER POLLUTION ON CATCHMENT SCALE

Agricultural land use in foreign parts of the river basins leaves an important mark on surface water quality measured at the Dutch border. More information about relationships between water quality and land use on catchment scale will contribute to a better understanding of in particular diffuse sources of transboundary river pollution. Various land uses will cause different water quality problems, which need specific measures to realize sustainable conservation of the environmental quality. For example, to protect transboundary river ecosystems against negative land use impacts, buffer zones which are partly or completely located in upstream countries may be needed.

Three catchment-orientated case studies were carried out to study the effects of agriculture as a non-point source of water pollution. Field maps of land use were digitized in order to determinate surface areas of each land use type and, secondly, to compute the nitrogen and phosphorus over-production per hectare. In the first phase of the case study, which is described in this paper, water quality aspects measured near the Dutch border are compared with upstream land use. Longitudinal water chemistry variations within the catchments related to land use will be studied in a second phase.

Case studies: preliminary results

Three catchments have been studied, all are situated in Pleistocene sand districts near the Dutch border. Except for some small forests, the land has been cultivated completely (Table 3). General characteristics per catchment are as follows:
(a) the river Dommel (number 13 in Fig. 2) after 30 km crosses the Belgium-Dutch border, where it has drained 13 454 ha of agricultural land. In this catchment also a complex of metallurgical industries has settled. Some point sources of zinc and cadmium contaminate a small tributary, which in turn pollutes the downstream system. Moderate levels of nutrients and extremely high heavy metal concentrations have been observed;
(b) the river Tongelreep (number 14 in Fig. 2), a small lowland stream parallel to the river Dommel, drains 8423 ha of mainly agricultural and partly forested land. Streamwater chemistry is characterized by moderate nutrient levels and relatively high heavy metal background levels;
(c) the river Boven-Slinge (number 48 in Fig. 2), a 5124 ha drainage area, is located in Germany and draining agricultural land to the eastern part of The Netherlands. Here high nitrogen and phosphorus and low metal concentrations have been monitored.

Discharge and water quality time series used in the framework of this study have been monitored during the period 1984-1989 by various Water

Table 3 Land use and water quality in three catchments.

catchment	Boven-Slinge	Tongelreep	Dommel
general:			
drained area (ha)	5124	8423	13 454
aver. discharge (l s^{-1})	630	940	2 075
land use percentages:			
pasture	29.1	33.7	43.7
maize	26.4	18.2	9.2
grain	24.1	0.1	3.2
other arable	3.1	3.4	3.3
forest	12.0	28.2	24.6
urban/industry	4.5	16.0	15.6
average solute levels at Dutch border:			
NH_4-N (mg l^{-1})	2.1	0.7	1.1
(NO_3+NO_2)-N (mg l^{-1})	9.5	2.9	3.3
TPO_4-P (mg l^{-1})	1.4	0.4	0.6
Zn-tot (µg l^{-1})	44	129	855
Pb-tot (µg l^{-1})	4.8	8.0	16.7
Cd-tot (µg l^{-1})	0.28	0.98	15.4
Cu-tot (µg l^{-1})	8.9	13.8	36.1

Water quality and discharge are based on 1984-1989, land use in 1990.

Boards, and by the principal author between 1988-1992. During these years, the average streamwater chemistry only had some minor changes. In general, nutrient concentrations in eastern districts increased and metal levels in the southern district decreased slightly. The land use is based on field mapping during 1990.

In the agriculturally orientated district of the Boven Slinge brooklet, intensive husbandry of cattle, pigs, and poultry is causing serious problems of soil, groundwater, and surface water contamination. Animal manure is being spread out over pasture, maize, and grain lands, accounting for 75% of the catchment area. Consequently, out of this drainage area annually a stream load of 230 tons of nitrogen is being transported across the border, corresponding to 56 kg N per hectare of cultivated land. Total nitrogen concentrations measured at the border in rivers Dommel and Tongelreep both are relatively low. The net over-production of nitrogen per hectare of cultivated land, based on stream nitrogen loads, is 37 and 24 kg N ha^{-1}, respectively. These proportions correspond with calculated differences in manure over-production based on data about total production of manure per municipality surface area. Differences, with respect to weighted (to surface area of cultivated land) mean annual over-production between the study areas, are not as large as stream concentrations of nitrogen imply.

Also highest values of weighted mean phosphate loads occur in the Boven Slinge catchment (6.8 kg P ha^{-1} year^{-1}), but are more or less comparable with calculated values of the river Dommel (5.2 kg P ha^{-1} year^{-1}). However, it is known that unpurified sewage water within the Dommel basin influences phosphate levels up to the Dutch border (Molenaar, 1992). Beside this, the extent of phosphate leaching is probably higher in the southern sand district (De Wit *et al.*, 1988).

The river Tongelreep is almost not influenced by point sources, but is located downwind of air pollution point sources in the Dommel catchment. Consequently, in this basin heavy metal values can be considered as a high unnatural background level caused by atmospheric deposition. It has been calculated that, based on the actual average stream concentration of 129 µg l^{-1}, 0.45 kg zinc ha^{-1} year^{-1} leaches towards the surface water, not taking into account the adsorption by soil layers and river sediments. Because the industries are settled in the Dommel basin, it is assumed that background levels are higher in this basin due to upwind location in the gradient of zinc deposition. This is demonstrated by higher concentrations in the source area. Taking into account a background level of zinc of 150 µg l^{-1}, it can be estimated that within the Dommel basin still 46 t year^{-1} is originating from point sources.

Case studies: future plans

In the second phase of this study the longitudinal water chemistry gradient form source area downstream to the Dutch border will be studied. Within each catchment area sub-basins have been defined. Relationships between measured concentrations of nutrients and agricultural land use will be studied with a cascade model at catchment level, using cumulative stream loads per sub-basin and data concerning manure production per land-use type. Finally, a quantification of total agricultural input to the river system will be made for the model catchments. Results may be useful in estimating the cumulative agricultural impact on national scale and to give more insight in natural background levels.

CONCLUSIONS

GIS has been a useful tool in calculating cumulative loads of pollutants and in investigating water quality/land use relationships both on national and regional (catchment) scales. On national scale, for the first time, the total contribution of regional river systems to transboundary transport of solutes could be calculated and visualized. It has been found that several water quality districts can be distinguished as a result of differences in land use and soil characteristics, although specific point sources of industrial and domestical pollution partly disturb this picture. Average levels of nutrients and especially of heavy metals in regional rivers and brooklets often exceed standard levels and levels measured in the river Rhine, consequently the cumulative transboundary load is relatively high.

On catchment scale, GIS has been used to map land use and supported load calculations which needed information about surface areas of sub-basins or land-use types. However, to achieve a better understanding of land use/water quality relationships on catchment level, especially with respect to longitudinal variations, more research supported by a catchment information system will be necessary.

It is clear that, for a policy on regional transboundary rivers to be successful, it has to be based on the river basin principle. A river basin analysis based on Information Systems of land use and of hydrological, hydrochemical, geohydrological and ecological features, will be of use to establish a basin-orientated, international and integral water management.

REFERENCES

CBS (1990) *Environmental statistics of The Netherlands*. Netherlands Central Bureau of Statistics, The Hague.

De Wit, N.H.S.M., Vissers, H.J.S.M. & Bleuten, W. (1989) The production and use of nitrate and phosphate in agriculture and their consequences on regional groundwater quality. *Soil Technology* **1**, 393-412.

Geochem-research (1992) Natural background levels of heavy metals and other trace elements in Dutch surface water. Ministry of Housing, Physical Planning and Environment, The Hague (in Dutch).

Heathwaite, A.L., Burt, T.P. & Trudgill, S.T. (1990) The effect of land use on nitrogen, phosphorus and suspended delivery to streams in a small catchment in Southwest England. Chapter 12. In: *Vegetation and Erosion* (ed. by J.B. Thornes). John Wiley & Sons Ltd., 161-177.

Ministry of Transport and Public Works (1989) *Water in The Netherlands: a time for action*. Summary of the National Policy Document on Water Management, The Hague.

Molenaar, A. (1992) Water quality of regional transboundary brooklets and rivers: a first attempt to determinate natural background values. Department of Environmental Studies, University of Utrecht, internal report (in Dutch).

Molenaar, A. & Bleuten, W. (1992) Regional transboundary rivers: the case of The Netherlands. *European Water Pollution Control* **2** (4), 25-30.

Van der Made, J.W. (1972) Hydrography of the Meuse catchment. H_2O. **5** (17), 356-362.

Flood simulation assisted by a GIS

I. MUZIK
Department of Civil Engineering, The University of Calgary, Calgary, Alberta, Canada T2N 1N4

C. CHANG
Hydrology Division, Water Resources Branch, Environment Canada, 75 Farquhar Street, Guelph, Ontario, Canada N1H 3N4

Abstract A microcomputer-based geographic information system (GIS) supporting hydrological simulation is described. The hydrologically-oriented GIS can automatically provide parameters required for prediction of flood hydrographs and physically-based synthetic flood frequency curves for ungauged watersheds. The hydrological model uses the Soil Conservation Service (SCS) runoff curve number (CN) method and a synthetic unit hydrograph to represent the physical transformation of rainfall into runoff. The standard SCS method was modified to account for randomness of key parameters whose probability distributions can be derived from regional rainfall and runoff data. The GIS and the hydrological theory and techniques, including the modified SCS method, derivation of a regional dimensionless unit hydrograph, and Monte Carlo simulation procedure, are described. Examples of results of hydrological analysis and simulation by the system implemented for an area of 11 000 km^2 in Alberta foothills are presented.

INTRODUCTION

Implementation of hydrological models requires handling of large quantities of various spatial data for computation of parameters associated with the models used, which can be laborious, repetitive, costly, and even error prone. These problems can be tackled by incorporating the spatial data in a geographic information system (GIS). A GIS has proven to be a potentially valuable organizing concept for spatial data in the field of water resources, and an efficient method of parameterizing and using models which require spatial data (Grayman, 1985). However, at present, few GISs are designed especially for hydrological purposes. Many parameters and procedures required for the implementation of various hydrological models are rarely included in existing standard GISs, severely limiting their applicability in hydrological design (Muzik & Pomeroy, 1990).

The paper describes results of a study having the following objectives:
(a) to develop a microcomputer - based GIS capable of supporting flood hydrograph simulation, and generation of synthetic flood frequency curves at ungaged sites;
(b) to develop necessary hydrological software for flood simulation using a lumped watershed model based on the SCS runoff curve method for abstractions, and a dimensionless unit hydrograph;
(c) assess the performance of the standard SCS method and modify it if

necessary to improve simulation results;
(d) assess the performance of a regional dimensionless unit hydrograph in flood simulation.

STUDY AREA

The study region runs generally in the north-south direction along the eastern slopes of Alberta's Rocky Mountains, as shown in Fig. 1. The total length of the region is about 330 km with an average width of 70 km. The topography of the region varies from rolling, forested foothills in the east to rugged, bare mountain ridges in the west. Soil in the area generally fall into the B or C hydrological soil group, according to the SCS classification (16). Runoff curve numbers determined from standard SCS tables and for antecedent moisture condition II, range from 65 to 85. The majority of the area receives some 300 mm to 400 mm of precipitation over the summer months, which is about 55% to 65% of the annual precipitation.

THE HYDROLOGICALLY-ORIENTED GIS

Organization, hardware, and software

The GIS of the present study has been created for the task of hydrological simulation. Figure 2 shows the organization of the GIS. The system consists of four major menus: (1) handling GIS database; (2) handling parameters of watersheds; (3) hydrological model; and (4) utilities. Each menu contains several sub-menus. Each menu or sub-menu has on-line help to assist users in manipulating data and operating the system. This GIS has been designed specially for IBM MS-DOS microcomputers and compatible. It was implemented on a Zenith 80386 microcomputer with a math co-processor linked to a digitizer, a graphic monitor, a printer, a plotter, and a modem (optional). All computer programs are written in Microsoft QuickBASIC version 4.5 programming language and compiled to executable files. The menu-driven software chains 15 programs together. Nevertheless, each program performing a specific task can be executed individually.

Data structure

The raster data structure was adopted in the present GIS because, generally, it is easier to perform overlay and update operations than with the vector data structure. In the raster method, the data structure consists of an array of orthogonal grid cells. Each grid cell is referenced by a row and a column number, and is assigned a number representing the type or value of the attribute of interest.

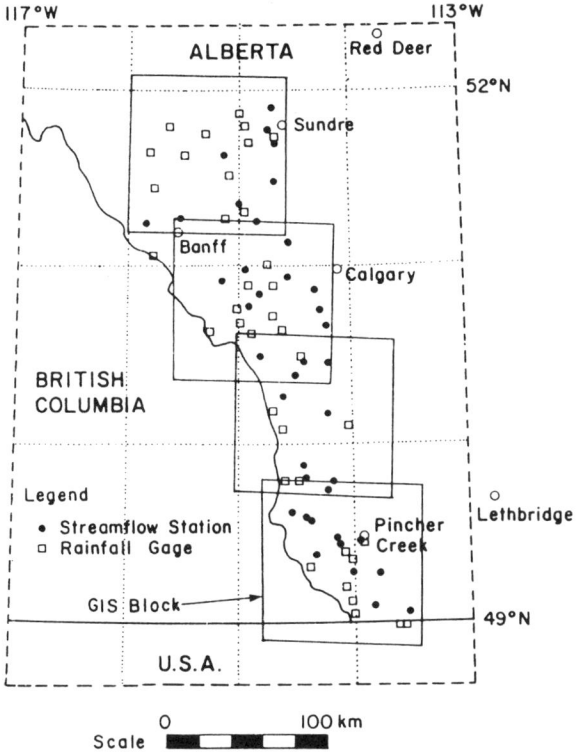

Fig. 1 Location of the study region.

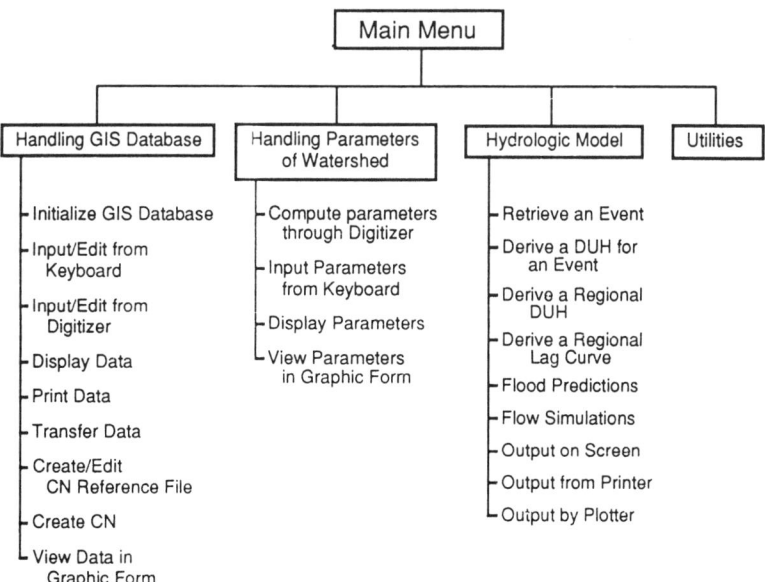

Fig. 2 Organization of the menu driven hydrologically-oriented GIS.

For the purpose of input, retrieval, analysis, and display of data, the area covered by the GIS is subdivided into 10x10 km blocks numbered in a similar sequence as on National Topographic Survey maps. The block numbers are used primarily to move large amounts of data between files. With the exception of rainfall data, which represent average values over 10x10 km grid squares, the blocks are further subdivided into 1x1 km grid cells. The (X,Y) coordinates at the SW corner of each grid square form the reference coordinates (I,J) for the square. Further, (I,J) ordinates at any point (X,Y) are equal to the rounded-down integer parts of the respective (X,Y) coordinates. When data are input into the GIS, the Y axis is aligned in the north-south direction and the origin is identified with the Canadian Universal Transverses Mercator grid to facilitate cross-referencing. Data are stored in random access files to ensure rapid retrieval for any part of a study region.

Data stored in the GIS database

To keep the size of the GIS database manageable within the microcomputer operating environment, the study region was divided into four 100x100 km areas as shown in Fig. 1. The GIS database was created within these four block for an area of about 11 000 km². Four types of spatial data were stored: (a) land cover and land use classification; (b) soil drainage classification; (c) SCS runoff curve number, CN; and (d) extreme rainfall statistics. Except for CN, the data were extracted manually from maps or plans and then input from the digitizer or the keyboard. Land cover and land use data were abstracted from composite forest cover series maps of 1:100 000 scale, and soil drainage data (hydrological soil groups A, B, C, and D) were acquired from ecological land classification maps of the same scale. The CN value for each cell was computed automatically by overlaying land cover and land use data with soil drainage data. The extreme rainfall statistics stored are the mean and the standard deviation for the Gumbel distribution of the rainfall depth for storm durations of 1, 2, 6, 12, and 24 hours extracted from the Rainfall Frequency Atlas of Canada (Environment Canada,1985). For rainfall durations of 48 and 72 hours, the statistics can be computed by multiplying the values of 24-hour rainfall by a factor of 1.35 and 1.5 respectively.

In addition to the spatial data, some derived regional hydrological information (details will be discussed later) is also stored in the GIS database. This includes a regional dimensionless unit hydrograph, a regional lag-time relationship, as well as probability distributions of storm duration, 5-day antecedent precipitation, and initial abstraction.

HYDROLOGICAL ANALYSIS AND SOFTWARE

The fifteen major programs comprising the hydrological and GIS software (Fig. 2) can simulate single floods, and by extension through Monte Carlo simulation, can derive synthetic flood frequency curves. This can be achieved at both gauged and ungauged basins by using regional hydrological parameters. To create the data base of regional parameters the hydrological software serves a dual purpose: (a) in analytical mode it enables derivation of regional parameters of the hydrological model from raw data, such as the observed discharge and precipitation data; (b) in predictive mode it is part of the flood simulation model.

The hydrological model which is used to convert excess rainfall into direct runoff is based on a dimensionless unit hydrograph (DUH) concept. The excess rainfall is determined by the standard SCS method, and the modified SCS method developed by the authors. Simulation results were used to evaluate: (a) the performance of the average dimensionless unit hydrograph, and (b) performance of the standard SCS method in comparison with the modified SCS method.

Dimensionless unit hydrograph

The Bureau of Reclamation (US Department of Interior, 1976) procedure of deriving a dimensionless unit hydrograph from a direct runoff hydrograph was used in this study. The procedure calls for derivation of unit hydrographs from observed rainfall and runoff events.

It was found that the computational effort could be greatly decreased, without much loss in the accuracy of simulation, by using a triangular unit hydrograph instead of a curved one. Sixty-one triangular unit hydrographs derived for a thirty-one basins in the study region from observed rainfall runoff events were non-dimensionalized and then averaged. This average, or regional dimensionless unit hydrograph, (DUH), was stored in GIS.

Standard SCS method

The Soil Conservation Service (US Department of Agriculture, 1972) procedure for computing abstractions from storm rainfall relates the depth of excess rainfall or direct runoff P_e to the total rainfall depth P, initial abstraction I_a, and potential maximum retention S by the following equation

$$P_e = \frac{(P - I_a)^2}{P - I_a + S} \tag{1}$$

An empirical relation was developed by the Soil Conservation Service to estimate I_a as follows

$$I_a = 0.2\ S \tag{2}$$

For convenience, the potential maximum retention S is converted into a parameter CN, having the range of values between 0 and 100, by the following equation

$$CN = \frac{25400}{S + 254} \qquad (3)$$

Values of CN have been tabulated by the Soil Conservation Service (US Department of Agriculture, 1972) on the basis of soil type and land use for three antecedent moisture conditions.

Modified SCS method

Although the standard CN tables provide helpful guidelines, the curve number is not a constant, but can vary from event to event. The variability in CN can not be explained by three discrete antecedent moisture conditions alone (Hjelmfelt, 1991). Moreover, the validity of conventional relationship between I_a and S, given by equation (2), has been questioned by several investigators (Bosznay, 1989). The following outlines a modified SCS CN method which has been developed by the authors for use with local rainfall and runoff data. The modified method takes the stochastic characteristics of parameters S and I_a into account by treating them as random variables.

Modifications to the standard SCS method consist of deriving the following relationships: (a) S - P5 relationship, where P5 is the five day antecedent precipitation, and (b) a probability distribution of the initial abstraction I_a. Details of the method are given by Chang (1992).

Rainfall analysis

For the purpose of Monte Carlo simulation, used in this study to derive synthetic flood frequency curves, it is necessary to know probability distributions of the five day antecedent precipitation P5, total duration of storm D_s, and total depth of rainfall P. Log-Pearson type III distribution was found to fit the variables P5 and D_s in the study region very well. The parameters of the probability distributions of P5 and D_s are stored in the GIS database.

SIMULATION PROCEDURE AND RESULTS

Having created the data base for a region, the software can be used for predicting either a single event flood hydrographs, and/or a synthetic flood frequency curve for any location of interest in the region. The simulation procedure can be broken down into the following steps:
(1)   place a topographic map on the digitizer;
(2)   input the map scale;

(3) digitize two points on the map through the digitizer cursor in order to orientate X and Y axes;
(4) digitize the basin boundary and the main watercourse;
(5) input the elevations of end points of the main stream.

At this point, the software automatically computes and displays the following watershed parameters: drainage area, L, Lca, So (parameters used to determine the time to peak of the DUH), the average CN, and the average rainfall statistics (mean and standard deviation of Gumbel distribution) for various storm durations.

(6) input the number of flood hydrographs to be generated, N;
(7) select the representative time distribution of storm. This may be user specified, or selected as one of the nine time distribution curves by Huff (1967). In Monte Carlo simulation the curve is selected randomly;
(8) choose either the standard or the modified SCS CN method to determine excess rainfall.

If a single hydrograph is simulated, the user determines parameters in steps 9 through 12, and 14. In Monte Carlo simulation the parameters are generated randomly from appropriate probability distributions, stored in the GIS.

(9) generate a random value of storm duration, D_s;
(10) given D_s, generate a random value of total rainfall, P;
(11) given P, compute the time distribution of the storm, i.e., compute the storm hyetograph.

If the modified SCS CN method was selected in step 8, the software continues in step 12. If the standard SCS CN method was selected, the software uses equation (3) and equation (2) to calculate the maximum potential retention S and the initial abstraction I_a, respectively; then goes to step 15.

(12) generate a random value of the 5-day antecedent precipitation, P5;
(13) given P5, determine S using the P5-S relationships;
(14) given P5, determine the appropriate I_a group and generate a random value of I_a;
(15) given the storm hyetograph, S, and I_a, compute the excess rainfall hyetograph by equation (1);
(16) determine the basin time to peak of the DUH by regional regression equation;
(17) synthetic unit hydrograph is computed by coupling the time to peak and the regional DUH;
(18) the predicted flood hydrograph is computed by the convolution of the excess rainfall hyetograph with the synthetic unit hydrograph;
(19) the peak discharge of the predicted flood hydrograph, Qp, is determined;

Single hydrograph simulation terminates by displaying and plotting the results. In Monte Carlo simulation steps 9 through 19 are repeated N times and terminated in step 20.

(20) flood frequency analysis of the generated Qp series is performed; the

synthetic flood frequency curve, observed data (if any) and the fitted probability distribution to observed data are displayed and plotted.

Flood frequency curve simulation

Synthetic flood frequency curves were generated by the Monte Carlo technique for twenty-seven watersheds in the study region, using the modified and standard SCS methods coupled with the regional DUH. Drainage areas of the simulated watersheds ranged between 64 and 820 km^2. The watersheds had between 10 and 55 years of observed natural flow records, with the mean record length of 26 years. Synthetic frequency curves, consisting of 5000 generated peak discharges, were compared with the empirical frequency curves of the watershed of annual maximum discharges, and with the log-Pearson III probability distributions fitted to the annual flood series. An example is shown in Fig. 3 for the Little Red Deer River near Water Valley.

The goodness of fit to the empirical distribution of observed discharges was judged by the χ^2-test. In case of the modified SCS method 21 out of 27, or 78%, of synthetic frequency curves passed the test at 95% confidence level. For the standard SCS method only 4 out of 13 (only 13 watersheds had complete soil information available to compute CN), or 31% of synthetic curves passed the test. This result indicates, in accordance with results obtained for single hydrograph simulation, that the modified SCS method provides much more accurate estimates of excess rainfall than the standard SCS method does.

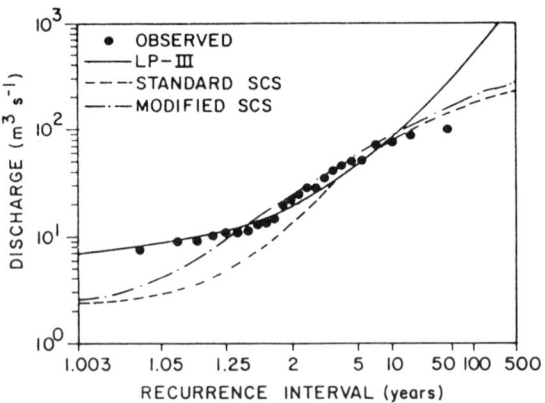

Fig. 3 Empirical flood frequency curve fitted by log-Pearson III distribution, and generated synthetic frequency curves for Little Red Deer River near Water Valley.

CONCLUSIONS

Main conclusions derived from the study are:
(a) once the GIS database has been created the time required for hydrological simulation is not a significant factor, freeing the modeller to

investigate different scenarios to arrive at the optimum solution;
(b) regionalization of hydrological parameters, which is promoted by the GIS, improves accuracy of simulation results;
(c) it is easy to update or modify the GIS database in order to study the impact of watershed changes, such as urbanization, on runoff;
(d) output in form of text, tables, graphs, thematic maps, etc. is easily produced;
(e) the method of derived distributions, or synthetic frequency curves of flood flows is physically based, thus allowing a judicial extrapolation of empirical frequency curves, and computation of frequency curves for changed watershed conditions;
(f) the lumped model used in the study is based on a modified SCS method for abstractions and a synthetic unit hydrograph. This model performed well in simulating flood frequency curves for most tested watersheds. It is sensitive enough to respond to relatively small variations in runoff conditions, thus it should be very useful in planning studies exploring impacts of various stages of urban development on flood flows;
(g) the modified SCS method for abstractions significantly improves flood hydrograph predictions in comparison with the standard SCS method. Application of the standard SCS method generally resulted in underestimation of peak flows in the study region;
(h) implementation of the SCS method and a regional dimensionless unit hydrograph requires considerable hydrological analyses of regional data, but the described software can be used in an "analytical mode" to simplify this task.

Acknowledgements This research was financially supported by a grant from Natural Sciences and Engineering Research Council of Canada.

REFERENCES

Bosznay, M. (1989) Generalization of SCS curve number method. *J. Irrigation and Drainage Engng* 115(1), 139-144.
Chang, C. (1992) Physically Based Flood Prediction Model Aided by a Geographic Information Systems, Ph.D. thesis, Department of Civil Engineering, The University of Calgary, Calgary, Alberta, Canada.
Environment Canada (1985) *Rainfall Frequency Atlas of Canada*. Canadian Climate Program, Canadian Government Publishing Centre, Ottawa, Ontario, Canada.
Grayman, W.M. (1985) Geographic and spatial data management and modeling. *Proc. ASCE Conf. on Computer Applications in Water Resources* (ed. by H.C. Torno), Buffalo, New York, USA, 50-57.
Hjelmfelt, A.T. (1991) Investigation of curve number procedure. *J. Hydraulic Engng* 117(6), 725-737.
Huff, F.A. (1967) Time distribution of rainfall in heavy storms. *Wat. Resour. Res.* 3 (4), Fourth Quarter, 1007-1019.
Muzik, I. & Pomeroy, S.J. (1990) A geographic information system for prediction of design flood hydrographs. *Canadian J. of Civil Engng* 17(6), 965-973.
US Department of Agriculture (1972) *National Engineering Handbook*. Soil Conservation Service, section 4, hydrology, United States Government Printing Office, Washington, D.C., USA.

US Department of Interior (1976) *Design of gravity dams*. Bureau of Reclamation, United States Government Printing Office, Washington, D.C., USA.

"APS on GIS": an operational forest hydrological GIS

D.W. VAN DER ZEL
Forestry Branch, Department of Water Affairs and Forestry, Private Bag X93, Pretoria, 0001, South Africa

F.V. RABE
Water Affairs Decision Support Service, Strategic Planning Directorate, Private Bag X313, Pretoria, 0001, South Africa

Abstract Due to limited water resources, legislation was introduced in South Africa in 1972 to control new afforestation of man made commercial timber plantations. After 20 years, 3700 permits for 900 000 hectares were granted and 400 permits for 130 000 hectares were refused. 1800 permits for 420 000 hectares were actually utilized.

INTRODUCTION

The Republic of South Africa (RSA) only receives an average annual rainfall of 444 mm. Successful establishment of man-made commercial timber plantations, from which timber and fibre needs can be satisfied, need a minimum of 750 mm rain per year. Less than 5% of the land does receive such minimum rainfall, but usually in the upper reaches of catchments. Runoff in South Africa is equivalent to only 7% of the rainfall so that it is clear that blanket afforestation would pose a threat to water resources. An Afforestation Permit System (APS) was therefore brought into being in 1972 by means of an amendment to the Forest Act. All land owners, state and private, in the RSA need a permit before any afforestation of new land can take place.

For more than two decades the APS has been in use. After 20 years, 3760 permits for 909 498 hectares have been issued, while 433 applications to afforest 132 779 hectares had to be refused. In practice, only 1100 permits were utilized by afforesting a new area of 420 391 hectares from 1972 to 1991. During the tenure of the system, the importance of accurate maps was increasingly realized. Special efforts were made to improve the plantation map and ultimately to incorporate it as part of the permit. On this map, areas which should remain unplanted, such as wetlands, riparian areas, and indigenous high forest, are shown, while on the permit several conditions (such as annual removal of stray trees outside the plantation area) are specified.

However, the continual increase in permits issued, the partial completion of most of the permits and the replacement of weed areas and non-commercial tree growths with commercial plantations made it more and more difficult, out of a viewpoint of water resources decision-making, to quantify and envisage the extent of the new afforestation. The dire need for spatial data was identified.

The "hand" system also became too time consuming, with an average time of three months between application and permit. Another disadvantage was the indirect link (through reports) with hydrological data available on computer and GIS. The system simply became outmoded.

A GIS programme was recently compiled to operationally use GIS in issuing permits under the APS (Afforestation Permit System). As a permit application is received, the information is keyed into GIS on a Sun Workstation. This Workstation operates in a TCP/IP Ethernet local area network (LAN) consisting of X-terminals, PC's emulating X-terminals, and SUN data applications and software servers.

A plantation map, indicating proposed afforestation, is captured on GIS as the basis to evaluate all possible influences on river flow, dam water input, water quantity, water quality, biodiversity, wetlands, scenic landscapes and protected indigenous forests. This system has recently become operational.

THE AVAILABLE GIS ENVIRONMENT

The planning and strategic sections of the Department of Water Affairs and Forestry (DWAF) have as goal the provision of an integrated and comprehensive range of professional and information related services and products, needed for facilitating and supporting decision-making in affairs related to water and forest management.

In order to achieve these goals, an extensive GIS structure has been developed, which forms the backbone of Planning Support Systems. ARC/INFO is used on an Ethernet LAN network, which consists of two Sun 690 servers, one SUN 370 server, one Prime server, nine workstations and 45 PC's. Presently there are 16 ARC/INFO workstation licenses and 73 PC ARC/INFO licences (in use at several locations in South Africa). Input devices include 6 digitising tables and a Screen 1000 scanner plus a V-track package to vectorize scanned data. Output is acquired through an electrostatic plotter, as well as a pen plotter. An Erdas system is being set up for image processing.

Planning Support Systems projects mainly consist of data capturing, land use management support, GIS support for other organizations, map production, surface analysis, presentation graphics and the integration of GIS with other decision-making tools such as expert systems, operational research, statistical information and remote sensing.

THE "APS ON GIS" PROJECT

A research project was commissioned by the Forestry Branch of DWAF with the Division of Forest Science and Technology of the Council for Scientific Industrial Research in South Africa (FORESTEK) to develop an operational Water Affairs Geographic Information System (WAGIS) for the Afforestation Permit System (APS). FORESTEK made use of two consulting firms, Infomet and Gims, to enable it to apply the most recent information technology systems in this first forestry GIS application.

The project involved designing and building the computerized component of "APS on GIS". The system would have a user-friendly menu-driven control program (to enable inexperienced field staff to input data), a data input sub-system, a modelling sub-system that will apply the APS criteria to a specific area and produce the resulting overlay, a generic query facility, and a reporting sub-system that will ultimately issue the permit plus map plus conditions. Three primary products were a logical, a functional and a technical specification.

The purpose of the logical specification of the project "APS on GIS" was to describe APS in terms of its different information dimensions: strategy, functions, data, organization, time, locality and operation, thus ultimately providing a basis from which the further design of the system could be derived.

The functional specification was then developed, outlining the system design in terms of hardware (SUN Sparc Station, A1 digitizer board, etc.), software (ARC/INFO), data (permit information, different RSA coverages, etc.), operations and personnel required to run the project.

Different datasets and procedures, menus and screens made up the technical specification, which denoted the system as ready to be programmed by the appointed programmers.

The system was then developed, producing program code by using the ARC/INFO system.

Data capture incorporated data capturing for the APS query system and the actual permit data itself. The query system consisted of the following:
(a) forest-economic map of southern Africa;
(b) forest regions of southern Africa;
(c) development region boundaries of southern Africa;
(d) major rivers and dams of the Republic of South Africa;
(e) catchment boundaries of the RSA;
(f) farm boundaries (shortly available for the whole country);
(g) magisterial districts of the country;
(h) provincial and international boundaries of southern Africa;
(i) afforestation permit information.

The system will also be able to display areas that have been afforested and the areas for which permits have been approved. Two days were found to be sufficient in training future users of the system.

There are multiple advantages to the "APS on GIS". One of its outstanding features is that eventually the complete scarce water resource decision-making procedure can be handled at the Sun Sparc Station and a permit and prescribing map produced at once.

The second advantage is the availability of all the historic data in spatial format. This will certainly provide perspective and better planning.

Another advantage is the provision of equally weighted input from all parties involved through their own GIS module inputs which can continually be updated. Finally this APS on GIS can also be linked to a total Forestry Industry on GIS spatial project.

In water resources decision-making, many parties require input or consideration. The basic "APS on GIS" programme is therefore written in such a way that it can accept several subprogrammes or modules into its decision-making process. If for example, the Natal Parks Board authorities require wetlands to be taken into consideration when evaluating permit applications in Natal, these authorities should provide the programme module for Natal wetlands. Such programme modules are expected to be developed and provided by three provincial nature conservation authorities, regional water affairs and/or irrigation board authorities, the central Agricultural Department and the central Department of the Environment. Such modules would include wetland areas, sensitive ecological areas, boundaries of irrigation board areas, national heritage sites, indigenous forests, etc. Expert programmes are also being developed to streamline uniform application of an unplanted riparian zone along all perennial streams and wetlands. The main product of FORESTEK's land use hydrology research will be a hydrological forest land management model. An advanced version, soon to be updated, already forms the basis of the APS. The ultimate model will eventually drive the APS towards balanced and sustainable land management and use.

THE ILLOVO CASE STUDY

The Illovo project was the pilot case study for "APS on GIS" and the cornerstone on which subsequent forestry permit projects are based.

The Illovo catchment area in the southern part of the Natal province formed the study area. It was ideal in the sense that it had a high level of forest and plantation activity. A high number of farms were also present. River and catchment activity could easily be monitored as the catchment bordered on the Indian Ocean.

A total of seven 1:50 000 maps were digitized, covering the whole of the catchment area. It consisted of three different coverages, namely original farm boundaries, plantations and natural forests, as well as primary, secondary, and tertiary catchment boundaries.

The historical (1977-1992) forest permit areas were then digitized onto these areas. Seeing that this was the first exercise of this sort quite a few problems were encountered. The main of these were due to the quality of the maps received from the farmers themselves. Most were hand drawn, lacking essential detail and without taking scale into account. To correctly locate and indicate these areas on a 1:50 000 map was next to impossible. The solution to this was for the regional offices to consult with the specific farmer. At present spatial data consists of the permit area before 1972, areas within the permit expanse actually planted and whether some areas have not been utilized as requested for. Permit particulars (alphanumerical data) were then linked to these areas. It consisted of the permit number, name and farmer or firm, name of the farm, postal and residential address, type of tree, application number, expiry date of the permit and other permit integral information.

To examine the functionality of satellite data, the afforested and natural forest coverage was overlayed with 1987 satellite forest imagery. Although the map forest data was of 1981 origin, the two sets corresponded considerably well. The use of satellite imagery and especially GPS (Global Positioning System) will be examined further and incorporated into "APS on GIS" for future use.

The Illovo case study was a valuable learning phase that contributed to the general outlay of the later "APS on GIS".

CONCLUSION

The Illovo Case Study has shown the variety of useful applications of an operational system such as "APS on GIS". We believe that we have only scratched the surface, and that the real diamonds will still be found when digging deeper. Water affairs' strategic planners envision an integrated catchment management system, whereby every land use and every land user will develop his water needs and his water use influences and will need to obtain his water licence with the consensus of the community. We strongly envision a system whereby the community considers and decides. It is, however, our duty to provide the facts, the mechanisms, the tools, the up-to-date technology: "APS on GIS" is a first attempt at such community involvement.

Spatially oriented surface water hydrological modelling and GIS

M. BRILLY, M. SMITH & A. VIDMAR
University of Ljubljana, Faculty for Architecture, Civil Engineering, and Survey, Hydraulics Department, 61000 Ljubljana, Slovenia

Abstract Geographic information systems (GIS) provide a convenient context for coupling hydrological models with spatial entities. Some of these entities are dictated by the internal data storage structure of the GIS, while others are based on the hydrological response of a geographic area. Much research has been directed into using GISs and existing hydrological models for input data preparation and presentation of final results. Most types of hydrological modelling cannot be fully implemented in a GIS context due to current inabilities to effectively handle the time dimension. Further research should consider the development of network functions to fully support hydrological analysis and a possible data structure which considers the time dimension.

INTRODUCTION

Definition of GIS

After more than 20 years of development, GISs have become common and almost indispensable tools used in construction, government, health, commerce, industry, politics, crime fighting, emergency services, and transportation. According to the Soil Conservation Service (SCS) of the US Department of Agriculture, a GIS can be defined as a system of hardware and software that allows the user to input, store, manipulate, analyze, and display geographic (map and image) data (SCS, 1992).

The definition of a GIS is so broad that it can include almost any of type of spatial data arrangement or suite of analysis capabilities. Yet, with any definition is the fundamental concept of dealing with spatial data. With this, we can say that the hydrological models developed in the early 1960's contained elements of GIS concepts. However, some may comment that these systems did not comprise a true GIS. Thus, the question aries as to what defines a true geographic information system. Moreover, the question remains as to what degree we can use a true GIS for solving water resource management problems.

GIS spatial primitives

In the Spatial Data Transfer Standard (SDTS) (US Geological Survey, 1992), spatial objects are classified and defined as the following:
(a) zero dimensional spatial objects - where one coordinate pair specifies location. These are points used for identifying point features and nodes

which link two or more chains. Points also serve as places where attribute values and names are attached to polygons;
(b) one-dimensional spatial objects - lines. These are line segments, strings (comprised of line segments), arcs (curves with mathematical expression), links (topological connection between two nodes), chains (a directed non-branching sequence of lines and arcs bounded by nodes), and rings (closure sequence of non-intersecting chains);
(c) two-dimensional spatial objects - areas. These can be with or without boundaries, polygons with inner non intersecting rings, and regular shaped pixels or grid cells.

The previously mentioned spatial primitives are aggregated into the following: digital images (pixels), grids of regular or almost regular cells (square, equilateral triangle, or regular hexagon), layers (an aerially distributed set of spatial data representing entity instances within one theme), rasters (a number of overlapping layers for the same grid or image), and graphs (all three types of objects which follow a set of topological rules).

Depending on the data structure, GIS can be categorized as raster-oriented (grid, image, layer, and raster), vector-oriented, and raster/vector-oriented. The accuracy of analysis in a raster-based GIS is dependent on the grid cell size. With vector-based systems, accuracy depends on the density of vectorized entities. GISs facilitate the manipulation and analysis of much more data than is contained on a single map.

GIS AND HYDROLOGICAL MODELLING

Linkages between GIS and hydrological models

Major linkages between GIS and hydrological models can be described as follows (Maidment, 1991):
(a) hydrological parameter determination. Here, GIS functions are used to more efficiently derive parameters needed to drive existing hydrological models such as the SCS family of models and the US Army Corps of Engineers HEC-1 program (Sasowsky & Gardner, 1992);
(b) GIS-hydrological model combinations. In this broad category, GISs are used to develop hydrological parameters from basic data sets. Results of spatial analyses are also developed as data layers and output using GIS functions. A large percentage of the GIS hydrological model research falls into this group. Johnson (1989) presents a GIS linked to a suite of hydrological models of various complexity;
(c) GIS and imbedded hydrological model. In this group, the hydrological model is an integral part of the GIS and uses the computing language of the host GIS to perform hydrological analyses.

This third category is the most limited in that GISs cannot yet effectively handle the fourth dimension (time) required by the majority of hydrological models (Maidment, 1991). However, some types of hydrological modelling can be performed, such as deriving average annual values of a variable such as pollutant loadings or annual flows (Maidment, 1991). Djokic (1991) developed an imbedded hydrological model - GIS linkage consisting of ARC/INFO and the Rational Method. These authors used the command language resident within ARC/INFO to analyze storm sewer capacities and network connectivity and were perhaps the first to effectively combine an expert system and GIS.

GIS and distributed parameter hydrological models

With their inherent ability to link many data layers, GISs have facilitated the use of distributed parameter hydrological models. Previously, these models had limited operational application due to tremendous data requirements. For example, one distributed parameter model required 22 pieces of information for each computational element (Young *et al.*, 1989). It is this very characteristics of distributed parameter hydrological models that often renders them inefficient for everyday operational hydrology. Yet, advances in computer technology have led to the implementation of these types of models in the microcomputer environment (Johnson, 1989). Soon, computing power will no longer constrain the use of these models (Beven, 1989).

While distributed parameter models are growing in applicability, Beven (1989) criticized most current models in this category and called them at best *lumped conceptual models at the grid scale*. His conclusion is based on the belief and common practice that sub-grid hydrological processes can be lumped together up to the scale of the data structure, say 250 by 250 m. His work points out that there is a danger when applying these models when the hydrological processes occurring at sub-grid scale are not well understood. Currently, there is no theory for lumping of sub-grid processes in hydrology.

A typical computational element for distributed parameter hydrological modelling is shown in Fig. 1 which illustrates one approach to quantifying the

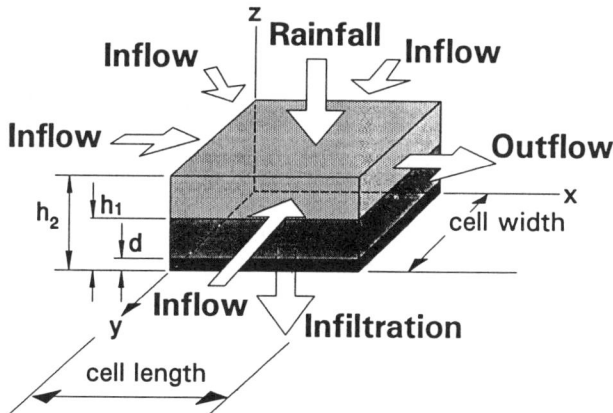

Fig. 1 Dimensions and processes used with computational element in distributed parameter hydrological modelling (adapted from Smith, 1992).

spatial variability of hydrological parameters. The processes of rainfall, infiltration, inflow, and runoff generation are computed for each element within each time step. A depth h_1 exists at the beginning of each time step and a depth h_2 at the end of each time step. A detention storage depth d is assigned depending on the land cover type. Depending on the complexity of the model, other processes can be incorporated such as those related to water quality modelling.

Partitioning a watershed into computational elements similar to that shown in Fig. 1, subsequently requires the determination of the proper hydrological computation sequence. Automated algorithms have been developed to derive this sequence for triangular irregular networks (TIN), digital elevation models (Gandoy-Bernasconi & Palacios-Velez, 1990) and for depressionless grid DEMs (Smith & Brilly, 1992). Figure 2 presents the procedure of Smith & Brilly. In their algorithm, raw optimal path densities described by Berry (1987) and Tomlin (1990) are ranked to produce the proper computational sequence as shown at the bottom of Fig. 2. This numbering scheme is similar to the link magnitude concept originally developed by Shreve (1967) for channel networks.

WATER RESOURCES MANAGEMENT INFORMATION SYSTEM

Introduction

A water resources management information system (WMIS) separate from a GIS is difficult to imagine. The flow, storage, and demand of water are spatially distributed and oriented. Point primitives (representing springs, waterous caves, sinks, etc.), line primitives (streams, pipes, etc.), and area primitives (watersheds, aquifers, etc.) are all aggregated into a complex three-dimensional system. Almost all features of sophisticated GISs are used for handling hydrological water resources information.

The basic structure of a WMIS is the stream system. Surface water flow is closely related to the stream network. Even on planar surfaces, water quickly forms small streams. The position and characteristics of the river network are crucial input descriptions for surface water modelling.

Network functions are powerful and necessary tools in geographic information systems. To this date, network models have largely been developed with a view towards transportation planning and routing. Shortest route and least cost path analyses are common examples of network functions. However, applications on network functions have been somewhat limited in the arena of water resources analysis. Problems arise during derivation of the stream/river network, assigning identifiers to segments, and other tasks. However, grid-based data structures combined with linear networks partitioned using dynamic segmentation techniques show promise as a spatial entity for hydrological modelling in a GIS context (Maidment, 1992a,b).

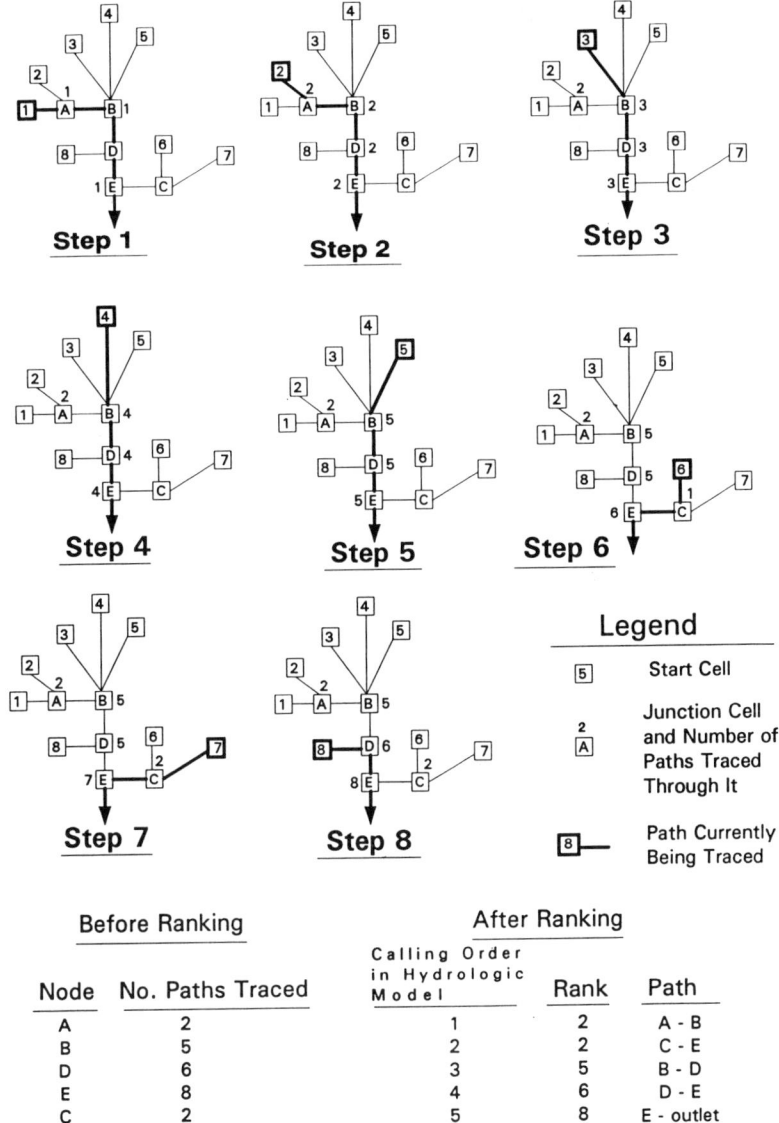

Fig. 2 Numbering and sequencing of computational elements for distributed parameter hydrological modelling (adapted from Smith, 1992).

Stream networks

For stream network coding, Strehler stream numbering is most commonly used (Garbrecht, 1988). In GIS-based applications, hydrologists are faced with the problem of stream coding using GIS formats while maintaining hydrological utility. Coding can be performed using universal identifiers useful for an entire country data base, or strictly for stream network operation purposes. Stream coding could also consider the spatial data associated with the adjacent areas.

For the WMIS for the Republic of Slovenia, a simple hydrographic coding system was developed (Vidmar, 1992). Coding is related to the downstream node of each river segment with the mouth of the river being the last node. Each stream is assigned an identifier equal to the coordinates of the mouth of the stream. Attributes are assigned according to the stream and the streams linked to it as shown in Fig. 3. An algorithm was derived for the automated determination of stream segments in the upstream and downstream directions of a chosen design point. In the near future, programs for the calculation of parameters for hydrological and hydraulic modelling will be incorporated into the WMIS.

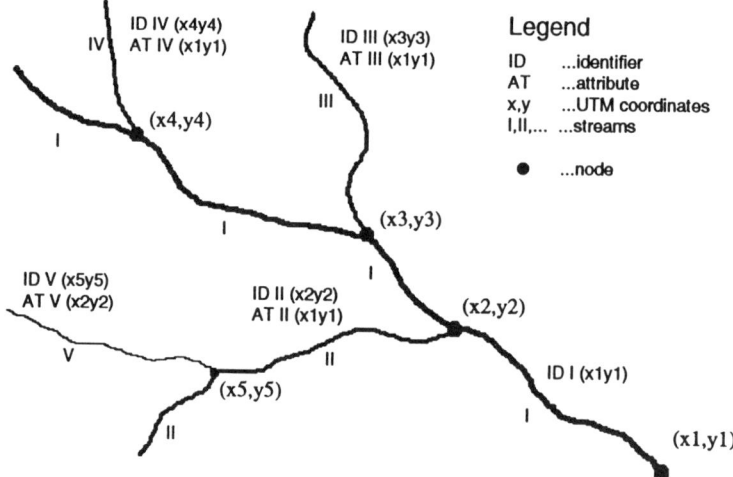

Fig. 3 Methodology for stream network coding for WMIS for Republic of Slovenia (Vidmar, 1992).

Using this system of coding, the stream network of Slovenia was delineated as shown in Fig. 4. There are 8000 streams are included in this data base. The system can potentially be expanded to include karst phenomena. In addition, possibilities exist to include algorithms for Strehler numbering, upstream and downstream distance calculations, stream-watershed linkage, watershed delineation for a specific stream point, and calculation of model parameters for hydrological and hydraulic modelling.

WMIS for the Republic of Slovenia

This project was started in 1988 and is still ongoing. WMIS is a developing system of software, hardware, and supporting services to handle scientific and administrative information requirements of the Water Industry of Slovenia. WMIS is an integrated system for data capture, storage, retrieval, analysis, and presentation. It presently includes the following data:
(a)   identified pollutant sources in 1986 and 1990;
(b)   administrative boundaries;

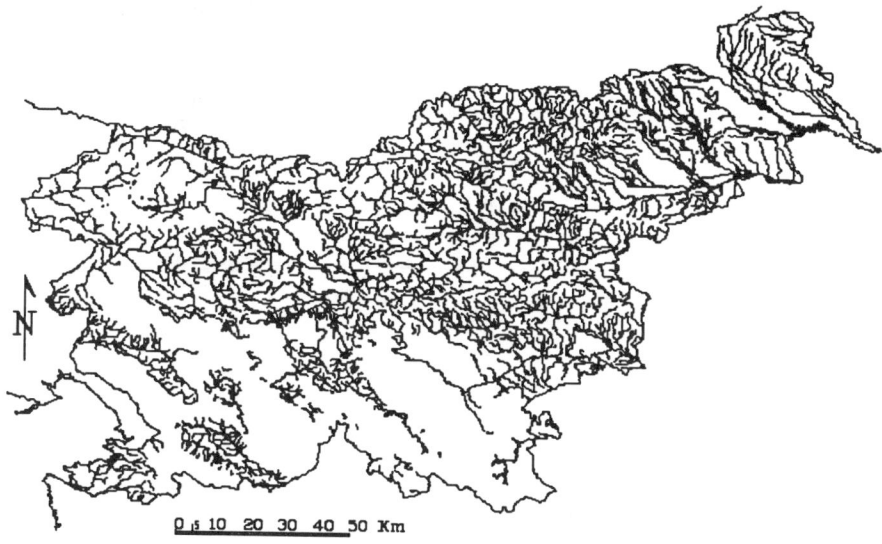

Fig. 4 Stream network in Slovenia.

(c) locations of water springs (Figs 5 and 6), caves, stream and rain gage stations, groundwater wells, piezometric stations, cities and settlements.
(d) stream network;
(e) Digital Terrain Model (DTM), a grid digital elevation model (DEM) with 100 by 100 m resolution.

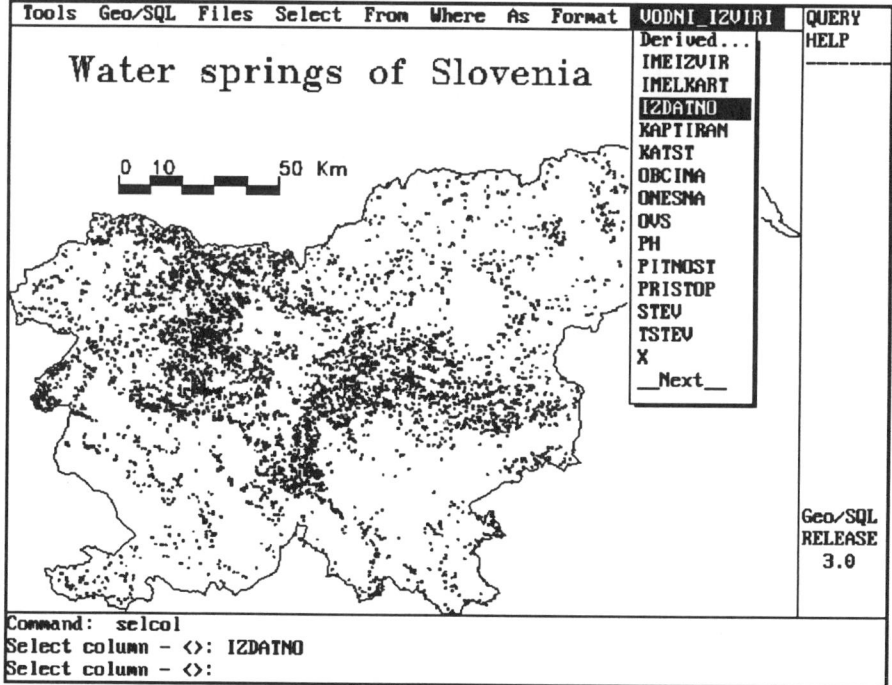

Fig. 5 Location of water springs in Slovenia.

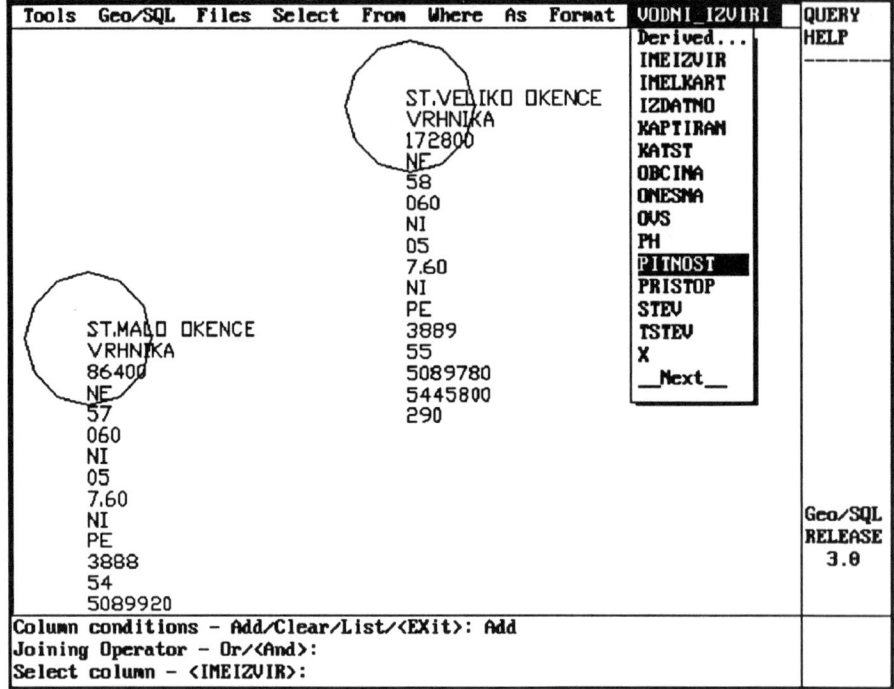

Fig. 6 Example of attribute data for springs using GEO-SQL GIS.

The WMIS for Slovenia is based on GEO/SQL software. In essence, GEO/SQL is a spatial database manager (SPDBM) which combines coordinate geometry (graphical information) such as points, lines, and polygons with attribute (descriptive) data, and is linked to AutoCAD.

GIS DATA STRUCTURES FOR HYDROLOGICAL MODELLING

An important consideration for GIS-based hydrological analysis is the selection of the data structure. Currently, GIS applications in this field typically contain some kind of digital representation of the terrain surface. Consequently, the data structure required to describe the hydrological processes must be compatible with the format of the terrain model. Three basic Digital Terrain Model (DTM) -hydrological model combinations have evolved. Most have developed from modelling applications to natural watersheds. These are:
(a)   Triangulated Irregular Network (TIN);
(b)   Stream Path / Contour (Moore & Grayson, 1991);
(c)   Grid / Raster (Johnson, 1989; Maidment, 1992a,b).

Moore *et al.* (1991) provide further discussion on the advantages and applications of each type of structure. It is clear that there seems to be no DTM format that is clearly superior for all tasks of analyzing a digital terrain model. Ebner & Eder (1992) describe hybrid DTMs where a grid DEM also contains

TIN elements to provide more details where necessary.

Rather than defining the data structure by the GIS data format or DTM structure, others are basing their hydrological analyses on spatial entities defined by hydrological response. Wood *et al.*, (1988) presented the concept of a representative elemental area (REA) for watershed modelling. While originally applied to groundwater analyses, these authors applied the REA concept to catchments. An REA is defined as a critical area of a watershed in which implicit continuum assumptions can be used without knowledge of the patterns of parameter values. In other words, an REA is a fundamental building block for catchment modelling. Sasowsky & Gardner (1991) applied this concept in a GIS application of a rangeland hydrological model. The authors found that for a watershed in Arizona, a partitioning consisting of 4th order streams produced results that were typically as good as a 2nd order partitioning. Thus, their work also pointed to the existence of an REA in catchment hydrology. A similar concept was described by Miller (1985), who partitioned a watershed according to hydrological response units (HRU). HRUs are defined as areas within a watershed which produce a similar amount of runoff per unit area when given equal amounts of rainfall. Other types of spatial primitives amenable to hydrological modelling were developed in Miller's work.

Mark (1978) stated that the "chief source of this (data) structure should be the phenomena in question, and not the problems, data, or machine considerations, as is often the case". In their proposal for a modelling strategy, Moore & Grayson (1991) argued that "there is a need for a method of partitioning landscapes into small areas where the hydrological processes and the soil, vegetation, and topographic characteristics can be considered uniform or at least can be characterized by simple relationships". Using this approach, these authors derived a contour-based partitioning based on the original work of Onstad & Brakensiek (1968).

Hydrological modelling is very often concerned with the movement of water through a stream system. Thus, some type of river network analysis is usually required to determine reach lengths and other parameters. However, certain channel characteristics required for flow routing cannot be derived directly from layered geographic information.

Other parameters such as channel lengths can be obtained directly from DTM analysis (Fairfield & Leymarie, 1991; Chorowicz *et al.*, 1992). Most commercial GISs contain some type of functions for network analysis. However, a large portion of these functions are typically geared to traffic analysis.

LIMITATIONS

Perhaps the biggest limitation to GIS and imbedded hydrological modelling is the current inability of GISs to model the time variability that is often required.

(Maidment, 1991). To accomplish a complete linkage between GIS and hydrological models, a GIS would be required to contain time-dependant data structures such that evolution through time of the spatial distribution of hydrological phenomena can be readily observed.

CONCLUSIONS AND RECOMMENDATIONS

Geographic information systems have facilitated the more efficient use of existing hydrological models. In addition, models have been recently developed that are based on the grid, TIN, or other data structures of commercial GISs. Current studies on watershed hydrological analysis in the context of a GIS shows that there are several basic data structures in use. Most often, these structures are defined first by the selection of DTM model, and then secondarily structured to efficiently fit a commercial GIS. Research is also pointing to the need to define the data structure based on the physics defining the hydrological response of watershed elements. Clearly, there is no structure that is superior considering DTM constraints and hydrological processes. GISs should be used as a platform for further research into spatial entities for DTM-hydrological modelling. Current research into GIS hydrological network analysis similar to that discussed in Maidment (1992b) should be continued.

REFERENCES

Berry, J.K. (1987) Fundamental operations in computer-assisted map analysis. *Int. J. of Geographic Information Systems* **1**(2), 119-136.
Beven, K. (1989) Changing Ideas in Hydrology, The case of physically-based models. *J. Hydrol.* **105**, 157-172.
Chorowicz, J., Ichoku, C., Riazanoff, S., Kim, Y. & Cervell, B. (1992) A combined algorithm for automated drainage network extraction. *Wat. Resour. Res.* **28**(5), 1293-1302.
Djokic, D. (1991) Urban stormwater drainage network assessment using an expert geographical information system. PhD dissertation, University of Texas at Austin.
Ebner, H. & Eder, K. (1992) State of the art in digital terrain modelling. *Proc. of Third European Conf. and Exhibition on Geographical Information Systems*, Munich, Germany, March 23-26, 682-690.
Fairfield, J. & Leymarie, P. (1991) Drainage networks from grid digital elevation models. *Wat. Resour. Res.* **27**(5), 709-717.
Gandoy-Bernasconi, W. & Palacios-Velez, O. (1990) Automated cascade numbering of unit elements in distributed hydrologic models. *J. Hydrol.* **112**, 375-393.
Garbrecht, J. (1988) Determination of the execution sequence of channel flow for cascade routing in a drainage network. *Hydrosoft* **1**(3), 129-138.
Johnson, L.E. (1989) MAPHYD: a digital mab-based hydrologic modelling system. *Photogrammetric Engng and Remote Sensing* **55**(6), 911-917.
Maidment, D.R. (1991) GIS and hydrological modeling. In: *Proc. of First Int. Symposium/Workshop on GIS and Environmental Modeling* (ed. by M.F. Goodchild, B.O. Parks & L.T. Steyaert), held in Boulder, Colorado, Sept., 1991, Oxford University Press, New York.
Maidment, D.R. (1992a) A grid-network procedure for hydrologic modeling. Report to Hydrologic Engineering Center, 14 October, US Army Corps of Engineers, Contract DACW05-92-P-1983.
Maidment, D.R. (1992b) Grid-based computation of runoff: a preliminary assessment. Report to Hydrologic Engineering Center, 7 August, US Army Corps of Engineers, Contract DACW05-92-P-1983.
Mark, D.M. (1978) Concepts of data structure for digital terrian models. In: *Digital Terrain Models (DTM)*, Symp. American Society of Photogrammetry, St. Louis, Mo., May.

Miller, S.W. (1985) A spatial data structure for hydrologic applications. *Geo-Processing* **2**, 385-408.

Moore, I.D. & Grayson, R.B. (1991) Terrain-based catchment partitioning and runoff prediction using vector elevation data. *Wat. Resour. Res.* **27**(6), 1177-1191.

Moore, I.D., Grayson, R.B. & Ladson, A.R. (1991) Digital terrain modelling: a review of hydrological, geomorphological, and biological applications. *Hydrol. Processes* **(5)**1, 1-30.

Onstad, C.A. & Brakensiek, D.L. (1968) Watershed simulation by stream path analogy. *Wat. Resour. Res.* **4**(2), 965-971.

Sasowsky, K.C. & Gardner, T.W. (1992) Watershed configuration and geographic information system parameterization for spur model hydrologic simulations. *Wat. Resour. Bulletin* **27**(1), 7-18.

SCS (1992) Geographic information system (GIS) and the Soil Conservation Service. US Department of Agriculture, Soil Conservation Service, Cartography and GIS Division, Washington, D.C.

Shreve, R.L. (1967) Infinite topologically random channel networks. *J. Geology* **75**(2), 178-186.

Smith, M.B. (1992) A distributed parameter hydrologic model for urban stormwater protection. PhD dissertation, Civil Engineering Dept. University of Ljubljana, Slovenia.

Smith, M.B. & Brilly, M. (1992) Grid element ordering for GIS-based overland flow modeling. *Photogrammetric Engng and Remote Sensing* **58**(5), 579-585.

Tomlin, C.D. (1990) *Geographic information systems and cartographic modeling*. Prentice Hall, Englewood Cliffs N.J.

US Geological Survey (1991) *SDTS User's Manual*. Reston, Virginia.

Vidmar, A. (1992) Mosnosti Uporabe GIS programsko orodje Geo/SQL pri upravljanju z vodnim bogastvom Republike Slovenie (Use of GIS GEO/SQL for water resources management in the Republic of Slovenia). Masters Degree thesis, Faculty of Architecture, Civil Engineering, and Survey, University of Ljubljana, Slovenia.

Wood., E.F., Sivapalan, M., Beven, K. & Band, L. (1988) Effects of spatial variability and scale with implications to hydrological modeling. *J. Hydrol.* **102**, 29-47.

Young, R.A., Onstad, C.A., Bosch, D.D. & Anderson, W.P. (1989) AGNPS: A nonpoint-source pollution model for evaluating agricultural watersheds. *J. Soil and Wat. Conservation*, March-April, 168-173.

Identification of heavy metal concentrations in surface waters through coupling of GIS and hydrochemical models

T.J. KERN & J.D. STEDNICK
Department of Earth Resources, College of Natural Resources, Colorado State University, Fort Collins, Colorado 80523, USA

Abstract The Upper Arkansas River Basin has been the site of land use activities for over a century. Past mining and metallurgical operations continue to adversely affect water quality and limit beneficial uses of water. Landscape elements (geology, soils, topography, hydrometry, aspect, vegetation, land use activity and streamflow generation mechanisms) were related to surface water chemistry by construction of chemical-hydrological response units (CHRUs). Chemical speciation models (MINTEQA2) identified metal species, since water quality standards are often specie specific. The CHRUs identified stream reaches that exceed water quality standards (or other user-input selection criterion) in space and time. The model identified heretofore unknown stream reaches that exceed standards. Further, the model allows evaluation of different mitigation scenarios in the basin as they affect downstream water quality.

INTRODUCTION

Human use of the Upper Arkansas River basin dates back over a century. Like other communities, local land uses (farming, ranching, mining and mineral processing, urbanization, and recreation) all have some impact on water quality. However, the entire Upper Arkansas River basin is also burdened with sins of the past. Given the abundance of mineral wealth, the area supported a booming industry based on resource extraction. These historic mining and metallurgical operations continue to adversely affect water quality by acid mine drainage and associated heavy metals. This acid mine drainage limits the use of basin waters for drinking water supplies, irrigation and livestock, fish and wildlife, and recreation. Waste piles, tailings deposits and heavy metal-laden sediments, smelter slag, and abandoned or inactive mine adit discharges all contribute to environmental limitations in the Upper Arkansas River basin (Moran & Wentz, 1974; Wentz, 1974; Kimball *et al.*, 1988; CWQCD, 1989; US EPA, 1989; US EPA, 1990).

In 1983 the Yak Tunnel, an abandoned mine drainage tunnel near Leadville, Colorado, discharged acid mine drainage waters laden with heavy metals. This episodic event resulted in a fish kill affecting 30 km of the Upper Arkansas River. Remediation of this acid mine drainage source continues today, a multimillion dollar program designed to monitor tunnel discharge and water quality, and remove heavy metals by precipitation in a wastewater treatment plant. However, the Leadville area is still of concern over 1300 other abandoned mines exist in the catchment. None of these are monitored or targeted for remediation. Currently, local authorities are not able to assess or predict which of these mines or drainage tunnels will become the next Yak Tunnel.

Regulatory authorities recognize that past resource extraction activities still affect surface water quality in the Upper Arkansas River catchment. The area has been the subject of 38 different water and soil chemistry studies over the past 20 years (US EPA, 1989; US EPA, 1990). Despite the efforts in the catchment, identification of potential heavy metal contamination is quite limited.

The biggest limitation seen in past work is in philosophy. The commitment of resources in the basin focuses on remediation rather than prevention. Linking water quality to landscape elements or potential sources of contamination is nonexistent. Studies identify potential water quality impacts throughout the watershed, but do not address possible cause, or sources of such contamination.

OBJECTIVES

The study objectives were to develop a new technology for the identification of water quality problems, one that identifies potential sources and addresses these sources before a substantive impact occurs. To accomplish this, we coupled a Geographic Information System (GIS) with hydrochemical software and synoptic water quality sampling and analysis. We defined three specific study objectives:

(a) to identify heavy metal specie concentrations in specified stream reaches;
(b) to couple landscape elements with water quality analyses; and
(c) to identify potential source areas of heavy metal contamination using these landscape element/water quality relations.

METHODOLOGY

The Upper Arkansas River basin is located in Lake County southwest of Denver, Colorado, United State of America. The catchment area is approximately 600 km^2, elevations range from 2700 m to 4200 m. Annual precipitation in the region varies substantially with elevation, averaging 47 cm over the entire watershed. The precipitation is divided between snow and early summer convectional thunderstorms (Fletcher, 1975). The cold snow zone has streamflow peaks as snowmelt in June, with lowflows in the winter months. Precipitation as rain contributes little to streamflow.

The Leadville Mining District sits at the intersection of the northeast trending Colorado mineral belt and the north-trending Mosquito-Weston fault system (Tweto & Sims, 1963). The geology of the region consists of a thick sequence of Pennsylvanian sediments overlying pre-Pennsylvanian Paleozoic rocks and Precambrian granitics. It is the Paleozoic structures that garner the most interest, with ore bodies containing gold, silver, copper, lead, molybdenum, and zinc (Emmons *et al.*, 1927). These mineral veins have produced over a half billion dollars worth of metals over the last 130 years. Other metals found in the area include cadmium, iron, and manganese.

We compiled existing surface water quality data from state and federal agencies (CWQCD, 1989; US EPA, 1989; US EPA, 1990). Surprisingly, the Upper Arkansas River water quality data base was limited in scope. Water quality data were often restricted to the main stem of the Arkansas River, complete chemical analyses were rare, and data accuracy was suspect.

We selected sample sites for basin water quality characterization based on known data, site access, land use activities, and personal communications. Stream reaches with elevated metal concentrations were sampled intensively to better identify metal sources. These 70 sites were sampled in January (lowflow conditions), and in May-June (spring snowmelt-highflow conditions). Since major decisions were to be based on the chemical results, we analyzed approximately 50% quality assurance samples. Chemical analyses of duplicate and standard samples show recovery of ±2%. A quality assurance project plan was prepared specifically for this study.

Complete chemical analyses were evaluated with a metal speciation model MINTEQA2 (US EPA, 1991) to identify metal species in solution for toxicological assessment. Metal speciation was, in general, independent of geological influences, and was regulated more by the surface water oxidation-reduction potential.

The water quality data base was input as tabular data to a geographic information system (GIS). The GIS allows visual spatial interpretation of water quality (metal concentrations) for the Upper Arkansas catchment in time and space.

RESULTS

Policy makers and land use managers have increasing needs to systematically evaluate and predict potential water quality changes from past, present, or future land use activities. To address this need we developed a method, the Chemical-Hydrologic Resource Information System (CHRIS), to predict water quality and assess land use management (or remediation) alternatives.

CHRIS is prompt-response spatial decision support system initially developed using ARC/INFO version 5.0.1 (ESRI, 1989) on a SUN 3/150 platform; the current configuration of the system uses more current hardware and software. CHRIS allows the user to store, analyze, manipulate, and output data based on specific user selected criteria. The system is designed for users who lack experience using computers or GIS. One of the principles guiding this project was that the GIS software used should be invisible to the viewer of the finished analysis.

Prior to the construction of this system, the only method of determining critical relationships among data was to manually sort different data files and correlate individual specie concentrations with map locations of potential controls. CHRIS automates this process. Data and outputs are easily displayed, and a land manager can quickly assess water quality in a specific stream reach.

GIS development

CHRIS uses both tabular and geographic data. The tabular data consists of aqueous chemistry and streamflow. The data used in this project represents a subset of the data available from federal and state agencies, with data relevance and validity determined by study team personnel. The files were transferred to the platform in flat-file format, then loaded into a relational data base.

USGS 1:24 000-scale Digital Elevation Model (DEM) files, obtained through the USDA Forest Service Geometric Service Center (Salt Lake City) were used to construct associated relief, aspect, and elevation information. Other geographic data consisted of regional hydrometry, physiography, geology, vegetation, soil, land use, transportation, and ownership mappings. Geological maps, many generated by CSU study team personnel, supplied the information to build the geological data base. The USDA Soil Conservation Service (SCS) data base did not have complete regional soils available in digital line graph (DLG) format. What soil maps were available were digitized.

United States Department of Interior Geological Survey (USGS) 1:24 000 topographic maps were used as the primary source of hydrometry, transportation, and land ownership. Vegetation, wetlands, updated transportation, and land uses were interpreted from infrared (IR) aerial photographs. Maps were digitized directly into the GIS, while aerial photographs were scanned and imported. This scanned data, in a raster format, had to be merged with the vector data. The raster data was converted to a vector polygon coverage, and polygons attributed based on the original IR color schemes. The structure of the system allows continuous field verification and update of vegetation and wetlands.

System construction

The GIS data base construction involved the following steps:
(a) input basin physiography, define hydrometry and distinct sub-basins;
(b) input landscape elements (geology, soils, vegetation, climate, land uses);
(c) link watershed physiography and controls to stream discharge measurements;
(d) build data base of water and sediment chemistry for distinct hydrological regimes;
(e) link chemistry to specific reach/sub-basin and relate to toxicity data base;
(f) link expected hydrological response to chemical data;
(g) define probable sub-basin loads and relate loads to landscape elements;
(h) incorporate other essential inputs (biomonitoring, aquatic life criteria);
(i) refine the model and repeat the process.

Our method of analysis began by breaking the watershed into distinct subunits, each with a characteristic hydrological and water chemistry component. These subunits, termed Chemical-Hydrologic Response Units (CHRUs), allowed us to model watershed response to both existing and projected conditions.

The idea of geographically distinguishing subunits is not new; the USGS developed a runoff modelling system based on hydrological response units (HRU) (Leavesley *et al.*, 1983). Our system builds upon this work, devising a method to define units based on expected chemical export.

The CHRU is initially defined by sub-basin boundaries. These distinct regions are inviolate. The CHRU cannot span more than one basin it must be confined to one drainage, with only one flow path allowed in each CHRU. Since the stream reach is the ultimate destination for attribute data, it can receive input only from areas hydrologically connected to it. The sub-basin delineation was expedited using a series of construction programs. The programs also added consistency to the delineation routine, regardless of the system operator, and allowed the user to quantify any input error.

The inputs came from maps with vastly different resolutions and scales. These differences were recorded in the coverage tolerance. Since the final product can be no more accurate than the inputs, care was taken to find maps at a scale that would maintain the integrity of the project. Large scale maps were the only source of some information (parts of the watershed were represented only by 1:100 000 geological map); this forced us to keep metadata files on each CHRU.

Identify stream reaches effected by heavy metals

The GIS and the water quality data bases, coupled with a front end program, enabled the user to visualize metal concentrations in space. The user selects a specific constituent (usually a metal) and a known concentration (standard, criteria, or other value of interest), and the program identifies stream segments those probably below the standard or criteria (low probability), and those predicted to exceed the standard (high probability) (Fig. 1). The water quality data bases include surface water hardness and alkalinity, two parameters necessary to determine the appropriate metal standard.

Prediction of metal contamination as related to landscape elements

After collecting hydrological, chemical, and landscape information, the GIS links the collected CHRU attributes, as control variables, to watershed response. These elements are linked through a series of relate operations to ASCII files. A separate file exists for each agency's results, as well as the method of analysis.

The hydrological contribution of a CHRU is dictated by the distribution of precipitation over time and space with ground cover and underlying soils. The hydrological model is a simple distributed parameter model using elevation, aspect, relief, vegetation, and precipitation. We developed

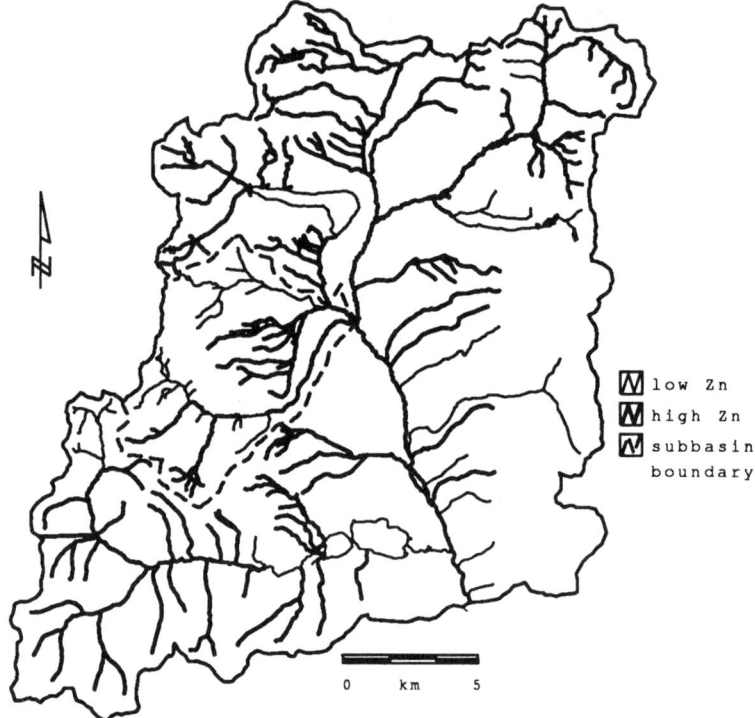

Fig. 1 Visualization of stream reaches for the probability of acute zinc toxicity (0.1 mg l^{-1}) for the Upper Arkansas River Basin, Colorado.

approximately 500 different CHRUs, with 30 distinct CHRU-types required to characterize the entire catchment area (Fig. 2).

As with the hydrological component, we used CHRU attributes to derive expected water chemistry contributions. We correlated field water quality sampling to landscape elements defining the CHRU; we used those elements to assign expected water quality to each unsampled CHRU. The chemical load from each CHRU was merged in a whole catchment model, this model output compared to the field data, and attribute-water chemistry relations refined.

We built a dynamic segmentation routine to construct the load portion of model. The segmentation routine provided the inputs needed to build a transport model. The stream segment identifier is linked to a series of attributes; the attributes can be summed along the length of the stream. Tributary inputs and flow/chemical impedances can be added to the mainstem attributes at any point in space and time for analysis of watershed transport patterns. Aqueous chemistry can be correlated with stream reach, evaluated for specie partitioning using MINTEQA2, linked to potential toxicity, and graphically displayed.

These routines make the model a dynamic system, inputs can be updated as changes in the watershed occur. Alteration to flow patterns (the addition of a flow modification or diversion, onset of spring snowmelt,...) are quickly added

Fig. 2 The 484 Chemical-Hydrologic Response Units (CHRUs) used to characterize water quality. The sub-basin delineation is Halfmoon Creek (see Figs 3 and 4).

to the data set, either as a permanent change or as a projected scenario. New water quality data, sampling locations, or attribute changes can easily be entered into the data base for real time evaluation of management options.

The land use manager can use the integrated system to view water quality trends, watch how stream chemistry changes with flow or season, or see how a proposed management action interacts with the temporal and spatial variability associated with the existing system. Proposed remediation options can be modelled as a simple routing change, and users can view the expected effect both on- and off-site (Figs 3 and 4).

Approach justification

To date, the most sophisticated use of GIS is as a decision support system for resource management. This requires complex spatial analysis, coupling the GIS data base with models designed to feed back to the GIS. The construction of this type of system can be complex and intimidating. We feel, however, that one must take such an approach to fully assess the potential controls on basin water quality. We are also aware that most agencies cannot build the same type of system without years of experience, experience that most managers do not

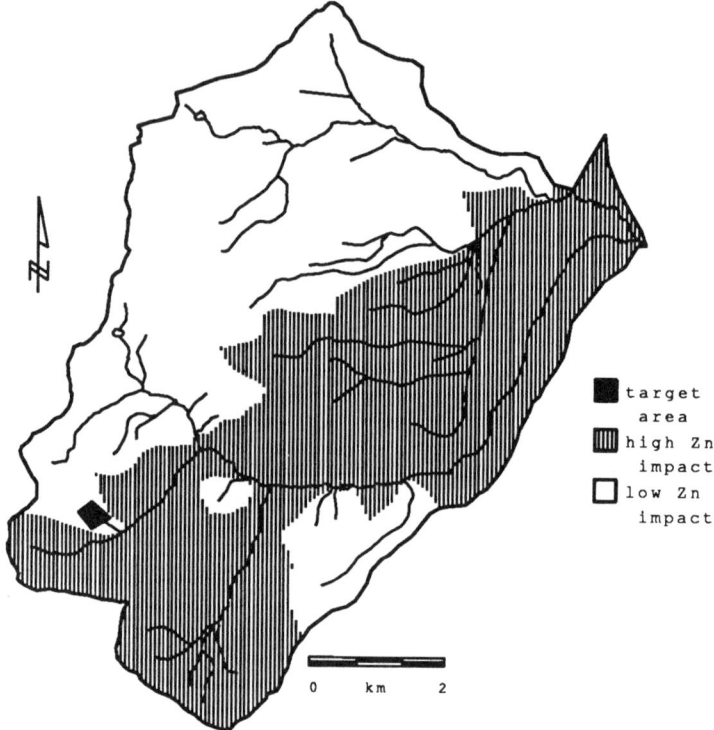

Fig. 3 Remediation with no apparent off-site benefits. Impacted area defined by acute zinc toxicity.

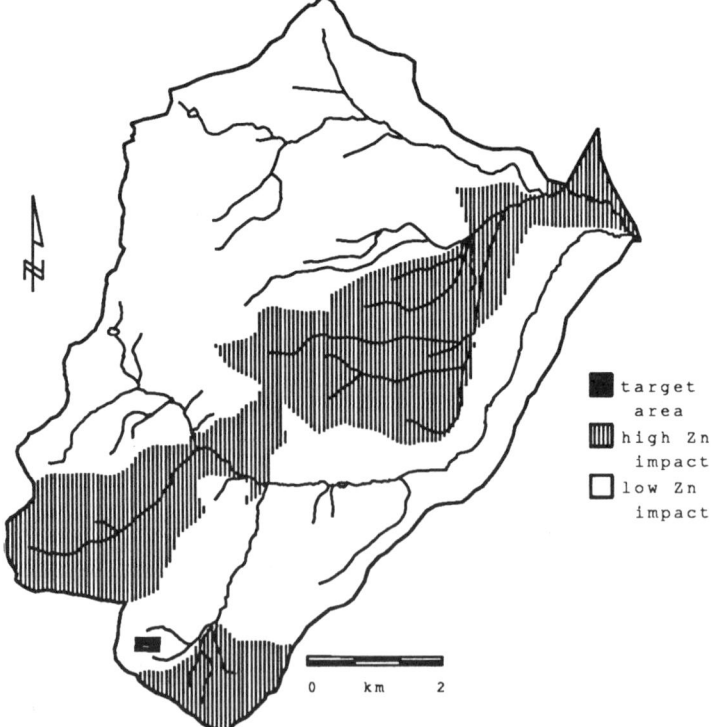

Fig. 4 Remediation effort resulting in significant improvement of off-site water quality, as identified by CHRU chemical load.

have and cannot afford. This model removes this experience requirement, giving the manager a tool to work with little or no background in GIS. As the manager's knowledge of the software matures, the model is flexible enough to allow user modification, accounting for regional nuances or adding new information.

Acknowledgements This research was made possible through a grant from the S.C. Johnson, Co. to the World Wildlife Fund, Inc. Washington, D.C.

REFERENCES

CWQCD (1989) Colorado nonpoint assessment report. Colorado Water Quality Control Division Report, November 1989.
Emmons, S.F., Irving, J.D. & Loughlin, G.F. (1927) Geology and ore deposits of the Leadville mining districts in Colorado. *US Geological Survey Professional Paper 148*.
ESRI (1989) *ARC/INFO, Version 5.0.1*. Environmental Systems Research Institute, Redlands, CA.
Fletcher, L.A. (1975) Soil Survey of Chaffee-Lake Area, Colorado: Parts of Chaffee and Lake Counties. US Department of Agriculture soil conservation Service, in cooperation with Colorado Agricultural Experimental Station, October 1975.
Kimball, B.A., Bencala, K.E. & McKnight, D. (1988) Research on metals in acid mine drainage in the Leadville, Colorado, area. In: *US Geological Survey Toxic Substances Hydrology Program* (ed. by G.E. Mallard & S.E. Ragone), Proc. technical meeting. Phoenix Arizona, September 26-30, 1988. *US Geological Survey Water-Resources Investigation Report 88-4420*, 65-70.
Leavesley, G.H., Litchty, R.W. Troutman, B.M. & Saindon, L.G. (1983) Precipitation-runoff modeling systems: Users Manual. *US Geological Survey Water Resource Investigations Rep. 83-4238*.
Moran, R.E. & Wentz, D.A. (1974) Effects of metal-mine drainage on water quality in selected areas of Colorado, 1972-1973. Colorado Water Conservation Board Water-Resource Circular No. 25.
Tweto, O. & Sims, P.K. (1963) Precambrian ancestry of the Colorado Mineral Belt. Geological Society of America Bulletin **74**, 991-1014.
US EPA (1989) Draft site management plan, appendix A, California Gulch, Leadville, Colorado. *US Environmental Protection Agency Document Control Number 4800-01-0106*, October 1989.
US EPA (1990) Surface water (spring runoff) and selected mine waste sampling, California Gulch, Leadville, Colorado. *US Environmental Protection Agency Document Control Number 4800-01-0387*, May 1990.
US EPA (1991) *MINTEQA2, Version 3.10*. Center for Exposure Assessment Modeling, Athens, GA.
Wentz, D.A. (1974) Effects of metal-mine drainage on the quality of streams in Colorado, 1971-1972. Colorado Water Conservation Board Water-Resource Circular No. 21.

8 Application of GIS in Groundwater Systems

Application of GIS for aquifer vulnerability evaluation

G. BARROCU
Department of Territorial Engineering, University of Cagliari, Faculty of Engineering, Piazza d'Armi, 09124 Cagliari, Italy

G. BIALLO
ESRI ITALIA S.p.A., Via Edoardo D'Onofrio 212, 00155 Rome, Italy

Abstract The current generation of Geographical Information Systems (GIS) are considered a powerful tool for representing aquifer vulnerability. Pollution danger scenarios, identified by integrating georeferenced, geological, hydrogeological, and soil use database, may be dynamically represented. By suitably combining updated database softcopies and hardcopies, integrated vulnerability maps are obtained and pollution fumes can be monitored and modelled to simulate and plan prevention and restoration actions.

INTRODUCTION

Methods generally adopted to evaluate aquifer vulnerability are necessarily schematic and affected by local conditions and scales of interest. Maps are mainly a static representation of a temporary situation and can quickly become obsolescent. Aquifer vulnerability maps are particularly complex, as they have to represent not only surface characteriscs but also underlying structures. In spatial monitoring of pollution at least several vertical sections should be used as well as 3D model representations made with continuously updated data bases.

The most advanced GIS are able to dynamically represent aquifer vulnerability data at various scales. In fact, they are speedy and flexible in analysing georeferenced raster and vectorial data from numerical maps, remote sensing, and field monitoring, and integrating these data in suitable formats. Therefore, GIS is a very efficient instrument for environment monitoring, dangerous area zoning, and regional planning.

VULNERABILITY

Aquifer *vulnerability* expresses the danger that a groundwater system might be polluted by addition of noxious substances, generally caused by accidental or intentional human activities. The term "aquifer" is here used to mean both the rock solid phase, at different saturation levels, and its fluid phase. In an aquifer system *pollution* affects mainly the top levels, but lower levels may also be damaged, especially by dense non-aqueous phase liquids (DNAPLs) such as chloride solvents percolating down the interspaces of improperly cemented wells.

Intrinsic vulnerability

The concept of *intrinsic or natural vulnerability* has been used for more than twenty years to indicate, from different points of view, the degree of danger of pollution and the conditions and characteristics that determine it (Albinet & Margat, 1970; Olmer & Rezac, 1974; Vrana, 1981; Civita, 1987; Bachmat & Collin, 1987; Foster S.S.D., 1987). Intrinsic vulnerability is controlled by (a) topography, (b) land cover, (c) climatology, (d) pedology, (e) hydrology, (f) unsaturated and saturated zone hydrogeology, (g) groundwater use and (h) land use.

Intrinsic vulnerability should be evaluated on a case-by-case basis, taking into consideration the physical and chemical characteristics of the pollutant (or of similar product families), the type of source (centre, diffuse), and the quantities, travel paths, and times of pollution (Andersen & Gosk, 1987; Bachmat & Collin, 1987).

Evaluation systems are very diverse, depending on physiography, the amount and quality of data, and the purposes for which the data have been obtained. Methodologies can be *universal*, for any physiographical situation, or *local*, for particular areas (Civita, 1990). In both methodologies three procedures may be adopted:

(a) in **homogeneous area zoning** vulnerability is generally evaluated for hydrogeological complexes by overlapping medium-to-large scale thematic maps suitable for large area coverages (Civita, 1990). The evaluation is qualitative and uses values prearranged for typical situations;

(b) **parametric systems** are based on a few parameters suitably selected for fixed value ranges and/or types. In *matrix systems* soil and aquifer classes and depth to water table are considered (Haertlé, 1983; Engelen, 1985; Carter *et al.*, 1987). *Rating systems* are all derived from the Legrand system (1964 and 1983). A score range for any parameter is subdivided in function of its variations. Any point and/or area is evaluated by summing its scores. Some authors emphasize the importance of chemical and physical soil characteristics, others bring hydrogeological and hydrological data into special prominence (Sotornikova & Vrba, 1987; Marcolongo & Pretto, 1987; Foster S.S.D., 1987). In *Point count system models* weight ranges are introduced for any basic parameter that is so amplified in a preordered way. DRASTIC (Depth to water, net Recharge, Aquifer media, Soil media, Topography, Impact of vadose zone, hydraulic conductivity of the aquifer) is the prototype of this system, conceived by the US Environmental Protection Agency (Aller *et al.*, 1983, 1987). Different hydrogeological situations are numerically evaluated by DRASTIC so that they can be compared, but vulnerability classes are not defined for thematic mapping (Civita, 1990). SINTACS, an automatic and integrated version of DRASTIC, has been tested in Italy (Civita, 1987, 1990). In *environmental evaluation* *systems* the impact of a danger centre on an aquifer and/or a groundwater development plan are evaluated by a hierarchy of pro and contrary environmental indicators (Pavoni *et al.*, 1972; Dee *et al.*, 1973). An intrinsic vulnerability index is determined with *evaluation numerical models*. For instance, Marcolongo & Pretto (1987) determine the vulnerability by taking into account the transfer time in the piston flow hypothesis, the hydraulic conductivity of the unsaturated zone, and the soil specific retention and infiltration capacities.

Integrated vulnerability

Integrated vulnerability implies the interaction between the intrinsic vulnerability of a hydrogeological system and the tential pollution centres. According to Legrand (1983), the more susceptible to pollution is a given aquifer, the higher is the impact potential of a danger centre towards it.

Pollution sources, depending on their type (and for the same type depending on their size), may be classified as:

(a) **point**, due to confined installations, include the legal or illegal use, stocking, and discharge of dangerous substances, such as liquid discharges in deep wells, landfills, stockfarms and connected structures (silos, fertilizer, hydrocarbon and pesticide stocks), underground reservoirs of harmful substances, and sewage systems. The danger is generally defined for specific pollutants, especially in complex hydrogeological conditions;

(b) **diffuse**, produced by irrational farming activities (using herbicides, manure, and fertilizers), urban settlements, and acid rainfall. They are represented by areas with different danger levels depending on the types and quantity of pollutants that could affect the aquifer.

In reality, aquifer regions that are already polluted (and the concentration levels of these pollutants) should also be considered when integrated vulnerability is evaluated. The integrated vulnerability of an aquifer will vary depending on the pollution already present in the aquifer.

Vulnerability and protection from pollution

Aquifer vulnerability and protection are complementary concepts. The former defines pollution dangers, the latter concerns the planning and implementation of safegard measures. They both require experimental observations and they both suggest methods of representing monitoring data. Monitoring is thus also a system of controlling integrated vulnerability.

Groundwater and soil quality monitoring should be part of ordinary management procedures, especially in critical regions such as protected zones, potential pollution centres, and coastal areas. To be effective, information from frequent updating ought to be represented with georeferenced databases. Monitoring is essential where restoration is to be planned.

Groundwater basin protection zones

The scientific principle of *protection zone* is based upon the fact, generally verified, that pollution danger declines as safety time increases. The *safe time* is the time necessary for a pollutant to reach the groundwater, and it depends on the different physical and chemical processes affecting flow. Inside protection zone perimeters, human activities are to be regulated or forbidden.

Where flow is not well known or definable, protection zones may not be the most effective method to tackle pollution dangers. Fixed perimeters may be difficult to alter, even when new technical information is provided, which suggests that the zone boundaries be modified.

VULNERABILITY REPRESENTATION METHODS

Vulnerability maps

The methods of protection zones and vulnerability maps, on different scales and at different precision levels, can be conveniently combined. Aquifer vulnerability maps are essential either in the first planning phase of preventing pollution disasters (danger identification, evaluation, and zoning, Foster H.D., 1980) or in later phases (disaster forecast, monitoring system planning, emergency planning, defence planning, effects mitigation, etc. Engelen, 1985; Civita, 1987; Aller *et al.*, 1987). Maps are classified as follows:
(a) *intrinsic vulnerability maps*, representing vulnerability variations with classes or with numerical ranges using various rating system criteria;
(b) *integrated vulnerability maps*, where the real groundwater quality situation in a given area with sufficient data is taken into proper account, together with real pollution danger sources and all other human elements of interest (Albinet & Margat, 1970; Vrana, 1977; Civita, 1990). Two approaches are taken:
 (i) one map accompanied by various contour line coverages, with a very detailed legend and notes;
 (ii) an atlas of maps on the same scale, representing different databases and partial and collateral elaborations, aimed at presenting all analytical information and leaving the synthesis to the user.

Aquifer vulnerability maps of many types have been produced, either orientative, suitable for assessing national or large regional scale problems, or special, for the solution of specific forecast and safegard problems.

Single hazard one purpose maps (Foster H.D., 1980) are used to zone the area variation of only one danger.

Integrated vulnerability maps are *single hazard multipurpose*, where the susceptibility of a soil-subsoil-aquifer system to pollution and real and potential danger sources are considered together. This methodology has been adopted by the Italian National Group of the National Research Council for the Defence against Hydrogeological Disasters (GNDCI VAZAR project) (Civita, 1990), and will be used in Italy according to the recent technical rules for the safegard of wellhead protection areas for human use. It partly follows the models by Aller *et al.* (1987), BRGM (1976), Carter *et al.* (1987) and Vrana (1977, 1984).

Vulnerability maps are typically obtained by manual overlapping of single information coverages (maps and/or transparent sheet) and through visual and personal interpretation to extract the essential information for final output. Homogeneous sets and subsets to which a class or a prearranged vulnerability value is attributed may be rapidly selected (Albinet & Margat, 1970; BRGM, 1976; Civita, 1990; Olmer & Rezac, 1974; etc.). The method is valid and universal for operational and schematic maps of large areas, but it is very stiff because a mean value has to be attributed even if certain parameters (depth to groundwater, permeability, recharge) are variable over limited areas. Maps are drawn on paper or plastic sheets and may be printed. Updating is costly and time-consuming as the maps must be redrawn.

Homogeneous set overlapping is also adopted in the last version of DRASTIC, integrated with an advanced GIS like ARC/INFO (Aller *et al.*, 1987). The territory is divided with a square regular grid (500 feet). In SINTACS square finite elements (500 m) are considered to which the scores of single parameters and the values of three different lines of weights interpreting the prevailing hydrogeological situation are attributed. Data processing is computerized, so that an intrinsic vulnerability map may be plotted with scores and weights according to six vulnerability degrees corresponding to as many percent SINTACS index ranges. Scales generally depend on the details to be represented. Data reliability varies considerably with investigation area.

Medium-to-large scale maps may more or less precisely indicate pollutant flow paths obtained by modelling (Kinzelbach, 1986; Bear & Verruijt, 1987).

The potential of GIS has recently been seen in preparing the prototype of an aquifer integrated vulnerability of Oristano (Sardinia) (Barrocu & Chiarini, 1989; Aru *et al.*, 1990; Barrocu, 1990). For that purpose, the same methodology used for mapping water resources quantity and quality parameters for the European Community Commission (ECC) CORINE-Water project has been adopted. Socio-economical, geological, hydrogeological, soil, and soil use databases have been assembled to define vulnerability classes and degrees for different aquifer combinations. Databases are consistent with ECC environmental information system guidelines. Aggregation, combination, and reattribution ARC/INFO functions have been repeatedly used with an AML program dedicated to developing required logical operations. The work was motivated on the one hand by the need to produce a map for the VAZAR project, and on the other by the desire to experiment with a methodology of general interest.

GIS in aquifer vulnerability analysis and representation

GIS is a software for acquiring, storing, elaborating and representing geographical data. The requirements of GIS are:
(a) data model flexibility, to allow system and database development;
(b) outward open architecture, and the possibility of accepting and transmitting data in different formats;
(c) interactive and multimedia interface between user and system;
(d) integrated management of alphanumerical, vectorial, raster, and graphical data, texts, and illustrations;
(e) land vectorial and raster representation, and possibility of switching from one system to the other;
(f) direct connections with several relational databases in parallel;
(g) connection with statistics elaboration packages of alphanumerical data;
(h) possibility of representing geographical data in 3D;
(i) possibility of developing interfaces, simulation, and representation models in the system with a suitable programming language.

Furthermore, a GIS for the development of an operative system for aquifer vulnerability evaluation should function to:
(j) acquire, elaborate and manage updated databases of territorial physical and hydrogeological characteristics and of potential and real pollution centres;
(k) manage and elaborate monitoring data;
(l) develop internal models of geographical and descriptive data analysis;
(m) connect external models;
(n) elaborate the zoning of aquifer vulnerability with techniques of overlay mapping, buffering, and continuous reclassifying;
(o) elaborate the planning map of necessary operations;
(p) monitor and represent alert situations and indicate any possible timely actions.

The different vulnerability configurations and the details of an aquifer in danger of being polluted may be dynamically obtained, from a GIS possessing the above-mentioned characteristics, in the form of transient 2D and 3D softcopies on a visual display unit and/or produced at the same time as hardcopies on maps and sections of suitable scale. For that purpose, different layers of information from updated databases are intersected depending on associated attributes in order to obtain synthesis layers of significant values. Softcopies may be decomposed in layers, so that different vulnerability factors may be analysed and evaluated.

Procedures have to be semiautomatic since in defining vulnerability degrees an expert hydrogeologist will need to evaluate how significant is information that has been acquired from interdisciplinary investigations, even if this information cannot easily be quantified. His interpretation is always needed, either when data processing begins and the first draft of the map for further elaboration is issued, or when a vulnerability map has to be updated and modified.

In Italy six degrees are conventionally used for aquifer vulnerability evaluation by the National Group for the Defence against Natural Catastrophies: BB (very low), B (low), M (medium), A (high). E (elevated), EE (very elevated).

Vulnerability degrees are attributed to polygons produced as noted above by an expert using a model, which can combine soil use, soil, and bedrock hydrogeological characteristics in any possible way. Aquifers are defined on the basis of hydrogeological and soil maps. Thus a vulnerability-aquifer coding is obtained. Groundwater flow directions may be considered to attribute the vulnerability degree of a polygon downstream of a pollutant. Its final vulnerability degree could be much higher than the level ascribed, only on the basis of vertical permeability. For that purpose, a digital model with productive wells and their drawdowns is generated to attribute an average water level to each polygon.

Final products are maps of aquifer horizons of interest, sections, and/or 3D representations, where different colours and symbols identify different vulnerability levels as a function of potential pollution sources, drainage, and topsoil characteristics. Colours, and symbols may be represented on the visual display unit with or without basic map coverage.

The main advantage of such a system is the possibility of digitizing thematic information as attributes of points, lines, and polygons in expandable databases. Thus, analysis results may be rapidly verified on the basis of new updated results. This flexibility is particularly useful when the output has to be ready in a short time and in nonstandard format.

When using digital data, the user is limited only by GIS system availability and mastery, and by costs, which have to be compared to the costs of traditional manual methods in terms of time, personnel, and required structures.

Emphasis is to be given to the fact that vulnerability maps are complex and cannot be automatically obtained by loading a GIS in a computer and striking keys according to screen menu instructions. Highly skilled and well trained operators are needed to develop the extensive possibilities of GIS.

Prospects

GIS can be essential system moduli to monitor danger situations and to aid in decision making. The most advanced configuration may be one in which updated data from monitoring networks, acquired from remote sensing and/or the field, are collected and elaborated for vulnerability mapping. The GIS user will be able to process data of varying characteristics, typology, and updating frequency in a homogeneous way, as needed. As confirmed by new applications in different sectors, geographical information may be treated with new generation GIS in vectorial format (topographical and thematic mapping), raster format (remote sensing data, digital elevation models, thematic data, etc.), alphanumerical data structured in relational databases, and interconnected with geographical data. Model potentialities are enlarged as matrix analysis applicable to raster data (overlay, reclassification, contour analysis, minimum cost surface, etc.) are combined with vectorial field base algorithms and mathematical and statistical analysis with suitably structured alphanumerical data .

Several applications have been experienced in aquifer vulnerability analysis and evaluation with ARC/INFO. A 2D and 3D system has been developed to represent aquifer configuration and chemical concentration in groundwater by analysing parameters surveyed with monitoring networks. Also, a dynamic model system has been obtained to represent flow vectors and pollutant concentration plumes in water bodies. Groundwater flow paths may be defined with vectors (links) applied to distort the map of an undisturbed medium. Thus, the rubber sheeting technique would act as an exact equivalent of real natural forces. A real and possible situation model may be generated by integrating these experiences in a vulnerability evaluation system. Its reliability will depend on the quantity and quality of information at disposal. A cost reduction can be foreseen as the technology takes hold, so that GIS will be increasingly adopted in minor projects.

A number of increasingly complex applications show that GIS are not only a technique but also an intelligent way of taking advantage of informatic instruments, involving human and institutional factors.

CONCLUSIONS

A GIS of area vulnerability is a very useful tool for zoning danger areas with no scale limitations, and for defining the seriousness of actual or potential disasters. GIS have been shown to be useful tools for combining quantity and quality to determine aquifer vulnerability degrees.

The actions to be decided in emergency conditions and in short, medium and long term prevention of pollution in groundwater bodies, and the consequences of these actions should be decided considering the spatial development of the phenomena.

Acknowledgements The present paper, no. 753 of the Italian National Group for the Defence against Hydrogeological Disasters (GNDCI), was carried out in the framework of the VAZAR special program of the National Research Council (U.O. 4.12, Department of Territorial Engineering, University of Cagliari, Sector of Engineering Geology and Applied Geophysics, Responsible: Prof. G. Barrocu).

REFERENCES

Albinet, M. & Margat, J. (1970) Cartographie de la vulnérabilité à la pollution des nappes d'eaux souterraines. *Bull. BRGM*, **2**, 3,4, 13-22.
Aller, L., Bennert, T., Lehr, J.H., Petty, R.J. & Hacket, G. (1987). DRASTIC: a standardized system for evaluating groundwater pollution potential using hydrogeologic settings. NWWA/EPA, Ser. **600**, 2-87.
Andersen, L.J. & Gosk, E. (1987) Applicability of vulnerability maps. In: *Vulnerability of soil and groundwater to pollutants* (ed. by W. Van Duijvenbooden & H.G. Van Waegeningh), Proc. Int. Conf., March 30-April 3, Noordwijk aan Zee, The Netherlands, Committee on Hydr. Research No. 38, 321-332.
Aru, A., Barbieri, G., Barrocu, G., Chiarini, E., Pani, F., Sanna, R.M., Uras, G. & Vernier, A. (1990) Applicazioni di cartografia automatica per la valutazione della vulnerabilità degli acquiferi di Oristano, Sardegna (Applications of Automatic mapping for the vulnerability evaluation of the aquifer system of Oristano). In: Protezione e Gestione delle acque sotterranee (Groundwater protection and management). (Proc. Symp. G.N.D.C.I.), *Marano sul Panaro*, Vol. 1, 41-60.
Bachmat, Y. & Collin, A. (1987) Mapping to assess groundwater vulnerability to pollution. In: *Vulnerability of soil and groundwater to pollutants* (ed. by W. Van Duijvenbooden & H.G. Van Waegeningh), Proc. Int. Conf., March 30-April 3, Noordwijk aan Zee, The Netherlands, Committee on Hydr. Research No. 38, 297-307.
Barrocu, G. & Chiarini, E. (1989) Using ARC/INFO to develop a map of aquifer vulnerability in Sardinia, Proc. the 4th Annual ESRI European ARC/INFO Conf., ROME.
Barrocu, G. (1990) Sistemi informativi geografici e cartografia della vulnerabilità come strumento di intervento per la difesa degli acquiferi GIS and vulnerability mapping as an action tool for the safeguard of aquifers). In: Protezione e Gestione delle acque sotterranee (Groundwater protection and management). (Proc. Symp. G.N.D.C.I.), *Marano sul Panaro*, Vol. 3, 87-103.
Bear, J. & Verruyt, A. (1987) Modelling groundwater flow and pollution. D. Reidl Publ., Dordrecht.
BRGM (1976) Cartes de la vulnérabilité à la pollution des eaux souterraines, 31 FF, 1/50,000. B.R.G.M., Orléans.
Carter, A.D., Palmer, R.C. & Monkhouse, R.A. (1987) Mapping the vulnerability of groundwater to pollution from agricultural practice, particularly with respect to nitrate. In: *Vulnerability of soil and groundwater to pollutants* (ed. by W. Van Duijvenbooden & H.G. Van Waegeningh), Proc. Int. Conf., March 30-April 3, Noordwijk aan Zee, The Netherlands, Committee on Hydr. Research No. 38, 333-342.
Civita, M. (1987) La previsione e la prevenzione del rischio d'inquinamento delle acquee sotterranee a livello regionale mediante le carte di vulnerabilità (Forecast and risk pollution prevention by regional vulnerability mapping). In: *Inquinamento delle acque sotterranee: previsione e prevenzione (Groundwater pollution: forecast and prevention)* (Proc. Symp., March 11, 1987). Mantova
Civita, M. (1990) Valutazione della vulnerabilità degli acquiferi. (Aquifer vulnerability evaluation) In: *Protezione e Gestione delle acque sotterranee (Groundwater protection and management)*. (Proc. Symp. G.N.D.C.I.), Marano sul Panaro, **3**, 39-86.
Dee, N., Baker, J., Drobny, N., Duke, K., Whitman, I. & Fahringer, D. (1983) An environmental evaluation system for water resource planning. *Wat. Resour. Res.* **9**(3), 523-535.
Engelen, G.B. (1985) Vulnerability and restoration aspects of groundwater systems in unconsolidated terrains in the Netherlands. I.A.H. (Proc. 18th Con.), 64-69.
Foster, H.D. (1980) Disaster planning. Springer Verlag, New York.
Foster, S.S.D. (1987) Fundamental concepts in aquifer vulnerability, pollution risk, and protection strategy. In: *Vulnerability of soil and groundwater to pollutants* (ed. by W. Van Duijvenbooden & H.G. Van Waegeningh), Proc. Int. Conf., March 30-April 3, Noordwijk aan Zee, The Netherlands, Committee on Hydr. Research No. 38, 69-86.
Haertlé, T. (1983) Method of working and employment of EDP during the preparation of groundwater vulnerability maps. In: *Groundwater in water resources planning,* IAHS Publ. no. 142, 1073-1085.
Kinzelbach, W. (1986) Groundwater modelling. Elsevier Sci. Publ., Amsterdam.
Legrand, H.E. (1983) System for evaluating the contamination potential of some waste sites. *J. Am. Wat. Wks Ass.* **56**(8), 959-974.
Marcolongo, B. & Pretto, L. (1987) Vulnerabilità degli acquiferi nella pianura a nord di Vicenza (Aquifer vulnerability in the plain to the north of Vicenza). Publ. GNDCI-CNR, 28, 1-13.
Olmer, M. & Rezac, B. (1974) Methodical principles of maps for protection of groundwater in Bohemia and Moravia, scale 1/200,000. *Mem. IAH* **10**(1), 105-107.
Pavoni, J.L., Hagerty, D.J. & Lee, R.E. (1972) Environmental impact evaluation of hazardous waste disposal in land. *Adv. Wat. Res.* **8**(6), 1091-1107.
Sotornikova, R. & Vrba, J. (1987) Some remarks on the concept of vulnerability maps. In: *Vulnerability of soil and groundwater to pollutants* (ed. by W. Van Duijvenbooden & H.G. Van Waegeningh), Proc. Int. Conf., March 30-April 3, Noordwijk aan Zee, The Netherlands, Committee on Hydr. Research No. 38,

471-475.

Vrana, M. (1977) Development of methods for the preparation of groundwater protection maps. *"Mem. IAH"* **13**(2), 22B-28B.

Vrana, M. (1984) Methodology for construction of groundwater protection maps. Lecture for UNESCO/UNEP Proj. PLCE-3/29, Moscow, Sept. 1981, In: *Hydrogeological principles of groundwater protection* (ed. by E. A. Kazlowsky), Vol. 1, UNESCO/UNEP, Moscow, 147-149.

Development and application of a groundwater model integrated in the GIS GRASS

O. BATELAAN, F. DE SMEDT & M.N. OTERO VALLE
Laboratory of Hydrology, Free University of Brussels, Pleinlaan 2, 1050 Brussels, Belgium

W. HUYBRECHTS
Institute of Nature Conservation, Kiewitdreef 3, 3500 Hasselt, Belgium

Abstract A one-layer and a multilayer regional groundwater flow model, based on the variable source area concept, have been developed and integrated at different levels in the GIS GRASS (Geographical Resources Analysis Support System). The models simulate quantitative groundwater recharge, discharge and groundwater elevation maps, that can be analyzed and compared with the GIS. Level and success of the integration in the GIS are discussed, as well as an application of the multilayer model to the nature reserve Walenforest (Belgium). It is shown that the groundwater flow is simulated in a very efficient way, yielding detailed insight into the regional and local flow systems, and division of the area in recharge/discharge zones. A succesful comparison in GIS is performed between the model results, soil mapping data and remote sensing.

INTRODUCTION

Progress in functionalities of Geographic Information Systems is primarily driven by the enormous developments in hardware components. The availability of more and more relatively cheap, highly computational personal workstations with large data storage, forces GIS to follow the hardware revolution. Striking for this revolution is that, nowadays, one MB of RAM is cheaper than two lines of source code (Plant, 1992). Logically, the development of GIS has shifted from vector based systems towards raster or hybrid systems, which were formerly considered inefficient.

In groundwater hydrology, model concepts increasingly deal with spatially distributed parameters, such as heterogeneous formation characteristics. Also, growing data sets for fine scales and realistic modelling of real world problems, demand more effective data management and analysis systems. Therefore, well established area discretization methods in groundwater modelling, like finite difference and finite elements, may be integrated efficiently in raster GIS.

The purpose of this study was to integrate a regional groundwater flow model in GRASS GIS. Two types of integration were developed:
(a) maximum integration with a simple one layer model, and
(b) minimum integration with a more complicated multilayer model.

GROUNDWATER MODEL CONCEPT

The regional groundwater flow model is based on the concept of "variable source area". In this concept infiltration areas are characterized by deep groundwater levels and recharge by effective precipitation. The natural groundwater flow from infiltration to discharge areas results in discharge areas characterized by maximum elevated groundwater levels, no recharge and groundwater seepage to wetlands and surface waters (Bronders, 1989). Any area can thus be divided in infiltration and discharge zones, connected to each other by local and regional groundwater systems.

In case of a three-layer system, the steady-state flow situation can be described by the following three features (items a through c):

(a) Horizontal flow in a phreatic aquifer

$$\frac{\partial}{\partial x}[K(h-z)\frac{\partial h}{\partial x}] + \frac{\partial}{\partial y}[K(h-z)\frac{\partial h}{\partial y}] + c(g-h) + N - P = Q \qquad (1)$$

where K = hydraulic conductivity of the aquifer [L T^{-1}],
  h = position of water table [L],
  z = elevation of base of aquifer [L],
  c = hydraulic resistance of underlying semi-confining layer [T^{-1}],
  g = piezometric level of the underlain semi-confined aquifer [L],
  N = effective precipitation [L T^{-1}],
  P = pumpage [L^3 T^{-1} L^{-2}], and
  Q = groundwater seepage [L^3 T^{-1} L^{-2}].

(b) Vertical flow through an aquitard

(c) Horizontal flow in a semi-confined aquifer

$$\frac{\partial}{\partial x}(T\frac{\partial g}{\partial x}) + \frac{\partial}{\partial y}(T\frac{\partial g}{\partial y}) + c(h-g) - P = 0 \qquad (2)$$

where T = transmissivity [L^2 T^{-1}].

In case of a one-layer system only, equation (1) is considered without the term c(g-h). Since no groundwater seepage exits in infiltration areas, these can be identified from equation (1) as the areas where the groundwater table is deeper than 0.5 m below the ground level. The latter accounts for an unsaturated soil zone. Consequently, the other areas in the study region are discharge areas, with the groundwater table equal to the ground level minus 0.5 m. The groundwater seepage in discharge areas can be calculated from equation (1).

DEVELOPMENT AND INTEGRATION OF CODES

The GIS GRASS (US Army Corps of Engineers, 1991) is a public domain hybride GIS with remote sensing analysis modules. It is written in the programming language C for UNIX-based computers. Its success lies in its powerful raster capabilities and an open-system concept, i.e. programming is possible at (a) command level in batch files with UNIX utilities or (b) C level with calls to GRASS library functions.

Based on above concepts, the one- and three-layer regional groundwater flow model have been integrated in GRASS GIS at different levels, using a finite difference numerical approach.

The one layer model has been programmed in C with maximum possible use of GRASS library functions. In this way, the model has been fully integrated as a GRASS command, with full GRASS functionalities and characteristics (e.g., zooming, masking, region definition, etc.). All model inputs and outputs are GRASS raster maps, and can be produced and changed by GRASS functions, such that highly spatially distributed data can be handled easily. Also output maps of groundwater levels, groundwater seepage rates, flow directions and velocities can be directly viewed, analyzed and compared with the available GRASS tools.

The one-layer model has been applied and tested in case of regional groundwater in Mid-Belgium (Bronders, 1989) and Munecas-Larecaja, Bolivia (Otero Valle, 1992).

The three layer model has been programmed in FORTRAN 77 and communicates by way of non-automatic transfer of adjusted format data with the GRASS GIS. It has been applied to the Walenforest region (Belgium), as will be discussed in the next paragraph.

Based on guidelines for comparing levels of integration of environmental models and GIS by Fedra (1991), the following conclusions can be drawn with respect to the development and implementation of codes:

(a) cost of development for the fully integrated model is considerably higher than for the FORTRAN program. This is mainly due to the GIS programming requirements for data structures, use of library functions, independency of input data size by use of segmentation routines, etc;

(b) ease and generality of use is clearly better for the fully integrated model, while the fortran code easily tends to become project dependent. The C code has a higher intrinsic independency caused by its integration in the GIS;

(c) power of the C model code is superior. Due to segmented input files, big size projects can be handled. However, the segmentation reduces the efficiency of computation considerably, and the present GRASS data structure (integer) is problematic for storing intermediate results;

(d) maintenance of code is lower for the C code than for the fortran code due to its integration.

Since the cost of development is a determining aspect of integration, it is

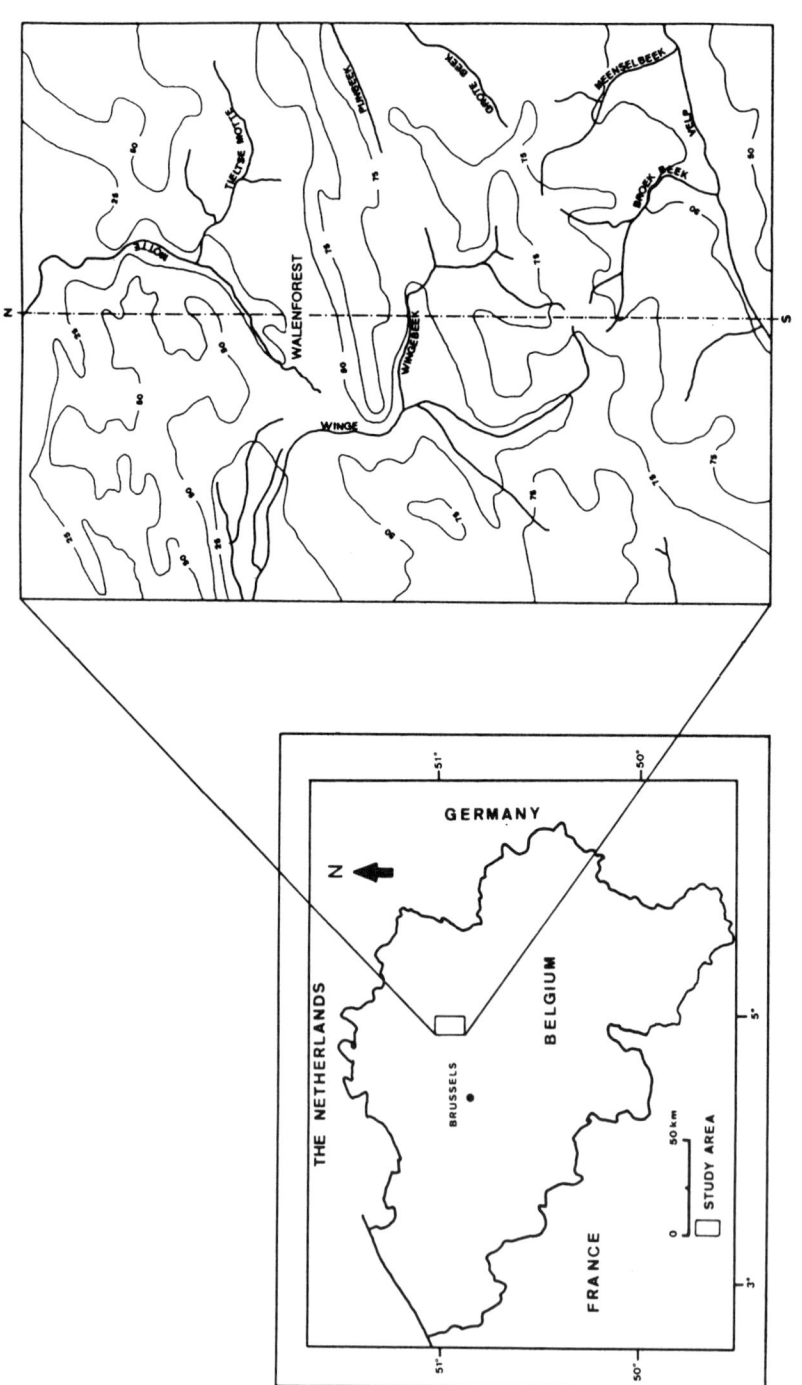

Fig. 1 Location of the Walenforest study area.

recommended that efficient methods of integration should be developed as much as possible using existing codes. Also, the use of powerful interface development tools or new programming techniques, like object oriented programming, could be very helpful.

APPLICATION TO WALENFOREST

Walenforest is one of the most important wet forest nature reserves in the northern part of Belgium (Fig. 1). It is located in the valley of the "Brede Motte" and highly dependent on groundwater seepage. An ecohydrological study is being performed in the forest since 1989, for protection and appropriate management of the reserve.

The hydrogeological situation can be schematized by a three layer system:

(a) a top phreatic aquifer consisting of loess or loamy colluvial deposits and a Pliocene formation, consisting of glauconite rich sand; the thickness of this aquifer ranges between 4 m in the south to about 95 m in the north of the study area (Fig. 2);

(b) a semiconfining 30 m thick layer, mainly consisting of plastic clay and sandy clay;

(c) a lower semiconfined aquifer with an impervious base consisting of 15 m coarse sands.

The hydraulic conductivity of the phreatic and semi-confined aquifer are respectively 11 and 7 m day^{-1} (Bronders, 1989). The hydraulic resistance of the

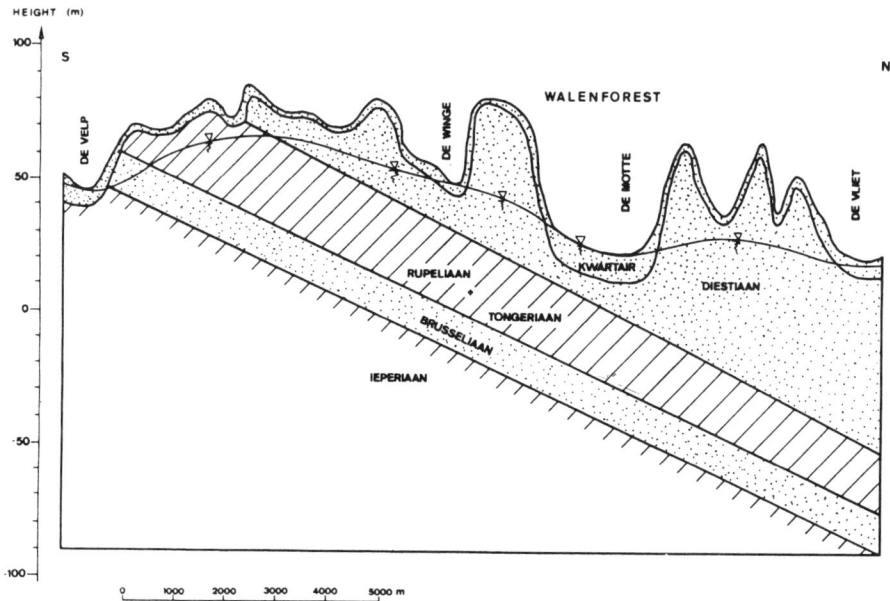

Fig. 2 Hydrogeological N-S profile through study area.

semi-confining layer was taken as 0.007 day^{-1}, giving a characteristic percolation time of 140 days. Effective precipitation was estimated as the difference between 800 mm of precipitation and about 510 mm of evapotranspiration per year.

The topography of the 192 km^2 area was obtained by digitalization of contour lines, subsequent rasterization and interpolation in a grid with cell size of 50 by 50 m, resulting in 76 800 cells.

The regional groundwater system was simulated using the three-layer model. The resulting system of 2 times 76 800 finite difference equations was solved iteratively by a four colour ordering (Hackbusch, 1985) and successive overrelaxation (Wang & Anderson, 1982).

RESULTS

The calculated results for the phreatic aquifer show that the Walenforest is a seepage area (Fig. 3), connected by a regional groundwater flow system to an infiltration area that is almost twice as large as would be deduced from the topography only. This is in terms of vulnerability and protection of the nature reserve a very important result.

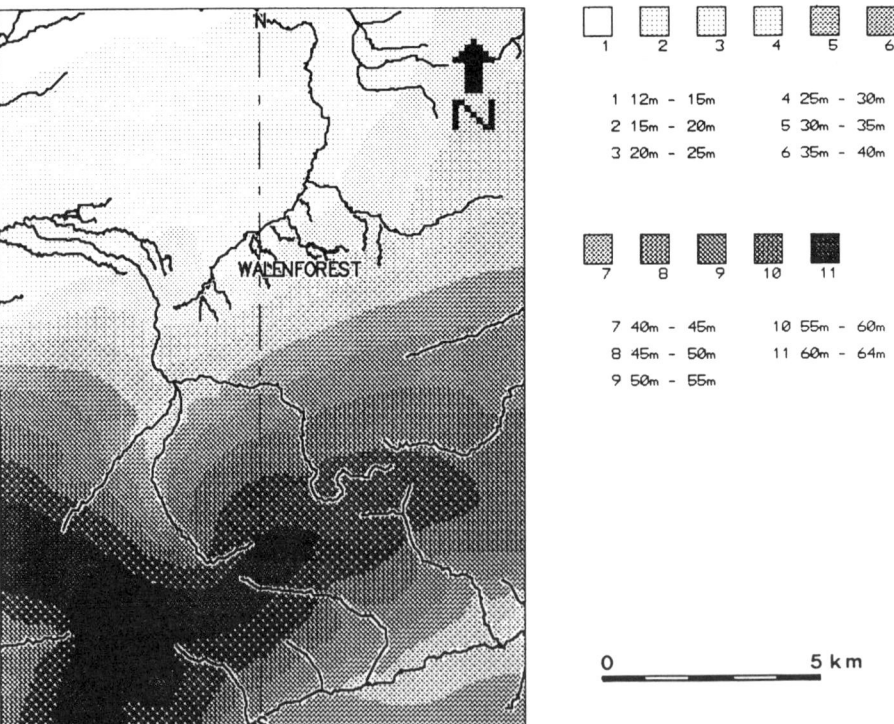

Fig. 3 Calculated phreatic water level overlain by river system.

Groundwater model integrated in the GIS GRASS 587

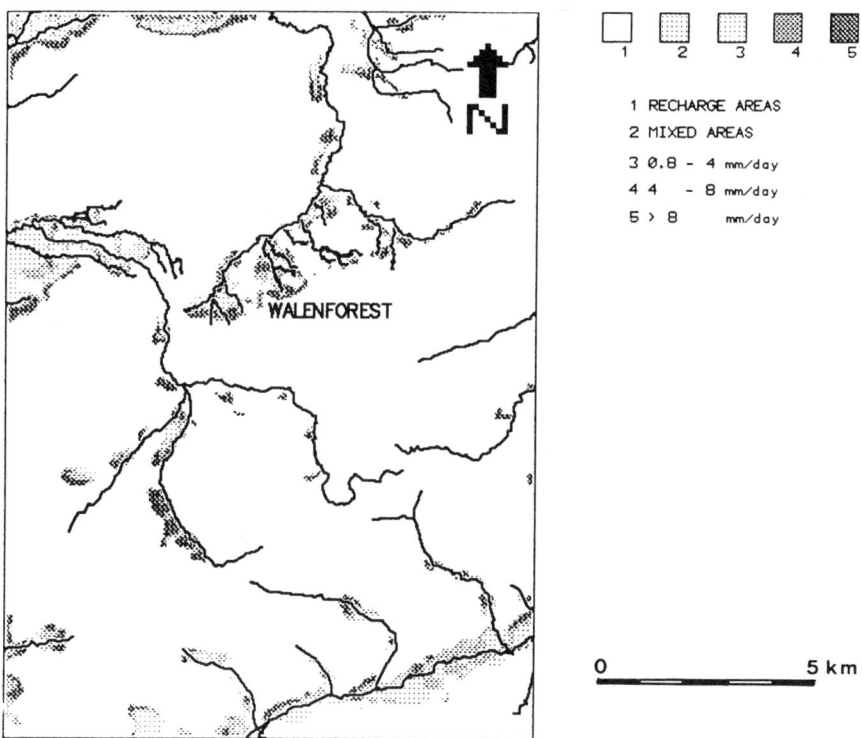

Fig. 4 Calculated seepage areas, overlain by river system.

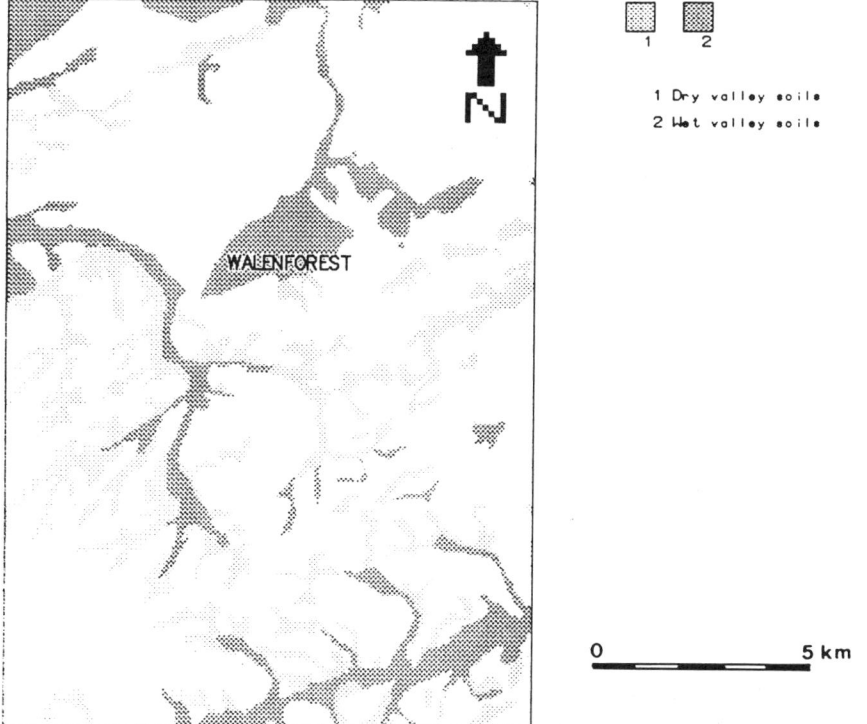

Fig. 5 Seepage areas derived from classified soil map of wet valley soils.

The calculated discharge areas (Fig. 4) are mostly located in valleys, but are seperated by infiltration areas, indicating that local groundwater systems are superposed on regional flow systems. This pattern is highly correlated with wet, often valuable biotopes, as indicated on the Biological Evaluation Map.

In Walenforest two important discharge zones emerge coinciding with the two most important drainage areas in the forest. Groundwater seepage in these areas is confirmed by ecohydrological field measurements, as mappings of seepage-dependent vegetation and measurements of changes in groundwater level and groundwater seepage rates. A good agreement is also obtained by comparing groundwater discharge zones with wet valley soils, as indicated on the soil classification map (Fig. 5).

For a slightly smaller study area, remote sensing information from Thematic Mapper band 3, 5 and 7 was used to perform a principal components analysis. The spectrum of the first component is a combination of three gaussian curves, indicating discharge, intermediate, and recharge areas (Bobba *et al.*, 1992). Classifying this spectrum with local terrain investigations (Sundara Rajan, 1992) resulted in a map with remote sensed discharge and recharge areas (Fig. 6), which compares favourably with the results obtained from the groundwater model.

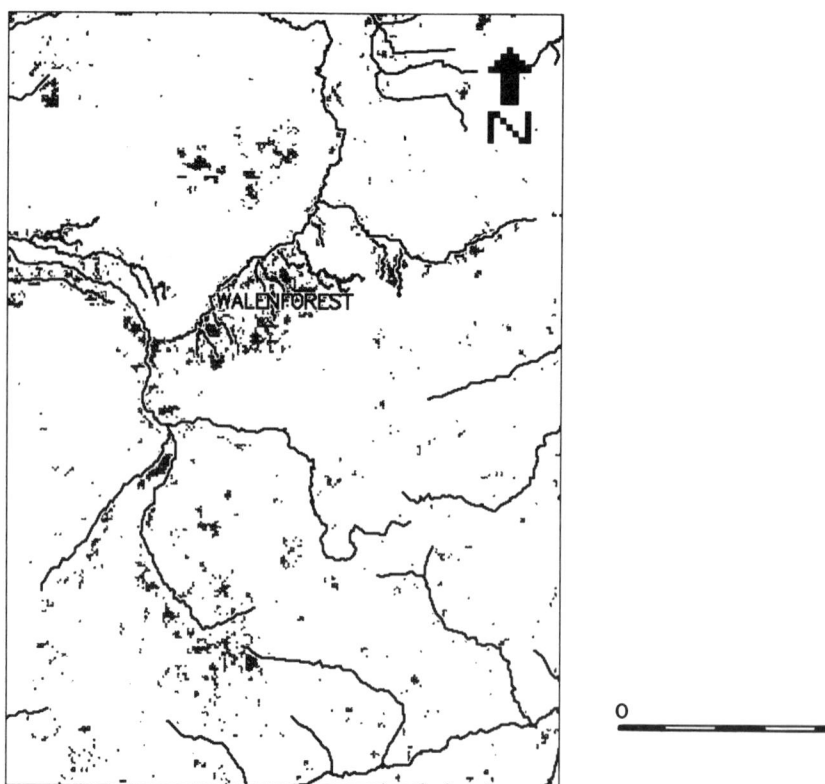

Fig. 6 Seepage areas obtained with principal components analysis of remote sensing data: Thematic Mapper bands 3, 5 and 7, overlain by river system.

CONCLUSIONS

Since cost of development is a major concern for integration of GIS and groundwater models, use should be made of interface building tools or of existing codes. In this study, GRASS GIS was succesfully linked with groundwater modelling codes, resulting in a more efficient data management. Also, the results could be more easily evaluated and checked with other information. The proposed model proves to be a valuable tool in delineating discharge and recharge zones and regional and local groundwater flow systems.

REFERENCES

Bobba, A.G., Bukata, R.P. & Jerome, J.H. (1992) Digitally processed satellite data as a tool in detecting potential groundwater flow systems. *J. Hydrol.* **131**, 25-62.

Bronders, J. (1989) Bijdrage tot de geohydrologie van Midden Belgie door middel van geostatische analyse en een numeriek model (Contribution to the geohydrology of Mid-Belgium by means of geostatistical analysis and a numerical model), Ph.D. thesis, Free University Brussels, Belgium.

Fedra, K. (1991) GIS and environmental modeling. *First Int. Conference/Workshop on Integrating Geographic Information Systems and Environmental Modeling,* Boulder, Colorado, USA.

Hackbusch, W. (1985) *Multi-grid methods and applications.* Springer-Verlag Berlin.

Otero Valle, M.N. (1992) Regional groundwater flow study using a two-dimensional mathematical model and GIS. MSc Thesis IUPHY, Free University Brussels, Belgium.

Plant, N. (1992) Software in crises. *Physics World* **9**, 43-47.

Sundara Rajan, C.V. (1992) Remote sensing as an aid to regional mathematical modelling of groundwater flow systems. MSc Thesis IUPHY, Free University Brussels, Belgium.

US Army Corps of Engineers (1991) *Geographic Resources Analysis Support System, version 4.0, user's reference manual.* Construction Engineering Research Laboratory, Champaign, USA.

Wang, H.F. & Anderson, M.P. (1982) *Introduction to groundwater modelling.* W.H. Freeman and Company, San Fransisco.

Use of volume modelling techniques to estimate agricultural chemical mass in groundwater, Minnesota, USA

WILLIAM A. BATTAGLIN

US Geological Survey, Water Resources Division, M.S. 406, Box 25046, Denver Federal Center, Lakewood, Colorado 80401, USA

Abstract The occurrence of agricultural chemicals such as nitrate or atrazine in groundwater beneath areas of agricultural activity can result from the leaching of those chemicals from the land surface. At the Rosholt Research Farm in Minnesota, three farming management systems were implemented on six test plots to determine if the management systems affected the movement of agricultural chemicals to groundwater. A method was developed that utilizes volume modelling computer software to estimate the mass of agricultural chemicals in groundwater beneath the test plots from sample data. Results demonstrate that: (1) farming management systems can affect the quality of groundwater; (2) techniques exist for estimating the mass of agricultural chemicals in the saturated zone; (3) there are advantages to visualizing all sample data for a given date in one graphic display; and (4) workstations available to the general user can now handle the large amounts of data, large number of calculations, and high resolution of graphics required to work effectively with three-dimensional data sets.

INTRODUCTION

The occurrence of agricultural chemicals such as nitrate or atrazine in surface water and groundwater is of growing concern to the general public. Recent investigations (Goolsby & et al., 1991, Thurman & et al., 1991) indicate that herbicides, in concentrations that exceed heath based limits, commonly occur during spring runoff and may persist though the summer in both small streams and large rivers in the midwestern United States. The US Environmental Protection Agency (EPA)(1992) summarized the results of several surveys of the occurrence of agricultural chemicals in groundwater. These studies indicated that 1.2 to 23% of the wells sampled in an individual study contained nitrate in concentrations that exceeded the drinking water standard of 10 mg l^{-1} (milligrams per liter). An estimated 118 000 tons of herbicides (Gianessi & Puffer, 1990), and an estimated 6.3 million tons of nitrogen fertilizer (US Environmental Protection Agency, 1990) are applied annually to cropland and pastureland in the midwestern United States.

At the Rosholt Research Farm in Westport, Minnesota, three farming management systems were implemented on six test plots (Fig. 1) by the University of Minnesota to determine if the systems affected the movement of agricultural chemicals to groundwater (Anderson & Stoner, 1989). The three farming management systems were implemented on the test plots as follows: (1) continuous cropping of corn using 72.6 kilograms of nitrogen as fertilizer and 454 grams of atrazine per acre on plots 1 and 3; (2) continuous cropping of

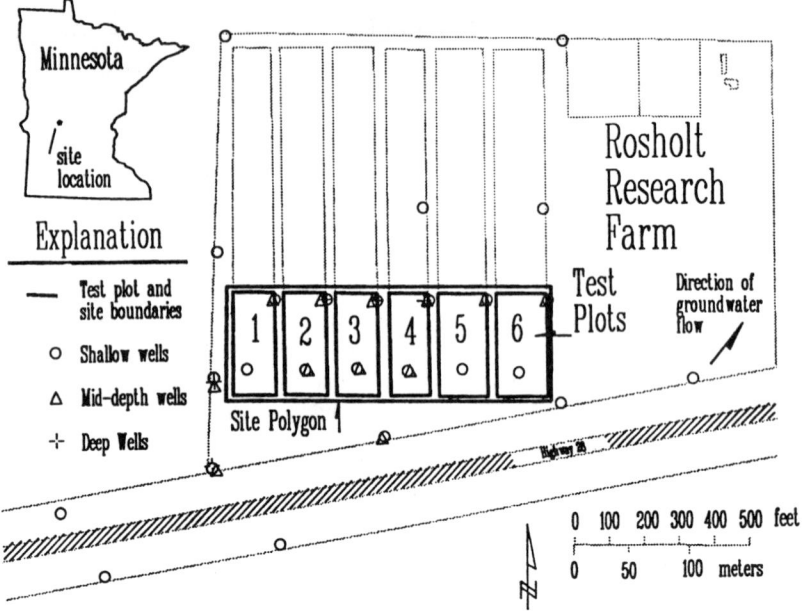

Fig. 1 Location of sampled wells, test plots, and site polygon at the Rosholt Research Farm, Minnesota, USA.

corn using 97.1 kilograms of nitrogen as fertilizer and 454 grams of atrazine per acre on plots 2 and 5; and (3) a corn-soybeans crop rotation sequence (corn on even years) with 72.6 kilograms of nitrogen as fertilizer and 454 grams of atrazine per acre applied on corn, and 6.8 kilograms of nitrogen as fertilizer per acre and no atrazine applied on soybeans on plots 4 and 6 (Anderson & Stoner, 1989). Rates of application for nitrogen as fertilizer on soybeans and atrazine on corn were estimated based on the average usage rates for Minnesota (Agricultural Statistics Board, 1991).

Other agricultural experiments (some that included that application of nitrogen and other agricultural chemicals) took place at the Rosholt farm in the area north (down gradient) of the six test plots, and other farming (not at the research farm) took place to the east of the site. Thus, concentrations of agricultural chemicals measured in monitoring wells to the north and east of the test plots (Fig. 1) may not solely reflect effects resulting from activities on the test plots. All monitoring wells were installed by the US Geological Survey. Sample collection and chemical analyses for nitrate and atrazine were done by both the US Geological Survey and the University of Minnesota.

The Rosholt Research Farm overlies an unconfined glacial outwash sand-plain aquifer with approximately 20 feet of saturated and 10 feet of unsaturated thickness. Aquifer hydraulic properties were estimated from analysis of data from a 31-hour aquifer test. Aquifer porosity and hydraulic conductivity were estimated to be 0.27, and 1250 ft day^{-1}, respectively (Anderson & Stoner, 1989).

Three-dimensional models were used to demonstrate: (1) a method for estimating the mass of agricultural chemicals in the saturated zone beneath the test plots, (2) how farming management systems affect shallow groundwater quality, and (3) insights gained by visualizing all sample data for a given date in one graphic display.

APPROACH

Estimating the mass of agricultural chemicals in the saturated zone at the Rosholt Research Farm provides a means for evaluating the effects on groundwater quality of the three farming management systems tested there. Interactive Volume Modelling (IVM) software was used to generate three-dimensional models depicting the concentration distributions of nitrate and atrazine (Interactive Volume Modelling software is a propriety package distributed by Dynamic Graphics Inc. Use of brand names in this paper is for identification purposes only and does not constitute endorsement by the US Geological Survey.). The three-dimensional grids are used to create display files for model visualization (Fig. 2), and to calculate the volumes required to estimate the mass of agricultural chemicals in the saturated zone.

Data from 43 observation wells were used to generate three-dimensional models depicting the concentration distributions of nitrate as nitrogen for 13 dates between July 1988 and May 1990, and of atrazine for 5 dates between April 1989 and September 1989. All models were bounded below by a surface representing the altitude of the base of the sandplain aquifer and bounded above by a water-table surface generated from water-level data collected on or near to the date of sample collection.

Agricultural chemical mass estimates were calculated by multiplying the volumes between concentration increments (1 mg l^{-1} increment for nitrate and 0.1 mg l^{-1} increment for atrazine) by the porosity of the aquifer and the average concentration of the volume, and then summing for all volumes in a given model. The region where volumes were calculated was limited laterally by the test plot boundary polygons (Fig. 1), and vertically by two-dimensional surfaces representing the altitude of the water-table (the upper model boundary), and the altitude of the base of the sand-plain aquifer (the lower model boundary). IVM software provides a means to develop three-dimensional gridded data sets and to calculate shell volumes.

GRIDDING AND VOLUMETRIC CALCULATIONS

Until recently, computers that were available to the general public were unable to handle the volume of data, number of calculations, or resolution of graphics required to work effectively with three-dimensional data sets. The advent of modern workstations has largely eliminated these constraints (Turner, 1990).

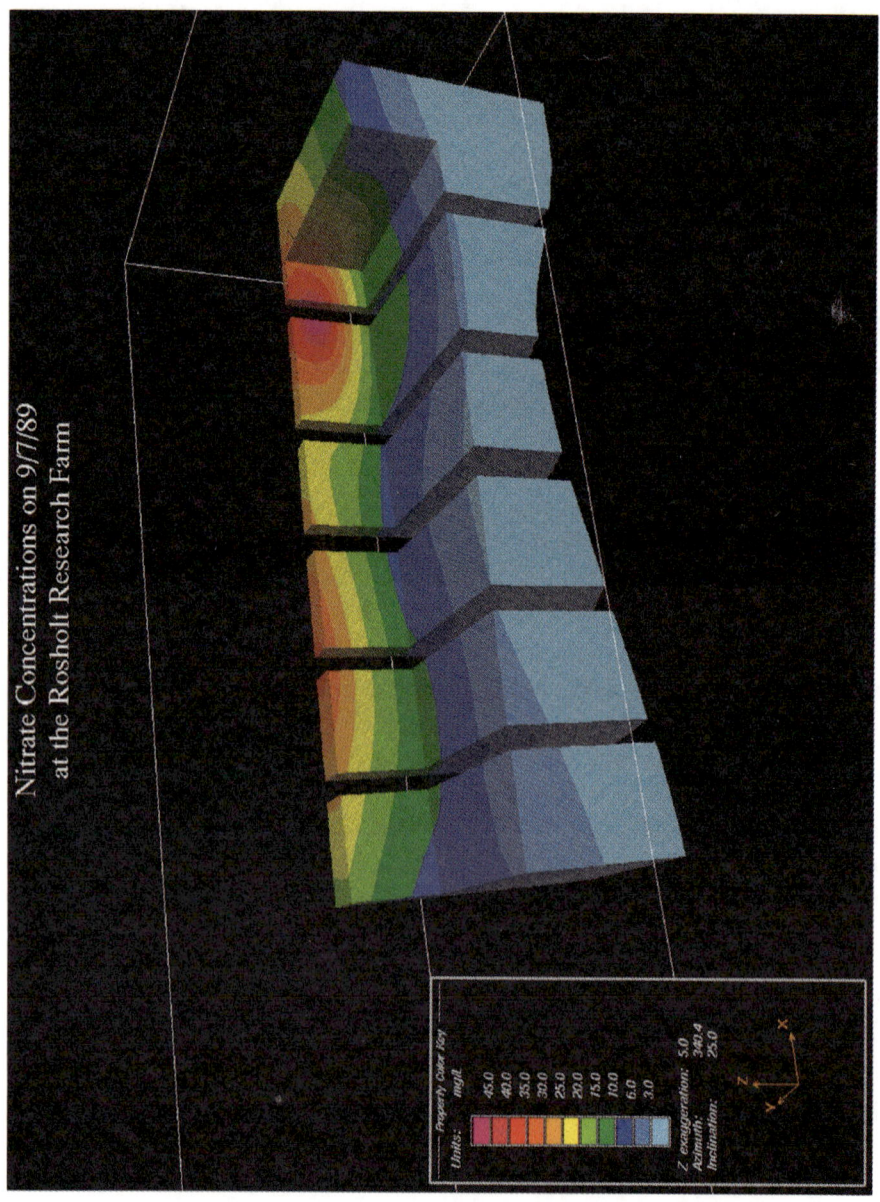

Fig. 2 Example of an IVM generated three-dimensional display of nitrate concentrations on September 9, 1989.

Flynn (1990) estimated that manipulation of three-dimensional solids requires 20 000 times the computing power required to manipulate two-dimensional wire-frame models.

Techniques for calculating the integral properties (volume, center of mass, etc.) of "designed" solid objects (e.g., a gear) are well established (Lee & Requicha, 1982), and have been available for industrial applications for over a decade. However, geologists and hydrologists are more concerned with "geo-objects" that are often highly irregular in shape, and may be "definition limited" such as a contaminant plume at or greater than a specified concentration value (Raper, 1989). The commonly used techniques for calculating integral properties of "designed" objects, such as constructive solid geometries, or primitive instancing, are more difficult to apply to complex "geo-objects." The accuracy of methods used for volume estimation are largely attributable to errors in approximating the solid and not to more traditional sources of numerical error such as roundoff, truncation, numerical cancellation or approximate integration formulas.

IVM software applies a minimum tension gridding algorithm to create three-dimensional grids of estimated concentration values from scattered data points. The minimum tension algorithm (also called minimum curvature) has been used by geoscientists to produce machine contoured maps of bathymetry (National Geophysical Data Center, 1988) and magnetic and gravity anomalies (Geological Society of America Committee for the gravity map of North America, 1987; Geological Society of America Committee for the magnetic map of North America, 1987) largely because of the algorithm's ability to generate results that are an adequate substitute to hand drafted maps (Briggs, 1974). The minimum tension gridding algorithm in its unaltered form honors the data at constrained points, but may have oscillation (loops or excessive curvature) between widely spaced points or in unconstrained areas (Smith & Wessel, 1990). These oscillations can be reduced by either relaxing the requirement that source data be honored exactly, or by relaxing the constraint that total curvature be minimized, by adding some tension to the splines (Smith & Wessel, 1990). IVM software may use one of these techniques, but the precise details of the IVM gridding algorithm are proprietary.

IVM software can be used to estimate the volume between two isovalue surfaces (shells or three-dimensional contours) from a previously defined gridded model. To calculate volumes, IVM first converts the gridded model into a boundary representation defining the two isovalue surfaces. Each cell in the gridded three-dimensional model is then divided into subcells, and the thickness between the isovalue surfaces at the centroid of each subcell is calculated. The thicknesses for each subcell with a positive thickness are then averaged and multiplied by the area of positive thickness within the grid cell to get the volume between the isovalue surfaces within each grid cell.

The term volume container refers to a geometric solid that defines the region in which a volume is calculated. IVM allows the user to specify constraints to the volume container. The volume container can be restricted by

2D surfaces above and below, or by maximum/ minimum values in the z direction. The container can be further constrained by polygons in the x-y plane that cut vertically through the solid. Yield factors and conversion factors can be incorporated into the volumetric calculations so that, for example, estimates for the volume of water in an aquifer can be made if the aquifer porosity is known (Paradis & Belcher, 1990).

RESULTS AND DISCUSSION

The estimated masses of nitrate as nitrogen and of atrazine are listed in Tables 1 and 2. Box-plots of estimated nitrate mass (Fig. 3) indicate that there is not a simple and consistent relation between the farming management systems and

Table 1 Estimates of nitrate mass in the saturated zone beneath the six test plots and the site polygon, at the Rosholt Research Farm, Minnesota.

Date of sampling	Mass of nitrate (kilograms) in the saturated zone beneath						
	Plot 1	Plot 2	Plot 3	Plot 4	Plot 5	Plot 6	Site polygon
07/21/88	35.0	26.4	18.4	13.8	16.8	13.4	169.0
08/17/88	29.7	31.9	22.6	18.6	36.0	26.9	225.7
04/18/89	35.6	23.5	15.5	17.6	28.4	22.4	193.5
05/03/89	37.7	31.2	21.0	17.1	25.1	26.8	215.7
05/25/89	38.4	32.8	22.4	16.0	23.4	23.2	211.8
07/10/89	40.6	36.7	23.3	14.9	21.5	20.4	213.6
08/09/89	34.7	33.6	26.4	17.5	23.2	21.4	213.5
09/07/89	36.5	40.0	34.4	27.5	53.6	46.1	322.2
09/13/89	44.5	55.2	46.8	30.1	61.8	49.1	385.6
10/11/89	49.7	50.0	45.2	31.6	62.0	49.4	389.9
11/30/89	36.1	31.8	38.7	24.1	26.6	39.0	266.7
03/22/90	30.9	22.2	23.8	18.6	26.1	26.5	203.7
05/03/90	35.6	28.9	22.8	21.9	39.8	33.4	246.9

Table 2 Estimates of atrazine mass in the saturated zone beneath the six text plots and the site polygon, at the Rosholt Research Farm, Minnesota.

Date of sampling	Mass of atrazine (grams) in the saturated zone beneath						
	Plot 1	Plot 2	Plot 3	Plot 4	Plot 5	Plot 6	Site polygon
04/18/89	0.0773	0.0692	0.0486	0.0186	0.0154	0.0058	0.350
05/03/89	0.0917	0.0871	0.0870	0.0535	0.0179	0.0090	0.518
05/25/89	0.0060	0.0055	0.0052	0.0043	0.0053	0.0057	0.044
07/07/89	0.0057	0.0056	0.0062	0.0065	0.0599	0.0804	0.260
09/13/89	0.2118	0.2356	0.1531	0.0755	0.0694	0.0394	1.086

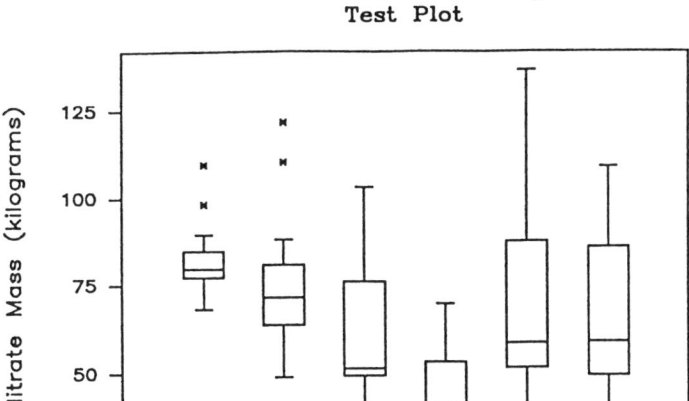

Fig. 3 Box-plots of nitrate mass estimates for the six test plots.

the nitrate mass occurring in the saturated zone beneath the test plots. In general, smaller nitrate masses occur beneath plots 3 and 4, and larger nitrate masses occur beneath plots 1, 2, 5, and 6. The nitrate mass estimates have been grouped by farming management system in Fig. 4, which indicates that the mass of nitrate is consistently smaller under plots using a corn-soybean rotation than under plots using continuous corn management systems. A Kruskal-Wallis test (Iman & Conover, 1983) on the nitrate mass estimates associated with the three farming management systems indicates that the null hypothesis of equal

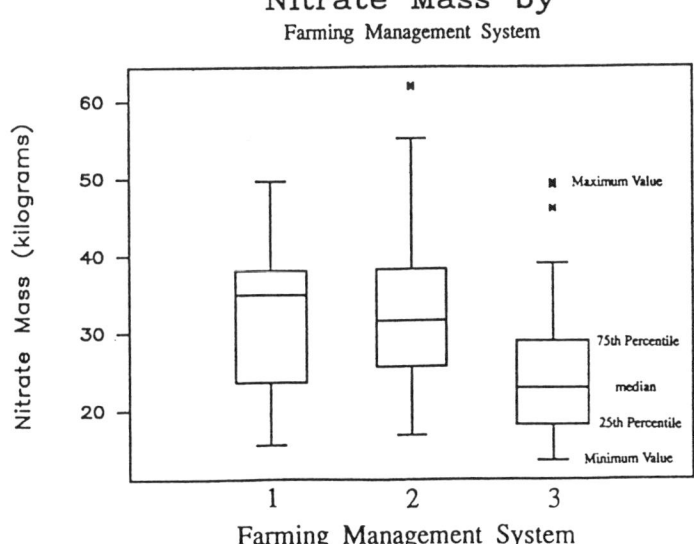

Fig. 4 Box-plots of nitrate mass estimates for the three farming management systems.

population means is rejected at the 0.05 significance level. The estimated mean nitrate masses associated with the three management systems listed in Table 3 indicates that estimated mean nitrate mass beneath test plots using farming management system 3 was 25% percent less than the estimated mean nitrate mass beneath test plots using farming management system 2, and 21% less that estimated mean nitrate mass beneath test plots using farming management system 1.

The results of the atrazine mass estimation procedure are more difficult to interpret due to the limited number of atrazine detections on any one sampling date. A KruskalWallis test on the atrazine mass estimates associated with the three management systems indicates that the population means are not significantly different at the 0.05 significance level. However, Table 3 indicates that the estimated mean mass of atrazine is smaller under plots using a corn-soybean rotation than under plots using continuous corn management systems.

The lack of stronger correlations between farming management systems and nitrate or atrazine mass in the saturated zone is in part the result of cross-contamination between the test plots (Fig. 2). This problem was identified by earlier researchers (Geoffery Delin, US Geological Survey, personal commun. 1992) and confirmed by visualizing the three-dimensional nitrate models generated with IVM. The cross-contamination results from groundwater flow that is not parallel to the orientation of the test plots and from the close proximity of the test plots to one another (Fig. 1) (Anderson & Stoner, 1989). Groundwater with relatively high concentrations of nitrate has moved down-gradient from test plot 2 to 3 and from test plot 5 to 6 (Fig. 2). Visualization of the nitrate models also shows that the concentration distribution of nitrate does not tend towards zero upgradient of the test plots indicating that there may be movement of nitrate laden water on to the site from off of the site (Fig. 2). The close proximity of other agricultural experiments has an unquantifiable effect

Table 3 Estimated mean nitrate and atrazine mass for the three farming management systems.

Farming management system	Mean nitrate mass (kilograms)	Mean atrazine mass (grams)
(1) Continuous Corn N applied at 72.6 kilograms per acre Atrazine applied at 454 grams per acre	32.55	0.063
(2) Continuous Corn N applied at 97.1 kilograms per acre Atrazine applied at 454 grams per acre	34.17	0.039
(3) Corn-Soybean Rotation N applied at 72.6 kilograms per acre to corn, 6.8 kilograms per acre to soybeans Atrazine applied at 454 grams per acre to corn 0 grams per acre to soybeans	25.66	0.014

on concentrations measured in some of the monitoring wells. The problem of cross contamination may be avoidable in future experiments if the test plots are oriented parallel to the direction of ground-water flow, separated by a greater distance, and isolated from other farming activities.

REFERENCES

Agricultural Statistics Board (1991) Agricultural Chemical Usage: 1990 Field Crops Summary. National Agricultural Statistics Service, US Department of Agriculture.
Anderson, H.W. & Stoner, J.D. (1989) Effects of controlled agricultural practices on water quality in a Minnesota sand-plain aquifer, US Geological Survey Open-filereport 89-267.
Briggs, I.C. (1974) Machine contouring using minimum curvature. *Geophysics* **39**(1), 39-48.
Flynn, J.J. (1990) 3-D computing geosciences update: hardware advances set the pace for software developers. *Geobyte* **5**(1), 33-36.
Geological Society of America Committee for the gravity map of North America (1987) *Gravity anomaly map of North America.* Washington, D.C.
Geological Society of America Committee for the magnetic map of North America (1987) *Magnetic anomaly map of North America.* Washington, D.C.
Gianessi, L.P. & Puffer, C. (1990) (revised 1991) Herbicide use in the United States: Washington, D.C., Resources for the Future, December (1990) (Revised April 1991).
Goolsby, D.A., Thurman, E.M. & Kolpin, D.W. (1991) Herbicides in streams: Midwestern United States, in Irrigation and Drainage. *Proc. Am. Society of Civil Engineers Conference,* Honolulu, HI., July 22-26, 17-23.
Iman, R.L. & Conover, W.J. (1983) A modern approach to statistics. John Wiley & Sons, Inc., New York.
Lee, Y.T. & Requicha, A.G. (1982) Algorithms for computing the volume and other intergral properties of solids. I. Known methods and open issues. *Communications of the ACM* **25**(9), 635-641.
National Geophysical Data Center (1988) ETOPO-5 Bathymetry/topography data. Data Announcement 88-MGG-02, National Oceanic and Atmospheric Administration, US Dept. Commerce.
Paradis, A.R. & Belcher,R.C. (1990) Interactive Volume Modeling: a new product for 3D mapping. *Geobyte* **5**(1), 42-44.
Raper, J.F. (1989) The 3-dimensional geoscientific mapping and modelling system: a conceptual design. In: *Three dimensional application in geographical information systems* (ed. by J.F. Raper), Taylor and Francis, London.
Smith, W.H.F. & Wessel, P. (1990) Gridding with continuous curvature splines in tension. *Geophysics* **55**(3), 293-305.
Thurman, E.M., Goolsby, D.A., Meyer, M.T. & Kolpin, D.W. (1991) Herbicides in surface water of the midwestern United States: The effect of the spring flush. *J. Environmental Science and Technology* **25**(10), 1794-1796.
Turner, A.K. (1990) Three-Dimensional GIS: Possibilities attract geoscientists in many industries, worldwide. *Geobyte* **5**(1), 31-32.
US Environmental Protection Agency (1990) *Country-level fertilizer sales data.* Office of Policy, Planning, and Evaluation (pm-221), Washington, D.C.
US Environmental Protection Agency (1992) *Another look: National Pesticide Survey Phase II report.* Washington, D.C.

Application of a geographic information system in analyzing the occurrence of atrazine in groundwater of the mid-continental United States

M. R. BURKART
Agricultural Research Service, National Soil Tilth Laboratory, 2150 Pammel Drive, Ames, Iowa 50010, USA

D. W. KOLPIN
US Geological Survey, Water Resources Division, Box 1230, Iowa City, Iowa 52244, USA

Abstract The US Geological Survey, US Department of Agriculture, and US Environmental Protection Agency are conducting research and regional assessments in support of policy alternatives intended to protect water resources from agricultural chemical contamination. The mid-continent was selected because of the intense row crop agriculture and associated herbicide application in this region. An application of a geographic information system is demonstrated for analyzing and comparing the distribution of estimated atrazine use to the detection rate of atrazine in groundwater. Understanding the relations between atrazine use and detection in groundwater is important in policy deliberations to protect water resources. Relational analyses between measures of chemical use and detection rate by natural resource units may provide insight into critical factors controlling the processes that result in groundwater contamination from agricultural chemicals.

INTRODUCTION

Increased yields for row crop agriculture in North America can, in part, be attributed to the increased use of herbicides. Herbicide use also has affected the quality of groundwater in many parts of the continent. Current research reflects the increasing concern over herbicide contamination of the Nation's water resources. The question of how agricultural practices can be modified to obtain a balance between production and water quality protection is currently being examined. Use of a geographic information system (GIS) facilitates the spatial integration and interpretation of numerous sources of information required when dealing with problems such as contamination of groundwater by herbicides.

The region of interest, the mid-continental United States, includes the states of Illinois, Indiana, Iowa, Kansas, Michigan, Minnesota, Missouri, Nebraska, North Dakota, Ohio, South Dakota, and Wisconsin (Fig. 1). This is an area of intense agricultural production where almost 60% of the nation's herbicides are applied (Gianessi & Puffer, 1990). This region is also where the US Geological Survey, US Department of Agriculture, and the US Environmental Protection Agency are planning and conducting multi-scale research to evaluate the effects of various farming practices on water resources (Burkart *et al.*, 1990).

Fig. 1 Location of wells from where atrazine analyses are available from WATSTORE (common reporting limit of 0.1 mg l^{-1}).

Purpose

The purpose of this paper is to demonstrate an application of a GIS to analyze the occurrence of atrazine in groundwater in the mid-continent. Atrazine was the chemical selected for this demonstration because it is the most extensively used herbicide in the region, has been continuously used for more than 30 years, and is the most frequently detected herbicide in groundwater. Because information is needed to answer both scientific and policy questions, there is a need to analyze the extent of contamination in as many contexts and scales as possible. A GIS provides for flexible methods of assessment and analysis. Using common geographic references, several layers of information can be combined and compared and the results studied from several perspectives.

Approach

The application of a GIS is demonstrated in the complex topic of atrazine contamination of groundwater. For this example, atrazine use estimates were selected to make spatial comparisons with well-specific detections of atrazine in groundwater. These end points will be analyzed at two scales, regional and subregional, using polygon coverages derived from classifications of soils and hydrogeological factors.

Atrazine use was obtained from county use estimates determined by Gianessi & Puffer (1990). The GIS was used to convert these estimates from political units into natural resource- based units. The total atrazine use in each natural resource polygon was divided by the polygon area to obtain atrazine use in units of mass per unit area so that polygons of different sizes could be compared.

Atrazine detection data were obtained from results of water-quality analyses in the US Geological Survey's National Water Data Storage and Retrieval System (WATSTORE). Latitude and longitude of the water-quality sampling locations were used to generate a GIS point coverage (Fig. 1). This point coverage was overlain onto polygon coverages to estimate the frequency of atrazine detection for each natural resource polygon. A detection of atrazine is defined as a concentration equal to or exceeding the laboratory reporting limit of 0.1 µg l^{-1}.

REGIONAL OCCURRENCE OF ATRAZINE

The estimated atrazine use (in kg km^{-2}) was reaggregated into "Major Land Resource Areas" (MLRA) (US Department of Agriculture, 1981) from the county-based estimates by summing the area-weighted atrazine use of all counties wholly or partly within the boundaries of each MLRA polygon. MLRA are large areas with similar soil, climate and vegetation attributes. A GIS display of this information shows the distribution of atrazine use in the mid-continent (Fig. 2). The working hypothesis for this demonstration is that atrazine use is proportional to atrazine detection. In other words, in areas where more atrazine is used to enhance crop production, atrazine will be detected more frequently in groundwater.

Detection in groundwater

To determine the spatial distribution of atrazine detection in groundwater resources of the mid- continental United States, approximately 3000 atrazine analyses collected from about 1700 wells sampled during 1979 to 1989 were

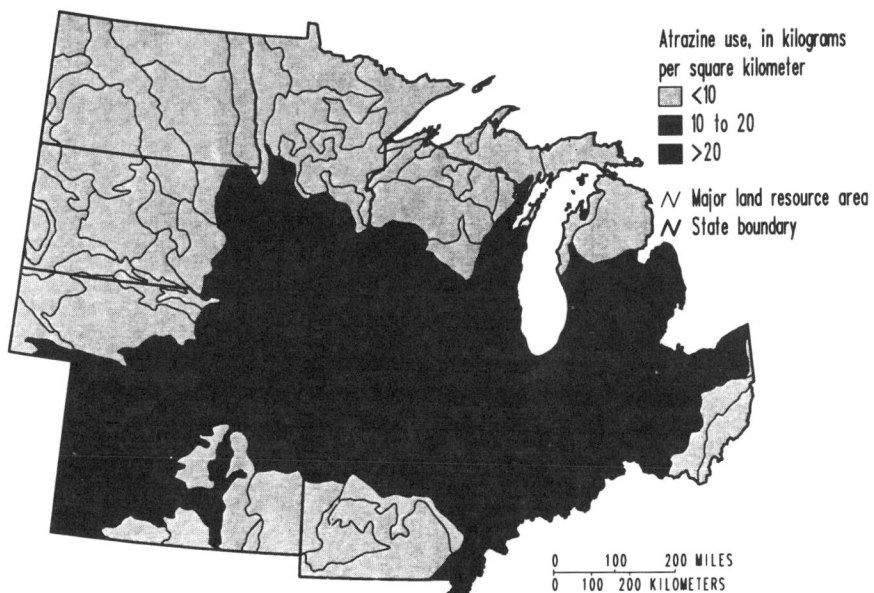

Fig. 2 Atrazine use aggregated by Major Land Resource Areas (MLRA).

retrieved from WATSTORE. A plot of the points in this coverage can provide a visual assessment of their spatial distribution. This can be used, for example, to determine if the data are evenly distributed within a particular MLRA, clustered, or thinly scattered. As can be seen in Fig. 1, these point data are not evenly distributed in all MLRA. Limiting the analysis to MLRA that include at least 10 wells tested for atrazine detection, a GIS display of the frequency of atrazine detections by natural resource units can quickly show where the smallest and largest atrazine detection frequencies occurred and where more information on atrazine detection is needed (Fig. 3).

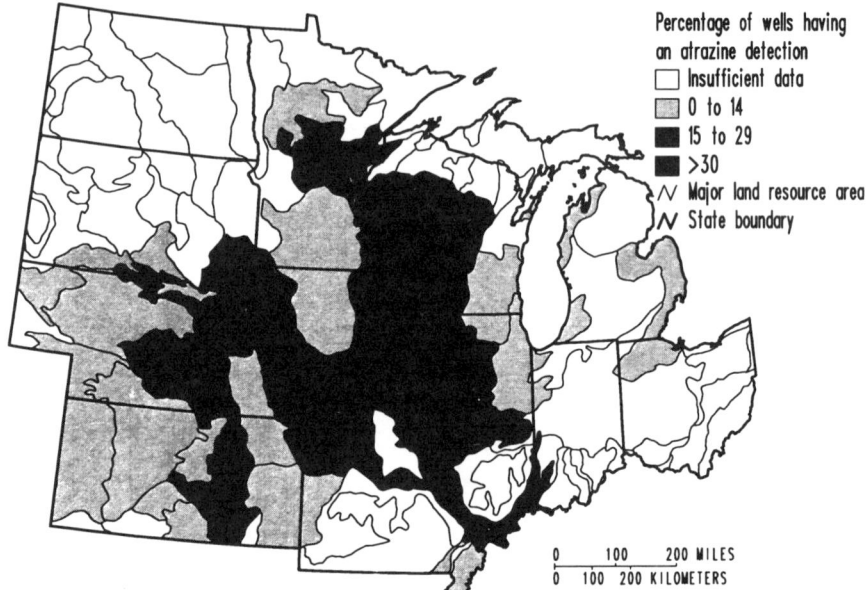

Fig. 3 Frequency of atrazine detections aggregated by Major Land Resource Areas (MLRA) (polygons with fewer than 10 wells tested for atrazine contained insufficient data).

Regional susceptibility

A visual comparison of Figs 2 and 3 gives a general perception as to where the working hypothesis is true. Specifically, a GIS can clearly display areas where large atrazine use corresponds to large frequencies of atrazine detection.

For this demonstration, the atrazine use rate for the entire mid-continent and the atrazine detection frequency for the entire region were standards used to define large and small use and detection frequencies. The atrazine use for the region (13.3 kg km^{-2}) was the quotient of the sum of all atrazine used and the area of the region. The regional atrazine detection frequency (26.2%) included all atrazine analyses available in the WATSTORE data base. Large use or detection values were those which exceeded these regional values and small use or detection was less than these values. The GIS can then be used to reselect and display the four permutations of atrazine use and detection. The permutations were ranked to produce the following index of hypothetical susceptibility to groundwater contamination by atrazine:

(a) small susceptibility (large use/small detection): Occurs where the relation appears to be opposite to the working hypothesis and large atrazine use does not correspond to large frequencies of atrazine contamination;

(b) intermediate susceptibility (small use/small detection): Occurs under conditions where the working hypothesis appears valid but the susceptibility is low because use is low;

(c) large susceptibility (large use/large detection): Occurs under conditions where the working hypothesis appears valid and atrazine detection is large;

(d) critical susceptibility (small use/large detection): Occurs under conditions where the relation appears to be opposite the working hypothesis and even small atrazine use corresponds to large frequencies of atrazine detection.

In Fig. 4, areas where atrazine detections were either greater or less than expected relative to atrazine use can be identified. The presence of larger frequencies of atrazine detection affords more opportunities for research on the causes of contamination. Conversely, the processes or factors that inhibit atrazine contamination can best be investigated in areas of large use but small detection frequencies. Also, if regulatory agencies adopt policies about a wide variety of agricultural practices designed to improve water quality, the areas of

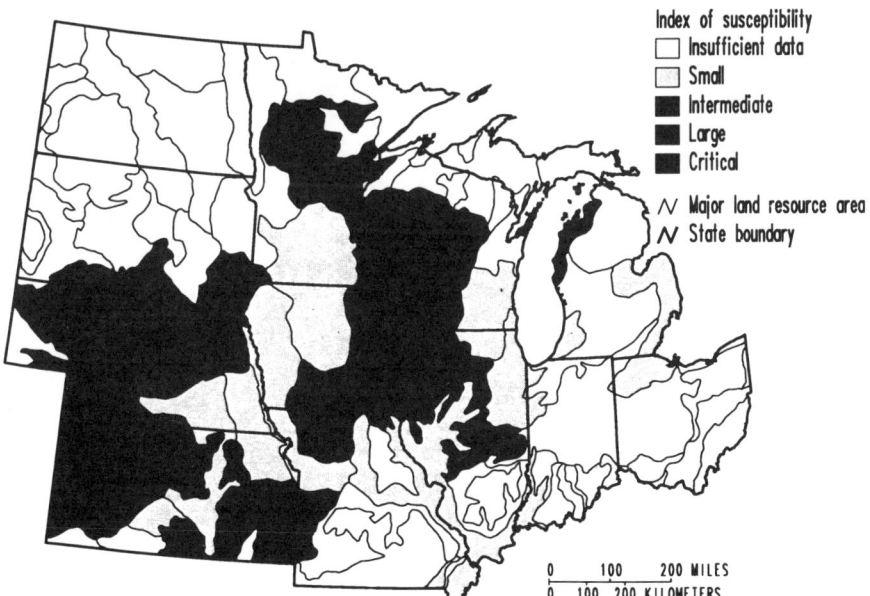

Fig. 4 Groundwater susceptibility to atrazine contamination based on detection frequency and atrazine use, displayed by Major Land Resource Areas (MLRA).

critical susceptibility may provide positive results in the shortest amount of time. If policy is to be limited only to a decrease in chemical use without considering other factors, success in improving water quality will most likely be in those areas where both the atrazine use and atrazine detections are large. Such policy may have little effect in the areas of small susceptibility because even a large use of atrazine does not appear to greatly increase the atrazine detection.

SUBREGIONAL ATRAZINE OCCURRENCE

More specific research and policy questions will require more detailed information at a larger scale to investigate the occurrence of atrazine in groundwater. The State of Iowa was selected as an example to illustrate the application of a GIS at this scale. Analysis was restricted to a subregion defined by political boundaries because detailed digital landscape data bases such as those used in this example are being developed using State rather than natural resource boundaries. Iowa was selected because it has the most densely and evenly distributed set of groundwater atrazine analyses in the mid-continent (1745 atrazine analyses from 817 wells in WATSTORE) and it has multiple landscape subdivisions available in digital geographic format. Eventually, these landscape data will be available for much of the mid-continent.

Susceptibility using state soil geographic data base

There are substantial variations in critical geological, hydrological and pedological factors that influence the potential for groundwater contamination within each MLRA. The "State Soil Geographic Data Base" (STATSGO) (US Department of Agriculture, 1991) includes a polygon coverage in which atrazine detection can be defined in relation to pedological factors at a scale considerably finer than, but using factors similar to those used for MLRA. Iowa is separated into 2032 polygons in 84 STATSGO units but only nine polygons in 7 MLRA units. Reaggregating the atrazine and atrazine detection data by STATSGO units shows more spatial variability in susceptibility to atrazine contamination than shown by the MLRA units alone (Fig. 5). An atrazine use rate for Iowa and an atrazine detection frequency for Iowa were standards used to define large and small use and detection values at the subregional scale. The atrazine use for Iowa (24 kg km^{-2}) was the quotient of the sum of all atrazine used in the State and the area of the State. The atrazine detection frequency for wells tested for atrazine in Iowa was 24%.

In general, the STATSGO units associated with major alluvial systems, represented by the network of dendritic patterns, have greater hypothetical susceptibility to atrazine contamination than immediately surrounding areas and are responsible for much of the appearance of local variability within the

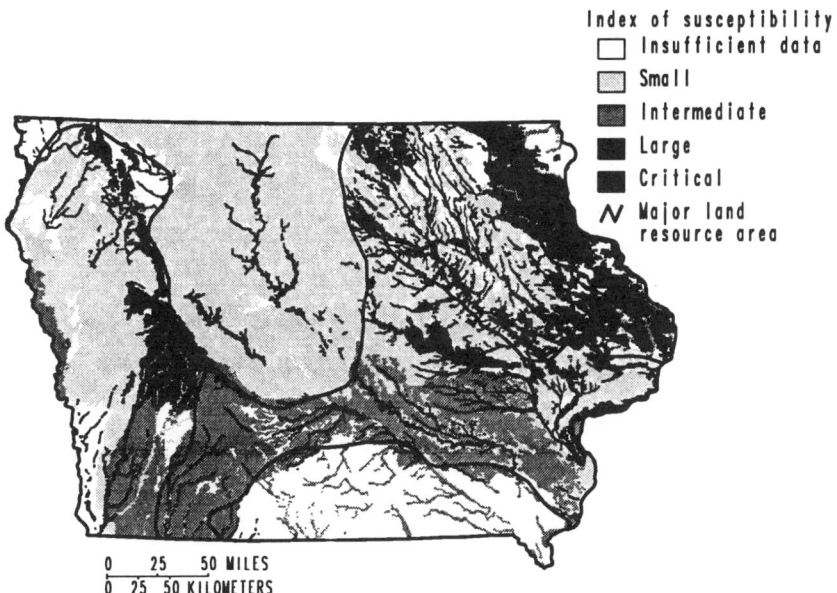

Fig. 5 Groundwater susceptibility to atrazine contamination for Iowa based on detection frequency and atrazine use, displayed by State Soil Geographic Data Base (STATSGO).

MLRA polygons. There are gross similarities in susceptibility between Figs 4 and 5 but the increased detail using STATSGO also reveals substantial differences. For example, the north-central part of the State is an area with the small susceptibility in both figures. However, in the southern part of the State, Fig. 4 shows an MLRA with an intermediate susceptibility, but this same MLRA in Fig. 5 shows intermediate susceptibility for only those STATSGO units associated with major alluvial systems. The remaining STATSGO polygons comprising this MLRA have insufficient information documenting atrazine detection. Coverages such as STATSGO can be used to isolate smaller areas of interest for conducting research or implementing management strategies.

Susceptibility using groundwater vulnerability regions

Another landscape perspective at the subregional scale is provided by the Iowa "Groundwater Vulnerability Regions Data Base" (GVR) (Hoyer & Hallberg, 1991). This coverage includes polygons with similar hydrogeological characteristics affecting the susceptibility of aquifers to contamination from surface applied sources. Iowa is separated into 3344 polygons using 9 GVR units. The pattern of hypothetical susceptibility index to atrazine contamination for the GVR (Fig. 6) has similarities and differences to the pattern determined for STATSGO. The differences were expected because the GVR units are explicitly defined by aquifer characteristics, whereas STATSGO uses soil

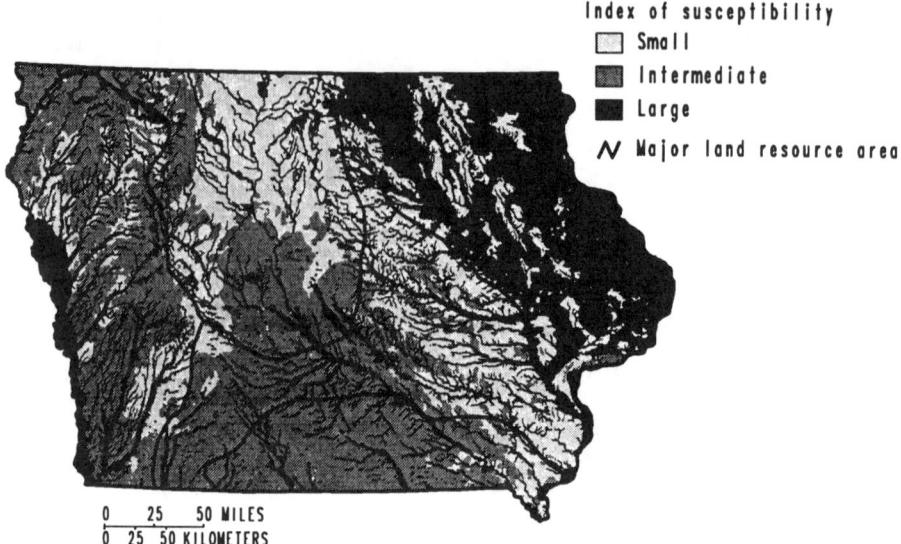

Fig. 6 Groundwater susceptibility to atrazine contamination for Iowa based on detection frequency and atrazine use, displayed by Groundwater Vulnerability Regions (GVR).

characteristics, which are only indirectly associated with aquifer properties. The GVR display clearly isolates alluvial aquifers and the karstic bedrock region of northeast Iowa as areas where large atrazine use corresponded to large atrazine detection. Independent, unpublished research by the authors, and other research (Detroy & Kuzniar, 1988; Libra *et al.*, 1991) has also shown these hydrogeological areas to be susceptible to sources of herbicide contamination resulting from application of agricultural chemicals.

DISCUSSION AND CONCLUSIONS

The value of using multiple polygon coverages for the type of spatial analysis demonstrated in this paper lies in the differing perspectives by which aggregated data can be analyzed. Each coverage will likely have similarities and differences in the distribution of susceptibility. These similarities and differences are useful to more realistically display the degree of confidence in conclusions drawn from the reaggregation of data. The consistent large to critical susceptibility shown for northeast Iowa at both regional and subregional scales is an example of multiple perspectives yielding similar conclusions. However, the larger susceptibility shown for alluvial aquifer systems in both the state Groundwater Vulnerability Regions Data Base (GVR) and the State Soil Geographic Data Base (STATSGO) could not be derived from the regional scale "Major Land Resource Area" (MLRA) display.

A direct comparison of the resolution attained by two scales of analysis (Fig. 7) shows that with an ideal distribution of water quality data, the

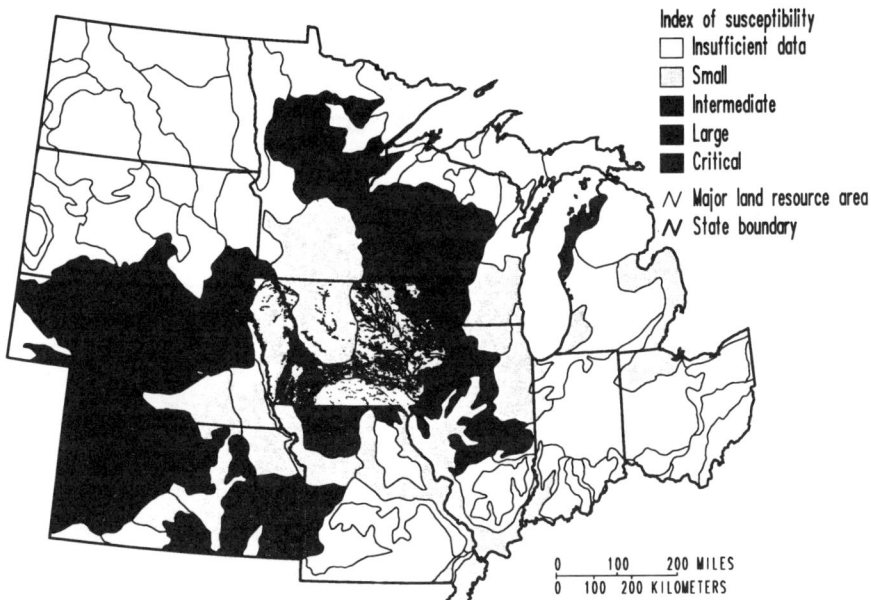

Fig. 7 Groundwater susceptibility comparing State Soil Geographic Data Base (STATSGO) of Iowa and Major Land Resource Areas (MLRA) of the mid-continent.

resolution shown for Iowa will be attainable when the STATSGO data base for the entire mid-continent is finalized. The STATSGO polygons also have attribute files associated with them which allow for extensive relational analysis of water quality and pedological data.

This type of spatial analysis presented in this paper is expected to lead to a better understanding of groundwater contamination from non-point sources of herbicides. STATSGO and GVR include many of the factors critical to the fate and transport of chemicals applied to the land-surface. With a better definition of the spatial distribution of contamination, it will be possible to concentrate research, monitoring, and policy efforts to maximize the information and results leading to protection of groundwater resources.

REFERENCES

Burkart, M.R., Onstad, C.A. & Bubenzer, G.D. (1990) Research on agrichemicals in water resources. *EOS, Trans. Am. Geophysical Union* **71**(29), 980-988.
Detroy, M.G. & Kuzniar, R.L. (1988) Occurrence and distribution of nitrate and herbicides in the Iowa River alluvial aquifer, Iowa-May 1984 to November, 1985. US Geological Survey Water-Resources Investigations Report 88-4117.
Gianessi, L.P. & Puffer, C.M. (1990) Herbicide use in the United States: Washington, D.C., Resources for the Future, [Inc.], Quality of the Environment Division. (Revised April 1991)
Hoyer, B.E. & Hallberg, G.R. (1991) Groundwater vulnerability regions of Iowa. Geological Survey Bureau, Iowa Dept. Natural Resources, Special Map Series 11.
Libra, R.D., Hallberg, G.R., Littke, J.P., Nations, B.K., Quade, D.J.& Rowden, R.D. (1991) Groundwater

monitoring in the big spring basin, 1988-1989. Iowa Department of Natural Resources, *Technical Information Series*, **29**.

US Department of Agriculture (1981) Land resource regions and major land resource areas of the United States. Agric. Handbook 296.

US Department of Agriculture (1991) State soil geographic data base (STATSGO) data users guide. Soil Conservation Service, Miscellaneous Publication Number 1492.

EGIS, a geohydrological information system

F. DECKERS
Institute of Applied Geoscience (IGG/TNO), P.O. Box 6012,-2600 JA Delft, The Netherlands

Abstract Although nowadays abundant software for geohydrological purposes exixts, the software environment of the geohydrologist often consists of a patchwork of packages featuring incompatible interfaces. As a result of this fragmentation, the full potential of the software cannot be exploited. A geohydrological information system can overcome this problem by providing the geohydrologist with an integrated working environment. In this paper, the requirements for such an information system are examined. Using the EGIS system - currently being developed by IGG/TNO as an example, technical solutions that permit to meet these requirements are discussed.

INTRODUCTION

During the past decades, the central groundwater institute in The Netherlands, the Institute of Applied Geosciences TNO (IGG), acquired considerable experience in the development of applications and large volume databases used for the proper management and analysis of geohydrological (point) data, collected from nationwide monitoring networks and geophysical surveys. During recent years, GIS have been applied by the institute to help in the assessment and analysis of geohydrological problems (Broers *et al.* 1990). Being confronted with the growing demand for integrated (ground)water management, the institute decided to develop a geohydrological information system, a tool able to provide support for this integrated approach. The design of this system which is called EGIS (Evaluation of Groundwater resources Information System) relied upon the experience acquired by numerous geohydrologists during geohydrological studies and projects carried out by the institute in The Netherlands and abroad. This paper examines the requirements for such a geohydrological information system and discusses how EGIS tries to comply to these requirements.

REQUIREMENTS FOR A GEOHYDROLOGICAL INFORMATION SYSTEM (Ghis)

A geohydrological information system should provide the necessary support for those activities that take place during a geohydrological project. In general, these projects have two main objectives; assessment of the groundwater resources in the area of interest, followed by management of these resources. During the assessment phase, quantitative and qualitative information about the groundwater resources in the project area, about the external factors influencing these resources and about the groundwater reservoir are acquired,

processed, interpreted and finally transformed into a conceptual geohydrological model of the area. This model should account for the spatial (3D) and temporal aspects (4D) of the groundwater system under study and will be used during the management phase to evaluate certain groundwater management scenarios. Figure 1 shows the activities in a geohydrological project that should be supported by a Ghis.In order to be able to support the projects outlined above, a geohydrological information system should provide :
(a) facilities to store and maintain all the geohydrological data collected in the area;
(b) functionality to derive and maintain a geohydrological model of the area under study;
(c) functionality to use this model in the management phase, e.g. to support simulation of the effect of management scenarios and provide answers to questions of the type - what if?

The software that is currently available in the field of geohydrology is still restricted to an agglomerate of independent programs showing often incompatible data- and user interfaces. The introduction of a geohy-drological information system will considerably improve the working environment of the geohydrologist. However, this will only be true if the Ghis overcomes the inconveniences of the current fragmented environment. A Ghis should therefore not only incorporate the required data and functionality, it should also be designed with three major design options in mind : integration, flexibility and support for multiple users.

The need for integration will exist on three levels; integration of data, integration of functionality and integration of user interface. Flexibility includes both scalability and adaptability. A Ghis is scalable when it is able to support local scale projects, but also regional and nationwide studies. Moreover it should be possible to tune the system in such a manner that all the topics that are really of interest to a certain problem can be covered but irrelevant topics are left out. Adaptability requires a system that can be customized in order to

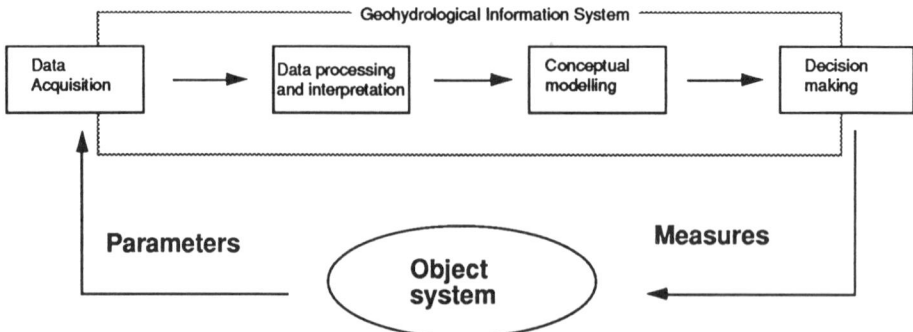

Fig. 1 Scheme of activities and information flow in a geohydrological project indicating the area supported by a Ghis. The geohydrological area of interest is indicated as object system.

adjust it to the specific needs of a particular project. Finally, because geohydrological studies are usually carried out by a team of geohydrologists, the Ghis should be configured as a fully fledged multi-user system.

EGIS

The EGIS system, currently under development at IGG/TNO, will be used by the institute in The Netherlands and in its projects abroad. For this purpose the system anticipates on being used as a support tool in different geohydrological projects throughout the world. The system contains two subsets (Fig. 2). The first subset is made up of range of general purpose packages for data presentation and processing, integrated with a database. The second subset extends this environment with a set of advanced geohydrological applications which cover either the processing and interpretation of a specific type of data (e.g. pumping test data, geoelectrical measurements) or focus on a specific type of analysis (e.g., time series analysis, numerical modelling). EGIS tries to comply to the requirements for a Ghis which have been stated above. The remaining sections of this paper have been used to point out how this will be accomplished.

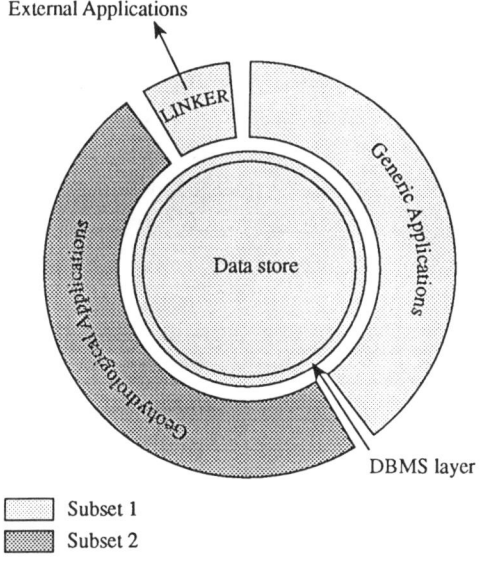

Fig. 2 Scheme of EGIS architecture showing the two different subsets.

The GIS component of EGIS

As a general guideline, the development of EGIS relies upon existing (commercial) software whenever appropriate. It has also been policy to restrict the use of such "off the shelf" components to a strict minimum. Increasing the number of components results in increased costs and introduces additional incompatibilities in user- and data interfaces, both hampering the integration effort. The pronounced spatial character of the data involved in geohydrological work makes it a perfect target for manipulation by a GIS. It should therefore not come as a surprise that a large part of the functionality of a Ghis can be filled in by incorporating such a GIS, what is actually being done in EGIS. In general, the use of GIS in (geo)hydrology has been concentrating on spatial analysis and to a lesser extent GIS have been used to prepare model input and to visualize modelling results. However, to be a valid candidate for incorporation in a geohydrological information system, a GIS should be able to support more functions :

(a) maintain the data stored in the system. Huge amounts of data may be required for projects on a regional and nationwide scale;
(b) act as the major interface with the user, by offering a spatial view on all the data stored;
(c) support the integration of the applications by providing them with a uniform data interface;
(d) support the definition and maintenance of the reservoir component of the geohydrological model;
(e) manage different versions of this reservoir model;
(f) provide the required spatial analysis functionality;
(g) support interfacing to groundwater models;

The items listed above will make it clear that a candidate GIS should offer more than just a set of spatial analysis functions acting on a set of spatial data structured in a proprietary format. In its context of Ghis component, the GIS will be used throughout an entire organization to manage and analyze all its corporate data. Moreover, it should be possible to incorporate the GIS in such a manner that the integration objective is not violated. The Smallworld GIS has been selected as GIS component of EGIS. This GIS comes with an object oriented (OO) development environment that has been used to implement the GIS core classes. This environment, which is a major asset in the adaptability of the GIS, has also been used to implement substantial parts of EGIS.

EGIS basic functionality

The data store The data store of EGIS is managed by two resources, the Smallworld GIS and a commercial SQL based RDBMS (Oracle). Although it is possible to store and maintain all the data - also non-spatial data - into the GIS data store, RDBMS systems still offer more powerful functionality for the maintenance of non-spatial data than the GIS has currently available. This is the main reason why a two-component architecture has been chosen. The data store has been loaded with a geohydrological data model, allowing storage and maintenance of the entire range of information that can be of interest to the field of geohydrological engineering, like reservoir characteristics, groundwater data (quality and quantity), groundwater production and use, groundwater systems and flow paths, well completion an external factors influencing the groundwater, like soil types, climatology and surface water.

The geohydrological model The geohydrological model that will be set up for a certain area of interest consists basically of a model describing the groundwater reservoir and a number of models (stochastical or deterministic) describing the temporal behaviour of the groundwater. In this paper only the groundwater reservoir part of the model will be considered.

A reservoir model that is set up and maintained using EGIS will be basically a 2.5D model. The term 2.5D is being used because only subsoil surfaces are modelled in 3D space, but complete geometrical and topological descriptions of volumes are still lacking. Although a full 3D model would be more appropriate and technology in the field of 3D GIS and even 4D GIS (O'Connail 1992) is emerging, this technology is still considered to be too immature to be incorporated with success into an operational system like EGIS. The data model implemented in EGIS incorporates objects modelling subsurface layers, subsurface layer interfaces and faults. In addition, a number of objects have been introduced to represent the topological relations between them. Subsoil surfaces are represented by conventional surface modelling objects like TINs or grids. The geometry of a reservoir model is represented in EGIS by means of a set of these modelled surfaces. Volumes are thought to be completely or partially enclosed by these modelled surfaces, the topology objects keep track of which surfaces enclose which volumes and vice versa. Functions like edit, close, split, assimilate etc. are defined upon the surfaces and volumes.

Next to the geometry model, separate property models can be set up for the different relevant geohydrological properties of the subsoil, for instance hydraulic conductivity. The properties can be related to the subsoil layers defined in the model. The spatial variation of these properties over the extent of a subsurface element are represented by means of contour maps and 3D attributes which can be used to associate properties to subsoil layers or subsoil surfaces.

Simulating management scenarios In order to predict the effects of management measures on the groundwater system, prediction tools are used. Because the (geo)hydrological problems encountered in groundwater management vary significantly from region to region, a geohydrological information system with one specific numerical model built in will be of limited use. It should be possible to accommodate a large number of these models which are often very specifically oriented towards one type of problem e.g. fresh-brackish groundwater interface.

EGIS makes use of a driver mechanism to allow flexibility in the accommodation of numerical groundwater models. A simulation will be defined as a high level object and is specified by means of a problem definition and a data set required for the simulation which includes a description of the target subsurface volume. In general, the data set will include parts of the 2.5D reservoir model, including both geometry and properties, and objects quantifying the factors influencing the groundwater in the area, that are to be taken into account by the simulation. Subsequently a numerical approach will be chosen (e.g., finite difference or finite elements) and a grid will be constructed using a grid editor/generator. This tool controls the layout of the grid, which should be conform the rules related to the numerical approach. Once this grid has been defined, the target volume represented by the aforementioned data set can be discretized. A uniform simulation definition has now been produced, which can be converted by the driver code into the input files required by a model code that is implementing the specified numerical approach. Using this three staged approach, incorporating a new model code into the system is reduced to the implementation of a driver definition, which specifies how the uniform simulation definition should be converted into its code-specific equivalent (Fig. 3).

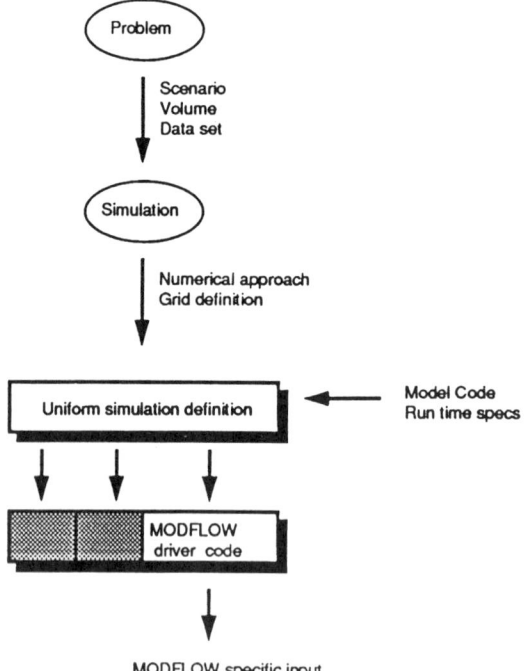

Fig. 3 Integration of numerical groundwater models in EGIS; scheme of the input flow path for MODFLOW.

EGIS basic characteristics

Integration Data integration is provided within EGIS through the concept of a single transparent data store. The data model that has been implemented in the data store allows storage of basic data e.g. groundwater levels or geophysical measurements, alongside processed or interpreted derivatives of this basic data in the form of 1D, 2D and 3D models and in the form of derived maps. Geohydrologists working on a project will process and interpret the basic measurements and store the result of their work back into the data store, where it can be accessed by other members of the organization, now or in the future. The major benefit of this single store mechanism, is that the geohydrologists can access all the relevant information about the area of interest during the same session.

Integration of functionality is dealt with on two levels; the data interface level and the user interface level. In EGIS, the access to the different physical databases is controlled and hidden by the virtual data store concept, implemented by the underlying GIS. An application interacts only with this virtual data store and hence accesses the stored data through a uniform interface. At the user interface level, the GIS is used to provide the user with a spatial viewport on the data, through which the data contents of the system can be visualized. Subsets of the data are selected using a general query engine allowing both spatial and non spatial query specifications. The different applications are on their turn activated upon these selected data sets. In this way a uniform interaction protocol is maintained for all the applications.

The different components of EGIS are presented to the user through a uniform interface based on a modern graphical user interface toolkit (Motif/OpenLook). The interaction with the user is based upon a number of standard interaction protocols. In this way the same look and feel is preserved over the entire system. This greatly improves the ease of use of the system and reduces the amount of time required to get acquainted with it. The object oriented environment that is available to the EGIS application programmers is used to maintain a large number of user interface objects, that are being re-used in the different application packages. In this way, the uniformity of the user interface can be maintained.

Scalability An operational version of EGIS can be packaged from a set of modules. These modules are covering different topics and contain a set of classes defined in the data model along with a number of applications operating on them. Using this assembly mechanism, the system can be tuned to accommodate only those modules that are of interest to a particular geohydrological project. The software components underlying EGIS are able to operate in different environments, varying from stand alone machines to heterogeneous networked environments where multiple users access the central data store. They guarantee the hardware independence of EGIS and its scalability to different operational sites.

Adaptability During the information analysis carried out for EGIS, it soon turned out to be impossible to produce a data model that will cover all the needs of any arbitrary geohydrological project. As the data model is actually at the core of any system, restructuring it can have severe consequences. In order to reduce the efforts related to such restructuring, it is important to produce a data model that anticipates on change. In EGIS, care has been taken to derive general data structures, that are modelling the general concepts behind the different types of data rather than their specific appearance. This has resulted in the definition of a set of core classes with associated functionality (behaviour) that can be redefined using the object oriented concepts of inheritance and polymorphism. These core classes are being used for the development of the system and remain available in the form of a toolbox that can be used when extensions to - or restructuring of parts of - the data model are required. Minor updates in the sense of adding some spatial or non-spatial attributes to a core object are possible rightaway by a number of customization hooks which have been built in order to allow dynamic modification of the data model. These updates can be effectuated without any programming effort. Generic functionality is available that can address such extensions.

Customizing the functionality and user interface of EGIS can be addressed on three levels:
(a) user level (styles, symbols, macro-commands, etc.);
(b) customizer level (generic packages, customization hooks);
(c) programmer level (object oriented toolbox and -language).

External applications can be linked into the EGIS system with the external linker module (Fig. 2). With the linker mechanism, activating an external application proceeds according to the same interaction protocol as used for the EGIS applications and the necessary data reformatting is hidden for the user. The user interface of the external application will be left unmodified.

Multiple users As mentioned before, many geohydrologists will co-operate in large scale geohydrological projects. It is extremely important that a system supporting those projects also supports multi-user operation. As a result, the data store should be able to handle multi-user access in a networked environment and the system should provide the necessary control mechanisms to support multi-user operation, including support for long transactions in the form of version management (Newell & Easterfield 1990). As these features are available in EGIS through the underlying software resources, they will not be discussed here.

CONCLUSION

A geohydrological information system improves the current working environment of the geohydrologist by providing support for his entire field of work. This covers the trajectory from data acquisition over conceptual modelling to final decision making. Although nowadays abundant software for geohydrological purposes is available, its fragmented character prevents exploiting it to its full benefits. A geohydrological information system can overcome this problem. To do this effectively, it should be based on three important design principles: integration, flexibility and providing support for multiple users. The EGIS system developed at IGG/TNO claims to be a geohydrological information system in the true sense. The system provides a high degree of data-, functionality- and user interface integration. Moreover sufficient mechanisms have been incorporated to make the system adaptable and to provide the necessary potential for future extensions.

REFERENCES

Broers, H.P., Peters, S.M.W. & Biesheuvel, A. (1990) Design of a groundwater monitoring network with GIS and remote sensing. *Proc. First European Conf. on GIS,* held in Amsterdam. EGIS Foundation, Faculty of Geographical Sciences, Utrecht, The Netherlands, 95-105.

Newell, R.G & Easterfield, M. (1990) Version management - the problem of the long transaction. *Mapping Awareness* **4**(2).

O'Conaill, M.A., Bell, S.B.M & Mason, D.C. (1992) Developing a prototype 4D GIS on a transputer array. *ITC Journal* **1992-1,** Special Issue on 3D GIS, Enschede, The Netherlands, 47-54.

The management of groundwater resources in the Salinas Valley, California, by computer

PHILIP HALL
Earthware of California, 30100 Town Center Drive 196, Laguna Niguel, California 92677, USA

MATTHEW ZIDAR
Monterey Water Resources Agency, 855 E. Laurel Drive, Salinas, California 93902-0903, USA

Abstract The Monterey Water Resource Agency is responsible for the management of water resources in the Salinas Valley, California which is approximately 100 miles (160 km) south of San Francisco. The area covers 4400 square miles (11 400 km^2) and groundwater is used to irrigate 210 000 acres, (850 km^2) which produces $1.4 billion per year in crop revenue. Groundwater also supplies drinking water for some 200 000 people in the valley. Five years of drought, various agrochemical pollution problems, and sea water intrusion along the coast has lead the Agency to implement a groundwater management plan utilizing computers.

INTRODUCTION

The valley has been broken down into four hydrogeological areas, the Upper valley, the Forebay, the Pressure and the East side aquifers, based on the recharge characteristics of each aquifer (California Department of Water Resources, 1946) as shown in Fig. 1. There are two main aquifers near the coast, the 180-foot aquifer and the 400-foot aquifer (Fig. 2). Some of the shallower aquifers are unconfined and subject to nitrate contamination, while some of the deeper aquifers near the coast are subject to sea water intrusion. There are approximately 10 000 water wells in the valley with capacities ranging from 500 to 2500 US gallons per minute (2 to 10 m^3 min^{-1}). The Agency intends to enter all well records into a database/Geographic Information System (GIS) thus enabling well locations to be mapped, hydrogeological cross sections to be created, wells to be sorted by producing aquifer and water quality and other parameters to be determined.

COMPUTER SYSTEMS

The Agency has a number of IBM PC computers used mainly for data entry and processing and an IBM RS 6000 workstation used to run ARC/INFO. PC terminals accessing ARC/INFO need to have at least 8MB RAM and a 486 CPU. Output goes to a number of printers or plotters. The hydrogeological software used by the agency is GEOBASE 6.0, a graphic relational database

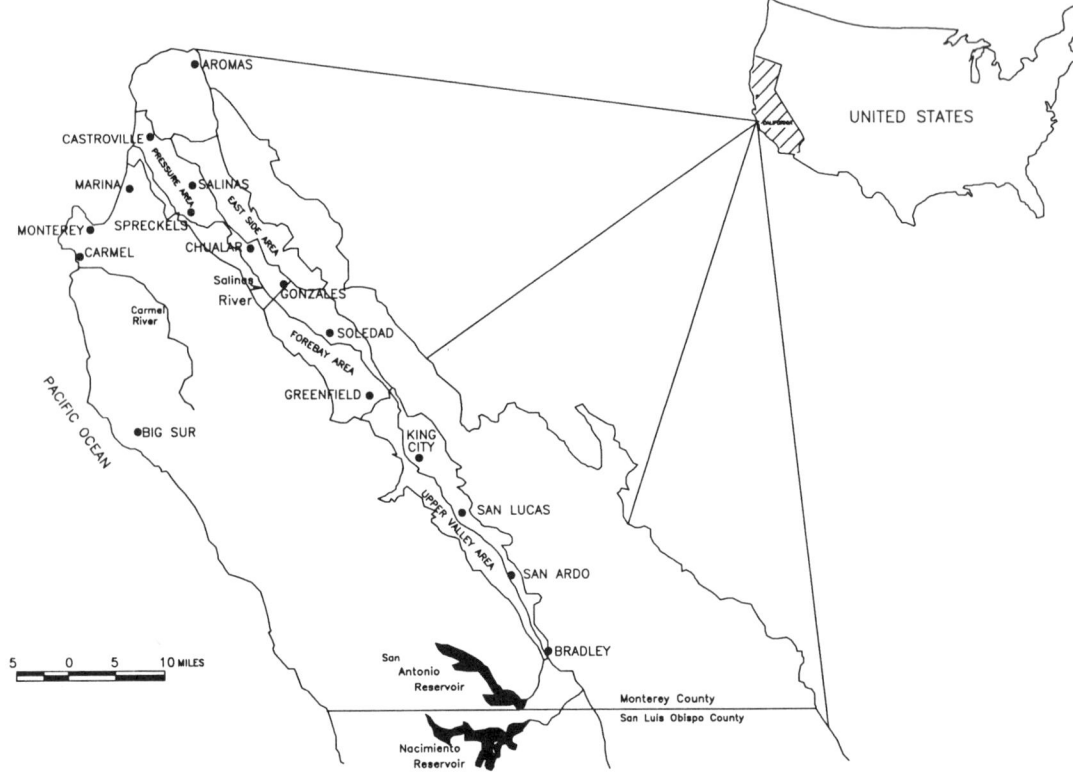

Fig. 1 General location map and hydrological sub areas of the Salinas Valley.

developed to process groundwater data. ORACLE will be the database engine on the workstation.

DATA ENTRY

Data entry is not a prestigious task and numerous checks have been built in to the programs to ease data entry and prevent typing errors. The employment of geology students to enter data from drillers logs has increased the accuracy of data interpretation. As a first step, water wells are located from descriptions found on the original driller's log. The well name is based on the location of the well using the Public Lands Survey Grid of Township, Range and Section. Coordinates may be calculated in UTM, Latitude/Longitude or the State Plane system from this grid. Coordinates can also be derived from maps using AutoCAD or ARC/INFO by placing the well, according to the physical description, on the drillers log. The coordinates can then be read off from the program. Field surveys using a Global Positioning System (GPS) have been obtained for more than 600 wells that are part of a routine monitoring program. Coordinates can be converted between Latitude/longitude, UTM or State Plane.

Well locations are displayed on AutoCAD .DXF basemap files using GEOBASE, AutoCAD or ARC/INFO. Lithology and well construction details are entered into GEOBASE. The standardized lithology names are entered by function keys, thus speeding up data entry and eliminating the inevitable typing errors. This limits the number of lithology names being used. The program also checks for continuity in depth intervals. The original drillers lithological description is added in a "comments" section.

Plotting up lithology and well construction details (Fig. 3) is the second step in the quality control process. Logs are checked against the written logs and compared against nearby logs for similarity. The positions of well screens are compared with lithology and checked against the written log. A visual output of each log is one of the best QA/QC control systems being used at this time. Hardcopy output is stored with the original written log. Geological cross sections are drawn on the computer and logs are selected from a map or index file. The logs are then displayed on the computer screen at their correct locations and elevations. A lithology is selected and highlighted on the borehole logs. A polygon is drawn with the aid of a mouse for a specified lithology and the angle of dip can be set. Well construction details and digitized electric logs can be shown alongside the borehole log with groundwater chemistry displayed as pie, bar or Stiff diagrams above each well. These features make it easier to correlate lithologies on the cross section. The cross sections can be displayed as fence diagrams.

Upon completion of the cross sections (Fig. 4), each borehole has hydrostratigraphic markers placed on it, typically including the top and bottom

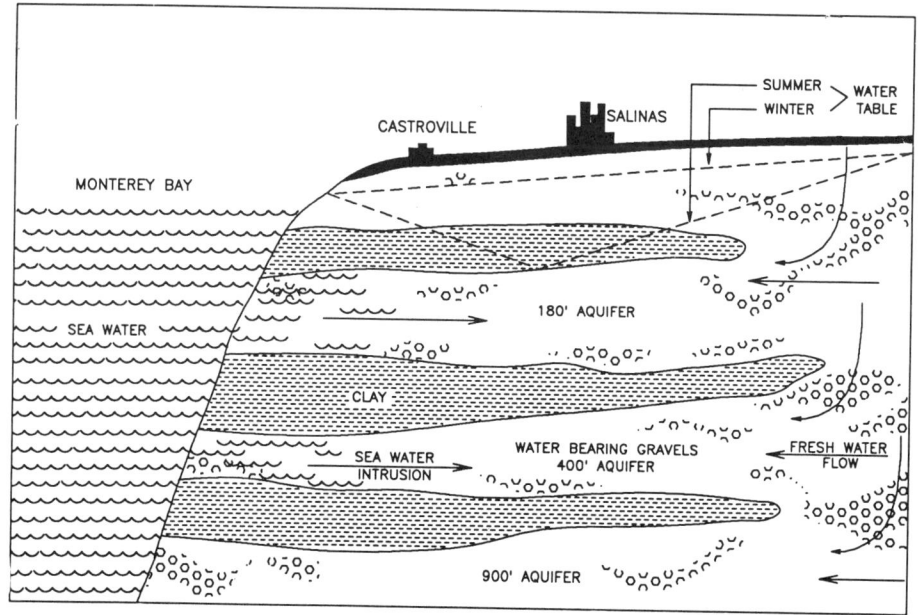

Fig. 2 Schematic hydrogeological cross section of aquifers and aquitards near the coast, Salinas Valley.

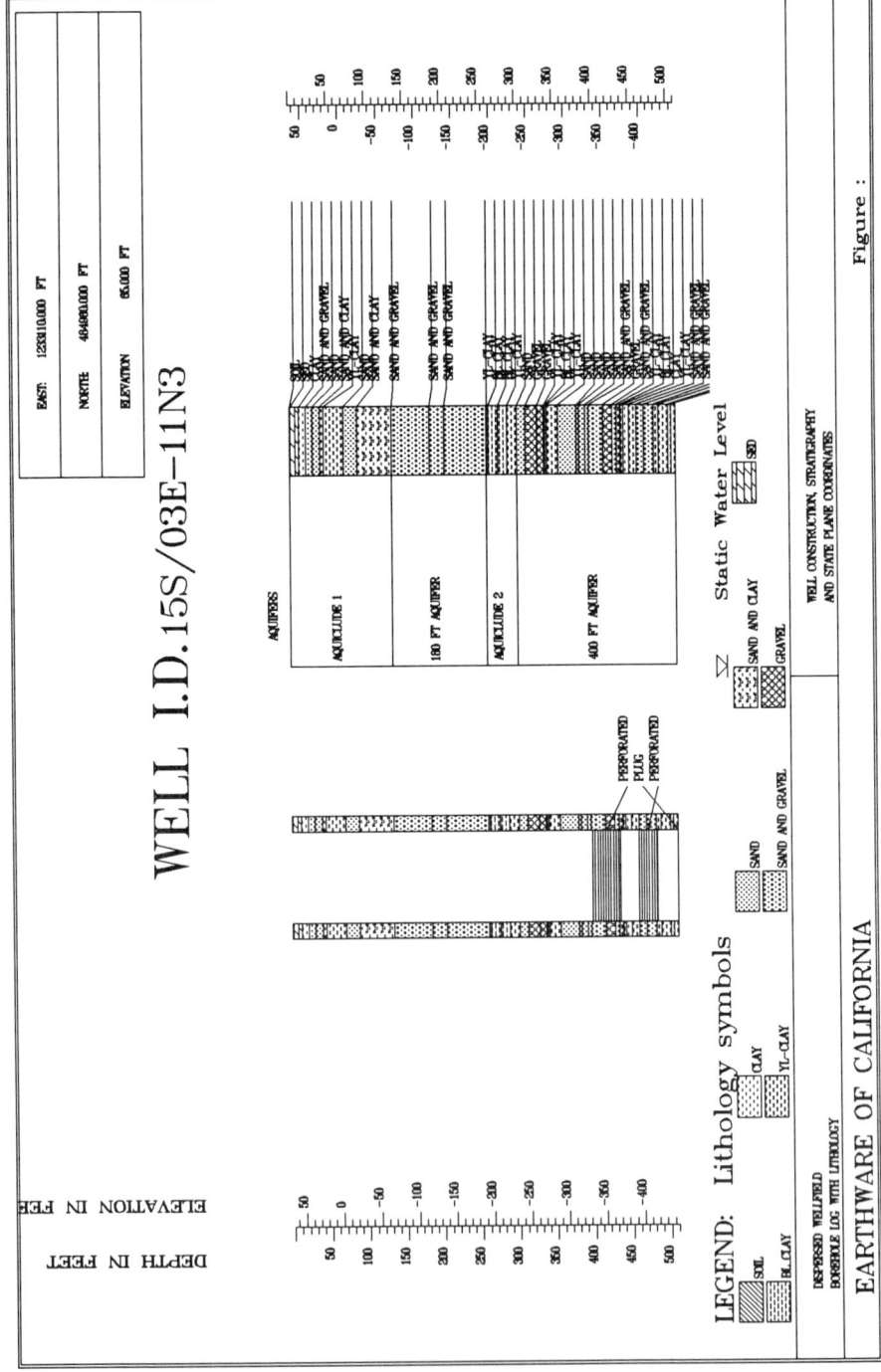

Fig. 3 Borehole log, well construction details and stratigraphy, from computer output.

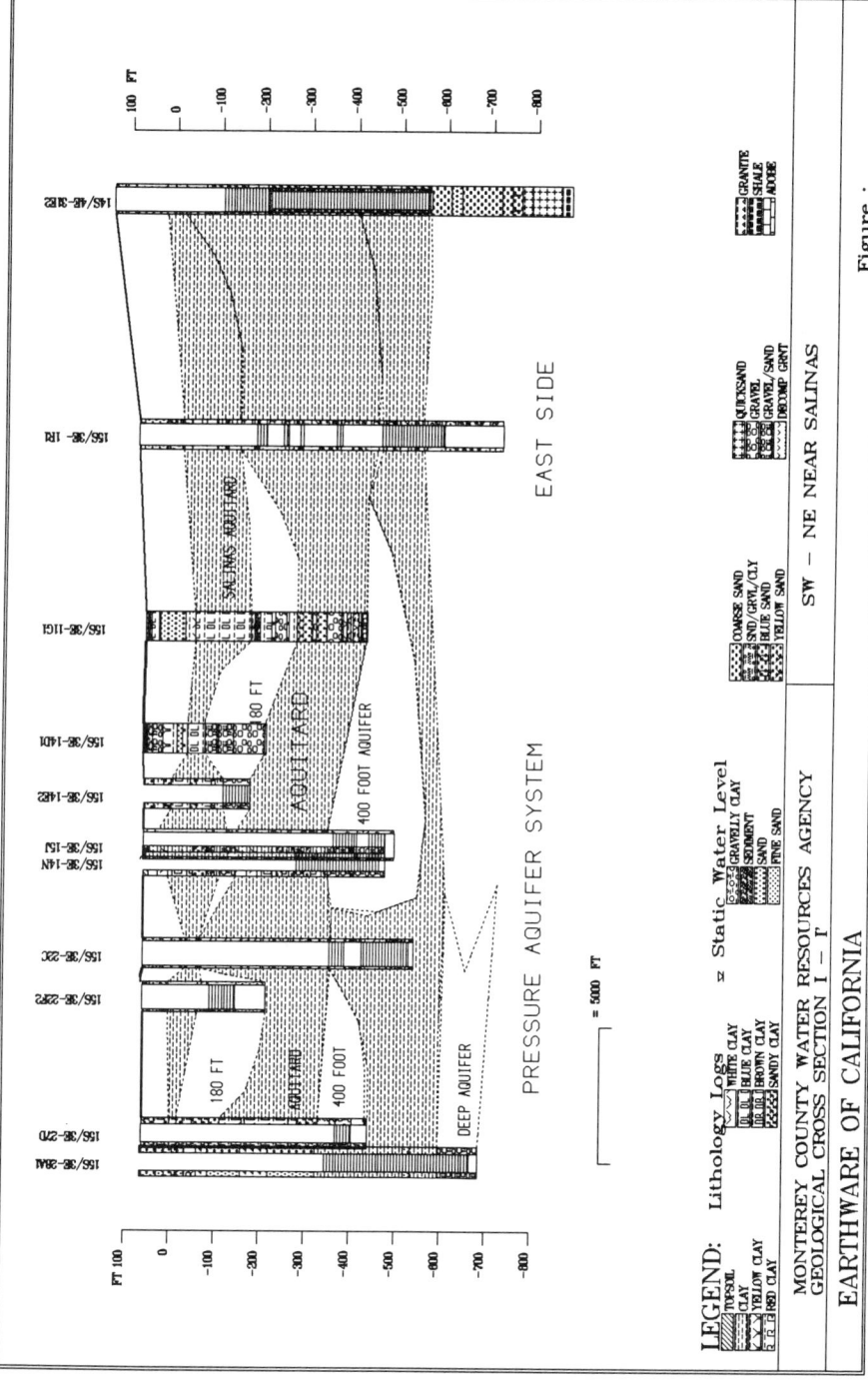

Fig. 4 Geological cross section across the Salinas Valley.

of the 180-foot and 400-foot aquifers and their related aquitards. This allows formation picks such as top, bottom or thickness to be mapped, contoured or displayed as a 3D wire diagram (Fig. 5). Contours can be displayed on top of base maps stored in the AutoCAD .DXF format or the contour drawing or data points exported to a GIS system and shown on top of selected maps or used for further spatial analysis.

Tne wells are categorized by the producing aquifer(s) thus allowing data to be sorted by aquifer. Chemistry maps of chloride, nitrate or other elements can be plotted for a specific aquifer for a given time interval. These data points will be contoured or displayed as a 3D wire diagram.

Logs of boreholes and well construction details will be displayed from ARC/INFO, providing that they are in a .DXF drawing format. This type of drawing typically takes up 1-2MB of disk space and as there are 10 000 records, this method is not recommended for display of well logs. Work is being carried out to generate well log diagrams from the database directly in ARC/INFO.

To date, more than 1000 records have been entered and checked. Most of these wells have periodic water level measurements and partial chemical analysese. More time has been spent on checking the locations of the wells than on any other part of the data entry process. The locations of the wells are important because of their high capacity which creates considerable impact on the surrounding groundwater system. The next phase will be to enter data for specific project areas on an "as-needed" basis.

BASE MAPS

The Agency began by using AutoCAD to prepare basemaps for the valley. US Geological Survey topographic maps were imported into AutoCAD from .DXF or .DLG files. It became apparent that AutoCAD would not be able to provide the capabilities required by the Agency. The US Bureau of Reclamation has contracted to supply the Agency with base map coverage in the ARC/INFO format. Maps being produced include topography, roads, rivers, property boundaries and farm use, thus enabling them to be used by other County agencies. Air photos are being used for the study of crop patterns for both present and past time studies.

HYDROGEOLOGICAL ANALYSIS

The ability to rapidly produce geological cross sections and to integrate water level and chemistry data allows the interpretations to be modified so that the sections can easily be changed. Producing contour and 3D wire surface maps

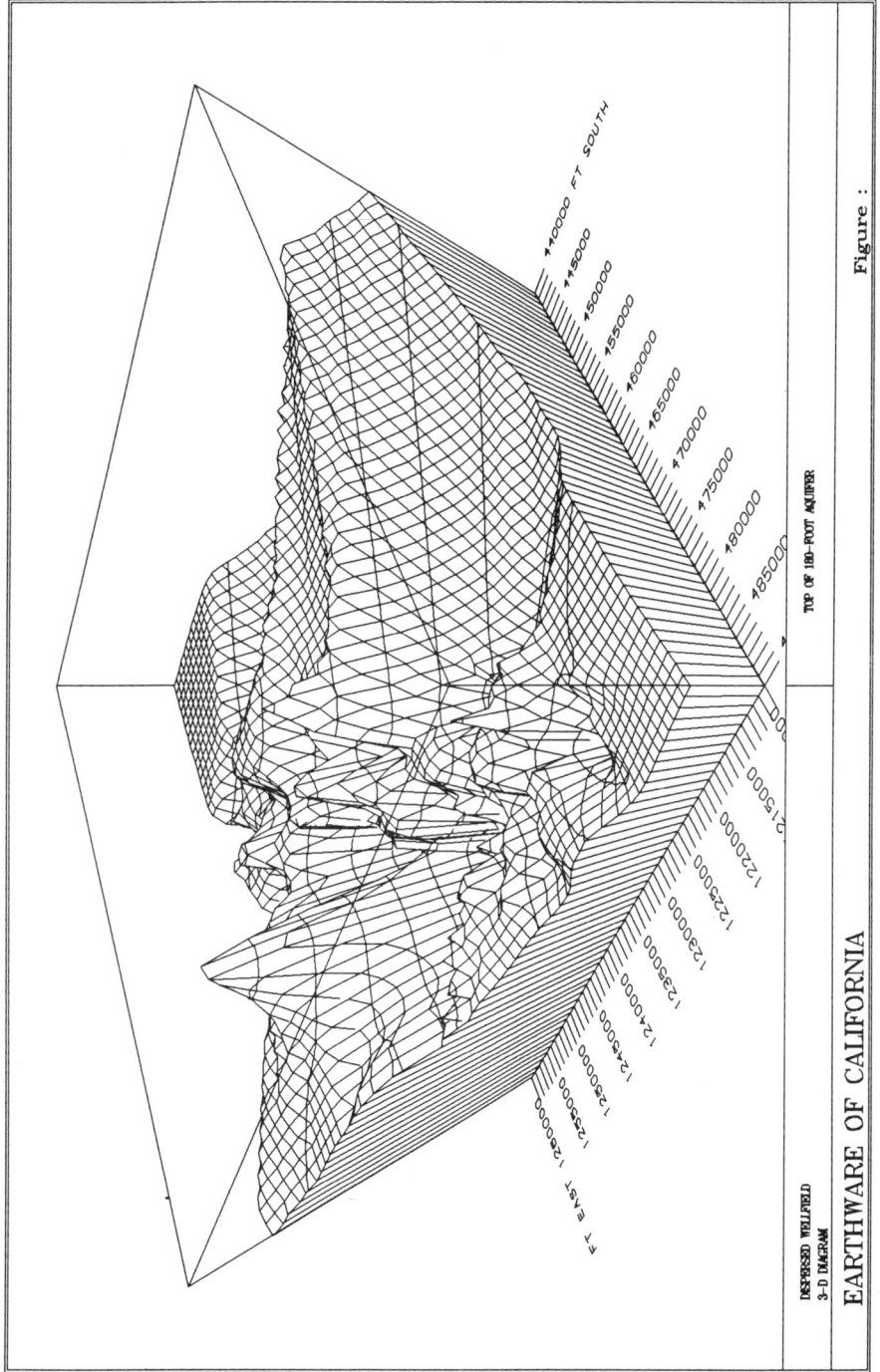

Fig. 5 Three-dimensional wire diagram of the top of the 180-foot aquifer.

further refines the interpretation of the hydrogeology. Use of the computer allows the addition of new records to be incorporated into cross sections and contour maps more easily than paper equivalents.

TIME SERIES DATA

Water levels for individual wells are measured on various schedules so that data is entered into the computer and hydrographs for selected wells are plotted automatically. The same data is used to generate water level and contour maps for specified areas and aquifers. Groundwater chemistry data can also be sorted by aquifers and plotted up as time series graphs or maps. These capabilities have been particularly useful in charting falling water levels due to drought and saltwater intrusion near the coastal areas.

Increased automation of data processing leaves staff time available for other tasks. Notable among these tasks is reviewing the reasons for monitoring data from existing monitor wells. A complete hydrogeological setting for each monitor well allows the effectiveness of each well to be assessed.

Pumping test data is taken from water level data files thus allowing water levels to be plotted with other data, such as pumping rate, precipitation or barometric pressure. The programs then analyse the pumping test and the aquifer coefficients can be used in groundwater flow models.

STATISTICAL ANALYSIS

Data processed in GEOBASE, such as stratigraphy and chemistry data, will be statistically analyzed by exporting the data to GEOPACK or GEO-EAS, two public domain programs available from the US EPA. Results will then be returned to GEOBASE for graphing. Time series data will be exported to spreadsheets or statistical analysis programs. Formation tops, thickness's, water levels and chemistry data will be contoured using either a triangulation technique or a kriging method. Processed data will be entered into the groundwater flow model MODFLOW, a US Geological Survey program. This is done via a graphics interface, making it easier for geologists to create models.

COMMUNICATION WITH OTHER DATA SOURCES

The Agency sends and receives data from several government agencies, including the US Geological Survey, US EPA, the National Weather Service and a number of State agencies. Most data is exchanged in an ASCII file format since there are currently no industry standards for data file structures or protocols for data exchange. Sort and capture routines are being developed to bring the required data into the programs being used by the Agency.

GROUNDWATER MODELS

The programs described thus far are used to help the Agency define its groundwater problems. Computer models are used to evaluate and compare management options by predicting effects of various courses of action. There are three levels of computer models currently being used.

Semi-analytical models are used for small projects such as designing pumping tests or assessing the impact of relatively small groundwater extractions on the surrounding areas and nearby wells. These models use simplified geological conditions and simulate boundaries with image wells. Only single layer aquifers can be simulated using these models. Output can be tabular or in the form of time-drawdown plots (linear, semi-log or log-log), contour maps of drawdown or water table elevation and 3D wire diagrams. Pilot studies using models to generate flowlines and capture zones around municipal drinking wells have been conducted (Hall & Zidar, 1993). The graphic results were useful in explaining impacts of proposed developments to general managment, politicians and members of the public.

A conjunctive-use finite element model has been developed for the whole valley. Individual cells are assigned aquifer characteristics. Water inputs and discharges are then assigned to each element and the model is run to produce water level contours for the whole valley and compute the water deficit/surplus.

For intermediate studies, such as the analysis of regional wellfields, the Agency proposes to use the US Geological Survey model MODFLOW. This can handle variable aquifer geology, multiple layers, river effects and variable discharge/recharge relationships. This model will be linked to the database via a graphics interface.

It is proposed that all of the models will be linked directly to the database. Thus changing input data arrays will not be a matter of merely changing abstract numbers, as so often happens. For example, changing transmissivity values could result in changing the aquifer thickness. If this is done through the database, the impact on the geology can readily be seen in cross sections and isopach contours.

THE ORACLE DATABASE

The workstation database will control the data being used. Processed data can be sent from other programs and raw data can be sorted and sent to programs as required. The database will control access to the data, limit the number of people who can change data and maintain a record log of data changes and updates. The database itself will not be capable of manipulating arrays of 3D data ie. converting borehole logs to cross sections or producing graphic output. A groundwater toolbox will be used to do the technical graphics processing. The database will also be used in the non hydrogeological data, such as, property ownership, land use and crop coverage.

The database will also be linked to the current well permitting process and involve several agencies, who are responsibile for installation, operation and public health. At the present time, water pumpage is not monitored for irrigation wells but it could be calculated from power consumption figures by relating various datasets.

SUMMARY

In summary, the complexities of managing groundwater resources in the Salinas Valley neccesitates the use of computer management. Not to do so would undermine the accuracy of data interpretation since the geologists would spend most of their time in the archaic methods of gathering and processing data rather than utilizing their talents of interpretation. The use of GIS systems enables more inter agency data exchange and thus make it easier to implement planning and management decisions requiring groundwater input. With computer management and the use of a customized database, data can easily and effectively be correlated, sorted and turned into graphic output. The net result is a more intensive and complete investigation in much less time. This allows resource people more time to deal with the "big picture" to analyze and solve problems that face the Salinas Valley not only today but those of the future.

REFERENCE

California Department of Water Resources (1946) Salinas Basin Investigation. Bulletin 52.

Mapping procedures for assessing groundwater vulnerability to nitrate and pesticides

G. SOKOL
Research Institute for Applied Knowledge Processing (FAW), University of Ulm, P.O. Box 2060, Helmholtzstrasse 16, D-7900 Ulm, Germany

Ch. LEIBUNDGUT
Department of Hydrology, University of Freiburg, Werderring 4, D-7800 Freiburg, Germany

K.P. SCHULZ
State Environmental Protection Agency of Baden-Württemberg, Griesbachstrasse 3, D-7500 Karlsruhe, Germany

W. WEINZIERL
State Geological Survey of Baden-Württemberg, Albertstrasse 5, D-7800 Freiburg, Germany

Abstract This paper focuses on methods for deriving maps of groundwater vulnerability to agricultural pollutants using GIS analyses and presentation tools. The mapping procedures applied are primarily based on an assessment of the filter capacity of the soil layers covering the aquifer. The area-distinguished groundwater recharge rate as one basic input parameter is estimated by a simple storage model. Nitrate leaching is estimated by calculating the frequency of nitrate transport below the effective root zone in times without vegetation cover. Pesticide leaching is assessed by a rating system containing rating matrices of binding, decay, volatilization and leachate. The modules are implemented within an ARC/INFO and ORACLE environment and have been made accessible with AML scripts and external programs. One of the basic objectives is to represent soil parameter heterogeneity both in the GIS database model and in the final vulnerability classification. In the end a variety of vulnerability maps for specific parameter combinations are produced. The GIS-based model environment thus is a flexible instrument to get an idea of groundwater vulnerability in the test areas.

INTRODUCTION

Contamination of groundwater by inorganic and organic substances has occurred in considerable parts of the State of Baden-Württemberg, as has been repeatedly detected by groundwater quality monitoring networks during the lasts years. Therefore the State Water Survey in Baden-Württemberg has intensified the monitoring efforts. In addition, it has started inventories of specific sources like waste disposals, liquid storage tanks, etc.

Besides groundwater quality data information on the soil filter capacity has to be added in order to be able to focus observation of groundwater quality on the aquifers most vulnerable. Such information is also needed for planning purposes, e.g. the optimal location of landfills or the prevention of spills from storage tanks.

A strict assessment of the vulnerability of an aquifer would have to be based on a description of migrant behaviour in the underground for a variety of substances. Although there exists a lot of transport models, they usually cannot be used for mapping. Mostly they are too detailed and therefore need a great variety of input data, which are only available for points or typical profiles. Hence they do not fit the needs of soil characterization on a regional scale.

For that reason simplified mapping approaches should be realized, based on routinely registered data and adequate to the scale of map in consideration. GIS mapping procedures are an effective tool to store, combine and process the different sources of information. In order to avoid misinterpretation a detailed description of the methods applied and its limits should always be given with the results when arguing with vulnerability maps. This paper deals mainly with this methodological aspect.

APPROACHES TO VULNERABILITY MAPPING

Usually, the vulnerability of groundwater is mapped area-wide and is not restricted to given catchment areas of (monitoring) wells. The applied mapping procedures are primarily based on the properties of the soil layers covering the aquifer. Methods of vulnerability mapping are determined mainly by the scale of the map and the homogeneity of the geological and pedological situation within the test area. For a meaningful interpretation it is suggested to map the test areas soil parameters at a scale not less than 1:50 000 in order to be able to interpret them meaningfully. Therefore, two pollutant-specific models describing the vulnerability of groundwater with respect to the agricultural pollutants nitrate and pesticides have been applied. The available input data are shown in Table 1.

ESTIMATION OF SPATIALLY DISTRIBUTED GROUNDWATER RECHARGE RATE

Pollutants from diffuse sources (agriculture, settlement and traffic areas) are primarily transported with the soil water flow. Therefore groundwater recharge rate is one important parameter for assessing the vulnerability of an aquifer. Disregarding lateral flow and transport, no pollution of the aquifer occurs without groundwater recharge. The groundwater recharge rate is estimated by a simple storage model. It calculates daily difference between rainfall and potential evapotranspiration using Haude's method, taking into account a reduction factor for the potential evapotranspiration if soil water content falls below 70% of the effective field capacity (Renger *et al.*, 1974). Furthermore, snow fall in winter is expressed by a storage constant and land-use is considered by evapotranspiration being higher on forest land. For the impervious areas a surface factor has been applied. The GIS-based model-process includes:

Table 1 Routinely registered data for the assessment of groundwater vulnerability.

DATA	PARAMETERS	TYPE OF DATA	SOURCE
Soil data	(effective) field capacity, cation exchange capacity, carbonate, clay and mould content, effective root zone, transmissivity, pH	mapping units, borehole data	GLA
Hydrogeological data	depth to water	time series of point data, isolines	GLA, LfU
Hydrological and climatic data	potential evapotraspirantion precipitation, temperature		DWD
land-use	agriculture, forest, settlement areas	maps, satellite data	LVA,....
Pollutant-specific data	absorption coefficient, decay rate		literature

GLA = State Geological Survey of Baden-Württemberg, LfU = State Environmental Protection Agency of Baden-Württemberg, DWD = German Metereological Survey, LVA = State Survey Office of Baden-Württemberg

(a) the interpolation of point values, e.g. precipitation and depth to groundwater;
(b) the intersection of all data layers (precipitation, effective field capacity, land-use, depth to water) in order to come to all occurring combinations of input parameters;
(c) the postprocessing, e.g., calculation of recharge rates for catchment areas or processing of data for groundwater modelling; and
(d) parameter variation and sensitivity analysis.

Water flow in the unsaturated zone is highly unsteady and complex. It is difficult to obtain information about concentrations in the aquifer, depending on the mass of the infiltrated substance. The estimated spatially distributed groundwater recharge allows only a comparison of different sites and does not indicate the actual pollution.

ESTIMATION OF NITRATE LEACHING

Nitrate leaching primarily occurs in times without vegetation cover. In these periods, nitrate is mostly transported tracer-like with the seepage water flow. If the nitrate remains in the effective root zone until the next vegetation period begins, it can be captured by plants again, otherwise it will be leached. For sites near to groundwater there might be a reverse nitrate dislocation due to capillary rise during summer. Therefore the model distinguishes between two approaches, one for sites far from groundwater surface, and one close to groundwater surface.

In order to estimate the nitrate leaching, a simple tracer transport model has been used. Dispersion and diffusion are not considered. Vulnerability classes are based on convective transport in times of no vegetation. The leaching factors are calculated as the quotient of groundwater recharge rate in times of no vegetation and field capacity of the effective root zone. For sites with small depth to groundwater, both the capillary rise in summer and different conditions for denitrification processes (i.e. anaerobic conditions) are considered. Therefore, a reduction factor for vulnerability classification is introduced. The model takes into account different crops, by varying the vegetation periods.

Since quantity of applied fertilizers is disregarded, this approach is not suitable for making statements about actual concentrations of nitrate in groundwater. It is rather used to compare different areas relatively. A preliminary validation of the model was performed by comparing the results with measured groundwater concentration of nitrate.

A RANKING SYSTEM FOR THE ASSESSMENT OF PESTICIDE LEACHING

In contrast to nitrate, pesticide leaching primarily depends on processes of adsorption and decay which are very complex, poorly known and substance-specific. Blume & Brümmer (1987) designed a ranking system in order to make rough approaches possible. It has been developed on a large database coming of field examinations in combination with laboratory analysis. The ranking system applied is based on three types of data: (1) site-specific data derived from soil mapping, (2) climatic and hydrological data, and (3) parameters describing substance behaviour depending on given soil properties. The applied ranking system has three components assigning values for:
(a) the prediction of binding;
(b) the prediction of decay and volatilization; and
(c) the prediction of leachate.

The rating process is described by rating matrices for these three components. Table 2 shows an example for the estimation of pesticide binding depending on mould content and soil texture respectively. Special site conditions, e.g. extreme pH-values or sites near to groundwater surface, are considered by additional values being added or subtracted. In the end a final classification of the summarized rating values in five groundwater hazard classes has been performed. The ranking system is based on values belonging to different specific site conditions. Therefore comparison of the ranking values can only be done with great caution (Blume & Brümmer, 1987). Nevertheless, an idea of the variability of pesticide behaviour for given site conditions and the sensitivity of pesticides to soil parameters will be obtained.

Table 2 Assessment of the potential binding of pesticides in soils dependent on mould content and soil texture respectively (Blume & Brümmer, 1987).

mould [%]	Diquat	Atrazin	Amitrol	Chloridazon	PCP	DDT
0-1	0	0	0	0.5	0.5	0.5
1-2	0	0.5	0.5	1	1	1
2-8	0.5	1	1	2	2	2
8-15	1	2	2	3	3	3
< 15	2	3	3	4	4	4

soil texture	Diquat	Atrazin	Amitrol	Chloridazon	PCP	DDT
S, Su2	0	0	0	0	0	0.5
Sl, St, Su, Us, U	0.5	0	0	0.5	0.5	1
Uls, Ul, Lu, Ls, Lt,Ts	1	0.5	0.5	1	1	1
Ts2, Tl	2	1	1	2	2	3
T	3	2	2	3	3	4

1 = (almost) not   2 = very low   3 = medium   4 = high   5 = very high

METHOD IMPLEMENTATION

The three modules of vulnerability mapping described have been implemented in a GIS-based model database environment using ARC/INFO and ORACLE. The interpolation, intersection and mapping procedure used for the three models are separate tools. The data processing is organized by ARC/INFO AML's, the macro language of the GIS, and by an external program for the calculation of the groundwater recharge rate. The database consists of different data types:

(a) hydrological and climatic time series stored as tables referring to each station;
(b) geometric data and their attributes stored in an area database; and
(c) substance-specific data for pesticides stored as rating matrices.

One of the basic objectives to be supported by the area database is to represent soil parameter heterogeneity and variability both in the GIS database model and in the final vulnerability classification. This is achieved by storing the relative frequency of each soil parameter for each distinguished soil map area in a separate table of our relational database system (Zwölfer & Vogl, 1989). Figure 1 shows the idea of parameter storage of field capacity (FK) and transmissivity (Kf). In order to minimize redundancy of data the database model distinguishes tables for the attachment of geometry and parameter group, for the attachment of parameter groups and parameter classes (e.g., FK of 0-130 mm) and finally a table with the relative frequencies (RH) of the parameter classes. This database model allows to distinguish classes and subclasses of soil types within a mapping unit, as they are usually registered. Furthermore, vulnerability processing can be performed for parameter distributions, outlining uncertainty of vulnerability classification.

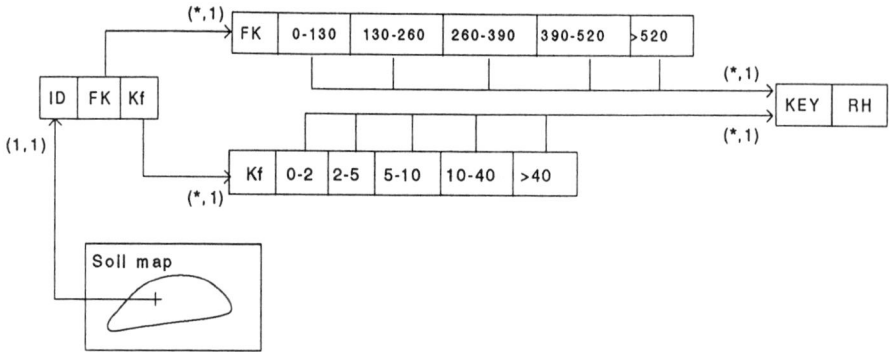

Fig. 1 ER-diagram of soil parameter storage.

APPLICATION

The methods have been implemented and tested in two areas of the Upper Rhine Valley. The estimation of groundwater recharge rate and nitrate leaching were implemented for an area near Karlsruhe (Kinzig-Murg-Rinne) at a scale 1:50 000. The pesticide leaching procedure was realized for the soil map of Mannheim-Northeast at a scale 1:25 000. The areas have similar site conditions. They are nearly plain and have a comparatively uncomplicated pedological and geological situation. Most of the agricultural areas under consideration are situated in a groove, about three kilometers wide. There we have a layer of about 2-3 m of clayey to silty materials covering the gravel beds of the latest glaciation. For the test area of Karlsruhe, the interpolation of climatic data for the assessment of groundwater recharge rate is based on isohyetes. Climate stations are assigned to the areas between isolines. A correction factor depending on the position of the climate station is required for the different stations. The land-use was mapped (Kaltenbach, 1988), the (effective) field capacity was derived from the soil map (Hummel, 1985) and depth to groundwater could be estimated from an existing hydrogeological map (State Geological Survey of Baden-Württemberg, 1988).

The parameters required for the estimation of pesticide leaching in the test area near Mannheim were derived from the soil map (Hummel, 1990), the spatially distributed groundwater recharge rate could be registered from an existing map (Weinzierl, 1990), and the land-use was derived from satellite data of Landsat-TM.

In addition, vulnerability mapping of nitrate leaching is regarded in more detail. With regard to usual agricultural land-uses, nitrate leaching is estimated for five different vegetation periods:

(a)   1 October - 31 March;
(b)   1 August - 31 May (grain);
(c)   1 September - 30 June (root crops);
(d)   1 August - 30 April (green land) ;
(e)   1 January - 30 June (intercropping, e.g.with green land).

Nitrate leaching is calculated as a quotient of groundwater recharge rate in these periods, and the averages of field capacity classes. For the period 1961-1986, frequencies of the depths of transport during winter time for all occurring combinations are obtained. Figure 2 shows the frequencies for a site with medium groundwater vulnerability. It allows to calculate the relative frequencies for situations where transport in winter, starting from the surface passes below the root zone. In our case, the relative frequency that nitrate is transported less than 0.7 m in the period from 1 October to 31 March is about 60% $(F(x \leq 0.7)) \approx 0.6$). These frequencies were calculated for all occurring site combinations.

Finally, for mapping purposes these frequencies of leachate are divided into five vulnerability classes. It is obvious that a lot of different vulnerability maps can be produced, e.g. 18 maps regarding five vegetation periods and three different soil depths. Comparing these maps we see that the sandy parts of the test area with high vulnerability are hardly influenced by parameter variation, whereas the sites of low and medium vulnerability are very sensitive to parameter variation. In conclusion, a comparison of the different maps has been performed interactively in order to outline sites where vegetation has a great influence on nitrate leaching and sites where the regarded soil body must be considered more carefully. Additionally, sites close to groundwater surface are marked and considered separately. Figure 3 gives an example for two situations: the potential nitrate leaching for root crops (Fig. 3a) and the potential for root crops and intercropping, e.g. with green land (Fig. 3b).

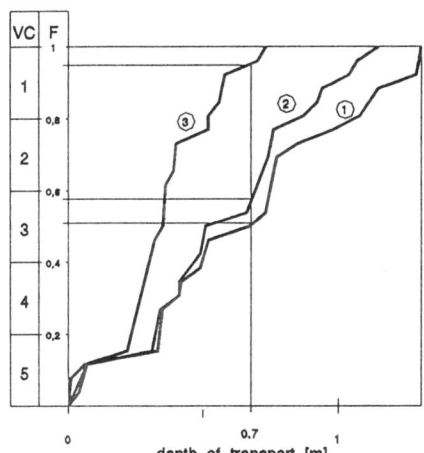

								VC	=	Vulnerability Class
								F	=	Cumulative relative frequencies of migration factors
								$F(x \leq 0.7)$	=	Relative frequency of a migration factor below 0.7 meter
								GWR	=	Groundwater recharge rate
								nFK	=	Effective field capacitity
								FK	=	Field capacity

	Soil type	Period	Averages					Frequency of leachate		
			GWR [mm/period]	nFK [mm]	FK [mm]	$\bar{x}$ [m]	s [m]	$F(x \leq 0.7)$	$F(x \leq 0.9)$ (mapped value)	$F(x \leq 1.0)$
1	Vega	1.9.-30.6	217			0.7	0.4	0.5	0.73	0.77
2		1.10.-31.3.	190	170	325	0.6	0.3	0.58	0.81	0.81
3		1.1.-31.3.	117			0.4	0.2	0.96	1	1

Fig. 2 Approximative distribution function of migration factors for different periods.

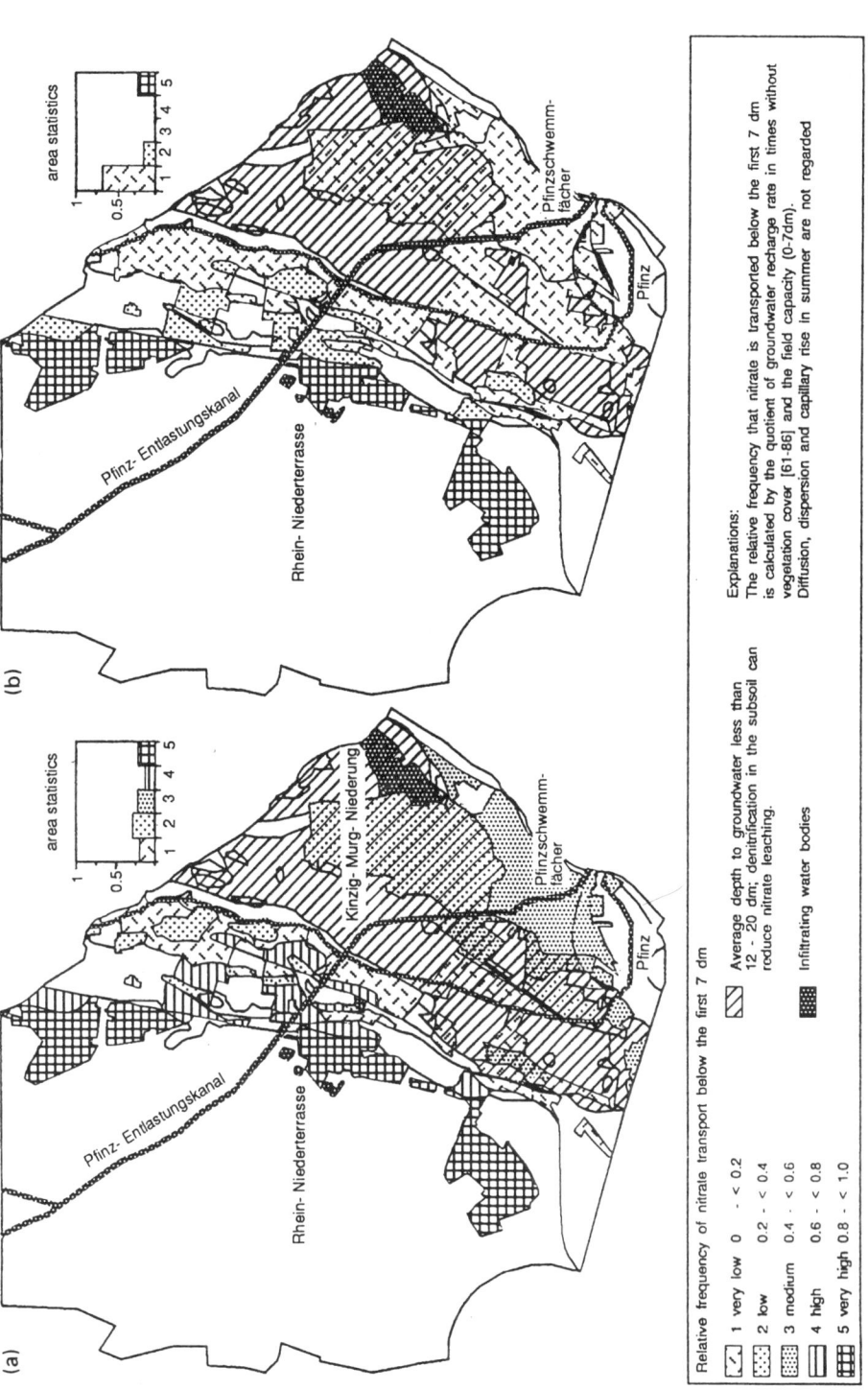

Fig. 3 Nitrate leaching from agricultural areas in the test area of Karlsruhe: (a) vegetation period 1 July - 31 August (root crops), (b) vegetation period 1 July - 31 December (root crops and intercropping).

CONCLUSION

A basic objective in our investigation was to outline the processes and the basic factors that control nitrate and pesticide leaching. The investigation is based on the assumption that surface water flow and lateral transport processes can be disregarded. Furthermore, only statements about potential leaching were aimed, regardless of the actual application of fertilizers and pesticides and actual concentrations in the aquifer. Especially pesticide-related processes are very complex. Hence statements with regard to spatial distribution should be made with great caution. Nevertheless, the described modules for vulnerability mapping could be successfully used for a reasonable outline of sensitivities of the aquifer in the test areas. The GIS-based data model allowed to interpret easily the effects of parameter variation.

In Baden-Württemberg, many aquifers which are important to water supply are situated in areas with similar site conditions (Upper Rhine Valley, Molasse Basin). Consequently, the approaches described above are feasible in many cases. Future research will focus on the validation of the methods using field data and extension of application in areas with different site conditions.

REFERENCES

Blume, H.-P. & Brümmer, G. (1987) Prognose des Verhaltens von Pflanzenbehandlungsmitteln in Böden mittels einfacher Feldmethoden (Prognosis of pesticide behaviour in soils, with simple field methods). *Landwirtschaftliche Forschung* **40**, 41-50.

Hummel, P. (1985) Bodenkarte von Baden-Württemberg 1:50 000 (Soil map of Baden-Württemberg 1:50 000). Geologisches Landesamt Baden-Württemberg, Freiburg im Breisgau.

Hummel, P. (1990) Bodenkarte von Baden-Württemberg 1:25 000, Blatt Nr. 6417 (Soil map of Baden-Württemberg 1:25 000, Sheet No. 6417). Mannheim-Nordost. Geologisches Landesamt Baden-Württemberg, Freiburg im Breisgau.

Kaltenbach, D. (1988) Eigenschaften und Nutzung von Böden in der Rheinebene nördlich von Karlsruhe (Characteristics and use of soils in the Rhine valley north of Karlsruhe). Diplomarbeit, Geographisches Institut der Westfälischen Wilhelms-Universität, Münster, Diplomarbeit (unpublished).

Renger, M., Strebel, O. & Giesel, W. (1974) Beurteilung bodenkundlicher, kulturtechnischer und hydrologischer Fragen mit Hilfe von klimatischer Wasserbilanz und bodenphysikalischen Kennwerten (Assessment of pedogenesis, agricultural and hydrological problems with climate water balance and physical soil parameters). 4.Bericht: Grundwasserneubildung, *Zeitschrift f. Kulturtechnik und Flurbereinigung* **15**, 353-366.

State Geological Survey of Baden-Wurttemberg (1988) Hydrogeologische Kartierung und Grundwasserbewirtschaftung im Raum Karlsruhe-Speyer (Hydrogeological mapping and groundwater management in area Karlsruhe-Speyer).

Weinzierl, W. (1990) Grundwasserneubildung aus Niederschlag - Bodenkarte von Baden-Württemberg 1:25000, Blatt Nr.6417, Mannheim-Nordost (Groundwater research from precipitation - Soil map of Baden-Württemberg, 1:25 000, Sheet No. 6417 North-East). Geologisches Landesamt Baden-Württemberg, Freiburg im Breisgau.

Zwölfer, F. & Vogl, W. (1989) Wichtige Faktoren des Bodenwasserhaushalts, Bodenbestandsaufnahme Baden-Württemberg, Auswertung der Bodenkarte 1:25000, Blatt Nr. 6417 (Important factors of soil water balance, soil inventory of Baden-Württemberg, 1:25 000, Sheet No. 6417). Geologisches Landesamt Baden-Württemberg, Freiburg im Breisgau.

Development of the GIS-based "RIVM National Groundwater Model for The Netherlands (LGM)"

R. LIESTE, K. KOVAR, J.G.W. VERLOUW & J.B.S. GAN
RIVM, National Institute of Public Health and Environmental Protection, Soil and Groundwater Research Laboratory, P.O. Box 1, 3720 BA Bilthoven, The Netherlands

Abstract A large scale groundwater model for The Netherlands was developed by coupling the finite element groundwater program package AQ-FEM with the geographical information system ARC/INFO. This paper describes the ARC/INFO data model, the program package AQ-FEM, and the interfacing between ARC/INFO and AQ-FEM. The interfacing is based on standardized ASCII data files. The parameters transferred via the interface are spatially distributed system parameters, well data and river data. Calculated phreatic groundwater heads in the top layer, and the seepage or infiltration flux between the top aquifer and the top layer, calculated for 10 study areas, were aggregated into single conjunctive maps stored in ARC/INFO, based on 1x1 km polygons. The conjunctive maps, covering 75% of The Netherlands' landsurface, were used for subsequent ecohydrological modelling (not considered in this paper) and for presentation purposes.

INTRODUCTION

By order of the Dutch Ministry of Housing, Physical Planning and Environmental Protection (VROM), the National Institute of Public Health and Environmental Protection (RIVM) began a study in 1990 at national scale to quantify the effects of future changes in groundwater abstractions (for drinking and industrial water supply) on the flora of terrestrial ecosystems. The resulting model is referred to as "RIVM National Groundwater Model for The Netherlands (Dutch abbreviation LGM)". Eight scenarios of possible future groundwater abstraction changes, based on the initial situation in 1988, were considered.

The volume of data to be handled required the use of a geographical information system. ARC/INFO has been chosen because it ensured an easy exchange of data with collaborating institutes in The Netherlands. The finite element program package AQ-FEM (Kovar, 1992) for the modelling of groundwater problems was chosen because of the spatial variability of the geohydrologic parameters, the existing expertise with finite element modelling at the RIVM, and the instantaneous availability of the software.

Groundwater modelling programs have their own characteristic input and output data structures. This inhibits direct access to database files. A standard ASCII file interface protocol was defined to ensure an easy and systematic exchange of data between the AQ-FEM modelling programs and the geographical information system ARC/INFO. Also, a set of interfacing programs was developed for AQ-FEM and ARC/INFO to transfer and reformat data (Lieste & Verlouw, 1992). In this way, the same procedure can always be followed in the GIS environment to make data available to the groundwater modelling program, while in the AQ-FEM program environment the interfacing routines have to be developed only once.

The storage of output data within a geographical information system facilitates the accessibility of the data for any other program, overlaying with existing data, and the display of a spatially distributed parameter.

DESCRIPTION OF THE "RIVM NATIONAL GROUNDWATER MODEL"

The groundwater model consisted of the finite element program package AQ-FEM and the geographical information system ARC/INFO, coupled by a series of interfacing programs. The finite element module was run on an ALLIANT minisuper computer with vector and parallel capabilities. The pre- and post-processors of the finite element program package were run on an IBM PS2/70, equipped with a mathematical co-processor (DOS or OS/2 operating systems). ARC/INFO was implemented on HP workstations with high level graphic capabilities (X-Windows). The hardware consisted of an HP 9000/832 server, a HP345 stand alone workstation, three HP340 diskless workstations running Unix System V, two HP425 workstations and two X-terminals. All hardware was connected to a network. The required disk storage capacity for the groundwater model amounted to approximately 500 MB.

The modelling approach of the "RIVM National Groundwater Model for The Netherlands" (Dutch abbreviation LGM) was based on a multi-aquifer system consisting of four aquifers, separated by semi-pervious layers (aquitards) (Fig. 1). The bottom aquifer is assumed to be underlain by an impervious layer (geohydrological base). The uppermost aquifer (aquifer 1) is covered by the so-called top layer.

The input values for AQ-FEM were obtained by interpolating the contour maps of the spatially distributed system parameters and from maps containing river and well data.

The credibility of maps representing the depth of the aquitard tops, and thickness of the aquitards was evaluated by calculating the thickness of all layers and checking for negative values. Similarly, the credibility of the calculated groundwater heads for the initial situation in 1988 was verified by evaluating the maps depicting the difference between calculated and observed groundwater heads. Both evaluations were carried out in ARC/INFO.

The modelling was carried out in 10 model areas, covering 75% of the landsurface area of The Netherlands (coastal areas excluded) (Pastoors, 1992a,b). The results for eight scenarios of the 10 overlapping model areas were aggregated into single conjunctive maps of 1x1 km polygons to be used for subsequent ecological modelling; either for presentation or for future use. The main steps of the model study are depicted in Fig. 2 and explained in the following section.

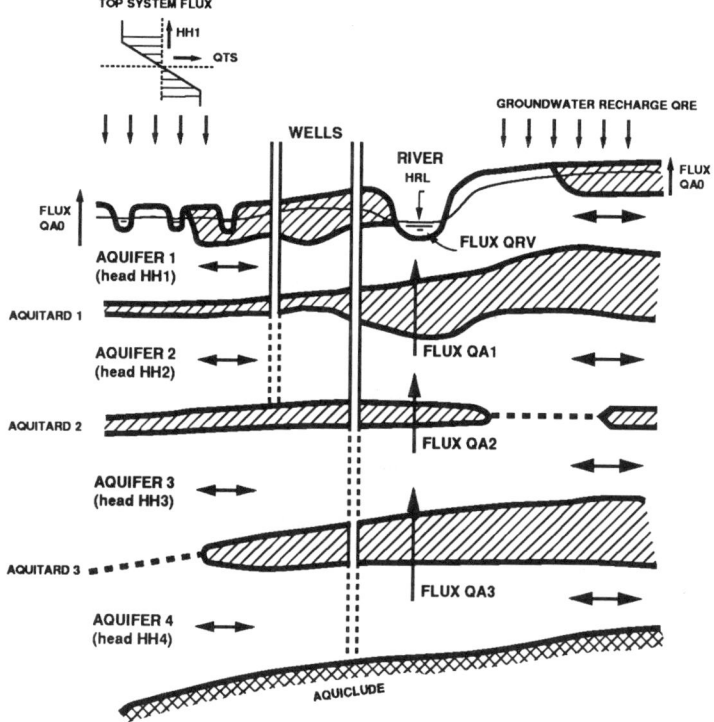

Fig. 1 Vertical cross section of geohydrological system.

Data in the geographical information system

Within a vector-based geographical information system, all spatially distributed parameters in x/y space required to define a multi-layered aquifer system can be represented in coverages (digital maps) by using points, lines or polygons.

The top and thickness of the aquitards, and the bottom of the system are represented by contour maps (line coverages). The elevation of the system top (i.e. ground level) was obtained from the point map (point coverage) of the locations of the groundwater head observation wells.

The hydraulic conductivity of the aquifers is represented by contour maps. The extent of the semi-pervious formations (aquitards) is represented by polygon maps; each polygon contains a code identifying a relationship for assessing the hydraulic resistance of the aquitard from the aquitard thickness, the depth of the aquitard below ground level and a constant parameter that depends on the lithology of the geological strata.

Polygon maps of the polders (areas with controlled surface water levels) were used to assess the so-called top system flux relation (flux QTS in Fig. 1). In addition, point maps of abstraction wells and observed groundwater heads were available, as well as line maps of rivers.

The elevation and depths are z-levels (vertical direction) with respect to the Amsterdam Ordnance Datum (NAP). The above listed data are referred to

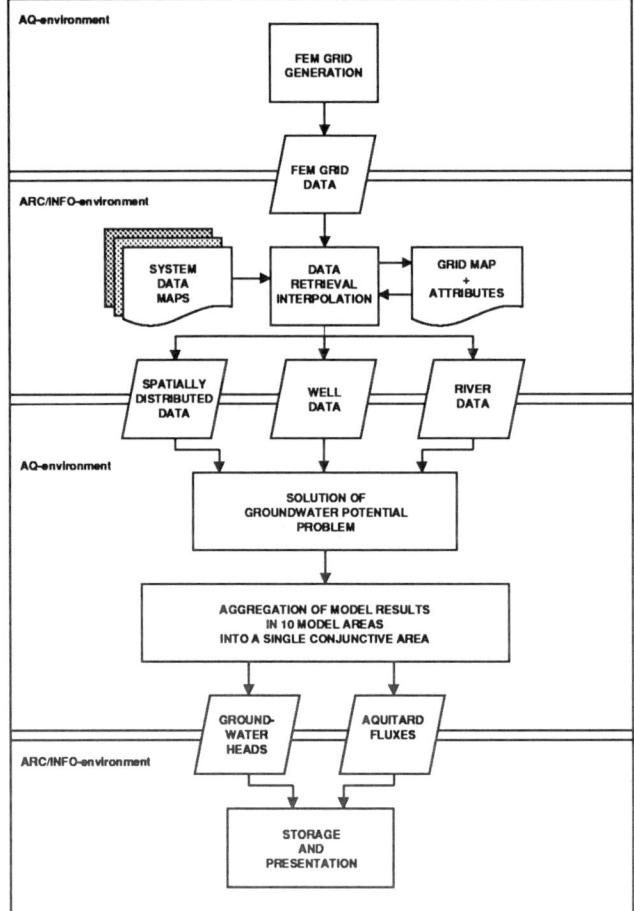

Fig. 2 Interface between program package AQ-FEM and ARC/INFO.

as "base data" and were obtained from collaborating institutes in The Netherlands. The adjusted data (error correction etc.) were stored as "project data" in the geographical information system.

Finite element program package AQ-FEM for groundwater modelling

The finite element approximation was applied to solve the differential equation describing the groundwater potential problem in a four-aquifer-system affected by groundwater abstraction by wells. The model areas were convex or concave polygons of arbitrary shape with triangular and quadrilateral shaped elements.

Figure 1 depicts schematically the geohydrological system used for the finite element simulation. The system consists of four aquifers separated by aquitards, underlain by an impervious base (geohydrological base). The top aquifer (aquifer 1) is covered by the so-called top system. The groundwater flow in the aquifers is assumed to be horizontal and two-dimensional, and in the aquitards vertical and one-dimensional.

The top aquifer is simultaneously affected by the groundwater recharge rate (variable QRE in Fig. 1) and the so-called top system flux relationship (variable QTS in Fig. 1). This top system flux relationship is specified by a series of linear line segments defining the relationship between the groundwater head in the top aquifer (variable HH1 in Fig. 1) and the flux rate (QTS). Both the groundwater recharge rate (QRE) and the top system flux (QTS) are spatially distributed parameters.

The top aquifer is hydraulically connected to the rivers. Rivers are defined by a sequence of points expressed by x,y-coordinates. The river parameter values vary linearly between the river points. The parameter values of river water level, river width, and river bottom drainage and infiltration resistance were obtained from the geographical information system. The well screens (groundwater abstraction or injection) are assumed to fully penetrate the aquifers. Prescribed groundwater heads were used as boundary condition along the periphery of the model area in the aquifers.

The main outputs of AQ-FEM are groundwater heads in the four aquifers (variables HH1 through HH4 in Fig. 1), the flux between the upper boundary of the groundwater system and the top aquifer (variable QA0 in Fig. 1), and the fluxes across the aquitards (variables QA1 through QA3 in Fig. 1).

Data exchange between ARC/INFO and AQ-FEM

Data stored as maps in the geographical information system ARC/INFO had to be converted into a different format for the finite element program package AQ-FEM. The geohydrological data are stored in ARC/INFO as (1) contour lines having a certain constant value, (2) points with certain attribute values, or (3) polygons with a certain value for the encompassed area. AQ-FEM requires all spatially distributed parameter values to be given at the nodes of the finite element grid, with the exception of well and river data. A complete procedure for data exchange was implemented in several controlling AML routines (ARC/INFO Macro Language) extended with FORTRAN programs (mainly to reformat data) using standardized ASCII files and standardized parameter names and file names. Only four different file types were needed to enable transfer all of the parameters:
(a) grid data file;
(b) spatially distributed parameter data file;
(c) well data file;
(d) river data file.

Grid data A finite element simulation for a given model area starts with the generation of a finite element grid (triangles and quadrilaterals) for that model area. The file with x,y coordinates of the finite element grid nodes is then converted into an equivalent ASCII file and transferred to the geographical information system.

Spatially distributed parameters The contour maps are interpolated to the x,y coordinates of the grid nodes. The results of the interpolation in the geographical information system are transferred via a standardized ASCII file to AQ-FEM, to serve as input.

Well data Well data are stored in the geographical information system in point coverages with several attributes, such as x,y-coordinates of the location, abstraction rate, number of well screens composing the well, and top and bottom depth for each well screen. AQ-FEM requires as input the x,y well location and well abstraction rate of wells per aquifer, assuming that wells fully penetrate the aquifers. Because the well data in the geographical information system are not related to the aquifers, the z-levels of the top and bottom of the aquifers and the hydraulic conductivity were interpolated from the contour maps to the location of the wells, and added as attributes to the well attribute file. The following well data were exported from the geographical information system to a standardized ASCII file:
(a)   x- and y-location of the well;
(b)   abstraction rate (total for all well screens composing the well);
(c)   number of well screens composing the well;
(d)   z-level of top and bottom for each well screen;
(e)   z-levels of the top and bottom for each of the four aquifers;
(f)   hydraulic conductivity for each of the four aquifers.

By means of a FORTRAN program, the abstraction rate of each well was allocated to each of the four aquifers, based on the portion of the well screen length occurring between top and bottom of an aquifer, and the hydraulic conductivity of the relevant aquifer. Hence, each well in the ARC/INFO database file is generally split up into several input wells for AQ-FEM. Allocation of the well abstraction rates was performed by a custom-made FORTRAN program because they tend to run significantly faster than equivalent INFO programs.

River data River data are stored in the geographical information system in a line coverage. A river consists of one or more lines (arcs), each line being delineated by a node (terminal point) at either side and traced by vertices (a set of ordered x,y coordinates) between the nodes. Within ARC/INFO line coverages, the attribute values refer to the whole line, each line having its specific value. Relevant attributes in the river database file are the river width, the river water level above the Amsterdam Ordnance Level (NAP), and the drainage and infiltration resistance of the river bottom. AQ-FEM requires a sequence of consecutive points as input for each river. Each point defining the river course is characterized by its x,y coordinates, river width, river water level, and hydraulic resistances of the river bottom. All input parameters are assumed to vary linearly between the adjacent x,y points. Unlike the attributes of the arcs, the x,y coordinates of arcs stored in an ARC/INFO coverage are not directly accessible via the usual tools. By creating an ASCII file by means of the export command, the x,y coordinates of the points (nodes, vertices) defining the river course became accessible, just as the attributes. Subsequently, the x,y coordinates of the river points and attribute values were extracted from the export file and converted to the standardized ASCII river file by a custom-written FORTRAN program. The ASCII file was used directly as an input for the AQ-FEM.

Transformations Before assigning the spatially distributed data to the nodes and elements of the finite element grid, the spatially distributed data had to be transformed to their final form and units required by AQ-FEM. Transmissivity of the aquifers was calculated from hydraulic conductivity and aquifer thickness, hydraulic resistance of the aquitards was derived from a code that identified a relationship between aquitard thickness, aquitard depth and a lithology-related parameter of an aquitard. Furthermore, units were converted into SI-units. After the solution of the groundwater potential problem, the groundwater heads in the four aquifers (variables HH1 through HH4 in Fig. 1) and vertical fluxes (variables QTS, QA1, QA2 and QA3 in Fig. 1) were available at the location of the finite element grid nodes. From these values, the phreatic groundwater head in the top layer, and the seepage or infiltration flux between the top aquifer and ground level (variable QA0 in Fig. 1) were derived and returned to the geographical information system.

Interpolation of contour maps and point maps in ARC/INFO

The depth of aquitard tops, the thickness of aquitards, and the hydraulic properties of the subsoil are assumed to be parameters continuously variable in x,y space. The distribution of these parameters is represented either by a set of contour lines, along which a constant value is given, or by a set of points at which a value is given, or by a combination of the two features. These contour and point maps were interpolated to obtain the parameter values at the location of the finite element grid nodes. Initially, the bivariate-quintic interpolation method of ARC/INFO was applied. For some of the maps, this method led to unexpected and unwanted results (for example, negative layer thickness) in areas containing relatively little data. Therefore, a linear interpolation method was used instead of the bivariate-quintic method. The interpolation was done by using a lattice (cell-based map).

Error verification

Databases of spatially distributed parameters are rarely error-free due to data-entry errors, or introduced by processes such as interpolation. Maps of the interpolated thickness and hydraulic conductivity were examined by colouring areas containing negative values or unrealistic values. The validity of the model was verified by comparing calculated groundwater heads with observed heads.

Coloured maps of the difference between calculated and observed heads gave an indication of the extent of the deviation.

Aggregation of multi-model results into single conjunctive maps

The results of the simulations, i.e. groundwater heads in the aquifers, fluxes across the aquitards, and the flux across the top layer were available at the nodes of the irregular finite element grids within each of the 10 model areas (Fig. 3). The model areas overlap each other partially. However, subsequent ecological modelling required groundwater heads in the top layer and the seepage flux across the top layer to be available on a 1x1 km grid. The first step in performing the data transformation was to calculate the changes of heads and fluxes with respect to the respective variables for the initial scenario (1988). Next, the changes calculated within each model area were interpolated to the 1x1 km grid covering the whole landsurface area of The Netherlands. Applying the principle of superposition, the results were added into one map of grid cells covering the conjunctive model area. Cells outside the hull of the model area were assigned a "no-data" value.

Fig. 3 Location of 10 model areas.

For the input to the subsequent ecological modelling, using, among other relations, the relation between phreatic water table and the terrestrial ecosystem, the groundwater heads in the upper aquifer (aquifer 1) were transformed into phreatic groundwater levels in the overlying top layer (below ground surface). An other important input variable for the ecological modelling is the seepage/infiltration flux across the top layer, i.e. the flux between the ground surface and the underlying saturated system. The calculated conjunctive

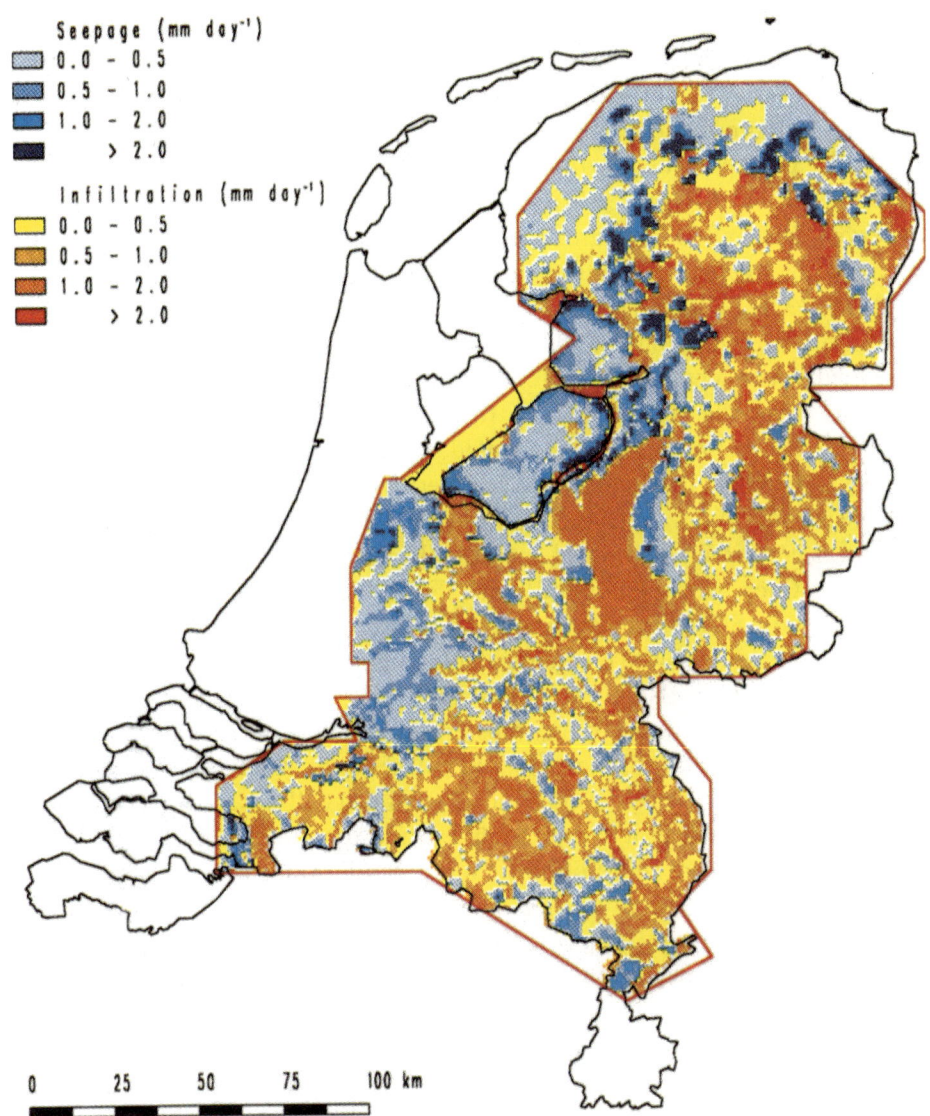

Fig. 4 Aggregated map of calculated seepage/infiltration flux between aquifer 1 (top aquifer) and ground surface. Calculation was carried out for 1988 (initial situation). The seepage flux is directed from aquifer 1 to the ground surface, the infiltration flux is in reversed direction.

map of the seepage/infiltration flux for 1988 (initial scenario, no changes in groundwater abstraction) is presented as an example in Fig. 4.

DISCUSSION

The geographical information system coupled with a finite element simulation program makes it possible to carry out efficiently the modelling of large areas characterized by a huge volume of spatially distributed data. Spatially distributed data can easily be entered into the geographical information system. Data can easily be verified or improved by using graphical tools of the geographical information system.

The tools provided by the geographical information system ARC/INFO offer the possibility to verify input data, to compare calculated data with observed data, and to compare in an easy manner the results of several model scenarios by overlaying maps of the relevant data. If another model has to be made in the area where the data is already available, only a new finite element grid has to be generated and the data discretization to be performed for that new grid. Even when grids differ, results within the same area are easy to compare using the geographical information system tools. In practice, the results for another model area or adapted grid in the same area can be obtained rapidly and without danger of introducing errors.

Transfer of data retrieved from ARC/INFO to serve as input for AQ-FEM was performed by a set of specifically developed AML routines. These AML routines were extended with FORTRAN programs to produce standardized ASCII files containing the reformatted data to be used by AQ-FEM. Results of the finite element calculations were returned to geographical information system, also via standard ASCII file interfaces. Values had to be interpolated in GIS as part of the transfer of input data. Because ARC/INFO's bivariate-quintic interpolation method could not be used in areas with little data, the interpolation was done by a lattice, i.e. a cell-based (linear) method. Another problem was a rather high calculation time when using INFO programs to perform more complicated calculations. As an example, the complicated assignment of well abstraction rates to either of the four aquifers was performed faster by exporting the data via an ASCII file and performing the calculations by means of a FORTRAN program rather then using an INFO program.

Model results can be reproduced significantly more reliable due to the consistent way in which data is stored and processed in the coupled geographical information system ARC/INFO and the finite element program AQ-FEM for the modelling of groundwater problems. This reproducibility of model results is a prerequisite to comply with the quality assurance standards.

FUTURE DEVELOPMENTS

Currently, in ARC/INFO, spatially distributed input parameters for the "RIVM National Groundwater Model for The Netherlands (LGM)" are represented by means of contour maps. The contour maps are based on irregularly spaced observations in x,y space. While creating a contour map (by interpolation between point values), it is important that the contouring interval is properly chosen. The reason for this is that in areas where the observed change of a parameter is smaller than the chosen contouring interval, the relevant parameter variation can not be represented. Therefore, to avoid interpolation errors, and thus to increase the accuracy of the model results, it is better to store the data at regularly spaced grids, instead of by means of contour maps. Fortunately, with the new ARC/INFO release 6.0, data can be stored and processed in grids by the GRID module; thus meeting the afore mentioned requirements. Interfacing techniques will handle the data stored in grid maps, instead of the data stored in contour maps.

The groundwater model will be extended by the implementation of the particle tracking calculation modules. The latter are a part of the AQ-FEM groundwater modelling program package. Particle tracking methods offer the possibility to study the (convective) transport of dissolved matter, to calculate recharge areas and to determine protection areas and travel times in saturated groundwater. A further improvement will be the implementation of the random walk method to perform calculations of the dissolved species transport, including the effect of hydrodynamic dispersion.

Acknowledgements Authors wish to thank their colleagues Toon Leijnse and Rien Pastoors, who have contributed substantially to the development of the model system described in this paper, as well as Melissa van der Poel for her contribution in checking the manuscript.

REFERENCES

Kovar, K. (1992) Groundwater Model for The Netherlands. Mathematical Model Development and User's Guide. *RIVM-report 714305002*, Bilthoven, The Netherlands.

Lieste, R. & Verlouw, J.G.W. (1992) Application of a Geographic Information System for the RIVM National Groundwater Model for The Netherlands (in Dutch). *RIVM-report 714305003*, Bilthoven, The Netherlands.

Pastoors, M.J.H. (1992a) RIVM National Groundwater Model for The Netherlands: Conceptual Model (in Dutch). *RIVM-report 714305004*, Bilthoven, The Netherlands.

Pastoors, M.J.H. (1992b) RIVM National Groundwater Model for The Netherlands: Model Results (in Dutch). *RIVM-report 714305005*, Bilthoven, The Netherlands.

Application of geographical information systems to support groundwater modelling

H.P. NACHTNEBEL, J. FÜRST & H. HOLZMANN
Institut für Wasserwirtschaft, Hydrologie und konstruktiven Wasserbau, Universität für Bodenkultur, Nussdorfer Lände 11, A-1190 Wien, Austria

Abstract The objective of this paper is to identify the user requirements addressed to geographical information systems (GIS) to support regional groundwater modelling. The paper summarizes some experiences gained from the application of GIS in water related management based on groundwater models. Five issues are discussed which refer to the exchange of information with a GIS, design of groundwater models, visualization and display of simulation results, analytical tools, and open system philosophy.

INTRODUCTION

A rational and efficient utilization of regional groundwater systems requires a reliable database, modelling tools to describe the spatial head distribution and flow pattern, a proper definition of goals and measures of performance to assess the efficiency of management alternatives, and a monitoring network to control groundwater pumping. In the last two decades numerous groundwater models have been developed from which many are reviewed in Bachmat *et al.*, (1980). Simulation runs of numerical models can describe various management alternatives, such as location of pumping wells, amount of groundwater pumping, impacts of modifications of the boundary conditions on the groundwater table and flow pattern, and the control of the transport of pollutants. The information gained from these simulations will assist in decision making related to regional groundwater modelling (Fürst *et al.*, 1993).

In the process of simulation and analysis, large data sets exhibiting quite distinct structures have to be manipulated. Because of the time consuming grid design and model setup procedure, often only a single geometric structure is developed to describe the groundwater flow by a discrete model. After calibrating a model, only a few management alternatives are evaluated, again because of the time consuming analysis of the spatial and temporal variability in the groundwater heads. It can be concluded that the effort to setup alternative grids and management schemes will be substantially reduced by:

(a) an easily accessible relational database;
(b) an interactive tool to administer and manipulate spatial data;
(c) a model base including different groundwater models for two- or three-dimensional problems, for describing flow and transport of solutes, or for multi-phase systems;
(d) utilities for the visualization of model results according to human perception;
(e) and tools for the analysis of the model outputs.

Geographical information systems provide some of these tools. Dependent on the software product there is a broad spectrum in the degree of user comfort and in the system's capabilities.

The experiences from several case studies (Fürst *et al.*, 1987; Nachtnebel *et al.*, 1991, 1992) in which different GIS (ARC/INFO and GRASS) were applied are summarized and user requirements are identified that should be covered by GIS. This paper discusses the following five issues: exchange of information to/from a GIS, design of a model, visualization of simulation results, analytical tools, and open system philosophy.

IMPORT AND EXPORT INTO/FROM A GIS

In this chapter three aspects are considered: The type of data that are needed in hydrological modelling, the exchange of information among different systems such as a relational database, and a GIS and the communication with the GIS. The database should handle:
(a)   time series data;
(b)   spatial information;
(c)   generally structured tables;
(d)   texts;
(e)   large bit mapped information.

The data are grouped into objects (Fig. 1), each composed of a block of general information characterizing uniquely the object, and a data block containing the time series or the text. To give an illustration, the general information includes, for instance, the name of an observation station, administrative region, coordinates of location, reliability of the station, type of measurement device.

Spatial data refer to points, arcs, areas and subspaces defined by separating three-dimensional boundaries. A typical example is the 3D characterization of the hydraulic conductivity of an aquifer.

Tables are important for easy handling of the hydrogeological data sets necessary for groundwater modelling. Especially when finite difference models are applied, the data sets are represented by a simple matrix type structure. Other tables might be required to describe the interaction between surface and groundwater system by a stage-infiltration relationship valid within a certain section of a river.

Texts constitute an important information which should be easily accessible during the modelling process. To give an example, the heterogeneity of an aquifer is often only verbally described.

The database integrated into the GIS is primarily designed to handle spatial information, either in vectorized form or in a grid-based structure. There is a deficit in the various GIS with respect to the handling of other than 2D spatial data sets. This implies that the support of 3D volumetric (solid) modelling (Raper, 1989; Flynn, 1990) should be improved and also that at least

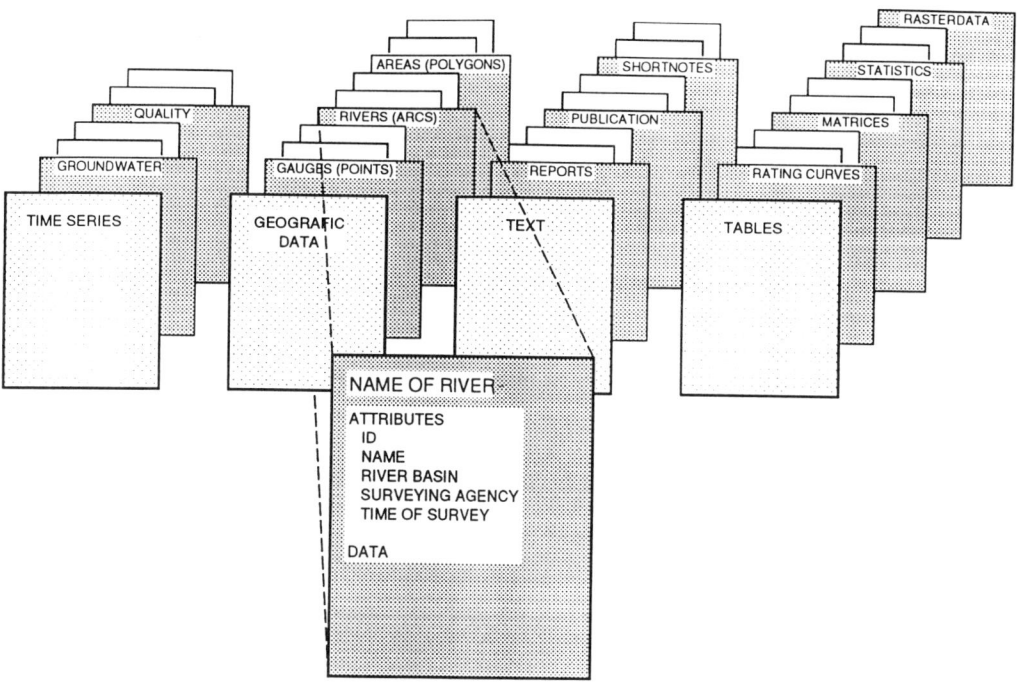

Fig. 1 Elements in the database.

time series data should be also integrated into the GIS database. In the long-term perspective, the most attractive approach is in the combination of an object oriented database with a GIS. Due to a lack of commercially available products another approach is in the combination of a GIS with a relational database (Nachtnebel *et al.*, 1991) supported by an interface that also handles predefined standard queries (Fig. 2). Due to the lack in the standardization of GIS databases the information cannot be fully communicated between an external database and the integrated GIS database.

The exchange of information with the GIS includes also the communication itself which is either supported by an interactive graphical or

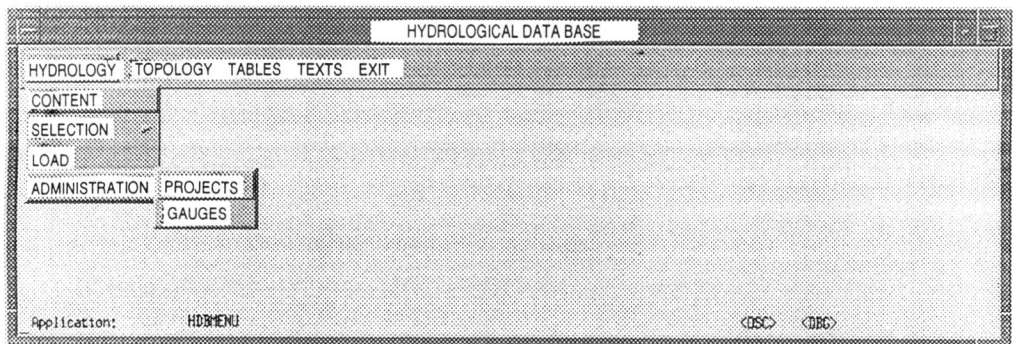

Fig. 2 Main menu of the interface to the database.

menu-driven user interface or which is based, as it is often the case, on a simple command structure. To facilitate the modelling of groundwater systems, an interface (Fig. 3) was developed integrating the system commands into macros which perform complex functions related to data selection and modelling. Some of the characteristics will be discussed in the subsequent chapters.

MODEL DESIGN

The model design requires at least the following functions:
(a) grid design under consideration of the observation network;
(b) definition of the boundaries and their hydrological characteristics;
(c) identification of sources and sink terms;
(d) interpolation and assignment of hydrological data to the grid cells.

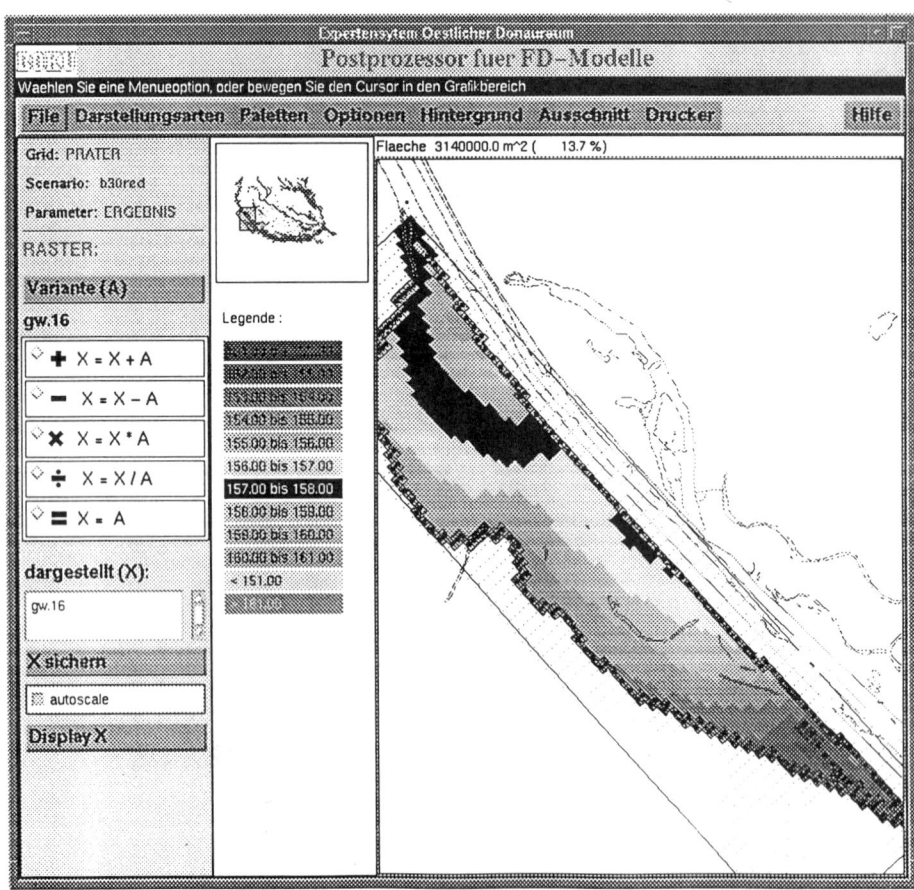

Fig. 3 Example of the interface to the groundwater modelling system. The figure illustrates the raster map display feature of the postprocessor for finite difference groundwater models (the user interface is in German).

In the case studies discussed in this paper, only 2D models (McDonald & Harbaugh, 1984; Voss, 1984) were applied. In Nachtnebel *et al.* (1991, 1992) a finite difference approach was adopted, while in Nachtnebel & Jawecki (1992) a finite element approach was applied. In the former case, a rectangular grid with an adjustable mesh size is arranged over the region under consideration of the main flow direction in the groundwater system. Raster-based systems facilitate this approach and subsequent GIS operations can be quickly performed. The disadvantage is that the grid extends uniformly over the whole region and that the minimum resolution has to be defined in the initial design phase. To avoid extremely large matrices subregions which require a more detailed spatial resolution can be discretized by local grid refinement. This methodology is only supported by vector-based systems.

When 3D models with several distinct layers in the aquifer are considered, a smooth and nonlinear gridding (Smith & Wessel, 1990) might be more appropriate to represent the physical system. From a modelling viewpoint, the locations of groundwater observation wells should coincide as good as possible with the centers of the grid cells. To achieve this goal, a rather detailed gridding would be necessary, resulting in a very large number of grid cells (McCullagh, 1989). Another approach is a finite element net, which provides more flexibility in the design process. From some applications it can be concluded that the GIS gridding algorithms do not fully satisfy hydrological requirements. Triangulation techniques should be able to handle internal break lines represented by vertices in the final grid, the outer boundary of which should not be restricted to the convex hull of the grid points.

The definition of the boundaries is related to the observation network and the hydrological conditions. A background map including the relevant information is useful to select the boundaries, which often follow the course of a major surface water body (Fig. 4). In this case, a Dirichlet type boundary is associated with the water table of the river which should be directly accessible as an object in the database. Similarly, the subterranean inflow from a watershed can be considered in the groundwater model design as a Neumann type boundary. A digital terrain model will substantially assist in drawing the boundary along a contour line directly on the display.

The top and bottom boundaries of geologic layers can be identified on the basis of hydrogeological informations obtained from borehole data or from spatial geophysical measurements.

In the next step, the physically justified boundaries have to be associated with the model grid. In principle, the same procedure as for the identification of sources and sinks will be applied. A graphical editor assists in (re)defining subsections of boundaries and in allocating wells to grid nodes.

In the case studies, the number of grid cells was between 950 and 2000. The average grid size ranged from 200x100 to 1000x1000 m for a regional study. Each grid cell requires a set of hydrogeological parameters such as:
(a) bottom layer of the aquifer;
(b) hydraulic conductivity values;

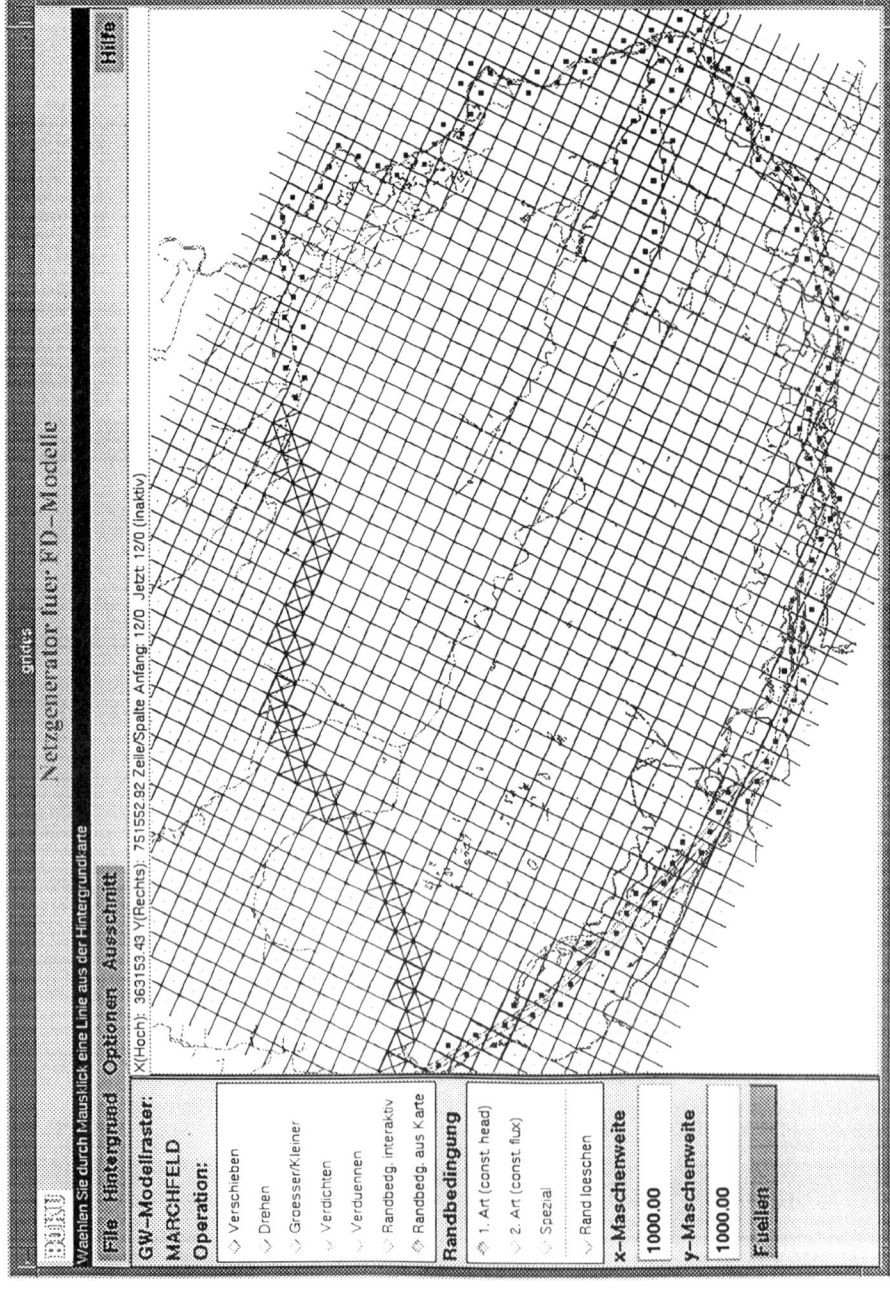

Fig. 4 Grid of a groundwater model with the surface water system in the background (the user interface is in German).

(c) storage coefficient;
(d) recharge rates;
(e) observed and calculated groundwater tables.

In comparison to the number of grid cells, all these parameters are obtained from only a few measurement locations. The point information has to be interpolated for each node and has to be averaged over the volume of the grid cell. To achieve this task, some advanced interpolation algorithms such as

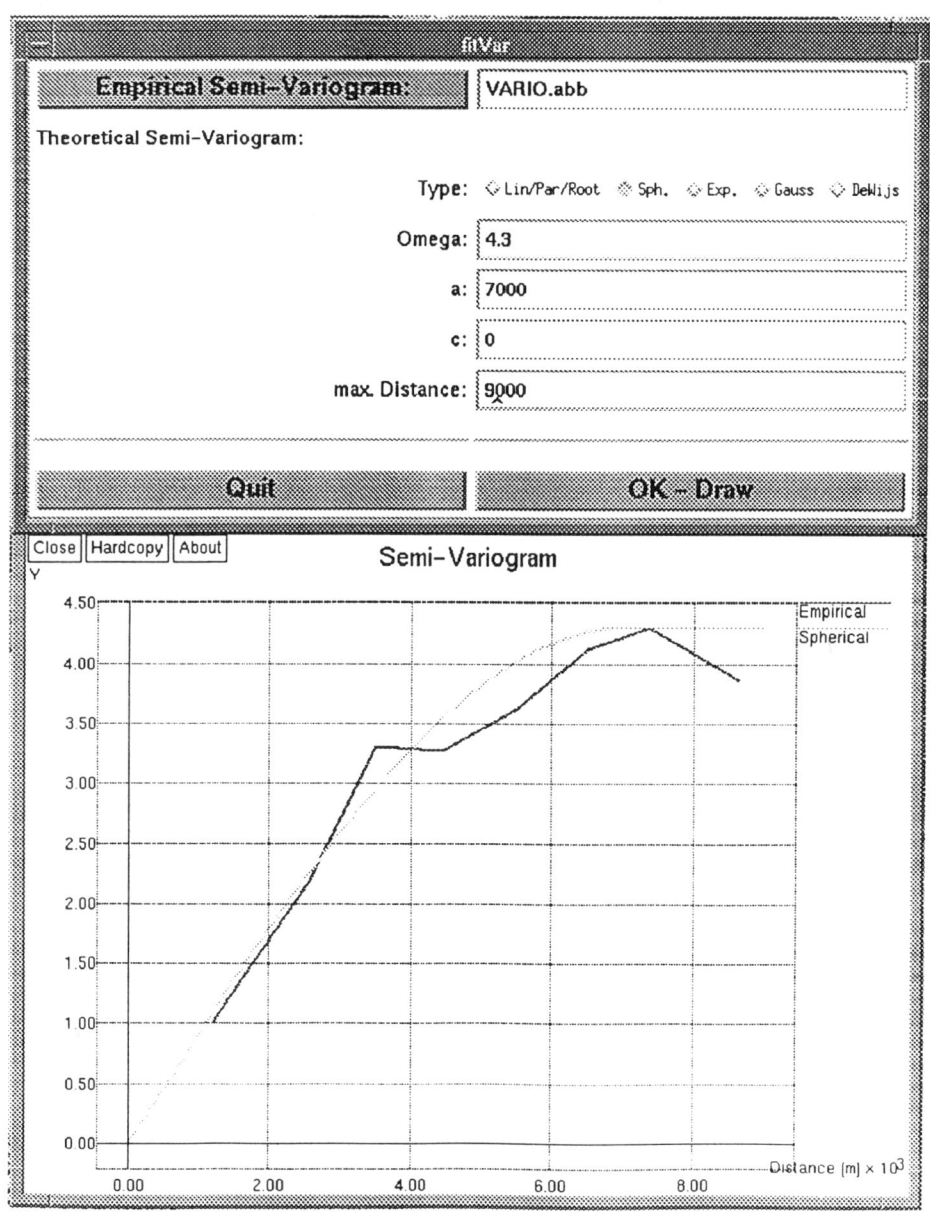

Fig. 5 Interface to the kriging routines - Fitting a theoretical semi-variogram to the empirical semi-variogram.

kriging, universal kriging are required. These tools are integrated only in a few GIS. However, recent attempts are made to include them in updated GIS versions. In the case studies described in this paper, user developed routines (Fig. 5) had to be combined with the GIS.

VISUALIZATION AND DISPLAY OF SIMULATION RESULTS

The groundwater table is observed at only a few locations while the model yields head data for every grid cell. Several techniques can be applied to compare observed and computed head values. Frequently, in the unsteady case, time series are compared for observation points which should be interactively identifiable by the modeller. To display the spatial pattern of the groundwater table, the observations have to be interpolated over the grid nodes. Subsequently, contour lines or colour-coded grid pattern can be displayed to compare, for instance, observed and simulated data. Contour lines, which can be generated by GIS routines, display the information about the groundwater system in a comprehensive way such that the modeller can compare two different groundwater tables simultaneously within the project area (Fig. 6). Further, conclusions with respect to the flow directions can be drawn. The difference in head values can also be displayed in a colour-coded grid pattern assisting in the identification of grid cells where the model describes the system unsatisfactorily. Many of the required operations are supported by a GIS.

To assist the modeller in the perception of the geographical context, the simultaneous display of several maps would be quite helpful. A scanned geographical map acting as a background information and an overlay of a transparent colour-coded grid map assist not only in the identification of unsatisfactory calibrated grid cells but also in their geographical allocation. It should be also mentioned in this context that the definition of colour families and the joint manipulation of foreground and background colours would facilitate hydrological modelling and analysis. Some PC based systems already provide such tools while more sophisticated GIS, which are based on workstations and on UNIX operating systems, exhibit some lacks in this respect.

ANALYTICAL TOOLS

From the hydrological viewpoint, there is a need for the generation of orthogonal nonlinear grids of head contours and stream lines. At least for steady-state situations, where stream lines are identical to the flow paths, this feature could eliminate the need for a particle tracking model. Some digital terrain models and related GIS routines can compute gradients of 3D surfaces. Applied to a groundwater surface, and combined with maps of hydraulic transmissivity, we can easily generate vectors of flow velocity and specific

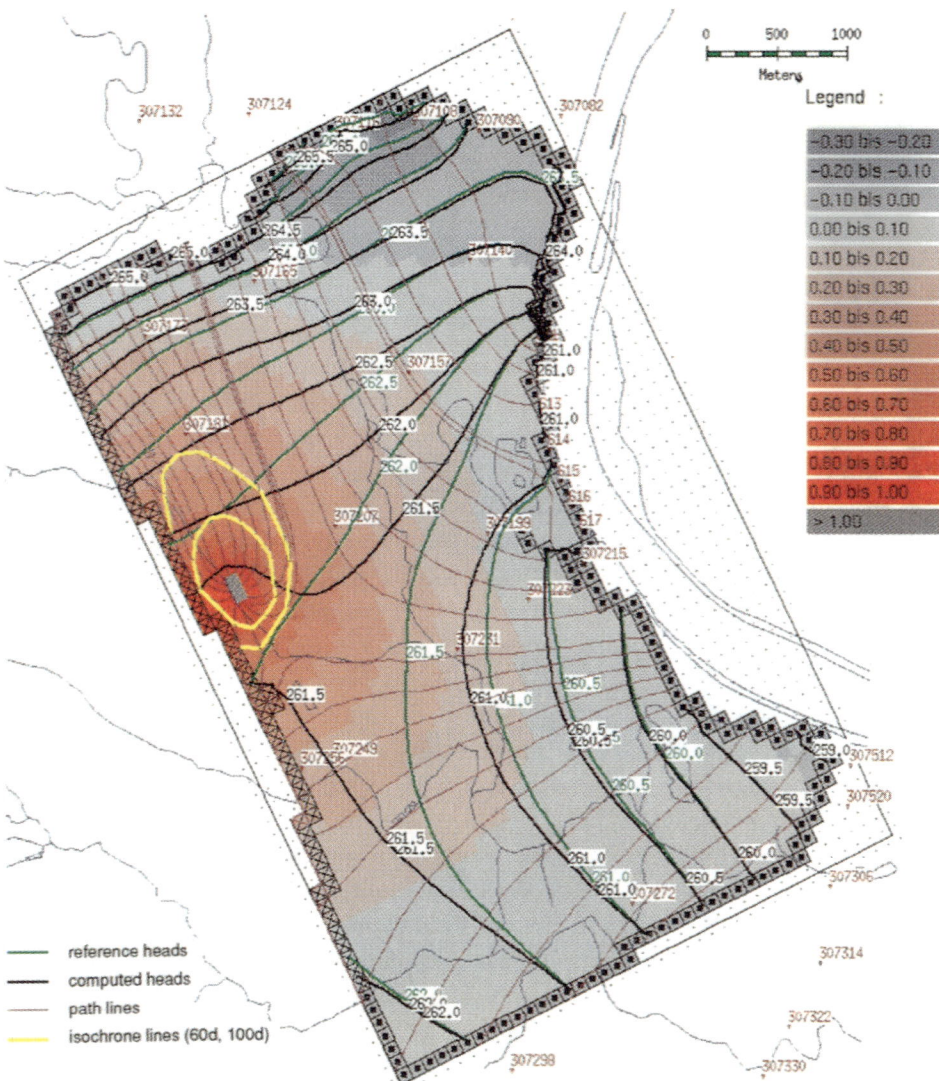

Fig. 6 Comparison of two computed groundwater heads by contour lines and a coloured grid map. The figure also shows pathlines and isochrones computed with a particle tracking model.

discharge at the grid points of a groundwater model.

Direct colour-based operations might assist in the analysis of simulation results. As an example, two hydrological variables, each represented by its coloured spatial pattern, have to be combined to yield new information about the system. First, the numerical range of each variable is expressed by a colour array and then these two arrays have to be mapped into a colour matrix. Obviously, the combination process which is described by the matrix is problem dependent. Booelan operations or tools for a user defineable combination based on colour families (Burger & Gillies, 1989) should improve the analysis of the system.

Also, the generation of transparent map displays would provide more insight into hydrological processes.

OPEN SYSTEMS

From the topics discussed above, it can be concluded that GIS can help substantially in groundwater modelling. However, the GIS has to communicate with a couple of other software tools. A GIS that should be efficiently applicable in groundwater modelling needs to be "open" on several levels:

(a) on the data level: GIS data structures should be well documented and easily accessible from external programs in both directions;
(b) on the functional level: GIS routines should be modular, so that individual routines can be selected for groundwater modelling purposes without the overhead of unneeded parts of the GIS;
(c) on the user interface level: To maintain a consistent user interface for all the modules of a "groundwater modelling workbench", the GIS should either provide a user interface development facility like a macro language that can also integrate external programs or it should be accessible by externally developed user interfaces;
(d) interdisciplinary openness: Coupling groundwater modelling and GIS makes groundwater related data and modelling results easily accessible to other disciplines in a workgroup environment (Hary & Nachtnebel, 1989). The GIS can become a communication platform, especially for environmental impact analysis of large water management projects (Fedra, 1991).

Experiences with ARC/INFO as a representative of vector-based, comprehensive, commercial GIS showed that easy data access via Import/Export routines to text files becomes slow, while direct access via subroutine libraries is computationally efficient but requires tedious programming. The modularity of ARC/INFO is good from the user's aspect, i.e. one can easily accomplish specific tasks and chain ARC/INFO routines and external programs together. However, one has to buy and install the complete ARC/INFO system. With ARC/INFO, a consistent, menu-based user interface can be developed with the Arc Macro Language AML, which is able to integrate ARC/INFO routines and external programs. Finally, as a widely used, general purpose, vector-based GIS ARC/INFO is an ideal communication platform for interdisciplinary, spatial analysis on different scales.

GRASS, as a representative of public domain, raster-based GIS provides easy and quick access to the data due to the simplicity of its raster-based data model. GRASS analysis tools are also organized as separate, independent programs that can be called from any higher level user interface system. The availability of source code further allows customization and adaptation of GRASS programs to specific requirements.

CONCLUSIONS

Experience with the application of GIS in several groundwater modelling projects shows several benefits:
(a) spatial data management and pre-processing is improved. These important steps are performed independently of a particular model code;
(b) the GIS aids in the model design process;
(c) GIS mapping facilities provide comprehensive and flexible visualization instruments for any type of groundwater model;
(d) the GIS can combine spatial information from different disciplines and thus become a communication platform especially for environmental impact analysis.

However, current software packages are not fully adequate for some aspects of groundwater modelling. They lack full 3D and solid modelling capabilities, and cannot efficiently handle time dependencies of spatial phenomena. Another weakness is in the communication with other software tools like RDBMS, visualization packages or geostatistics.

REFERENCES

Bachmat, Y., Bredehoeft, J., Andrews, B., Holtz, D. & Sebastian, S. (1980) Groundwater management: The use of numerical models. *Water Resources Monographs,* Vol. 5, AGU, Washington D.C., USA.
Burger, P. & Gillies, D. (1989) *Interactive computer graphics.* Addison Wesley Publishing Company.
Fedra, K. (1991) A computer based approach to environmental impact assessment. IIASA Research Report RR-91-13, IIASA, Laxenburg, Austria.
Flynn, J.J. (1990) 3D computing geoscience update: hardware advances set the pace for software developers. *Geobyte* 5(1), 33-36.
Fürst, J., Haider, S. & Nachtnebel, H.P. (1987) ARC-INFO as a tool in groundwater systems analysis. *Proc. of the ESRI Users Conference,* Kranzberg, Bavaria, Germany, 7/1-7/12.
Fürst, J., Girstmair, G. & Nachtnebel, H.P. (1993) Application of GIS in Decision Support Systems for groundwater management. In: *Application of GIS in Hydrology and Water Resources* (ed. by K. Kovar & H.P. Nachtnebel), Proc. Int. Conf. HydroGIS 93, 19-22 April, Vienna, Austria, IAHS Publ. No. 211.
Hary, N. & Nachtnebel, H.P. (eds) (1989) Ecosystem case study hydropower scheme Altenwoerth. Publication of the Austrian Man and Biosphere Program, Vol. 14, Universitätsverlag Wagner, Innsbruck, Austria (in German).
McCullagh, M.J. (1989) Power to the People! PC and Workstation Mapping and Database Systems. In: *Digital Geologic and Geographic Information Systems* (ed. by J.N. Van Driel & J.C. Davis), Vol. 10, Short Course in Geology, AGU, Washington DC, USA.
McDonald, M.G. & Harbaugh, A.W. (1988) A modular three-dimensional finite-difference ground-water flow model. US Geological Survey. Techniques of Water-Resources Investigations of the United States Geological Survey, Book 6, Chapter A1.
Nachtnebel, H.P. (1993) Interactive modelling of groundwater systems utilizing a GIS. In: *Decision Support Techniques for Integrated Water Resources Management* (ed. by J.J. Bogardi & H.P. Nachtnebel), Cambridge Univ. Press, UK (to appear).
Nachtnebel, H.P., Fürst, J. & Holzmann, H. (1991) *Expert system for groundwater management in the Marchfeld Region, Austria.* Final Report to the Ministry of Agriculture and Forestry, Vienna, Austria (in German).

Nachtnebel, H.P., Fürst, J. & Holzmann, H. (1992) GIS applications in water related management in Austria. *Proc. Int. Workshop on GIS in Water Management*, Technical University Budapest, 28-30 September.

Nachtnebel, H.P. & Jawecki, A. (1992) *Groundwater model for the Leibnitzer Feld, Styria, Austria*. Final Report to the Provincial Government of Styria, Graz, Austria (in German).

Raper, J.F. (1989) *3D application in geographical information systems*. Taylor and Francis, London, UK.

Smith, W.H.F. & Wessel, P. (1990) Gridding with continuous curvature splines in tension. *Geophysics* 55(3), 293-305.

Voss, C.I. (1984) A finite element simulation model for saturated-unsaturated fluid-density-dependent groundwater flow with energy transport or chemically reactive single-species solute transport. USGS Water Investigations Report 84-4369, US Geological Survey, Reston, VA, USA.

Groundwater modelling using GIS at the Amsterdam water supply

T.N. OLSTHOORN, P.T.W.J. KAMPS & W.J. DROESEN
Amsterdam Water Supply, Vogelenzangseweg 21, 2114 BA Vogelenzang, The Netherlands

Abstract The Amsterdam Water Supply carried out a three-year study to optimize the interests of water abstraction and nature protection in the dune area south of Zandvoort, Netherlands. It was foreseen that the study would necessitate intensive interchange of spatial information between hydrologists and ecologists. This led to the acquisition of a GIS (GENAMAP). This paper focuses on the data handling performed for the groundwater model (MODFLOW), for which input has been generated as much as feasible by means of the GIS. The groundwater model results were analyzed by GIS and further used via a DTM in ecological models. The use of GIS proved beneficial both from the view of model building and usage as well as for mutual use of tools and data by various disciplines.

INTRODUCTION

The Amsterdam Water Supply uses a 35 km^2 dune area along the North-Sea coast (Fig. 1) for groundwater recovery (since 1853) and artificial groundwater recharge (since 1957). The area serves both as a groundwater resource and a nature reserve.

Total production of drinking water from this area equals about 67 million m^3 year^{-1} of which about 10 million m^3 consists of groundwater from natural recharge and 57 million m^3 stems from purified River Rhine water after having passed the dune sands.

For a number of years, national and provincial policy insist on hydrological restoration of the dune area, in order to regain original high-valued groundwater-dependent vegetation and moist conditions.

The aim was to develop a set of hydrological scenarios for the area and to predict their ecological consequences. To do this, a detailed hydrological model was needed, as well as a series of ecological models, joined by a digital terrain model (DTM). Since this project involves extended spatial investigations and modelling by both hydrologists and ecologists including the exchange of spatial data, it was decided to make a GIS the project basis.

BUILDING THE GROUNDWATER MODEL

In this hydro-ecological project emphasis was placed on ecological predictions. As a consequence detailed modelling of the phreatic groundwater surface was critical. Input files for the model would be generated as directly as possible from GIS maps.

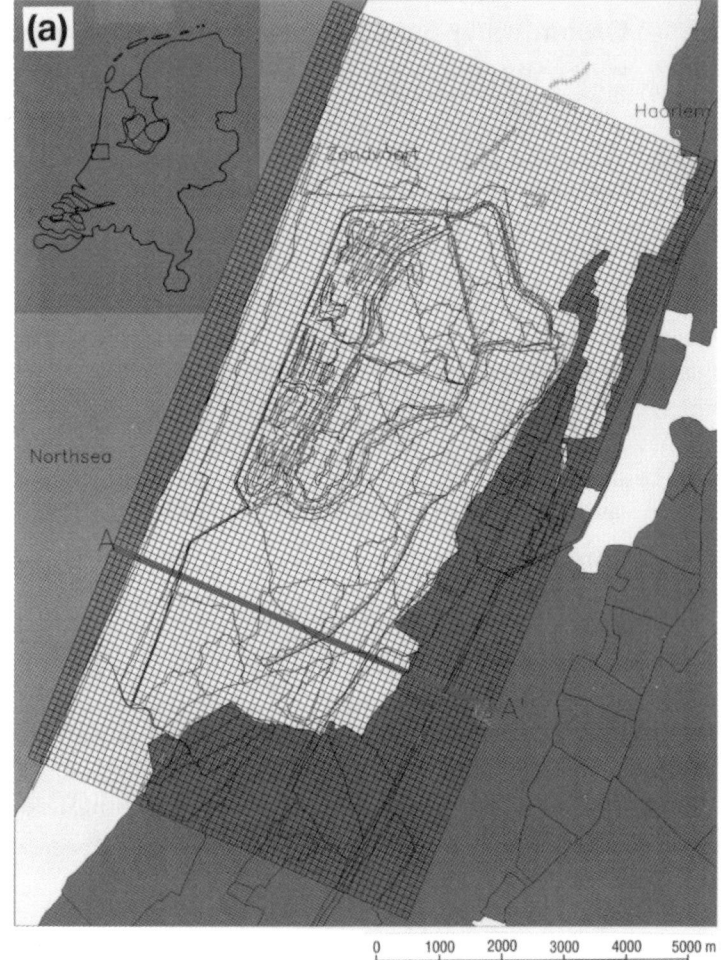

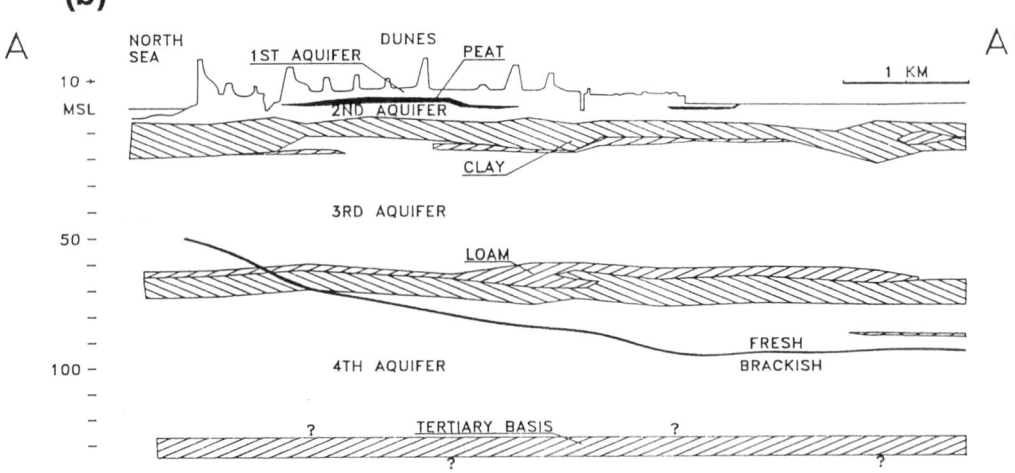

Fig. 1 (a) The model grid (100x100 m grid cells) for the dune area of the Amsterdam Water Supply; (b) a schematized W-E cross section through the line AA', showing the fresh-salt water interface and the three semi-pervious layers separating the four aquifers.

Geological information

The geological cross section in Fig. 1 shows the geological layers in a digitized form, made up of three semi-pervious layers dividing four aquifers. The uppermost aquifer is phreatic, the second one is partially semi-confined and partially phreatic, namely where the uppermost semi-pervious layer, the peat layer is absent. The second semi-pervious layer consists mainly of clay and the third of loam. The base of the fourth aquifer, and also the basement of the geohydrological system, are the clayey layers of the tertiary formations.

Available geological maps (Stuyfzand, 1988; Anonymous, 1989), for the semi-pervious layers, were used as basic information. These maps consist of contours of thickness-classes and indicate specific geological origin (Fig. 2). Further they have point data of their bottom elevation. The digitized maps are discussed below.

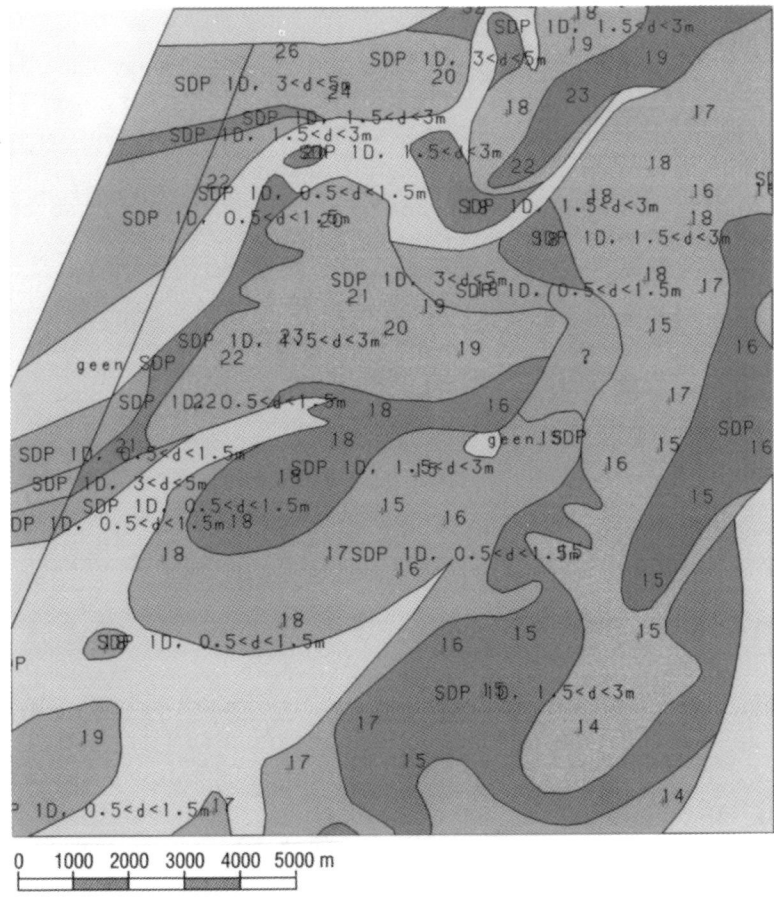

Fig. 2 Original map of second aquitard (clay) showing thickness, origin classes and point elevations of its lower plane.

The 150 m mean sea level plane has been taken as the hydrological basis, since there are too few boreholes sunk as deep as the tertiary formations. In the model the interface between fresh and saline water is taken as the actual basis (see cross section in Fig. 1). A map of this interface is available.

To obtain a full area coverage of the layer underside, the elevation points were interpolated by kriging on a 100 m grid using the Geo-Ease geo-statistical program. Due to layer thickness on the original maps being given as classes (such as "3 m<d<10 m") an average class-thickness was used as a rough estimate of actual thickness. Using these "average" thicknesses and the interpolated bottom of the semi-pervious layers, their top surface was estimated.

The thickness of the aquifers was obtained by subtracting the planes of the bottom of the overlaying and the top of the underlying aquitards, with the exception of the lowest aquifer where the top of the tertiary formations was assumed at 150 m below sea level, according to the few bore logs available. The actual base of the fourth aquifer during model runs is the digitized fresh-salt water interface.

Figure 3a shows part of the digitized peat map. The peat formed in ancient beach plains between beach embankments. It was covered by several m of wind-blown sand and compacted. This yielded a shallow (on average 0.5 m thick) layer of very low permeability and great hydrological importance (Anonymous, 1989). Where peat is present, there may be several meters of head difference. The elevated beach embankments are largely free of peat, consequently elongated gaps remain in the peat layer. These gaps connect the first and second aquifers and are of utmost importance to the local phreatic groundwater head. Consequently, the location and extent of these gaps and the margins of the peat layer is of utmost importance for the model.

The above approach was accepted as the most practical. Refinements are planned as new maps become available. The GIS and the automated procedures to generate model input from GIS-files should guarantee an easy update of model input with new data at any time. Hydraulic properties of the various layers, not being basic data, are estimated by model calibration.

Hydrological infrastructure

The hydrological infrastructure consists of maps of wells, ditches, drains, canals in the dune area and polders and other surface-water courses in the wider surroundings according to the table below. The name of each infrastructural unit is used as an identifier to link other attributes to them. Polders are areas having a maintained fixed winter and summer head respectively.

Choosing a mesh for the model

Due to the high hydraulic resistance of even a thin peat layer, an accurate

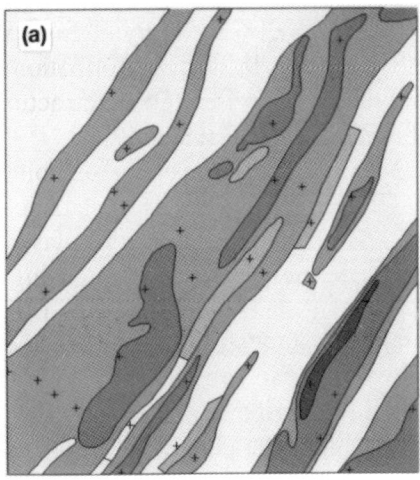
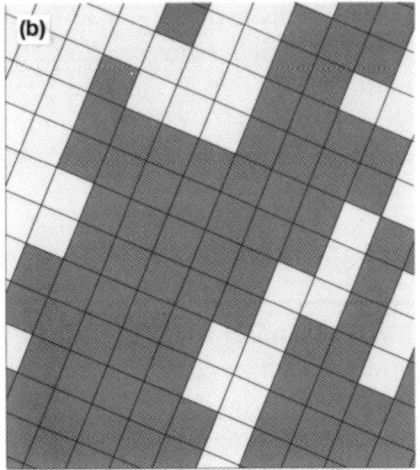
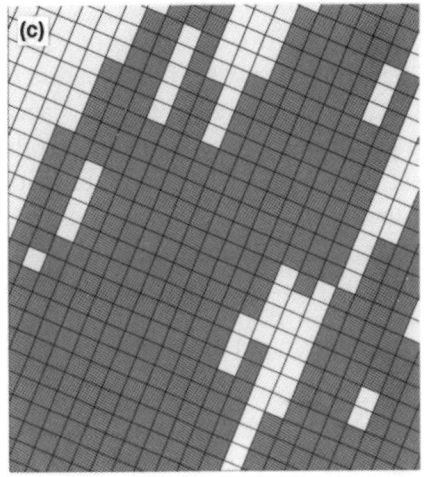

Fig. 3 (a) Part of the digitized peat layer; (b) overlay of Fig. 3a with a 400×400 m grid; (c) overlay of Fig. 3a with a 200×200 m grid; (d) overlay of Fig. 3a with the 100×100 m grid, used by the model (for grid see Fig. 1).

representation of the margins of the layer and any gaps in it (Fig. 3a) have more importance than mere thickness. There is no way to obtain a valid average for the hydraulic resistance of a model cell that is only partly covered by peat. So each model cell had to become peat or no peat. Hence, the model cells have to be small enough to properly represent the margins of the peat layer.

The thematic peat map has been overlain with three networks of square cells of different sizes (see Figs 3b to 3d), while a cell was made a "peat cell" if more than 50% of its area was peat covered. Clearly only the network with the smallest cells (100x100 m) can represent the peat layer well enough and was therefore chosen for the model.

Building model input

Maps of aquifer transmissivities and aquitard resistances cannot be created directly from geology. The logical approach would be to combine maps of the type of material on the one hand with layer thicknesses on the other, using pumping tests as a guide to derive proper values for the permeability of the materials involved, while later adjusting them by model calibration.

In the absence of independent data on spatial variation of aquitard or aquifer material Olsthoorn & Kusse (1986) in their self-programmed GIS-alike environment applied a single permeability per distinct geological formation. For the time being, we have simplified their approach in that we calculated the spatial transmissivity from the multiplication of a model-calibrated single-valued aquifer-specific permeability and the aquifer's spatially varying thickness.

The aquitards have been contoured into thickness classes (Figs 2 and 3a). We have applied a model-calibrated single-valued aquitard-specific permeability and multiplied its inverse by its average class thickness to obtain a map of the aquitard resistance.

Transmissivities and resistances for the model were obtained by first exporting their thicknesses from GIS into an ASCII file and then using a FORTRAN coded program to join thickness and aquifer/aquitard permeabilities. This procedure was kept outside the GIS in order to be able to perform automatic Monte-Carlo calibration with the model.

The thematic maps of the hydraulic infrastructure (drains etc.) were overlain on the model's network. In this way the surface area of infiltration ditches, and recovery canals per cell were obtained as well as the length of drain per cell. The resulting ASCII files are read by a FORTRAN coded program together with a parameter-value file to obtain hydraulic model input as required by the model.

Wells are point features. So simply attributing a well to its model cell leads to an apparent shift of the location of wells that happen to fall near a cell border. Therefore a circular buffer with a radius of 25% of the cell size around each well was used before overlying the network. The well buffer surface area in a cell was used to attribute well abstraction to model cells.

Net precipitation was derived from the combination of a map of the actual vegetation and the weekly measured outflow of four 25x25 m lysimeters. These lysimeters are operated by the Provincial Waterworks of North-Holland since 1940 (Wind, 1958) and located in the dune area near Castricum and covered with bare sand, natural dune vegetation, mainly bush (sea buckthorn), pedunculate oak and black pine, respectively.

Boundary conditions

The two uppermost aquifers contain many fixed-head cells such as the levels of the North-Sea, polders, canals, infiltration ponds and ditches. Furthermore, the abstraction of drains in the phreatic aquifers and the abstraction rates of wells are fixed-flow boundary conditions to the model. On the west and east border of the model, the aquifers are connected to a fixed head via a given resistance, thus compensating for the limited size of the model.

Some canals have a water level below the peat layer. This is of special significance to the boundary condition of the aquifer above the peat layer. While the given head of the aquifer below the peat layer is set equal to the water level of the canal, the head in the aquifer above the peat layer is set equal to the bottom of the peat layer at the canal borders.

MODEL CALIBRATION

Optimization of aquifer and aquitard permeabilities was done using a Monte-Carlo method. Latin Hypercube Sampling (Peck *et al.*, 1988) was used to generate parameter sets to run the model. A FORTRAN-coded program reads the exported ASCII-format GIS file with aquifer and aquitard thicknesses. It further reads the file with the sampled parameter sets and generates a model input file for each run. After each run another program reads model output together with measured data and weights and generates the weighted sum of squared differences between model and measured heads, adding this result to the file of the parameter sets.

After all (600) model runs, the parameter-set file is sorted, yielding a list with the best parameter sets at its top together with their weighted error.

The average of the best five parameter sets was further used, while the best say 5% parameter sets could be accepted as plausible sets to be run the model with in order to obtain a feel of the uncertainty of the model.

USE OF THE MODEL AND ITS RESULTS

The model was used for over a dozen water management scenarios for the southern part of the Amsterdam Dune Area. In order to analyze the results by means of GIS, the output of the MODFLOW-model was transformed by a FORTRAN-coded program into an ASCII table with 27 columns, one record for each model cell, containing the entire model output. This file is read into GIS, combined with the model network and used to present and analyze the results in the form of maps. The maps are in the from of model cells (area features), coloured according to a specified legend (Fig. 4). The large number of model cells guarantees a clear picture. These pictures were overlain by topographical features in GIS for easy inspection.

USE OF RESULTS BY OTHER DISCIPLINES

The predicted groundwater heads are only one step in the chain of models used in the project. For the ecological models not the heads as such, but their

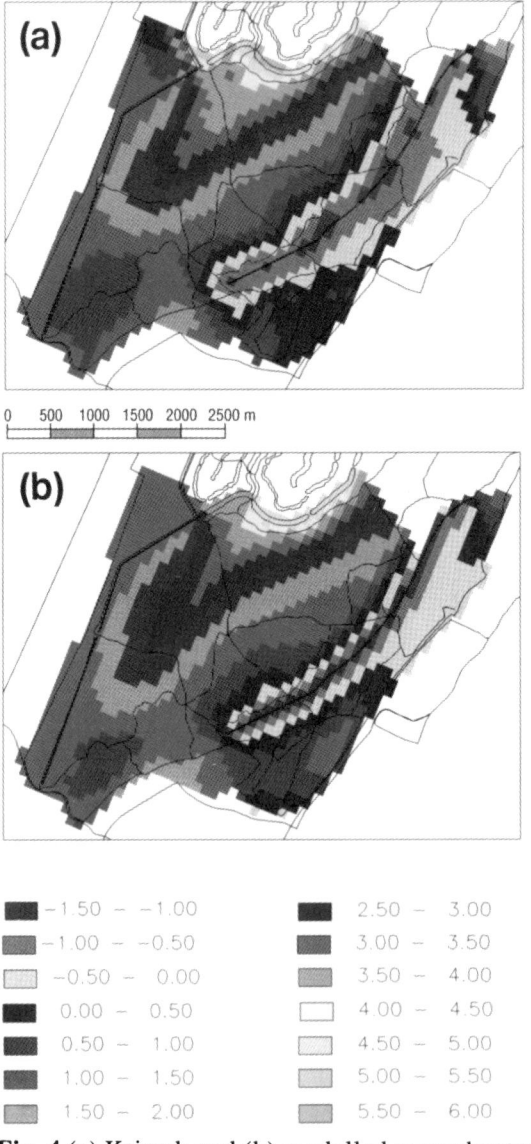

Fig. 4 (a) Kriged; and (b) modelled groundwater heads in the second aquifer (average values for 1974-1990). The groundwater heads are in meters above the Dutch Ordnance level (NAP).

distance below ground surface was needed. For this, a DTM (digital terrain model) has been made, using contoured ground-elevation maps together with more precise local elevation measurements. The spatial interpolation was done by the TIN module, from which a 10x10 m ground elevation grid was generated. A similar 10x10 m grid was obtained for the groundwater heads, by bilinear interpolation of the 100x100 m grid. The resulting depth was then divided into classes ("dry", "moist", "wet" and "open water") and the maps were coloured accordingly (Fig. 5).

The resulting digital maps were further used by the ecologists, who combined them with soil maps, soil acidity, soil humus and trophic level maps to predict spatial vegetation development under the various scenarios (Droesen *et al.*,1993).

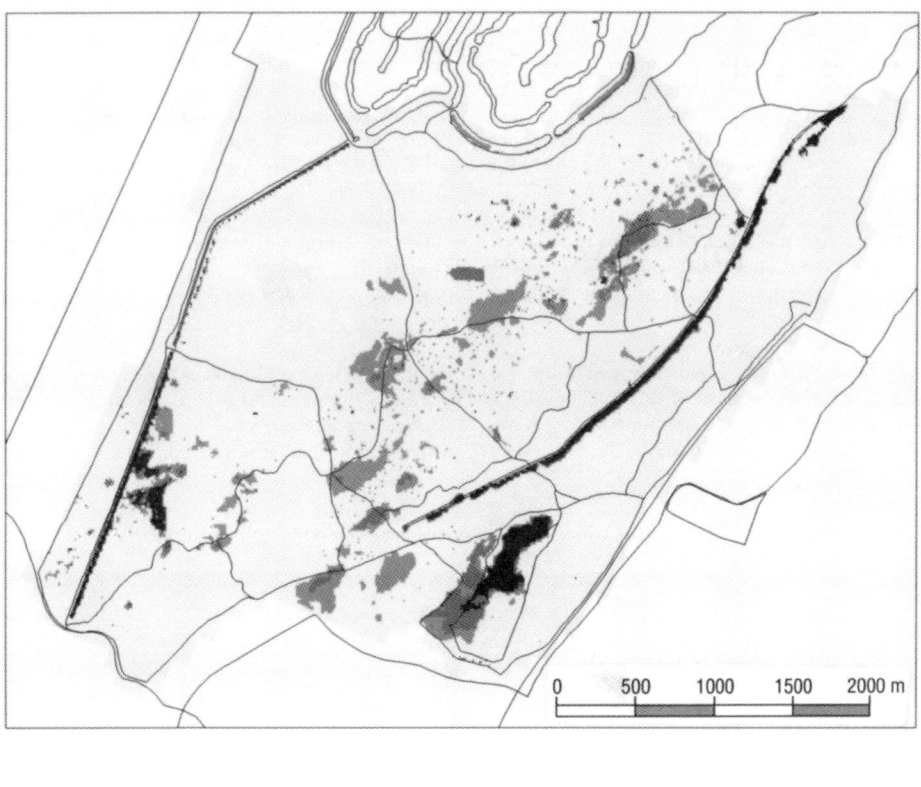

Fig. 5 Depths of groundwater table below ground surface, expressed as four moisture classes (case when no groundwater is abstracted from the southern part of dune area).

SOME BENEFITS OF GIS

One of the main benefits of the use of GIS is the possibility for the data exchange between the different research groups (groundwater heads by ecologists, and vegetation data by hydrologists). GIS thus works as a highly functional market place for readily available compatible spatial data, encouraging the exchange of tools, procedures, methods and data between disciplines.

The formalized automated use of spatial data guarantees reproducibility, which is critical when it comes to updating model input with new basic data, and preparation of input for other kinds of models.

The combination of GIS and models has uncovered errors in geological maps that could then be updated by field research hence improving both the basic data and the final model results.

REFERENCES

Anonymous (1989) Het voorkomen van veen- en humeuze lagen in en rondom het infiltratiegebied van Gemeentewaterleidingen Amsterdam (Presence of peat and humus layers in the Amsterdam Dune Area). Rijksgeologische Dienst (The Netherlands' Geological Survey), Report BP 10815.

Droesen, W.J. & Geelen, L.H.W.T. (1993) Application of fuzzy sets in ecohydrological expert modelling. In: *Application of GIS in Hydrology and Water Resources* (ed. by K. Kovar and H.P. Nachtnebel), Proc. Int. Conf. HydroGIS 93, Vienna, 19-22 April, IAHS Publ. No. 211.

Olsthoorn, T.N. & Kusse, A.A.M. (1986) Aanpak en gegevensverwerking bij grondwatermodellering, toegepast bij onderzoek naar de invloed van de Duitse bruinkoolwinning (Data handling in groundwater model study). H_2O **19**(16), 367-373.

Peck, A., Gorelick, S.M., De Marsily, G., Foster, S. & Kovalevsky, V. (1988) *Consequences of spatial variability in aquifer properties and data limitations for groundwater modelling practice.* IAHS Publ. No. 175.

Stuyfzand, P.J. (1987) Hydrochemie en hydrologie van duinen en aangrenzende polders tussen Noordwijk en Zandvoort aan Zee (Hydrochemistry and hydrology of dunes and adjacent polders between Noordwijk and Zandvoort aan Zee). KIWA SWE, Nieuwegein, SWE 87.007.

Wind, R. (1958) The Lysimeters in the Netherlands. Committee on Hydrological Research TNO, The Hague.

Evaluation of a new integrated software platform for data storage and analysis pertinent to hazardous waste site investigations

M.A. SEN & B. KELK
British Geological Survey, Keyworth, Nottinghamshire, NG12 5GG, UK

Abstract The British Geological Survey (BGS) is collaborating with the Intergraph Corporation in the evaluation and development of a new suite of software, built upon a standard GIS, but extended to include functionality for geological and hydrogeological modelling. This has been done to a large extent by integrating existing separate applications into the same environment. The paper describes the software components and some of the projects where BGS has made use of the software in order to evaluate how well it aids the workflow of a site investigation.

INTRODUCTION

The British Geological Survey has a long history of hydrogeological investigations, having been working on groundwater resource issues since the 1930s and waste management projects since the late 1960s. As a computer system with a broad range of earth science applications allied to a relational database and good graphics has great potential for many environmental applications, the British Geological Survey agreed to assist the Intergraph Corporation in the evaluation and development of a major new package, and has been actively involved in this since mid-1991. The software involved is the *Environmental Resource Management Applications System (ERMA)*. This paper does not attempt to describe all the possible uses and functionality of the software but concentrates on issues which have arisen whilst applying it to BGS environmental assessment projects.

As with all new software, the version that has been evaluated is already out-of-date. Many changes have been made since BGS first became involved and many more will have been incorporated by the time this paper is published. The initial agreement with Intergraph still has a further 18 months to run during which time BGS will continue its evaluation.

In the following text the software packages or modules referred to are all Intergraph products unless specifically identified otherwise. All the figures were photographed from an Interpro workstation.

DESCRIPTION OF SYSTEM

The Intergraph *Environmental Resource Management Applications System*, now incorporated in Intergraph's *ERMA Site Restoration System*, was described at the prototype stage as "a system designed to manage and help analyze data collected and created during the characterization, risk assessment and remedial design phases of site assessment and remediation. Fundamental to that design is the broad requirement of providing a core environment for data management, data verification and data presentation that is easy to use and a set of application modules that can be added to that environment as the task dictates" (Michael & Pearson, 1990).

Whilst the ERMA suite of software extends into the design of any remediation, and hence into civil engineering, these modules have not been evaluated by BGS and are not discussed here.

The ERMA environment in the British Geological Survey

The ERMA suite of software is running on an Interpro 6280 which is linked by a local area network (Ethernet) to a range of VAX mini-computers, a Convex, Sun servers and workstations, and numerous Interpro workstations and PCs. Networked output devices include Versatec colour electrostatic plotters. I^{2}S systems are available for image analysis.

Central to the ERMA software is the internal relational database; the purchaser has the option of Informix, Oracle or Ingres. BGS requested Oracle, as the factual databases of BGS are largely managed in that system. In addition to GIS and CAD packages, the ERMA suite also includes modules for terrain modelling, posting database attributes to graphics, borehole correlation, creation of geological cross-sections, manipulation of raster data, chart plotting, complex statistics, geological surface generation, and a groundwater flow model.

EVALUATION

The evaluation of ERMA carried out by BGS has, to the time of writing, concentrated on three sites which together provide a variety of data and workflows. Two of the sites studied were part of on-going projects, thus providing experience of real-time pressures on the workflow. To describe the main aspects of the ERMA system and of the present BGS evaluation, the following text will concentrate on one site, Wolvey Villa Farm, near Coventry, UK.

Wolvey Villa Farm

The lagoons at Villa Farm had previously been chosen by BGS for investigation as an example of the potentially most unfavourable landfill scenario. Wastes were being introduced directly into the saturated zone of an intergranular aquifer where the potentially beneficial effects of an unsaturated zone were absent, and there were no records of the waste material for the early years of its use. A total of 84 boreholes were drilled on the site from which geological logs, water table and flow parameters and chemical concentrations were measured, but not all parameters in all boreholes. Two-dimensional groundwater flow and conservative and reactive transport modelling were carried out to see if the measured pollution plume concentrations could be predicted from the measured hydrogeological parameters (Mackay *et al.*, 1993).

For the ERMA evaluation it was not feasible to repeat the original scientific study which extended over several years, but a simplified workflow was followed from data gathering to groundwater flow modelling.

In the initial stages of the evaluation the data were limited to surface geography including topographic contours, a surface geology map, and a small number of boreholes. The geographical information was in the form of aerial photographs at about 1:20 000 scale and the Ordnance Survey 1:10 000 scale topographic map. The geological and topographic maps were in paper form. In the future these maps will increasingly be available in digital format but the conversion of paper information will remain important.

The topographic maps were scanned to provide a reference backdrop for other information and helped to check the accuracy of, for example, borehole locations. Some features like the contours, surface geological boundaries and a subset of the stream, road and field boundary locations were digitized so that they could be further manipulated. Other types of data have been successfully imported in other projects, for example, satellite images from the I^{2}S image analysis system as byte files, maps digitized in AutoCAD (Autodesk), and gridded magnetic and gravitational anomaly data as text files.

Surface characterization

Using the CAD module, MicroStation, and MGE Terrain Modeler, the topographic contours were converted into a 3D file and then to a 2.5D (Raper & Kelk, 1991) terrain model. The surface geology having been digitized was passed into the GIS modules MGE and MGA, where the topology was created and attributed. The resulting lithostratigraphic units could be patterned and projected from two dimensions onto the terrain model. Figure 1 shows a screen photograph of the resulting surface (with a factor of 10 vertical exaggeration) with some of the streams, roads and field boundaries also projected onto the surface (the two lower surfaces are described in the next section). At this stage, a significant effort had been expended in digitizing to produce a picture which is useful for presentation to the layman but conveys no more information to an experienced hydrogeologist than the paper surface geology map. However, for projects where there is a large amount of varied two-dimensional information, or where presentations to the public are required, the effort to convert the data into digital format is worthwhile. Even if the only output is 2D maps, these can be edited easily and different data sets selectively superimposed on each other.

Geological characterization

This stage demanded the incorporation of the borehole data. The geological data available from the existing boreholes was lithology and stratigraphy against depth. Some hydrogeological and chemical data were also recorded from them. In order that this information could be used effectively it had to be incorporated into a database. One aspect of ERMA which saves a considerable amount of setting up time is that a database structure of approximately 80 tables is already available. These covered all the categories of information needed for this project.

The basic positional information for boreholes or sample locations are stored in a well table. Other information, eg lithological logs and chemical concentrations, have their own tables which are linked to the well table and reference their entries to the depth or depth interval within the borehole. The geographical location of some of the boreholes already existed in the central BGS borehole database (in Oracle) and could be accessed by the local area network using the inbuilt Relational Interface System (RIS), but the location of others was entered into the database by digitizing from paper maps. The depths to the top and bottom of the main aquifer were in a text file. The rest was entered into spreadsheet files on a personal computer, transferred as text files to the ERMA workstation, and loaded using a database loading application.

The first part of the geological analysis was to generate correlation panels in order to check the interpretation of the borehole geology. Figure 2 shows an example; the wells to be used were selected by drawing a fence around their positions posted onto a basemap (illustrated in the top right corner), and then after various parameters for the style of the panel were selected it was automatically generated by the EPSECT module from the database information. The lower half of the figure shows the whole panel with colour coded patterns corresponding to the main lithologies: orange lines joining the recorded top and bottom depths of the aquifer and a dashed grey line where the water level has been measured. As can be seen in the left two wells the bottom of the aquifer corresponds to the bottom of the sand (brown coloured), but the two wells to the right of these appear to need a reinterpretation of the aquifer extent. The stratigraphic tops in the database can be interactively modified from the graphics. At the top left of the figure is a close-up of one borehole showing that more detailed lithological descriptions have also been posted from the database at the relevant depth interval. Where graphic logs of boreholes already exist in paper form these may be scanned, scaled as required and used as references to the correlation panels.

Following correlation of the borehole geology, 2.5D surfaces of the top and bottom of the aquifer were produced. These are illustrated in Fig. 1 but displaced vertically for the sake of clarity. In interactive use the ability to change viewpoint and section these surfaces helps the user to understand the structure of the aquifer. The ability to post the borehole locations on the surface(s) helps to trace any anomalous shapes to the data from which they

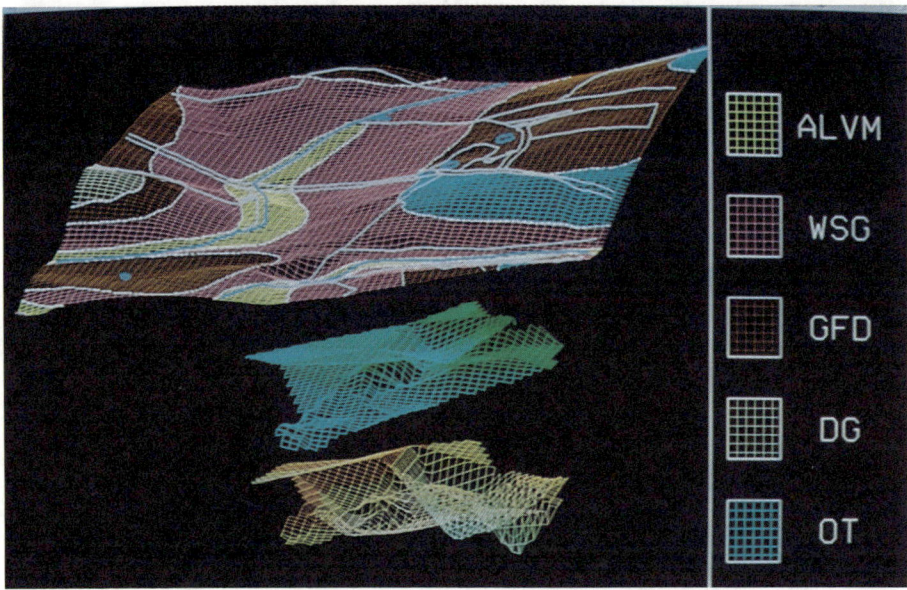

Fig. 1 Digital elevation model of ground surface overlain by surface geology, with top and bottom surfaces of the aquifer below. Vertical exaggeration is x 10. Aquifer surfaces displaced vertically for clarity.

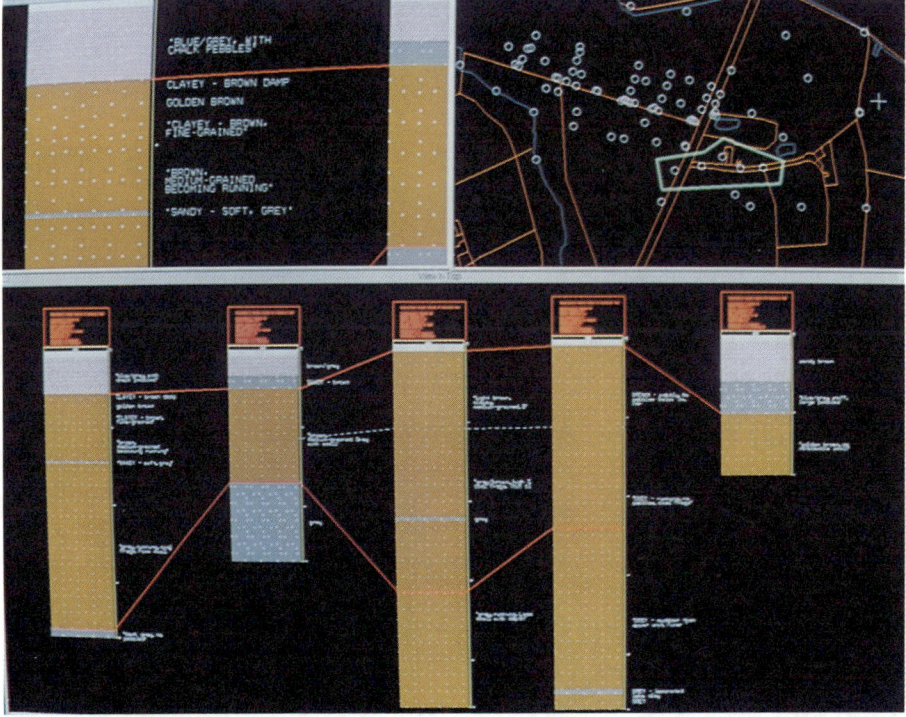

Fig. 2 Correlation panel generated by selection of boreholes from a basemap (top right), showing main lithology (coloured patterns), top and bottom of aquifer (orange lines), water level (dashed grey line) and lithological descriptions.

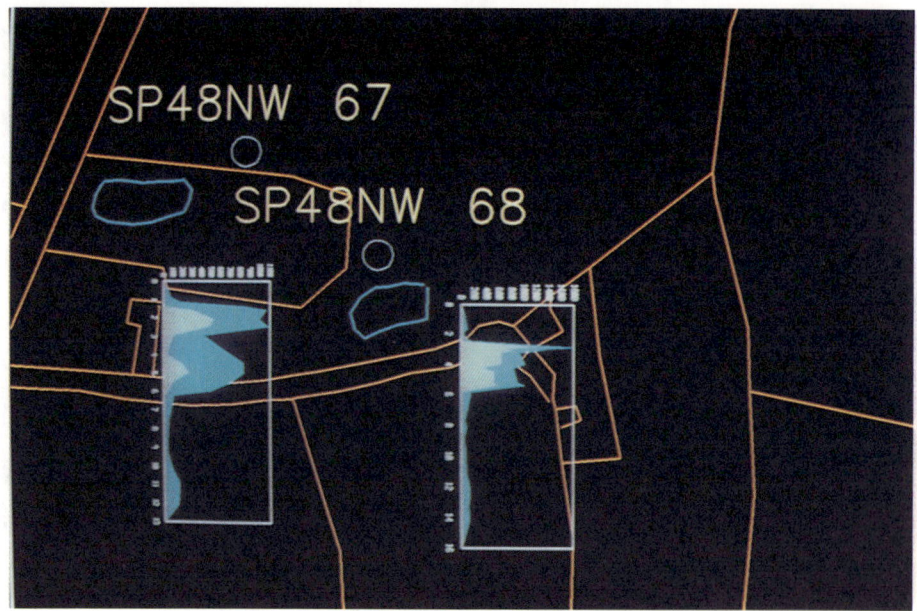

Fig. 3 Depth profiles of chloride (blue) and phenol (white) posted next to the boreholes in which they were measured.

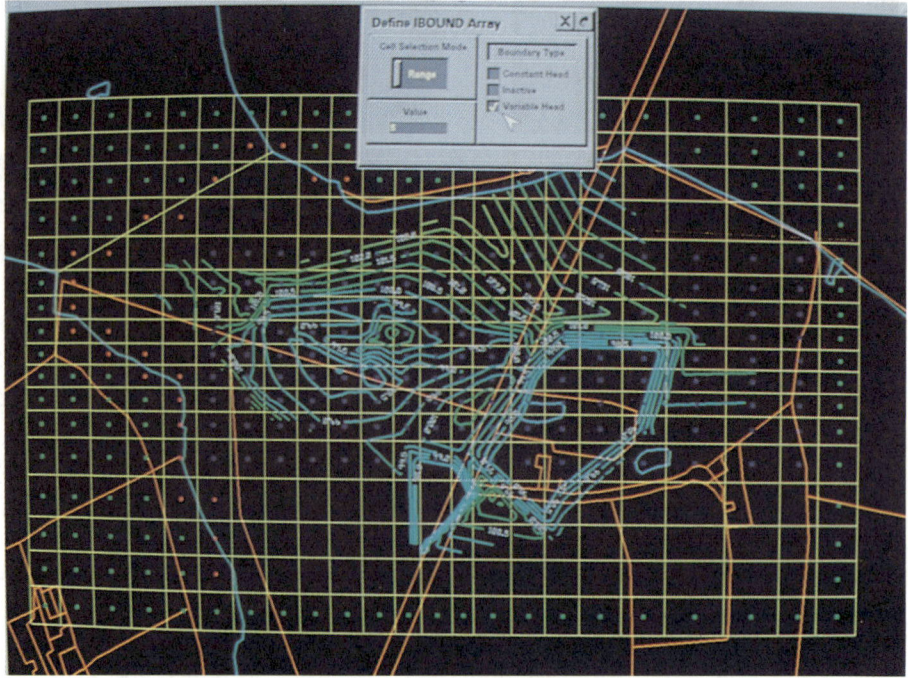

Fig. 4 Setting MODFLOW grid parameters graphically.

originated. The same information can also be presented, of course, in the form of two-dimensional contour maps, or be used to calculate enclosed volumes. Where the geology is complex or significantly faulted the user may wish to take advantage of the interface to Radian Corporation's surface modelling package, CPS-3, which also runs on the same platform.

To emulate the actual project, the remaining boreholes were added to the database, correlated and surfaces for the aquifer top and bottom were generated.

Other types of data may be too sparse or otherwise unsuitable for contouring but can be presented in other ways. Pie charts, histograms and scatter plots can all be generated from data in the database and posted at suitable locations on a map. Figure 3, for example, shows depth profiles of the concentrations of chloride (blue) and phenol (white) posted next to the boreholes in which they were measured.

Hydrogeological characterization

The third step of the analysis was some initial groundwater flow modelling. The software platform includes an implementation of the US Geological Survey's finite difference groundwater flow code MODFLOW (McDonald & Harbaugh, 1988) with a forms interface. The most important part of the interface is the ability to create the model grid graphically with reference to existing map information. Figure 4 shows a grid drawn with reference to a map of the site where the type of model cell is being specified. The colour of the dots in the rectangular cells shows whether they have been set up as inactive cells outside the model boundary, cells with a specified constant head, or variable head cells. The cell types were set up by graphically selecting a cell or range of cells. Behind the grid a contour map of the aquifer bottom surface has been plotted so that values for this parameter in the model can be specified for each cell while referencing the contoured value in the same location. When the planned feature of automatically interpolating from a gridded surface of, for example, hydraulic conductivity values, onto the model cell locations is implemented this will further speed up the initial, time-consuming task of setting model parameters. After a model run, the output head values were transferred to the terrain modelling package for contouring and display. It is planned to include in the near future the US Geological Survey's MODPATH code for contaminant transport modelling.

COMMENT

Overall, there is no doubt that a single environment for the integration of database, GIS, various earth science applications, groundwater modelling and interfaces to a number of third-party products will bring advantages to many environmental projects.

Data in a host of formats-paper maps, text files, satellite images in byte format, graphic borehole logs and air photographs as scanned images, digitized maps in DXF, Autocad or MicroStation formats, data from networked external relational database systems-have all been input to the ERMA system. The easy combination of raster and vector files provided valuable flexibility.

The history of ERMA is that it has incorporated several individual but major application packages into the one environment so there is a major learning curve in becoming familiar with even an adequate subset of them. Furthermore, the transfer of data from one part to another, at this stage in its development, is not always as straightforward as it might be. The evaluation has demonstrated that such integration demands that considerable attention be paid to:

(a) the ease of moving data and results between modules;
(b) the flexibility of workflow through the system;
(c) uniformity of the user interface (to ease the learning curve).

There is insufficient space here to deal in detail with all the above but one example will illustrate something of the case for flexibility of workflow.

One of the processes where the integration of database with GIS and surface modelling packages yields benefits is that, for example, the positions of the penetrations of particular stratigraphic units by boreholes can be posted into a three-dimensional graphics file by an application which automatically retrieves the information from the database records and makes the appropriate calculations. The posted symbols are linked by the GIS to the database so that the relevant records can be retrieved graphically or queried spatially. The (x,y,z) coordinates of these graphic symbols can be read by the CPS-3 surface

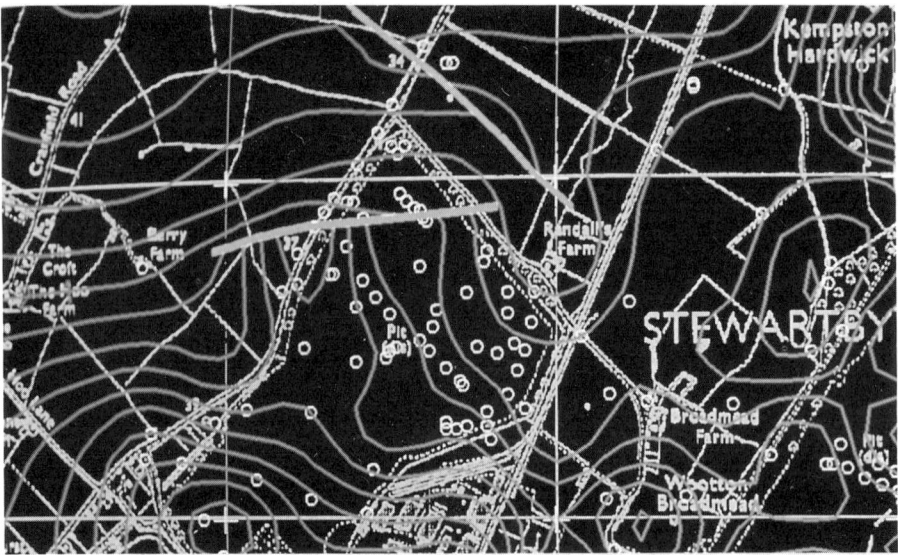

Fig. 5 Contours of the top surface of an aquifer unit plotted together with borehole and fault locations over a scanned topographic map.

modelling package and gridded. The contours of the resulting model can then be posted in the same graphics file. If there appears to be some unexpected structure in these contours the nearby boreholes which have determined the structure can be seen and their database records retrieved easily to check whether they have been correctly entered. Figure 5 shows part of an area where this process of generating surfaces from the database records was iterated several times allowing re-examination of the original data in the light of successive results. The integration undoubtedly made this process easier to do and probably contributed to a more accurate end result.

However, although this particular workflow proceeded smoothly, not all work sequences are necessarily so well catered for. For example, in earlier versions of the software a restricted set of database information, such as stratigraphic unit penetrations and water table contacts, could be posted into graphic files whereas concentrations of a particular chemical species could not. This arose because the original database posting application was designed for hydrocarbons exploration where environmental data was not needed. The situation has considerably improved during the period of evaluation by BGS but the type of problem illustrates how software originally designed to function in a particular well-defined environment may need increased flexibility when moved into a broader application area.

The existence of a default, but customisable, database structure, which covered all the types of data required for the projects used in the evaluation, is a very useful feature. For the user to design a database to manage the variety of data commonly encountered in hydrogeological investigations and remediation projects would be a complex and lengthy task.

An important feature not so far mentioned is the ease of generation of plots and formatted text reports, essential both for the investigating scientist/engineer and client.

At the time of writing, the major omissions from the ERMA suite are a contaminant transport code and a toolkit for the creation and visualization of the true three-dimensional geological spatial model. It is the authors' understanding that by the time of publication, the MODPATH code and an initial 3D visualization module will be incorporated.

Acknowledgements The authors publish by permission of the Director, British Geological Survey, NERC, UK. The reference to any proprietary hardware or software does not imply any BGS endorsement of those products. The authors acknowledge the invaluable assistance of many colleagues in BGS and the Intergraph Corporation, in particular C. Gray, R.C. Michael, I. Nixon and G. Stephens.

REFERENCES

Mackay, R., Riley, M.S. & Williams, G.M. (1993) Simulation of contaminant plume development in the shallow sand aquifer below the Villa Farm lagoons. *Water Research* (in press).
McDonald, M.G. & Harbaugh, A.W. (1988) A modular three-dimensional finite-difference ground-water flow model. *Techniques of Water-Resources Investigations of the US Geological Survey*, Book 6, Chapter A1.
Michael, R.C. & Pearson, M. (1990) The Environmental Resource Management and Assessment System (unpublished report).
Raper, J.F. & Kelk, B. (1991) Three-dimensional GIS. In: *Geographical Information Systems: principles and applications* (ed. by D.J. Maguire, M.F. Goodchild & D.W. Rhind), Vol. 1, Longman, London, 299-317.

Application of water balance model to western Saudi Arabia and use of GIS in future

ALI Ü. SORMAN
King Abdulaziz University, Department of HWR, Jeddah, Saudi Arabia

YAKUP BASMACI
Water and Sewage Department, Madinah, Saudi Arabia

KAMIL EREN
Surveying and Cadastral Department, Ministry of Municipal and Rural Affairs (MOMRA), Riyadh, Saudi Arabia

Abstract Annual water balance of representative wadis on the Western Saudi Arabia was studied by mathematical integration of a one-dimensional model on the basis of hydrometeorological data, topographical maps and remote sensing images. The model proposed by Eagleson (1979) is modified for the Southwestern region of the Kingdom to introduce a transient flow component as moisture storage in the unsaturated zone. The statistical storm parameters and soil characteristics are considered. Functional parameters for surface flow evapotranspiration, retention and recharge are determined. Spatial variability in land use, soil formation, and vegetation are derived from maps and remote sensing. Thus aerial variation of all the sensitive input model parameters are considered in order to improve estimates of surface runoff and groundwater recharge. A summary is provided for the use of GIS in future for the Kingdom of Saudi Arabia.

INTRODUCTION

This paper discusses an application of groundwater recharge model of annual water balance to some representative basins in the Western Saudi Arabia. The model, based on simplified physics of the cycle elements, formulates the water budget at a land-atmosphere interface on a probabilistic basis. It has the ability and capability of providing physical insight into the dynamic coupling of climate-soil vegetation systems over the wadis at several locations.

The concept of water mass conservation and change of energy equation enables us to express infiltration, exfiltration, percolation to groundwater, and capillary rise of the water table during rainy season. Uncertainty is introduced through the probability function of independent storm variables. The distribution functions for dependent variables of the mass balance equation derived by Eagleson (1978, 1979) are determined in terms of the physical parameters of soil, vegetation, and depth of water table.

Estimation of annual recharge is repeated for various combinations of model input parameters. Some of the model parameters are found sensitive and are determined by using topographical maps and remote sensing images to account for their actual variability.

BACKGROUND

The Western Saudi Arabia has three administrative regions as shown in Fig. 1, namely Northern (VI), Central (IX) and Southwest (II and III). The wadis located in Regions II and III are composed of two geological formations: rocks and alluvial wad channels. But these in Regions VI and IX, in addition to precambrian rocks and alluvial deposits consist of Tertiary and Quaternary basaltic formations called sub-basaltic alluvials (Sogreah,1968, 1970).

Rainfall and surface runoff directly infiltrate from alluvial channels to produce direct recharge to shallow aquifers as it occurs in the southwestern region. But there is a subsurface inflow through buried valleys from neighboring wadis in the basaltic formations, in addition to direct flow, which produce an indirect recharge to the deeper groundwater aquifer formation in Central and Northwestern regions (Bayumi, 1992).

MODEL DESCRIPTION

The annual recharge depth is related to the atmospheric-soil-vegetation system and can be determined using the concept of conservation of water mass and energy. The governing equation is expressed by Eagleson (1978) as

$$E[P_A] - E[R_{SA}] = E[E_{TA}] - E[E_{rA}] + E[R_{GA}] \tag{1}$$

Where E [ ] denotes expected value and the terms in the brackets are annual precipitation $[P_A]$, surface runoff $[R_{SA}]$, total evapotranspiration $[E_{TA}]$, evaporation from surface retention $[E_{rA}]$, and groundwater recharge $[R_{GA}]$.

The equation can be made dimensionless either in terms of annual precipitation (P_A) or mean annual precipitation (m_{pA}) as

$$1 = \frac{E[R_{SA}]}{E[P_A]} + \frac{E[E_{TA}] - E[E_{rA}]}{E[P_A]} + \frac{E[R_{GA}]}{E[P_A]} \tag{2}$$

This equation is valid under the condition that

$$E[R_{SA}] - E[E_{rA}] \geq 0 \tag{3}$$

Otherwise, the balance equation is expressed by

$$1 = \frac{E[E_{PA}] \, J(E)}{E[P_A]} + \frac{E[R_{GA}]}{E[P_A]} \tag{4}$$

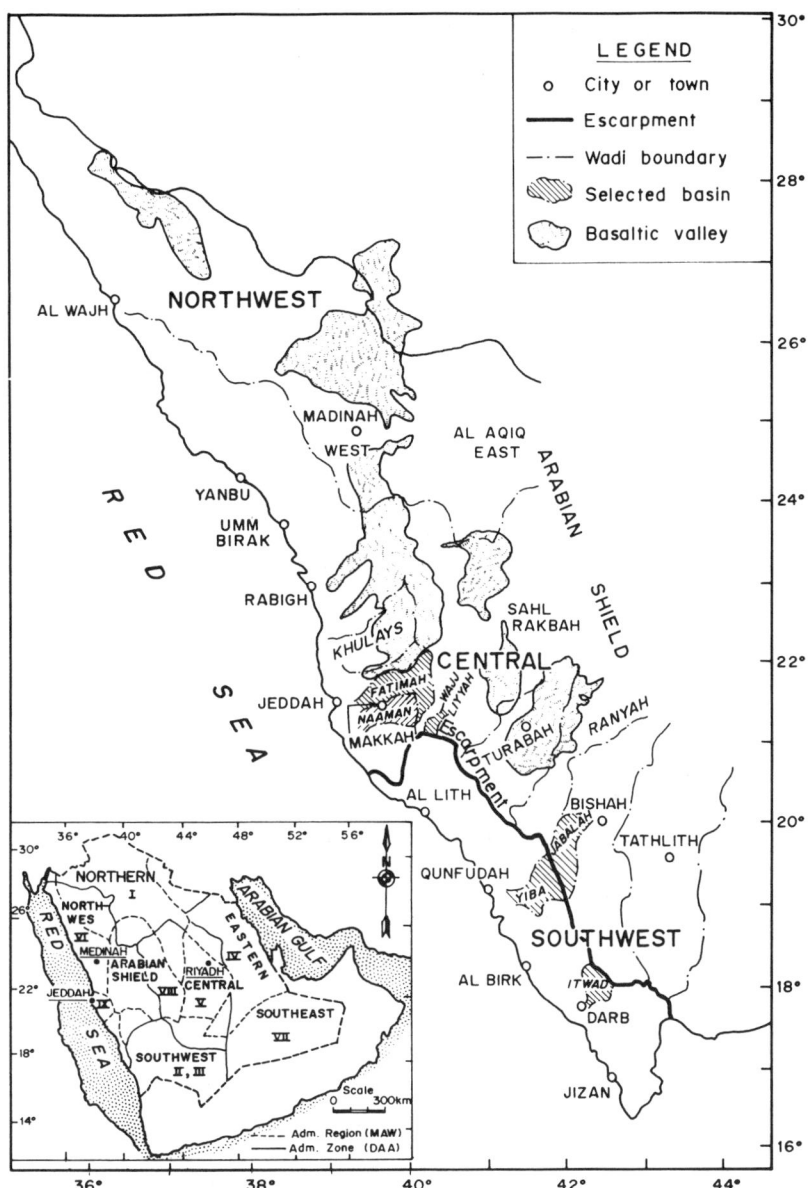

Fig. 1 Location map and representative wadis in western Saudi Arabia.

The original equation is applicable when there is a surface runoff and surface retention such as in the SW part of the Kingdom of Saudi Arabia. Otherwise, the model may be simplified for wadis which have no runoff and retention depth as is the case in most of the basins within the central zone and NW part of the Western Saudi Arabia. In some cases, a modified model would be more appropriate to consider the replenishment of the soil moisture.

The functional parameters playing role in the balance equation are expressed by Eagleson (1978) as follows:

$$E[E_{GA}] = m_\tau \, k(1) \, s_o^c - T \, W \tag{5}$$

$$E[E_{PA}] = 365 \, e_p \quad \text{or} \quad (m_v-1) \, m_{tb} \, e_p \tag{6}$$

$$J(E) = 1 - (1+\sqrt{2}\,e) \, e^{-E} + (2E)^{0.5} \, \Gamma(3/2, E) \tag{7}$$

and

$$E = (2\beta \, n_e \, k(1) \, \Psi(1) \, \phi_e(d) / \pi \, m \, e_p^2) \, s_o^{(d+2)} \tag{8}$$

Various terms in these formulaes are:

c	=	pore disconnection index,
d	=	soil diffusivity index,
E	=	exfiltration parameter,
e_p	=	mean annual potential evaporation rate,
E_{PA}	=	annual potential evapotranspiration,
$J(E, M_o, k_v)$	=	evapotranspiration function,
$k(1)$	=	saturated hydraulic conductivity,
k_v	=	plant coefficient,
m	=	pore size index,
m_τ	=	mean length of rainy season,
M_o	=	vegetal canopy,
m_{tb}	=	time between storms,
m_v	=	mean number of storms,
n_e	=	effective porosity,
s_o	=	soil moisture concentration,
T	=	time in years,
W	=	capillary rise velocity,
β	=	$1/m_{tb}$,
$\Gamma()$	=	Gamma function (GM),
$\phi_e(d)$	=	exfiltration diffusion parameter,
$\Psi(1)$	=	soil suction head.

When the groundwater recharge is normalized by division to the mean annual precipitation, inverse of the function is expressed in terms of m_{PA}

$$\text{Prob}(R_{GA}/m_{PA} < z) = \text{Prob}[PA/m_{PA} < f^{-1}(z)] \tag{9}$$

Then, the probabilistic rainfall input can be tied to the probability of recharge over the mean annual precipitation depth being less than a value of z which varies between 0.0 and 3.0.

MODEL APPLICATION WITH REMOTE SENSING

It is quite difficult to estimate from land survey representative soil and hydraulic parameters of the model, which would resemble the areal and temporal variability. But, one may get an estimate of climatic parameters using long and reliable recording station information. The wadis have different geological features and heterogenous soil formations. A reliable value of pore size index (m) for sand and fractured formation are not reported in the literature. For soils with well developed structure, the value of index is less than two but the areal average value must be weighted for sand, basalt and fractured rocks all together. A limited number of laboratory soil analyses from representative wadis and early model calibration studies has provided us an initial and rough estimate for pore size index (m), saturation conductivity (k), and porosity (n). Although, it has been reported by Sorman & Abdulrazzak (1993) that saturation hydraulic conductivity can be expressed in terms of average annual precipitation or runoff coefficient, so far, the studies have indicated a weak correlation between k(1) and m. The effective porosity n_e is determined from various analyses and field survey and found to be 10 to 15% corresponding to alluvium and basaltic valley formation. A study of the variability in hydraulic conductivity, in relation to rainfall depth and runoff coefficient for the selected wadis (see Sorman & Abdulrazzak, 1993) provides us with a mathematical relationship between the saturation hydraulic conductivity and average annual precipitation or runoff coefficient.

Some of the sensitive input parameters of the model are determined for their areal variation using remote sensing maps. Landsat map imageries at two different scales, collected and corrected for radiometric and geometric relations, have been obtained by King Abdulaziz City for science and Technology (KACST) from Riyadh Center and used in the research. Types of land use, soil formation and vegetation index values are numerically expressed to get an estimate for aerial average pore size index parameter for sand and fractured formations over the two selected catchments. Also the areal percentage coverage of alluvials, basaltic and fractural rocks are determined to weigh the mean values for inputing into the model.

The early model runs for calibration studies are done with those values and then the values are improved with the additional information extracted from laboratory analysis of soils. The location of soil samples are marked on the maps to represent various soil characteristics. The analysis of samples provides additional information for soil saturation and porosity. The data obtained from maps and laboratory experiments are confirmed with the field survey results.

USE OF GIS IN FUTURE MODEL APPLICATION

The GIS has basically three major components (McLaughlin & Nichols, 1987);

digital topographical maps (DTM), geodetical control and cadastral overlay. Since DTM is of special importance in this paper, this section is provided to show how to use GIS in future model applications. For that purpose, Table 1 is presented to summarize commonly-used map scales in Saudi Arabia along with intervals of spot heights and to show the current scheme of base mapping.

Surveying and Cadastral Department (SCD) of Ministry of Municipal and Rural Affairs (MOMRA) in Riyadh has initiated a pilot work on digital mapping of the Madinah region at a scale of 1:25 000 and presented in Fig. 2. Similar work is under preparation for the Makkah region to the scales of 1:1000. The digital map is checked with 1:50 000 topographical map for the consistency.

The map such as presented in Fig. 2 could be more effectively used in similar studies of profiles, heights, and slopes in future model studies in order to obtain the input parameters associated with the morphology and topography. Also Military Survey Department is in the process of the stereo-compilation of the whole country of the Kingdom at 1:250 000 scale. So that, in the very near future, Saudi Arabia will be equipped with digital topographical map database at both scales.

SCD of the same Ministry has also started for large-scale mapping process using a software (TRIM, 1987) in order to reach his objectives for non-uniformly spaced terrain data which is the most commonly encountered case in the Kingdom. Some of these objectives are:
(a) generate a digitized terrain model (DTM);
(b) analyze topographical features of the terrain using topographical maps in digitized form;
(c) obtain elevation read-outs; and
(d) edit the model for interactivity.

As these studies become more easily accessible to the scientist through communication systems by modems, it would be possible to generate a set of 3D triangular planes for any study area to display and analyze digitized data. As a result, regularly spaced data matrix or grid system would be superimposed on the irregular network (geodetical or hydrological) to calculate the morphological and topographical features of the wadi systems which will be used as input in model studies.

Table 1 Map scales, contour interval and spot heights in Saudi Arabia.

Scale	Contour interval (m)	Spot height interval (m)	Features
1:1000	1	50	primary/rectangular/settlements and dev. areas
1:2500	2	100	derived/rectangular/settlements and dev. areas
1:10 000	10	500	primary/rectangular/immediate periphery
1:25 000	20	1000	primary/geographical/settlements
1:50 000	-	-	primary/geographical/kingdomwide
1:250 000	-	-	derived/geographical/kingdomwide

Fig. 2 Digital map to the NE of Madinah.

RESULTS

When the critical model input parameters are obtained and storm characteristics are determined as presented in Table 2, then the problem becomes the determination of other parameters of the model, and functional forms using the mathematic expressions gives in model descriptions and presented in the same Table 2.

Two representative wadis are selected to show their computed and measured model input parameters. From the two administrative regions mentioned earlier in the West as Naaman and southwest Tabalah, two stations are considered to represent the wadi Tabalah located at the upstream, B221 and downstream B004 and only one station J204 for Wadi Naaman in Central region.

Table 2 Input and derived model parameters with process variables.

Type	Title	Symbol	(Unit)	Wadi Naaman J204	Wadi Tabalah B221	Wadi Tabalah B004
Climatic Parameters	Mean annual prec.	m_{PA}	(cm)	21.80	28.14	10.37
	Mean no. of storms	m_v		18.7	26.8	7.2
	Mean rainy season	m_τ	(days)	123.6	198.7	63.7
	Potential evap.	e_P	(cm day^{-1})	0.44	1.0	1.2
	Annual evap	E_{PA}	(cm)[1]	51-159.9	173-365	68-438
	Mean time between storms	m_{tb}	(days)	6.62	6.7	9.1
	Inverse of m_{tb}	β	(day^{-1})	0.151	0.149	0.110
Soil Parameters	Total porosity	n_t		0.35	0.30	0.30
	Effective porosity	n_e [2]		0.20	0.15	0.15
	Sat. hyd. conductivity	$k(1)$ [2]	(cm day^{-1})	(10-20)	(50-200)	(100-300)
	Intrinsic permeability	k	(cm^2)	5.2×10^{-10}	5.9×10^{-9}	2.65×10^{-8}
	Soil moist concent.	s_o		0.3	0.4	0.46
Derived	Pore size index	m [2]		(1-2)	(0.30-0.50)	(0.20-0.25)
	Pore discon. index	$c=(2/m+3)$		5.0	9.7	11.7
	Soil diffusivity	$d=(2+1/m)$		3.0	5.33	6.35
	Sat. soil matrix	$\psi(1)$	(cm)	180.0	3.53	2.50
	Pore shape par.	$\Phi(m)=5(\frac{2+m}{m})$		15	38.3	48.5
	Exfilt. diffusivity	$\phi_e(d)$		0.126	0.081	0.039
Functions	Evap effectiveness	E		0.018	0.0033	0.0003
	Dimensionless evap.	$J(E)$		0.129	0.07	0.02

				(cm)	(%)	(cm)	(%)	(cm)	(%)
Process Variables	Surface runoff	RSA (cm)		-	-	2.11	7.5	0.31	3.0
	Annual evapotr.	ETA (cm)		20.6	94.5	19.53	69.0	4.91	47.4
	Surface retention	ERA (cm)		-	-	3.75	13.3	1.40	13.4
	Annual recharge	RGA (cm)		1.207	5.5	2.45	8.7	0.715	6.9

(1) $E_{PA} = 365 \, e_p$ or $(m_v-1) \, m_{tb} \, e_p$
(2) Sensitive model parameters

The mass balance equation is solved for the determination of groundwater recharge once the functional process elements are physically computed using the respective equations for surface flow, evapotranspiration, retention and others where the sum becomes equal to the mean annual precipitation. A modified model version is used for wadi Tabalah to take into account the soil moisture depletion as shown in Table 2.

In order to get representative values for aerially distributed model parameters such as vegetation, land cover, soil and rock formation, Landsat images and topographical maps at a scale of 1:250 000 and 1:50 000 were used to get areal average values for inputting to the model.

It was found out that the weighted average values of pore size index m, saturation hydraulic conductivity k(1), porosity n_e are the most sensitive parameters to which the hydrological process variables (runoff, retention, evaporation, soil moisture concentration and groundwater recharge) are highly dependent on.

Model studies provided the following conclusive results which are summarized below:

(a) the estimated annual recharge can vary around 5 to 9% of the annual mean precipitation;
(b) the surface runoff is considered to be 3 to 7.5% for the southwest wadis; no runoff and surface retention are considered for the central and northwest wadis;
(c) the highest percentage of estimate for the hydraulic process was the potential evapotranspiration which ranged between 47 and 94.5%;
(d) the dimensionless exfiltration function J(E) was found within a range of 2% for the inland wadis, 7% for the escarpment and 12.9% for the wadis located in the central part of the Western Saudi Arabia;
(e) the initial soil moisture concentration is determined around field capacity of volumetric moisture content ranging (9 ≈ 14%). These values represent the temporal and spatial average of surface boundary layer at the time of starting of a storm but might change within the interstorm period of rainy season.

Acknowledgements The authors acknowledge the sponsorship received from the University and Agencies. They are also grateful for the use of the facilities which made it possible to publish this paper.

REFERENCES

Bayumi, H.M. (1992) Groundwater resources of the northern part of Harat Rahat, Saudi Arabia. King Abdulaziz University, Earth Science Faculty, Jeddah.
Eagleson, P.S. (1978) Climate soil and vegetation No.1-7. *Wat. Resour. Res.* **14**(5), 705-776.
Eagleson, P.S. (1979) The annual water balance. *Hydraulic Div. Proc. ASCE* **105**(8), 923-941.
McLaughlin, J.D. & Nicholas, S.E. (1987) Parcel-Based Land Information Systems. Lecture Notes in Digital Mapping and Land Information, Canadian Institute of Surveying and Mapping.
Sogreah (1968, 1970) Water and agricultural development studies for areas V and VI. Ministry of Agriculture and Water, Riyadh, Saudi Arabia.
Sorman, A.U. & Abdulrazzak, M.J. (1993) Flood hydrograph estimation for engaged wadis in Saudi Arabia. *J. Wat. Resour. Planning and Management* **119**(1).
TRIM (1987) TRIM User's Guide (8.8, rev. 2). Intergraph Corporation, Huntsville, Alabama, USA.

LOOK FOR BARCODE ←